Finite Element Analysis of Reinforced Concrete Structures

Proceedings of the Seminar sponsored
by the Japan Society for the Promotion of Science
and the U.S. National Science Foundation

Tokyo, Japan
May 21-24, 1985

Approved for publication by the
Structural Division of the
American Society of Civil Engineers

Edited by Christian Meyer, Columbia University, New York
and
Hajime Okamura, University of Tokyo

Published by the
American Society of Civil Engineers
345 East 47th Street
New York, New York 10017-2398

ABSTRACT

The 40 papers included in this publication were presented at the Joint U.S.-Japan Seminar on Finite Element Analysis of Reinforced Concrete. Two state of the art reviews discuss the role of finite element analysis of reinforced concrete structures, and finite element analysis of reinforced concrete in Japan. Specific topics include material models, cracking of concrete and fracture models, shear behavior, models for bond and reinforcement, thermal analysis, models for dynamic and cyclic loads, and numerical, computational and verification aspects of finite element analysis. Practical applications of finite element analysis are reviewed. Future development and prospects are also discussed.

Library of Congress Cataloging-in-Publication Data

Finite element analysis of reinforced concrete structures.

Includes index.
1. Reinforced concrete construction—Congresses. 2. Finite element method—Congresses. I. Meyer, Christian, 1943- . II. Okamura, Hajime, 1938- . III. Nihon Gakujutsu Shinkokai. IV. National Science Foundation.
TA683.2.F46 1986 624.1′8341′01515353 86-17279
ISBN 0-87262-549-4

Cover photo courtesy of
Kajima Institute of Construction Technology, Kajima Corporation

Library of Congress Catalog Card No.: 86-17279
ISBN 0-87262-549-4
Manufactured in the United States of America.

PREFACE

The 40 papers contained in this volume represent the efforts of two committees: the Committee on Finite Element Analysis of Reinforced Concrete Structures of ASCE's Structural Division, and the Research Committee on Finite Element Analysis of RC Structures of the Japan Concrete Institute.

The ASCE Committee was formed in 1978 as a Task Committee, with the objective of writing a state-of-the-art report on the subject, which was published by ASCE in 1982. In 1981, the Committee was elevated to a standing committee and set out to expand its contacts with researchers outside North America, mostly through corresponding memberships. While communication with European researchers was generally excellent, it was felt that similarly fruitful exchanges of ideas with Japanese engineers were hampered by the language barrier, and this could be best overcome by holding a Joint U.S.-Japan Seminar on the subject. Traditionally, this barrier has not been completely mutual. Whereas Japanese researchers are often sufficiently proficient in English in order to keep up with developments published in the English literature, American researchers are typically limited to those Japanese accounts that are published in English, although the recent efforts of, for example, the Japan Concrete Institute to translate many important technical papers and to publish the Transactions in English are widely appreciated.

In 1983, the two committee chairmen, Christian Meyer of Columbia University and Hajime Okamura of the University of Tokyo, started to plan for the seminar and in May 1985, the two committees met in Tokyo for a most beneficial exchange of ideas. Most of the papers presented in Tokyo were preprinted by JCI in two volumes of preliminary proceedings. Following the seminar, all papers were subjected to the standard ASCE peer review process and underwent extensive editing in appreciation for the Japanese authors' efforts to publish their work in English. All papers are eligible for discussion in the ASCE Journal of Structural Engineering, and they are eligible for ASCE awards.

The editors believe that the work contained in this volume represents the current state of the art in finite element analysis of reinforced concrete. It should be of interest to academic researchers because it is testimony of the considerable progress that this relatively new technology has made in the last ten years. This book should be of equal importance to practitioners because this new analysis tool is not anymore limited to problems of academic interest, but is increasingly being used to solve complex problems in engineering practice.

The topical categorization of papers, following to some degree the definitions of individual seminar sessions, has a certain degree of arbitrariness because many of the subjects listed are closely interrelated. Practical applications, for example, can be found in most of the contributions.

Special thanks are due to the two funding agencies, without whose generous support the seminar and this proceedings volume would not be possible. The Japan Society for the Promotion of Science and the U.S. National Science Foundation shared the funding for this seminar under the U.S.-Japan Cooperative Science Program.

The seminar benefitted from the contributions of three European representatives—Professors Johan Blaauwendraad from the Netherlands, Herbert Mang from Austria and Gerhard Mehlhorn from Germany—who were able to obtain funding from their own representative government agencies.

December, 1985

Christian Meyer
Columbia University, New York

Hajime Okamura
University of Tokyo

CONTENTS

THE ROLE OF FINITE ELEMENT ANALYSIS OF REINFORCED CONCRETE STRUCTURES

William C. Schnobrich[1], M.ASCE

Abstract: The roles of the tensile strength smeared crack models and the fracture mechanics smeared and discrete crack models are discussed. The tensile strength model is seen as an appropriate tool to study global behavior of complex surface structures like plates and shells. For these types of structures, multiple cracks do exist, while there behavior is not sensitive to the precise geometry of cracking. A hypar gabled roof and a cooling tower represent two examples for which this concrete model should produce engineering solutions. Fracture mechanics models are suggested for problems where one or at most a few cracks dominate and behavior of the model depends strongly on the extent of cracking. A number of transverse shear problems are suggested that seem to demand the fracture mechanics approach.

1. INTRODUCTION

It has been 20 years now since the first introduction of a concrete material model into a finite element program in order to produce a system capable of analytically researching the behavior of reinforced concrete members [21]. This first model included a cracking and bond simulation capability so that the stresses in the vicinity of the crack could be computed. This led to the formulation of the discrete crack model. Soon thereafter a second and basically different approach was begun to assist in the design of the containment structure for a gas cooled nuclear reactor [24]. This second approach looked at a problem in a more global sense. It was necessary to model a large segment of the concrete containment vessel, its reinforcement and liner in order to study the overall behavior of this structure to overload. The concrete model, therefore, had to locate zones of cracking and how crack development affected the overall behavior of the structure. The consequence of these demands led to the development of a smeared crack model.

Subsequent studies were undertaken predominantly to study overall behavior of special types of structures such as reinforced concrete plates and shells. Therefore, in these follow-on studies,

[1]Professor of Civil Engineering, University of Illinois at Urbana-Champaign, 208 N. Romine, Urbana, IL 61801.

emphasis was placed on the development of the smeared crack model. The objectives of these studies were to determine the basic failure mechanisms for complex plate and shell structural systems [9] [13] [15]. The basic failure mechanisms for a number of these structures were not well defined. Physical experiments to determine the sequence of formation and the locations of the failure modes were not easy to conduct. The scaling of these structures made the construction of physical models of the size to fit into the laboratories very difficult, demanding and expensive. Some of the attempts to actually construct reduced scale models proved to be unsatisfactory. This prompted the numerical experiments. A number of these early studies using concrete modeling incorporating smeared cracking within a layered plate and shell element produced reasonable pictures of the nonlinear behavior of those structures.

With increased activity in concrete modeling by finite elements additional researchers became involved. Soon questions were raised regarding the correctness of the simple concrete smeared crack models [3]. The major point that was raised related to the independence of the solution relative to the grid size used, a concept known as objectivity. If the width of the crack zone in a smeared crack model diminishes with mesh refinement, stresses in the elements preceding the crack undergo an increase, thereby triggering a premature crack propagation when the cracking phenomenon is governed solely by a strength level criterion. When projected to infinitesimally small element sizes, this should produce a self propagating crack requiring negligible energy. However, the stress fields computed from low order finite elements, constant strain triangles, etc., often are somewhat irregular and require some form of interpretation or averaging in order to produce realistic stress fields. Some of these lower order elements due to their propensity to not respond to sharp gradients properly, can, therefore, confuse the cracking situation.

For many of the now commonly used elements there is a tendency to spread or distribute the cracking over a wider area than a single element or integration point. Thus, the localizing or focusing behavior that leads to the breakdown of the objectivity of the solution does not develop. The fault of the method then lies in those problems where one or two isolated cracks do control the behavior, and the depth of penetration of these cracks is important to the overall result. Here localization of the crack is necessary if the solution is to focus on a few cracks.

Often many elements with their limited assumed displacement fields have preferential stress responses so that for a given size mesh the make up of the stress field is biased toward those dominating stresses. This behavior is sometimes termed shear locking and membrane locking. With mesh refinement these stress distortions disappear. This does mean what may at first appear to be a failure of objectivity may be a consequence of the element formulation rather than the material model.

For those problems where localizing behavior is important, there is cause for changing from a simple strength criterion for evaluating crack propagation to another criterion such as one incorporating fracture mechanics concepts [4][26]. Significant strides in this direction have been made in recent years by several investigators [4][14]. Problems controlled by dominant shear cracks are examples for which problems can arise and the fracture approach may have to be considered.

Questions have also been raised regarding the consistency of the results obtained using adjustments of the material properties [9][13] as employed by the early investigators. On the basis of theoretical arguments, it can be hypothesized that the minimum element size should be restricted to roughly three times the diameter of the maximum size aggregate if consistency is to be maintained. However, within the competition performed on the Collins tests [25], one of the four best performances was achieved by such an altered material stiffness model. In view of the homogeneity of the stress field only a single element was used, and the element size was unspecified.

It is this writer's belief that the engineering profession has need for both the simple material property adjustment models and fracture mechanics models in both their blunt and discrete crack forms. Overall structural behavior of some complex systems requires only that the overall stiffnesses of the various constituents reflect the average conditions of the various regions. Problems such as the determination of the strength of cooling towers under wind loading, the collapse mechanism of a hypar roof with edge beams, the influence of the weight of the edge and valley beams on the strengths of these roofs, the behavior of frame wall systems in an earthquake environment, etc., can be studied with these simple cracking strength models with their changing material property matrix.

On the other hand some problems result in a behavior which is controlled by a single dominant crack. This would be the case, for example, for the transverse shear behavior in the spherical end closures of cylindrical vessels in offshore structures like Condeep. It is anticipated that this behavior is markedly different from that present in a beam because of the existence of the multiaxial stress field in which the transverse shear acts. Any adaptation of empirical equations present in most design codes is inadequate to define the failure levels for these structures in this mode. Most shear problems are prime candidates for the dominant crack mode. These problems are then in a group for which the fracture mechanics form of the smeared crack approach seems most appropriate.

2. PLATES AND SHELL PROBLEMS

Some stiffened reinforced concrete plates and many reinforced concrete shells are complex structure whose failure modes are not well defined. When faced with decisions such as amount and placement of reinforcement, the influences of cracking, nonlinear concrete behavior, creep, etc. upon strength, cracking conditions, deflections, and buckling levels for these shells, the designer has little published information from which to draw upon. This is in contrast to the vast body of literature on structural elements for standard frame systems.

Because of the interest by the aerospace industry in the modeling of shell structures, this was an early area of application of the finite element method. Investigators have been seeking the ultimate in a general shell element ever since this early beginning. However, after 20 plus years, the search still goes on.

Because of the difference in in-plane and out-of-plane stiffnesses, many shell elements can experience phenomena known as "shear locking" and "membrane locking" [5]. Stress fields can overestimate some stresses at the expense of others. One well known example of this is the pinched cylinder, Fig. (1). When Lagrangian nine node shell elements are used in a fully integrated mode, large discrepancies can occur, Fig. (2). Convergence can be slow. One remedy for these locking conditions has been the use of uniform "under integration" for many of the isoparametric numerically integrated elements. Thus, one element currently in frequent use, the nine node Lagrangian element, is integrated with 2 x 2 mesh of Gaussian points rather than the fully integrated 3 x 3 mesh. This remedy, however, results in a side effect about as deleterious as the original malady, the appearance of spurious zero energy modes [23], that is the element is capable of deforming into a shape which has zero strains at the 2 x 2 integration points. Several remedies for the suppression of these spurious modes now exist [5][20]. Fortunately, these remedies are relatively easy to apply. Consensus is now that the Lagrangian nine node shell element [5] with 2 x 2 integration is currently the "best" element to achieve accuracy on a shell problem.

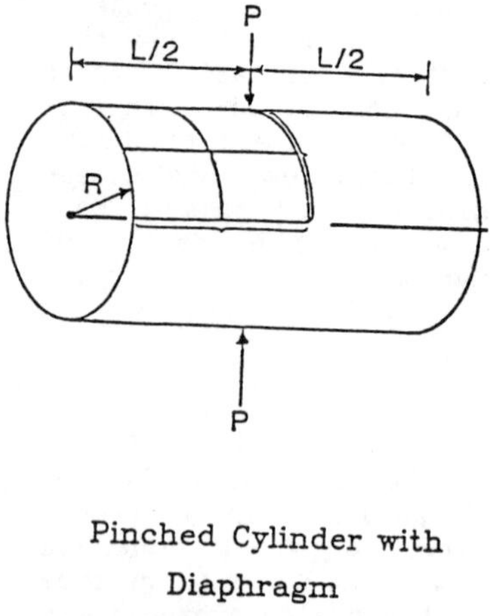

Fig. 1 Pinched Cylinder with Diaphragm

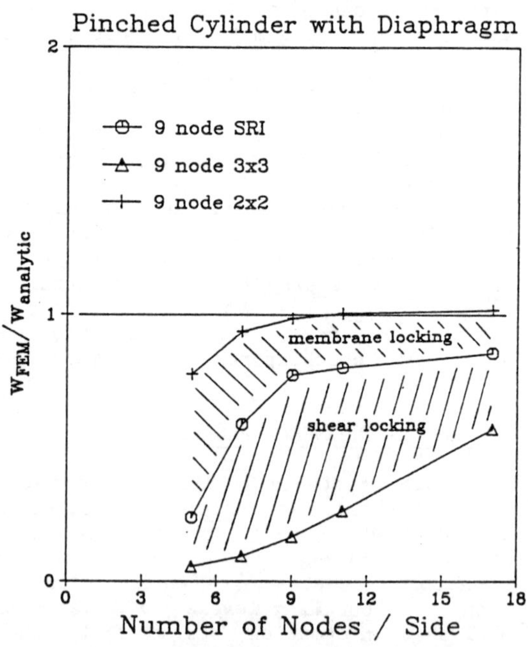

Fig. 2 Membrane and Shear Locking for Pinched Cylinder

This nine-noded Lagrangian element has been prepared in a layered format by Chan [7] and by Milford [20]. Reinforcing steel is placed as smeared orthotropic layers. Concrete behavior is specified by whatever material model one wants to use. The objective in the preparation of these models is in the determination of the failure mechanisms of shell structural systems. Crack patterns for these structures are quite extensive as the failure loads are approached. Use of a simple strength criterion for crack determination is the proper material model to use for the stated objectives. As observed by Gupta [11], the material model must be capable of producing the limit state condition for the shell.

The sequence of events that develop as the applied loading increases follows the pattern described next. The first discussion pertains to membrane behavior. Initially as the load is applied, cracks form approximately normal to the direction of maximum principal strain. The amount and the orientation of the reinforcement has only a negligible effect on both the behavior of the element prior to cracking and on the crack orientation. The angle that the crack makes with the x-axis is given by

$$\mathrm{Tan}(2\theta) = \frac{2N_{xy}}{N_x - N_y} \tag{1}$$

In a conventional smeared crack model, once the crack forms it is generally assumed that the direction θ is fixed and remains unchanged in subsequent analysis. This crack may close and a second system may form, but with restrictions on the direction relative to the first crack. If the region of cracking is in a zone that is isotropically reinforced, this sequence is not unreasonable. However, in a nonisotropically reinforced region as the loading is further increased and the reinforcement yields, usually one direction then the other these initially formed cracks become less prominent and new cracks are formed. This is possible due to the ability of the cracks to transfer shear. The change in crack direction and the consequential change in direction of the maximum panel stiffness are clearly observed in the experiments of Vecchio and Collins [25].

A lower bound to the ultimate load that the panel can sustain, considering failure by yielding of both layers of reinforcement, can be obtained from the equilibrium of a cracked element as shown in Fig. (3). This approach, often referred to as the "Principle of Minimum Resistance" or the limit state, Gupta [11], yields

$$N_{xy} = \sqrt{(N'_x - N_x)(N'_y - N_y)} \tag{2}$$

and

$$\mathrm{Tan}(\theta) = \frac{N'_x - N_x}{N_{xy}} = \frac{N_{xy}}{N'_y - N_y} \tag{3}$$

in which N_x, N_y, and N_{xy} are the components of the applied in-plane loading, N'_x and N'_y are the reinforcement capacities in the x and y directions respectively, and θ is the crack orientation. In this approach it is implied that the crack orientation at failure is independent of the initial crack orientation. The crack direction in this limit state is governed by the equilibrium condition, Eq. (3). In general this direction will not coincide with that given by Eq. (1). In arriving at Eqs. (2) and (3), and in Fig. (3), it is assumed that the crack is normal to the maximum principal concrete stress and that no tension from the concrete exists across the crack.

As noted, failure of reinforced concrete panels subject to biaxial tension-compression and/or shear loading is not limited to yielding of both layers of reinforcement, but a compressive failure can also occur. Considering failure in which the concrete crushes in compression preceded by one layer of reinforcement yielding (constrained shear), equilibrium of the cracked element in Fig. (3) results in

$$N_{xy} = \sqrt{(N'_x - N_x)\{N'_c - (N'_x - N_x)\}} \tag{4}$$

and

$$\mathrm{Tan}(\theta) = \frac{N'_x - N_x}{N_{xy}} = \frac{N_{xy}}{N'_c - (N'_x - N_x)} \tag{5}$$

where N'_c is the concrete compressive force at failure, accounting for any degradation of the concrete compressive strength [25]. In Eqs. (4) and (5) it is assumed that the x-direction reinforcement yields. Provided that the concrete compressive stress at failure is known, these equations provide a lower bound to the actual failure load, Braestrup and Nielsen [6].

It appears desirable that any numerical technique should approximate the changing crack direction effect observed in non-isotropically reinforced sections. However, this has not been

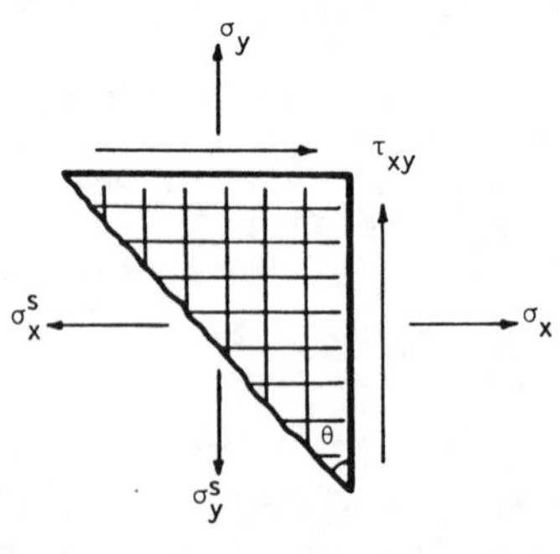

Fig. 3 Free Body Diagram of Cracked RC Element

accounted for consistently in the literature. With the exception of Cope, et al. [8] and Gupta [11], the crack direction is assumed fixed once it forms and while it remains open. Retaining some shear stiffness for singly cracked concrete allows for subsequent cracking, but the direction of the second crack is limited by the shear stiffness employed for doubly cracked concrete, ASCE [2].

Analytical methods which do not retain any shear stiffness for doubly cracked concrete can still allow for nonorthogonal cracking, Hand, et al. [13]. Restricting secondary cracking to form orthogonal to the primary crack may violate the cracking criterion within the element. Retaining some shear stiffness for doubly cracked concrete requires the formation of orthogonal cracks, Lin [14], unless the material axes are rotated upon formation of the second crack to bisect the crack directions, Abdel Rahman [1], ASCE [2]. Using this latter approach it is also necessary to limit the formation of subsequent cracks to above a specific divergence angle to the primary crack, say 30 degrees.

3. CRACKING MODEL

The type of constitutive model suggested for use in shell analysis is based on an incrementally linear smeared crack element model together with a material model that assumes that the concrete principal stresses and strains coincide.

Progressive cracking, or changes in the crack orientation, are accounted for in the cracking model by assuming that the crack direction is always normal to the direction of the maximum principal strain. In contrast to the model used by Hand, et al. [13] and Lin [15], the material axes are not fixed after formation of the initial crack, but their orientation is determined as the direction of the maximum principal strain at the beginning of each iteration.

For intact concrete, the concrete contribution to the incremental tangent stiffness matrix in the material coordinates is

$$\{\Delta \underline{\sigma}\} = \frac{1}{1-\nu^2}\begin{bmatrix} E_1 & \nu\sqrt{E_1 E_2} & 0 \\ \nu\sqrt{E_1 E_2} & E_2 & 0 \\ 0 & 0 & \frac{1}{2}(E_1+E_2-2\nu\sqrt{E_1 E_2}) \end{bmatrix}\{\Delta\varepsilon\} \quad (6)$$

where $\Delta\sigma_i$ (i=1,2) is the stress increment in the direction of the current principal strain, $\Delta\varepsilon_i$ is the corresponding strain increment, E_i is the tangent modulus and ν is Poissons ratio.

A maximum principal stress criterion is used to determine failure in tension. The constitutive relation for cracked concrete is then obtained by setting the tangent modulus, say E_1, to zero in Eq. (6),

$$\{\Delta\underline{\sigma}\} = \begin{bmatrix} 0 & 0 & 0 \\ 0 & E_2 & 0 \\ 0 & 0 & \beta_s E_2 \end{bmatrix} \{\Delta\underline{\varepsilon}\} \tag{7}$$

The shear retention factor, β_s in Eq. (7), is introduced to account for the effective shear modulus along the cracks due to shear friction and dowel action. Although it is implied that there is no shear along the current crack at the moment of its formation, this does not exclude the presence of a shear stress on previously formed cracks nor along this crack later in the development. In fact, it is this shear stress which will cause subsequent cracking and effectively produce changes in the crack orientation.

In developing a numerical algorithm, Gupta [11] obtained the rotating crack tangent stiffness matrix as the sum of the conventional fixed crack tangent stiffness matrix, such as Eq. (7), plus a contribution which reflects possible changes in the crack direction. This can be obtained as follows

$$\Delta\underline{\sigma} = \underline{K}\Delta\underline{\varepsilon} + \frac{\partial\underline{\sigma}}{\partial\theta} \{E\}\Delta\theta$$

where (8)

$$\Delta\theta = -\frac{\cos^2 2\theta}{2(E_x - E_y)} [\tan(2\theta) \quad -\tan(2\theta) \quad -1]\{\Delta\varepsilon\}$$

Gupta used these equations to obtain an incremental stress-strain relationship:

$$\{\Delta\sigma\} = [K^+]\{\Delta\varepsilon\}$$

where (9)

$$[K^+] = [K] + [G]$$

in which $\underline{K}$ is given by Eq. (7), and $\underline{G}$ represents the possible changes in the crack direction and is given by

$$[G] = -\frac{\sigma_2}{2\sqrt{(E_x - E_y)^2 + \gamma_{xy}^2}} \begin{bmatrix} \sin^2 2\theta & -\sin^2 2\theta & -\sin 2\theta\cos 2\theta \\ & \sin^2 2\theta & \sin 2\theta\cos 2\theta \\ \text{Sym.} & & \cos^2 2\theta \end{bmatrix} \tag{10}$$

An expression for the rotating crack tangent stiffness matrix, assuming the concrete to be linearly elastic and neglecting the tensile strength of the concrete (and tension stiffening) was given by Gupta [11]. This early expression can easily be extended to overcome these limitations using Eq. (8).

While Eq. (10) is theoretically correct (within the assumptions made), in keeping with the iterative nature of a nonlinear finite element approach, any suitable incremental tangent stiffness matrix can be used. It is therefore possible to neglect the rotating crack contribution $\underline{G}$, provided that the changing crack direction is adequately accounted for in the pertinent parts of the solution procedure, particularly in selecting the material axes and transforming from the material to the global or element axes.

In Milford's investigation [20] it was found that neglecting the rotating crack stiffness matrix [G] only rarely increased the number of iterations and did not introduce any numerical instabilities.

4. TENSION STIFFENING

The inclusion of a realistic tension stiffening model is very important when analyzing lightly reinforced sections such as those present in shells which are subject to in-plane loading. This is also true for structures which are endangered by stability failures. An example of the significance of tension stiffening to the modeling of instability problems is shown in Fig. (4). In Fig. (5), the factor β characterizes the length of the descending branch of the concrete stress-strain curve.

Tension softening which many researchers are now investigating should not be confused with tension stiffening. Tension softening is a local material softening observed during strain controlled experiments. It is a property of plain concrete. Tension stiffening is a global property of reinforced concrete even though it manifests itself as a softening phenomenon. It is a consequence of the presence of steel within the concrete and the fact that we are using average material properties over a finite domain to establish the stiffness characteristics of that domain. Although tension

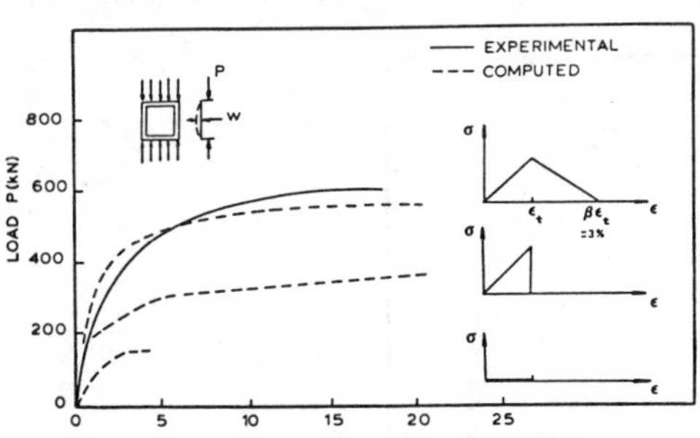

Fig. 4 Effect of Tension Stiffening on Calculated Response

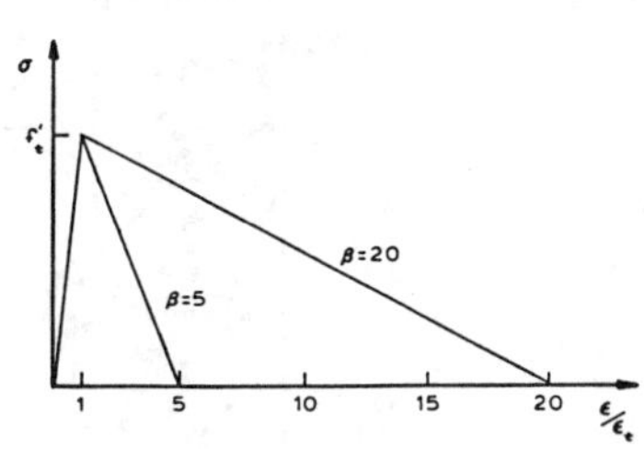

Fig. 5 Tension Stiffening Model

stiffening is frequently assigned to the concrete, it can be handled by a change in the modulus assigned to the steel, Gilbert and Warner [10]. In view of the strain levels associated with tension stiffening it usually is the more dominant effect.

Different values of β have been used in the literature: Lin [15] adopted a value of about 6, Gilbert and Warner [10] used 10. Mehlhorn [19] used $\beta\varepsilon_t$ equal to 2.0$\mu\varepsilon$ and Cope, et al. [8], set $\beta\varepsilon_t$ to 1.5$\mu\varepsilon$. There is, however, little experimental evidence from which definitive values of β can be determined. It should also be noted that an upper bound to $\beta\varepsilon_t$ must be set so that the tension stiffening does not artificially increase the total stress in the direction of any yielded reinforcement. A conservative procedure to insure that this does not occur is to limit the upper bound of $\beta\varepsilon_t$ equal to the yield strain of the reinforcement.

Limited experimental results have been obtained from the experiments of Vecchio and Collins [25] who report reasonably high tension stiffening effects "even for strains one hundred times greater than the cracking strain". Admittedly, these results were obtained from relatively heavily reinforced specimens, with a reinforcement percentage of 1.875% in at least one direction. In Fig. (6) the results obtained by Vecchio and Collins are reproduced, together with assumed unloading concrete stress-strain curves corresponding to $\beta\varepsilon_t$ equal to 2.0$\mu\varepsilon$ ($\beta \approx 20$). From Fig. (6) it appears that β=20 is sufficiently conservative. Consequently, that value was used as the default value in Milford's investigation.

5. MODELING OF FLEXURAL BEHAVIOR

The algorithm presented previously can be extended to model flexural behavior as well as combined in-plane plus flexural behavior. This is accomplished by incorporating the solution algorithm with a layered model approach such as described by Hand [13]. In this approach the concrete is divided up into a number of "layers" through the thickness, with each layer assumed to be in a state of plane stress. The only new wrinkle is the rotating crack concept of Cope [8] and Gupta [12]. The rotating crack concept is

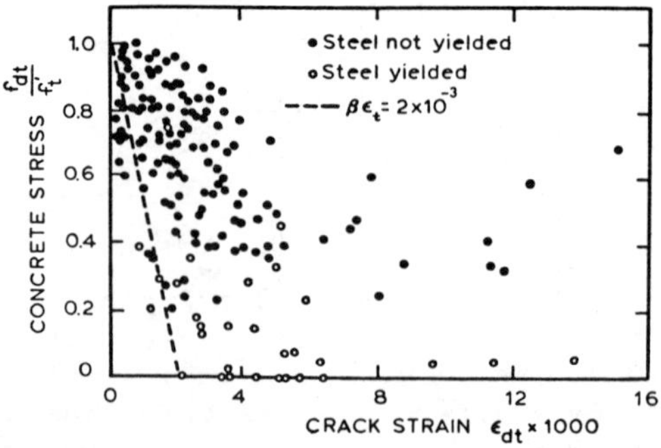

Fig. 6 Measured Tension Stiffening from Collins' Panels

applied to each layer individually. Thus, crack orientation can change from layer to layer. Milford demonstrated that this model produced yield line orientations and failure loads which were much more consistent with experimental results and yield line theory than was achievable with earlier fix crack models. This was particularly true for sections which had reinforcing meshes which were particularly anisotropic. One example demonstrating the increased accuracy of these models is shown in Fig. (7) were the results of a rotating crack model are compared with experimental values from a statically determinate plate specimen loaded in flexure. Also shown on the figure are the fixed crack results from Hand.

In view of the agreement of the computed results with the corresponding experimental quantities, the suitability of this type of smeared crack model for use in examining the global response of reinforced concrete plates and shells seems well grounded in spite of any size dependence the grid may have. It appears that the onset of crack localization does not occur until late in the grid refinement, and the precise initiation, location and extent of cracking is not critical to the determination of overall behavior. Flexural problems are less sensitive than membrane problems to this as well as other effects such as tension stiffening.

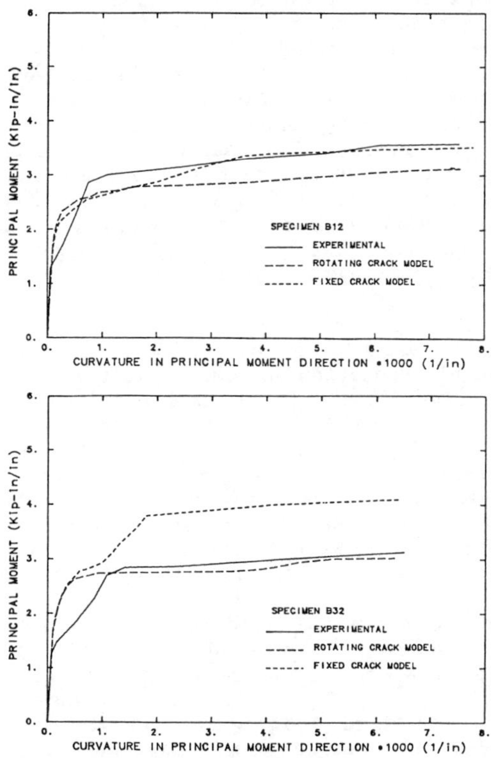

Fig. 7 Moment Curvature Relationships Using Rotating Crack Model

6. ROOF SHELLS AND COOLING TOWERS

Two examples of hyperbolic shells that have been numerically loaded to failure in order to determine their nonlinear behavior are selected to demonstrate the capabilities of the method. Chan [7] is the latest investigator to study hypar roof shells. His model is a layered nine-node Lagrangian element. The concrete model is the orthotropic model of Darwin. Tension stiffening is included by increasing the steel modulus as proposed by Gilbert [10]. An eight point integration rule is used in lieu of a reduced two by two rule in an effort to minimize the development of spurious zero energy modes. The hypar shown in Fig. (8) was studied to determine its mode of failure and particularly the role played by the creep deflections and their influence on the ultimate load carrying capacity. Because deflections resulting from creep might be influential in establishing the failure load, nonlinear geometric effects are included by Chan in the model. One reason for studying this shell is that several failures of this type of hypar have occurred. One failure was a complete collapse.

Figure (9) shows the significant variations in total loading the structure can sustain as computed by Chan under several possible loading conditions which include sustained versus instantaneous loads. The drastic reductions in live load capacity as a consequence of creep is extremely unsettling. The notion that creep does not reduce the failure load unless of course it triggers a buckling failure, seems to be violated. For this structure the gable rib serves as the main compression member. It normally is shallow enough that its dead weight must be locally supported by the shell. For the gable roof the central region of the shell is somewhat flat unless a conscious effort is made so that the quadrants intersect at some larger angle. This will also produce a slope to the gable rib. Otherwise, the rib must support itself through flexure in the central region while relying on the shell as an elastic foundation only near the ends.

In view of the results the question must be raised: What can prompt such a sizeable reduction in load capacity? A possible answer is that the creep deflection of the shell flattens a wide central region of the shell so that some of the support the shell provides the rib disappears. The increase in span the rib must cover increases the flexural stresses in the rib so when the axial stresses increase with applied surcharge, the reduced interaction capacities are exceeded, and failure results. To simulate this type of failure should not require a precise definition of crack geometry, and a basic smeared crack model should be sufficient for the study.

The second example demonstrating the type of problem for which the global concrete model might be used is the failure behavior of a cooling tower when loaded by extreme wind. Again, this class of shell has suffered failures in the form of collapse of some actual towers during severe wind loadings. Several investigators have applied finite element analyses to study the failure of these towers. The most extensive studies are those of Mang, et al. [18] and Milford [20]. Both investigators used layered shell elements,

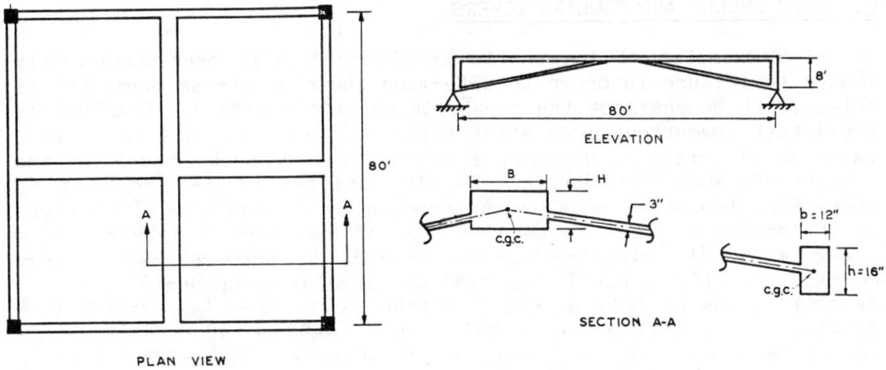

Fig. 8a Gable Roof Geometry (from Ref. 7)

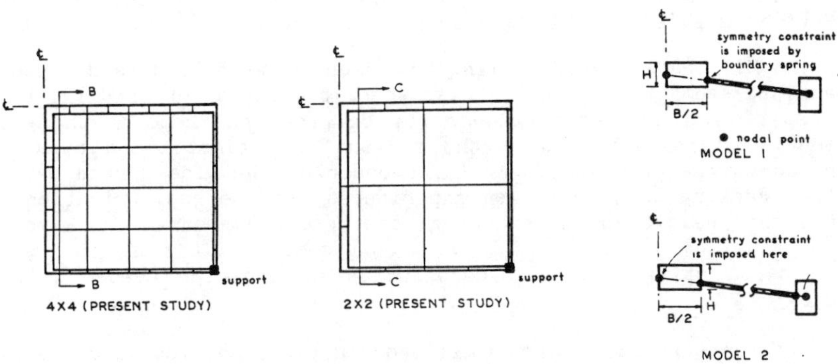

Fig. 8b Gable H.P. Finite Element Grid (from Ref. 7)

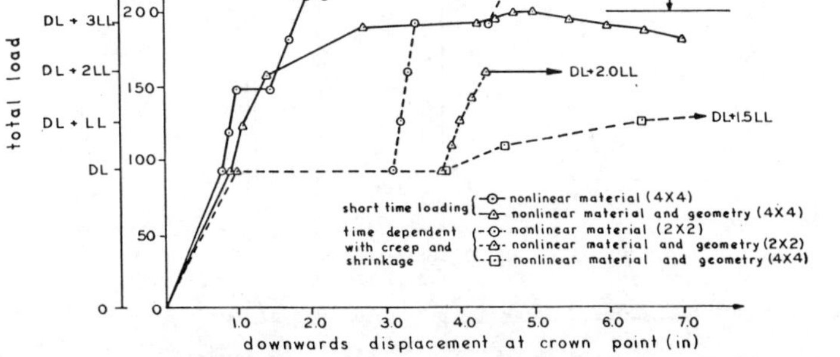

Fig. 9 Load Deflection Curves for Gabled H.P. Shell under Instantaneous or Sustained Load (from Ref. 7)

specified tensile strength, smeared crack approach and tension stiffening. Milford used the nine node Lagrangian shell element with two by two reduced integration. To suppress the zero energy modes an additional fully integrated element with a soft modulus was placed over the last shell element. The geometry of the tower that was investigated is shown in Fig. (10).

In the Milford study the material model used for concrete is based on a biaxial orthotropic material model of the form proposed by Liu [16] together with a modified form of the failure envelope proposed by Darwin [9]. Crack initiation is controlled by a maximum principal stress criterion. With the initiation of cracking the concrete is modeled using the "rotating crack model" in which the direction of the crack does not remain fixed after its formation but is assumed to remain normal to the direction of maximum principal strain. Cracks defined by this model are not cracks in the strict sense but rather are notational cracks defining an average crack direction as described in Section 3.

The Milford study of cooling towers under wind load is discussed because it seems to indicate a case where the manner of handling tension stiffening has an influence not only on the shape of the load deflection curve but also on the maximum or ultimate load capacity ascribed to the structure. Again, because of the wide spread nature of the cracking that the tower experiences and the redistribution of forces that result as the cracking develops, the smeared crack model can be considered adequate. In view of the sensitivity of results to the tension stiffening this description of behavior needs further study.

Tension stiffening in the Milford study was modeled as a gradual unloading of the concrete stress-strain curve in tension, as shown in Fig. (5). In this figure β is a measure of the length of the

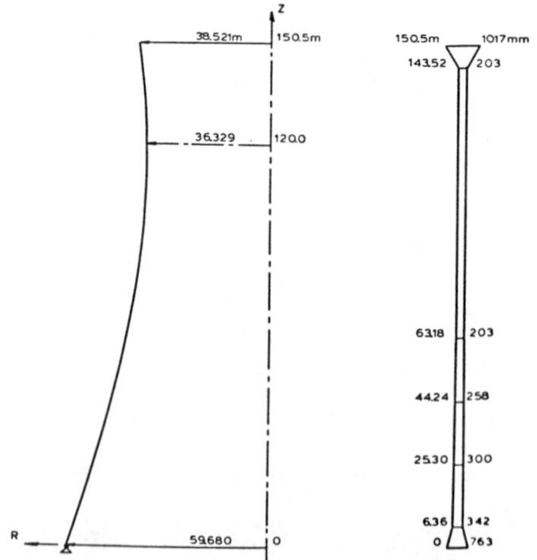

Fig. 10 Geometry of Cooling Tower

unloading part of the stress-strain curve and ε_t is the strain at which cracking occurs. The decision as to what values to assign to the β quantity currently is somewhat arbitrary. There is little experimental data available on which to quantify tension stiffening in particular for lightly reinforced sections such as one frequently finds in shell structures. Some measurements were made by Vecchio and Collins (in their in-plane or membrane test specimens). These measurements indicate significant tension stiffening effects. Their results are shown in Fig. (6) together with the value selected by Milford with $\beta\varepsilon_t$ = 2.0$\mu\varepsilon$ ($\beta \approx 20$ for f_t = 440 psi (3.0 MPa)). As sketched on the figure, the 2.0$\mu\varepsilon$ value seems to represent a lower limit to the experimental results. It must be kept in mind, however, that these tests involved for relatively heavily reinforced specimens. Their steel percentages were 1.875%. Ideally, the tension stiffening used must be related to the reinforcement quantities present. It must be a function of both bar size and spacing in each direction. At present there is insufficient experimental information available to convert a specified concrete strength and steel arrangement into a numerical value for tension stiffening.

Milford studied the same tower geometry as Mang. This tower is shown in Fig. (10). Steel reinforcement is slightly less than that used by Mang. Circumferential reinforcement is controlled throughout by the requirement of 0.35% minimum. Meridional reinforcement in the upper one third of the tower is also 0.35%. The steel in the lower two thirds is shown on Fig. (11). These percentages are substantially less than those employed in the Collins tests.

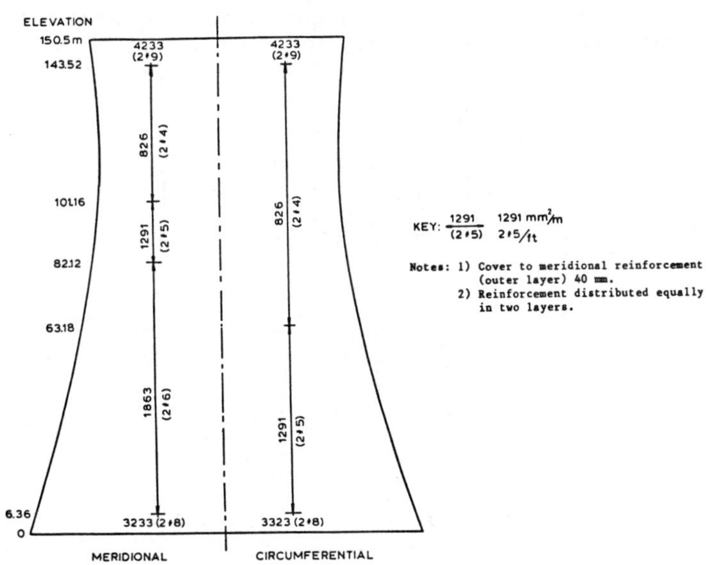

Fig. 11 Reinforcement for Cooling Towers

Over part of the shell the steel capacity is less than the cracking strength of the concrete. Milford assumed two concrete strengths, a normal concrete with 440 psi (3 MPa) as its tensile strength and a weak concrete with f_t = 220 psi (1.5 MPa). A tensile strength of 440 psi (3 MPa) is typical of that which would be obtained from a modulus of rupture test and, therefore, as such it is applicable to sections subjected to the high stress gradients present in flexure. Cooling towers are predominantly experiencing a membrane stress field with its uniform stress gradient through the thickness. Any gradient that does exist is in-plane. Hence, a tensile strength obtained from a split cylinder test should be more representative of the controlling value. The split cylinder test results are typically 60 to 80% of those of a modulus of rupture test. Thus, the lower value of 220 psi (1.5 MPa) is not greatly out of line with what should be the stress level at which cracking is initiated.

The grid used involves ten elements around the circumference and nine in the meridional direction, Fig. (12). A hinged base was assumed. As the actions of interest are up in the body of the shell, this assumption has no influence on the overall result observed. The ASCE-ACI winding loading was used.

Four different cases were studied by Milford. All but Tower II had a geometrical nonlinear formulation. Tower I allowed for a full tension stiffening effect (β=20). Tower III used a low value of stiffening (β=5). Tower IV used the low tensile strength of f_t = 220 psi (1.5 MPa). The same reinforcing was used in all four towers.

The load deflection plots computed for these four cases are shown in Fig. (13). The deflections are taken at elevation z = 331 ft (101 m), which is about the point of maximum. The ultimate load for these towers varies from 2.10 for Tower I to 1.35 for Tower IV. Zones of cracking for Tower I are shown in Fig. (14). This crack

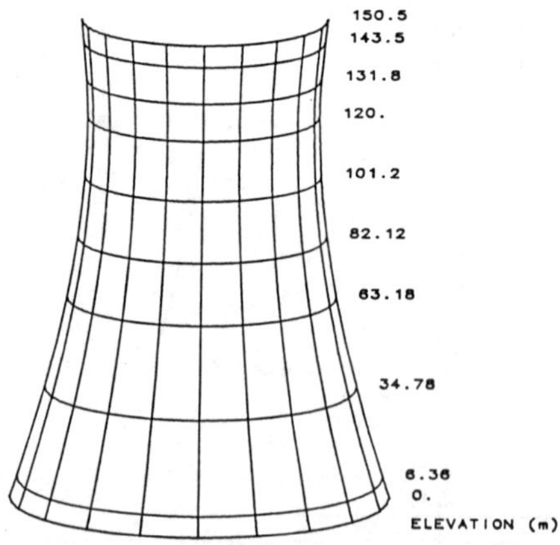

Fig. 12 Finite Element Mesh for Cooling Tower

pattern is for a load near failure. The membrane nature of the meridional cracks and the flexural nature of the circumferential cracks can be seen. The load factor at which cracking initiates is approximately 1.45 for the first three towers and about 1.0 for Tower IV.

The role of tension stiffening in the numerical solution can be seen by examining the results predicted for Tower III. Reducing the tension stiffening in comparison with that for Tower I significantly reduces the ultimate load. This difference in behavior between Towers I and II seems to indicate that contrary to the popularly held belief the calculated failure load of a structure may be dependent upon the tension stiffening adopted. The high tension stiffening model performed as a ductile shell while the low value resulted in a brittle shell.

Because reduced integration is being used, the loss of any integration point as a consequence of its reaching the capacity of

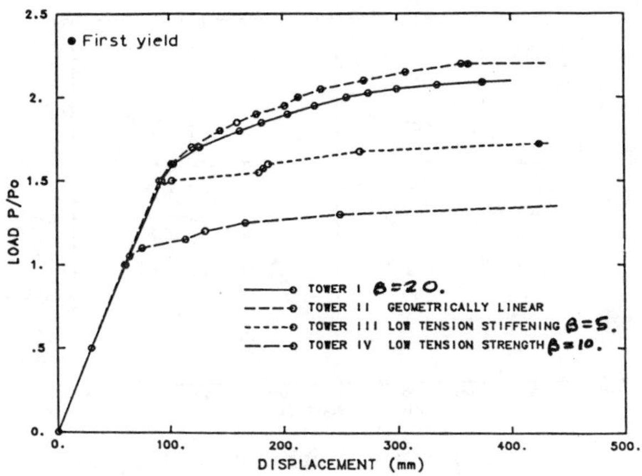

Fig. 13 Load Displacement Curves for Cooling Tower

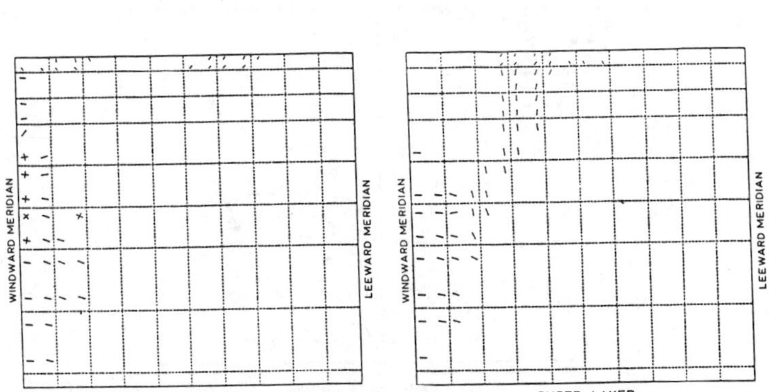

Fig. 14 Crack Patterns for Tower I

the point can result in additional zero energy modes being present in the element. To suppress this occurrence some small incremental stiffness was retained in each direction. This was done to minimize the possibility that the predicted failure is numerical, not real.

The difference in behavior between Tower I and Tower II can be explained by examining the variation in the meridional force for various integration points around the circumference, at the level at which cracking first occurs (z = 188 ft (57.2 m)). These results are shown in Figs. (15) and (16) for Towers I and III respectively. The corresponding variations in the meridional strain are shown in Figs. (17) and (18) respectively. The somewhat erratic behavior in Fig. (16) is due to the fact that the meridional forces are not necessarily recorded immediately prior to cracking. As one must expect, Fig. (16) indicates a rapid redistribution of stresses as cracking occurs. This is accompanied by an accelerated change in the reinforcement strain which, when combined with the rapidly spreading

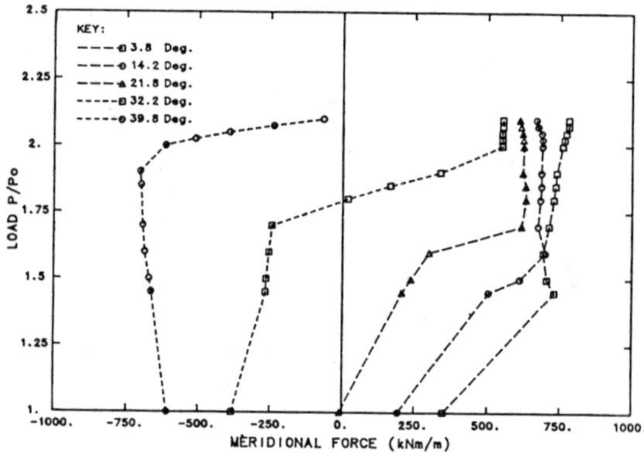

Fig. 15 Meridional Force Distribution for Tower I

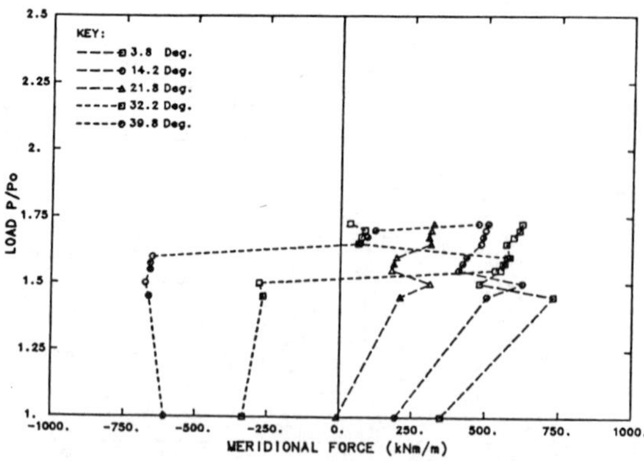

Fig. 16 Meridional Force Distribution for Tower III

zone of cracking, results in the large displacements observed in Fig. (13) at little increase in load. It is clear that the reinforcement is ineffective in that it cannot carry all the stress being redistributed by the concrete. The behavior of Tower IV is similar to Tower III.

The figures for Tower I show that there is a gradual redistribution of stress from the concrete to the reinforcement. Figure (19) indicates that the total meridional force remains nearly constant with increasing load up to the point at which the reinforcement yields. This behavior must be questioned, and is due to the significant portion of the total meridional force still carried by the cracked concrete through tension stiffening.

It is also interesting to note that the initiation of yielding of the reinforcement for both Towers I and II begins at the same elevation at which cracking is initiated (z = 188 ft (57.2 m)). However, though Tower III begins to crack at the same elevation, the first yielding of the reinforcement occurs at an elevation of 273 ft

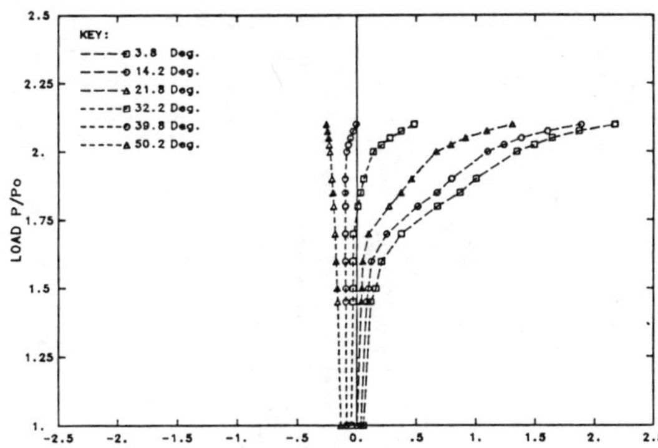

Fig. 17 Meridional Strain Distribution for Tower I

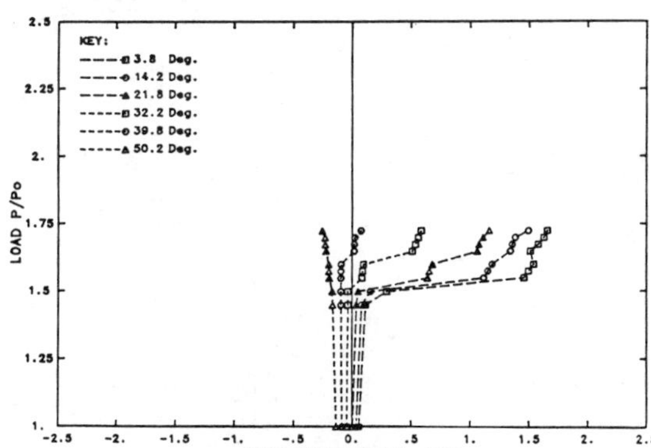

Fig. 18 Meridional Strain Distribution for Tower III

(86.2 m). This is caused by the ultimate capacity of the reinforcement being substantially less than the cracking strength of the section at this elevation and by the rapid distribution of the meridional force from the concrete to the reinforcement as a result of the low tension stiffening used. Initially the tower responds to the wind load by the meridional forces forming a tension-compression (T-C) couple within the windward face of the tower to balance the overturning component of the wind. The inability of Tower III (and Tower IV) to compensate for any loss of these meridional forces due to cracking by a shift in the neutral axis can be seen in Fig. (19). In this figure, the distribution of the meridional force and meridional strain around the circumference at the elevation at which yielding first occurs, is shown for increasing load. A small change in the neutral axis position is observed for γ = 1.725, but this is clearly not sufficient to prevent failure by excessive displacements resulting from straining of the windward meridian.

In a similar investigation, Peters [22] also observed only small changes in the neutral axis position with increasing load. The inability of the neutral axis to shift appears to be related to the predominant membrane behavior of the cooling tower, in which the circumferential distribution of the N_ϕ meridional force is controlled by the circumferential distribution of the wind pressure. The position of the neutral axis is, therefore, effectively set by the circumferential pressure distribution.

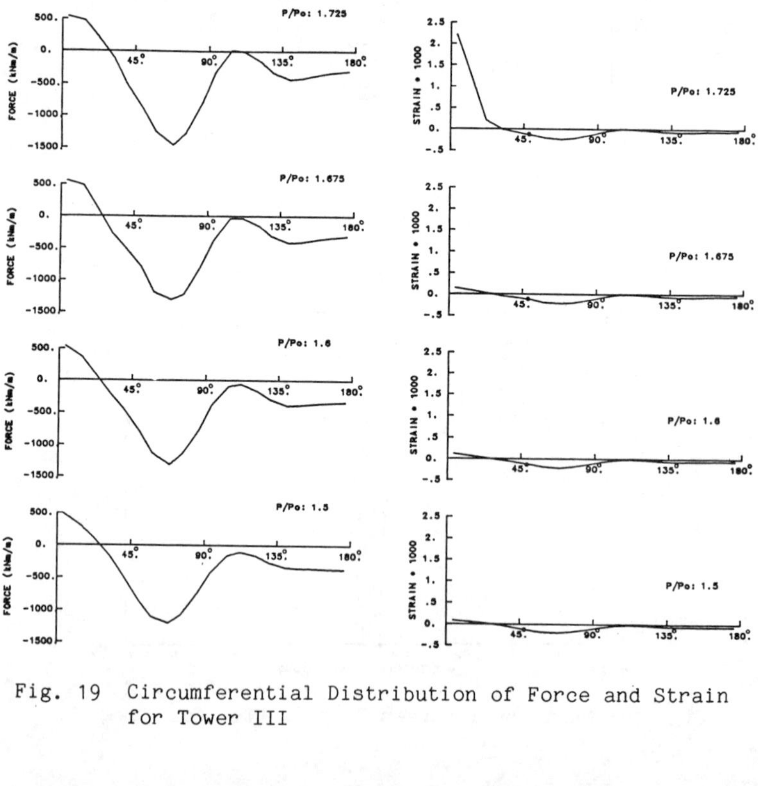

Fig. 19 Circumferential Distribution of Force and Strain for Tower III

The influence of the tensile strength on the failure load of the cooling tower is also shown in Fig. (13). It is, however, recognized that a tensile strength in excess of 220 psi (1.5 MPa) could be expected in an actual tower (except possibly at construction joints). For Tower IV it is seen that the required load factor of 1.52 is not achieved. First yielding of the reinforcement occurred at an elevation of 345 ft (105.2 m), at which location the reinforcement capacity is less than the tensile capacity of the concrete and at a maximum displacement exceeding 540 mm.

7. SUMMARY AND CONCLUSIONS

Two classes of problems were discussed. In one group of problems the local behavior is of prime interest. This group represents problems which are mostly research oriented but can develop within a design environment (such as transverse shear capacity at the dome-cylinder intersection of an offshore structure). Within this class typically one dominant crack develops. To study this behavior a fracture mechanics approach using a smeared crack model such as Bazant's is recommended. However, additional material properties such as the fracture energy for three dimensional conditions are needed before such models can be implemented.

The second class of problems for which a reinforced concrete finite element approach is needed stems from the determination of the failure loads for complex plate and shell systems. This class of problems is viewed in a more global sense. The simple tensile strength smeared crack model is appropriate to solve such problems. Multiple crack systems are anticipated. It is desired to determine the limit mechanism. Several examples of analyses following this approach were described.

Several conclusions on the models to use for the global design problem show that:

1. The simple strength criterion with a smeared crack model is adequate to study global behavior of many structural problems.

2. The rotating crack model is necessary if anisotropic steel percentages are used.

3. Tension stiffening can have a significant influence on the ultimate load capacity for some special structures with the tensile strength of the concrete which exceeds the strength of the steel used, and the zones of critical response are limited by the loading.

4. Study of creep in reinforced concrete shells using finite elements can demonstrate cases where creep deflections alter the mode of failure of the shell.

REFERENCES

1. ABDEL RAHMAN, H. H., HINTON, E. and HUQ, M. M., "Nonlinear Finite Element Analysis of Reinforced Concrete Slab and Slab-Beam Structures," Numerical Methods for Nonlinear Problem, Part I, Proc. of the Int. Conf. held at the University College of Swansea (U.K.), Pineridge Press, Swansea, September 1980, pp. 493-502.

2. ASCE Task Committee on Concrete and Masonry Structures, "State of the Art Report on Finite Element Analysis of Reinforced Concrete," ASCE, 1981.

3. BAZANT, Z. P., "Advances in Deformation and Failure Modes for Concrete," Introductory Report, IABSE Colloquium on Advanced Mechanics of Reinforced Concrete, Delft, June 1981, pp. 9-39.

4. BAZANT, Z. P. and OH, B. H., "Crack Band Theory for Fracture of Concrete," Materials and Structures, Rilem, Paris, Vol. 16, 1983, pp. 155-177.

5. BELYTSCHKO, T., STOLARSKI, H., LIU, W., CARPENTER, N. and ONG, J., "Stress Projection for Membrane and Shear Locking in Shell Finite Elements," Computer Methods for Applied Mechanics and Engineering, 1985.

6. BRAESTRUP, M. W. and NIELSEN, M. P., "Plastic Design Methods of Analysis and Design," in "Handbook of Structural Concrete," F. K. Kong, et al., Eds., McGraw-Hill, 1983.

7. CHAN, E. C., "Nonlinear Geometric, Material and Time Dependent Analysis of Reinforced Concrete Shells with Edge Beams," Ph. D. Thesis, UC-SESM 82-8, University of California at Berkeley, Berkeley, California, August 1982.

8. COPE, R. J., RAO, P. V., CLARK, L. A. and NORRIS, P., "Modeling of Reinforced Concrete Behavior for Finite Element Analysis of Bridge Slabs," Numerical Methods for Nonlinear Problem, Part I, Proc. of the Int. Conf. held at the University College of Swansea (U.K.), Pineridge Press, Swansea, September 1980, pp. 457-470.

9. DARWIN, D. and PECKNOLD, D. A., "Inelastic Model for Cyclic Biaxial Loading for Reinforced Concrete," Civil Engineering Studies, SRS No. 409, University of Illinois, Urbana, Illinois, July 1974.

10. GILBERT, R. I. and WARNER, R. F., "Tension Stiffening in Reinforced Concrete Slabs," Journal of the Structural Division, ASCE, Vol. 104, No. ST12, Proc. Paper 14211, December 1978, pp. 1885-1900.

11. GUPTA, A. K. and HABIBOLLAH, A., "Changing Crack Direction in Reinforced Concrete Analysis," Report, Civil Engineering

Department, North Carolina State University, Raleigh, North Carolina, January 1982.

12. HABIBOLLAH, A. and GUPTA, A., "Membrane Reinforcement in Concrete Shells Design versus Nonlinear Behavior," Report, North Carolina State University, Raleigh, North Carolina, January 1985.

13. HAND, F. R., PECKNOLD, D. A. and SCHNOBRICH, W. C., "A Layered Finite Element Nonlinear Analysis of Reinforced Concrete Plates and Shells," Civil Engineering Studies, SRS No. 389, University of Illinois, Urbana, Illinois, August 1972.

14. HILLERBORG, A., MODEER, M. and PETERSSON, P., "Analysis of Crack Formation and Crack Growth in Concrete by Means of Fracture Mechanics and Finite Elements," Cement and Concrete Research, Vol. 6, No. 6, pp. 773-782.

15. LIN, C. S. and SCORDELIS, A., "Nonlinear Analysis of RC Shells of General Form," Journal of the Structural Division, ASCE, Vol. 101, No. ST3, Proc. Paper 11164, March 1975, pp. 523-538.

16. LIU, T.C.Y., NILSON, A. H. and SLATE, F. O., "Biaxial Stress-Strain Relations for Concrete," J. Struct. Div., ASCE, Vol. 98, No. ST5, May 1972, pp. 1025-1034.

17. MAEKAWA, K. and OKAMURA, H., "The Deformational Behavior and Concrete Using the Elasto-Plastic and Fracture Model," Journal of the Faculty of Engineering, The University of Tokyo, Vol XXXVII, No. 2, 1983, pp. 253-328.

18. MANG, H. A. and FLOEGL, H., "Tension-Stiffening Concept for Reinforced Concrete Surface Structures," Final Report, IABSE Colloquium on Advanced Mechanics of Reinforced Concrete, Delft, June 1981, pp. 351-369.

19. MELHORN, G., "A Calculation for Reinforced Concrete Beams under Bending and Torsion using Three-Dimensional Finite Elements," IABSE Colloquium on Advanced Mechanics of Reinforced Concrete, Delft, Final Report, June 1981, pp. 591-601.

20. MILFORD, R. V. and SCHNOBRICH, W. C., "Nonlinear Behavior of Reinforced Concrete Cooling Towers," Civil Engineering Studies, SRS No. 514, University of Illinois, Urbana, Illinois, May 1984.

21. NGO, D. and SCORDELIS, A. C., "Finite Element Analysis of Reinforced Concrete Beams," Journal of the American Concrete Institute, Vol. 64, No. 3, March 1967, pp. 152-163.

22. PETERS, H. J., "Tragreserven in Flächentragwerken aus Stahlbeton," in Vorträge Betontag 1981, Hamburg Deutscher Beton-Verein E. V.

23. PARISCH, H., "A Critical Survey of the 9-Node Degenerated Shell Element with Special Emphasis on Thin Shell Application and

Reduced Integration," Comp. Meth. Appl. Mech. Eng., Vol. 20, 1979, pp. 323-350.

24. RASHID, Y. R., "Analysis of Prestressed Concrete Pressure Vessels," Nucl. Eng. Design, Vol. 7, No. 4, April 1968, pp. 334-344.

25. VECCHIO, F. and COLLINS, M. P., "The Response of Reinforced Concrete to In-plane Shear and Normal Stresses," Publication No. 82-03, University of Toronto, Toronto, Canada, March 1982.

26. WILLAM, K. J., "Experimental and Computational Aspects of Concrete Fracture," Computer Aided Analysis and Design of Concrete Structures, Damjanic, F., Hinton, E. Owen, D. R., Eds., Pineridge Press 1984.

FINITE ELEMENT ANALYSIS OF REINFORCED CONCRETE IN JAPAN

by

Norio Inoue[1] and Hiroshi Noguchi[2], M. ASCE

ABSTRACT

A general review is presented herein of the recent nonlinear analyses of reinforced concrete structures in Japan, by applying the finite element method (FEM). The authors (30) already made similar reviews in which attention was paid to the analytical method, and so this paper presents the current state and the future problems to be studied, with emphasis on the application to structures. In the description the structures are classified into colums, beams, column-beam joints and nuclear power plant structures because they have specific structural features respectively.

1. INTRODUCTION

Many analytical studies of reinforced concrete members or structures have been done recently by the Finite Element Method due to the development of computers and analytical methods based on fundamental experimental data. When the general view of the studies is observed many cases are found in which experimental results were investigated analytically and the specific problems for reinforced concrete were made clear by the feedback between experiments and analyses. This approach should clarify the unknowns which might be impossible to investigate only with experimental studies. Further for the buildings of nuclear power plants or offshore structures, analytical studies are sometimes considered to be more favorable because the structures are so large and complicated, hence, experimental studies are very difficult at an adequate scale. Such problems are also seen in the case where the number of necessary test specimens is excessively large in accordance with the variety of loading and boundary conditions. Currently this approach is still supplemental but it should become possible to make up for experimental studies and reduce the number of necessary test specimens when the accuracy of the analyses increases in the future. From the view of structural design the current design formulae are established empirically from test results obtained for various members such as columns, beams or shear walls. This sometimes causes discrepancies when a formula is used in an inapplicable range or a formula is applied as a matter of convenience to a structure which does

---

1 Senior Research Engr., Muto Institute of Structural Mechanics, Kajima Corporation, Tokyo, Japan

2 Associate Prof. of Architectural Engr., Chiba Univ., Chiba, Japan

not belong to any member under classification. For such members FEM is considered to be useful to clarify their behavior, but this approach is still thought of as being supplemental.

From this background active movements have appeared to rearrange the present status and to look over the future prospects of the FEM. In Europe IABSE held a colloquium on "Advanced Mechanics of Reinforced Concrete" at Delft in 1981 (9) and in the United States the ASCE published the "State-of-the-Art-Report" (4). In Japan the Japan Concrete Institute held the colloquium on "Shear Analysis of RC Structures" in 1982 and 1983 (44), where the fundamental problems of reinforced concrete were discussed especially on analytical procedures. In the colloquium the authors (30) presented a review of the analytical approaches by FEM and pointed out future problems concerning each basic item as follows: (a) stress-strain relation of concrete under multi-axial stresses, (b) behavior of concrete after peak stress (c) cracking and its propagation, (d) internal shear transfer, (e) modelling of reinforcing bars, (f) bond between reinforcing bars and concrete, (g) dowel action of reinforcing bars.

In this paper the application of FEM is reviewed in accordance with various members such as columns, beams, beam-column joints, shear walls and structures of nuclear power plants. These members have distinct characteristics form one another in the size, space and arrangement of bars, the effect of bond and the dimensions of space in which calculations are performed. Consequently it seems wise to represent respective items to which applications are made so that the problems may be clarified. In the presentation herein only short time loads are considered excluding creep, thermal loads and impact loads. The subjects are selected from among those that are comparatively new and have distinct characteristics.

2. COLUMN

A column contains comparatively large reinforcing bars which are arranged rather sparsely. From this specific feature the characteristics of bars including the effect of interaction with surrounding concrete, is shown to be remarkably well, hence the bond behavior and the dowel action become important. From an analytical view each bar can be modelled individually and the bond behavior can be represented easily by bond links which have assumed bond-slip relationships. The failure modes of columns are classified into the bending-type, the shear-type, the bond-type etc. and depend on the shear span ratio and the quantity of reinforcement. But it is very difficult for only one model to simulate all of the different failure modes. As to the loading condition a column is subjected to not only lateral but also axial forces and, besides, the direction of stresses change in accordance with the increase of lateral forces. This sometimes causes difficulty so that some additional means for modelling of cracking or its propagation are necessary, or in some cases the questions have to be left undecided. Furthermore in case of analyzing a column with axial forces, which is deformed in two horizontal directions such as in circular displacements, an analytical method is required which can treat cyclic loads in three dimensions. In this model it is worthwhile to note that the concrete confined by hoops is in a 3-dimensional stress condition, and for its evaluation the 3-dimensional constitutive law should be applied.

Nomura et al. (36) defined the yield and failure of concrete by the Mohr-Coulomb formula and represented reinforcement by axial elements. Bond was modelled by bond links, in which the strength of the bond was reduced corresponding with the distance from a crack surface. Columns under cyclic loads were analyzed and the influence of axial forces on the shape of hysteresis was presented. In the results the area within a hysteretic loop was comparatively small (Fig. 1). The assumption of the hysteretic characteristic should be reconsidered. Furthermore from the study (37) in which the pullout of main bars at the end was modelled by spring elements, the influence of the pullout was pointed out to be very important, because the story deflection increased 35% to 25% when this effect was considered. In considering that, Okamura et al. (3) represented the bond behavior of reinforcement embedded in massive concrete from the strain of reinforcement vs. slip relation. This procedure seems to be effective for modelling anchorages.

Noguchi et al. (31) represented concrete by an orthogonally anisotropic model and modelled reinforcement by two-dimensional elements for expressing the dowel effect (Fig. 2). Bond was represented by bond links in which the detailed characteristics of bond were incorporated, such as the difference of their properties after the bar yielding or the attainment of the ultimate bond strength. Further variable properties of bond were assumed for the upper and lower portion, and the deterioration of bond near a crack surface was given consideration. From the analysis results the fact was pointed out that the assumed bond characteristics severely affected the distribution of the strain in the main bars, the bond stress and the bond slip. In this study, detailed analyses were carried out on the basis of the precise properties of each material, although crack locations were predetermined to represent them

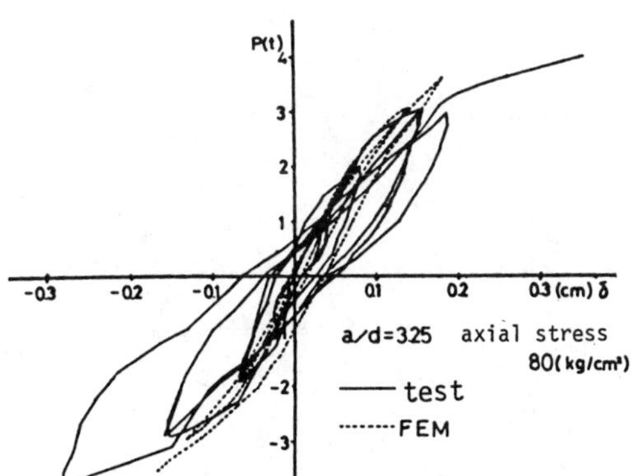

Fig.1 Cyclic load - deflection diagram (36)

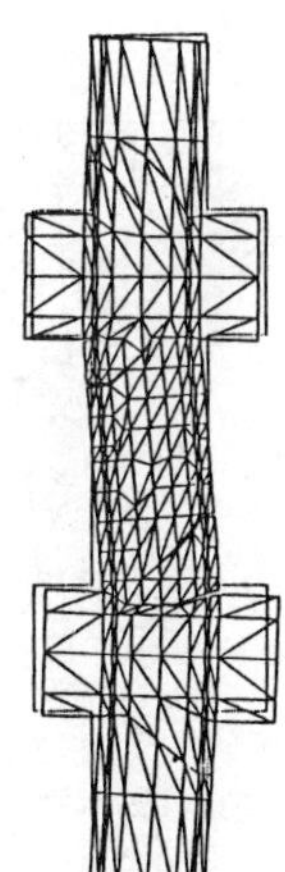

Fig.2 Deformation of column (31)

by crack links. This shows that FEM is useful approach to clarify the influence of each factor on the behavior of reinforced concrete members.

Watanabe (55) investigated the shear failure mechanism of columns caused by splitting bond failure, by considering the ultimate bond stress obtained by experiments for splitting bond failure. In the analytical model the stress-strain relation of concrete was represented by a simple bi-linear curve, and reinforcement was modelled by axial elements. Bond was represented by bond links and cracks were modelled by crack links whose locations were established beforehand (Fig. 3). Furthermore, from the obtained stress distribution in the concrete a simplified failure mechanism was obtained, and a macroscopic model which could represent its failure more adequately was established. In this study a microscopic model like FEM was used to create a macroscopic model. This approach seems to be the future trend.

Muto and Inoue et al. (27) proposed a nonlinear three-dimensional model in which core concrete confined by hoops was represented by three-dimensional elements based on plasticity theory, and reinforcement and bond were modelled by axial elements and bond links, respectively. In this model the hysteretic characteristics were assumed for each element. This approach was applied to analyze columns under cyclic lateral forces in one direction and to those which were deformed in two horizontal directions (14)(49)(Fig. 4). The results showed that their three-dimensional behavior was simulated well. When analyzing a column under two-directional lateral forces, a three-dimensional FEM becomes necessary to investigate the effect of shear forces which cannot be estimated by a fiber model, although this analysis is comparatively difficult because the stresses change every moment in accordance with the hysteresis loops. But in the results of the shear-deflection relationship, the area surrounded by the hysteretic loop was comparatively small. This tendency was also seen in Nomura's analysis (36), so the assumption of the hysteretic characteristics requires reconsideration.

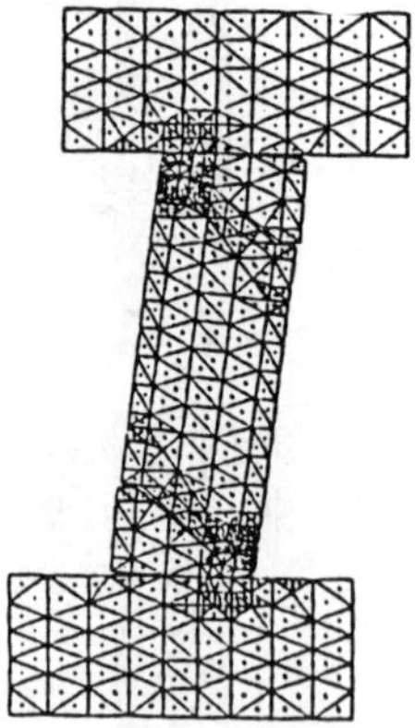

Fig.3 Deformation of column (55)

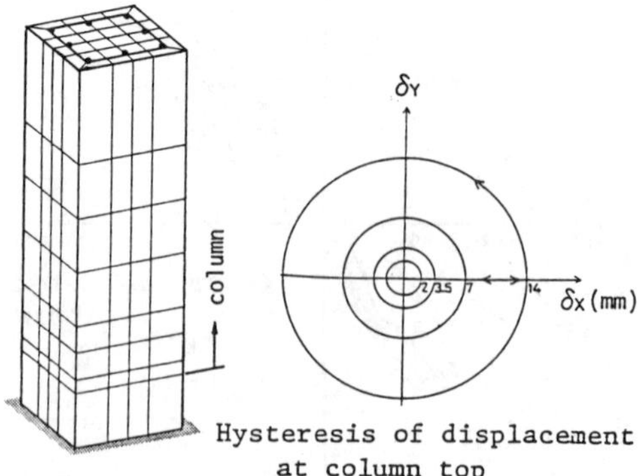

Fig.4 3-Dimensional model of column (14)(49)

A column is a member which carries vertical loads. Its failure can lead directly to collapse of the building and so it is very important to investigate its ductility. Especially the influence of cyclic loads should be clarified. But in many cases analyses by FEM were carried out for monotonic loading only up to adequate strength. This tendency of loading below the strength is seen also in the analyses of cyclic loading. Further the influence of axial forces on the shear capacity of a column has not been cleared by experiments, and the ductility after the peak loads is still not understood very well. From now on it is hoped that proposal will be made for an analytical model in which the ductility of a column after peak loads can be evaluated with consideration for the effect of axial forces and in which the influence of cyclic loading can be discussed.

3. BEAM

A beam is a fundamental member which is often analyzed by FEM to clarify the mechanism of shear resistance in reinforced concrete. Analytical models are almost the same as for columns, from the structural characteristics. The authors (30) have already made a presentation on the detailed review of beams in a past colloquium, so in this report only research conducted after that review is presented.

Noguchi et al. (29) have tested simply-supported beams without stirrups, in which cracks of different depths were installed beforehand, and measured the shear resistance of the elements after the occurrence of inclined cracks, contributed by concrete, dowel action of reinforcement and aggregate interlock. In the analytical study by FEM the mechanism of each factor was presented and the obtained shear stress well explained the experimental results in the resistance ratio (Fig. 5). This approach seems effective to clarify the resistance role of each element in reinforced concrete by both experiments and analyses.

Kokusho and Hayashi et al. (21) performed the test of a beam under bending and shearing forces with prior installed cracks to investigate the bond-slip characteristic of a cracked member. From the analytical results by FEM in which the bond characteristic obtained by tests was assumed for bond links (Fig. 6), it was pointed out that the bond-slip relation should be varied with the distance from a crack surface to simulate the distribution of the strain of longitudinal bars and bond

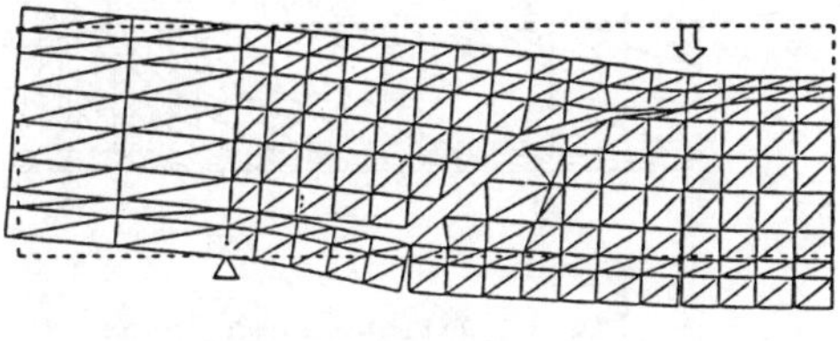

Fig.5 Deformation of beam (29)

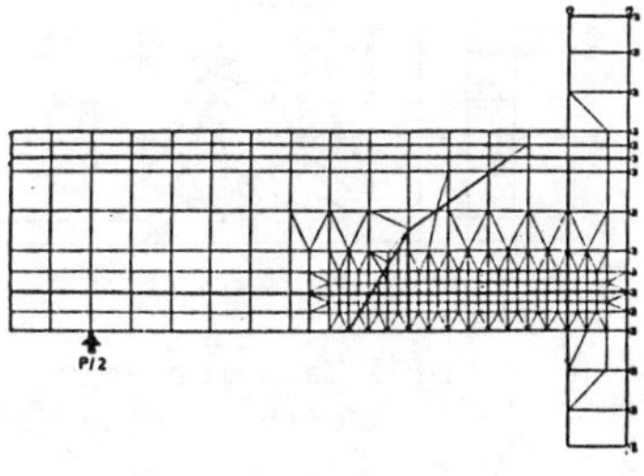

Fig.6 Mesh layout of beam (21)

stresses. In this approach one dominant factor for the behavior of reinforced concrete was picked out and investigated by both experiments and analyses. This is a favorable approach expected in the future.

Fukuhara (6) made parametric analyses of beams with prior cracks by an elastic model to grasp their mechanical behavior in the vicinity of inclined cracks which occurred at the ultimate stage under shear forces and to gain the fundamental data for proposing experimental formulae for ultimate shear strength (Fig. 7). Obtained results were studied quantitatively in stresses and crack width. This elastic analysis seems to be useful when considering the fact that many specimens have to be analyzed to obtain the formulae of ultimate strength.

Niwa (28) analyzed deep beams by nonlinear FEM assuming smeared cracks in order to clarify the mechanism of their shear resistance (Fig. 8). Some parameters which affected the shear capacity were picked out and their influence on the distribution of compressive stresses in the diagonal concrete struts was grasped quantitatively. Finally from these results a formula for the ultimate shear strength for deep beams was proposed. This approach, in which a design formula was obtained analytically by parametric calculations, is highly significant to connect FEM with design.

Kokusho et al. (20) analyzed beams with an opening to study the effect of the reinforcement around it (Fig. 9). In this study two models, with and without bond, were analyzed. Results showed that bond

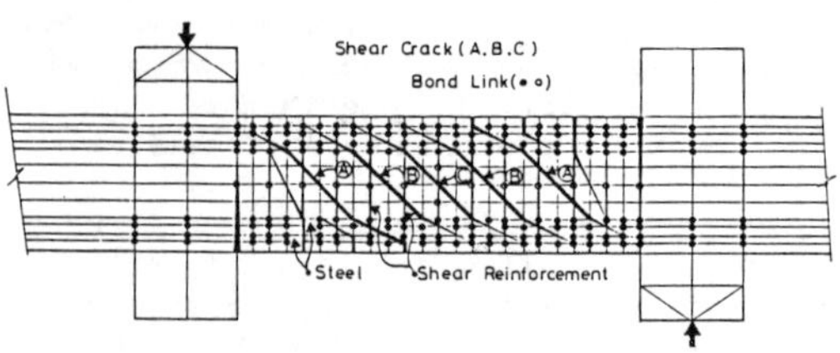

Fig.7 Mesh layout of beam (6)

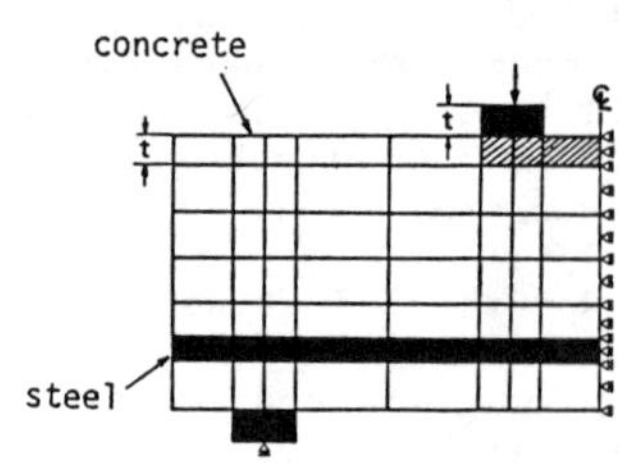

Fig.8 Mesh layout of deep beam (26)

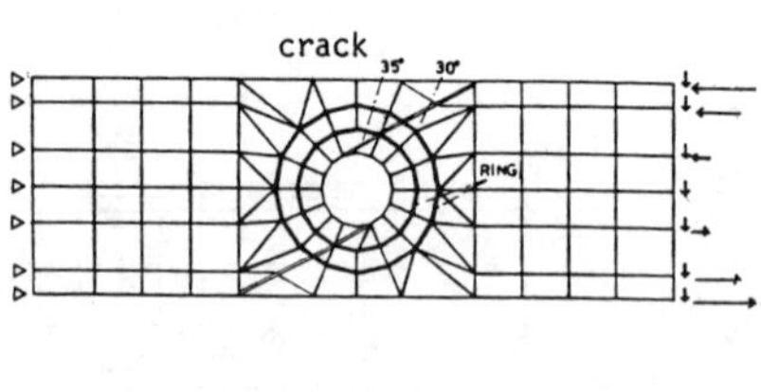

Fig.9 Mesh layout of beam with opening (20)

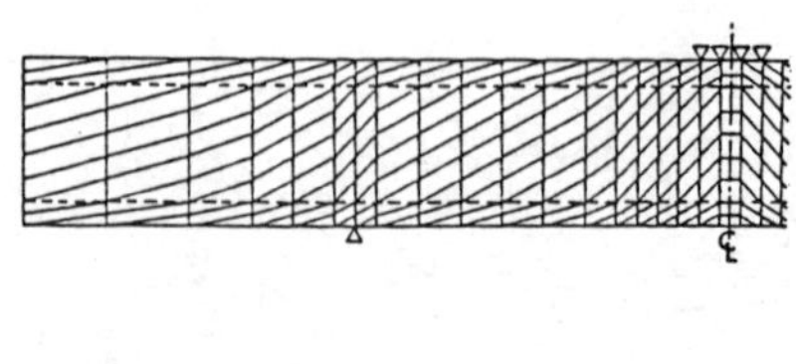

Fig.10 Oblique mesh layout of beam (39)

had a large influence on the behavior of X-shaped reinforcement but it did not influence that of ring-shaped reinforcement. This is an example to show that certain fundamental behavior can be grasped by a simple approach like this elastic analysis.

Obata and Katsumizu (39) pointed out that the element mesh layout affected the strength, deformation, crack distribution and stress of longitudinal reinforcement in the analysis of a simply-supported beam. In this study variations were considered, such as the fineness or coarseness of the mesh size and the orthogonal or oblique division (Fig. 10).

FEM analyses make it possible to grasp the internal stress and strain distributions of a beam which cannot be obtained easily by experiments. Consequently it is expected that the various failure mechanisms of a beam corresponding with the quantity of lateral reinforcement and the shear span ratio is clarified by FEM. Furthermore, by investigating the resistance ratio of each element quantitatively and comparing it with the results from a macroscopic model, it is possible that the current experimental formulae may be substantiated analytically and a more reasonable design formula can be proposed by utilizing the results of parametric studies by FEM. Regarding the method of modelling, the degree of precision in the adopted method should be considered in accordance with how it will affect the analytical results.

4. BEAM-COLUMN JOINT

In the design of reinforced concrete frames the stresses in a beam-column joint can be large if the ultimate strength of beams and columns is increased. Consequently the necessity of its study has increased but not many analytical studies exist as yet. This is because it is very difficult to model the nonlinear behavior such as the cracking and crushing of concrete under a complicated high stress condition, and the bond slip of longitudinal reinforcement in a beam at the joint as the bond deteriorates.

Shimodaira et al. (46) analyzed cross-shaped joints under low stresses (Fig. 11). In this model the concrete was assumed to be elastic. Cracking was represented by separating the adjoining nodal points, and the bond was modelled by bi-linear springs. The results showed good agreement with experimental values in the distribution of stress and strain, cracking and its propagation at the joint.

Otsuki et al. (18) analyzed joints in statically indeterminate frames by a nonlinear model to investigate the influence of various assumptions like the mesh layout, the stress-strain relation and failure criteria of concrete and the bond-slip relation. In this study inclined cracks were neglected at the joint because the assumed failure mode was restricted to yielding of the beam and bond failure of main bars (Fig. 12).

Ohwada (40) analyzed joints elastically by 3-dimensional FEM to investigate the influence of parameters such as the shape of joints, the existence of orthogonal beams and the width of beams. Recently in another research (41) the behavior after cracking was studied elastically with a model, incorporating reinforcement and bond, in which cracks were installed beforehand (Fig. 13). In considering that the effect of orthogonal beams may decrease when they are subjected to cyclic loads up to their yield moment, such a study of nonlinear behavior is desirable.

Kamimura et al. (54) investigated by three-dimensional elastic analysis the influence of the eccentricity of beams joined eccentrically to a column.

Noguchi et al. (33) analyzed cross-shaped joints under cyclic loads, paying attention to the constitutive relation of concrete, the opening and closing of a crack and the bond property. Results indicated that the deterioration of hysteretic characteristics was influenced by the

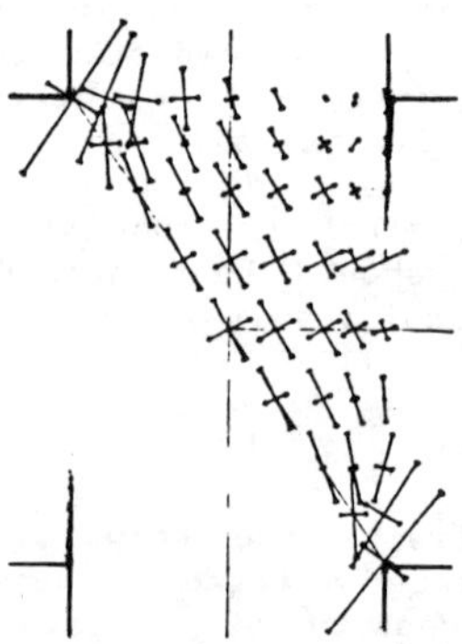

Fig.11 Principal stress distribution in beam-column joint (46)

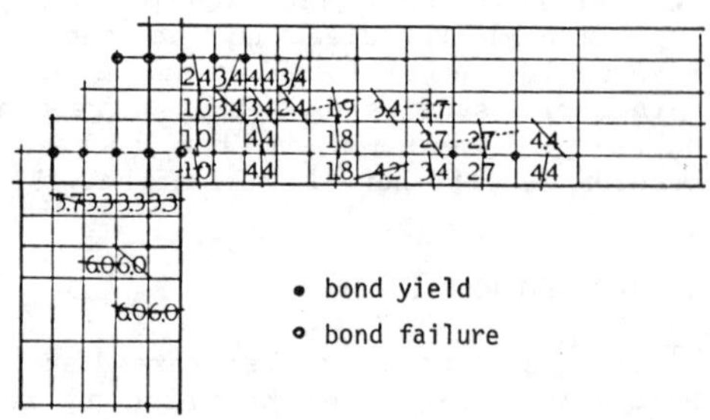

Fig.12 Distribution of crack in beam and column (18)

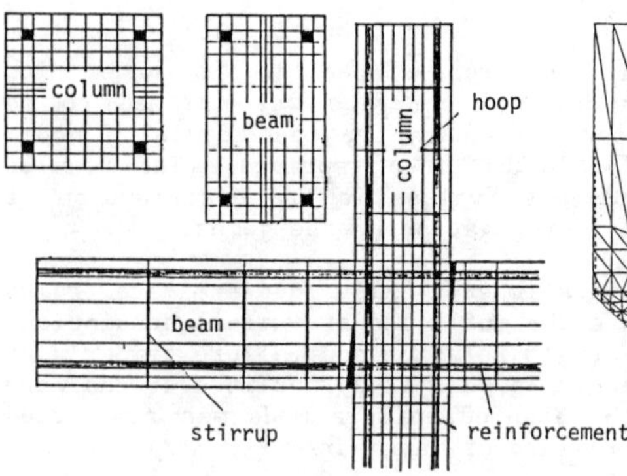

Fig.13 3-Dimensional mesh layout of beam-column joint (40)

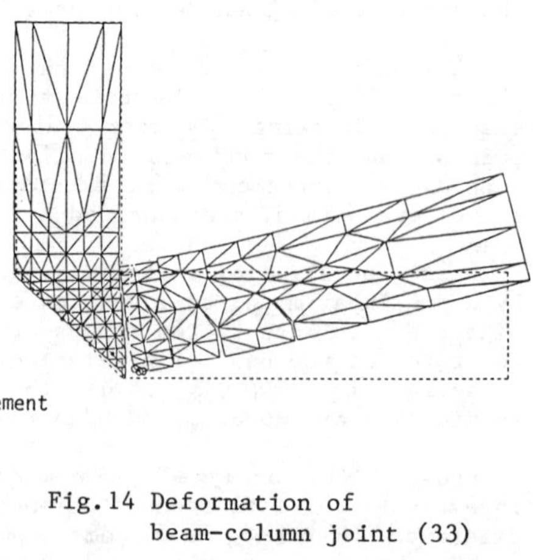

Fig.14 Deformation of beam-column joint (33)

shear failure of concrete and the deterioration of bond of the beam main bar in the joint (Fig. 14).

Hayashi and Kokusho et al. (7) analyzed a joint by applying assumed bond characteristics equal to those obtained from an experiment in which the bond property of beam reinforcement was measured within the joint. The results showed good agreement with the experimental ones. Especially it is an interesting indication that the deterioration of bond near a crack surface and the axial stress near the surface of a column were influential in expressing the bond properties of a beam.

Until now the number of analytical examples on beam-column joints is small. Hereafter it is expected that the mechanism of shear resistance will be clarified by investigating analytically the change of internal stress distribution and the deterioration of bond, and that the analysis can be useful for establishing a design method for joints. Especially the following problems seem to be important: the correlation between the bond deterioration of beam reinforcement and the shear failure of a joint, the effect of lateral reinforcement in a joint, the change of the mechanism of shear failure under cyclic loads after the bending yield of beams.

5. SHEAR WALL

Generally in a shear wall a large number of reinforcing bars are placed. Due to this feature it is rather difficult to model each reinforcing bars individually and to represent bond by bond links. Consequently in many cases the average behavior of a shear wall was studied by using a distributed representation of the reinforcement. In this modelling, there are cases where perfect bond is considered, but better results were obtained when evaluation was made of tension stiffening for the bond behavior. But for a detailed investigation of bar arrangement or bond behavior, the modelling methods used in the analysis of a column or a beam, in which a bar is modelled by an axial element and bond by bond links, are sometimes adopted. Such modelling methods retain certain specific features that correspond with the subjects of the investigation, such as to evaluate experimental results of the total behavior of a shear wall, or the microscopic behavior of the crack space and the bond characteristics. In the study to design a shear wall there are many problems to be investigated, such as the adequate arrangement of reinforcing bars around an opening, the behavior of the joints in a precast wall or infilled wall, the stiffness and strength of partial walls (wing, wainscot and/or hanging portion of wall). When these subjects were analyzed, the most suitable modelling method was proposed for evaluating the most noted points and the studies corresponding with its aim were executed.

Mochizuki et al. (24) represented the nonlinear behavior of concrete by the Darwin-Pecknold method, and assumed uniform distribution of reinforcement and perfect bond. By using this method shear walls with surrounding frames, under cyclic pure shear forces, were analyzed, in which a nondimensional hysteretic loop was assumed from the experimental results of stress vs. crack width relation (Fig. 15). This showed a simple method to simulate the cyclic behavior but some problems remain to be solved such as generality of the hysteretic loop obtained from the specific test specimens, and the applicability to a shear wall in which the yield of concrete in compression is dominant.

Shirai et al. (47) assumed plastic theory for concrete and uniform distribution of reinforcement. Cracks were represented by the smeared model but the proposed procedure was able to quantitatively evaluate tension stiffening, crack spacing and width, aggregate interlock and dowel action by using the bond theory which was based on the results of pullout tests under two axial stresses considering compression perpendicular to the direction of reinforcement. The application to a shear wall of three stories and one span showed good results (Fig. 16). This approach is useful to evaluate each factor in detail although cracks are modelled by smearing.

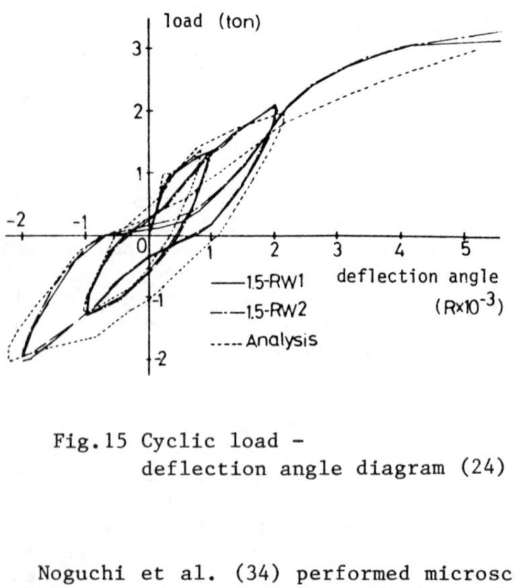

Fig.15 Cyclic load - deflection angle diagram (24)

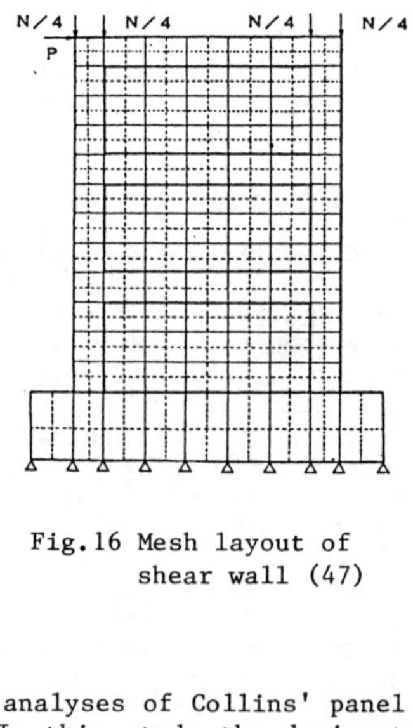

Fig.16 Mesh layout of shear wall (47)

Noguchi et al. (34) performed microscopic analyses of Collins' panel tests by a discrete crack model (Fig. 17). In this study the dominant factors, such as strength of concrete, reinforcement ratio, yield strength of reinforcement and loading method, were selected and the shear stress vs. shear strain relation of panels was compared in correspondence with each parameter. Finally, by evaluating the influence of each parameter, formulae to evaluate the maximum shear capacity were proposed. Here not only qualitative but also quantitative analyses were executed instead of experimental studies in which the difficulty of conducting the many tests corresponding to each parameter was recognized. This approach showed the usefulness of FEM. However, panels with deformed bars should be investigated in the future because the bond characteristic might show some differences from the Collins tests, in which welded wire meshes were used. Inoue and Koshika et al. (13) analyzed a reinforced concrete panel under pure shear by a smeared crack model (Fig. 18). In this study reinforcement was modelled by axial elements and the properties of bond links were assumed with consideration for the change of bond strength with the distance from a crack surface. Furthermore, the mesh size for the concrete was kept smaller than the crack spacing. From the results it was concluded that in order to simulate the crack pattern the bond strength near a crack surface should be decreased. In considering that these studies of panel tests were conducted mainly to grasp the total macroscopic behavior and to obtain the constitutive relation for a reinforced concrete panel, the purpose of a microscopic analysis by FEM can be said to be the identification of the dominant factors, which cannot be evaluated well only by experiments, and to substantiate the constitutive relationship analytically.

Inoue et al. (15) (50) presented a method in which the principal stress-strain relationship, proposed by Collins and Vecchio on the basis

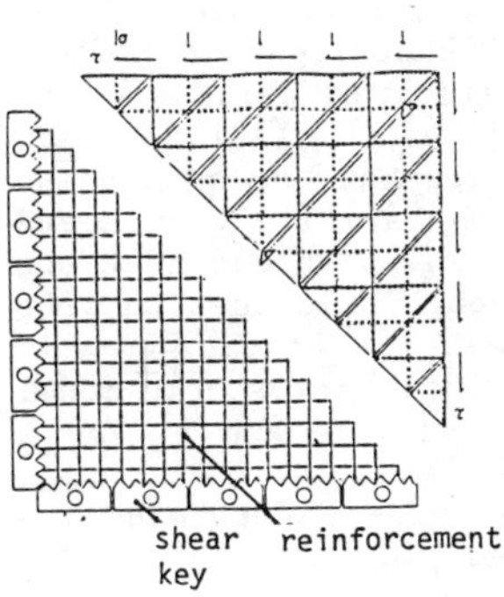

Fig.17 Reinforced concrete panel and mesh layout (34)

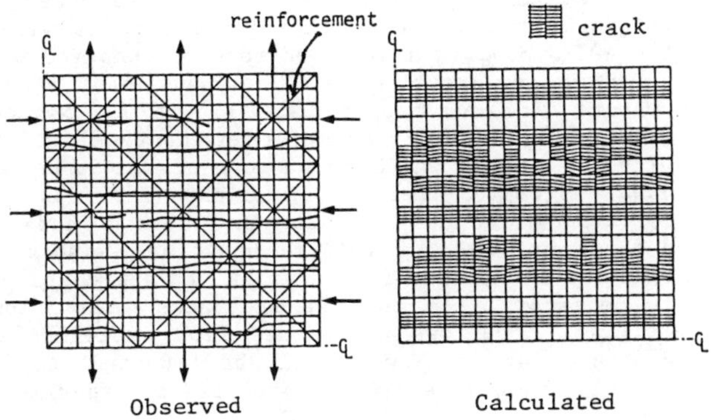

Fig.18 Crack pattern of reinforced concrete panel (13)

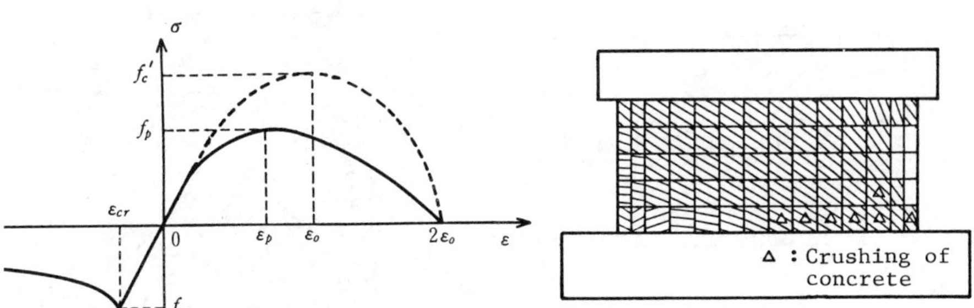

Assumed stress - strain curve of cracked concrete

Crack pattern

Fig.19 FEM analysis of shear wall based on Collins' theory (15)(50)

of reinforced concrete panel tests, was assumed in compression-tension, and the orthogonally anisotropic model was assumed in compression-compression. In the applications to shear walls with different shear span ratios and reinforcement ratios, it was shown that this method could reproduce the yielding propagation and ultimate strength corresponding with each parameter (Fig. 19). This approach required comparatively less assumptions because it was more macroscopic on the basis of the test results in which reinforced concrete panels were considered to be the minimum elements. Consequently in conjunction with the iterative calculation procedure using secant modulus, the determination of results was simple and this approach is thought to be useful for design purposes. However, it is very important to know the applicability of the test results on which this approach is based.

As to the study of openings, Seya et al. (45) analyzed shear walls with an opening by a model in which the plastic theory was assumed for concrete and reinforcement was modelled by axial elements. From the obtained stress distribution and crack pattern it was concluded that the analyses can reproduce ultimate failure modes observed in experiments. Sotomura et al. (26) analyzed shear walls with many small openings by the ADINA program and showed results compared with experiments regarding shear vs. deformation relation, the loads at initial cracking and the feature of crack propagation (Fig. 20). Noguchi et al. (32) analyzed shear walls with different arrangements of reinforcement for an opening with a model in which reinforcement was modelled by axial elements and bond was represented by bond links (Fig. 21). Results showed that the difference in the reinforcing method had much influence on the deformation and the failure behavior of shear walls. It can be said that for a microscopic study like this a corresponding microscopic model is necessary. In this study cracks were assumed beforehand by crack links at locations determined from experimental results. But from a design point of view it will be important to be able to determine the location of cracks without preliminary experiments.

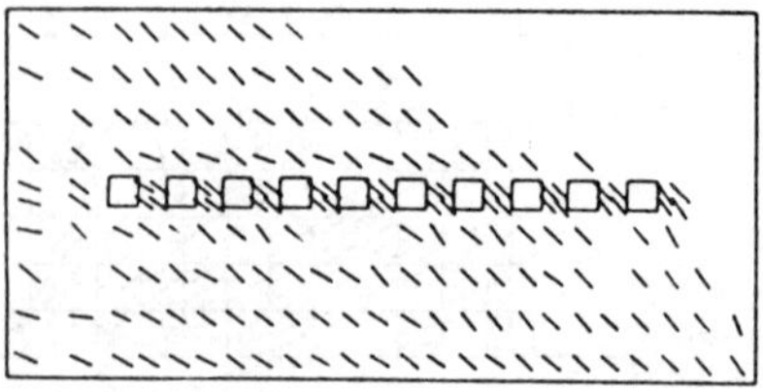

Fig.20 Crack pattern of shear wall with many openings (26)

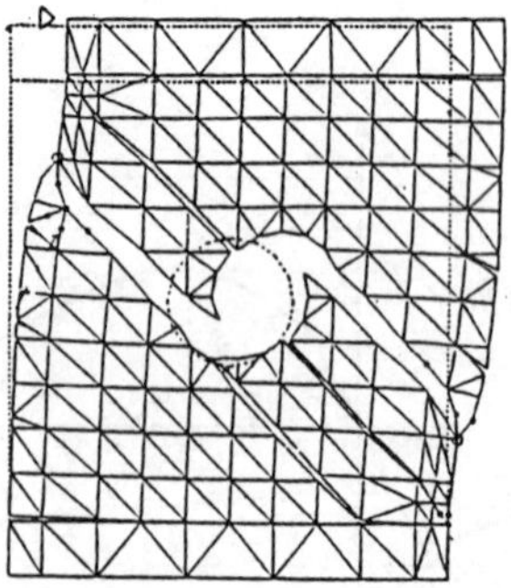

Fig.21 Deformation of shear wall with opening (32)

In a study of concrete joints, Kokusho et al. (19) analyzed an infilled shear wall which was placed with dowel reinforcement to strengthen an existing building (Fig. 22). In this analysis the joints were modelled from the experimental results of joints with dowel reinforcement using spring elements which were mutually orthogonal. From the analytical results in which the resistance ratio of dowel and concrete was estimated, it was concluded that the behavior of an infilled shear wall could be grasped if the property of the joint was given. Suenage et al. (51) represented the joint by an anisotropic finite element considering the friction, bond and dowel action to analyze a precast shear wall with horizontal joints. The application to a walled frame well represented the load vs. deformation relation and the nonlinear behavior. Higashi et al. (8) represented the vertical and horizontal joints of a precast shear wall by bond link elements, and determined the horizontal stiffness from shear tests and the vertical stiffness from uni-axial compression tests. Noguchi and Maruta (35) estimated the stiffness and the failure criteria for the spring elements which joined the frame and the infilled shear wall on the basis of the latest test data for dowels and the analytical results presented by Kokusho et al. This application showed that the failure process of joints had much influence on the overall behavior of a shear wall, determined in the study of the dowel effect on the ultimate strength and deformability of the infilled shear wall (Fig. 23).

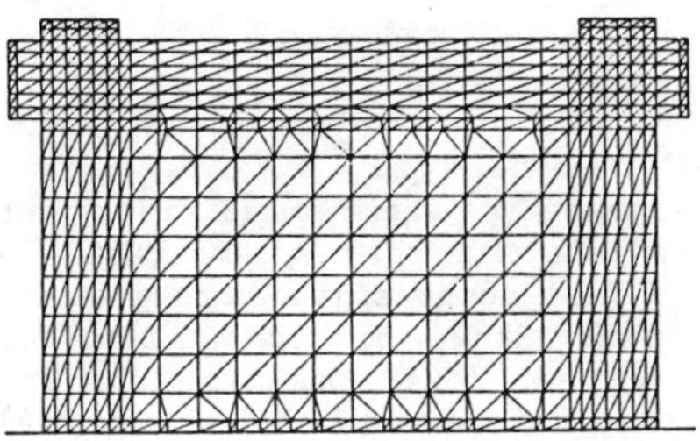

Fig.22 Mesh layout of infilled shear wall (19)

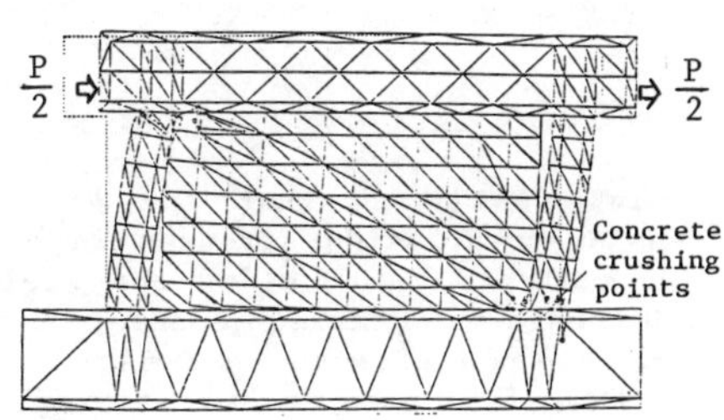

Fig.23 Deformation of infilled shear wall (35)

Nomura et al. (38) studied the effect of partial walls which were placed in a frame (Fig. 24). In the analysis the nonlinear behavior of concrete and reinforcement was evaluated by assuming an equivalent elastic stiffness. Here the initial stiffness, the yield strength and the ultimate strength were compared with the current design formulae to investigate their applicability. Aizawa et al. (1) analyzed walled frames with openings of different shapes and sizes to obtain the equivalent opening ratio and presented the applicable limit of the modelling by frames. These studies investigated the applicability of the design formulae by parametric analyses and should be extended as the analytical accuracy is improved.

Furthermore, Ono and Adachi et al. (42) investigated the resistance ratio of the shear wall panel and its surrounding frame in a multi-storied shear wall by a discrete crack model (Fig. 25). Kano et al. (17) studied the same problem by the smeared crack model. As to the calculation approach, Arai (2) proposed an iterative method for monotonically increasing loading and Mitsukawa and Baba (23) pointed out the usefulness of a generalized code with many options of modelling.

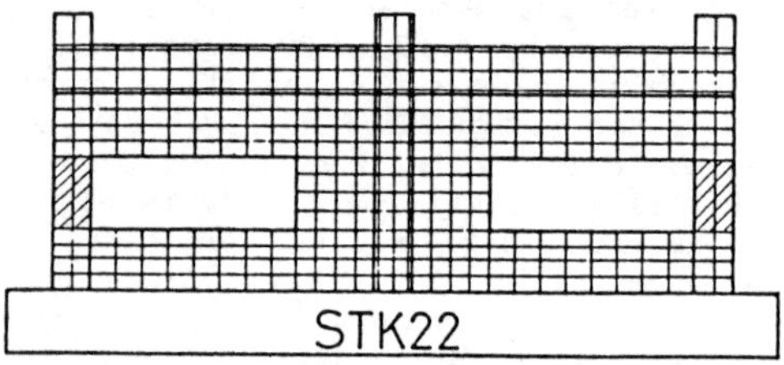

Fig.24 Mesh layout of shear wall (38)

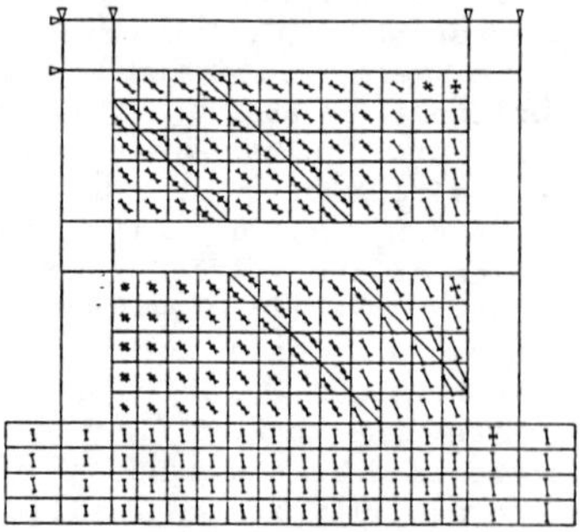

Fig.25 Principal stress distribution of shear wall (42)

In considering that the greater part of a shear wall is under tension-compression stresses in shear failure, it is very important in the analysis to consider the decline of the compressive strength and stiffness in cracked concrete, which was shown in the reinforced concrete panel tests by Collins. Ueda et al. (52), Maekawa (22) and Morikawa (25) used methods to decrease them beforehand and Noguchi (34) represented their decline by assuming the characteristic under bi-axial tension-compression stresses without deteriorating the concrete strength itself. Inoue et al. (15)(50) assumed the stress-strain relation proposed by Collins as it was. This problem should be clarified urgently. About the shear slip failure observed in experiments some studies have been started like Ueda's proposition (52) in which the failue was represented by assuming the limit criteria of shear strain. But this is a future problem and should be investigated to clarify its mechanism and the means to represent it in analyses. Furthermore there are problems to be studied such as the change of ultimate shear strength with the confinement of the surrounding frame and the influence of the loading methods such as push, push-pull and distributed loading.

6. NUCLEAR POWER PLANT STRUCTURES

As for nuclear power plant structures there are box-type and cylinder-type shear walls, and prestressed concrete reactor vessles

(PCRV) among others. In analyses they should be handled as three-dimensional structures because they are subjected to out-of-plane forces besides in-plane forces. The loading conditions are very complicated. A cylinder-type shear wall under internal pressures or thermal loads can be analyzed as an axi-symmetric problem and can be handled like a two-dimensional problem by using solid plane elements for concrete and axial elements for reinforcement. But when subjected to non-axisymmetric forces like earthquake loads the entire structure must be considered and then in many cases it is analyzed by a model in which the out-of-plane bending is represented by layered elements, reinforcement is replaced by equivalent plates and smeared cracks are assumed. When subjected to earthquake loads after cracking by internal pressures, the direction of principal stresses changes greatly and new cracks occur in an inclined direction to the previous cracks. So it becomes very difficult to represent the cracks by the generally used method in which the orthogonally anisotropic model is assumed after cracking. Also, it becomes a basically complicated problem because the shear transfer between cracks is dominant. For these reasons there are few studies for such loading conditions.

Isobata (16) proposed an orthogonally anisotropic model by introducing parameters concerning Young's modulus and Poisson's ratio to analyze three-dimensional structures like a prestressed concrete pressure vessel (PCPV) as a three-dimensional axisymmetric problem (Fig. 26). The application to the PCPV under internal pressures showed that not only the deformation but also the crack propagation could be grasped in a three-dimensional structure with thick cross sections and that the results simulated the experimental crack pattern.

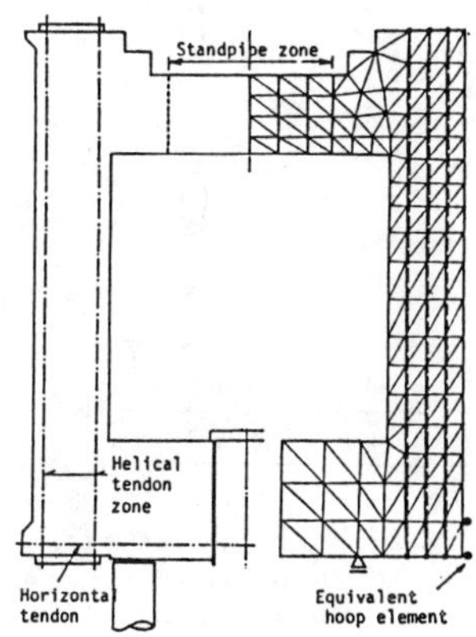

Fig.26 Model of PCPV (16)

Miyashita and Morikawa et al. (53) analyzed a reinforced concrete containment vessel under internal pressure by a three-dimensional axisymmetric model in which the constitutive relation of concrete was defined by the plastic theory, and reinforcement and bond were represented (Fig. 27). This is the same approach as for the two-dimensional problem and makes it possible to investigate the deformation and the propagation of nonlinearity in a structure of revolution with thick cross sections like Isobata's model.

Imoto (10) proposed an analytical method using three dimensional solid elements for three dimensional structures with complicated shapes like a PCRV. As to the yield condition of concrete a cracking criterion reflecting the effect of the intermediate principal stress for tension, and a generalized Drucker-Prager's criterion for compression were adopted to precisely represent the three dimensional stress state. The application to a PCRV under internal pressures showed a good coincidence with the observed deformation and yielding propagation (Fig. 28). In the future the clarification of the applicability of the stress strain relation of plain concrete under triaxial stresses is desired although this is a problem for experiments.

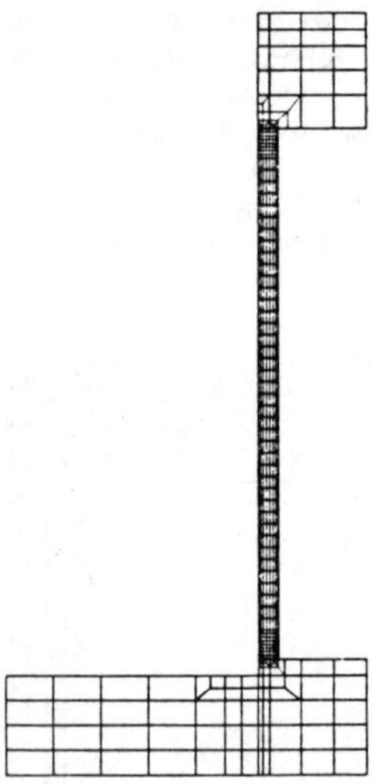

Fig.27 Mesh layout of reinforced concrete containment vessel (53)

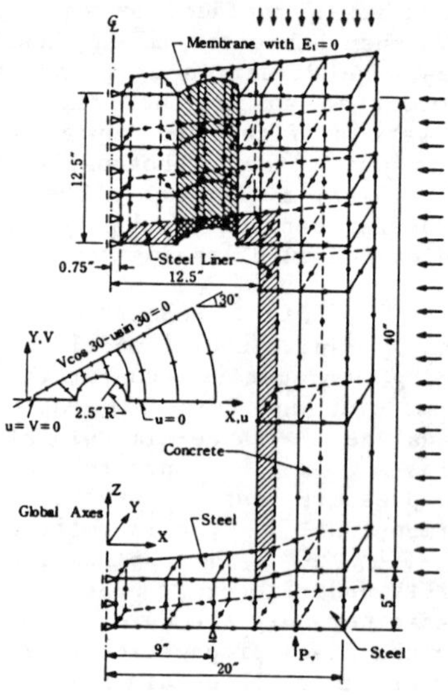

Fig.28 3-Dimensional model of PCRV (10)

Ouchi et al. (43) analyzed a cooling tower under horizontal forces by a layered model assuming that the in-plane stiffness of concrete was obtained from plastic theory and reinforcement was replaced by equivalent plates (Fig. 29). The results simulated the experimental ones for deformations, the crack pattern and strains of the reinforcement. But it was pointed out that the observed shear failure at the throat of the cooling tower could not be predicted from the analysis although its possibility was shown. The clarification of this failure is a very important problem pertaining to shear walls in general.

Fujita et al. (5) analyzed a cylindrical shear wall under internal pressure and a box-type shear wall under lateral forces by the layered model using the orthogonally anisotropic theory proposed by Isobata (Fig. 30). Results showed good agreement with the observed ones for deformations, the stress distribution and the strain distribution in the reinforcement.

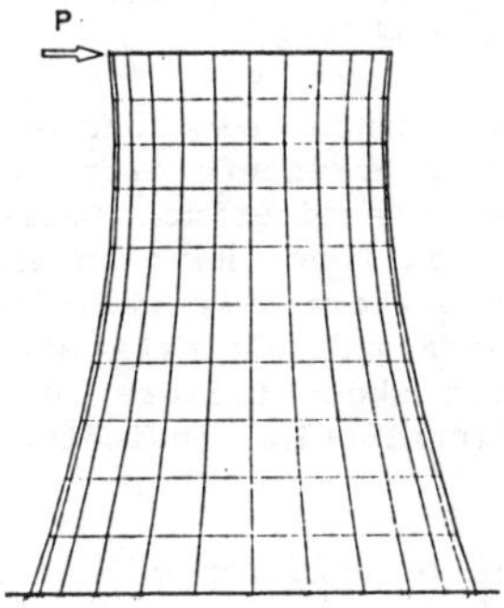

Fig.29 Model of cooling tower (43)

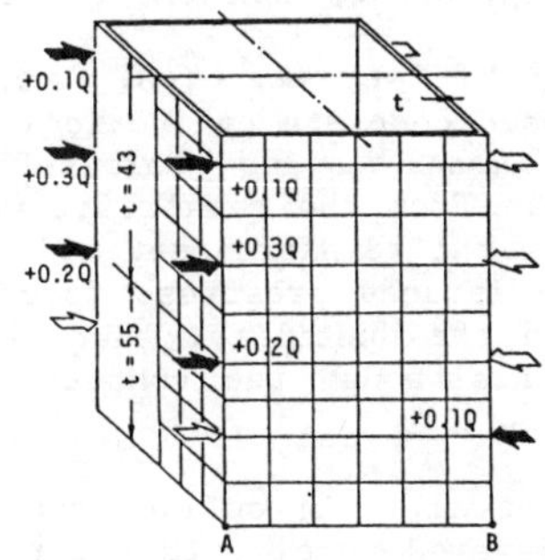

Fig.30 Model of box-type shear wall (5)

Imoto et al. (11) analyzed a prestressed concrete containment vessel (PCCV) using Ouchi's method and investigated the method of estimating the bending and shear stiffness for lumped mass models, which are widely used in earthquake response analyses (Fig. 31). This was conducted by separating the rotation, the bending deformation and the shear deformation from the FEM results to obtain each nonlinear characteristic. This approach seems to be very useful for obtaining the necessary load-deflection curve in executing dynamic analyses.

To obtain the load-deflection curve of box-type shear walls, Inada et al. (12) made parametric studies with Fujita's layered model using the orthogonally anisotropic theory and investigated the influence of each parameter such as the reducing ratio of shear rigidity, the shear span ratio, the reinforcement ratio, the axial stress and the loading method (Fig. 32). Furthermore the shear stress vs. shear strain relation, which was obtained by separating the bending and shear deformation from analytical results, was simplified to tri-linear curves and compared

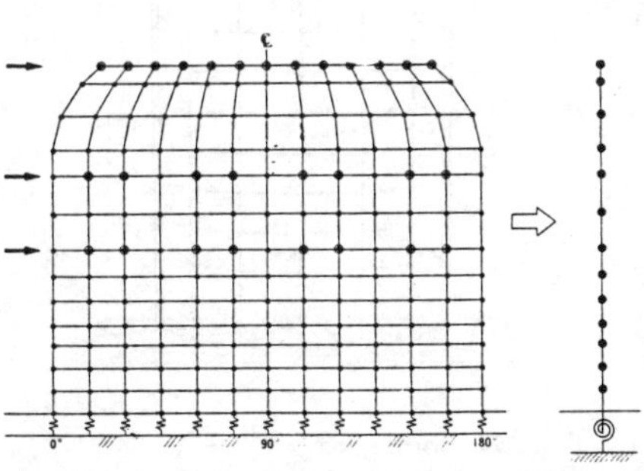

FEM model Lumped mass model

Fig.31 FEM model and lumped mass model of PCCV (11)

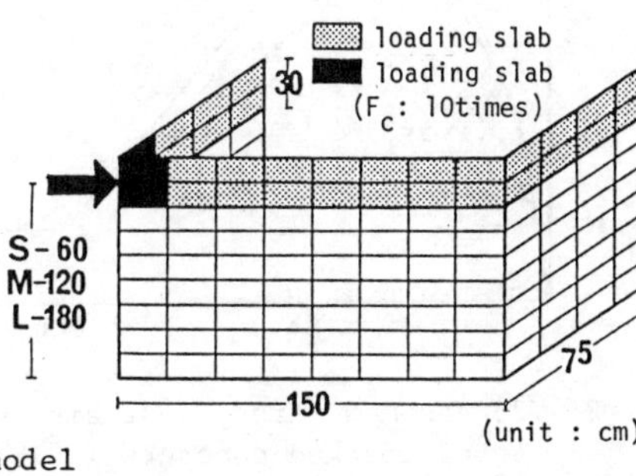

Fig.32 Model of box-type shear wall (12)

with the current formulae. In considering that there are not many experiments of box-type shear walls and that it is difficult to generalize the load-deflection curve only with test data, this is a useful approach to complement the insufficient data.

Shiraishi et al. (48) analyzed a reinforced concrete and a prestressed concrete cylindrical shell under horizontal forces with the layered model using Darwin-Pecknold's constitutive relation for concrete. From the results it was concluded that the deformation, the strain in the reinforcement and the crack pattern were well grasped and that the induced prestress forces increased the ultimate strength and decreased the ductility. But the observed shear failure could not be simulated although the computed shear distribution indicated it as a possibility.

Ueda et al. (52) defined the constitutive relation of concrete used in the layered model with reference to the experimental results of the reinforced concrete panel tested by Collins (Fig. 33). That is, it was assumed that the stiffness and strength of concrete parallel to a crack after cracking were reduced in the tension-compression region and that the stiffness of concrete dropped to zero when the shear deformation reached the assumed limit value. The former represented the deterioration of concrete after cracking and the latter represented the shear slip failure by shear strain. These assumptions are very important and should be clarified urgently as a general problem of shear walls, in considering that the greater portion of the wall is in the tension-compression stress domain when shear failure occurs. The applications to a cylindrical shell and a box-type shear wall showed good agreement with the experimental results for the load vs. deformation relation and the failure pattern.

Morikawa (25) analyzed reinforced concrete cylindrical shells under lateral forces with internal pressure or thermal loads using the layered model assuming plastic theory for concrete (Fig. 34). To represent this

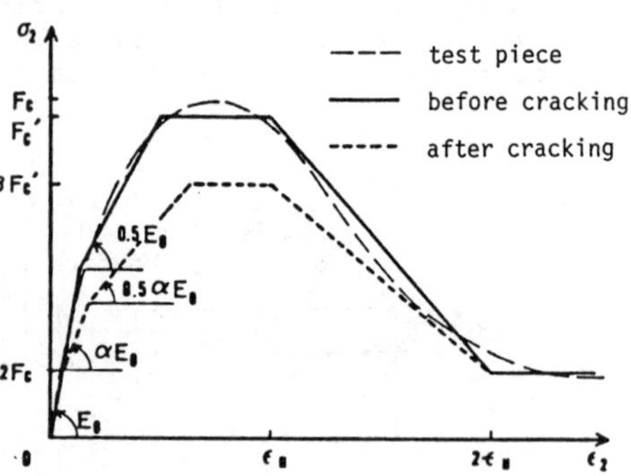

Fig.33 Assumed stress - strain curve of cracked concrete (52)

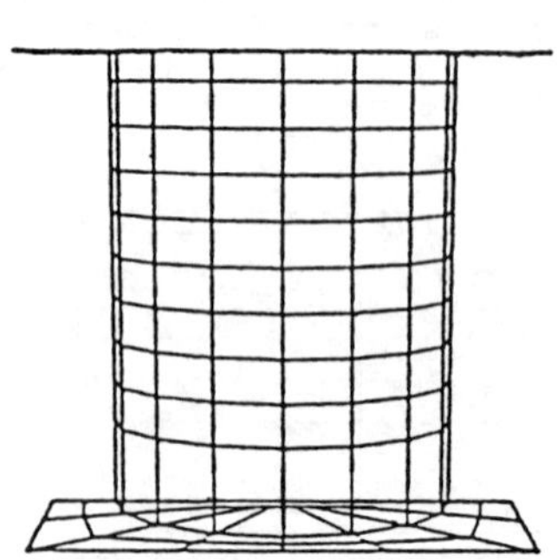

Fig.34 Model of cylindrical shell (25)

complex loading condition, it was proposed that the aggregate interlock after cracking in two directions was modelled precisely, assuming the equivalent shear rigidity and that the previous cracks in two directions closed and an inclined crack occurred newly when the principal stresses in the inclined direction reached the assumed cracking criterion. The results indicated good representation of the experimental behavior although the shear slip failure could not be simulated. This approach seems to be useful when the loading direction changes greatly.

The structures of a nuclear power plant are generally very stiff, but in many cases dynamic studies are needed due to their importance. In such analyses an adequate load-deflection curve has to be assumed and certain suitable formulae based on experimental studies are used. But until now there are not enough experimental data available. Therefore, FEM analyses raise expectations to fill the insufficient points. Furthermore, because the structures have in general very complicated shapes and are under complicated loading combinations due to their functions, the current design formulae might not be applied adequately in some cases. These reasons make the necessity for a FEM analysis very high. Especially in considering that the safety assessment has become necessary during the nonlinear behavior in the worst earthquakes, besides an allowable design, the FEM application for these structures may be the most popular and useful. From the experimental view, because the actual structures are so huge, the test specimens become very small, comparatively. In many cases the scale is reduced to approximately 1/30. Consequently the scale effect should be studied by analyses on the basis of the material properties besides experimental investigations.

7. CLOSING REMARKS

In this paper a description was presented of respective structural specific features. As for the future direction to proceed, it depends on each researcher's opinion whether the development of computer program codes being applied to each structural feature is a better approach or not. But the authors think it is preferable to solve structural members, such as colums, beams, beam-column joints and shear walls, by one program, because they are made from the composite material of reinforcement and concrete with not such difference in size and they could be analyzed by the same model if it was assumed on the basis of the true material property. Furthermore, such a possibility for FEM analyses is its strong point over macroscopic analyses. However, as each member has its own structural feature it seems effective to use a general code in which a suitable option can be selected corresponding with the following points; the discrete crack or the smeared crack, modelling reinforcement by an individual element or by a distributed element, representing the dowel action and the aggregate interlock by a discrete model or by distributed model. Furthermore, when calculating the whole structure it will be necessary to develop a more macroscopic model on the basis of the test results of fundamental members like the Collins panel tests or the analytical results obtained by microscopic FEM analyses. In these analyses it is desirable that better modelling be proposed on the basis of the latest knowledge concerning the dominant factors in the nonlinear behavior of reinforced concrete. Then FEM analyses can contribute to the clarification of actual phenomena and be applied to the actual design by promoting their usefulness.

Finally in considering that similar analytical results are obtained in some cases even though researchers use rather different models individually, it is desirable that they should exchange their knowledge and make a comparison with each other without keeping it on personal basis. Such opportunities are desirable not only domestically but also on a world-wide basis.

REFERENCES

1. Aizawa, S., Tanno, Y., Tohata, Y., Ueda, M., Hayami, H., Yamaguchi, I., "Study on Nonlinear Behavior of Frame with Shear Wall," Proc. of Annual Conv., AIJ, Oct. 1982. pp. 1529-1532 (in Japanese)
2. Arai, Y., "Nonlinear Analysis of a Shear Wall-Frame Systems of Reinforced Concrete," Trans. of AIJ, No. 265, Feb. 1978, pp. 41-49 (in Japanese)
3. Chou, L., Niwa, J. and Okamura, H., "Bond Model of Steel Embedded in Massive Concrete," Proc. of JCI 2nd Colloquium on Shear Analysis of RC Structures, Oct. 1983, pp. 45-52 (in Japanese)
4. Committee on Concrete and Masonry Structures, "State-of-the-Art Report on Finite Element Analysis of Reinforced Concrete," ASCE, 1982
5. Fujita, F., Isobata, O., Kawamata, S. and Tsuboi, Y., "Finite Element Elasto-Plastic Analysis of Thin Walled Structures of Reinforced Concrete as Applied to Reactor Facilities," 6SMIRT, 1981, M 1/3
6. Fukuhara, M., "Examination of Behavior of RC Beam Near Inclined Crack Using FEM Linear Analysis," Proc. of JCI 2nd Colloquium on Shear Analysis of RC Structures, Oct. 1983, pp. 111-118 (in Japanese)
7. Hayashi, S., Kokusho, S. and Morimoto, T., "Experimental Study of Bond Behavior of Reinforcing Bars through the Interior of Beam-Column Joint," Report of the Research Laboratory of Engineering Materials, Tokyo Institute of Technology, No. 9, 1984, pp. 283-297
8. Higashi, Y., Iso, K. and Tachi, M., "Studies on Shear Behavior of Precast Reinforced Concrete Shear Wall-Part 5," Proc. of Annual Conv., Sep. 1983, pp. 1839-1840 (in Japanese)
9. IABSE, "Final Report on Advanced Mechanics of Reinforced Concrete," IABSE Colloquium Delft 1981
10. Imoto, K., "An Analytical Method for Three-Dimensional Nonlinear Responses of Prestressed Concrete Nuclear Reactor Vessels," 6SMIRT, 1981 H3/7
11. Imoto, K., Matsumoto, T. and Tomura, H., "Nonlinear Analysis of Prestressed Concrete Containment Vessel," Proc. of Annual Conv., AIJ, Oct. 1982, pp. 945-946 (in Japanese)
12. Inada, Y., Ogawa, Y. and Saruta, M., "Parametric Study on the Seismic Behavior of 3-Dimensional Shear Walls for the Reinforced Concrete Reactor Building Using Nonlinear FEM Analysis," Proc. of annual Conv., Sep. 1983, (in Japanese)
13. Inoue, N., Koshika, N., Suzuki, N. and Hayami, Y., "Nonlinear Analysis of Reinforced Concrete Panel Subjected to In-Plane Shear by the Finite Element Method," Transaction of JCI, 1984, pp. 485-492

14. Inoue, N., "3-Dimensional Non-linear Analysis of Reinforced Concrete Column under Biaxial Lateral Forces by Finite Element Method," Trans. of AIJ, No. 350, Apr. 1985, pp. 67-76 (in Japanese)
15. Inoue, N., Koshika, N., Suzuki, N. and Hayami, Y., "Nonlinear Analysis Method of Reinforced Concrete Based on Collins' Theory," Proc. of JCI 7th Conference, 1985, pp. 405-408 (in Japanese)
16. Isobata, O., "On an Anisotropic Constitutive Model of Concrete for the FEM Nonlinear Stress Analysis," Trans. of AIJ, No. 265, March 1978, pp. 11-18 (in Japanese)
17. Kano, Y., Takagi, H. and Morinaga, K., "Experimental Study on Strength and Deformation of 3-storied Small Reinforced Concrete Shear Wall-Part 2," Proc. of Annual Conv., AIJ, Sep. 1981, pp. 1653-1654 (in Japanese)
18. Kimura, H., Otsuki, K. and Obata, M., "Nonlinear Analysis of Frame with Infinite Uniform Span by FEM-Part 1," Proc. of Annual Conv., AIJ, Oct. 1970, pp. 771-772 (in Japanese)
19. Kokusho, S., Hayashi, S. and Aya, M., "Study on the Effect of Joint to the Behavior of Infilled Reinforced Concrete Shear Wall," Proc. of Kanto District Symp., AIJ, 1978, pp. 161-164 (in Japanese)
20. Kokusho, S., Hayashi, S., Shinke, H., Iwakura, T. and Honda, Y., "Experimental Study on Reinforced Concrete Beam with Opening," Proc. of Annual Conv., AIJ, Sep. 1981, pp. 1505-1506 (in Japanese)
21. Kokusho, S., Hayashi, S., Kobayashi, K. and Yoshida, H., "Fundamental Study on Bond between Deformed Bar and Concrete -Part 3," Proc. Annual Conv., AIJ, Oct. 1982, pp. 1329-1330 (in Japanese)
22. Maekawa, K., "Constitutive Law of Concrete Based on the Elasto-Plastic and Fracture Theory," Proc. of JCI 2nd Colloquium on Shear Analysis of RC Structures, Oct. 1983, pp. 1-8 (in Japanese)
23. Mitsukawa, E. and Baba, K., "Study on Hysteretic Characteristic of Shear Wall," Proc. of Kanto District Sym., AIJ, 1978, pp. 169-176 (in Japanese)
24. Mochizuki, S. and Kawabe, S., "Experiment of Shear Wall with Surrounding Frame under Reversed Cyclic Pure Shear-Part 3," Proc. of Annual Conv., AIJ, Sep. 1980, pp. 1619-1620 (in Japanese)
25. Morikawa, H., "Nonlinear Analysis of Reinforced Concrete Cylindrical Shell Subjected to Internal Pressure and/or Thermal Loads with Seismic Loads," Dec. 1984, pp. 165-172
26. Murakado, Y., Sotomura, K., Yoshizaki, S. and Ezaki, T., "Strength and Deformation of Reinforced Concrete Shear Wall with Many Small Openings," Proc. of Annual Conv., AIJ, Sep. 1981, pp. 1627-1628 (in Japanese)
27. Muto, K., Sugano, T., Miyashita, T. and Inoue, N., "3-Dimensional Nonlinear Analysis of Reinforced Concrete Columns," Final Report on Advanced Mechanics of Reinforced Concrete, IABSE Colloquium, Delft, 1981, pp. 381-396
28. Niwa, J., "Equation for Shear Strength of Reinforced Concrete Deep Beams Based on FEM Analysis," Proc. of JCI 2nd Colloquium on Shear Analysis of RC Structures, Oct. 1983, pp. 119-126 (in Japanese)
29. Noguchi, H. and Maruta, M., "Nonlinear Finite Element Analysis of Shear Resistance Mechanism of Reinforced Concrete Beams," Proceedings of JCI Colloquium on Shear Analysis of RC Structures, June 1982, pp. 99-106 (in Japanese)
30. Noguchi, H. and Inoue, N., "Analytical Techniques of Shear in Reinforced Concrete Structures by Finite Element Method," Proceedings of JCI Colloquium on Shear Analysis of RC Structures,

JCI, Sep. 1983, pp. 57-92

31. Noguchi, H. and Iizuka, S., "Nonlinear Analysis of the Effect of the Bond Characteristics on the Mechanical Behavior of R/C Short Columns," Proc. Annual Conv., AIJ, Sep. 1983, pp. 2105-2106 (in Japanese)
32. Noguchi, H. and Maruta, M., "Nonlinear Analysis of the Detailing of Reinforcement in R/C Shear Walls with Opening," Proc. of Annual Conv., AIJ, Sep. 1983, pp. 2059-2060 (in Japanese)
33. Noguchi, H. and Naganuma, K., "Nonlinear Analysis of Reinforced Concrete Beam-Column Joints under Reversed Cyclic Loading," Proc. of JCI 2nd Colloquium on Shear Analysis of RC Structures, Oct. 1983, pp. 139-146 (in Japanese)
34. Noguchi, H. and Naganuma, K., "Nonlinear Parametric Analysis of Reinforced Concrete Panels," Proc. of JCI 2nd Colloquium on Shear Analysis of RC Structures, Oct. 1983, pp. 147-154 (in Japanese)
35. Noguchi, H. and Maruta, M., "Nonlinear Analysis of Strength and Ductility of Infilled RC Shear Walls," Proc. of JCI 6th Conference, 1984, pp. 745-748 (in Japanese)
36. Nomura, S., Sato, K. and Ono, H., "Analysis of Hysteretic Behavior of Reinforced Concrete Member by FEM," Proc. Annual Conv., AIJ, Sep. 1978, pp. 1469-1470 (in Japanese)
37. Nomura, S., Sato, K. and Ono, H., "Study on Effect of Bond Characteristic to Hysteretic Behavior of Reinforced Concrete Column," Proc. of Annual Conv., AIJ, Sep. 1980, pp. 1799-1800 (in Japanese)
38. Nomura, S., Sato, K., Zama, Y. and Yamada, T., "Nonlinear Behavior of Reinforced Concrete Structure with Partial Wall," Proc. of Kanto District Sym., AIJ, 1980, pp. 161-168 (in Japanese)
39. Obata, M. and Katsumizu, Y., "Nonlinear Analysis of Reinforced Concrete Beam by the Finite Element Method," Proc. of Annual Conv., AIJ, Sep. 1983, pp. 1781-1782 (in Japanese)
40. Ohwada, Y., "Three-Dimensional Stress Analysis of Reinforced Concrete Beam-Column Joints under Seismic Stress," Proc. of JCI Colloquium on Shear Analysis of RC Structures, June 1982, pp. 107-114 (in Japanese)
41. Ohwada, Y., "Stress Analysis of Reinforced Concrete Beam-Column Joint by Three-Dimensional FEM-Part 2," Proc. of Kanto District Symp., AIJ, 1984, pp. 193-196 (in Japanese)
42. Ono, A., Adachi, H., Nakanishi, M. and Tanaka, N., "Study on Aseismic Capacity of Reinforced Concrete Shear Wall-Part 15," Proc. of Annual Conv., AIJ, Oct. 1976, pp. 1617-1618
43. Ouchi, H., Koike, K., Ito, M. and Takeda, T., "Experimental and Analytical Study on Inelastic Behavior of the RC Cooling Tower Model Subjected to Horizontal Loads," Proc. of JSCE, No. 266, Oct. 1977, pp. 39-59 (in Japanese)
44. Research Committee on Shear Strength of Reinforced Concrete Structures, "Proc. of JCI Colloquium on Shear Analysis of RC Structures," JCI, June 1982 (in Japanese), Sep. 1983, Oct. 1983 (in Japanese)
45. Seya, Y. and Matsui, G., "Study of Stress and Displacement of Shear Wall with Opening-Part 2," Trans. of AIJ, No. 301, March 1981, pp. 101-105 (in Japanese)
46. Shimodaira, H. and Higashi, Y., "Analysis of Reinforced Concrete Beam-Column Joint by FEM," Proc. of Annual Conv., AIJ, Nov. 1971, pp. 843-844 (in Japanese)

47. Shirai, N. and Sato, T., "Bond-Cracking Model for Reinforced Concrete," Proc. of JCI 2nd Colloquium on Shear Analysis of RC Structures, Oct. 1983, pp. 9-16 (in Japanese)
48. Shiraishi, I., Machida, N. and Nakamura, Y., "Experimental and Analytical Study on Prestressed Concret Cylindrical Wall Subjected to Lateral Force," Proc. of Annual Conv., AIJ, Sep. 1983, pp. 1537-1538 (in Japanese)
49. Sugano, T., Inoue, N., Koshika, N., Hironaka, Y. and Hayami, Y., "3-Dimensional Nonlinear Analysis of Reinforced Concrete Columns under Repeated Bending and Shearing Forces," Proc. of JCI 2nd Colloquium on Shear Analysis of RC Structures, Oct. 1983, pp. 87-96 (in Japanese)
50. Sugano, T., Inoue, N., Koshika, N. and Suzuki, N., "Nonlinear Parametric Study of Reinforced Concrete Shear Wall," Proc. of JCI 7th Conference, 1985, pp. 409-412 (in Japanese)
51. Suenaga, Y. and Halim, J.K., "2-Dimensional Nonlinear Stress Analysis of Precast Reinforced Concrete Shear Wall," Proc. of Annual Conv., AIJ, Oct. 1982, pp. 1705-1706 (in Japanese)
52. Ueda, M., Seya, H. and Kei, T., "Nonlinear Analysis of RC Cylinder and Box Type Wall Subjected to Shear Force," Proc. of JCI 2nd Colloquium on Shear Analysis of RC Structures, Oct. 1983, pp. 163-170 (in Japanese)
53. Uchida, T., Miyashita, T., Endo, A. and Morikawa, H., "Model Experiment of Reinforced Concrete Containment Vessel for Nuclear Power Plant under Accidental Internal Pressure and Lateral Force-Part 3," Proc. of Annual Conv., AIJ, Sep. 1978, pp. 1831-1832
54. Umemura, H., Yamada, D., Matsuda, O. and Kamimura, T., "Experimental Study on Reinforced Concrete Beam-Column Joint with Eccentric Wall Girder-Part 7," Proc. of Annual Conv., AIJ, Oct. 1982, pp. 1661-1662 (in Japanese)
55. Watanabe, F., "Mechanism of Shear Failure of R/C Column due to Bond Splitting," Proc. of JCI 2nd Colloquium on Shear Analysis of RC Structures, Oct. 1983, pp. 61-68 (in Japanese)

ON CONSTITUTIVE MODELING OF CONCRETE MATERIALS

W. F. Chen [1], M.ASCE and E. Yamaguchi [2]

1. INTRODUCTION

The rapid advancement of computer technology made the usage of a sophisticated constitutive law possible. Since then, the constitutive modeling of concrete has been an attractive subject to many researchers. Various models have been proposed [13,14] and today the behavior of concrete up to failure can be pursued quite well. However, the simulation of post-failure behavior is still under development. Since some portions of a structural system may reach this range prior to the overall failure, the post-failure behavior cannot be neglected in the simulation study of the response of some concrete structures to loads. Thus, this subject has received a wide attention and is currently being studied very actively [2-6,15,21,36,42,45,46].

In this paper, the fundamental mechanism of concrete is discussed first. In conjunction with the modeling, often the emphasis is placed on the experimental data in terms of stress-strain relationship. It is possible to develop a model without resort to microscopic events. However, the essential factor of mechanical properties is definitely the reflection of material microscopic behavior. Microscopic observations can serve to reason out the stress-strain response and to capture its fundamental characteristics. Hence the study using microscopic observations is of great interest to model the behavior of concrete. To this end, in the following section, special attention is paid to the microscopic mechanism in various loading circumstances rather than to the stress-strain relationship. The study of the stress-strain relationship is also an important issue to quantify the behavior of concrete, and a brief summary on this aspect can be found elsewhere [13,14]. Against this background, together with the theory of solid mechanics, the existing models are reviewed. For the practical application, the models suitable for use in computation, particularly, finite element analysis, are of great importance, and thus only such models are considered. All the discussions are confined to time-independent mechanical properties. Besides, the degradation of elastic moduli in unloading or reloading in the post-failure range is outside the scope of this paper due to the lack of microscopic observations of this phenomenon.

1 Professor and Head of Structural Engineering, School of Civil Engineering, Purdue University, West Lafayette, IN 47907.

2 Research Assistant, School of Civil Engineering, Purdue University, West Lafayette, IN 47907.

2. BASIC MECHANISM OF CONCRETE

Experiments are carried out under a particular loading condition such as uniaxial compression and the results are usually filed according to the loading situation. Therefore in this section, the discussion is organized by loading case.

The study of rock is often helpful in connection with the work on concrete since their mechanical properties are quite similar to each other. Therefore, some researches on rocks are also referred to in the discussions that follow.

2.1 UNIAXIAL COMPRESSION

For a heterogeneous brittle material like concrete, cracking is a dominant factor for its behavior. The development of microcrackings is closely related to the characteristics of stress-strain relation. In the uniaxial compression case, a comprehensive study has been made by Hsu et al. [25].

Three distinct stages are observed in the behavior of concrete. Even before any loads are applied, bond cracks (micro-cracks at the interface between coarse aggregate and mortar) exist. These cracks reveal no extension until the applied load reaches about 30% of the peak value. In this stage, the stress-strain relation is linear. In the next stage, bond cracks begin to increase in length, width and number. The stress-strain curve begins to deviate from a straight line. At about 70 to 90% of the ultimate load, mortar cracks (cracks through mortar) increase noticeably and start bridging between bond cracks to form a continuous crack pattern. This load is associated with that of the minimum volumetric strain. Typical cracking maps are shown in Figs. 1(a) to (c) in which cracks are depicted as heavy lines.

Shah et al. [38] have shown the interesting result that the cement paste itself does not expand under the compression load. The paste specimen continues to consolidate up to failure. Only when it is mixed up with aggregates, volumetric expansion is observed. Aggregate-paste bond strength as well as aggregate size appears to be an essential factor for this inelastic behavior. This indicates that the composite nature of concrete is primarily responsible for the volume dilatation and related to inelastic behavior. In addition, their work has revealed that the load at which the stress-volumetric strain curve deviates from linearity is related to a significant increase in microcrackings at the aggregate-paste interface (bond cracks) and that the stress at which volume begins to increase is related to a noticeable increase of microcracks through the matrix (mortar cracks) and a sharp increase in the lengths of continuous cross-linked microcracks. These observations agree very well with those of Hsu et al. mentioned above.

Based on the experimental observations, Hsu et al. [25] and Krishnaswamy [26] have explained the mechanical process of concrete to failure as follows. Bond cracks alone cannot cause failure since they

(a) Prior to Loading

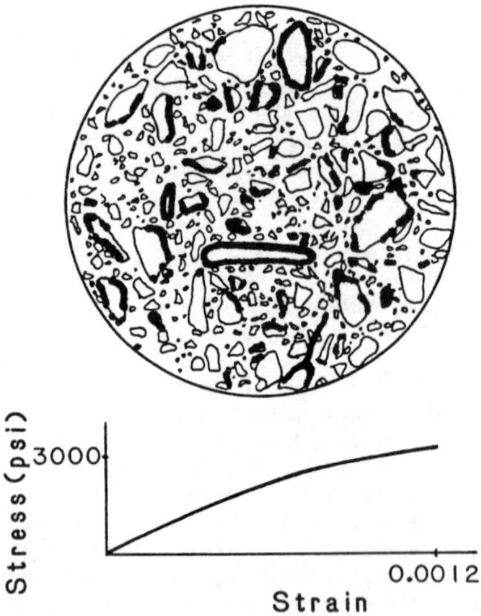

(b) Prior to Peak Stress

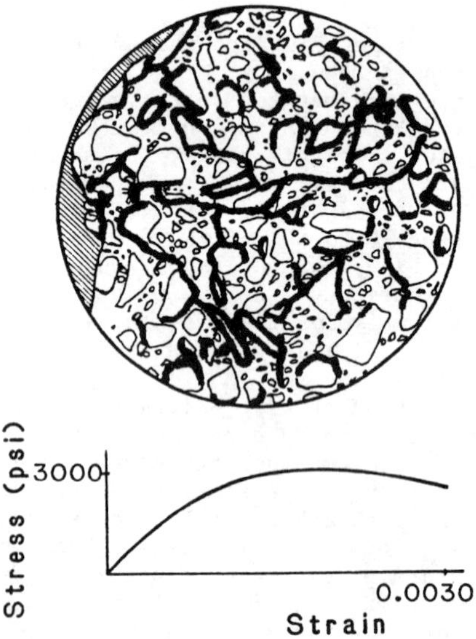

(c) Post Peak Stress

Fig. 1 Cracking Maps Under Uniaxial Compressive Loading
(Hsu et al. 1963)

are separated from one another. Failure occurs only when there are sufficient bond cracks interconnected with mortar cracks. The development of continuous crack patterns does not lead to the immediate loss of load-carrying capacity because concrete at this stage behaves as a highly redundant structure. As successive load paths become inoperative through bond cracking, alternative load paths (either entirely through mortar, or partly through mortar and partly through aggregate) continue to be created and become available for carrying the additional load. As the number of load paths decreases, the intensity of stress and hence the magnitude of strain on the remaining paths increase at a faster rate than the external load. When an extensive continuous crack pattern has developed and the load paths have been reduced considerably, the load-carrying capacity of the concrete decreases and at this stage the stress-strain curve begins to descend.

This explanation implies that the strain-softening behavior is attributed to the extension of continuous cracks, which may be regarded as the reduction of effective cross-sectional area. In fact, Brady et al. [7] have succeeded in explaining the strain-softening characteristics of brittle rocks by the reduction of unfractured area, postulating a maximum true stress which does not change with the increasing axial deformation. This approach can also reason out the degradation of elastic moduli in unloading or reloading in the post-peak region.

From these discussions, it is clear that the descending branch of a stress-strain curve of concrete can no longer be treated as a material property. Rather, it has to be referred to as a structural property. This statement is strongly supported by the work of Van Mier [41], presented in Fig. 2. It has shown that in pre-peak region, stress-strain curves are identical regardless of the heights of specimens, however, that in the post-peak region, they vary considerably.

Thus, three characteristic stages have to be considered in modeling the behavior of concrete, i.e., linear elastic stage, inelastic stage and localized stage. It is noted here that in a uniaxial compression test the failure mode is axial splitting running parallel to the applied load. The process to reach the axial splitting (tensile failure) is depicted in Fig. 3. Hour-glass mode is often encountered in an experiment, but this is probably due to the frictional confinement between the bearing plates and the concrete specimen. Therefore it should not be considered as a real material behavior.

2.2 MULTIAXIAL LOADING

The loading circumstances up to moderate confinement are considered. Moderate confinement refers to the stress state for which the octahedral shear stress, τ_{oct}, is large enough to cause the brittle fracture. Compared with the former case, the crack propagation process is not studied well in concrete materials. However, inferring from the qualitative similarity of stress-strain relation between under uniaxial compression and under multiaxial loading, again, three stages may be required for modeling the behavior of concrete. The major difference

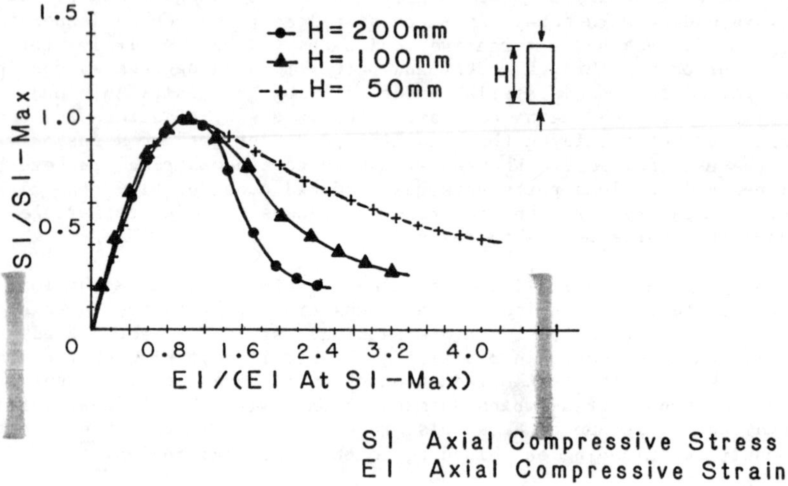

Fig. 2. Influence of the Specimen Height on Uniaxial Stress-Strain Curve (Van Mier, 1984)

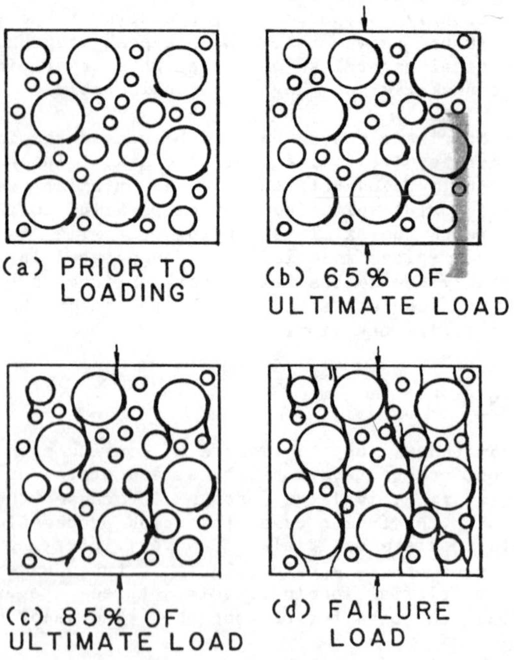

Fig. 3 Crack Pattern Obtained from X-ray Examination (Uniaxial Loading) (Liu et al. 1972)

from uniaxial compression case arises in the post-failure behavior. Macroscopic observations show that there exist two failure modes. One of them is axial splitting failure which is the same as observed under uniaxial compression and the other is shear failure. In spite of the name, shear failure, the microscopic study on rock by Hallbauer et al. [20] has indicated that this failure is more likely the product of tensile crackings because the fracture plane is produced by stepwise joining at the existing microcracks (the areas of high crack densities). The essential difference is that shear failure is formed by the interaction of small flaws and the tension cracks emanating from these flaws, whereas axial splitting is caused by the growth of cracks extending from the large flaws, according to Horii et al. [24]. Wong's work [48] is also noteworthy. He worked on the faulting mechanism in Westerly granite under the compressive pressure by scanning electron microscope observations. His work has revealed that there exists the strain localization almost immediately after shear failure occurs. In other words, the deformation in the post-shear failure sample is localized. This implies that strain-softening behavior after shear failure may not be a material property, but a structural property as is the case for axial splitting failure.

Horii and Nemat-Nasser [24,31] have shown that in the presence of lateral confinement, compressive loading initiates shear failure while in its absence or in the presence of tensile stress normal to the direction of maximum compression, axial splitting failure takes place. They have estimated also the orientation of shear failure plane analytically. However, their method is not so simple for practical application. On top of it, the systematic experimental data on the failure modes is insufficient. Therefore the prediction of shear failure direction may not be made with a good accuracy. Observe the substantial scatter in the experiment data of Westerly granite by Wawersik et al. [43], Fig. 4, although there is probably the trend that the gradient of fracture angles is greater in low confinement.

2.3 MISCELLANY

Lastly, the cases under high confinement and under simple tension are briefly mentioned. High confinement is defined to be the stress state for which the octahedral shear stress, τ_{oct}, is kept so small that concrete does not collapse. If the applied loads are large enough under this condition, concrete behaves like a ductile material without collapse. This corresponds to the fact that the failure surface is not closed. As an example, Willam-Warnke five-parameter failure model with experimental results is presented in Fig. 5, in which ρ stands for the distance from the hydrostatic axis in principal stress space ($\rho = \sqrt{2J_2}$), f_c' a uniaxial compressive strength and σ_{oct} octahedral normal stress component. In rock materials, Mogi [30] has obtained the brittle-ductile boundary from the experimental data. Nemat-Nasser et al. [32] have demonstrated this transition phenomenon from brittle fracture to ductile flow analytically . Their results are presented in Fig. 6 in which it is clearly observed that under high confinement a crack length (l_t) stops growing while a plastic length (l_p) continues to

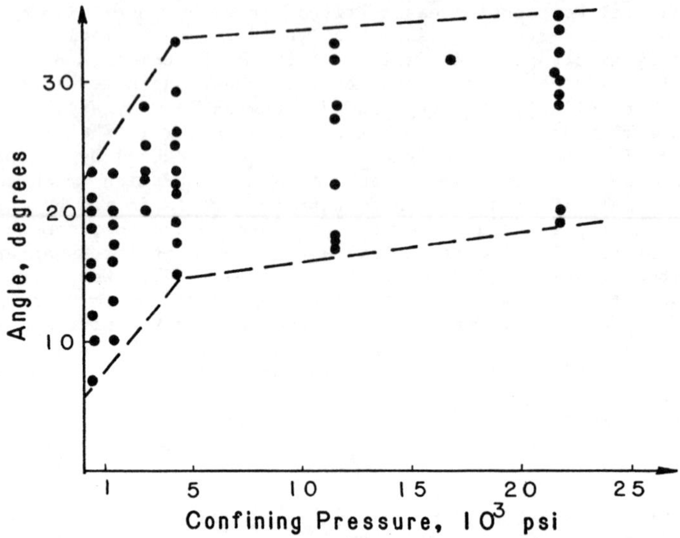

Fig. 4 Orientation of Macroscopic Failure Zone Relative to the Direction of Greatest Compression in Westerly Granite (Wawersik et al. 1971)

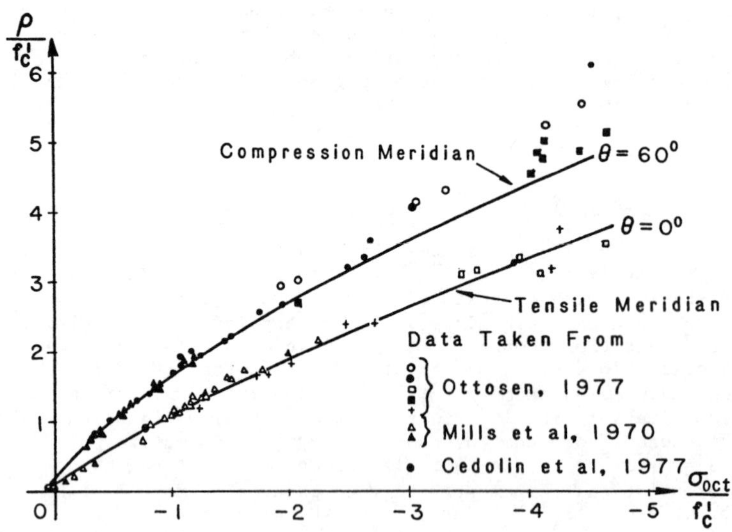

Fig. 5 Willam-Warnke Five Parameter Failure Model (Han et al. 1985)

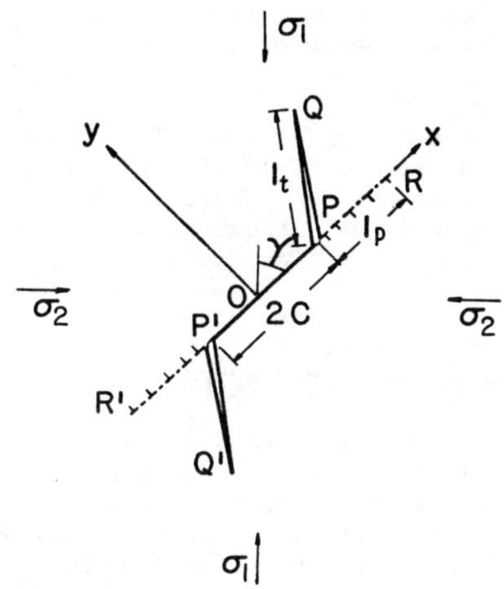

(a) Straight cracks PQ and P'Q' and plastic zones PR and P'R' at the tips of a pre-existing flaw PP'

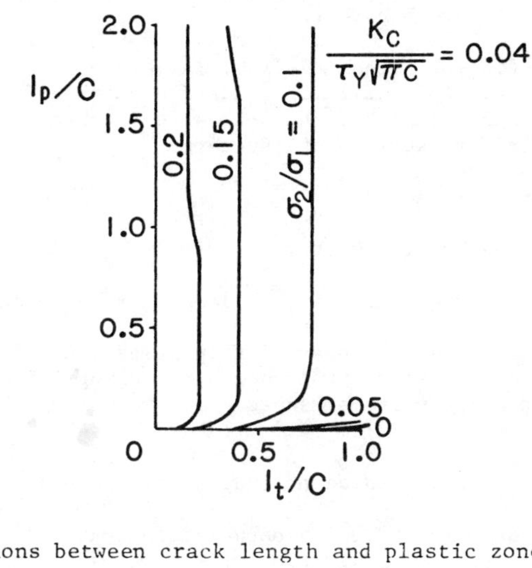

(b) Relations between crack length and plastic zone size

Fig. 6 Transition from Brittle Fracture to Ductile Flow (Nemat - Nasser et al., 1985) increase. Their results agree well with Mogi´s work. In this loading condition, strain-softening behavior is not observed.

Under a simple tension test, strain-softening behavior is reported and obviously tensile cracking occurs. This strain-softening behavior may be attributed to the reduction of effective cross-sectional area and thus not a material property. Characteristically, in this case, stress-strain relation is almost linear up to a peak stress point, Fig. 7. Hence inelastic behavior is missing.

3. MODELING OF THE BEHAVIOR OF CONCRETE

The models developed for a general stress state are preferred to those applicable only to a specific loading condition, since any concrete element in an actual structure is more likely to be subject to complex multiaxial stress state. In general, the behavior of concrete changes, once the failure occurs. Particularly after failure, the stress starts decreasing as the strain increases. This makes a sharp contrast to the pre-failure behavior for which Drucker´s stability postulate holds. To this end, the review of existing models is made in a different manner from that in the previous section. Namely, the discussion is divided into two parts; one is for pre-failure behavior and the other for post-failure behavior.

In conjunction with the modeling, failure criterion is an important issue. In fact, the failure criterion has been studied quite well. Experimental results indicate the following features in principal stress space for the failure surface of concrete.

In the deviatoric plane:

1. The failure curve is smooth.

2. The failure curve is convex, at least for compressive stresses.

3. The failure curve is nearly triangular for tensile and small compressive stresses and becomes increasing bulged (more circular) for higher compressive stresses.

In the meridian plane:

1. The failure curves depend on hydrostatic component of stress, σ_m.

2. The meridians are curved, smooth and convex.

3. The ratio ρ_t/ρ_c is less than one, where indexes t and c correspond to the tensile and the compressive meridian, respectively.

4. The ratio ρ_t/ρ_c increases with increasing hydrostatic pressures. It has a value of about 0.5 near the Π - plane and reaches a high value of about 0.8 near the hydrostatic pressure $\zeta/f_c' = -7$ where ζ is a distance from the origin along the hydrostatic axis in the

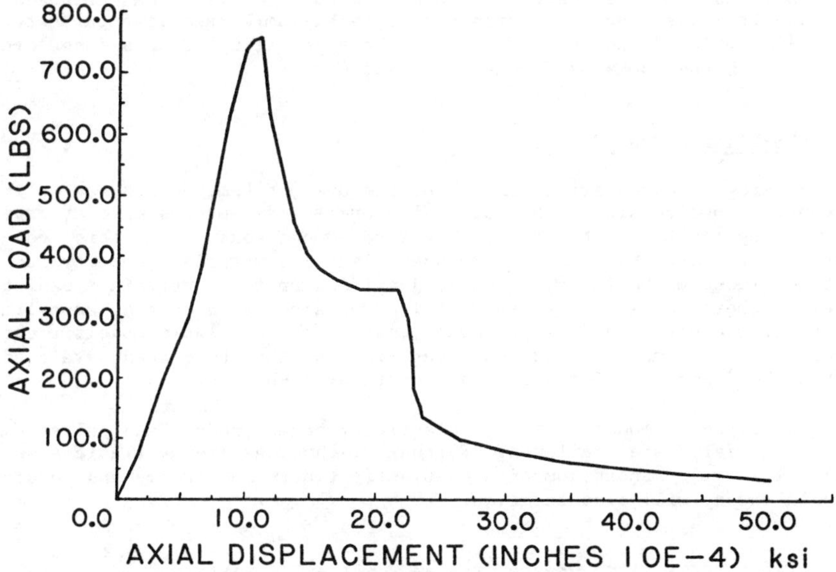

Fig. 7 Load-elongation Diagram of Mortar Specimen (Cylinder Size 5.40 x 11.43 cm) (Willam et al. 1984)

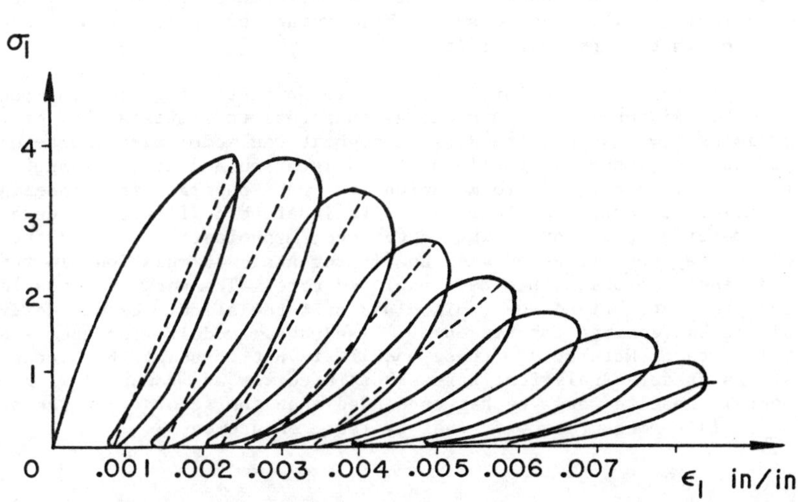

Fig. 8 Stress-Strain Curve under Cyclic Uniaxial Compressive Loading (Sinha et al. 1964)

principal stress space ($\zeta = I_1 / \sqrt{3} = \sqrt{3}\sigma_{oct}$).

5. Pure hydrostatic loading cannot cause failure.

Various failure models describing these features have been proposed. One of the most sophisticated models is Willam-Warnke five-parameter model [47] which is shown in Fig. 5. More detailed discussions are available in the books by Chen et al. [13,14].

3.1 PRE-FAILURE BEHAVIOR

Heterogeneous materials like concrete are not homogeneous at all. However, if one considers the material property in a macroscopic average sense, they may be treated as if they were homogeneous. In this context, microcrackings can be imbedded also in a material property such that continuum mechanics approach is justified up to a certain loading level. There exist some interesting research based on the composite nature of concrete [33,34]. However, most of the available models treat concrete as a simple, homogeneous material and they have been developed within the framework of traditional continuum mechanics.

Two types of models, i.e., elasticity-based model and elastic-plastic model, are reviewed. Further discussions are available elsewhere [1,13,14]. Short comments on plastic-fracturing theory and endochronic theory will also be given.

3.1.1 ELASTICITY-BASED MODELS

Linear isotropic elasticity is a good approximation for the behavior of concrete under the tensile type of force, since in this situation a linear relation is observed almost up to failure. This is also a good model for the material in the initial stage of other loading condition. It becomes poorer, however, as the applied loads increase. Nonlinear elastic model improves this drawback drastically. This can be achieved by using the theory of hyperelasticity or that of hypoelasticity.

The hyperelastic formulation is the one starting from the postulate of the existence of strain energy such that the satisfaction of thermodynamics laws is guaranteed. Although it can model many characteristics of concrete such as nonlinearity, dilatation and stress-induced anisotropy, this class of formulation is not popular for modeling the behavior of concrete, because of its inherent difficulty in determining the material constants. The latter one, hypoelasticity-based modeling, which may be regarded as an incremental formulation in the time-independent problem, has been explored more. The work of Gerstle [18] can be categorized by this class of formulation. He has derived the simple expressions for tangent bulk and shear moduli from the experimental data. Naturally, these moduli are not constant, but functions of stress state. Analytical stress-strain curves are compared with experimental results and he has concluded that the accuracy of the proposed model is commensurate with that of test measurements.

However, the hypoelastic type of formulation encounters an inherent difficulty in the construction of its constitutive relations. In the nonlinear range, concrete becomes anisotropic, even if the initially linear behavior is isotropic. This anisotropy implies that the behaviors in the principal-stress directions are different from each other and that the principal axes of stress and strain are different, introducing coupling between normal stresses and shear strains. This means that for a general triaxial state of stress, 21 material moduli have to be defined for each point of the material-loading path - an insurmountable task for practical application.

The models mentioned so far are primarily directed toward monotonic (not necessarily proportional) loading regimes wherein no distinction need be made between loading and unloading behavior. To account for the unloading behavior, the most common approach has been to arbitrarily introduce a loading criterion and to use different stress-strain relations for loading and unloading behaviors. The popular model in this class is the one known as variable moduli models and used extensively for soils. The defect of this model is visible at or near the neutral loading. Due to the inconsistency of the formulation, discontinuity or jump in strains may appear between loading and unloading behaviors near the neutral loading path. This difficulty can be overcome by the application of plasticity concept.

3.1.2 PLASTICITY-BASED MODELS The more phenomena one tries to capture, the more material constants one may have to deal with as long as the general technique used in the elasticity-based modeling is followed. In plasticity-based modeling, the number of unknowns is significantly reduced, which is credited to the postulated rules for inviscid elastic-plastic materials with work-hardening. The application of plasticity concept, based on the dislocation theory of metals, to concrete is criticized in that the inelastic behavior predicted by a plasticity-based model is not accompanied by the degradation of elastic moduli, i.e., the decrease of the unloading stiffness. This degradation is sure to be observed in experiments. A typical stress-strain relationship under a cyclic loading is presented in Fig. 8. This phenomenon may be modeled by the plastic-fracturing model [5]. However, it is noted that this degradation is not crucial in the pre-failure region while it is eminent in the post-failure region. Therefore, the application of the plasticity concept in the pre-failure region is not eliminated by the criticism mentioned above. Recalling the evidence of microscopic observation, the degradation of elastic moduli is found to be closely related to the extension of mortar cracks.

One of the best-known plasticity-based models is the one by Chen et al. [11] in which the associated flow rule is applied. For simplicity, this rule is frequently used [8,15] and the model generally yields a good agreement with experimental results. The major deviation, however, is observed in the volumetric strains. Volume expansion tends to be overestimated. Therefore, the application of non-associated flow rule has been introduced [9,21,22,42]. It has succeeded in controlling the volumetric strains, but much more computational efforts are required in finite element analysis since the stiffness matrix is no longer symmetric.

Another problem at hand is the shape of loading surface. Often it is assumed that the loading surfaces are similar to the shape of failure surface for simplicity [8,11,15,42]. Consequently, the loading surface is open toward the hydrostatically compressive direction. The typical characteristics are found in Fig. 9. However, it seems unrealistic that a material like concrete shows a linearly elastic response along a certain stress path even under rather large magnitude of loads. In fact, some experimental results have indicated a closed elastic limit [27]. The work of Han et al. [21,22] is one of the earliest efforts employing a closed loading surface. The meridians of this loading surface are shown in Fig. 10. Evolution of the loading surface is also presented. It uses a nonuniform hardening rule and not an isotropic hardening rule adopted usually. Stankowski et al. [40] have also used a closed loading surface, although their formulation itself is different from the plasticity-based model.

Once these factors are determined, the development of a constitutive model follows immediately the standard formulation in plasticity theory, which can be found in many books, for example the book by Chen [13]. In general, the plasticity-based models with non-associated flow rule can describe the stress-strain relationship fairly well. A typical result of such modeling is presented in Fig. 11.

3.1.3 PLASTIC-FRACTURING THEORY AND ENDOCHRONIC THEORY In the plastic-fracturing theory the inelastic behavior is attributed to two sources, plastic slip and microcrackings (or faulting). The former is formulated by the conventional plasticity concept while the latter is modeled by using the potential function in strain space. This second concept was initiated by Dougill [16,17]. Very good reproduction of the nonlinear multiaxial behavior of concrete has been reported by Bazant et al. [5]. However, much complexities in defining two loading functions are inevitable.

The endochronic theory was first applied to concrete by Bazant et al. [3]. The basic concept in this theory is that of intrinsic time, playing a similar role of effective plastic strain in the plasticity theory. The intrinsic time is the non-decreasing scalar variable representing the evolution of the measure of irreversible damage. This model can cover many phenomena like nonlinear behavior, inelastic volume dilatancy, hydrostatic pressure sensitivity, etc. This can be achieved, however, only at the expense of greater complexity and increasing number of material parameters.

3.2 POST-FAILURE BEHAVIOR

Brittle material like concrete fails in two different modes, tensile cracking (axial splitting) failure and macroscopic shear failure. These two modes are quite different in terms of post-failure behavior.

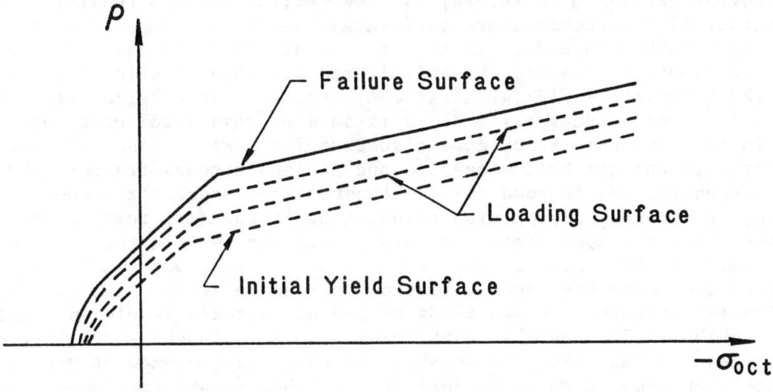

Fig. 9 Chen-Chen Model (Chen et al. 1975)

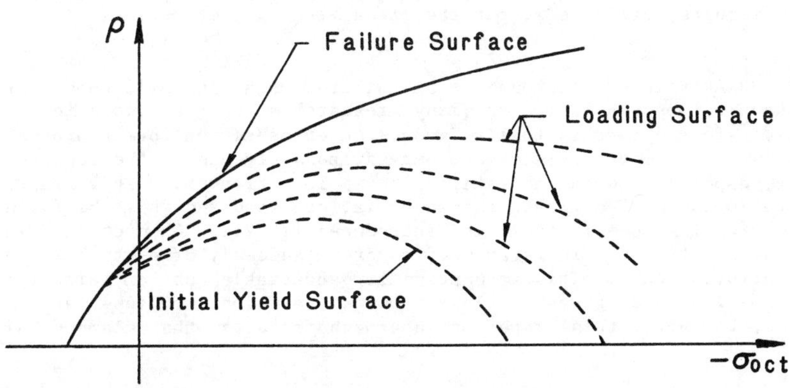

Fig. 10 A Nonuniform Hardening Plasticity Model (Han et al. 1985)

In tensile cracking, the resistance drops to zero as the crack propagates whereas in shear failure some resistance due to friction can be expected. Besides, as pointed out earlier, the formation (or propagation) process of a failure zone is different in each of these two modes. Prior to shear failure, a high-density crack zone is observed. It is this region which leads to macroscopic shear failure. On the other hand, the extension of an individual crack is a dominant factor in tensile cracking failure. Therefore, in the following, discussions are made separately for these two failure modes.

3.2.1 TENSILE CRACKING (AXIAL SPLITTING) FAILURE The degradation of strength due to cracks is modeled in several ways. The most popular method is to treat the cracking as distributed cracks on the continuum level, i.e., the cracks are smeared out in a continuous fashion. This is the easiest method in numerical implementation, especially when the direction of crack propagation is unknown. Naturally, this method is associated with the modification of elastic constants. In the simplest form, some elastic moduli are set equal to zero when the collapse (crack) occurs. However, such a simple application using the failure criterion in terms of strength leads to the unobjective result in finite element analysis [2]. As the finite elements are made smaller, the tensile stress in the element just ahead of the crack becomes larger such that the crack grows at a lower load level. This superficial dependence on the chosen finite element size has been overcome through the introduction of fracture mechanics concept by Bazant et al. [4]. Instead of strength failure criterion, they have employed the energy criterion which is a common practice in fracture mechanics. The applicability of linear fracture mechanics to concrete is confined to a large scale structure [44], however, the nonlinear fracture theory for concrete developed by Bazant et al. [6] can be applied also to a small structure. The constitutive model thus obtained has a branch of gradual stress drop. This strain-softening portion represents the fracture process zone as well as the crack propagation in an element. The softening modulus varies in accordance with the finite element size so that size effect diminishes. Some results obtained by this approach are shown in Fig. 12 where d_a is a maximum aggregate size, P_o failure load based on strength criterion applied to uncracked cross section and P_{max} a maximum applied load. As is observed, the nonlinear theory reproduces experimental results fairly well, but the linear theory does not.

3.2.2 SHEAR FAILURE Although shear failure mode is well-known from macroscopic observation, not many researches on the modeling have clearly referred to this failure mode. Under nearly uniform hydrostatic stresses, crushing failure is observed, possibly due to the rupture of mortar, especially when the material has a low strength. This rupture failure is found also within the shear failure band and thus the failure termed crushing mode is sometimes introduced [15,21]. Chen et al. [15] have assumed that the resistance of concrete suddenly drops to zero when this failure occurs. This assumption is acceptable as a fair first approximation and gives a lower bound solution. Unless carefully treated, however, this type of approach creates the element size

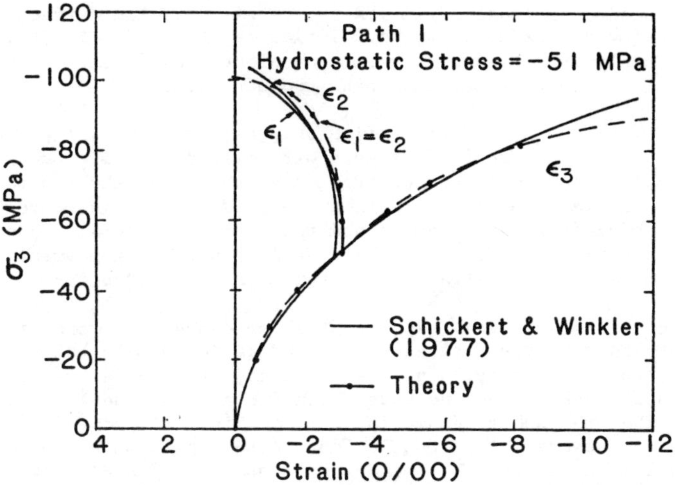

Fig. 11 Results of the Plasticity-Based Model for Triaxial Nonproportional Loading (Han et al. 1985)

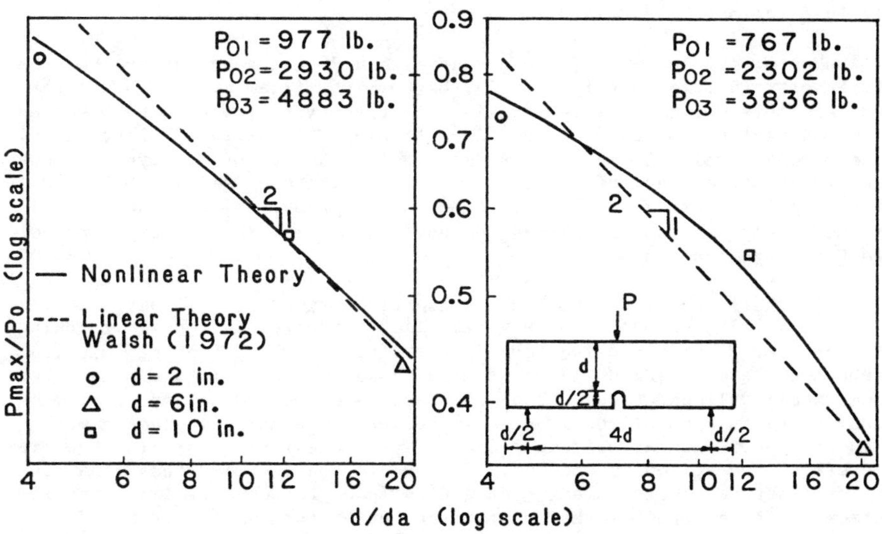

Fig. 12 Results of the Fracture Mechanics Approach on Notched Beams of Various Sizes (Bazant et al. 1981)

dependency mentioned previously. As its counterpart, if the perfectly plastic flow is allowed, the upper bound solution is obtained. In this case, a limit analysis serves as a very convenient tool [12,13]. The actual behavior lies somewhere between these two approaches.

Bazant et al. [5] and Han et al. [21] have obtained the results in strain-softening range in good agreement with available experimental data. Their formulations are different from each other, but the introduction of a potential function in strain space is common. A conventional plasticity concept accompanies it and thus two potential functions have been involved. In spite of their good results, a fundamental question has to be raised. Is strain-softening a material property?

As discussed earlier, the experimental evidence of uniaxial compression test indicates that strain-softening is not a material property, but a structural property. In the essence, the stress-strain curve in the post-peak range depends significantly on the specimen height, while in the pre-peak range it is not. This is for tensile cracking failure. However, this conclusion can be applied also to the shear failure case if one recalls the experimental evidence that a localized deformation is observed after shear failure. The similar discussion has been made by Hegemier et al. [23]. Also from the different aspect, the same conclusion has been derived by Willam et al. [45].

If this is the case, the stress-strain relationship in the softening range is merely a nominal property. Then, the results of Bazant et al. [5] as well as those of Han et al. [21] are not physically meaningful since their models yield the stress-strain relationship without resort to specimen size. Any other efforts along this line are similarly questionable.

Hence, the load-displacement curve should be traced rather than the stress-strain relationship. It is the work of Willam et al [45,46] that strikes this point. Their model is one of the very few efforts based on the understanding that a strain-softening is a structural property. Their model, called composite damage model, consists of two types. One is for tensile cracking and the other for frictional slip. Their model for tensile cracking was not reviewed in the previous section (3.2.1) since their approach follows a unified framework for both types of post-failure behaviors.

In their model, damage zone is clearly defined as illustrated in Fig. 13. It embodies highly localized fracture (tensile cracking failure) or distributed fracture (shear failure). Different material behaviors are assumed in a damage zone and in an intact zone. For numerical implementation, these two phases are smeared out in an element and thus this model comes up as an equivalent continuum element with a linear elastic-linear softening law. The immediate task is how to determine the softening modulus. Through the parametric study, the constant fracture energy release concept is found to serve well in tensile cracking case whereas the constant softening modulus is recommended in shear failure case. Then they have concluded that frictional slip type fracture takes place in cementitious materials in a far more distributed fashion than tensile cracking [46]. This conclusion is expected from the aforementioned experimental results. This new approach, however, is still a first order approximation for the description of a post-failure behavior. Further refinement is needed, for instance, inelastic dissipation mechanism must be taken into account. The distinction of two failure modes is a crucial matter. Mohr-Coulomb failure criterion with tension cutoff is adopted to identify the failure mode in their study. However, this criterion is rather rough. It may predict the shear failure under the biaxial loading of tension and compression as long as the maximum tensile strength is not reached. This is not justified by the experimental evidence usually reported.

4. CONCLUDING REMARKS

The discussion in Sec. 2 indicates that the behavior of concrete can be classified into three stages.

i) Linear Elastic Behavior

ii) Inelastic Behavior

iii) Localized Behavior

Strain-softening is a structural property so that it cannot be considered as a stage of material behavior. A simple tension test does not reveal an inelastic behavior and concrete under high confinement shows no strain-softening behavior. Under high confinement, plastic deformation becomes dominant after some crack extension. This behavior should be in the position parallel to localized behavior. However, in terms of stress-strain relationship no clear-cut difference between inelastic behavior (due to the extension of microcracks) and plastic deformation (due to plastic slip) has been made. In fact, the experimental data under high confinement is insufficient. Therefore, no further refinement need be made for the purpose of modeling. These three stages are seen in a stress-strain relationship (Fig. 14), although the boundaries are not always obvious. Besides, they are classified by the degree of crack extension as described earlier. Hence, it is essential that all these three stages be imbedded in the constitutive modeling to pursue a simulation of the complete behavior.

Due to the presence of a large number of constituents, it is impractical to analyze a heterogeneous material like concrete by retaining the exact geometry of all the constituents. From the macroscopic viewpoint, concrete material is usually modeled as an isotropic homogeneous material since constituents are expected to lie randomly. Thus at a low stress level, a linear isotropic elasticity model is accepted exclusively.

In the next stage, inelastic behavior, the best compromise between accuracy and complexity may be attained by a plasticity-based model. This model can capture many essential features of concrete fairly well.

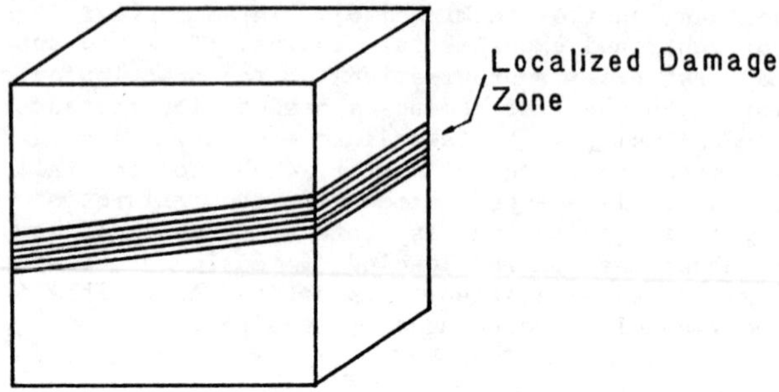

Fig. 13 Composite Damage Model (Willam et al. 1984)

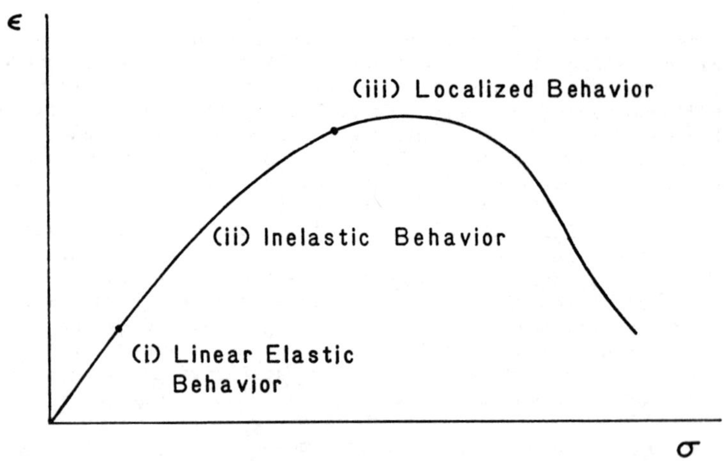

Fig. 14 Typical Stress-Strain Curve

Sophisticated models like the plasticity-fracturing model and endochronic model, can yield better predictions. However, they have to deal with greater complexity and more material parameters. Considering the scatter of experimental data (see the interesting report by Gerstle et al. [19], the high accuracy is not very urgent. Rather, simplification would be preferred.

In the localized behavior, the stress-strain relationship is a nominal property. Therefore, the effort to trace the stress-strain relationship is ruled out unless the structural effect is reflected by some means. The fracture mechanics approach has been applied successfully to the tensile cracking case. On the other hand, the model for the post-shear failure behavior is still under development, although the composite damage model has opened a new insight. The behavior of concrete under general loading is an important issue. However, the extension of the composite damage model, developed for plane strain condition, to three dimensional stress state, is not an easy task, due to the lack of experimental data. For example, the orientation of a shear failure plane is not defined well. The experimental work on the failure mode is not adequate. The experimental data on the ductile behavior of concrete is also insufficient. Further, the microscopic observations of concrete under the cyclic loading have not been reported much, especially in the post-failure range, so that the discussion here for localized behavior is associated just with monotonic loading. To explore a more comprehensive model, these experimental data are urgently needed.

5. REFERENCES

1. ASCE Committee on Concrete and Masonry Structures, A State-of-the-Art Report on Finite Element Analysis of Reinforced Concrete, Task Committee on Finite Element Analysis of Reinforced Concrete Structures, ASCE Spec. Publ., New York, 1981.

2. Bazant, Z.P., Instability, Ductility and Size Effect in Strain-Softening Concrete, J. Eng. Mech. Div., ASCE, Vol. 102, EM2, April 1976, pp. 331-344.

3. Bazant, Z.P. and Bhat, P.D., Endochronic Theory of Inelasticity and Failure of Concrete, J. Eng. Mech. Div., ASCE, Vol. 102, EM4, August 1976, pp. 701-722.

4. Bazant, Z.P. and Cedolin, L., Blunt Crack Band Propagation in Finite Element Analysis, J. Eng. Mech. Div., ASCE, Vol. 105, EM2, April 1979, pp. 297-315.

5. Bazant, Z.P. and Kim, S., Plastic-Fracturing Theory for Concrete, J. Eng. Mech. Div., ASCE, Vol. 105, EM3, June 1979, pp. 407-428.

6. Bazant, Z.P. and Oh, B.H., Concrete Fracture via Stress-Strain Relations, Report No. 81-10/665C, Center for Concrete and Geomaterials, The Technological Institute, Northwestern University,

Evanston, IL, 1981.

7. Brady, B.T., Duvall, W.I. and Horino, F.G., An Experimental Determination of the True Uniaxial Stress-Strain Behavior of Brittle Rock, Rock Mechanics 5, 1973, pp. 107-120.

8. Buyukozturk, O., Nonlinear Analysis of Reinforced Concrete Structures, Computers and Structures, Vol. 7, 1977, pp. 149-156.

9. Buyukozturk, O., A Constitutive Model for Concrete in Compression, Proc. 3rd Eng. Mech. Div. Spec. Conf., ASCE, Austin, TX, 1979.

10. Cedolin, L., Crutzen, Y.R.J. and Dei Poli, S., Triaxial Stress-Strain Relationship for Concrete, J. Eng. Mech. Div., ASCE, Vol. 103, EM3, June 1977, pp. 423-439.

11. Chen, A.C.T. and Chen, W.F., Constitutive Relations for Concrete, J. Eng. Mech. Div., ASCE, Vol. 101, EM4, August 1975, pp. 465-481.

12. Chen, W.F., Limit Analysis and Soil Plasticity, Elsevier, Amsterdam, 1975.

13. Chen, W.F., Plasticity in Reinforced Concrete, McGraw-Hill, New York, 1982.

14. Chen, W.F. and Saleeb, A.F., Constitutive Equations for Engineering Materials, Vol. 1 - Elasticity and Modeling, Wiley-Interscience, New York, 1982.

15. Chen, W.F. and Suzuki, H., Constitutive Models for Concrete, Computers and Structures, Vol. 12, 1980, pp. 23-32.

16. Dougill, J.W., Some Remarks on Path Independence in the Small in Plasticity, Q. of Applied Math., Vol. 32, October 1975, pp. 233-243.

17. Dougill, J.W., On Stable Progressively Fracturing Solids, Zeitschriftfuer Angewandte Mathematik und Physik (ZAMP), Vol. 27, 1976, pp. 423-437.

18. Gerstle, K.H., Simple Formulation of Biaxial Concrete Behavior, ACI Journal, Vol. 78, 1981, pp. 62-68.

19. Gerstle, K.H., Aschl, H., Bellotti, R., Bertacchi, P., Kotsovos, M.D., Ko, H.Y., Linse, D., Newman, J.B., Rossi, P., Schickert, G., Taylor, M.A., Traina, L.A., Winkler, H. and Zimmerman, R.M., Behavior of Concrete under Multiaxial Stress States, J. Eng. Mech. Div., ASCE, Vol. 106, EM6, December 1980, pp. 1383-1403.

20. Hallbauer, D.K., Wagner, H. and Cook, N.G.W., Some Observations Concerning the Macroscopic and Mechanical Behavior of Quartzite Specimens in Stiff, Triaxial Compression Tests; Int. J. Rock Mech. Min. Sci & Geomech. Abstr., Vol. 10, 1973, pp. 713-726.

21. Han, D.J. and Chen, W.F., Constitutive Modeling in Analysis of Concrete Structures, J. Eng. Mech. Div., ASCE, Vol. 112, 1986. (To appear).

22. Han, D.J. and Chen, W.F., A Nonuniform Hardening Plasticity Model for Concrete Materials, J. Mech. of Materials, Vol. 4, No. 4, December 1985.

23. Hegemier, G.A. and Read, H.E., On Some Outstanding Issues Pertaining to the Nonlinear Static and Dynamic Response of Rock and Concrete, Presented at Workshop on Inelastic Deformation and Failure Modes, Northwestern University, Evanston, IL, November 18-21, 1984.

24. Horii, H. and Nemat-Nasser, S., Compression-Induced Micro-Crack Growth in Brittle Solids: Axial Splitting and Shear Failure, J. Geophys. Res., Vol. 90, pp.3105-3125, 1985.

25. Hsu, T.T.C., Slate, F.O., Sturman, G.M. and Winter, G., Microcracking of Plain Concrete and the Shape of the Stress-Strain Curve, ACI Journal, Vol. 60, 1963, pp. 209-223.

26. Krishnaswamy, K.T., Microcracking of Plain Concrete Under Uniaxial Compressive Loading, Indian Concrete Journal, Vol. 66, April 1969, pp. 143-145.

27. Launay, P. and Gachon, H., Strain and Ultimate Strength of Concrete Under Triaxial Stresses, Special Publ., Sp-34, ACI, 1, 1970, pp. 269-282.

28. Liu, T.C.Y., Nilson, A.H. and Slate, F.O., Stress-Strain Response and Fracture of Concrete in Uniaxial and Biaxial Compression, ACI Journal, Vol. 69, 1972, pp. 291-295.

29. Mills, L.L. and Zimmerman, R.M., Compressive Strength of Plain Concrete Under Multiaxial Loading Conditions, ACI Journal, Vol. 67, October 1970, pp. 802-807.

30. Mogi, K., Pressure Dependence of Rock Strength and Transition from Brittle Fracture to Ductile Flow, Bulletin of the Earthquake Research Institute, Vol. 44, 1966, pp. 215-232.

31. Nemat-Nasser, S. and Horii, H., Compression-Induced Nonplanar Crack Extension with Application to Splitting, Exfoliation and Rockburst, J. Geophys. Res., Vol. 87, 1982, pp. 6805-6821.

32. Nemat-Nasser, S. and Horii, H., Micro-Mechanics of Fracture and Failure of Geo-Materials in Compression, Advances in Fracture Research, Proc. of the ICF Sixth Int. Conf. on Fracture, New Delhi, India, 4-10 December 1984, Pergamon Press, 1985, pp. 515-524.

33. Ortiz, M., A Constitutive Theory for the Inelastic Behavior of Concrete, J. Mech. of Materials, 1985 (To appear).

34. Ortiz, M. and Popov, E.P., Plain Concrete as a Composite Material, Journal of Mechanics of Materials, Vol. 1, 1982, pp. 139-150.

35. Ottosen, N.S., A Failure Criterion for Concrete, J. Eng. Mech. Div., ASCE, Vol. 103, EM4, August 1977, pp. 527-536.

36. Reinhardt, H.W., Fracture Mechanics of an Elastic Softening Material Like Concrete, Heron, Vol. 29, No. 2, 1984.

37. Schickert, G. and Winkler, H., Results of Test Concerning Strength and Strain of Concrete Subjected to Multiaxial Compressive Stress, Deutscher Ausschuss Fur Stahlbeton, Heft 277, Berlin, 1977.

38. Shah, S.P. and Chandra, S., Critical Stress, Volume Change, and Microcracking of Concrete, ACI Journal, Vol. 65, 1968, pp. 770-781.

39. Sinha, B.P., Gerstle, K.H. and Tulin, L.G., Stress-Strain Relations for Concrete under Cyclic Loading, ACI Journal, Vol. 61, 1964, pp. 195-211.

40. Stankowski, T. and Gerstle, K.H., Behavior of Concrete Under Multiaxial Load Histories, RILEM-CEB-CNRS, International Conference on Concrete Under Multiaxial Conditions, Vol. 1, Presses de l'Universite' Paul Sebatier, Toulouse, France, May 1984, pp. 124-132.

41. Van Mier, J.G.M., Complete Stress-Strain Behavior and Damaging Status of Concrete under Multiaxial Conditions, RILEM-CEB-CNRS, Int. Conf. on Concrete under Multiaxial Conditions, Vol. 1, Presses de l'Universite' Paul Sabatier, Toulouse, France, May 1984.

42. Vermeer, P.A. and De Borst, R., Non-Associated Plasticity for Soils, Concrete and Rock, Heron, Vol. 29, No. 3, 1984.

43. Wawersik, W.R. and Brace, W.F., Post-Failure of a Granite and Diabase, Rock Mechanics 3, 1971, pp. 61-85.

44. Walsh, P.F., Fracture of Plain Concrete, Indian Concrete Journal, Vol. 46, November 1972, pp. 469,470 and 476.

45. Willam, K.J., Sture, S., Bicanic, N., Christensen, J. and Hurlburt, B., Identification of Strain-Softening Properties and Computational Predictions of Localized Fracture, Report No. 8404, Department of Civil, Environmental and Architectural Engineering, Univ. of Colorado, Boulder, CO, March 1984.

46. Willam, K.J., Bicanic, N. and Sture, S., Constitutive and Computational Aspects of Strain-Softening and Localization in Solids, Presented at ASME-WAM '84 Symposium on Constitutive Equations: Micro, Macro and Computational Aspects, New Orleans, December 10-14, 1984.

47. Willam, K.J. and Warnke, E.P., Constitutive Model for the Triaxial Behavior of Concrete, Int. Assoc. of Bridge and Structural Engineers, Seminar on Concrete Structure subjected to Triaxial Stresses, Paper III-1, Bergamo, Italy, May 1974.

48. Wong, T.F., Micromechanics of Faulting in Westerly Granite, Int. J. Rock Mech. Min. Sci. & Geomech. Abstr., Vol. 19, 1982, pp. 49-64.

ELASTICITY AND PLASTICITY OF CONCRETE UNDER PRINCIPAL STRESS ROTATION

Koichi Maekawa * Baolu Li **

Investigating the deformational behavior of concrete subjected to general loading paths, the authors carried loading tests including the principal axis rotation of stresses. According to the test results, the anisotropic change of normal and shear stiffnesses was quantitatively investigated and correctly simulated by constitutive laws of the concrete elasticity. plasticity and fracture. However, under high stress conditions, disagreement between the principal plastic flow direction and that of the principal stress was observed, and the effect of axis turning on the plastic flow rate was confirmed from experimental results. These facts are considered characteristic of concrete, but can be neglected in the mathematical formulation if time dependent plasticity would be taken into account.

1. INTRODUCTION

Constitutive equations have been recognized as one of the basic mechanical theories used in FEM analyses of reinforced concrete, and the necessity for the correct understanding of the behavior of reinforced concrete has promoted the research on constitutive laws from experimental and theoretical points of view. However, we should understand the fact that the proposed concrete models were checked by experimental results obtained in multi-axial loading tests carried out with limited loading paths. Generally speaking, proportional loading paths without principal stress rotation have been popular, and parameters corresponding to time were implicitly included in the stress or strain speed and often represented by the indistinct terminology "Short-Term Loading". In meeting the many-sided demands on the functions of concrete structures, path-dependent characteristics of concrete and applicability of constitutive laws should be widely investigated.

The main subject of this paper is the deformational behavior of concrete under conditions where the principal stress axis continuously and intermittently rotates. In the case of 2-dimensional reinforced concrete members, the rotation of principal stress directions is usually caused by the anisotropic stiffness changes of concrete such as cracking and by the existence of reinforcement. In some cases, failure of concrete elements subjected to the axis rotation is considered to

* Assistant Lecturer, Dept. of Civil Eng., University of Tokyo, Hongo, 7-3-1, Bunkyo-Ku, Tokyo, 113 JAPAN

** Graduate Student, Dept. of Civil Eng., University of Tokyo, Hongo, 7-3-1, Bunkyo-Ku, Tokyo, 113 JAPAN cause the shear failure of RC beams. Taking these cases into account, the behavior of concrete under axis rotation should be accurately predicted. In many cases, concrete constitutive models indirectly assume some rules or physical models corresponding to the stress rotation paths.

In order to collect the fundamental data for discussing the overall deformational mechanics of concrete, the authors carried out biaxial loading tests with histories of stress axes rotation. Based on the experimental results, the elastic and plastic behavior of concrete was quantitatively investigated and the effect of the axis rotation on the rate of plastic flow and its direction were discussed.

2. EXPERIMENT

The experiments consisted of two types of loading tests,
(A) biaxial compression-tension and,
(B) uniaxial compression with a history of principal axis rotation.

(A) Compression-Shear Loading Test (continuous rotation)

Principal rotation paths under biaxial compression-tension stresses were introduced by applying the compressive force and the torsional moment to cylindrical shell specimens [4,5,7] (diameter D=320mm, height h=950mm, thickness W=35mm and area A=35160mm², See Fig.1). The torsional moment was transmitted to concrete by the steel arm attached to the specimen with an epoxy adhesive. In order to eliminate the friction between the steel arm and the compression-loading apparatus, ball bearings, silicon grease and teflon sheets were placed between them. Shear and normal stresses in the global coordinates as defined in Fig.1 are calculated from the compressive force C (negative) and the torsional moment T (negative) as

$$\sigma_{xx}=0, \quad \sigma_{yy}=C/A, \quad \sigma_{xy}=T/(WD^2\pi) \tag{1}$$

The loss of shear stress due to friction was not greater than 0.05Mpa.

(B) Repeated Uniaxal Loading Test (cyclic rotation) [1]

The loading paths including the principal axis rotation were reproduced by the following method.

(1) Uniaxial compressive stress was monotonically applied to a certain strain level and unloaded completely. (The first stage.)

(2) A concrete piece was cut off the concrete plate (200x200x50mm) loaded in the first stage and re-shaped as shown in Fig.2.

(3) Uniaxial compressive stress was applied to the cut-off concrete piece (The second stage). In this case, the direction of the principal stress was different from that in the first stage. Then, let ''γ'' denote the rotational angle as shown in Fig.2.

In the cutting process, concrete plates were split statically with round steel bars as shown in Fig.3. Four faces of cut-off concrete plates were recapped with super-high-early portland cement mortar whose uniaxial compressive strength was almost equal to the strength of concrete used. The compressive strength of concrete was 20-30Mpa. In case of the compression-shear test, the normal and shear stress rates were 0.2 and 0.15Mpa/sec, respectively. In the uniaxial rotation test (B), 0.2Mpa/sec was used as the principal compressive stress rate in the first and second stages. Therefore, we should note that the delayed elastic and plastic strains are included in the overall total strains under the adopted time paths.

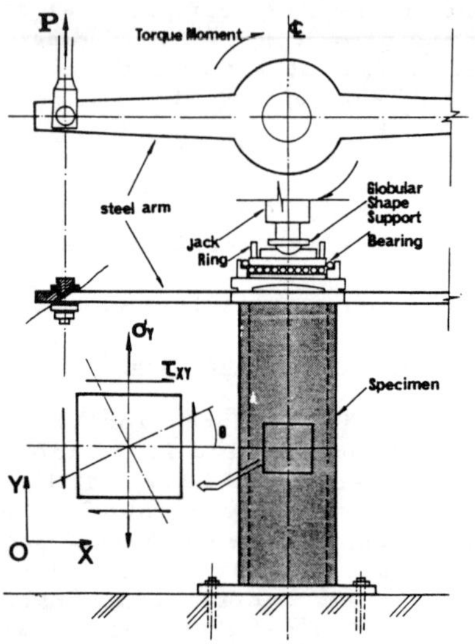

Fig.1 Set-up of compression-shear tests.

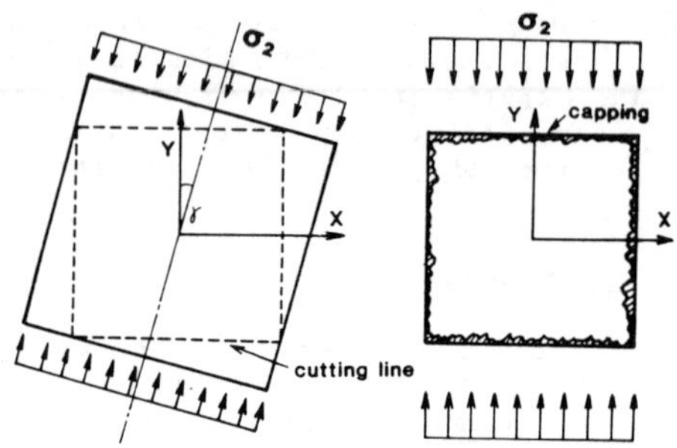

Fig.2 Uniaxial rotation tests

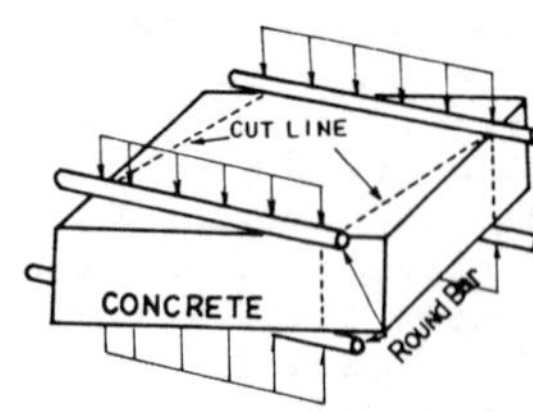

Fig.3 Cutting of concrete plate

3. ANISOTROPIC DEFORMATIONAL CHARACTERISTICS UNDER PRINCIPAL STRESS ROTATION

Stress-strain diagrams are shown in Fig.4 where the shear stress varied monotonically ($d\sigma_{xy}\neq 0$) with the principal stress rotation under various normal compressive stress levels ($d\sigma_{yy}=0$). Let fc and ε_o denote the uniaxial compressive strength (negative) and corresponding compressive strain (negative). In this type of loading path in Fig.4, the principal stresses σ_1, σ_2 are calculated by

$$\sigma_1 = \sigma_{yy}/2 + \sqrt{\sigma_{yy}^2/4 + \sigma_{xy}^2}\,, \quad \sigma_2 = \sigma_{yy}/2 - \sqrt{\sigma_{yy}^2/4 + \sigma_{xy}^2} \tag{2-1}$$

The directional angle between the maximum principal stress 'H_s' and the x-coordinate axis is expressed using the function H_f as

$$H_s = H_f(\sigma_{ij}), \quad H_f(\sigma_{ij}) = 0.5\tan^{-1}(-2\sigma_{xy}/\sigma_{yy}) \tag{2-2}$$

where the counterclockwise direction is defined positive. For the convenience of describing the relation between incremental stresses and strains, the authors utilize the following compliances.

$$\begin{bmatrix} d\varepsilon_{xx} \\ d\varepsilon_{yy} \\ d\varepsilon_{xy} \end{bmatrix} = \begin{bmatrix} \phi_{11} & \phi_{12} & \phi_{13} \\ \phi_{21} & \phi_{22} & \phi_{23} \\ \phi_{31} & \phi_{32} & \phi_{33} \end{bmatrix} \begin{bmatrix} d\sigma_{xx} \\ d\sigma_{yy} \\ d\sigma_{xy} \end{bmatrix} \tag{3}$$

Organizing the concepts related to the axis rotation, it would be beneficial to understand and verify the idealization of simple and representative popular models for concrete using the experimental data in Fig.4 as an example.

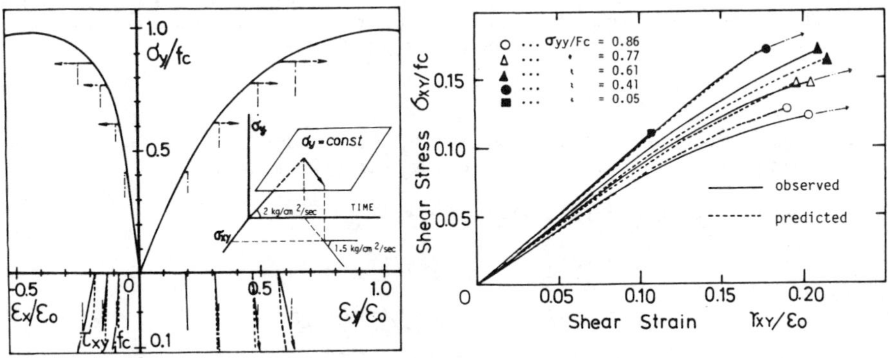

Fig.4 Loading paths and corresponding total stress-strain diagrams.

1) Nonlinear Isotropic Elasticity

This type of models [6,10,11] are recognized as the nonlinear modification of the well-known linear isotropic elasticity model and are formulated or generally reformed as

$$\varepsilon_m = K_{in}(I_1)\sigma_m, \qquad e_{ij} = G_{in}(J_2)s_{ij} \tag{4}$$

where, $\sigma_m=\sigma_{kk}/3$, $\varepsilon_m=\varepsilon_{kk}/3$ (mean stress, strain), $e_{ij}=\varepsilon_{ij}-\delta_{ij}\varepsilon_m$, $s_{ij}=\sigma_{ij}-\delta_{ij}\sigma_m$ (deviatoric strain and stress), $I_1=3\sigma_m$, $J_2=0.5\cdot s_{ij}\cdot s_{ij}$ (the first and second invariants of stress). K_{in} and G_{in} are the explicit functions of I_1 and J_2. The compliances are respectively calculated by differentiating Eq.(4) as

$$\phi_{13}=s_{xx}s_{xy}(dG_{in}/dJ_2), \qquad \phi_{23}=s_{yy}s_{xy}(dG_{in}/dJ_2),$$

$$\phi_{33}=G_{in}(J_2)+(dG_{in}/dJ_2)s_{xy}{}^2 \tag{5}$$

Eq.(4) is described in the global coordinate system where the y-axis agrees with the compressive normal stress direction and σ_{xy} is equal to zero at the beginning of the principal stress rotation.

According to this model for concrete, the compliance ϕ_{33} in Eq.(5), where J_2 is almost constant under high stress levels during axis rotational paths, increases rapidly and nonlinearly due to the shear stress increment ($d\sigma_{xy}\neq 0$). On the other hand, the increment of the volumetric strain $d\varepsilon_m$ is prescribed zero because of the fixed mean stress. The ratio of the compliance ϕ_{13} to ϕ_{23} does not vary during the axis rotation.

2) Elasto-Plastic Model [2,6]

The plasticity criterion described by stress tensors considers this type of loading to be irreversible. Accordingly, the compliances defined by Eq.(3) are calculated as the sum of the elastic and plastic components. Let us consider the plasticity model including the plastic potential function P formulated by the first and the second invariants of stresses, I_1 and J_2. The vector of plastic flow is expressed parallel to the normal and outer vector which stands on the plastic potential surface, where P is constant as follows.

$$d\varepsilon_{pij}=dg(\partial P/\partial \sigma_{ij}) \tag{6}$$

where g is the scalar value defined by the hardening rule in the frame of plasticity. We get compliances defined in Eq.(3) as

$$\begin{aligned}\phi_{13} &= (PI_1+PJ_2\cdot s_{xx})(\partial g/\partial s_{xy})\\ \phi_{23} &= (PI_1+PJ_2\cdot s_{yy})(\partial g/\partial s_{xy})\\ \phi_{33} &= 1/(2Go) + PJ_2\cdot s_{xy}(\partial g/\partial s_{xy})\end{aligned} \tag{7}$$

where, Go is the initial shear stiffness and PI_1, PJ_2 are the partial differential coefficients of the plastic potential function P by I_1 and J_2.

The shear compliance ϕ_{33} does not agree with that in case of the linear isotropic model in general, but at the beginning of this axis rotation produced by the compression-shear tests, ϕ_{33} becomes constant elastic value because s_{xy} is equal to zero. As the shear stress s_{xy} increases, the shear compliance does not vary so rapidly but changes in a linear fashion because PJ_2 changes very little in these rotation paths under high stress levels. The volumetric strain increment appears when PI_1 has a non-zero value. Generally speaking, however, the ratio of ϕ_{13} to ϕ_{23} is not constant in this loading program.

3) Orthogonally Anisotropic Model

At the beginning of the principal stress rotation, this type of modeling [3,6,13] specifies the concrete tangential stiffness as

$$\begin{bmatrix} d\sigma_{xx}\\ d\sigma_{yy}\\ d\sigma_{xy}\end{bmatrix} = \frac{1}{1-\mu^2}\begin{bmatrix} E_1 & \mu\, E_1E_2 & 0\\ \mu\, E_1E_2 & E_2 & 0\\ 0 & 0 & (E_1+E_2-2\mu\, E_1E_2)/4\end{bmatrix}\begin{bmatrix} d\varepsilon_{xx}\\ d\varepsilon_{yy}\\ d\varepsilon_{xy}\end{bmatrix} \tag{8}$$

where, $d\sigma_{xy} \neq 0$ and $d\sigma_{xx} = d\sigma_{yy} = 0$. Accordingly, the compliances ϕ_{13}, ϕ_{23} and the volumetric strain increment are explicitly equal to zero.

As mentioned above, the qualitative actions under the principal axis rotation are predicted in different manners. Accordingly, it is significant to verify the basic concept as to the axis rotation path by the comparison with the experimental data reported in Fig.4. The tangential shear and normal stiffnesses $G(=1/2\phi_{33})$, $E(=1/\phi_{22})$ normalized by the initial stiffnesses at each compressive level (σ_{yy}/fc, uniaxial compression) are shown in Fig.5. We can see the anisotropic decrease of both stiffnesses according to the increase of stress level. However, the shear stiffness does not drop so remarkably as the compressive normal stiffness decreases. In other words, concrete under the compressive stress is more rigid in the shear direction of deformation. The prediction made by models (1) and (3) are qualitatively reasonable as concerns the change of shear and normal stiffnesses, but model(2) estimates the shear stiffness to be constant.

In order to discuss the qualitative fitness of concrete modeling, the change of the shear compliance with rotation axis as shown in Fig.6 must be the object for checking. When less shear stress is applied to concrete, ϕ_{33} varies in accordance with the increase of the shear stress in a linear fashion and furthermore, its slope $\partial\phi_{33}/\partial s_{xy}$ tends to become greater according to the increase of stress level σ_{yy}/fc. With reference to this deformational behavior, model (2) provides the best qualitative estimate. However, the shear compliance ϕ_{33} increases nonlinearly when greater shear stress is applied to concrete, then the prediction by model (1) is qualitatively appropriate.

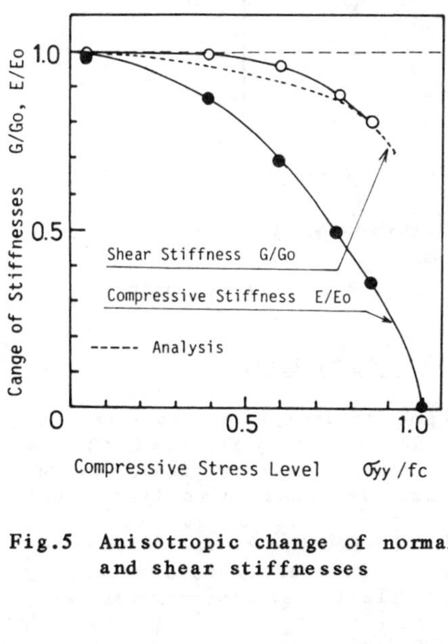

Fig.5 Anisotropic change of normal and shear stiffnesses

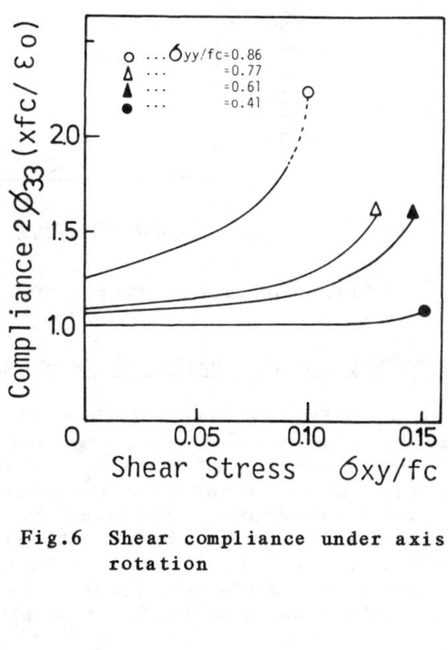

Fig.6 Shear compliance under axis rotation

The volumetric strain increase due to shear stresses is often defined as shear dilatancy and recognized as one of the characteristic properties of geomaterials including concrete. But, the dilatancy under axis rotational paths had not been obtained quantitatively. Accordingly, it may be reasonable to use the experimentally obtained shear dilatancy shown in Fig.7 to verify the constitutive models. The volumetric strain increment was not observed under low stress conditions where predictions according to models (1)-(3) coincide with the actual phenomenon. However, at greater stresses, the finite volumetric expansion occurs due to the introduction of the shear stress as shown in Fig.7. Among the constitutive laws above, model (2) has the capacity to reproduce this behavior.

Each concrete constitutive model has physically appropriate concepts for concrete deformation, but their assumptions are not sufficient. If we would try to formulate the complicated overall response of concrete directly from experimental data, the mathematical descriptions would become complex. Taking this situation into account, the authors adopted the approach to decompose the total strains into the reversible and irreversible deformation for simulating the path-dependency of the concrete mechanics.

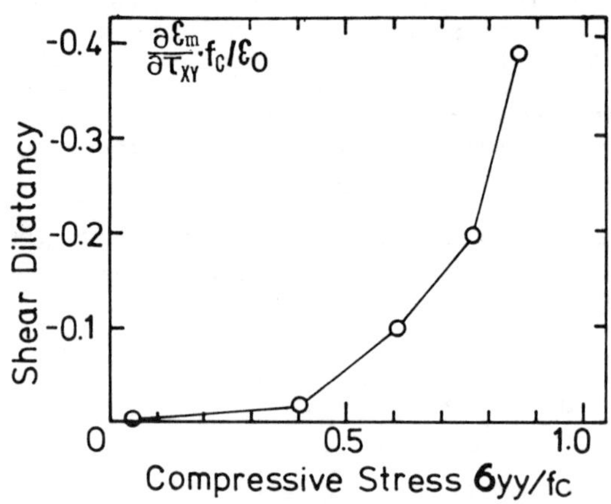

Fig.7 Volumetric strain change under axis rotation paths

4. REVERSIBLE DEFORMATION UNDER PRINCIPAL STRESS ROTATION

We define the irreversible (plastic) strain ε_{pij} as the total one when stresses would be instantaneously and completely released, and the reversible (elastic) strain ε_{eij} which is explicitly determined by the updated stresses and damage conditions. In order to analyze total strain increments introduced by the principal stress rotation, unloading paths were added to the monotonic loading histories as shown in Fig.8. The plastic strain increments were obtained directly by measuring the difference between the two plastic strains corresponding to Points A and B in Fig.8, respectively.

According to the definition, the reversible shear and normal strain increments were separated from total strain increments as illustrated in Fig.9, where we can find an interesting fact that the reversible shear strain increment accounts for the greater part of the total strain when the shear stress increases under fixed normal stresses ($d\sigma_{yy}=0, d\sigma_{xx}=0$). On the other hand, increments of the reversible normal strains are very small under the principal stress revolution. In other words, the tangential shear stiffness in the irreversible process is equal to the elastic stiffness of concrete. Moreover, the elastic shear dilatancy does not appear, and only the elastic shear strain changes when a shear stress increment is introduced. Then, we can interpret this behavior as the isotropic action of elasticity and understand that the principal direction of the reversible strain increment agrees with the principal one of the stress increment.

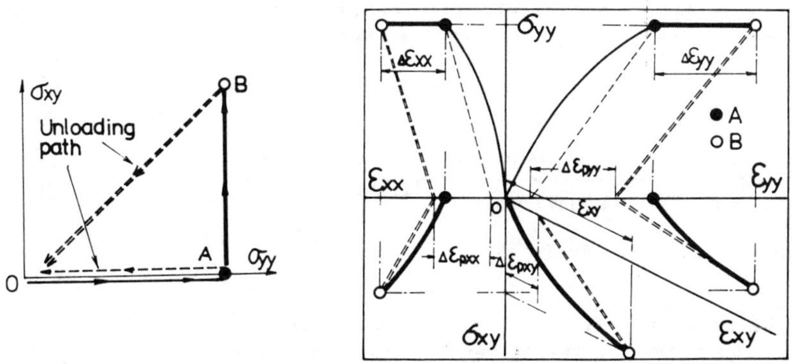

Fig.8 Unloading paths and corresponding strains

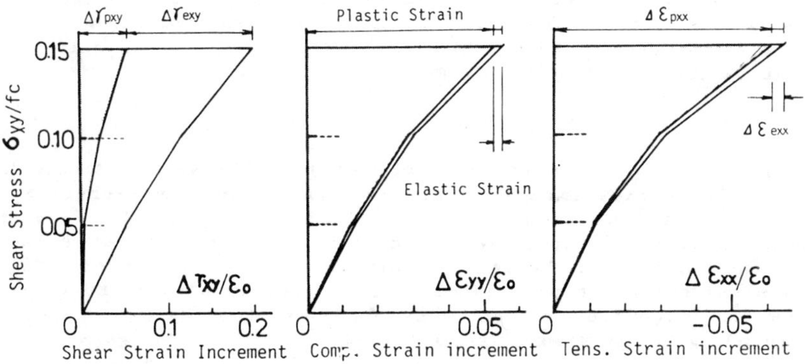

Fig.9 Separated strain components under the axis rotation ($\sigma_{yy}/fc = 0.77$ in experiments. γ_{xy} is a engineering shear strain and equal to $2\varepsilon_{xy}$.)

The agreement of the overall shear stiffness with the elastic one is certainly simulated by the elasto-plastic constitutive model (2) as discussed in Chapter 3. The predicted shear (elastic) stiffness, according to this model, is constant and equal to the initial stiffness. But the actual shear stiffness is not constant but changes and depends on paths as shown in Fig.5. Therefore, we must take the decrease of the elastic shear stiffness into account.

Under biaxial compression-tension stress states where the axes of principal stresses are fixed, the stiffness matrix relating principal stresses and reversible principal strains was verified to be isotropic [1]. Furthermore, it may be reasonable to neglect the elastic dilatancy effect from experimental data as shown in Fig.9. Accordingly, we can assume the isotropic and symmetric constitutive equation for elastic response. Then, the change of elasticity coefficients must be considered as shown in Fig.5. The authors represent the change of elastic coefficients with a parameter K and define the elastic constitutive equation [1] as

$$\sigma_{ij} = DE_{ijkl}\varepsilon_{ekl}$$

$$DE_{ijkl} = \frac{\mu(K)KEo}{(1+\mu(K))(1-2\mu(K))}\delta_{ij}\delta_{kl} + \frac{KEo}{2(1+\mu(K))}(\delta_{ik}\delta_{jl}+\delta_{il}\delta_{jk})$$

$$\text{or} \quad \{\sigma\} = [DE(K)]\{\varepsilon_e\} \tag{9}$$

where, Eo is the initial stiffness and the scalar value K is the condition parameter which represents the degree of fracture in concrete and defined as fracture parameter. $\mu(K)$ is the isotropic Poisson's ratio and varies according to the past loading history. The value of K is equal to or less than unity when concrete is idealized as an initially undamaged continuum. For checking the applicability of Eq.(9), the authors introduced the biaxial unloading paths to concrete shells as shown in Fig.10. From the strain response corresponding to the release of compressive normal stress, the assumed shear stiffness G_{cal} can be calculated by Eq.(10) as

$$G_{cal} = E_{se}/(1+\mu_{se}) \tag{10}$$

where, E_{se} and μ_{se} are unloading (elastic) stiffness and Poisson's ratio, obtained experimentally as shown in Fig.10. The actual and assumed shear stiffnesses G, G_{cal} are illustrated in Fig.11. From an engineering point of view, the two shear moduli correlate with each other. The volumetric change of deformation was not observed. This result means the isotropic response of reversible strains to stress increments. Accordingly, we can use Eq.(9) in this type of loading path including the principal axis rotation.

The concept of fracture is represented by the loss of the effective area which resists external actions and was investigated by some researchers [1,8,9,12]. Under compression-tension 2D conditions, the value of K was reported [1] as

$$K = \exp(-0.73Emax(1-\exp(-1.25Emax))) \tag{11-1}$$
$$E = \int (\partial F/\partial \varepsilon_{eij}) d\varepsilon_{ij}, \quad F = \sqrt{(0.62\overline{eo})^2 + (0.98\overline{ro})^2}/\varepsilon o$$
$$\overline{eo} = (\varepsilon_{exx} + \varepsilon_{eyy})/\sqrt{2}, \quad \overline{ro} = \sqrt{2}\sqrt{(\varepsilon_{exx} - \varepsilon_{eyy})^2/4 + \varepsilon_{exy}^2}$$

The effective isotropic Poisson's ratio can be obtained as a unique function of K as

$$\mu(K) = \mu o \quad \text{(initial Poisson's ratio)} \qquad 0.8 \leq K$$
$$\mu(K) = \mu o(-3.75K+4) \qquad 0.4 \leq K < 0.8 \tag{11-2}$$
$$\mu(K) = 2.5\mu o \qquad K < 0.4$$

where, E is defined as the equivalent total strain and Emax is the maximum value of E under past loading histories. The irreversible process is defined as Emax=E. The K value is expressed in terms of strain tensors.

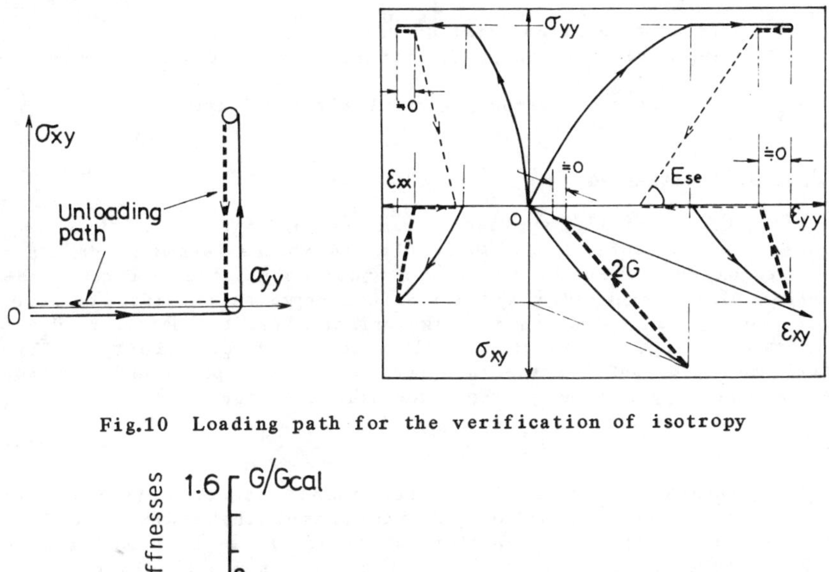

Fig.10 Loading path for the verification of isotropy

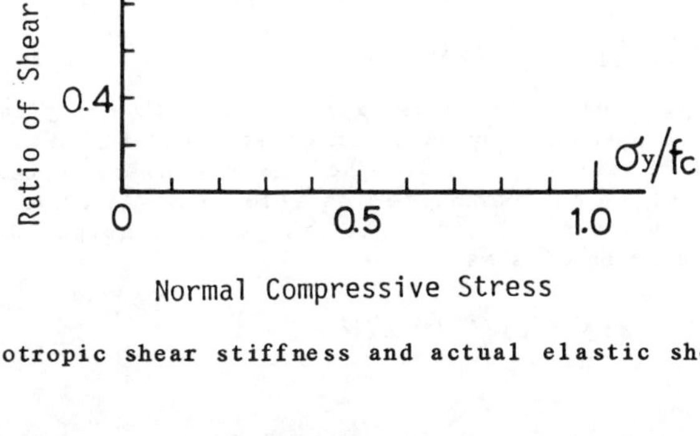

Fig.11. Isotropic shear stiffness and actual elastic shear stiffness.

5. IRREVERSIBLE DEFORMATION UNDER PRINCIPAL STRESS ROTATION

The elastic strain expresses the reversible deformation, and the corresponding constitutive laws were verified to be isotropic and path-independent when fracture does not proceed. Accordingly, the separated irreversible plastic strain represents the path-dependent nonlinearity as discussed in Chapter 3. Therefore, the movement of the plastic strain should be exactly investigated. In case of the 2-dimensional stress analysis, the proportion and absolute value of plastic increments $d\varepsilon_{pxx}$, $d\varepsilon_{pyy}$, $d\varepsilon_{pxy}$ must be predicted under arbitrary strain paths. Within these experiments, ε_{zz} was almost the same as ε_{xx}, and taking a more intuitive understanding of the plastic flow into account, the authors express the experimentally obtained plastic flow as

$$\begin{bmatrix} d\varepsilon_{pxx} \\ d\varepsilon_{pyy} \\ d\varepsilon_{pxy} \end{bmatrix} = [C(H_p)] \begin{bmatrix} d\varepsilon_{p1} \\ \\ d\varepsilon_{p2} \end{bmatrix} \tag{12}$$

where, [C] : coordinate transformation matrix.

H_p : angle between the principal axis 1 and the x-coordinate, $= H_f(d\varepsilon_{pij})$.

$d\varepsilon_{p1}, d\varepsilon_{p2}$: maximum and minimum principal plastic increments.

5.1 Principal Direction

When shear stress is applied to concrete and the axes of principal stresses start to rotate, the shear plastic strain increment does not appear as discussed in Chapter 4. In other words, the value of the principal direction H_p of plastic strain increments is equal to zero, that is to say, it coincides with the principal stress direction H_s as the normality rule predicts in the theory of plasticity. This directional agreement holds within the series of experiments reported here. Accordingly for the present, the authors assume

$$H_p = H_s \tag{13}$$

As mentioned above, it is not reasonable or convenient to use stress tensors for expressing the irreversible behavior including strain-softening [1]. It is preferable to redefine the hypothesis of Eq.(13) in terms of strain tensors. Then we can use the isotropy of elasticity where the principal directions of stresses agree with those of elastic strains as

$$H_s = H_f(\varepsilon_{eij}) = H_f(d\varepsilon_{pij}) \tag{14}$$

In the local coordinates whose axes agree with the principal axes of elastic strains, shear components of elastic and incremental plastic strain tensors become zero, i.e., the shear tensorial component of $\varepsilon_{eij} + d\varepsilon_{pij}$ in the local coordinates is also equal to zero. The sum of the elastic strain and plastic flow has the same principal direction as that of the elastic strains as

$$H_f(\varepsilon_{eij} + d\varepsilon_{pij}) = H_f(\varepsilon_{eij}) \tag{15}$$

Developing Eq.(15) in the form of total differentiation, we get

$$(\partial H_f/\partial \varepsilon_{eij})d\varepsilon_{pij}=0 \tag{16}$$

The plastic incremental vector stands normal to the envelope where H_f is constant. Under the irreversible process, the change of stress direction dH_s is expressed to be the same as that of the elastic strain direction. Using the isotropy assumption for the reversible strain and Eq.(16), we can get

$$dH_s = (\partial H_f/\partial \varepsilon_{eij})d\varepsilon_{eij} = (\partial H_f/\partial \varepsilon_{eij})(d\varepsilon_{eij}+d\varepsilon_{pij}) = (\partial Hf/\partial \varepsilon_{eij})d\varepsilon_{ij} \tag{17}$$

Eq.(17) indicates the constitutive equation which controls the principal directions of stresses and strains, and should be verified by another type of loading paths including the axis rotation. The experiment (B) in Chapter 2 is appropriate to check Eq.(17) because we can set arbitrary directions of stress rotation and furthermore, the loading paths which this type of experiments can produce were not utilized directly in deriving Eq.(17). The loading paths and corresponding straining are expressed in the coordinates illustrated in Fig.2 and shown in Table 1 and Fig.12. Within the reversible process under the second loading after rotation, the shear strains hardly change in any cases, that is to say, the principal direction of elastic deformation agrees with that of the total stress. This experimental fact supports the isotropy and symmetry of the stiffness matrix described by Eq.(9).

The disagreement between the principal stress direction H_s and H_{sc}, the one assumed by Eq.(14), is shown in Fig.13. When the equivalent strain E exceeds Emax, which the concrete was reached in the first loading, the plastic strain changes and nonlinearlity appears as shown in Fig.12. But the deviation of the calculated principal direction from the experimental one is negligible when the stress levels of the first stage are not as high as Emax=0.95. Critically speaking, when concrete is compressed to a greater level, the hypothesis in Eq.(17) equivalent to Eq.(14) does not hold, as illustrated in the case of Emax=0.95 in Fig.13. The plastic flow direction slightly tilts from the principal stress axis and the plastic deformation proceeds taking the mode as shown in Fig.14.

Table 2. Loading program of principal axis rotation tests Type (B) [1].

Exp. No.	γ(deg)	Emax	fc(Mpa)	ε_o(micro)
RT1	15	0.46	-27.4	-2100
RT2	25	0.41	-27.4	-2180
RT3	45	0.46	-28.0	-2200
RT4	65	0.46	-28.0	-2200
RT5	15	0.58	-26.9	-2100
RT6	25	0.56	-28.0	-2200
RT7	45	0.66	-26.9	-2200
RT8	65	0.55	-27.7	-2200
RT9	21	0.95	-34.4	-2350
RT10	28.4	0.85	-35.1	-2350
RT11	45	0.95	-34.0	-2350

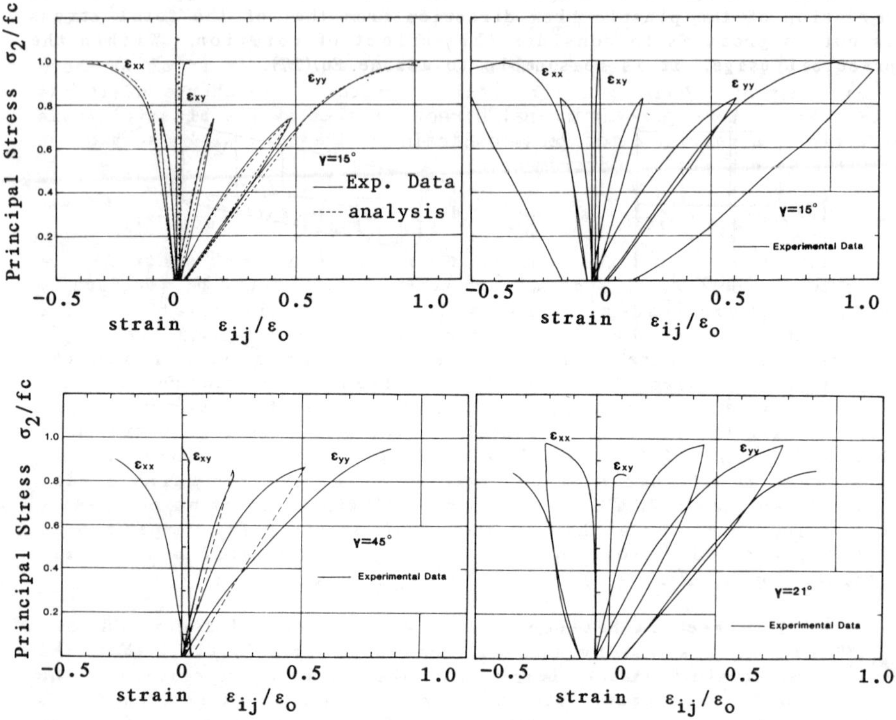

Fig.12 Stress-strain relations under axis rotation paths in Type (B)

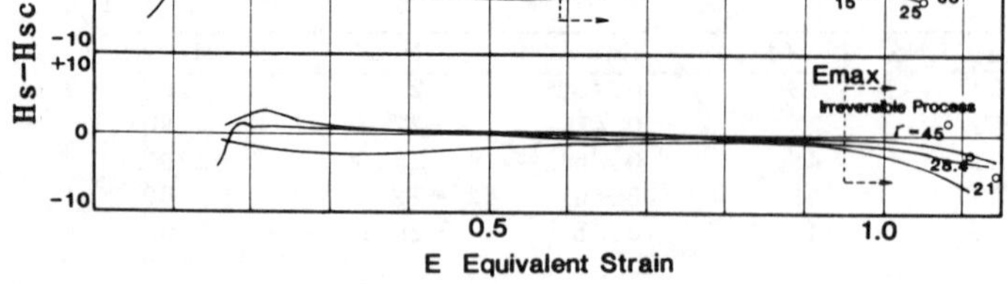

Fig.13 Principal directions of stress and strain

Similar instances were reported as to the delayed plasticity in high stress creep tests [7]. This may be caused partly because of slip and dilatancy on the microcracking plane introduced in the first loading. Damaged concrete with distributed microcracking is considered to be in a transient condition to discrete cracking. Unfortunately, we do not have enough data to idealize this directional disagreement. But deviation of the plastic flow direction from that of the total stress is not so great as to consider this effect of rotation. Within the practical usage, it is reasonable to assume Eq.(17).

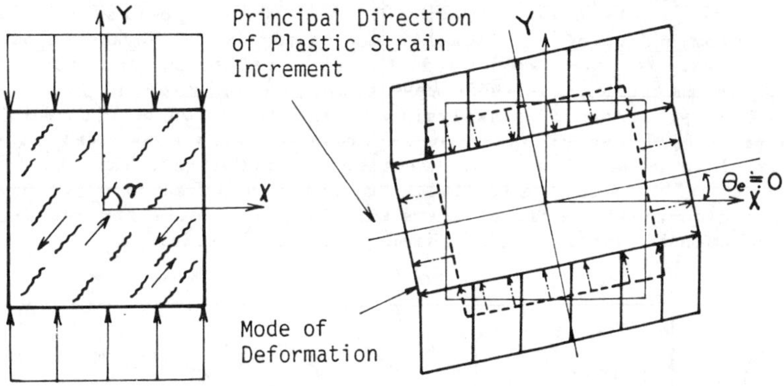

Fig.14 Principal stress axis and plastic flow direction

5.2 Directional Vector of Plasticity in Principal Space

It is necessary and sufficient to discuss the plastic flow rate in principal space because the principal direction is defined by Eq.(17) in advance. The authors adopt the unit directional vector of the plastic flow as

$$\{PF\}=(d\varepsilon_{p1},\ d\varepsilon_{p2})^T/\sqrt{d\varepsilon_{p1}^2 + d\varepsilon_{p2}^2} \tag{18}$$

where, the notation { } indicates a column vector. The vector {PF}, experimentally obtained by compression-shear tests, is shown in the principal stress space, Fig.15. The directional tangent of {PF} at the beginning of the axis rotation is calculated as the ratio of the normal strain increment $d\varepsilon_{xx}$ to $d\varepsilon_{yy}$, and the corresponding stress state is uniaxial. Even if the direction of the principal stress vector (σ_1,σ_2) is fixed, the corresponding direction of {PF} is not constant but varies according to the absolute stress level. Vector {PF} tilts to the principal tensile direction when the applied stress reaches higher levels as shown in Fig.15. The existence of the principal tensile stress accelerates this tendency very much. This behavior is not peculiar to the case of the axis turning but characteristic of concrete under compression-tension stress states [1], where some mathematical expresions have been proposed, based on multi-axial tests without axis rotation.

Specifying the effect of axis rotation on the principal plastic direction, let us compare the unit flow vector {PFc} obtained experimentally under fixed principal stress axes with {PF} obtained by experiment (B) containing rotational paths in Chapter 2. Only for this purpose, the authors carried out verification tests where biaxial compression-tension stresses without rotation were applied to concrete panels with the same time-principal stress history as calculated by Eq.(2). The experiment was executed in the same manner as the biaxial loading tests including unloading paths (See Fig.8).

The directions of vector {PFc} are shown in Fig.15. Within this case, some difference between {PF} and {PFc} appears, however, no particular effect of rotational paths is observed. We should note that in general, the precision of differetial values of directly obtained data drops compared with the accuracy of non-processed data. Accordingly, we can adopt rules as to the direction of plastic flow in the principal space, based on experiments with non-rotating stress paths. This conclusion is useful when we investigate the plastic flow direction from experiments. Maekawa and Okamura [1] executed compression-tension loading tests and obtained models for the direction of incremental strains under fixed principal directions.

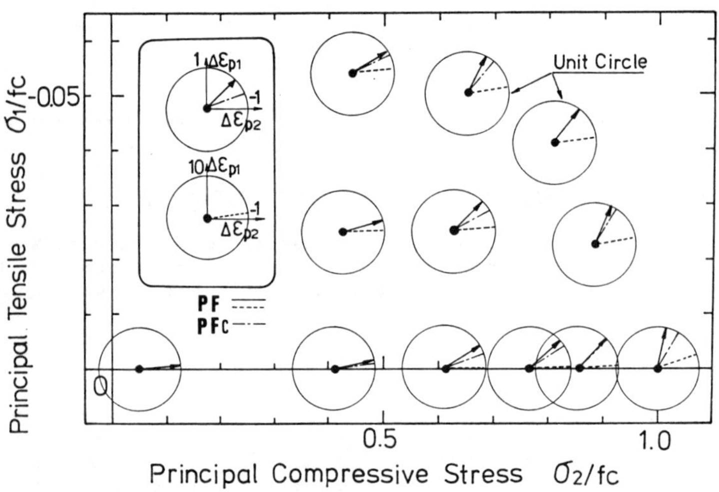

Fig.15 Direction of the unit plastic flow vector {PF} and calculated vector {PFc} on the principal strain space.
(Dotted lines indicate {PF} on the local coordinate ($d\varepsilon_{p1}/10, d\varepsilon_{p2}$) which coincides with the stress coordinate scale.)

5.3 Plastic Flow Rate

The formulation of the length of the plastic incremental vector completes the system of constitutive laws related to the plasticity. In case of strain-hardening plasticity, the length of the plastic incremental vector is defined by the hardening rule in terms of stresses. Within the strain-hardening conditions, plastic deformation flows in accordance with the increase of stresses, but on the other hand, the plastic flow appears even when stresses appear to decrease in the strain-softening region of concrete. Therefore, the mathematical expression would not be simple if stress tensors were used as one of the main parameters for describing plastic flow rates. However the irreversible process of plastic deformation is judged by the criterion using elastic strains even in the strain-softening area. Taking these situations into consideration, the following rule for the length of the plastic incremental vector is proposed as

$$dEp=(\partial F/\partial \varepsilon_{eij})d\varepsilon_{pij}=\{\partial F/\partial \varepsilon_e\}^T\{d\varepsilon_p\} \quad \text{(equivalent plastic strain)} \quad (19)$$

where $\{\partial F/\partial \varepsilon_e\}$ is the column vector having the terms $\partial F/\partial \varepsilon_{eij}$. The equivalent plastic strain Ep represents the length of the plastic flow vector and was verified by compression-tension loading paths "without axis rotation" [1] as

$$dEp=(1 - \exp(-0.35E))dE = H\ dE \qquad \text{when } E=Emax \qquad (20)$$

The experimentally obtained plastic flow rate H=dEp/dE under rotation paths is plotted in Fig.16 normalized by the value predicted by Eq.(20). As shown in Fig.16, the axis rotation of principal stresses has an effect on the appearent plastic flow rate. Especially under high stress condition, the principal axis rotation effectively accelerates the plasticity.

Investigating the influence of the principal axis rotation on the plasticity, let us consider the difference of loading conditions as to time. In case of the compression-shear tests, the shear stress rate was equal to 0.15Mpa/sec as discussed in Chapter 2, but the loading rate of principal stresses was not greater than 0.015Mpa/sec (nearly 1/10 - 1/50 times as high). However, the plastic flow rate by Eq.(20) was derived from loading tests with the principal stress rate 0.2Mpa/sec. Accordingly, the greater time path-dependent plasticity was included within the measured strains under high stress conditions. Then, using test results obtained in Section 5.2, the authors obtained the plastic flow rate Ept experimentally under the no-rotation paths where the principal stress rate agreed with the compression-shear loading . The actual plastic flow rate normalized by Ept is shown in Fig.16 where the delayed plasticity may be eliminated. We can recognize the greater effect of the time parameter on the plastic flow rate at higher stress levels. In this loading condition, the accelerated plastic flow rate due to the rotation of stress axes is almost explained by the difference of loading speeds.

However, there is some influence of the axis rotation on the plasticity which cannot be explained by the delayed plasticity (creep). In order to understand the effect of rotation on the plastic flow rate, let us consider the extreme case, that is to say, "orthogonal rotation". The adopted loading path and corresponding equivalent plastic strain are shown in Fig.17. When concrete is strained to a higher level, excessive plastic flow owing to the orthogonally rotating path was observed. The non-isotropic property of the plastic flow rate cannot be expressed when the hardening formulation such as Eq.(20) is an explicit function of tensor invariants. It is imagined that the tensile plastic strain easily flows due to the compressive stress normal to the distributed microcracking surface until enough recontact between cracking faces is ensured. The accelerated plastic flow subjected to rotation histories is considered a substantial property of concrete. But there exist little data for representing this behavior. It is planned to execute a systematically arranged series of experiments for this purpose.

In case of 2-dimensional reinforced concrete structural members, however, the produced rotational angle is not so great actually as to take their additional flow rate into account. When concrete is biaxially loaded until Emax=unity, the plastic flow rate may be described mathematically by invariants of strains which are independent of the coordinate transformation.

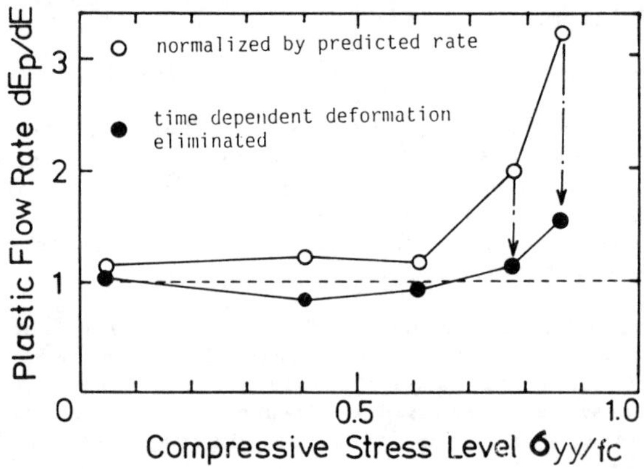

Fig.16 Plastic flow rate under axis rotation and no rotation paths

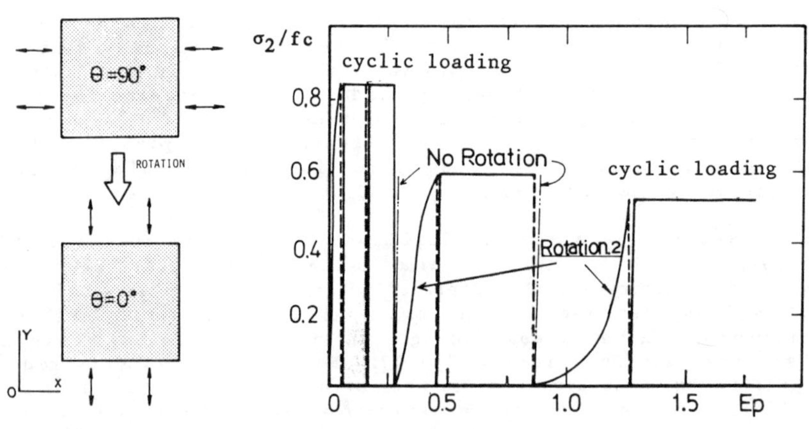

Fig.17 Plastic flow rate under orthogonally rotated stress paths. (Cyclic uniaxial loading test)

6. TOTAL STRESS-STRAIN RELATIONSHIP AND TIME-PATH DEPENDENT DEFORMATION

The deformational behavior is expressed by a combination of elastic and plastic strains. The total stress-strain relationship (combination) is mathematically formulated by the simultaneous solution of constitutive equations concerning the elasticity including the condition parameter K and the plasticity. The increment of the total strain is the sum of elastic and plastic components as

$$\{d\varepsilon\} = \{d\varepsilon_e\} + \{d\varepsilon_p\} \tag{21}$$

where $\{d\varepsilon\}$ is the incremental strain vector having the terms $d\varepsilon_{ij}$. The corresponding stress increment is expressed by the elastic strain increment and the rate of fracture dK, differentiating Eq.(9) as

$$\{d\sigma\} = [DE(K)]\{d\varepsilon_e\} + [dDE/dK]\{\varepsilon_e\}dK \tag{22}$$

where [ ] indicates a square matrix. The rate of fracture is the function of elastic strains and described by the total differentiation of Eq.(11-1) as

$$dK = (dK/dE)\{\partial F/\partial \varepsilon_e\}^T\{d\varepsilon\} \tag{23}$$

where dK becomes zero when the equivalent strain E is less than Emax in the reversible process. The plastic strain increment $d\varepsilon_{pij}$ is obtained in terms of total strain increments by solving Eqs.(12),(19) as

$$\{d\varepsilon_p\} = [C(H_p)]\{PF\}dm \tag{24}$$

where dm represents the rate of the plastic deformation and can be determined by the constitutive law which defines the equivalent plastic strain Ep. Then, the value of dm is determined by substituting Eq.(24) into Eq.(19) and solving simultaneously with Eq.(20), we can get

$$dm = \frac{H\,\{\partial F/\partial \varepsilon_e\}^T\{d\varepsilon\}}{\{\partial F/\partial \varepsilon_e\}^T[C(H_p)]\{PF\}} \tag{25}$$

Substituting Eq.(25) into Eq.(24),

$$\{d\varepsilon_p\} = \frac{H}{\{\partial F/\partial \varepsilon_e\}^T[C(H_p)]\{PF\}}\,[C(H_p)]\{PF\}\{\partial F/\partial \varepsilon_e\}^T\,\{d\varepsilon\} \tag{26}$$

Substituting Eq.(26) into Eq.(21), we can obtain the explicit expression of the elastic strain increment as

$$\{d\varepsilon_e\} = \left([I] - \frac{H}{\{\partial F/\partial \varepsilon_e\}^T[C(H_p)]\{PF\}}\,[C(H_p)]\{PF\}\{\partial F/\partial \varepsilon_e\}^T\right)\{d\varepsilon\} \tag{27}$$

Substituting Eq.(23) and Eq.(27) into Eq.(22),

$$\{d\sigma\}=([DE(K)] - \frac{H}{\{\partial F/\partial \varepsilon_e\}^T[C(H_p)]\{PF\}}[DE(K)][C(H_p)]\{PF\}\{\partial F/\partial \varepsilon_e\}^T)\{d\varepsilon\}$$

$$+[dDE(K)/dK]\{\varepsilon_e\}(dK/dE)\{\partial F/\partial \varepsilon_e\}^T\{d\varepsilon\} \tag{28}$$

Eq.(28) is the derived constitutive equation applicable to the principal stress rotation paths. The calculated shear stress-strain relationship and the shear stiffness G $(=1/2\emptyset_{33})$ by Eq.(28) are shown in Fig.4 and Fig.5. As the material coefficients were experimentally verified in each chapter, the calculated values certainly agree with experimental data with reasonable accuracy.

However, the material constants under the axis rotation were verified within the compression-tension stress states only. More general stress conditions should be checked. Furthermore, as mentioned in chapter 1, the experimentally obtained strain components include time-path dependent strains. These strain components are often expressed as delayed elasticity and plasticity (creep). However, the authors defined the plastic flow rate derived from experiments under constant principal stress rates. Accordingly, the analytical results of normal strains ε_{xx}, ε_{yy} as shown in Fig.4 were estimated to be much less than the actual response. Taking the greater rates of plasticity and fracture under high stress or strain levels into account, the strict definition of time dependent deformation and fracture should be formulated when we are looking for wide applicability of concrete modeling.

7. CONCLUSIONS

Loading tests including stress axis rotation were carried out for the experimental verification of constitutive laws for concrete subjected to continuous rotation of principal axes under compression-tension stress states. The authors obtained the anisotropic behavior of concrete from the experiments as follows.

When shear stress is applied to concrete under fixed normal stresses (principal axis rotates), the normal stiffness parallel to the principal stress axis and the shear stiffness deteriorate with an increase of the compressive stress level. However, the shear stiffness does not drop so remarkably as the normal stiffness decreases in appearance. Understanding this overall deformational anisotropy, the authors introduced unloading paths, and it was made clear that the shear plastic strain does not flow, but normal plastic strain increments account for nearly 100 per cent of the normal total strain increments parallel to the principal stresses when concrete experiences axis rotational paths. Taking these experimental results into account, the anisotropic behavior of concrete was explained by a combination of elasticity and plasticiy.

The relationship between elastic strains and stresses was verified to be isotropic, but two material coefficients used in formulating isotropy change in accordance with damage conditions of concrete. These are the fracture parameter and isotropic elastic Poisson's ratio. The fracture parameter can be predicted by the function of strain paths. The idealization of the fracture concept makes it possible to simulate the path-dependent change of the shear stiffness with reasonable accuracy.

Strictly speaking, the principal direction of plastic strain increments does not coincide with that of the updated stress direction. The mechanical slip and dilatancy along microcracking introduced is considered to cause the disagreement of both axes. However, it is sufficient to assume that two principal directions coincide with each other when concrete is regarded a continuous body.

In the compression-shear loading paths under high stress states, the authors observed accelerated plastic flow rates due to the axis rotation. Because of the restrictions of the experimental technique, shear and normal stress rates of 0.15 and 0.20Mpa/sec were adopted, but the principal stress rate became less than 0.015Mpa/sec (1/10 - 1/50 times the stress component speeds). The fact is that the apparent effect of the axis rotation is practically caused by the difference of time paths. Accordingly, we should use with caution constitutive laws without time parameter when we analyze RC structures where the direction of stress increments is greatly different from the updated stress direction. It is indispensable to formulate the time path-dependent deformation under high stress levels when we are looking for wide applicability of constitutive laws for concrete.

ACKNOWLEDGMENT

This research was sponsored by Grant-in-Aid for scientific research No.58350122 from Japan Ministry of Education to the Technological University of Nagaoka in 1984. The authors wish to thank Dr. H.Okamura, University of Tokyo, for his advice and express their gratitude to Mr. Y.Nakamura and Mr. M. Odagawa, Technological University of Nagaoka, for their help in the experiments.

REFERENCES

[1] Maekawa,K. and Okamura,H. : The Deformational Behavior and Constitutive Equation of Concrete Using the Elasto-Plastic and Fracture Model, Journal of Faculty of Engineering, the University of Tokyo(B), Vol.XXXVII, No.2, 1983, pp.253-pp.328.

[2] Chen,W.F. : Plasticity in Reinforced Concrete, McGraw-Hill Co., 1982.

[3] Darwin,D. and Pecknold,D.A. : Nonlinear Biaxial Stress-Strain Law for Concrete, Journal of Engineering Mechanics Division, ASCE, Vol.103, No.EM2 April 1977, pp.229-pp.421.

[4] Bresler,B. and Pister,K.S. : Strength of Concrete Under Combined Stresses, Journal of ACI, Sept. 1958, pp.321-pp.345.

[5] Okajima,T. : Fracture Criterion of Concrete under Combined Stresses (Comp-ression-Torque, Tension-Torque), Proceedings of AIJ, No.178, Dec. 1970 pp.1-pp.10. (in Japanese)

[6] Task Committee on Finite Element Analysis of Reinforced Concrete Structure of Structural division, ASCE : State-of-the-art Report on Finite Element Analysis of Reinforced Concrete, 1982.

[7] Li,B. and Maekawa,K. : Deformational Behavior of Concrete under Principal Stress Rotation, Proc. of JCI Colloquium on Finite Element Analysis of RC Structures, Dec. 1984, pp.29-pp.34 (in Japanese).

[8] Bazant,Z.P. and Kim,S. : Plastic Fracturing Theory for Concrete, Proc. of ASCE, No.EM3, June, 1979.

[9] Dougill,J.W. : Application and extensions of elasticity and plasticity theory, William Prager Symposium on Mechanics of Geomaterials, Rocks, Concretes, Solids, Northwestern University, 1983.

[10] Kupfer,H.B. and Gerstle,K.H. : Behavior of Concrete Under Biaxial Stresses ASCE Journal, No.EM4, Aug. 1973, pp.853-pp.866.

[11] Kotsovos,M.D. and Newman,J.B. : A mathematical description of the defor-mational behavior of concrete under complex loading, Magazine of Concrete Research, Vol.31, No.107, June, 1979, pp.77-pp.90.

[12] Ishikawa,M., Yoshikawa,H. and Tanabe,T. : The Nonlinear Constitutive Equation Based on the Damage Tensor and Its Application to Finite Element Method, Proc. of JCI Colloquium on Finite Element Analysis of RC Structures Dec. 1984 (in Japanese)

[13] Isobata, O. : On an anisotropic Constitutive Model of Concrete for the FEM Nonlinear Stress Analysis, Trans. of AIJ, No.265, March, 1978, pp.11-pp.18

THE CONSTITUTIVE MODEL IN TERMS OF DAMAGE TENSOR

Masami ISHIKAWA* Hiromichi YOSHIKAWA** Tada-aki TANABE***

SYNOPSIS

The damage concept was proposed originally by Kachanov in 1958 and developed to damage tensor and applied to the creep problem of metals by Murakami and Ohno in 1980 [4].
In this study, the damage concept is applied to construct a constitutive model for concrete. In the beginning, damage tensor is defined following Murakami and Ohno's derivation ([5], [8]), then it is developed to a three dimensional damage field where the damage increase in one direction is treated to affect the damage rate in the other two orthogonal directions. From them, 3-D damage strain is defined to allow for a stress analysis to be treated as an initial strain problem. It is shown then, that the failure process from initial softening to ultimate failure expresses successfully both the uniaxial loading history and multi-axial loading history. The proposed model was compared with Kupfer and Gerstle's model ([2], [3]) under biaxial loading, and a good agreement was shown between these two models. Stress-strain histories for uniaxial cyclic loading were also treated and reasonable expression was obtained.

1. INTRODUCTION

The constitutive models to describe the nonlinear behavior of concrete materials presently available can be classified roughly into two groups: a nonlinear elastic model and an elasto-plastic model. In the elasto-plastic theory, concrete materials are treated elastically, until their stress states reach the assumed yielding criteria. When a material is stressed beyond the yielding criterion, its behavior is

* Research engineer at the Technological Research Center of Tokyu Construction Co., Ltd. 2-13-9 Miyazaki, Miyamae-ku, Kawasaki City, 213 Japan

** Research engineer at the Technical Research Institute of Hazama-Gumi, Ltd. Yono City, Saitama Pref., 338, Japan

*** Professor in the Civil Engineering Department, Nagoya University. Furo-cho, Chikusa-ku, Nagoya City, 464, Japan described according to the associated flow rule or the non-associated flow rule based on the plasticity theory. Thus the elasto-plastic theory divides the behavior of materials into two states, before yielding and after yielding.

However, the stress-strain relation of concrete shows apparent nonlinearity from low stress levels, and concrete is brittle and little yielding takes place. Therefore, the determination of the yield point to be adopted in analysis, is not so obvious.

On the other hand, in a nonlinear elastic model, a comparatively simple and direct expression of the nonlinearity of concrete is adopted. This is easy to understand compared with the elasto-plastic theory. However, the theory does not have a phenomenological basis. It only describes the phenomenon mathematically which was observed from experiments.

The damage mechanics which were based on the damage concept were proposed by Kachanov in 1958. He applied this theory to the creep analysis of metals under uniaxial tension. Murakami and Ohno ([5], [8]) show that the material damage of creep can be represented by a second rank symmetrical tensor. This tensor is called damage tensor which estimates the decreasing ratio of the effective area three-dimensionally in the body. Kyoya and Ichikawa et al. [6] applied the damage concept to the analysis of discontinuous rock mass.

Herein, the authors attempt to introduce the damage concept into the nonlinear analysis of concrete materials, which is fundamentally different from both a plastic model and from a nonlinear elastic model. In accordance with the damage theory, internal micro-cracks and structural defects of concrete caused by loading are modeled as a continuous defect field. This defect field is described by a damage tensor which can take into account the direction of micro-cracks or micro-defects. Through the damage tensor, the relation between the nominal stress tensor and effective stress tensor is derived. Damage strain is then defined as the increase in strain caused by the reduction of the effective net area resisting the stress.

The damage theory is thus introduced into the Finite Element Method as an initial strain problem. The procedure enables one to describe the nonlinear behavior of concrete as brittle and non- or little yielding material, from the state of a low strain level to a failure state through the softening range.

2. DEFINITION OF DAMAGE TENSOR

Microscopic cracks occur around aggregates in concrete when it is loaded. These microscopic cracks grow with an increase of load level. When the concrete reaches a critical state, they link up with each other and grow rapidly. Nonlinear behavior of concrete may be attributed to the growth and linkage of these cracks, i.e., the development of a damage area within the concrete.

Generally, damage may be classified into two types: initial damage such as microscopic cracks or micro-defects, which exist in concrete before loading and the extending damage which is caused by external loading. The latter damage will be discussed in this paper.

Consider a body with damage which depends upon stress intensity. As shown in Fig. 1, we define the damage ϕ_1 as the damaged area per unit area of the section perpendicular to the X_1 axis. Now we assume the damage increment $d\phi_1$, which is observed in the section perpendicular to the X_1 axis as a result of the stress increment $d\sigma_1$ in the direction of X_1. It is assumed that the damage increment $d\phi_i$ has the effect to decrease the net resistance area. In general, the damage increment $d\phi_i$ may depend upon stress level σ_i, stress increment $d\sigma_i$, hysteresis of stress K_i, and initial damage ϕ_{oi}. It can be expressed by the following equation.

$$d\phi_i = f\,(\sigma_i,\ d\sigma_i,\ K_i,\ \phi_{oi},\ \text{etc.}) \tag{1}$$

We assume that the damage $d\sigma_1$ which was observed in the section perpendicular to the X_1 axis has some influence on the other two sections perpendicular to axes X_2 and X_3. This influence can be expressed as the damage influence ratios ζ_{21} and ζ_{31}. They are the ratios of the damage increment $d\phi_1$ in the direction of X_1 to the damage increment which was generated by the influence of $d\phi_1$ in the direction of X_2 and X_3. Therefore the damage increment which includes the mutual influence of each damage can be written as follows.

$$d\Omega_i = \zeta_{ij}\, d\phi_j \quad (i = 1,\ 2,\ 3,\quad j = 1,\ 2,\ 3) \quad \begin{cases} \zeta_{ij} = 1 \text{ for } i = j \\ \zeta_{ij} < 0 \text{ for } i \neq j \end{cases} \tag{2}$$

Explicitly,

$$\begin{Bmatrix} d\Omega_1 \\ d\Omega_2 \\ d\Omega_3 \end{Bmatrix} = \begin{bmatrix} \zeta_{11} & \zeta_{12} & \zeta_{13} \\ \zeta_{21} & \zeta_{22} & \zeta_{23} \\ \zeta_{31} & \zeta_{32} & \zeta_{33} \end{bmatrix} \begin{Bmatrix} d\phi_1 \\ d\phi_2 \\ d\phi_3 \end{Bmatrix} .$$

The infinitesimal tetrahedron OABC is taken within a body as shown in Fig. 2. The area of face ABC is designated, S, and let ν be the unit vector along the external normal direction on face S. e_i is a set of unit base vectors in the local coordinates. The area vector **S** can be defined as the product of the vector e_i and a scalar which is equal to the area of face ABC.

$$\mathbf{S} = S\nu = S_i e_i \quad (i = 1,\ 2,\ 3) \tag{3}$$

where S_i can be expressed as,

$$S_i = S(\nu \cdot e_i) \ . \tag{4}$$

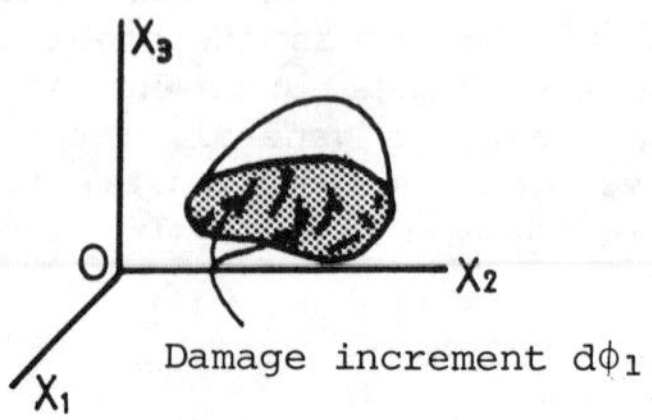

Fig. 1 A body with damage

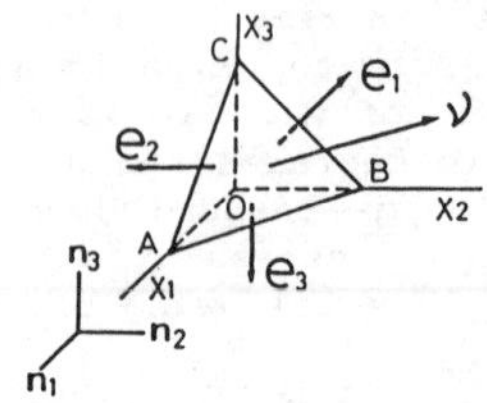

Fig. 2 A infinitesimal tetrahedron

As shown in Fig. 3, let us consider the case that the damage increment $d\phi_3$ is observed only at the section perpendicular to the X_3 axis as a result of a load increment in the direction of X_3. The damage increment $d\phi_3$ is assumed to be distributed over each plane perpendicular to X_3. A similar tetrahedron is taken in this damaged body. The area of ΔAOB of the tetrahedron OABC may be transformed into ΔA^*OB^* by the damage increment $d\phi_3$, as shown in Fig. 4. Introducing the damage influence ratios ζ_{13} and ζ_{23}, the areas ΔAOC and ΔBOC of this tetrahedron may also be transformed into ΔA^*OC^* and ΔB^*OC^*, respectively. The outward normal vector ν is reduced to ν^*. S_1, S_2 and S_3 are transformed into S^*_1, S^*_2, and S^*_3 as follows, respectively.

$$
\begin{aligned}
S_1^* &= (1 - \zeta_{13} d\phi_3)\, S_1 \\
S_2^* &= (1 - \zeta_{23} d\phi_3)\, S_2 \\
S_3^* &= (1 - \quad d\phi_3)\, S_3
\end{aligned}
\tag{5}
$$

Accordingly, the effective area vector $\mathbf{S}^*$ is

$$
\begin{aligned}
\mathbf{S}^* &= S^* (\nu^* \cdot e_i)\, e_i \\
&= (1 - \zeta_{13} d\phi_3)\, S_1 e_1 + (1 - \zeta_{23} d\phi_3)\, S_2 e_2 + (1 - d\phi_3)\, S_3 e_3 \, .
\end{aligned}
\tag{6}
$$

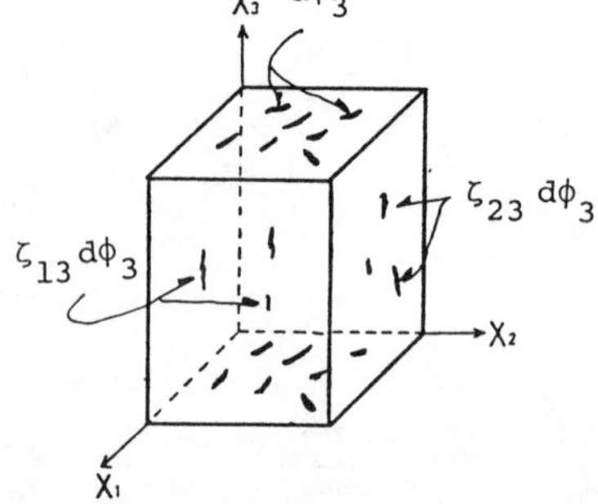

Fig. 3 A body with damage $d\phi_3$

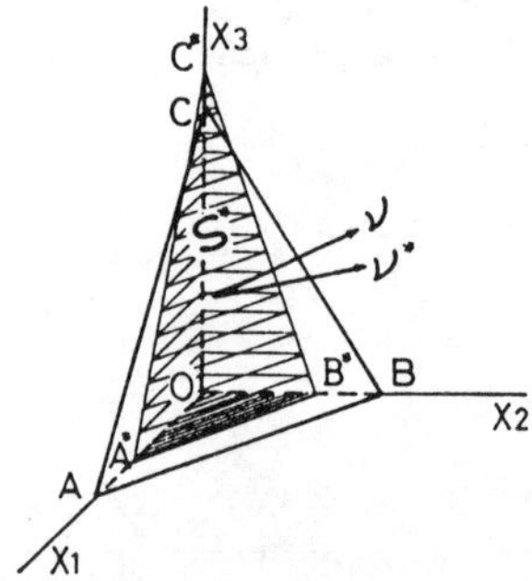

Fig. 4 Effective area of the tetrahedron

Under the multiaxial state, the effective area vector S^* can be given the following form.

$$S^* = (1 - d\Omega_1)\, S_1 n_1 + (1 - d\Omega_2)\, S_2 n_2 + (1 - d\Omega_3)\, S_3 n_3 \qquad (7)$$

$$= (I - d\Omega)\, S\nu$$

where I is a unit tensor, and $d\Omega$ is the second rank damage tensor, that can be defined as follows.

$$d\Omega = d\Omega_1\, (n_1 \otimes n_1) + d\Omega_2\, (n_2 \otimes n_2) + d\Omega_3\, (n_3 \otimes n_3) \qquad (8)$$

where n_1, n_2 and n_3 are unit base vectors in global coordinates. The summation of every step until the N-th step is

$$\Omega_N = \sum_{i=1}^{N} d\Omega_i \; . \qquad (9)$$

In the following discussion the reason for defining the damage tensor as in the above equation will be clarified.

The resultant force vector acting on the face ABC in Fig. 4 can be represented by S{T}, noting that T is traction. This resultant force vector is related to the stress tensor $[\sigma]$ as follows.

$$S\{T\} = [\sigma]\{S\nu\} \qquad (10)$$

In the body with damage, the resultant force vector acting on the effective area A*B*C*, is expressed as in Fig. 4.

$$S^*\{T^*\} = [\sigma^*]\{S^*\nu^*\} \qquad (11)$$

Since the equilibrium should be satisfied between the two states, before and after the damage,

$$S\{T\} = S^*\{T^*\} \qquad (12)$$

where T* is a traction acting on A*B*C*. In the view of the relation of Eq. (11), Eq. (12) can be rewritten as

$$[\sigma]\{S\nu\} = [\sigma^*]\{S^*\nu^*\} \qquad (13)$$

$$= [\sigma^*]([I] - [\Omega])\{S\nu\}$$

where $[\sigma^*]$ is a stress tensor. It can be defined as follows.

$$[\sigma^*] = [\sigma]([I] - [\Omega])^{-1} \qquad (14)$$

As can be seen, the nominal stress tensor $[\sigma]$ was transformed into the effective stress tensor $[\sigma^*]$ by the influence of damage, and it should be noted that the effective stress tensor generally is a non-symmetrical tensor of secondary rank. Once the symmetric tensor is obtained, it can be expressed in the form of a vector. If a symmetric tensor is needed, we can make it symmetric in the following way.

$$[A]_{sym} = \frac{1}{2}([A]^T + [A]) = \begin{bmatrix} A_{11} & A_{12} & A_{13} \\ A_{21} & A_{22} & A_{23} \\ A_{31} & A_{32} & A_{33} \end{bmatrix} \rightarrow \quad \{A\} = \begin{Bmatrix} A_{11} \\ A_{22} \\ A_{33} \\ A_{12} \\ A_{23} \\ A_{31} \end{Bmatrix} \tag{15}$$

The nominal stress tensor can be rewritten in vector form, considering Eq. (14).

$$[\sigma] = [\sigma^*]([I] - [\Omega]), \ \{\sigma\} = ([I] - [\Omega^*])\{\sigma^*\}$$

where,

$$[\Omega^*] = \begin{bmatrix} [\Omega]^T & 0 & 0 \\ 0 & [\Omega]^T & 0 \\ 0 & 0 & [\Omega]^T \end{bmatrix} . \tag{16}$$

Note that $[\Omega^*]$ is a matrix of 9 × 9. To apply this matrix $[\Omega^*]$ to F.E.M. code, it can be reduced to $[\Omega^*]_{sym}$ to be matrix of 6 × 6 by Eq. (16), considering symmetry of stress tensor.

$$\{\sigma\}_{sym} = ([I] - [\Omega^*])_{sym} \{\sigma^*\}_{sym} . \tag{17}$$

3. DEFINITION OF DAMAGE STRAIN

The initial elasticity is assumed to keep strain constant in the net resistant area. In other words, the strain $[\varepsilon]$ corresponding to effective stress $[\sigma^*]$ acting on the intact part can be estimated through the initial elastic modulus. Referring to Fig. 5, the nominal stress corresponding to $[\sigma^*]$ is $[\sigma]$, and the strain $[\varepsilon_e]$ corresponding to the nominal stress may also be obtained by use of the initial elasticity. It is shown in Fig. 5 as ε_e. The difference between these two strains $[\varepsilon]$ and $[\varepsilon_e]$ can be defined as damage strain $\{\varepsilon_f\}$.

$$\begin{aligned} \{\varepsilon_f\} &= \{\varepsilon\} - \{\varepsilon_e\} \\ &= [Do]^{-1} \{\sigma^*\} - [Do]^{-1} \{\sigma\} \\ &= [Do]^{-1}(\{\sigma^*\} - ([I] - [\Omega^*])\{\sigma\}) \\ &= [Do]^{-1} [\Omega^*][Do] \{\varepsilon\} \end{aligned} \tag{18}$$

where [Do] is the matrix of initial elasticity.

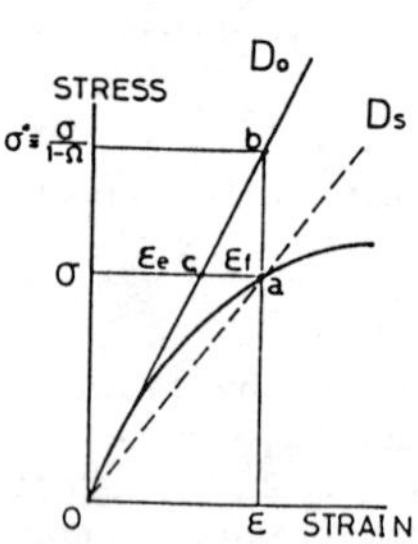

Fig. 5 Relation between total strainand damage strain

Next we discuss the application of the damage tensor to the constitutive equations for nonlinear analysis. When we introduce the damage tensor into the stress-strain relation, the following three kinds of application may be considered. The first kind is to associate the damage tensor with strain. Stress can be expressed by the following equation, referring to the point c in Fig. 5.

$$\{\sigma\} = [D_o](\{\varepsilon\} - \{\varepsilon_f\}) = [D_o]\ (\{\varepsilon\} - [D_o]^{-1}[\Omega^\star][D_o]\{\varepsilon\}) \tag{19}$$

This equation corresponds to the concept of elasto-plastic theory in which the strain is defined with two components, elastic and plastic. It is also possible to associate the damage tensor with the [D] - matrix, as the second kind. The material stiffness can decrease as the result of damage. In this method the matrix of initial elasiticity is changed according to the damage level.

$$\{\sigma\} = ([I] - [\Omega^*])[Do]\{\varepsilon\} = [D_s]\{\varepsilon\} \tag{20}$$

This equation corresponds to nonlinear elastic theory or the tangential method, and it is illustrated as D_S in Fig. 5. The last kind is to associate the damage with stress. That is the method to transform the nominal stress into effective stress, referring to point b in Fig. 5. It can be developed into an initial stress method in the F.E.M code.

$$[\sigma^*] = [\sigma]([I] - [\Omega])^{-1} \rightarrow \{\sigma^*\}_{sym} \tag{21}$$

Note that the equation is described in tensorial form.

4. NUMERICAL EXAMINATION

4.1 Nonlinear Behavior with Strain Softening

Let us consider a member in a uniaxial stress state. The definition of damage strain ε_f in Eq. (18) in general form turns out to be the following equation in the uniaxial state.

$$\varepsilon_f = \Omega\varepsilon \tag{22}$$

Ω is the damage parameter (or damage scalar) in the uniaxial state, and the strain ε is called total strain in the following discussion. The total strain ε can be expressed as the sum of the elastic and damaged strain,

$$\varepsilon = \varepsilon_e + \varepsilon_f \tag{23}$$

Assuming that

$$\Omega = \frac{1}{2\varepsilon_o}\varepsilon \tag{24}$$

the damage strain in the uniaxial state is expressed as

$$\varepsilon_f = \frac{1}{2\varepsilon_o}\varepsilon^2 \quad . \tag{25}$$

The stress-strain relation obtained from substituting the Eq. (25) into Eq. (19), coincides with CEB-FIP code which is the equation of parabolic approximation, i.e.

$$\frac{\sigma}{f_c} = \{2\left(\frac{\varepsilon}{\varepsilon_o}\right) - \left(\frac{\varepsilon}{\varepsilon_o}\right)^2\} \tag{26}$$

where f_c and ε_o are the uniaxial compressive strength and the strain at the peak stress, respectively. Fig. 6 shows the stress-strain curve simulated by using Eq. (25). The solid line shows the nominal stress-strain curve and the dotted line shows the effective stress-strain curve. Fig. 7 shows the comparison of experimental results by Kupfer et al. in 1969, [2] with the model. Fig. 7 shows a good agreement between the two.

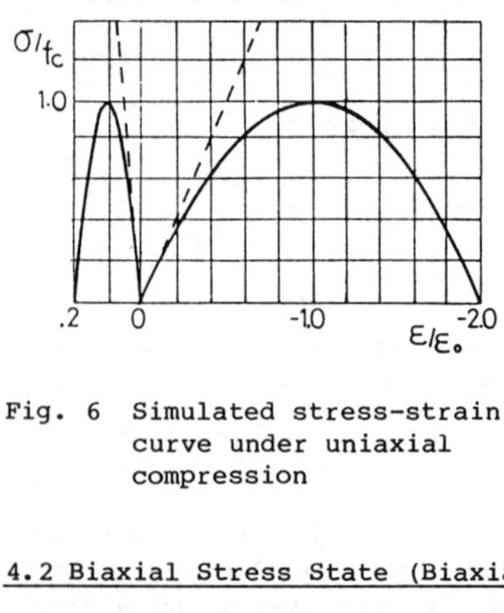

Fig. 6 Simulated stress-strain curve under uniaxial compression

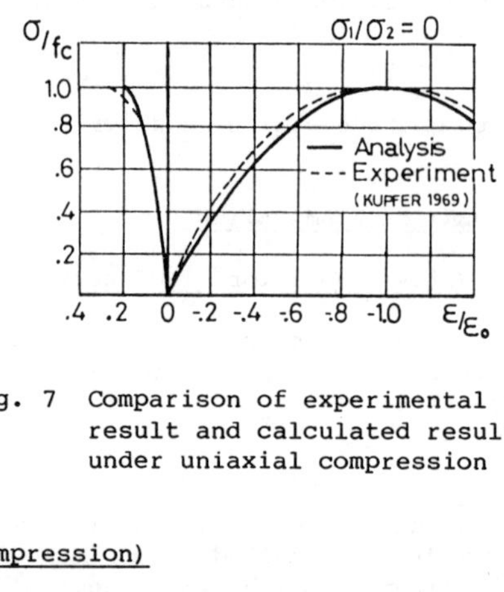

Fig. 7 Comparison of experimental result and calculated result under uniaxial compression

4.2 Biaxial Stress State (Biaxial Compression)

To describe the biaxial stress behavior of concrete, we use the damage influence ratio in Eq. (2). Kupfer et al. [2] concluded that the strength of concrete subjected to biaxial compression is higher than the uniaxial strength. Such a phenomenon may be explained that the extending damage is confined by the effect of each orthogonal stress. Therefore, we assume that the damage influence ratio is negative in the biaxial compression state. Fig. 8 shows that stress-strain curve obtained by the use of this model under proportional loading with various principal stress ratios. Kupfer et al. [3] suggested that the elasticity of material varies with bulk modulus K and shear modulus G. Kupfer's model can be classified as one of the nonlinear elastic models, and the stress-strain curve under the proportional loading is shown in Fig. 9. These two models are comparable to each other and show similar nonlinear characteristics though they are derived from fundamentally different models. Fig. 10 shows the comparison of experimental results by Kupfer and Gerstle in 1969, with calculated results of damage model, when the principal stress ratio is 0.52.

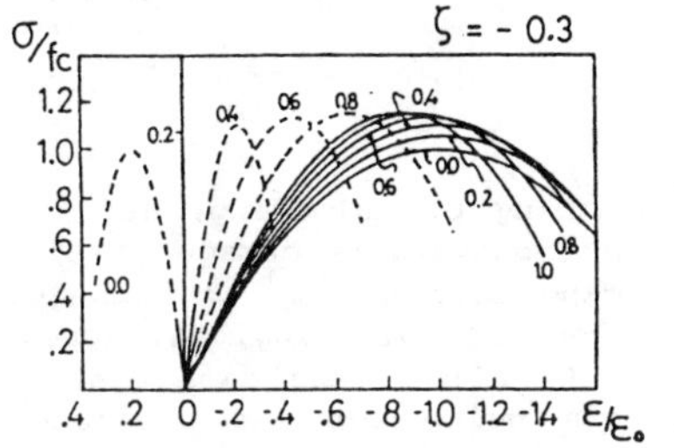

Fig. 8 Biaxial compression curve of the proposed model

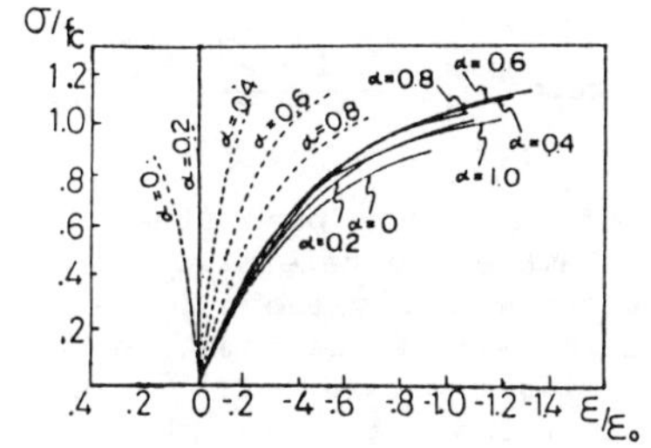

Fig. 9 Biaxial compression curve of Kupfer's model

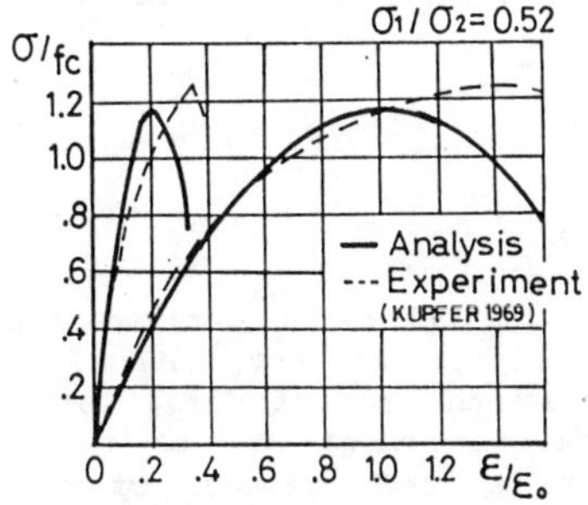

Fig. 10 Comparison of experimental result and calculated result under biaxial compression

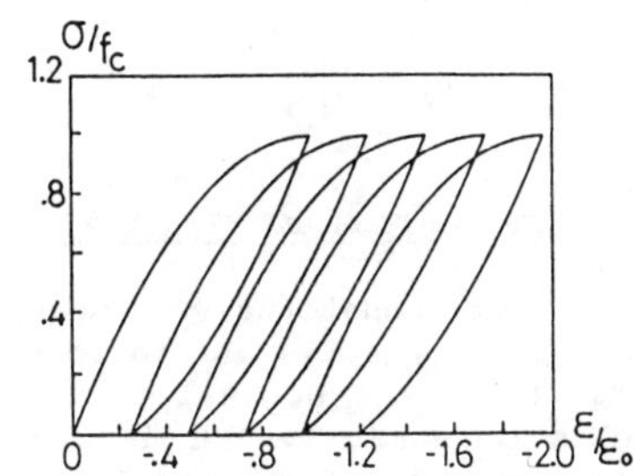

Fig. 11 Unloading and Reloading model

4.3 Unloading and Reloading Example

The behavior of concrete subjected to cyclic loading is simulated with the damage model as shown in Fig. 11. Eq. (25) is used to estimate the initial slope. At unloading branch and reloading branch, the definition of damage parameter Ω_{unl} and Ω_{rel} are as follows.

At unloading,

$$\Omega_{unl.} = 2 - \frac{\varepsilon_0}{2(\varepsilon_p - \varepsilon)} - \frac{\varepsilon_p - \varepsilon}{2\varepsilon_0}\xi_1 \quad , \qquad (27\text{-a})$$

at reloading,

$$\Omega_{rel.} = \frac{\varepsilon - \varepsilon_{resd.}}{2\varepsilon_0}\xi_2 - \frac{\varepsilon - \varepsilon_{resd.}}{2\varepsilon_0\varepsilon}\varepsilon_{resd.}\cdot\xi_2 - \frac{\varepsilon_{resd.}}{\varepsilon}, \tag{27-b}$$

$(0 \leqq \xi_1 \leqq 1,\ 0 \leqq \xi_2 \leqq 1)$

where ε_{resd} is the residual strain defined as

$$\varepsilon_{resd.} = (1 - \frac{1 - \sqrt{1 - \xi_1}}{\xi_1})\varepsilon_p, \tag{27-c}$$

and ε_p is the peak strain at the beginning of unloading, and ξ_1 is a parameter to determine the magnitude of recoverable damage strain. It assumes the value 1.0 for the reloading branch. If ξ_1 is equal to zero, all of the damage strain is kept during unloading, and appears as residual strain. If ξ_1 is equal to 1.0, all of the damage strain is recovered. By selecting an appropriate value for the parameter ξ_1, the model can describe the recoverable strain. The parameter ξ_2, determines the slope of the reloading branch. The damage parameter is defined as in Eq. (28) when the stress reaches the peak stress.

$$\Omega = 1 - \frac{\varepsilon_o}{2\varepsilon}\xi_1 \qquad (\xi_1 = 0.9) \tag{28}$$

5. APPLICATION TO FINITE ELEMENT METHOD

The constitutive equation Eq. (19) which is derived from the damage concept, is introduced into Finite Element Method. As in Eq. (23), the nominal strain is given by the sum of damage strain and elastic strain which is obtained from the nominal stress through the initial elastic modulus. Considering the damage strain, the constitutive equation can be written in matrix form,

$$\{\sigma\} = [D_o](\{\varepsilon\} - \{\varepsilon_f\}). \tag{29}$$

Using the principle of virtual work,

$$\{F\} = \int_V [B]^T [D_o][B]\, dv\, \{u\} - \int_V [B]^T [D_o]\{\varepsilon_f\}\, dv \tag{30}$$

where $\{F\}$ is the load vector and $\int_V [B]^T [D_o]\{\varepsilon\}$ dv is the equivalent load vector corresponding to damage strain. Eq. (30) means that it is possible to analyze the nonlinear behavior of concrete as an initial strain problem.

As another procedure, if the damage strain is given as a function of total strain from Eq. (18), $\{\varepsilon_f\}$ can be written as

$$\{\varepsilon_f\} = [D_o]^{-1}[\Omega^*]_{sym}[D_o][B]\{u\}. \tag{31}$$

Eq. (30) can be developed according to the following procedure.

$$\begin{aligned}\{F\} &= \int_V [B]^T [D_o] [B]\, dv \{u\} - \int_V [B]^T [\Omega^*]_{sym} [D_o][B]\, dv \{u\} \\ &= \int_V [B]^T ([I] - [\Omega^*]_{sym}) [D_o][B]\, dv \{u\} \qquad (32) \\ &= \int_V [B]^T [D_s][B]\, dv \{u\}\end{aligned}$$

where [I] is unit matrix. [D] is the constitutive equation for damaged material,

$$[D_s] = ([I] - [\Omega^*]_{sym}) [D_o] \qquad (33)$$

6. CONCLUSION

The damage concept is applied to constitutive models for concrete material, and several forms of its application were discussed. As an example, the damage tensor Ω is made to correspond to the parabolic stress-strain equation proposed by the CEB-FIP code. The damage influence ratio ζ_{ij} is also made to reflect the behavior of concrete in multi-axial stress states.

The present study is only the first step of utilizing the damage concept to concrete material and further study is necessary to make a comparative study of how the damage tensor of the second rank corresponds phenomenologically to experimental results including the investigation of the problem of the tensorial invariance restriction.

7. REFERENCES

[1] F.H. Wittman ed., "Fracture Mechanics of Concrete," Elsevier, 1983.

[2] H. Kupfer, H. K. Hilsdorf, and H. Rusch, "Behavior of Concrete Under Biaxail Stress," Journal ACI Vol. 66, No. 8, August 1969, pp. 656-666.

[3] H. Kupfer and K. Gerstle, "Behavior of Concrete Under Biaxial Stresses," ASCE (EM4), August, 1973, pp. 853-866.

[4] S. Murakami, "Damage Mechanics - Continuum Mechanics Approach to Damage and Fracture of Materials," Journal of the Society of Material Science, JAPAN Vol. 31, 340 1982. (in Japanese)

[5] S. Murakami and N. Ohno, "A Continuum Theory of Creep and Creep Damage," International Union of Theoretical and Applied Mechanics, Creep in Structures, Proc. of 3rd Symposium, Leicester, UK, September 1980.

[6] T. Kyoya, Y. Ichikawa et al., "On Damage Mechanics Model for Rock and Rock Mass," The 17th JAPAN National Conference on Soil Mechanics and Foundation Engineering, June, 1982 (in Japanese)

[7] T. Kyoya," An Elast-Plastic Damage Mechanics Model for Rock and Rock Mass." Master of Engineering Thesis. University of Nagoya, 1983 (in Japanese)

[8] S. Murakami and N. Ohno, "Tensorial Expression of Creep Damage Multi-axial State of Stress," Transactions of the JAPAN Society of Mecnanical Engineers, 46, 409, 1980 (in Japanese)

Invariant Formulation of Constitutive Relations for Concrete

Kurt H. Gerstle, F.A.S.C.E.[1] and Niels S. Ottosen[2]

ABSTRACT

A simple constitutive relation for concrete is outlined which appears capable of predicting strains under axisymmetric stress histories. Test results are cited which document the existence of coupling between volumetric and distortional concrete response. Loading-unloading behavior is shown which appears to contradict that usually assumed for concrete. Comparison with test results indicates good correlation between prediction and observation.

A critique of the method indicates several limitations and a more general approach is outlined which extends the applicability of the method.

INTRODUCTION

The need for the quantitative determination of concrete properties arises from the requirements of the analyst. It is therefore important to present the experimental results in such a way that they can be used in the analysis of the behavior and strength of structures. It is the purpose of this paper to formulate the observed behavior of concrete for use in such analyses.

Many theories and formulations have been applied to describe concrete strength and behavior [8,7,1], ranging from the very simple to the very involved. It is not the purpose of our study to discuss the pros and cons of different theories; rather, we will aim at a reasonable compromise between reality and simplicity. Too much of the former may lead to intractable solutions of such complexity, or requiring so many material parameters, as to be unsuitable for the structural engineer; too much of the latter may lead to meaningless predictions.

The vast scatter of the test results should be considered in deciding on the desired accuracy of the formulation. The expense of an "exact" formulation may be unwarranted in a statistical material such as concrete in which there may be a large variation of behavior

---

[1] Prof. of Civil Engineering, University of Colorado, Box 428, Boulder, Colorado 80309

[2] Research Engineer, Riso National Laboratory, Roskilde, Denmark.

and strength from one specimen to the next, or between specimen and structural concrete.

The relation of test conditions to structure conditions is important but highly uncertain. Test specimen type and boundary conditions have a great influence on the observed test results [3], but it is by no means certain which conditions are best able to make the concrete "feel as if it were part of the structure". Little attention has been paid to such matters in the past.

A unified formulation of concrete response and strength which is applicable to all possible stress states is highly desirable since uni- and biaxial cases should be considered nothing more than special cases of the more general triaxial stress state. Because in the past uni-, bi-, and triaxial testing was often undertaken on separate occasions using different apparatus, different analytical representations resulted for each case. Our test program [13] contained all of these special cases and may thus form the basis for a unified formulation; this is attempted in the following.

SIMPLE FORMULATION

The proposed formulation of concrete behavior is based on a constitutive relation between stress and strain invariants, as provided by the octahedral representation. Accordingly, the procedure consists of three steps:

1. Transformation of given principal stress increments into their octahedral equivalents.
2. Application of constitutive relations to obtain octahedral strain increments.
3. Re-transformation of the octahedral strain increments into principal strain increments.

In the following, these steps are concisely outlined.

Octahedral Stress and Strain

The octahedral normal, or mean normal, stress σ_0, and octahedral shear, or deviatoric, stress τ_0, in terms of the principal stresses σ_1, σ_2, σ_3, are given by

$$\sigma_0 = \frac{1}{3}(\sigma_1 + \sigma_2 + \sigma_3) = \frac{1}{3}\sigma_{ii} \;, \; i = 1,2,3 \qquad (1)$$

$$\begin{aligned} \tau_0 &= \frac{1}{3}\sqrt{(\sigma_1-\sigma_2)^2 + (\sigma_2-\sigma_3)^2 + (\sigma_3-\sigma_1)^2} \\ &= \sqrt{\frac{1}{3}}\;\sqrt{S_{ij}\,S_{ji}} \;, \end{aligned} \qquad (2)$$

in which $S_{ij} = (\sigma_{ij} - \sigma_0 \cdot \delta_{ij})$ are the deviator stresses.

Similarly, the octahedral normal, or volumetric, strain ε_0, and the octahedral shear, or distortional, strain γ_0, are related to the principal strains ε_1, ε_2, ε_3, by

$$\varepsilon_0 = \frac{1}{3}(\varepsilon_1 + \varepsilon_2 + \varepsilon_3) = \frac{1}{3}\varepsilon_{ii} \tag{3}$$

$$\begin{aligned}\gamma_0 &= \frac{1}{3}\sqrt{(\varepsilon_1-\varepsilon_2)^2 + (\varepsilon_2-\varepsilon_3)^2 + (\varepsilon_3-\varepsilon_1)^2} \\ &= \sqrt{\frac{1}{3}}\ \sqrt{e_{ij}\,e_{ji}}\ ,\end{aligned} \tag{4}$$

in which $e_{ij} = (\varepsilon_{ij} - \varepsilon_0 \cdot \delta_{ij})$ are the deviator strains.

Octahedral Stress and Strain Increments

For general cases of non-proportional loading, the increments of the octahedral stresses and strains are related to the principal stress and strain increments by

$$\Delta\sigma_0 = \frac{1}{3}\Delta\sigma_{ii} = \frac{1}{3}(\Delta\sigma_1 + \Delta\sigma_2 + \Delta\sigma_3) \tag{5}$$

$$\Delta\tau_0 = \frac{1}{3}\ \frac{s_{ij}\Delta s_{ij}}{\tau_0} = \frac{1}{3\tau_0}\ (s_{11}\Delta s_{11} + s_{22}\Delta s_{22} + s_{33}\Delta s_{33}) \tag{6}$$

$$\Delta\varepsilon_0 = \frac{1}{3}\Delta\varepsilon_{ii} = \frac{1}{3}(\Delta\varepsilon_1 + \Delta\varepsilon_2 + \Delta\varepsilon_3) \tag{7}$$

$$\Delta\gamma_0 = \frac{1}{3}\ \frac{e_{ij}\Delta e_{ij}}{\gamma_0} = \frac{1}{3\gamma_0}\ (e_{11}\Delta e_{11} + e_{22}\Delta e_{22} + e_{33}\Delta e_{33}) \tag{8}$$

Eqs. 6 and 8 show that because of their nonlinear nature, the octahedral shear stress and strain increments depend not only on the stress and strain increments, but also on the current state of stress and strain.

Octahedral Stress-Strain Relations

In order to predict strains in a nonlinear material under arbitrary, non-proportional load histories, we seek to relate the octahedral stress and strain increments in the following fashion:

$$\begin{Bmatrix}\Delta\varepsilon_0 \\ \Delta\gamma_0\end{Bmatrix} = \begin{bmatrix}\dfrac{1}{3K} & \dfrac{1}{H} \\ \dfrac{1}{Y} & \dfrac{1}{2G}\end{bmatrix}\begin{Bmatrix}\Delta\sigma_0 \\ \Delta\tau_0\end{Bmatrix} \tag{9}$$

The moduli K, G, H, and Y are tangent values which may depend on stress, strain, and load history. Their values are to be determined from tests. In general, this constitutive matrix will be non-symmetric.

For classical elasticity, the values of the off-diagonal terms, which couple volumetric and deviatoric behavior, are zero, leading to infinitely large values of the moduli H and Y. In our test results, to be presented in the following, we demonstrate the existence of this coupling for concrete.

It should be noted that the two octahedral strain increments obtained from Eq. 9 are insufficient for determination of the three principal strain increments desired. One additional condition will be needed to solve for these three unknowns in the general case. However, for axisymmetric cases in which $\Delta\sigma_2 = \Delta\sigma_3$, $\Delta\varepsilon_2 = \Delta\varepsilon_3$, we have only two unknowns and Eq. 9 is sufficient.

DETERMINATION OF MODULI

Reported test data on concrete stiffness vary widely [4]. In view of these uncertainties, simple formulations are adopted here; as new, and more reliable, information becomes available, these moduli can be improved. Compressive stresses and strains will be considered positive in the following.

The Bulk Modulus K

Fig. 1 [12] shows a typical volumetric stress-strain curve from a cube test. This behavior can be represented by a linear decrease of the bulk modulus K, of limited range:

$$K = K_0\ (1 - C \cdot \sigma_0) \geq C'\ K_0 \tag{10}$$

From our tests, the initial bulk modulus $K_0 = 2.1 \times 10^3$ ksi (14.5×10^3MPa), C = .16, and C' = .14.

The Shear Modulus G

Fig. 2a shows mean deviatoric (τ_0-γ_0) stress-strain curves from tests under various biaxial proportional loadings [5]. We approximate these curves by the exponential expression

$$\tau_0 = \tau_{ou}\ [1 - e^{-\frac{2G_0}{\tau_{ou}}\gamma_0}] \tag{11}$$

in which G_0 is the initial shear modulus, and τ_{ou} is the octahedral shear strength, to be determined from a suitable strength envelope.

Differentiating, and substituting, we obtain the tangent shear modulus as function of the stress level:

$$G = G_0\ [1 - \frac{\tau_0}{\tau_{ou}}]\ , \tag{12}$$

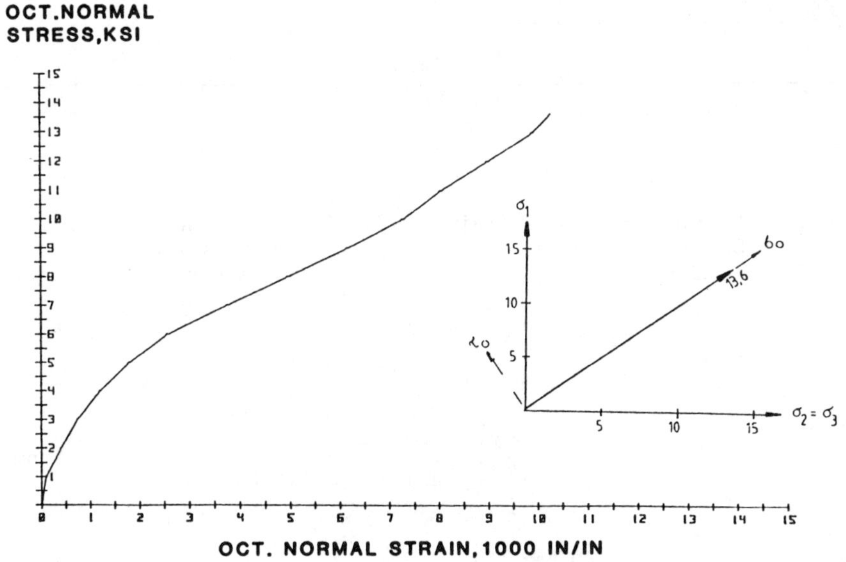

Fig. 1 Octahedral Normal Stress Versus Strain

in which G_0 itself is a function of the hydrostatic stress σ_0:

$$G_0 = G_{e1} - C_g \cdot \sigma_0 \geq C'_g \cdot G_{e1} ,$$

in which G_{e1} is the initial shear modulus corresponding to σ_0 = 0. In our tests, G_{e1} = 1.7 x 10^3 ksi (11.7 x 10^3 MPa), C_g = .25, and C'_g = .47.

The deviatoric shear strength τ_{ou} can be defined in several different ways, as shown in Fig. 3. τ_{ou1} is the triaxial compression (or tension) strength corresponding to the current value of σ_0. τ_{ou2} is defined by the direction of the current stress increment extended; and τ_{ou3} is defined by the radial line from the origin through the current stress point, extended. For a given stress path, such as the one shown dashed, all three definitions will converge to the same value at failure, but the shear stiffness G along the load path, and therefore the strains at failure, will be different.

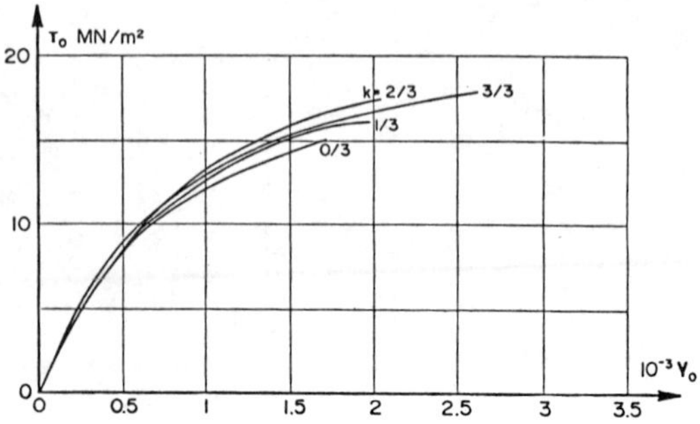

Fig. 2a. Deviatoric Stress-Strain Curves for Biaxial Loadings

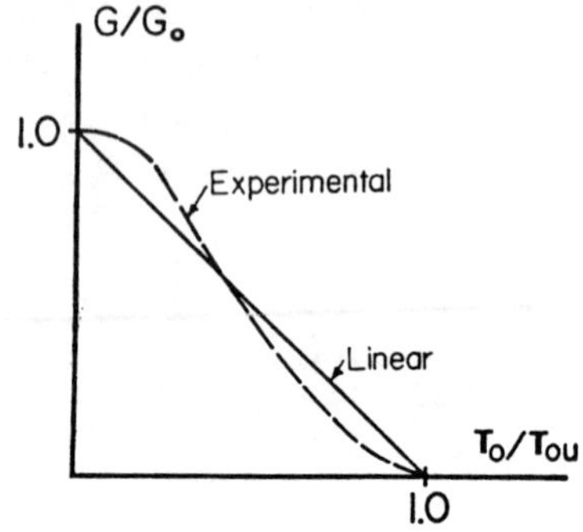

Fig. 2b. Variation of Shear Modulus with Octahedral Shear Stress

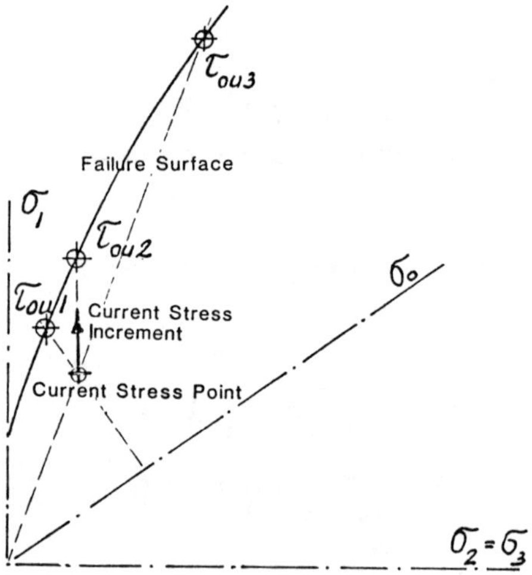

Fig. 3. Definitions of Shear Strength

In our calculations, the definition $\tau_{ou} = \tau_{ou1}$ was adopted; it should be noted that this might lead to difficulties for the special case of uniaxial loading.

Eq. 12 shows that the assumption of an exponential τ_0-γ_0 curve leads to a shear modulus which decreases linearly from its initial value G_0 for $\tau_0 = 0$ to zero at failure, as shown in Fig. 2b. As such, it appears to be the simplest variation of modulus possible; it contains no linear range. It requires only the initial shear modulus G_0 and the strength as discussed earlier for quantification. The fact that compressive failure τ_{ou} coincides with G = 0 indicates a ductile failure in accordance with the principles of perfect plasticity.

The Coupling Modulus H

Volume contraction occurs during pure stress deviation until the stresses are close to failure. This coupling between deviator stress and volume strain increases with the hydrostatic stress σ_0. This effect is accounted for by the modulus H in Eq.9. The coupling modulus H was given in Ref.[6] as

$$H = (1.45 + \frac{6.3}{\sigma_0 - 1.45}) \times 10^3 \text{ Ksi}, \quad \sigma_0 > 1.45 \text{ Ksi}$$

$$[H = (10 + \frac{300}{\sigma_0 - 10}) \times 10^3 \text{ MPa}, \quad \sigma_0 > 10 \text{ MPa}] \tag{13}$$

For information about this coupling behavior during unloading and reloading, several tests were carried out; Fig. 4b represents the τ_0-ε_0 relations for the three deviatoric loading-unloading histories shown in Fig. 4a [12]. The observed volume change during unloading was deemed negligible, suggesting elastic unloading behavior in the sense of a zero off-diagonal term 1/H in Eq. 9. Accordingly, the modulus H is set infinitely large during un- and reloading in the computations.

The Coupling Modulus Y

Away from the hydrostatic axis, distortion can occur in the material even under pure hydrostatic stress increments. This phenomenon is accounted for by the coupling modulus Y in Eq. 9. In Ref. [11], the variation of this modulus, based on preliminary information, was adopted as

$$Y = \frac{12000}{\tau_0^2} \times 10^3 \text{ ksi}$$

$$(Y = \frac{3942}{\tau_0^2} \times 10^6 \text{ MPa}) \tag{14}$$

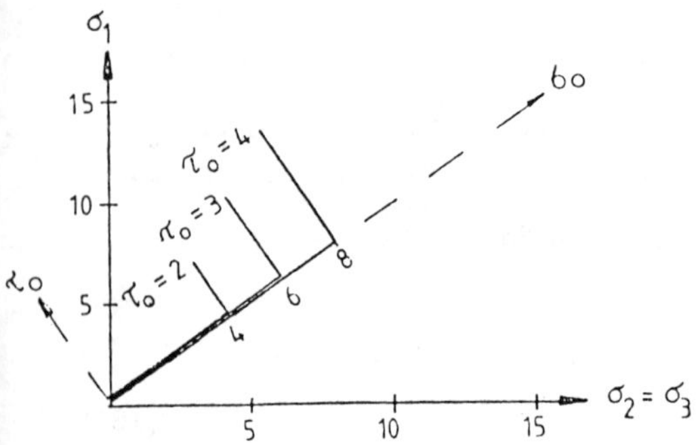

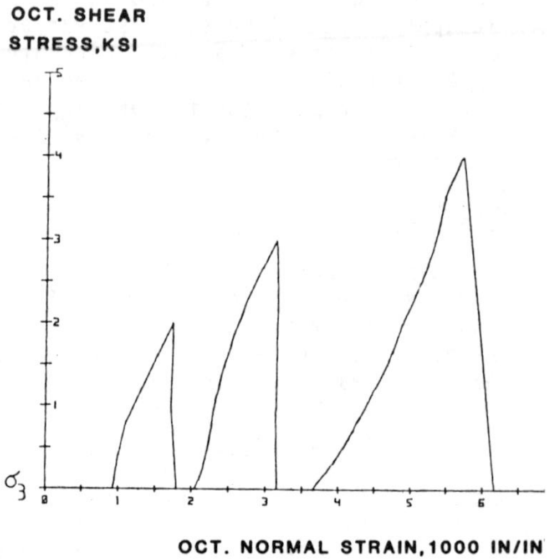

Fig. 4a. Stress Histories for Three Tests

Fig. 4b. Octahedral Shear Stress Versus Octahedral Normal Strain for Three Tests

that is, coupling vanishes for stress increments along the hydrostatic axis, which might be expected by symmetry. In order to explore Y-coupling under un- and reloading, the three tests shown in Fig. 5a were carried out. They consisted of hydrostatic pre-loading to σ_0 = 8 ksi, unloading to 4 ksi, triaxially compressive stress deviations to τ_0 = 1,2, and 3 ksi, followed by repeated cycles of hydrostatic increments between σ_0 = 4 ksi and σ_0 = 8 ksi. The measured curves of σ_0 versus γ_0 shown in Fig. 5b indicate that Y-coupling might reasonably be neglected during un- and reloading, suggesting elastic behavior corresponding to the off-diagonal term 1/Y equal to zero. In computations, this is accounted for by setting the value of Y-modulus very large during un- and reloading.

Loading Surface and Flow Rule

Virgin loading, unloading, and reloading are clearly defined in proportional loading, but this is not so in more complex stress histories. Fig. 6a [14] shows a load history consisting of several successive deviatoric excursions within one deviator plane σ_0 = constant. The measured principal stress-strain curves of Fig. 6b indicate that during any one of these excursions, only that principal stress component which increases beyond its previously attained highest value results in a nonlinear strain increment. Un-, and

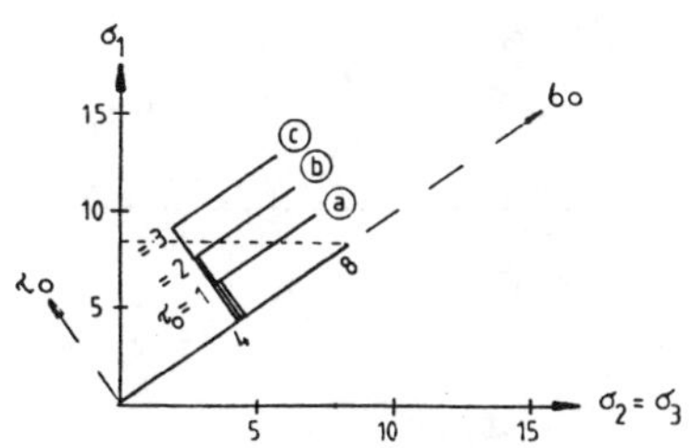

Fig. 5a. Stress Histories for Three Tests

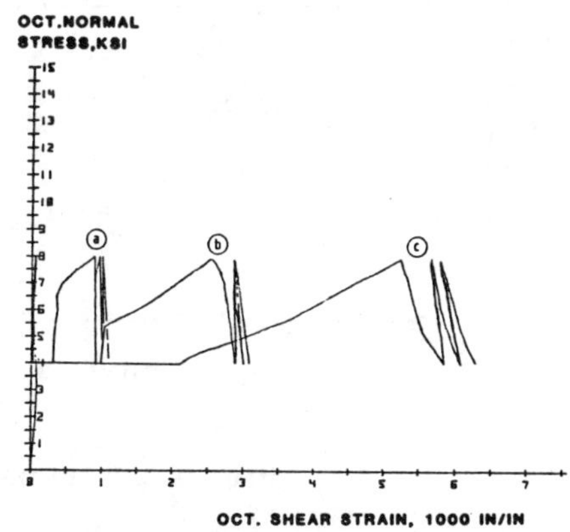

Fig . 5b. Octahedral Normal Stress versus Octahedral Shear Strain for Three Tests

reloading up to the previously attained highest principal stress value proceed essentially elastically. These observations are confirmed by the stress-strain curves of Fig. 5b. The highest stress σ_1 previously attained during the preloading to σ_0 = 8 ksi, and shown by a horizontal dashed line in Fig. 5a, determines the onset of nonlinear strains during Tests a and b.

It should be noted that this reloading behavior is quite different from that which would have been predicted by classical plasticity theory [2], according to which the current loading surface would be a circle of radius τ_{0max}, shown dashed in Fig. 6a. According to this criterion, all loadings following Excursion 1 would be purely elastic, since they are contained within the circular loading surface defined by Loading 1.

The previous discussion suggests that elastic and inelastic (or "plastic") strain components ε^E and ε^P can be separated; the former follows Hooke's law, the latter occurs only for principal stresses in excess of their previous highest value. The total response is then governed by the variable moduli defined in Eq. 9.

Geometrically, these observations can be represented in principal stress space by a loading surface in the shape of a cartesian "box"

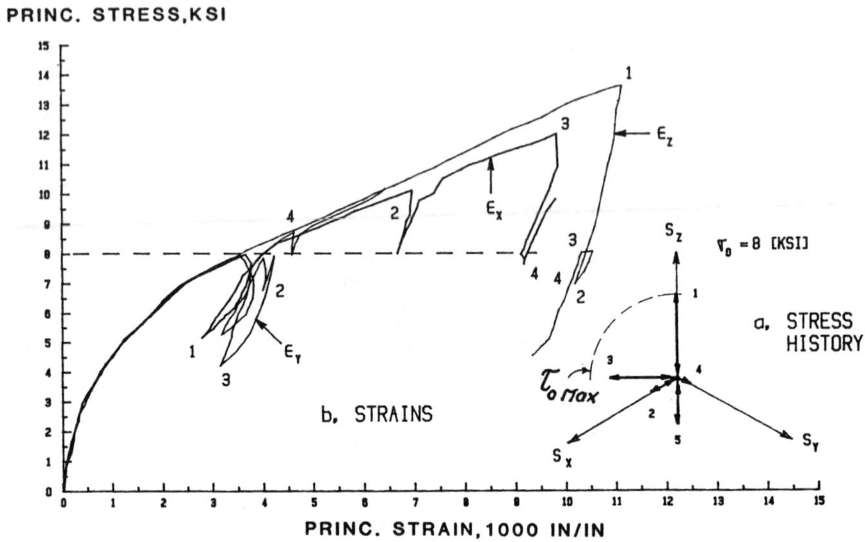

Fig. 6. Multiaxial Stress History

which expands under virgin loading. Its side dimensions are defined by the highest attained principal stress components; if the current load increment causes increase of σ_i, the stress path will translate ("push") the loading surface with it; a moving loading surface defines virgin loading. Any subsequent load history confined within the current location of the loading surface will consist of elastic unloading and reloading.

In the projection of the principal stress space into the Rendulic plane for which $\sigma_2 = \sigma_3$, the loading surface will appear as shown in Fig. 7a. In the deviatoric plane σ_0 = constant, the trace through the loading surface appears triangular, as in Fig. 7b.

If we associate a plastic strain space with the principal stress space, as shown in Fig. 7 by the axis labeling, then we can represent the plastic strain increment discussed earlier by a vector attached to the current stress point. Since the direction of this vector, according to earlier observation, must be parallel to the increasing principal stress, it must therefore be normal to the loading surface at the current stress point, as shown in Fig. 7a. This vector can be decomposed into two components: that parallel to the hydrostatic axis ε_0 represents the inelastic volume change, and that parallel to the deviator axis γ_0 represents the inelastic distortion of the plastic strain vector.

The postulated loading surface is not affine to the failure surface. Further, the rectangular loading surface will always predict

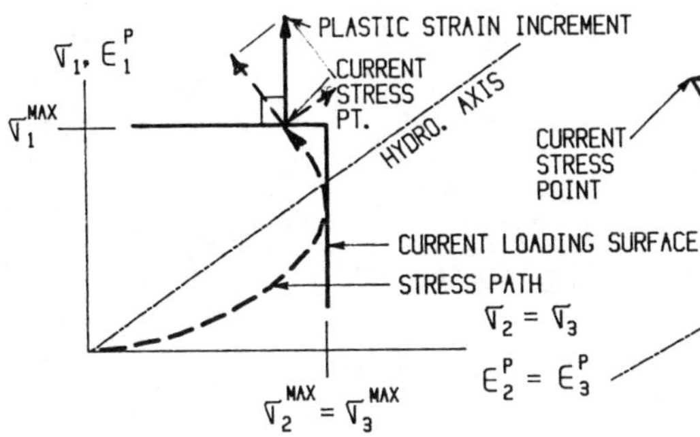

Fig. 7a. Loading Surface in Rendulic Plane

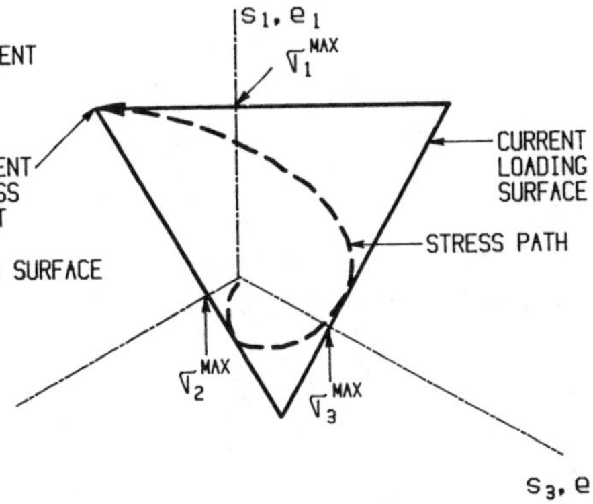

Fig. 7b. Loading Surface in Deviatoric Plane

volume contraction, which is in conflict with actual concrete behavior near failure. We account for this by introducing a transition [14] which rotates the plastic strain vector near failure to make it normal to the failure surface, thus reproducing the nonlinear unloading strains and resulting volume expansion observed at concrete failure.

Rational Derivation of Coupling Moduli

It appears that the role of the coupling moduli H and Y is nothing more than to nudge the strain increment vector in the proper direction normal to the loading surface. Conversely, knowing this proper direction, we can determine the values of the coupling moduli in a rational manner. In the Rendulic plane, three cases of a virgin loading are possible:

Case 1: $\sigma_1 = \sigma_1^{max}$, $\sigma_2 = \sigma_3 < \sigma_3^{max}$, $d\sigma_1 > 0$: Plastic strain increment vector is parallel to Axis 1.

Case 2: $\sigma_1 = \sigma_1^{max}$, $\sigma_2 = \sigma_3 = \sigma_3^{max}$, $d\sigma_1 > 0$: $d\sigma_2 = d\sigma_3 > 0$: Corner loading.

Case 3: $\sigma_2 = \sigma_3 = \sigma_3^{max}$, $\sigma_1 < \sigma_1^{max}$, $d\sigma_2 = d\sigma_3 > 0$: Plastic strain increment vector is parallel to Axis 3.

To indicate the approach, we will only consider Case 1 in detail: We resolve the plastic strain increment vector $\Delta\varepsilon^P$ into its two components $\Delta\varepsilon_0^P$ and $\Delta\gamma_0^P$ as shown dashed in Fig. 7a. Their ratio is found from the geometry of this figure as

$$\frac{\Delta\varepsilon_0^P}{\Delta\gamma_0^P} = \frac{1}{\sqrt{2}} \tag{15}$$

We calibrate this vector by considering, first, a purely triaxially compressive stress deviation, during which

$$\Delta\gamma_0 = \Delta\gamma_0^E + \Delta\gamma_0^P = \frac{\Delta\tau_0}{2G_0} + \Delta\gamma_0^P = \frac{\Delta\tau_0}{2G} \quad . \tag{16}$$

and

$$\Delta\varepsilon_0 = \Delta\varepsilon_0^E + \Delta\varepsilon_0^P = 0 + \Delta\varepsilon_0^P = \frac{\Delta\tau_0}{H} \quad . \tag{17}$$

Solving Eqs. 15, 16, and 17 we find

$$H = 2\sqrt{2} \cdot G \cdot \frac{1}{1 - G/G_0} \tag{18}$$

The magnitude of H depends on the value of the shear modulus.

Next, we consider a hydrostatic stress increment involving Case 1 virgin loading, during which

$$\Delta\gamma_0 = \Delta\gamma_0^E + \Delta\gamma_0^P = 0 + \Delta\gamma_0^P = \frac{\Delta\sigma_0}{Y} \quad . \tag{19}$$

and

$$\Delta\varepsilon_0 = \Delta\varepsilon_0^E + \Delta\varepsilon_0^P = \frac{\Delta\sigma_0}{2K_0} + \Delta\varepsilon_0^P = \frac{\Delta\sigma_0}{3K} \quad . \tag{20}$$

Solving Eqs. 15, 19 and 20, we find

$$Y = \frac{3}{\sqrt{2}} \cdot K \cdot \frac{1}{1 - K/K_0} \tag{21}$$

The coupling modulus Y depends on the value of the bulk modulus.

For Case 3, calculations similar to those for Case 1 can be carried out. For the corner loading, Case 2, non-uniqueness prevails, just as in the case of Tresca plasticity. In the absence of further information, we have resorted to a definition of the coupling moduli as specified by Eqs. 13 and 14.

EXPERIMENTAL VERIFICATION

The analysis which has been presented was programmed for microcomputer, and used to calculate the strains for various applied stress histories, to be compared with those measured in tests. Of the multitude of load histories tested [13,14,10] and calculated

[11,14,10], only one, within the triaxially compressive region, will be considered here.

The only material constants required for the concrete used were the easily obtained initial bulk modulus $K_0 = 2.1 \times 10^3$ ksi, and shear modulus $G = 1.7 \times 10^3$ ksi, and the failure surface in the triaxial compression plane over the relevant range of σ_0, given by

$$\tau_{0u} = .549 + 1.079\ \sigma_0 - .051\ \sigma_0^2\ . \tag{22}$$

Fig. 8a shows a complex load history involving virgin loading, unloading and reloading. In Fig. 8b, the solid curve shows experimentally determined strains, while the chain-curve shows the analytical prediction. The agreement between the calculated and measured strains is satisfactory.

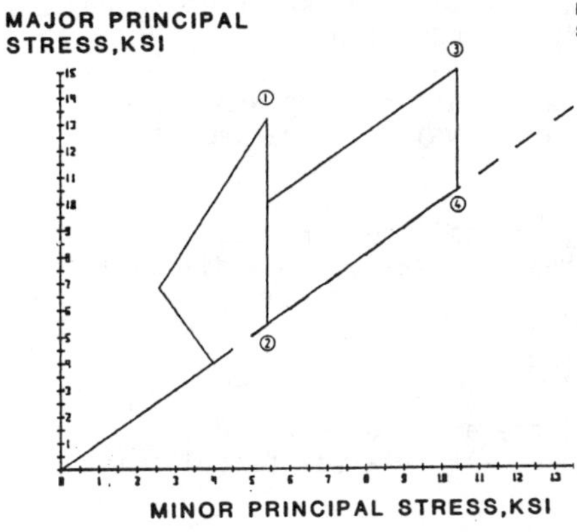

Fig. 8a. Axisymmetric Stress History

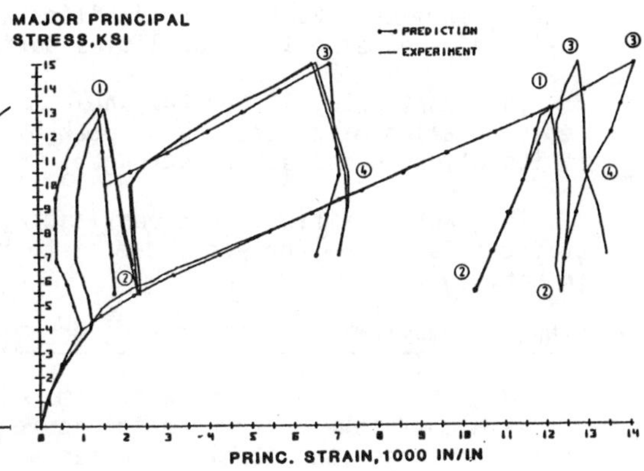

Fig. 8b. Principal Strains for Axisymmetric Test

A CRITIQUE OF THE "SIMPLE FORMULATION"

While the analysis which has been outlined appears to be able to predict concrete strains under axisymmetric stress histories reasonably closely, and with relative simplicity, it has several severe drawbacks:

1. The restriction of the constitutive Eqs. 9 to two invariant quantities does not, by itself, permit the determination of the three independent principal strain increments, nor, in general, of the six

cartesian strain components. (This defect did not appear in our presentation because we restricted ourselves to the axisymmetric case for which $\Delta\varepsilon_2 = \Delta\varepsilon_3$, and where the principal stress and strain directions are fixed, thus, only two unknown strains were to be determined).

2. Circular stress paths [10] within one deviatoric plane correspond to $\Delta\sigma_0 = 0$, $\Delta\tau_0 = 0$, and the strains cannot therefore be captured by Eq. 9. These paths depend on the third stress invariant which is not included in the Simple Formulation.

3. The assumption of linearly-elastic strains under decreasing components of principal stress which has been observed in Fig. 6 and postulated in the theory becomes invalid near failure. While a correction for this is included in the Simple Formulation [8], this device must be considered purely empirical.

4. The assumption of the rectangular loading surface as in Fig. 7 leads to discontinuity of strains, depending on whether a neutral loading path is approached from within, or without, the current loading surface. Such discontinuities are physically implausible and may lead to numerical instabilities during computation.

5. The adopted Eq. 12 for the shear modulus variation does not permit representation of concrete post-peak behavior which may be important in its effect on structure strength.

6. The lack of affinity between loading and failure surfaces can lead to inconsistencies for some load histories near failure, as discussed in Ref. [9].

GENERAL INVARIANT CONSTITUTIVE FORMULATION

Recent work [9], in view of some of the above enumerated concerns, has taken a fundamental approach. It relates the following invariant stress and strain increments:

$$\begin{Bmatrix} \Delta\sigma_0 \\ \Delta\tau_0 \end{Bmatrix} = \begin{bmatrix} A_1 & A_2 & A_3 \\ B_1 & B_2 & B_3 \end{bmatrix} \begin{Bmatrix} \Delta\varepsilon_0 \\ \Delta e \\ \Delta w \end{Bmatrix} \tag{23}$$

in which the two new strain invariant increments Δe and Δw are defined by

$$\Delta e = \frac{s_{ij}\ \Delta e_{ij}}{3\tau_0}$$

$$\Delta w = \frac{s_{ij}\ s_{jk}\ \Delta e_{ki}}{3\tau_0{}^2}$$

It can be shown that A_3 and B_3 represent the influence of direction of the distortion or "angle of similarity" [2], that is, the third strain invariant.

For the axisymmetric case $\varepsilon_2 = \varepsilon_3$, considered earlier, Δe becomes identical to $\Delta\gamma_0$; for this special case, which is fixed in direction within the deviatoric plane, $A_3 = B_3 = 0$, and Eq. 23 becomes identical with Eq. 9.

As demonstrated in Ref. [9], the formulation of the problem in terms of Eq. 23 implies three significant advantages:

Firstly, it is possible to reformulate the invariant formulation, Eq. 23, so that the general tangential stiffness relation between all six stress and strain increments can be constructed. This is necessary for use in finite element analysis.

Secondly, this general incremental stress-strain relation can be augmented by consistent loading-unloading criteria which fulfill the continuity requirement touched upon in Point 4, above. The proposed formulation separates volumetric and deviatoric loadings, among which the latter becomes affine to the failure surface to remove the inconsistency listed under Point 6, above.

Thirdly, the formulation can be extended to cover the post-peak region.

The constitutive theory as sketched above is of surprising generality. As shown in Ref. [9], it can be specialized by appropriate choice of the constitutive functions to a number of well-known constitutive relations including associated and non-associated plasticity, hypoelasticity, and plastic-fracturing theory.

Current work involves the calibration of the material functions, and comparison with test results. Future studies will incorporate this formulation in a finite element code for improved prediction of the behavior and strength of concrete structures.

REFERENCES

1. Bazant, Z.P., and Bhat, P.D., "Endochronic Theory for Inelasticity and Failure of Concrete," Trans. A.S.C.E., Jnl. E.M.D., No. EM4, August 1976.

2. Chen, W.F., "Plasticity in Reinforced Concrete," McGraw-Hill Book Co., New York, 1982.

3. Gerstle, K.H., Linse, D., et al., "Strength of Concrete under Multiaxial Stress States," Proc. McHenry Symposium, Publication SP-55, A.C.I., 1978.

4. Gerstle, K.H., Aschl, H., et al., "Behavior of Concrete under Multiaxial Stress States," Trans. A.S.C.E., Jnl. E.M.D., No. EM6, December 1980.

5. Gerstle, K.H., "Simple Formulation of Biaxial Concrete Behavior," Jnl. A.C.I., Jan.-Feb. 1981.

6. Gerstle, K.H., "Simple Formulation of Triaxial Concrete Behavior," Jnl. A.C.I. Sept.-Oct. 1981.

7. Kotsovos, M.D., and Newman, J.B., "Generalized Stress-Strain Relations for Concrete, Trans, A.S.C.E., Jnl. E.M.D., No. EM4, August 1978.

8. Ottosen, N.S., "A Failure Criterion for Concrete," Trans. A.S.C.E., Jnl. E.M.D., No. #M4, August 1977.

9. Ottosen, N.S., "Theoretical Framework for Modelling the Behavior of Frictional Materials," submitted for publication in Int. Jnl. of Solids and Structures, 1985.

10. Scavuzzo, R., "Behavior of Concrete under Multiaxial Load Histories," M.S. Thesis, C.E.A.E. Dept., University of Colorado, 1982.

11. Scavuzzo, R., Cornelius, V.H., Gerstle, K.H., Ko. H.Y., Stankowski, T., "Simple Formulation of Concrete Response to Multiaxial Load Cycles," Proc. Int. Conf. on Constitutive Relations, Univ. of Arizona, Tucson, January 1983.

12. Stankowski, T., "Concrete under Multiaxial Load Histories," M.S. Thesis, C.E.A.E. Dept., University of Colorado, Boulder, February 1983.

13. Stankowski, T., Scavuzzo, R., Gerstle, K.H., and Ko, H.Y., "Tests of Concrete under Multiaxial Load Histories," Report, Dept. of Civil, Env., and Arch. Engin., University of Colorado, Boulder, 1983.

14. Stankowski, T., and Gerstle, K.H., "Simple Formulation of Concrete under Multiaxial Load Histories," Jnl. A.C.I., March-April 1985.

FRACTURE MECHANICS AND STRAIN-SOFTENING OF CONCRETE

Zdeněk P. Bažant,[1] F.ASCE

ABSTRACT: Finite element analysis of failure of certain types of concrete structures requires modeling of distributed cracking. Propagation of failure zones with distributed cracking must be treated from the viewpoint of fracture mechanics if the calculation should be objective and properly converge as the mesh is refined. After a brief review of the crack band model, it is shown that this model is applicable to both localized and nonlocalized cracking, while the strain-softening fracture models based on line cracks are applicable only to localized cracking. The article subsequently expounds a novel approach to strain-softening— a nonlocal formulation in which stress at a point depends on the entire strain field from a neighborhood of the point. Application of the theory is demonstrated by finite element analysis of waves in strain-softening materials. Furthermore, the formulation of strain-softening constitutive relations on the basis of the microplane model is briefly outlined, and the size effect is discussed. Finally, the structural size effect is analyzed and a generalized form of the size effect law is presented.

INTRODUCTION

Heterogeneous brittle materials such as concretes or rocks fail by progressive fracturing distributed over a finite-size zone within the material. In the continuum approximation, this zone is characterized by strain-softening, i.e., a stress-strain relation in which the maximum principal stress decreases at increasing strain. The purpose of the present article is to present an overview of some recent results obtained at Northwestern University in the use of strain-softening material models for the description of fracture propagation in concrete structures.

CRACK BAND MODEL

Due to the distributed nature of microcracking, as well as the fact that the path of a final crack is generally not smooth but highly tortuous, it is not unrealistic to model cracking by means of stress-strain relations characterized by strain-softening in which the major principal tensile stress is gradually reduced to zero at increasing strain (Fig. 1). This approach is particularly convenient for finite element analysis, since it necessitates merely an adjustment in the incremental stiffness matrix of the finite element.

In the form of sudden cracking, this approach was introduced in 1967 by Rashid (34,49). For strain-softening analysis of fracture this ap-

[1]Professor of Civil Engineering and Director, Center for Concrete and Geomaterials, Northwestern University, Evanston, Illinois 60201, USA.

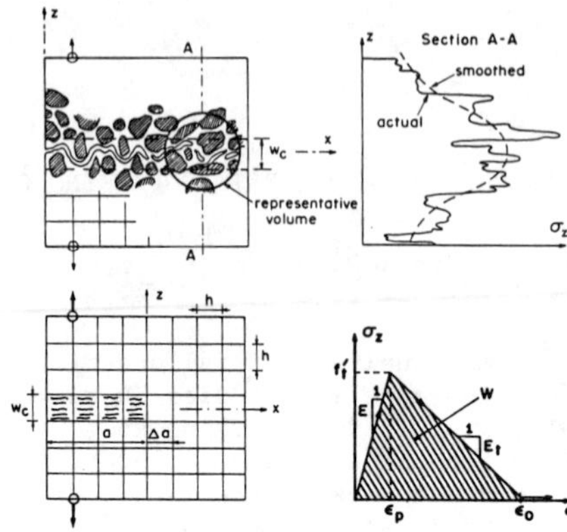

Figure 1. Fracture in a heterogeneous aggregate material, statistical scatter of stress, crack band model, and strain-softening stress-strain diagram

proach was extended in Refs. 5, 24, in which the strain-softening triaxial stress-strain relation was formulated as:

$$\underset{\sim}{\varepsilon} = \underset{\sim}{D}\underset{\sim}{\sigma} + \underset{\sim}{\xi}. \tag{1}$$

Here $\underset{\sim}{\varepsilon}$ and $\underset{\sim}{\sigma}$ are the column matrices of the components of strain and stress, $\underset{\sim}{D}$ is the 6×6 matrix of variable secant elastic compliances and $\underset{\sim}{\xi} = (\xi_{11}, \xi_{22}, \xi_{33}, 0, 0, 0)^T$, where superscript T denotes a transpose and the numerical subscripts refer to cartesian coordinates x_i $(i = 1,2,3)$; $\underset{\sim}{\xi}$ is the column matrix of additional smeared-out strains due to cracking. The normal stresses are assumed to be uniquely related to their associated cracking strains (Fig. 1), i.e.,

$$\sigma_{11} = C(\xi_{11})\xi_{11}, \quad \sigma_{22} = C(\xi_{22})\xi_{22}, \quad \sigma_{33} = C(\xi_{33})\xi_{33} \tag{2}$$

in which C is the secant modulus for cracking strain. C reduces to zero at very large cracking strain and may be calibrated from direct tensile test data (33,35,44,45,50,51). Different algebraic relations are used for unloading. In the simplest form of the model it is assumed that cracking can occur only in three mutually orthogonal directions which are known in advance and do not rotate against the material once the cracking starts. For details of the procedure, see Ref. 24. Cracking in all directions can be described by the microplane model (see Eq. 29).

In the classical formulation of finite element analysis of concrete structures, as employed in the current large computer programs, a strain-softening stress-strain relation, usually one that exhibits a sudden stress drop, is used indiscriminately for an arbitrarily chosen finite element size. It has been demonstrated, however, that this approach is inconsistent, unobjective with regard to the analyst's choice of the element size. It can yield greatly different results for different mesh sizes (6,9,16-18). For failures of the brittle type, the load causing further fracture extension converges to zero as the mesh size tends to zero and so does the total energy consumed by failure of the structure. These are physically unrealistic results (4,16,20).

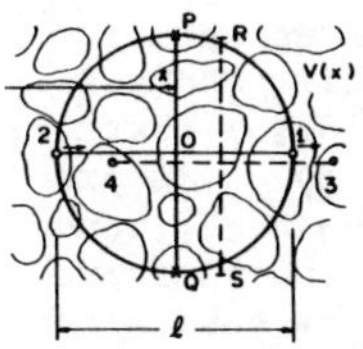

Figure 2. Characteristic volume of a heterogeneous material

It has been shown that these incorrect features can be eliminated by using an energy criterion rather than the stress-strain relation as the primary condition for the extension of the cracked band of finite elements. Considering a crack band of a single-element width, consisting of finite elements with a uniform strain distribution across the band, the fracture energy, i.e., the energy consumed per unit extension of the crack band (and per unit thickness), may be expressed as

$$G_f = w_c \int \sigma_{33} d\varepsilon_{33} = \frac{w_c}{2} f_t'^2 \left(\frac{1}{E_0} - \frac{1}{E_t}\right) \quad (3)$$

in which w_c is the width of the crack band front, which must be considered as a material property (4), σ_{33} is the normal stress across the central plane of the crack band, f_t' is the direct tensile strength of the material, E_0 is the initial elastic Young's modulus, and E_t is the mean downward slope of the strain-softening segment of the stress-strain diagram, which is negative; w_c may be approximately considered as $3d_a$ where d_a = maximum aggregate size.

It was shown (24) that the crack-band finite element model yields results which agree with all essential experimental evidence from fracture tests, both the maximum load data and the R-curves (9,55). However if the cracking at the failure front tends to localize into a single-element width, a behavior which fracture specimens are designed to ex-

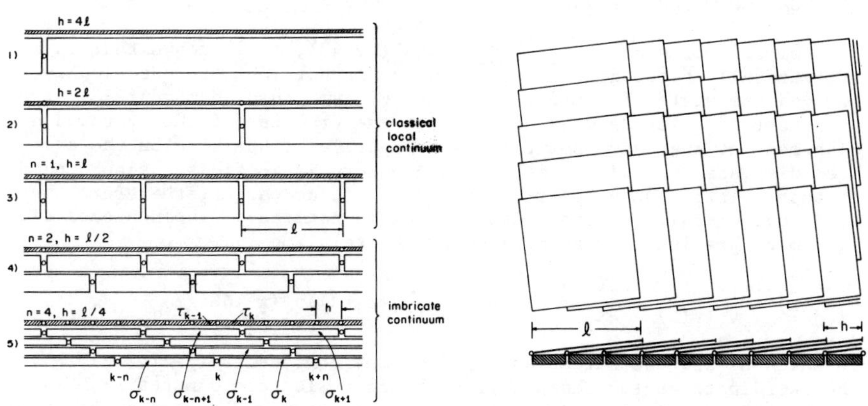

Figure 3. Illustration of mesh refinement for a one-dimensional imbricate continuum (left) and of element arrangement for a two-dimensional imbricate continuum (right) (the elements are slightly rotated for the purpose of illustration)

hibit, then the use of the correct element size $h = w_c$ is essential. For large structures, such elements are impracticably small, and a larger element size needs to be used. It was shown (16-18,24) that this leads to consistent results if the value of fracture energy G_f is preserved. This can be achieved by first adjusting the value of E_t, and if $E_t \rightarrow -\infty$ (vertical drop), by adjusting also f_t' while always satisfying Eq. 3, in which w_c is replaced by h. So, one must use some equivalent strain-softening slope and some equivalent tensile strength, depending on the element size, if consistent results should be achieved.

On the other hand, one may also imagine the width of the crack band to be reduced to zero, and if the strain-softening slope E_t is adjusted so as to keep the correct value of G_f, the model in the limit becomes equivalent to the use of the stress-displacement relation on the centerline of fracture. This limiting case of the crack band model coincides with the model of Hillerborg et al. (36,44).

The adjustment of constitutive properties in order to achieve consistent results for different mesh sizes is, however, disconcerting from the fundamental, mathematical viewpoint. It appears as if the finite element model was approximating a continuum that is actually not defined. Obviously, in the crack band model we cannot say we are approximating the solution for a classical, local continuum because for such a continuum the crack band in fact can localize into a layer of zero thickness, while an adjustment of material properties is not permitted.

The source of the difficulty appears to be (7,19) the assumption that we deal with a local continuum, in which the stress at a certain point is a function of the strain at the same point. We now focus attention on the development of a new type of continuum for which the aforementioned mathematical difficulties do not arise. At the same time we gain with this new type of continuum a means to resolve in detail the distributions of the averaged stress and strain throughout the cracking zone.

NONLOCAL CONTINUUM APPROACH

From the works of Kröner and others (39-43), it is known that in a statistically heterogeneous medium which is not in a macroscopically homogeneous state of strain, the averaged (smoothed, homogenized) stress at a certain point depends not only on the gradient of the averaged displacement at that same point (local properties), but also on the averaged displacements within a certain characteristic finite neighborhood of that point. The properties of such a medium cannot, therefore, be said to be local, and the medium is called nonlocal. The nonlocal displacement gradient may be defined by the relation

$$D_i u_j(\underset{\sim}{x}) = \frac{1}{V} \int_{V(\underset{\sim}{x})} \frac{\partial u_j(\underset{\sim}{x}')}{\partial x_i} \, dV' = \frac{1}{V} \int_{S(\underset{\sim}{x})} u_j(\underset{\sim}{x}') n_i(\underset{\sim}{x}') dS' \tag{4}$$

in which u_j are the cartesian displacement components ($j = 1,2,3$), $\underset{\sim}{x}$ is the coordinate vector characterized by cartesian coordinates x_i, $V(\underset{\sim}{x})$ is the characteristic volume of the material (Fig. 2) centered at point $\underset{\sim}{x}$, $S(x)$ is the surface of this volume, $n_i(\underset{\sim}{x}')$ is the unit normal of this surface at point $\underset{\sim}{x}'$, and D_i is the gradient averaging operator. The surface integral in Eq. 4 follows from the volume integral by application of the Gauss integral theorem. More generally, a weighting function can be introduced in Eq. 4. Using the gradient averaging operator, the mean strains may be defined as

$$\bar{\varepsilon}_{ij} = \tfrac{1}{2}(D_i u_j + D_j u_i). \tag{5}$$

In previous works dealing with nonlocal continua it has been generally assumed that the continuum equation of motion may be written as

$$\frac{\partial}{\partial x_j} \bar{C}_{ijkm}(\bar{\underset{\sim}{\varepsilon}})\, D_m u_k = \rho \ddot{u}_i \tag{6}$$

in which C_{ijkm} are the secant elastic moduli which, in general, depend on the mean strain, ρ is the mass density, and superior dots refer to time derivatives. It is found, however, that Eq. 6 is incapable of describing a strain-softening continuum. It always leads to unstable response as soon as strain-softening begins. The difficulty has been traced to the asymmetry of these equations which consists in the combination of partial derivatives $\partial/\partial x_j$ with the gradient averaging operator D_m. This nonsymmetry gives rise to nonsymmetric finite element matrices even if C_{ijkm} are constant, i.e., if the medium is elastic. Such a nonsymmetry is certainly an objectional property.

For this reason, a systematic derivation of the continuum equation of motion on the basis of Eq. 4 was attempted, using the calculus of variations. It was found (19,7)that the proper form of the continuum equation of motion is

$$(1-\gamma) D_j \bar{C}_{ijkm}(\bar{\underset{\sim}{\varepsilon}}) D_m u_k + \gamma \frac{\partial}{\partial x_j} C_{ijkm}(\underset{\sim}{\varepsilon}) \frac{\partial}{\partial x_m} u_k = \rho \ddot{u}_i \tag{7}$$

in which γ is an empirical coefficient between 0 and 1, and C_{ijkm} are the local secant moduli. In contrast to Eq. 6, each term of the last equation has a symmetric structure, and consequently, discretization by finite elements leads to symmetric stiffness matrices if the elastic moduli $\bar{C}_{ijkm}$ and C_{ijkm} are symmetric. To prevent instabilities which involve periodic zero-energy deformation modes, it is necessary that $\gamma > 0$. However, if γ is close to 0 there is excessive noise in the numerical solution, and to avoid it, γ should not be chosen less than about 0.1, as experience shows (19).

Eq. 7 can also be written in the form

$$(1-\gamma) D_j \sigma_{ij} + \gamma\, \tau_{ij,j} = \rho\, \ddot{u}_i \tag{8}$$

in which the subscript following a comma denotes a partial derivative and

$$\sigma_{ij} = \bar{C}_{ijkm}(\bar{\underset{\sim}{\varepsilon}})\bar{\varepsilon}_{km} = \bar{C}_{ijkm}(\bar{\underset{\sim}{\varepsilon}}) D_m u_k \tag{9}$$

$$\tau_{ij} = C_{ijkm}\, \varepsilon_{km} = C_{ijkm}\, u_{k,m}. \tag{10}$$

τ_{ij} are the usual, local stresses, and σ_{ij} are the stresses characterizing the stress state in the entire representative volume of the material, called the broad-range stresses (7,19).

When the continuum defined by Eq. 7 or 8 is discretized by finite elements the size of which is smaller than the size ℓ of the representative volume, one obtains a system of overlapping (or imbricated) finite elements visualized in Fig. 3. Therefore, the present type of nonlocal continuum has been called imbricate. The finite elements keep a constant size ℓ as the mesh is refined, and the number of finite elements crossing a given point is inversely proportional to the mesh size, while the cross section of these elements diminishes so that all imbricated elements have the same total cross section for any mesh size. It can also be shown that the limiting case of the finite difference equations describing such an imbricated system of finite elements is the differential equation in Eq. 7 or 8 (7,19). If the finite element size h is larger than the characteristic length ℓ, then the finite element model of the imbricate continuum becomes identical to that for the classical local continuum.

To assure convergence and stability, the local stress-strain relations (Eq. 10) may not exhibit strain-softening, or else unstable response and spurious sensitivity to mesh size, along with incorrect convergence, may be obtained. The strain-softening properties must be described solely by the broad-range stress-strain relation in Eq. 9.

EXAMPLES: WAVES IN STRAIN-SOFTENING MATERIALS

To demonstrate that the concept of imbricate nonlocal continuum yields convergent finite element solutions for problems with a finite-size strain-softening zone, we will now describe two examples of one-dimensional waves. As the first example, Fig. 4 reproduces some of the results of explicit dynamic finite element calculations from Ref. 19, in which wave propagation in a strain-softening bar of length ℓ was analyzed. The characteristic length ℓ for nonlocal solution is assumed to be 1/5 of bar length L. Both ends of the bar are subjected to a constant outward velocity v_0 beginning at time $t = 0$. This loading produces step waves of strain propagating inward. When these waves meet at midlength, the strain suddenly increases and strain-softening ensues.

If this problem is analyzed with the usual finite element method for local continuum, it is found that strain-softening always localizes into a single finite element. Thus, the width of the strain-softening zone reduces to zero as the element mesh is refined. Since the energy dissipation per unit volume equals the area under the stress-strain diagram, which is finite, the energy W consumed by failure decreases with decreasing mesh size and approaches zero as the mesh size tends to zero (Fig. 5), which is unrealistic from the physical point of view. Moreover, the finite element model of local continuum exhibits a discontinuous dependence of response on the prescribed end velocities as well as on the slope E_t of the strain-softening branch (12). The solution, however, converges to a unique exact solution (12) and the convergence is quadratic.

By contrast, for the present imbricate continuum, the solution of wave propagation in the strain-softening bar exhibits correct convergence, with a strain-softening zone of a finite size in the limit (Fig. 4, right column). Also, the energy consumed by failure in the bar converges to a finite value, as shown in Fig. 5.

As another example, we now briefly describe the solution for an implosive (converging) spherical wave, a problem which was proposed by T. Belytschko and was originally solved in full detail in Ref. 28. Although this problem is also one-dimensional it involves three-dimensional stress and strain states. Like for the bar problem, strain-softening is produced inside the body but there is an important difference —waves

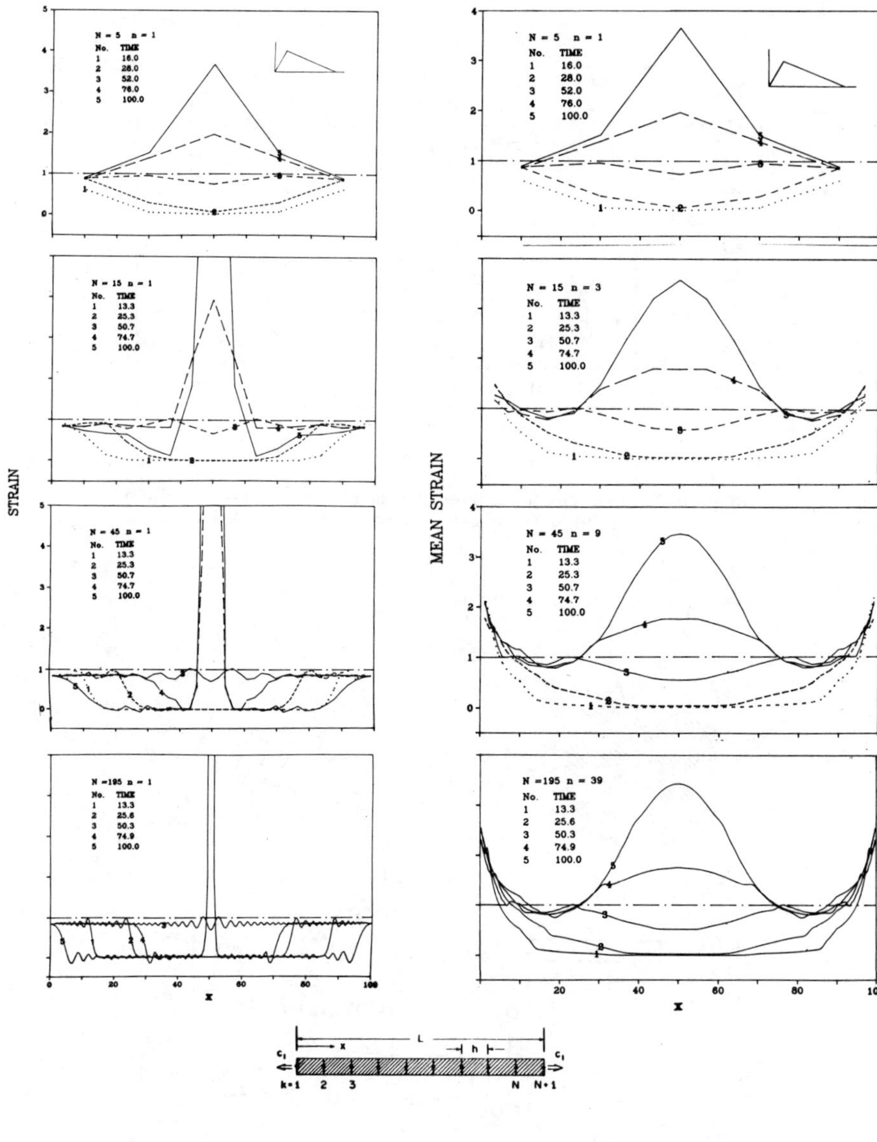

Figure 4. Numerical results of Bažant, Chang and Belytschko (19) for wave propagation in a strain-softening bar, obtained with different numbers N of identical elements (left column - local continuum, right column - imbricate nonlocal continuum)

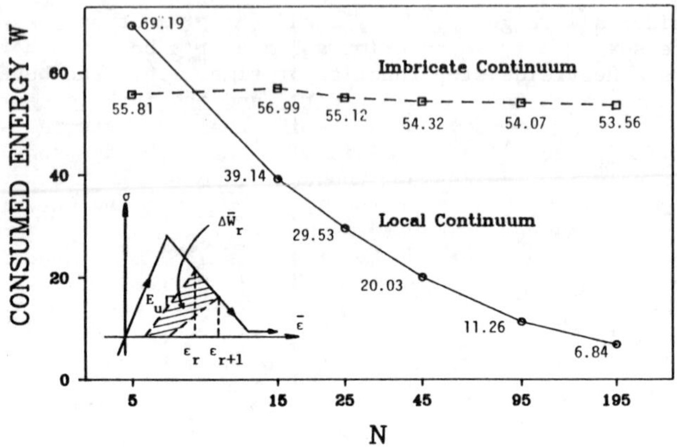

Figure 5. Energy consumed by failure caused by wave propagation in a one-dimensional bar

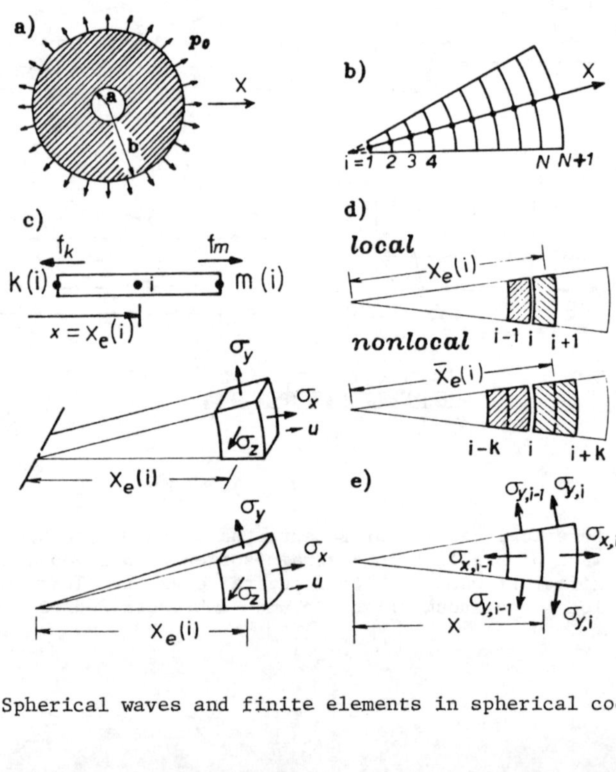

Figure 6. Spherical waves and finite elements in spherical coordinates

can pass through the points of strain-softening even if a local continuum is assumed.

We consider a wave generated by a sudden application of a constant uniform pressure p at the exterior surface of a hollow sphere. The pressure is a Heaviside step function of time, i.e., the boundary condition at $x = b$ (Fig. 6a) is $\sigma_x = p_0 H(t)$, and the interior surface is load-free, i.e., $\sigma_x = 0$ at $x = a$ (a, b = internal and external surface radii, x = radial coordinate). Initially, at $t = 0$, the body is at rest. The elastic solution, as is well known (1), consists of a wave with a step front of strain whose magnitude is growing as the wave front propagates toward the center. Therefore, at a certain time t_1 (and at $x = x_1$) the strain at the wavefront reaches the elastic limit, and if the material has strain-softening properties, strain-softening is thus produced in the interior of the solid by a one-directional wave, while for uniaxial waves (planar wavefront) strain-softening can be produced in the interior only if waves of opposite directions meet.

The governing differential equations in spherical coordinates are:

$$\varepsilon_x = u_{,x}, \quad \varepsilon_y = \frac{u}{x}, \quad \varepsilon_z = \varepsilon_y \tag{11}$$

$$\varepsilon = \varepsilon_x + 2\varepsilon_y, \quad e_x = \frac{2}{3}(\varepsilon_x - \varepsilon_y), \quad e_y = \frac{1}{3}(\varepsilon_y - \varepsilon_x) \tag{12}$$

$$\bar{\varepsilon}_x, \bar{\varepsilon}_y, \bar{\varepsilon}, \bar{e}_x, \bar{e}_y = \text{the means of } \varepsilon_x, \varepsilon_y, \varepsilon, e_x, e_y \text{ analogous to Eqs. 11 and 16} \tag{13}$$

$$\sigma_{x,t} = \bar{K}\,\bar{\varepsilon}_{,t} + 2\bar{G}\,\bar{e}_{x,t}, \quad \sigma_{y,t} = \bar{K}\,\bar{\varepsilon}_{,t} + 2\bar{G}\,\bar{e}_{y,t} \tag{14}$$

$$\tau_{x,t} = K\,\varepsilon_{,t} + 2G\,e_{x,t}, \quad \tau_{y,t} = K\,\varepsilon_{,t} + 2G\,e_{y,t} \tag{15}$$

$$\bar{\sigma}_x(x) = \int_{x-\ell/2}^{x+\ell/2} \sigma_x(x+s)w(s)ds, \quad \bar{\sigma}_y(x) = \int_{x-\ell/2}^{x+\ell/2} \sigma_y(x+s)w(s)ds \tag{16}$$

$$S_x = (1-\gamma)\bar{\sigma}_x + \gamma\tau_x, \quad S_y = (1-\gamma)\bar{\sigma}_y + \gamma\tau_y \tag{17}$$

$$S_{x,x} + \frac{2}{x}(S_x - S_y) = \rho u_{,tt}\,. \tag{18}$$

Here ℓ = characteristic length of material = size of representative volume, u = radial displacement, x = radial coordinate (Fig. 6a); ε_x, ε_y = radial and circumferential normal strains (local); ε, e_x, e_y = volumetric strain (= volume change) and deviatoric strains (local), τ_x, τ_y = local radial and circumferential normal strains; σ_x, σ_y = broad-range radial and circumferential normal stress; S_x, S_y = total radial and circumferential normal stresses, K, G, $\bar{K}$, $\bar{G}$ = local and broad-range (nonlocal) bulk and shear moduli. In Eq. 18, isotropy of the material is assumed. The shear moduli G and $\bar{G}$ are assumed to be constant, while K and $\bar{K}$ depend on ε and $\bar{\varepsilon}$, respectively; $K > 0$ but $\bar{K}$ may become negative, which represents strain-softening. Eq. 18 represents the differential equations of motion of the usual form (1).

For elastic spherical wave, the present problem has a closed-form solution:

$$u = -\frac{b^3 p_0}{4Gr^2}\{1 - e^{\xi\tau}[\frac{\xi}{\omega}(\frac{2r}{b} - 1)\sin\omega\tau + \cos\omega\tau]\} \quad \text{for } \tau > 0 \tag{19}$$

as may be checked by substituting in Eqs. 11-18 with $\gamma = 1$, and $K =$ constant; and

$$\tau = t - (b-r)/c_1, \quad \xi = 2c_2^2/(bC_1), \quad \omega = \xi[(c_1/c_2)^2 - 1]^{\frac{1}{2}},$$
$$c_1 = [3K(1-\nu)/(1+\nu)\rho]^{\frac{1}{2}}, \quad c_2 = (G/\rho)^{\frac{1}{2}}, \quad \nu = (3K-2G)/(6K+2G). \tag{20}$$

For the purpose of numerical solution, a one-dimensional mesh with constant step h is used (Fig. 6b). All finite elements are two-node elements with a linear distribution of displacement u, integrated with a single numerical integration point at the element center. This is true of both the local elements of length h and the imbricate elements of length $\ell = nh$. Mass is lumped in the nodes. The radial nodal forces are referred to a radial conical segment of the sphere such that its cross section area at a unit distance from the center of coordinates is 1. The computational algorithm is similar to the usual explicit algorithm and is as follows:

1. Read a, b, ℓ, n, Δt (time step), N_e, $\bar{N}_e$, N_k, N_t (numbers of all local and imbricate elements, of all modes and of time steps), γ, and p_0. Generate arrays $\hat{k}(i)$, $\hat{m}(i)$, $\ell_e(i)$, $x_e(i)$, $\bar{k}(i)$, $\bar{m}(i)$, $\bar{\ell}_e(i)$, $\bar{x}_e(i)$ giving the number of the left and right nodes of the ith local or imbricate element, its length, and its coordinate at the center of element. Also generate externally applied nodal forces $f^{ext}_{k,r}$. Initialize as zero the values (for $r=1$) of $v_{k,r}$, $m_{k,r}$, $\tau_{xk,r}$, $\tau_{yk,r}$, $\sigma_{yk,r}$, $\max\varepsilon_k$, $\max\bar{\varepsilon}_k$, $\tau_{\max\varepsilon,k}$, $\sigma_{\max\varepsilon,k}$, for all $k=1,\ldots,N_e$ (local elements) and $k=1,\ldots,\bar{N}_e$ (imbricate elements).

2. DO 8, $r = 2,\ldots,N_t$.

3. Initialize nodal forces $F_k = 0$, $f_k = 0$ (for all nodes, $k = 1,\ldots,N_k$).

4. DO 5, $i = 1,\ldots,N_e$ (local elements).

5. Set $k=k(i)$, $m=m(i)$, $x=x_e(i)$, and evaluate $\Delta\varepsilon_{xi,r} = (v_{m,r}-v_{k,r})\Delta t/h$, $\varepsilon_{xi,r} = \varepsilon_{xi,r-1} + \Delta\varepsilon_{xi,r}$, $\Delta\varepsilon_{yi,r} = (v_{m,r}+v_{k,r})\Delta t/2x$, $\varepsilon_{yi,r} = \varepsilon_{yi,r-1} + \Delta\varepsilon_{yi,r}$, and $\Delta\varepsilon_{i,r} = \Delta\varepsilon_{xi,r} + 2\Delta\varepsilon_{yi,r}$. Then call a subroutine which determines the incremental tangential moduli $K_{i,r}$, $G_{i,r}$ for the local elements from their strains $\varepsilon_{xi,r}$, $\varepsilon_{yi,r}$, $\varepsilon_{i,r}$ and also decides whether virgin loading, unloading or reloading applies. Then calculate $\Delta\tau_{xi,r} = K_{i,r}\Delta\varepsilon_{i,r} + 2G_{i,r}e_{xi,r}$, $\Delta\tau_{yi,r} = K_{i,r}\Delta\varepsilon_{i,r} + 2G_{i,r}e_{yi,r}$, $\tau_{xi,r-1} + \Delta\tau_{xi,r}$, $\tau_{yi,r} = \tau_{yi,r-1} + \Delta\tau_{yi,r}$. Then calculate all the local nodal forces at element nodes k and m (Fig. 6c);

$$\Delta f_k = -\gamma[\tau_{xi,r}x^2 - \tau_{yi,r}(x - \frac{h}{2})h], \quad \Delta f_m = \gamma[\tau_{xi,r}x^2 + \tau_{yi,r}(x+\frac{h}{2})h]. \tag{21}$$

These forces are accumulated at each node, $f_k \leftarrow f_k + \Delta f_k$, $f_m \leftarrow f_m + \Delta f_m$.

6. DO 7, $i = 1,\ldots,\bar{N}_e$ (imbricate elements).

7. Set $k=\bar{k}(i)$, $m=\bar{m}(i)$, $\ell=\bar{\ell}_e(i)$, $x=\bar{x}_e(i)$ and evaluate $\Delta\bar{\varepsilon}_{xi,r} = (v_{m,r}-v_{k,r})\Delta t/\ell$, $\bar{\varepsilon}_{xi,r} = \bar{\varepsilon}_{xi,r-1} + \Delta\bar{\varepsilon}_{xi,r}$, $\Delta\bar{\varepsilon}_{yi,r} = (v_{m,r}+v_{k,r})\Delta t/2x$, $\bar{\varepsilon}_{yi,r} = \bar{\varepsilon}_{yi,r-1} + \Delta\bar{\varepsilon}_{yi,r}$, and $\Delta\bar{\varepsilon}_{i,r} = \Delta\bar{\varepsilon}_{xi,r} + 2\Delta\bar{\varepsilon}_{y,r}$. Then call a subroutine which determines the incremental (tangential) moduli $\bar{K}_{i,r}$, $\bar{G}_{i,r}$ for the imbricate elements from their mean strains $\bar{\varepsilon}_{xi,r}$, $\bar{\varepsilon}_{yi,r}$, $\bar{\varepsilon}_{i,r}$

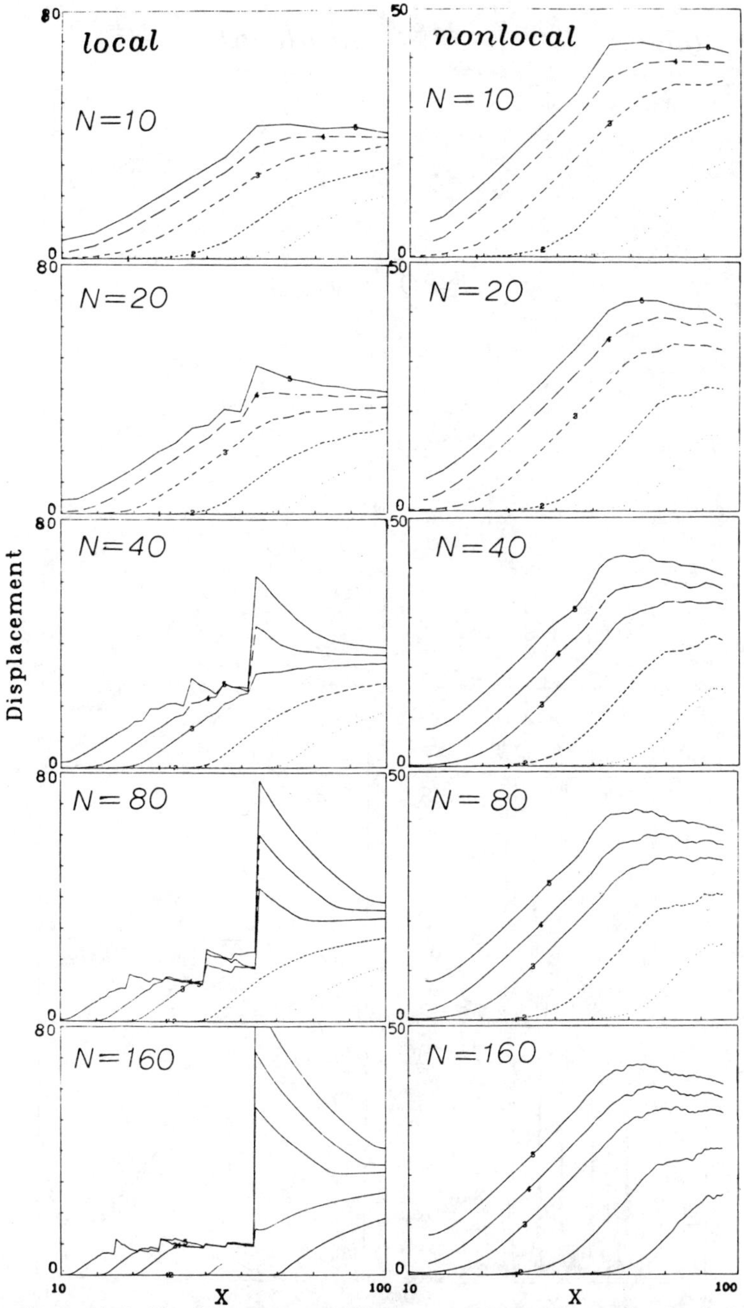

Figure 7. Radial displacement profiles at various times obtained for various finite element subdivisions (N = number of elements)

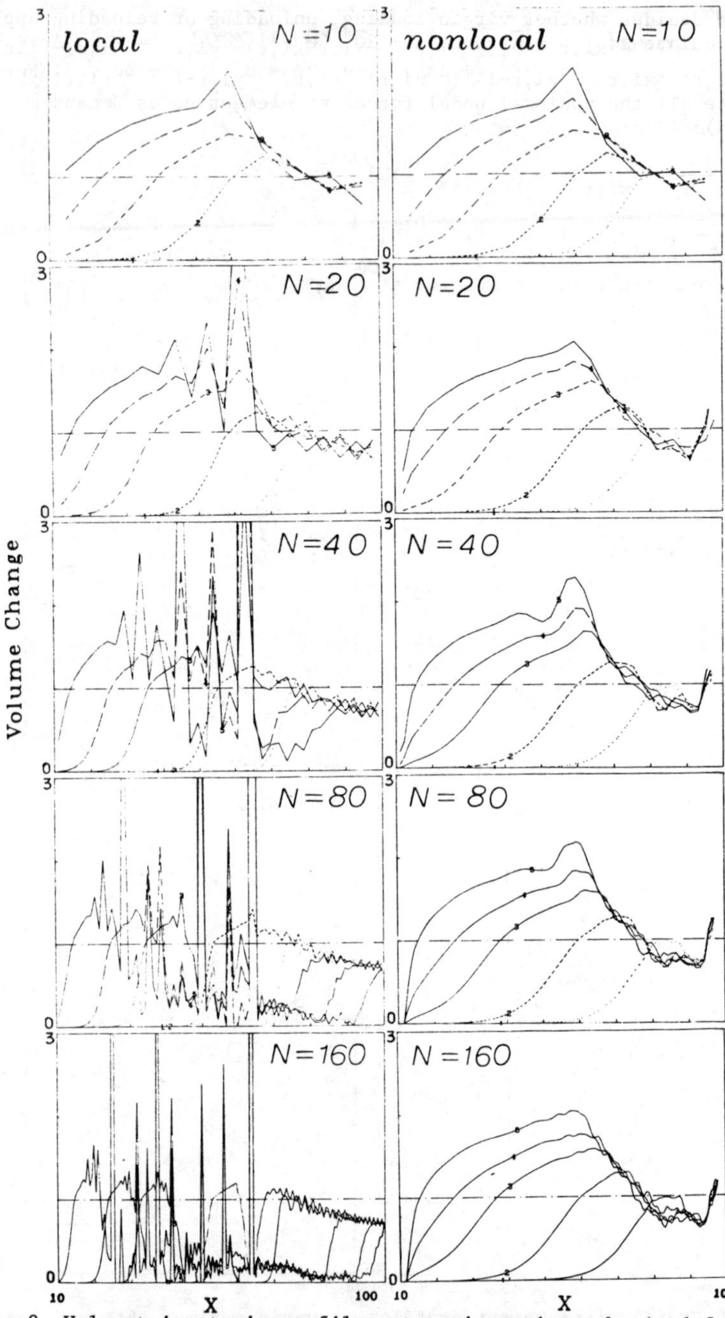

Figure 8. Volumetric strain profiles at various times obtained for various finite element subdivisions (N = number of elements)

and also decides whether virgin loading, unloading or reloading applies. Then calculate $\Delta\sigma_{xi,r} = \bar{K}_{i,r}\Delta\varepsilon_{i,r} + 2\bar{G}_{i,r}e_{xi,r}$, $\Delta\sigma_{yi,r} = \bar{K}_{i,r}\Delta\varepsilon_{i,r} + 2\bar{G}_{i,r}e_{yi,r}$, $\sigma_{xi,r} = \sigma_{xi,r-1} + \Delta\sigma_{xi,r}$, $\sigma_{yi,r} = \sigma_{yi,r-1} + \Delta\sigma_{yi,r}$. Further calculate all the nonlocal nodal forces at element nodes k and m (Fig. 6d):

$$\Delta F_k = -\frac{1-\gamma}{n}[\sigma_{xi,r}x^2 - \sigma_{yi,r}(x-\frac{h}{2})h],$$
$$\Delta F_m = \frac{1-\gamma}{2}[\sigma_{xi,r}x^2 + \sigma_{yi,r}(x+\frac{h}{2})h]. \tag{22}$$

These forces are then accumulated at each node: $F_k \leftarrow F_k + \Delta F_k$, $F_m \leftarrow F_m + \Delta F_m$.

7. DO 8, $k = 1,\ldots,N_k$ (all nodes).

8. Calculate $v_{k,r} = v_{k,r-1} + (F_k + f_e + f^{ext}_{k,r})\Delta t/(\rho h x^2)$ and $u_{k,r} = u_{k,r-1} + \Delta t\, v_{k,r}$.

For more detailed explanations, a similar algorithm for uniaxial stress wave given in Ref. 19 may be consulted. It may be checked that summing all the nodal forces (according to Eqs. 21-22) which act on one node yields a second-order discrete approximation of the continuum equation of motion, Eq. 18.

Figs. 7-8 show the numerical results obtained with the foregoing algorithm for the special case of volumetric strain-softening (28). The shear moduli G and $\bar{G}$ are assumed to be negligibly small (10^{-6}), and the bulk behavior follows the bilinear total stress-strain diagram in Fig. 9, characterized by elastic bulk moduli $\bar{K} = \bar{K}_0 = 1$, strain $\varepsilon_p = 1$ at

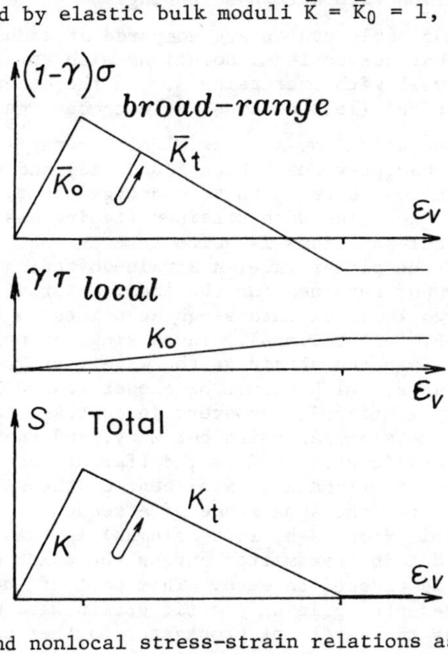

Figure 9. Local and nonlocal stress-strain relations assumed in examples

peak stress, and $\varepsilon_f = 5$ at the end of softening. This diagram is obtained by assuming that the local behavior is elastic-ideal plastic, and that the broad-range behavior (Fig. 9) exhibits oversoftening, the purpose of which is to achieve that, for the total behavior, the strain-softening terminates at exactly zero stress (19). The dimensions are $a = 10$, $b = 100$, $L = b - a = 90$. The use of coefficient γ is required for stability. For one-dimensional planar waves, it was shown (16) that stability is assured for $\gamma > 0$. However, γ-values too close to 0 are not possible for numerical solutions since they lead to an ill-conditioned equation system and produce excessive noise. For one-dimensional planar waves, the value $\gamma = 0.1$ gave noise-free response. The same value sufficed for the spherical wave ($\gamma = 0.1$), in order to obtain a response with noise-free appearance. The value of the applied surface pressure is chosen as $p_0 = 0.708$. For this boundary condition, the wave propagating from the outer surface remains elastic until the wavefront reaches 30% of the thickness $b - a$.

The nonlocal inelastic finite element program was first verified by running with it the solution for the special case of an elastic local material, for which one may set $\varepsilon_p = 10^8$, $\varepsilon_f = 10^9$, $\ell = h =$ mesh step (ℓ is variable here). The convergence with increasing numbers of elements (N = 10, 20, 40, 80, 160) was good (28). The results for the spherical wave converge (quadratically) to the exact solution given by Eq. 19 (28). Subsequently, the problem was solved (28) for a nonlocal continuum ($\varepsilon_p = 1$, $\varepsilon_f = 5$) with the characteristic length $\ell = L/10 = 9 =$ constant. The results for progressively refined meshes (N = 10, 20, 40, 80, 160) are shown in Figs. 7 and 8. For comparison, the solution was also run for a local continuum ($\ell = h = L/N =$ variable) having the same stress-strain relation with strain-softening.

The local and nonlocal results are compared in each figure. It is seen (Figs. 7-8) that the nonlocal solutions with strain-softening (right columns) converge well with increasing N. On the other hand, the corresponding local solutions (left columns) do not converge at all.

While an exact solution for a uniaxial planar wave in a local strain-softening material has previously been found (12) and the corresponding finite element solutions were shown to converge to it, for the present case no exact local solution with strain-softening has been found. The behavior of the local solutions is quite interesting. Similar to the exact solution for the planar wave, a strain-softening zone of finite size is apparently not obtained for the local material, and strain-softening appears to localize into singular points on the x-axis (the spikes in Figs. 7-8, left columns). These singular points begin to appear, as expected, when the strain at the wavefront grows to reach the strain-softening range, which cannot be closer than 0.3L from the outer surface, as already mentioned. However, in contrast to the planar wave, there is not only one singular point but many, and they appear at different locations for different N (i.e., different mesh refinements). The impression is that of chaos (reminiscent of the situation in turbulence). The reason for the appearance of a sequence of strain-softening points (the spikes in Figs. 7-8, left columns) is that part of the strain step at the wavefront is transmitted across the point of strain-softening before the stress is reduced to zero. This part of the wave which has crept through represents again an elastic strain wave with a step front, which then grows as the wavefront propagates further until the elastic limit ϵ_p is reached again. Then the situation repeats itself.

A similar strain-softening problem has also been solved for cylindrical waves (28). A static two-dimensional imbricate finite element analysis of a fracture specimen is reported in Ref. 47.

To sum up, the numerical results for the spherical waves (28), as well as the result for planar waves (19), show that the imbricate nonlocal continuum is a mathematically sound model for strain-softening.

COMPARISON WITH STRAIN-SOFTENING FRACTURE MODELS OF HILLERBORG, WILLAM AND OTHERS

Let us now turn attention again to the crack band model, which represents a special case of the imbricate nonlocal approach when the element size is not less than ℓ. The crack band model (5,9,16,18,24), for which the basic ideas of a fixed width of a strain-softening crack band were proposed in 1974 on the basis of stability analysis (4), has three independent material parameters: fracture energy G_f, tensile strength f_t', and crack band front width w_c. The value of the strain-softening modulus E_t then follows from Eq. 3. When a curved shape of the strain-softening segment of the diagram is used, E_t should be interpreted as the mean slope of this segment. The shape of the strain-softening segment must of course be specified, too.

Although the optimum fit of the entire data set was obtained when $w_c \approx 3d_a$ for concrete (24) and $w_c \approx 5d_a$ for rock, it was found that within the relatively broad range $d_a \leq w_c \leq 10d_a$ (where d_a = maximum size of aggregate or grain) the effect of w_c in the fits of fracture test data (localized cracking) is weak, and almost equally good fits of the test data can be obtained with various values of w_c within this range. Another reason for choosing the value of $w_c = 3d_a$ for concrete was the fact that the corresponding E_t value seems to represent well the mean slope of the descending branch for some direct tensile tests in which the specimen appeared to be in an approximately homogeneous state of strain.

Therefore, as far as the analysis of localized fracture is concerned, only two independent parameters of the crack band model, namely f_t' and G_f, are important, same as for the fictitious crack model of Hillerborg et al. (36,44,45) and the composite damage model of Willam, Sture et al. (53,54). Although these two models have not yet been systematically compared with the basic fracture test data from the literature, they should be capable of representing these data equally well as the crack band model since they are essentially equivalent to it. Each of them is also characterized by two independent material parameters G_f and f_t' to be found from fracture test data.

Hillerborg's model (36,44,45), which represents an adaptation to concrete of the formulations of Barrenblatt, Dugdale, Knauss, Kfouri, Rice and others for ductile fracture of metals (2,30,38,37), considers a line crack with a sharp tip and assumes a stress-displacement relation for this crack to be the basic material characteristic. If this relation is approximated as linear, the relative displacement across the line crack is $\delta = (\sigma - f_t')/C_f$ where C_f is the slope; $C_f < 0$. The fracture energy is then represented by the area under this strain-softening stress-displacement diagram, i.e., $G_f = -f_t'^2/2C_f$, from which C_f =

$-f_t'^2/2G_f$. Equating this expression for G_f to that according to the crack band model we have $G_f = -f_t'^2/2C_f = w_c(E^{-1} + E_t^{-1})f_t'^2/2$, from which we obtain the equivalence condition

$$\frac{1}{C_f} = -w_c\left(\frac{1}{E} - \frac{1}{E_t}\right). \tag{23}$$

The same relation may be obtained by application of Rice's J-integral, or by considering uniaxial stress and equating the stress displacement relations for both models for a bar element of length $L_e \geq w_c$. For the crack band model, segment w_c of the length L_e exhibits strain-softening, while the rest of the bar element undergoes elastic unloading, as may be concluded from stability analysis (4,5). For Hillerborg's model, only one line crack is allowed within the bar length L_e, according to stability conditions. Thus, the declining portion of the $\sigma - \delta$ relationship is described according to these two models as

$$\delta(\sigma) = w_c\left(\frac{f_t' - \sigma}{-E_t} + \frac{f_t'}{E}\right) + (L_e - w_c)\frac{\sigma}{E} = \frac{f_t' - \sigma}{-C_t} + L_e\frac{\sigma}{E} \tag{24}$$

($E_t < 0$, $C_t < 0$), and it may be checked that this equation is satisfied for any σ if Eq. 23 holds true.

The composite damage (or composite fracture) model of Willam, Sture et al. (53,54), which represents an adaptation of Pietruszczak and Mróz's model for strain-softening in shear (46), assumes a line crack (or a narrow cracking strip) to be embedded within the finite element and the additional strain due to cracking to be added to that due to elastic deformation of the element. For uniaxial stress in a finite element of length L_e in a bar, one obtains for this element the same stress-displacement relationship as Eq. 24, which again confirms the equivalence of this model.

We may thus conclude that the crack band model, the fictitious crack model of Hillerborg et al., and the composite damage (or composite fracture) model of Willam et al. are equivalent for uniaxial stress, except that continuum concept in the crack band model makes no sense for $L_e < w_c$; but neither of the other models can be applied for this case because the continuum description is no longer admissible if $L_e < w_c$. For problems with localized fracture and arbitrary geometry, these three models are not exactly equivalent, but they are nearly equivalent, with differences that are probably too small to be clearly detected from test results. This is because, according to St. Venant's principle, only the total additional relative displacement due to cracking at the fracture front matters for the energy release rate from the entire body.

It has been claimed (53,54) that the composite damage (composite fracture) model and the fictitious crack model differ from the crack band model by having only two rather than three independent material constants, G_f and f_t'. However, this claim is unjustified because w_c, the third independent constant in the crack band model, has almost no effect in fitting fracture test data, as already pointed out, and can be only determined if the softening modulus E_t is specified independently, which was done in Ref. 23 on the basis of direct tensile tests.

It has been also claimed that the composite damage model and the fictitious crack model permit the use of finite elements of any size, L_e,

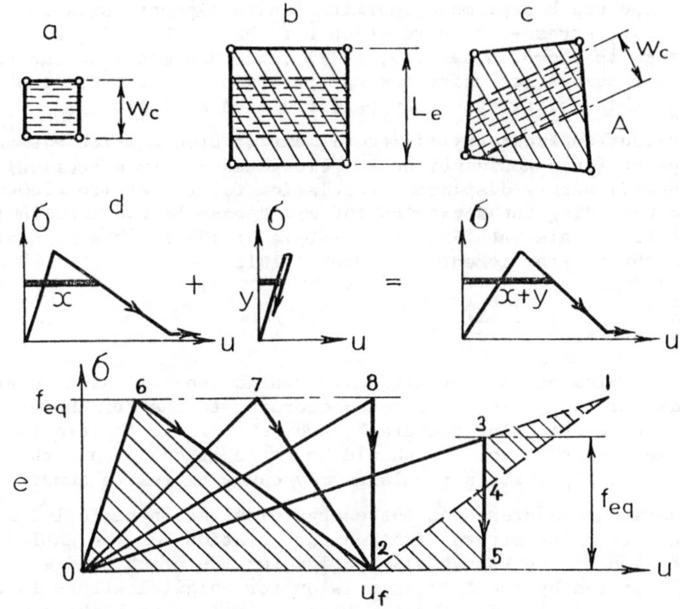

Figure 10. Crack band model when element size is increased (a-c), determination of overall stress-strain relation for finite element (d) and resulting stress-displacement relations for elements of various sizes

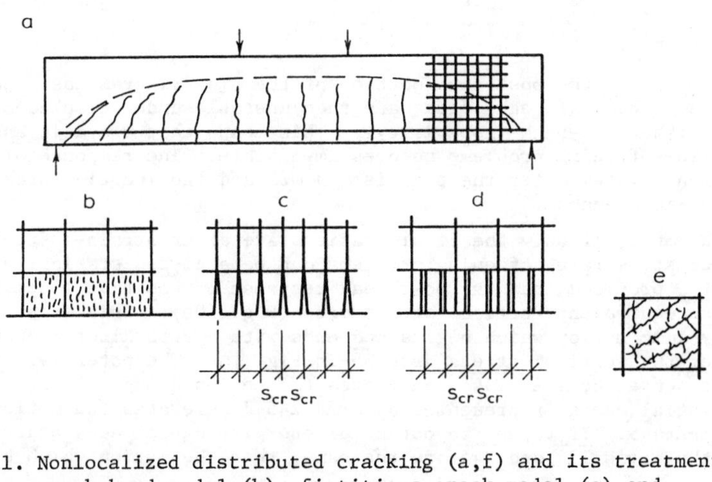

Figure 11. Nonlocalized distributed cracking (a,f) and its treatment by crack band model (b), fictitious crack model (c) and composite damage model (d)

whereas the crack band model is restricted to finite elements of size $L_e = w_c$. This is not true, however. As an approximation involving some error, the crack band model permits finite elements of sizes $L_e > w_c$ if the average stress-strain relation for the cracked finite element is modified (as shown in Refs. 9,16,17), and this modification turns out to yield the same stress displacement relation as the other two models. In general this modification may be formulated as follows.

Consider an arbitrary finite element of area A which has one integration point (and should not be too elongated in any direction); Fig. 10c. The overall stress-displacement relation for the entire element is obtained by adding the strain on the crack band of width w_c to the strain in the finite element (Fig. 10c). This yields for the mean strain tensor in the finite element the general rule

$$\underset{\sim}{\varepsilon} = \frac{w_c}{\sqrt{A}}\,\underset{\sim}{\varepsilon}_{soft} + \left(1 - \frac{w_c}{\sqrt{A}}\right)\underset{\sim}{\varepsilon}_{el} \tag{25}$$

where the element is approximately treated the same as a square element of side $\sqrt{A}$; $\underset{\sim}{\varepsilon}_{soft}$ is the strain according to the strain-softening constitutive relation for the crack band, and $\underset{\sim}{\varepsilon}_{el}$ is the elastic strain for the same stress value. It should be noted, though, that the use of an element size L_e that is too large may cause excessive numerical error.

Equation 25 assures that the energy dissipation at full cracking, as obtained from the stress-displacement relation for the entire element, is the same as the energy dissipation for the crack band w_c. This may be illustrated by the $\delta(\sigma)$ expression for uniaxial stress in Eq. 24. For various element lengths L_e, these relations give the curves in Fig. 10e. While the peak stress point in the stress-displacement relation moves to the right as the element length L_e increases, the end point with $\sigma = 0$ remains fixed because the area under this diagram, representing the fracture energy, must remain constant. The declining portion of the $\delta(\sigma)$ diagram has the inverse slope $d\delta/d\sigma = (w_c/E_t) + (L_e - w_c)/E$; it becomes steeper as L_e increases (Fig. 10e) and becomes vertical when $L_e = L_{cr}$,

$$L_{cr} = w_c\left(1 + \frac{E}{-E_t}\right) = \frac{2G_f E}{f_t'^2} \qquad \left(\frac{d\sigma}{d\delta} \to \infty\right). \tag{26}$$

For $L_e > L_{cr}$, the post-peak portion of the $\delta(\sigma)$ diagram has a positive slope. The post-peak states are then unstable under displacement control (the snap-back instability). This means that a static analysis of strain-softening problems becomes impossible. The response of the finite element system after the peak is *dynamic* and the inertia forces must be taken into account.

However, if only the final static state after strain-softening is of interest, strain-softening problems can be analyzed statically even for $L > L_{cr}$ provided that the post-peak response of positive slope is replaced by an equivalent vertical stress drop (Fig. 10e). This drop represents a dynamic motion which begins and ends with a zero kinetic energy. Consider the vertical stress drop $\overline{35}$ in Fig. 10e. The potential energy represented by area $\overline{134}$ accelerates the motion along segment $\overline{34}$, and the potential energy represented by area $\overline{245}$ decelerates the motion along segment $\overline{45}$. If these two potential energies cancel each other, i.e., if the areas $\overline{134}$ and $\overline{245}$ are the same, then the motion which begins with an equilibrium state at point 3 will end with an equilibrium state (zero kinetic energy) at point 5. On the other hand, if the areas $\overline{134}$ and $\overline{245}$ were unequal, the final state at point 5 would not be an equilibrium state and a static analysis would then be incorrect.

We may conclude that, for the purpose of static analysis, in which some finite elements are larger than L_{cr}, the reverse post-peak behavior with positive slope can be replaced by a vertical stress drop for which the areas $\overline{134}$ and $\overline{245}$ in Fig. 3e are equal. This means that the area $\overline{035}$ under the entire diagram with the vertical stress drop must be equal to the area $\overline{012}$ under the actual stress-displacement diagram for the finite element of length L_e, which in turn equals the fracture energy G_f. Thus, the strength limit at the beginning of the equivalent vertical stress drop, which has been called the equivalent strength f'_{eq} (9,16,18), must satisfy the condition $f'_{eq}\ \delta_0/2 = G_f$ where $\delta_0 = L_e f'_{eq}/E$. This yields for the equivalent strength at $L_e > L_{cr}$ the expression

$$f'_{eq} = \sqrt{\frac{2G_f E}{L_e}} = f'_t \sqrt{w_c\left(1 - \frac{E}{E_t}\right)}\ \frac{1}{\sqrt{L_e}} \ . \tag{27}$$

This is the same expression as that deduced before in a different manner (9,16,18). We see that, for $L_e > L_{cr}$, the strength limit used in the calculation must be decreased in inverse proportion to $\sqrt{L_e}$.

When the vertical stress drop is used, and the mesh is not too crude, the results of analysis are very close to those of linear elastic fracture mechanics. The essential nonlinear characteristics of the response are caused by strain-softening, and so the strain-softening behavior cannot be analyzed with large elements that require a vertical stress drop, unless the structure is so large that linear elastic fracture analysis is applicable.

In the preceding we considered only the situations where the front of distributed cracking is known to localize to its minimum admissible width (w_c in the crack band model, zero in Hillerborg's and Willam's models). When typical fracture specimens are analyzed by finite elements, we know from testing experience that such complete localization will occur. Fracture test specimens are of course designed to obtain just that (localized fracture). In general situations, though, it is not known whether complete localization of the cracking front will occur, and in many practical situations it will not. For example, large, nonlocalized zones of distributed cracking may develop if they are stabilized by compression zones ahead of the cracking front, by reinforcement, or by boundary restraint (Fig. 11). Whether or not complete localization occurs must be decided on the basis of stability criteria, which are discussed in detail in Ref. 10.

The crack band model is ideally suited to handle also the situations of nonlocalized cracking. No additional material parameters are needed for this purpose because a stress-strain relation is used.

On the other hand, further extensions are needed for Hillerborg's fictitious crack model and Willam-Sture's composite damage model in order to handle nonlocalized cracking. An array of line cracks must be introduced for these models, either as a system of interelement line cracks or as a system of cracks with two or more cracks (or cracking strips) embedded in each finite element (see Fig. 10). These line cracks cannot have arbitrary spacing s_{cr}, or else the analysis would be unobjective; any desired results could be obtained depending on the analyst's choice of s_{cr}. This is illustrated by the example in Fig. 12 which was solved at Northwestern University by P. Pfeiffer and is similar to an example presented before by Crisfield (29). The solution is given for a reinforced bar of length 10 in., concrete cross section $A_c = 2$ in.^{2}, steel cross section $A_s = 0.1$ in.^{2}, elastic moduli of steel and concrete $E_c = 4 \times 10^6$ psi, $E_s = 30 \times 10^6$ psi, and straight-line softening diagram of concrete with $E_t = -E_c/8$. The stress displacement relation is assumed so that it gives the same displacement over an element of length 2 in. (Eq. 23). It may be checked by stability criteria (4) that in this bar a uniformly distributed strain-softening is stable and no localization occurs. Three meshes of element sizes 4:2:1 are used, and the calculated load displacement diagrams (Fig. 12) are identical for the crack band model. However, for Hillerborg's and Willam's models, in which one line crack or crack strip per element (or per node) is implied, the results are very different (Fig. 12), illustrating that these models are for nonlocalized cracking unobjective with regard to the choice of element size.

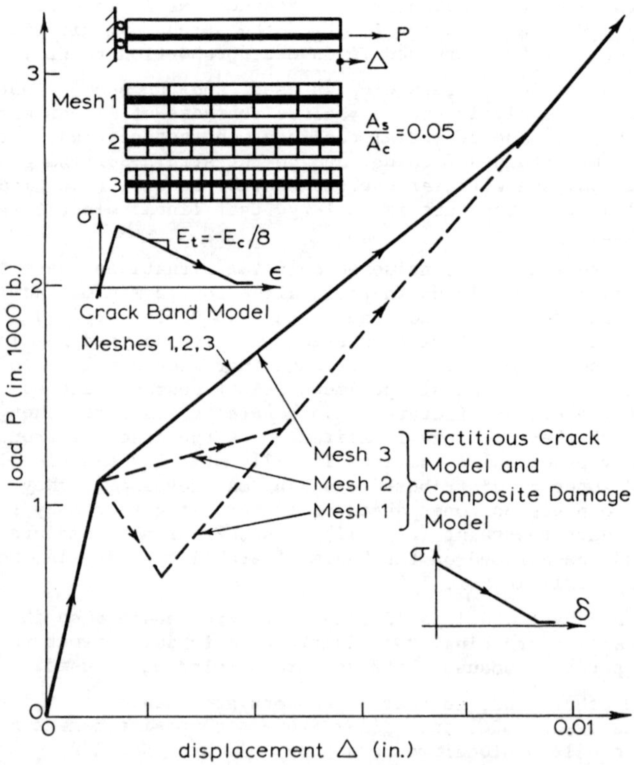

Figure 12. Finite element results of Pfeiffer and Bažant for three different meshes, for concrete bar with 5% reinforcement (nonlocalized cracking)

Thus we again see that the previous claims that these models with line cracks have one independent material parameter less than the crack band model (i.e., two vs. three) are unjustified. These models with line cracks require a characteristic length such as spacing s_{cr} as the *third* material parameter in order to be capable of handling any situations more general than the fracture specimens, which are designed so that the cracking is certain to localize.

It may be concluded that no strain-softening fracture model can be objective with regard to mesh size unless it involves some characteristic length as a material property. The situation is summarized in Table 1.

Table 1. Attainment of Objectivity with Respect to Element Size

Model	Localized Cracking	Nonlocalized Cracking
1. Classical Smeared Cracking	NO	YES
2. Fictitious Crack Model	YES	NO
3. Composite Damage Model	YES	NO
4. Crack Band Model	YES	YES

Aside from the foregoing limitation, Hillerborg's and Willam-Sture's models with line cracks do not seem very suitable for handling the situations where cracking orientation varies over the space (Fig. 11f). These models require proper crack spacing s_{cr} to be objective, but what is the proper spacing if the cracks are nonparallel? This must be expected in general situations. The question how to handle such situations apparently has not yet been addressed, and would have to be addressed if these models should be implemented in a general finite element code. A strain-softening constitutive relation appears to be the only practically manageable way to deal with nonlocalized cracking whose orientation is spatially variable (Fig. 11f).

Other behaviors which have been modeled with the crack band approach but appear difficult for the models based on line cracks are: 1) cracking in more than one direction at a point and 2) cracking process where the prevailing crack direction at one point changes its orientation during the loading process (for example, when partial cracks form in one direction, and later, after the principal stress has rotated, further cracks form in another direction).

CONSTITUTIVE RELATIONS FOR STRAIN-SOFTENING

In the analysis of many practical situations, including all fracture tests, the principal stress direction in the fracture process zone remains constant during fracturing. Triaxial strain-softening can then be introduced in the form of Eqs. 1 and 2. However, especially in dynamics, it is often necessary to describe progressive formation of fracture during which the principal stress directions rotate. In such a case, Eqs. 1-2 are inadequate. A satisfactory formulation can be obtained with an analog of the slip theory of plasticity (3,52), which has been called the microplane model (6,9,25). In this model it is assumed that the strain on a plane of any inclination consists of the resolved components of one and the same macroscopic strain tensor ε_{ij}. Using the condition of equal energy dissipation when calculated in terms of the stresses and strains on all such planes and in terms of the macroscopic stress and strain tensors, one may obtain (6,9,25) the stress-strain relation

$$d\sigma_{ij} = D^{c}_{ijkm}\, d\varepsilon_{ij} \tag{28}$$

in which

$$D^{c}_{ijkm} = \int_0^{2\pi} \int_0^{\pi/2} n_i n_j n_k n_m\, F'(e_n) \sin\phi\, d\theta d\phi. \tag{29}$$

This equation superimposes contributions to inelastic stress relaxations from the planes of all directions within the material, defined by spherical coordinates θ and ϕ; n_i are the direction cosines for all such directions, and $F(e_n)$ is a function characterizing the constitutive properties and representing the stress-strain relation for one particular microplane within the material; $e_n = n_i n_j \varepsilon_{ij}$ = normal strain on a plane with direction cosines n_i. The integral in Eq. 24 must be evaluated numerically. Efficient numerical integration formulas for integration on the surface of a sphere are given by Stroud (48), and some further formulas which are superior in certain situations are given in Ref. 26. It has been demonstrated that the microplane model allows description of tensile strain-softening under general stress or strain histories and always leads to a reduction of stress to zero at sufficiently large tensile strain (26,9).

SIZE EFFECT LAW AND ITS GENERALIZATION

From the practical viewpoint, the imbricate nonlocal continuum model, as well as the special case of crack band model, differs from the usual local finite element models principally by the size effect. The structural size effect, a salient aspect of fracture mechanics, is observed when geometrically similar structures of different characteristic dimensions d are compared. It can be described in terms of the nominal stress at failure, defined as $\sigma_N = P/bd$ where P = load at failure (maximum load) and b = structure thickness. While according to the strength or yield criteria used in plastic limit design or elastic allowable stress design, σ_N is independent of structure size d, in fracture mechanics σ_N decreases as the structure size increases. This is because fracture mechanics is based on energy criteria for failure.

On the basis of an approximate but apparently quite reasonable hypothesis that the energy release caused by fracture is a function of both the fracture length and the area traversed by the fracture process zone, it can be shown (8) by dimensional analysis and similitude arguments that, for geometrically similar structures or specimens,

$$\sigma_N = Bf'_t\left(1 + \frac{d}{\lambda_0 d_a}\right)^{-\frac{1}{2}} \tag{30}$$

in which f'_t is the direct tensile strength of concrete, d_a is the maximum aggregate size, and d, λ_0 are empirical parameters characterizing the shape of the structure or specimen. According to this size effect law, the plot of log σ_N vs. log (d/d_a) represents a gradual transition from the strength criterion (i.e., σ_N proportional to strength f'_t) to the failure criterion of the classical, linear elastic fracture mechanics (i.e., σ_N proportional to $d^{-\frac{1}{2}}$). This size effect law is verified, within the limits of inevitable statistical scatter, by all available Mode I fracture tests of concrete and mortar. The size effect law in Eq. 26 applies not only to cleavage (Mode I) fractures but also to shear (Mode II) fractures (27). Moreover, this size effect law has also been shown applicable to the diagonal shear failure of longitudinally reinforced beams without stirrups, unprestressed (23) or prestressed (13), punching shear of slabs (15), failure of unreinforced pipes (14), and is probably applicable to all the so-called brittle failures of reinforced concrete structures. Evaluation of fracture tests for different specimen sizes on the basis of Eq. 26 offers the simplest means for determining the fracture energy and other nonlinear fracture parameters.

When the usual, local finite element codes are applied to similar structures with similar meshes, the results exhibit no size effect. On the other hand, when the nonlocal model (or the special case of crack band model) is used, the results conform to the size effect law in Eq. 26, as experience confirms.

The size effect law in Eq. 30 was originally derived by dimensional analysis and similitude arguments from the following fundamental hypothesis:

The total potential energy release W caused by fracture in a given structure is a function of both

1. The length of the fracture a; and
2. The area of the cracked zone, $nd_a a$.

nd_a is a material constant representing the effective width of the micro-cracking zone at the fracture front, and d_a is the maximum aggregate size.

The original derivation (8), which used the nondimensional variables $\alpha_1 = a/d$ and $\alpha_2 = nd_a a/d^2$, tacitly implied one further hypothesis, namely that W as a function of the cracked zone area can be linearized. This is because otherwise the functions α_1 and α_2 used in Ref. (8) would not be size-independent as parameter α_2 depends on the relative size d_a/d when a/d is constant. This fact, as well as the need for the additional linearization assumption, was independently discovered by J. Planas and M. Elices (47).

We will now show a different derivation of the size effect law which involves linearization only as an asymptotic approximation that is always admissible, and which at the same time shows a more general form for the size effect law. Instead of the aforementioned nondimensional variables α_1 and α_2, we now choose as nondimensional variables:

$$\theta_1 = \frac{a}{d}, \quad \theta_2 = \frac{nd_a a}{ad} = \frac{nd_a}{d} . \tag{31}$$

We consider geometrically similar structures of different sizes d, which are characterized by a constant ratio a/d. Thus, parameter θ_1 is size-independent, while parameter θ_2 characterizes the size. As before (8), the potential energy release W of any structure may always be written in the form

$$W = \frac{1}{2E_c}\left(\frac{P}{bd}\right)^2 bd^2\ F(\theta_1, \theta_2, \varepsilon_i) \tag{32}$$

in which E_c is the Young's elastic modulus of concrete, b is the thickness of the structure, P is the maximum load (i.e., the failure load in a load-controlled test), F is a certain function of nondimensional parameters, and ξ_i are parameters characterizing the structure shape (geometry) which are constant when geometrically similar structures are considered. The nominal stress at failure may be defined as $\sigma_N = P/bd$. Then, substituting $P = \sigma_N bd$ and Eq. 32 into the well known fracture equilibrium relation

$$\frac{\partial W}{\partial a} = G_f b \tag{33}$$

where G_f is the fracture energy, we obtain

$$\frac{1}{2E_c}\frac{\sigma_N^2 b^2 d^2}{b}\frac{\partial F}{\partial \theta_1}\frac{1}{d} = G_b b \tag{34}$$

from which

$$\sigma_N = (2G_f E_c)^{\frac{1}{2}}\left[\frac{\partial F(\theta_1,\theta_2,\xi_i)}{\partial \theta_1} d\right]^{-\frac{1}{2}}. \tag{35}$$

In this formulation, the fracture energy itself is considered to be size-independent, being regarded as the final value of the R-curve. According to the blunt crack band theory, $G_f = nd_a(1 - E_c/E_t)f_t'^2/2E_c$, in which f_t' = direct tensile strength and E_t = mean slope of the strain-softening portion of the stress-strain diagram = strain-softening modulus. After substitution for G_f, Eq. 35 becomes

$$\sigma_N = \left[\frac{nd_a}{d}\left(1 - \frac{E_c}{E_t}\right)\right]^{\frac{1}{2}} f_t'\left[\frac{\partial F(\theta_1,\theta_2,\xi_i)}{\partial \theta_1}\right]^{-\frac{1}{2}}. \tag{36}$$

We may now choose the state $\theta_2 = 0$, which corresponds to an infinitely large structure, $d/d_a \to \infty$, as the reference state, and expand $\partial F/\partial \theta_1$ as a function of θ_2 in Taylor series. However, it is preferable to expand, more generally, the function $g(\eta) = (\partial F/\partial \theta_1)^r$ where $\eta = \theta_2^r$ and exponent r $(r > 0)$ is an empirical constant to be chosen so that the size effect could be described with the minimum number of terms in the series expansion. Thus we may set

$$g(\eta) = \left[\frac{\partial F(\theta_1,\theta_2,\xi_i)}{\partial \theta_1}\right]^r = c_0 + c_1\eta + c_2\eta^2 + c_3\eta^3 + c_4\eta^4 + \cdots \tag{37}$$

in which

$$\begin{aligned} c_0 &= \left[\left(\frac{\partial F}{\partial \theta_1}\right)^r\right]_{\theta_2=0}, \quad c_1 = \frac{1}{1!}\left[\frac{\partial}{\partial \eta}\left(\frac{\partial F}{\partial \theta_1}\right)^r\right]_{\theta_2=0}, \\ c_2 &= \frac{1}{2!}\left[\frac{\partial^2}{\partial \eta^2}\left(\frac{\partial F}{\partial \theta_1}\right)^r\right]_{\theta_2=0}, \quad c_3 = \frac{1}{3!}\left[\frac{\partial^2}{\partial \eta^2}\left(\frac{\partial F}{\partial \theta_1}\right)^r\right]_{\theta_2=0}, \cdots \end{aligned} \tag{38}$$

Now, introducing $\partial F/\partial \theta_1 = [g(\eta)]^{1/r}$ into Eq. 36 and substituting Eq. 37 for $g(\eta)$, we can rearrange the resulting expression to the form

$$\sigma_N = Bf_t'\left[\left(\frac{\lambda_0}{\lambda}\right)^{-r} + 1 + \left(\frac{\lambda_1}{\lambda}\right)^r + \left(\frac{\lambda_2}{\lambda}\right)^{2r} + \left(\frac{\lambda_3}{\lambda}\right)^{3r} + \cdots\right]^{-1/2r} \tag{39}$$

in which $\lambda = d/d_a$ = relative size of the structure, and

$$B = c_1^{-1/2r}(1 - E_c/E_t)^{\frac{1}{2}}, \qquad \lambda_0 = n(c_1/c_0)^{1/r},$$
$$\lambda_1 = n(c_2/c_1)^{1/r}, \qquad \lambda_2 = n(c_3/c_1)^{1/2r}, \qquad \lambda_3 = n(c_4/c_1)^{1/3r}, \ldots \tag{40}$$

Constants B, λ_0, λ_1, must normally be determined from maximum load data for geometrically similar specimens of different sizes. These constants are different for each specimen geometry. Note also that they are size-independent, since they depend on the derivatives of $F(\theta_1, \theta_2, \xi_i)$ at a fixed value of size parameter θ_2.

Eq. 39 is the most general asymptotic approximation for the size effect. However, full generality is preserved even if we substitute $r = 1$ because any smooth function $g(\eta)$ in Eq. 37 can be approximated by a polynomial with any desired degree of accuracy. The usefulness of coefficient r consists in the fact, found by experience, that if we use a certain optimal value of r we can describe any given shape of the size effect curve with much fewer terms than if we used $r = 1$. It is more effective to vary r than to include more terms in the series expansion.

If the structure is very large ($d/d_a \to \infty$), the first term in the brackets in Eq. 39 dominates over all other terms, which may be neglected. Then Eq. 39 reduces to

$$\sigma_N = Bf_t'(\lambda_0 \, d_a/d)^{\frac{1}{2}} = \text{const.} \times d^{-\frac{1}{2}}. \tag{41}$$

This is the size effect typical of all solutions according to linear elastic fracture mechanics.

Eq. 41 represents the zero-th order asymptotic approximation of the size effect. The first-order approximation is obtained by retaining the first two terms of the series in Eq. 41 and truncating the rest:

$$\sigma_N = Bf_t'\left[1 + \left(\frac{d}{\lambda_0 d}\right)^r\right]^{-1/2r}. \tag{42}$$

This formula, which may be called the generalized size effect law since for $r = 1$ it reduces to the originally proposed approximate size effect law (Eq. 30), can closely represent various size effect curves which may be obtained by finite element analysis for various shapes of the softening diagram (11). It is hardly conceivable that an asymptotic approximation of higher than first order might be needed, provided that the optimum value of r for given data is used. However, if r were fixed as 1, then higher order approximations would be needed to closely describe certain finite element results for σ_N.

The size of the smallest castable concrete structure is a certain fixed multiple of the maximum aggregate size, say $d = 3d_a$. Substituting this into Eq. 42 or Eq. 39, we find that

$$\sigma_N = B_0 f_t' \tag{43}$$

in which B_0 is a constant for a particular geometry of the structure. Eq. 43 coincides with the failure criterion used in limit analysis (plasticity) as well as in the elastic allowable stress design. This criterion, as is well known, exhibits no size effect.

The size effect law in Eq. 30 or Eq. 42 offers the simplest means of determining the fracture parameters of concrete, particularly the fracture energy. The parameters of Eq. 42 may be determined from given σ_N data by using a nonlinear optimization subroutine. However, they can be also determined by linear regressions, since Eq. 42 can be algebraically transformed to the linear relation:

$$Y = AX + C, \text{ with } X = d/d_a, \quad Y = \sigma_N^{-2r}, \quad A = (Bf_t')^{-2r}, \quad C = A/\lambda_0. \tag{44}$$

Choosing the value of r, we may plot Y versus X, and then determine the slope A, the Y-intercept C, and the coefficient of variation ω for the deviations from the regression line. Repeating the regression for various values of r (e.g., 1, 0.8, 0.6, 0.4), we may consider ω as a function of r and determine r so as to minimize ω. From the slope A of the regression line for optimum r one can then easily determine the fracture energy G_f, as shown in Ref. 22. Also $Bf_t' = A^{2r}$ and $\lambda_0 = A/C$.

While the exponent r can be determined from finite element results for a given strain-softening constitutive relation (and a given characteristic length), the existing test data are insufficient to determine r and equally good data fits are possible with different r-values. The reasons are that the random scatter of test results is too high and the size range is not broad enough. We may therefore use $r = 1$ (Eq. 30) until better test data become available. For the typical range of applications, the value of r is in fact unimportant. With different values of r, different values of the fracture energy are obtained from the analyses of given test data, but they all yield approximately the same failure predictions in the practical range of structure sizes such as d/d_a = 3 to 100. Thus the value of r does not matter unless extrapolation to extremely large structure sizes is intended.

CONCLUSION

The concept of imbricate nonlocal continuum, along with strain-softening stress-strain relations, allows a mathematically consistent and realistic description of progressive distributed cracking of concrete structures. When fine resolution inside the strain-softening zone is not needed, this approach reduces to the crack band theory. While the existing models based on the line crack concept are unobjective and inapplicable for nonlocalized cracking, the present nonlocal formulation, along with the crack band model as its special case, is objective and applicable for both localized and nonlocalized cracking. Examples of one-dimensional planar and spherical waves demonstrate that stable strain-softening regions of finite size can be modeled. General strain-softening constitutive relations can be obtained by the microplane model approach. The results of nonlocal finite element models conform to the size effect law. A generalized size effect law is derived by asymptotic analysis.

Acknowledgement: Partial financial support under AFOSR Grant No. 83-0009 to Northwestern University is gratefully acknowledged. Mary Hill is thanked for her exemplary secretarial assistance.

APPENDIX I. REFERENCES

1. Achenbach, J.D., Wave Propagation in Elastic Solids, North Holland Publ. Co., Amsterdam, 1973.
2. Barrenblatt, G.I., "The Formation of Equilibrium Cracks During Brittle Fracture, General Ideas and Hypothesis, Axially Symmetric Cracks," Prikladnaya Matematika i Mekhanika, Vol. 23, No. 3, 1959, 434-444.
3. Batdorf, S.B., and Budianski, B., "A Mathematical Theory of Plasticity Based on the Concept of Slip," NACA Technical Note TN 1871, Apr. 1949.
4. Bažant, Z.P., "Instability, Ductility and Size Effect in Strain-Softening Concrete," Journal of the Engineering Mechanics Division ASCE, Vol. 102, Apr. 1976, 331-344; Discussions, Vol. 103, 357-358, 775-777, Vol. 104, 501-502, based on Structural Engineering Report No. 74-8/640, Northwestern University, August 1974, and private communication to K. Gerstle, S. Sture and A. Ingraffea in Boulder, Colorado, June 5, 1974.
5. Bažant, Z.P., "Crack Band Model for Fracture of Geomaterials," Proc. 4th Intern. Conf. on Numerical Methods in Geomechanics, held in Edmonton, Alberta, Canada, June 1982, ed. by Z. Eisenstein, Vol. 3, 1137-1152.
6. Bažant, Z.P., "Fracture in Concrete and Reinforced Concrete," Chap. 13 in Mechanics of Geomaterials: Rocks, Concretes, Soils, John Wiley and Sons, Chichester-New York, 1985 (Proc. of IUTAM W. Prager Symposium held at Northwestern University in Sept. 1983), pp. 259-304.
7. Bažant, Z.P., "Imbricate Continuum and Its Variational Derivation," Jour. of Eng. Mech. ASCE, Vol. 111, Dec. 1984, 1693-1712.
8. Bažant, Z.P., "Size Effect in Blunt Fracture: Concrete, Rock, Metal," J. of Engineering Mechanics ASCE, Vol. 110, Apr. 1984, 518-535.
9. Bažant, Z.P., "Mechanics of Fracture and Progressive Cracking in Concrete Structures," Chap. 1 in Fracture Mechanics Applied to Concrete Structures, ed. by G.C. Sih, Martinus Nijhoff Publishers B.V., The Hague, Netherlands, 1985.
10. Bažant, Z.P., "Distributed Cracking and Nonlocal Continuum," Preprints, Europe-US Symposium on "Finite Element Methods for Nonlinear Problems," ed. by P. Bergan, Norwegian Institute of Technology, Trondheim, Aug. 1985, pp. II.2-1...II. 2-25.
11. Bažant, Z.P., "Comment on Hillerborg's Comparison of Size Effect Law with Fictitious Crack Model," Dei Poli Anniversary Volume, ed. by L. Cedolin, Politecnico di Milano, Oct. 1985.
12. Bažant, Z.P., Belytschko, T.B., "Wave Propagation in a Strain-Softening Bar: Exact Solution," J. of Engineering Mechanics ASCE, Vol. 111, No. 3, March 1985.
13. Bažant, Z.P., and Cao, Z., "Size Effect in Shear Failure of Prestressed Concrete Beams," ACI Journal, in press.
14. Bažant, Z.P., and Cao, Z., "Size Effect in Brittle Failure of Unreinforced Pipes," ACI Journal, in press.

15. Bažant, Z.P., and Cao, Z., "Size Effect in Punching Shear Failure of Slabs," Report No. 85-8/4285, Center for Concrete and Geomaterials, Northwestern University, Evanston, Illinois, May 1985.

16. Bažant, Z.P., Cedolin, L., "Blunt Crack Band Propagation in Finite Element Analysis," Journal of the Engineering Mechanics Division, ASCE, Vol. 105, No. EM2, April 1979, 207-315.

17. Bažant, Z.P., Cedolin, L., "Fracture Mechanics of Reinforced Concrete," Journal of the Engineering Mechanics Division, ASCE, Vol. 106, No. EM6, Dec. 1980, 1287-1306; with Discussion, Vol. 108, 1982, pp. 464-471.

18. Bažant, Z.P., Cedolin, L., "Finite Element Modeling of Crack Band Propagation," Journal of Structural Engineering, ASCE, Vol. 109, No. ST2, Feb. 1983, 69-92.

19. Bažant, Z.P., Chang, T.P., Belytschko, T.B., "Continuum Theory for Strain-Softening," Report No. 83-11/428c, Center for Concrete and Geomaterials, Northwestern University, Evanston, Ill., Nov. 1983; also Journ. Eng. Mech. ASCE, Vol. 111, Dec. 1984, 1666-1692.

20. Bažant, Z.P., and Chang, T.P., "Imbricate Finite Element Analysis of Strain-Softening Solids," Report No. 85-8/428, Center for Concrete and Geomaterials, Northwestern University, Evanston, Ill., Aug. 1985.

21. Bažant, Z.P., and Gambarova, R.G., "Crack Shear in Concrete: Crack Band Microplane Model," J. of Engineering Mechanics ASCE, Vol. 110, Sept. 1984, 2015-2035.

22. Bažant, Z.P., Kim, J.-K., and Pfeiffer, P., "Nonlinear Fracture Properties from Size Effect Tests," J. of Structural Eng., ASCE, in press.

23. Bažant, Z.P., and Kim, J.-K., "Size Effect in Shear Failure of Longitudinally Reinforced Beams," ACI Journal, Vol. 81, 1984, 456-468.

24. Bažant, Z.P., Oh, B.H., "Crack Band Theory for Fracture of Concrete," Materials and Structures (RILEM, Paris), Vol. 16, 1983, 155-177.

25. Bažant, Z.P., Oh, B.H., "Microplane Model for Fracture Analysis of Concrete Structures," Proc. Symp. on the Interaction of Nonnuclear Munitions with Structures, U.S. Air Force Academy, Colorado Springs, May, 1983, 49-55.

26. Bažant, Z.P., and Oh, B.H., "Microplane Model for Progressive Fracture of Concrete and Rock," J. of Engineering Mechanics ASCE, Vol. 111, 1985, pp. 559-582.

27. Bažant, Z. P., and Pfeiffer, P. D., "Tests of Shear Fracture and Strain-Softening in Concrete," Proc. 2nd Symp. on "The Interaction of Non-Nuclear Munitions with Structures," Panama City Beach, Florida, Apr. 1985, pp. 254-264; and further extension in Matériaux et Constructions (RILEM, Paris), in press.

28. Belytschko, T.B., Bažant, Z.P., Hyun, V.V., Chang, T.P., "Strain-Softening Materials and Finite Element Solutions," Computers and Structures, in press.

29. Crisfield, M.A., "Difficulties with Current Numerical Models for Reinforced Concrete and Some Tentative Solutions," Proc., Intern. Conf. on "Computer Aided Analysis and Design of Concrete Structures," held in Split, Yugoslavia, Sept. 1984, ed. by F. Damjanić, E. Hinton et al., publ. by Pineridge Press, Swansea, U.K., 331-358.

30. Dugdale, D.S., "Yielding of Steel Sheets Containing Slits," J. of Mech. Phys. Solids, Vol. 8, 1960, 100-108.

31. Eringen, A.C., Edelen, D.G.B., "On Nonlocal Elasticity," Int. J. Engng. Science, Vol. 10, 1972, 233-248.

32. Eringen, A.C., Ari, N., "Nonlocal Stress Field at Griffith Crack," Cryst. Latt. Def. and Amorph. Mat., Vol. 10, 1983, 33-38.

33. Evans, R.H., Marathe, M.S., "Microcracking and Stress-Strain Curves for Concrete in Tension", Materials and Structures (RILEM, Paris), No. 1, Jan.-Feb., 1968, 61-64.

34. "Finite Element Analysis of Reinforced Concrete," ASCE State-of-the-Art Report prepared by a Task Committee chaired by A. Nilson, ASCE, New York, 1982.

35. Heilmann, H.G., Hildsdorf, H.K., Finsterwalder, K., "Festigkeit und Verformung von Beton unter Zugspannungen," Deutscher Ausschuss für Stahlbeton, Heft 203, W. Ernst & Sohn, West Berlin, 1969.

36. Hillerborg, A., Modéer, M., Petersson, P.E., "Analysis of Crack Formation and Crack Growth in Concrete by Means of Fracture Mechanics and Finite Elements," Cement & Concrete Research, Vol. 6, 1976, 773-782.

37. Kfouri, A.P., and Rice, J.R., "Elastic-Plastic Separation Energy Rate for Crack Advance in Finite Growth Steps," in Fracture 1977 (Proc. of 4th Intern. Conf. on Fracture, Waterloo, Ontario, June 1977), ed. by D.M.R. Taplin, Univ. of Waterloo Press, 1977, Vol. 1, 43-59.

38. Knauss, W.C., "On the Steady Propagation of a Crack in a Viscoelastic Plastic Solid," J. of App. Mech., ASME, Vol. 41, No. 1, 1974, 234-248.

39. Kröner, E., "Elasticity Theory of Materials with Long-Range Cohesive Forces," Int. J. Solids Structures, Vol. 3, 1967, 731-742.

40. Kröner, E., "Interrelations between Various Branches of Continuum Mechanics," Mechanics of Generalized Continua, ed. by E. Kröner, Springer-Verlag, 1968, 330-340.

41. Krumhansl, J.A., "Some Considerations of the Relation between Solid State Physics and Generalized Continuum Mechanics," Mechanics of Generalized Continua, ed. by E. Kröner, Springer-Verlag, 1968, 298-331.

42. Kunin, I.A., "The Theory of Elastic Media with Microstructure and the Theory of Dislocations," Mechanics of Generalized Continua, ed. by E. Kröner, Springer-Verlag, 1968, 321-328.

43. Levin, V.M., "The Relation between Mathematical Expectation of Stress and Strain Tensors in Elastic Microheterogeneous Media," Prikladnaya Mathematika i Mekhanika, Vol. 35, 1971, 694-701 (in Russian).

44. Petersson, E., "Fracture Energy of Concrete: Method of Determination," Cement & Concrete Research, Vol. 10, 1980, 78-89, and Fracture Energy of Concrete, Practical Performance and Experimental Results," Cement and Concrete Research, Vol. 10, 1980, 91-101.

45. Petersson, P.C., "Crack Growth and Development of Fracture Zones in

Plain Concrete and Similar Materials," Doctoral Dissertation, Lund Institute of Technology, Lund, Sweden, Dec. 1981.

46. Pietruszczak, S., Mróz, Z., "Finite Element Analysis of Deformation of Strain-Softening Materials," Intern. J. of Num. Meth. in Engng., Vol. 17, 1981, 327-334.

47. Planas, J., and Elices, M., Private Communication to Z.P. Bažant, Dept. of Material Science, ETS Universidad Politecnica de Madrid, Spain, March 1985.

48. Stroud, A.H., Approximate Calculation of Multiple Integrals, Prentice Hall, Englewood Cliffs 1971, 296-302.

49. Rashid, Y.R., "Analysis of Prestressed Concrete Pressure Vessels," Nuclear Engng. & Design, Vol. 7, No. 4, April 1968, 334-344.

50. Reinhardt, H.W., Cornelissen, E., "Post-Peak Cyclic Behavior of Concrete in Uniaxial Tension and Alternating Tensile and Compressive Loading," Cement & Concrete Research, Vol. 14, 1984, 263-270.

51. Rüsch, H., Hilsdorf, H., "Deformation Characteristics of Concrete under Axial Tension," Vorunterschungen, Bericht Nr. 44, Munich, May 1963.

52. Taylor, G.I., "Plastic Strain in Metals," Jour. of the Institution of Metals (London), Vol. 62, 1938, 307-324.

53. Willam, K.J., Bicanić, N., Sture, S., "Constitutive and Computational Aspects of Strain-Softening and Localization in Solids," Proc. Sym. on Constitutive Equations, Micro, Macro & Computational Aspects, K.J. Willam, ed., held at ASME Winter Annual Meeting, New Orleans, Dec. 1984, Am. Soc. of Mech. Engrs., New York, 1984.

54. Willam, K.J., Hurlbut, B., Sture, S., "Experimental, Constitutive and Computational Aspects of Concrete Failure," Preprints, U.S.-Japan Seminar on Finite Element Analysis of Reinforced Concretes, Tokyo, May 1985, 149-172.

55. Wittmann, F.H. (editor), Fracture Mechanics of Concrete, Elsevier, Netherlands, 1983.

Analysis of "Shear Fracture" Tests of Concrete Beams

Anthony R. Ingraffea, M. ASCE* and Malcolm J. Panthaki, A.M. ASCE**

Abstract

Recent tests by Bazant and Pfeiffer (1985) have purported to show the existence of shear fracture in plain concrete beams. This behavior is in contrast to that observed in similar specimens tested by Arrea and Ingraffea (1981). It is here shown by analysis, both classical elasticity and finite element, that the specimens tested by Bazant and Pfeiffer (1985) fractured in a manner quite similar to that which occurs in a splitting tensile strength test. It is concluded that, although shear fracture can possibly occur under certain conditions, tensile and not shear fracture occurred in their specimens.

Introduction

The influence of shear on the process of cracking in concrete structures is of great practical importance. A limiting case of this influence occurs when a pre-existing crack is loaded by placing it in a plane of principal shear stress. In the terminology of fracture mechanics, this is called Mode II loading.

The results of Mode II fracture initiation tests on plain concrete beams were reported by Arrea and Ingraffea in 1981. Tests were performed on specimens in the configuration shown in Figure 1. This loading arrangement places the notch in a state of nearly pure Mode II. In all nine tests performed response was like that shown in Figure 2. Fracture initiated at the notch-tip in a direction approximately normal to the local principal tensile stress, Figure 2c. Fracture continued in a direction which minimized the Mode II component, with the fracture plane describing a curvilinear path as shown in Figure 2b. Finite element analyses of the tests were reported in References (ANSI/ASTM C496-71) and (Arrea and Ingraffea, 1981). The same type of testing was later performed on 38 specimens of limestone and granite (Ingraffea, 1981) with the same behavior patterns as shown in Figure 2.

One of the principal results of all of those tests and analyses was that "shear fracture" did not occur. Although the initial notch was

*Associate Professor, Department of Structural Engineering, Hollister Hall, Cornell University, Ithaca, NY 14853.

**Graduate Student, Department of Structural Engineering, Hollister Hall, Cornell University, Ithaca, NY 14853.

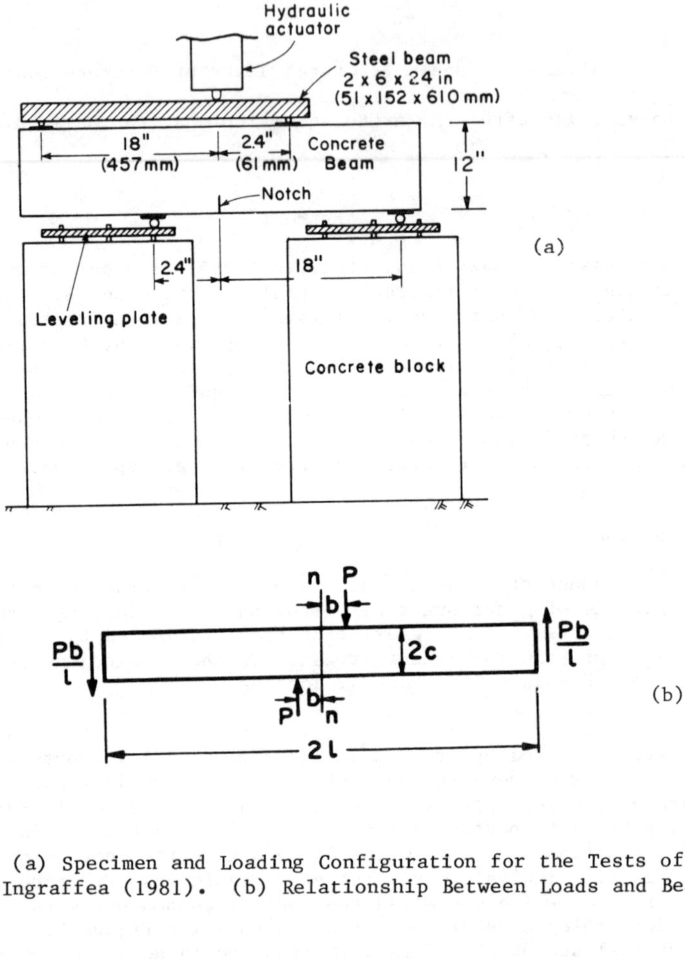

Figure 1. (a) Specimen and Loading Configuration for the Tests of Arrea and Ingraffea (1981). (b) Relationship Between Loads and Beam Geometry.

loaded in almost pure Mode II, cracking was not co-planar with the notch. The crack propagated not in the plane of principal shear stress, but rather in the direction always normal to the local principal tensile stress, that is, in Mode I.

Recently, Bazant and Pfeiffer (1985) performed similar tests but reported dissimilar results. Their specimen and loading arrangement is shown in Figure 3. Comparison with Figure 1 shows that Bazant and Pfeiffer (1985) added a second starter-notch, and moved the major loads, those closer to the notches, much closer to the centerline of the specimen. The latter change was an attempt to create a narrow zone of intense shear stress. In contrast to the behavior shown in Figure 2, Bazant and Pfeiffer (1985) reported that a fracture plane connected the two notch tips in their tests. Further, and most

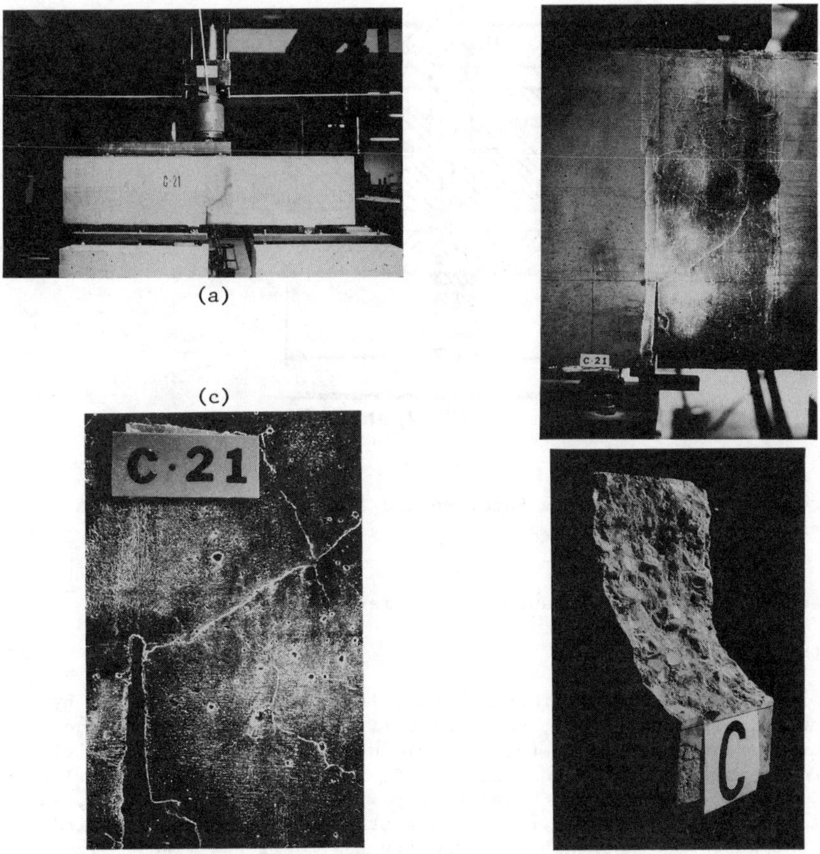

Figure 2. Results of Test by Arrea and Ingraffea (1981).
(a) Visualization of Tensile Crack Using Water Impregnation.
(b) Crack Visualization Using Fluorescent Particle Penetrant.
(c) Close-up Around Notch Tip. (d) Final Fracture Surface.

importantly, they claimed that since this plane is one of high shear stress, the observed phenomenon was a "shear fracture."

The purpose of this paper is to show that this claim cannot be supported by analysis. It will be shown that the fracture plane observed in the Bazant and Pfeiffer (1985) tests is not the plane of principal shear, but rather is nearly perpendicular to high principal tensile stresses. It will also be shown that the shear stresses in the region between the notch tips are much lower than those predicted by elementary theory of bending. In fact, we will attempt to show that those tests, rather than measuring Mode II critical fracture energy, were an indirect measure of tensile strength.

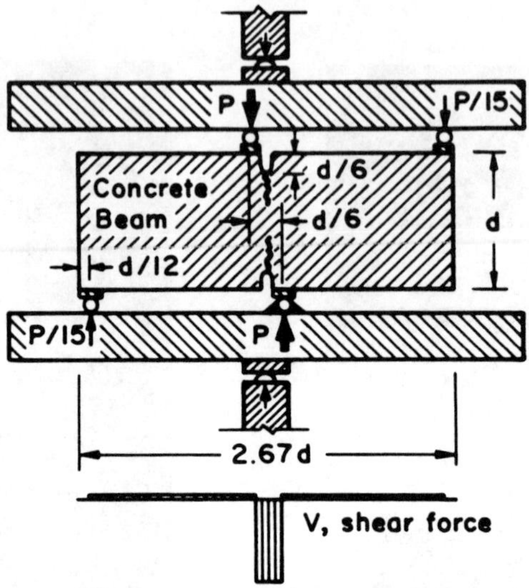

Figure 3. Mode II Fracture Specimen and Loading Apparatus. From Bazant and Pfeiffer (1985).

Stress Analysis of the Bazant and Pfeiffer Specimens

Finite element analyses

Before reporting on a fracture analysis of the specimens used by Bazant and Pfeiffer (1985), we present first the results of a linear finite element analysis of their specimens. Bazant and Pfeiffer (1985) tested specimens with depths, d, of 1.5, 3, 6, and 12 inches (38.1, 76.2, 152.4, and 304.8 mm, respectively). Specimen thickness, t, was constant at 1.5 inches (38.1 mm). All other dimensions were scaled with the depth so that only one geometry had to be analyzed. Figure 4 shows 3 finite element meshes used in a convergence study for the 12 inch deep specimen. Quadratic displacement, isoparametric elements were used throughout. The Finite Element Fracture Analysis Program (FEFAP) (Ingraffea and Saouma, 1985; Ingraffea et al., 1984) was used to perform all finite element analyses reported here.

The stress state in the region between the notch-tips is of paramount interest. Figure 5 shows major and minor principal stress contours in this region. The photographs of stress contours reproduced here have been taken from a color graphic display device. They have been reduced to a degree that some small text is not easily legible; however, it is the nature of the stress distribution that we wish to convey with these photographs. Quantitative information is contained in other figures.

The most important observation here is that Figure 5a indicates that the major principal stress in this region is tensile, and that

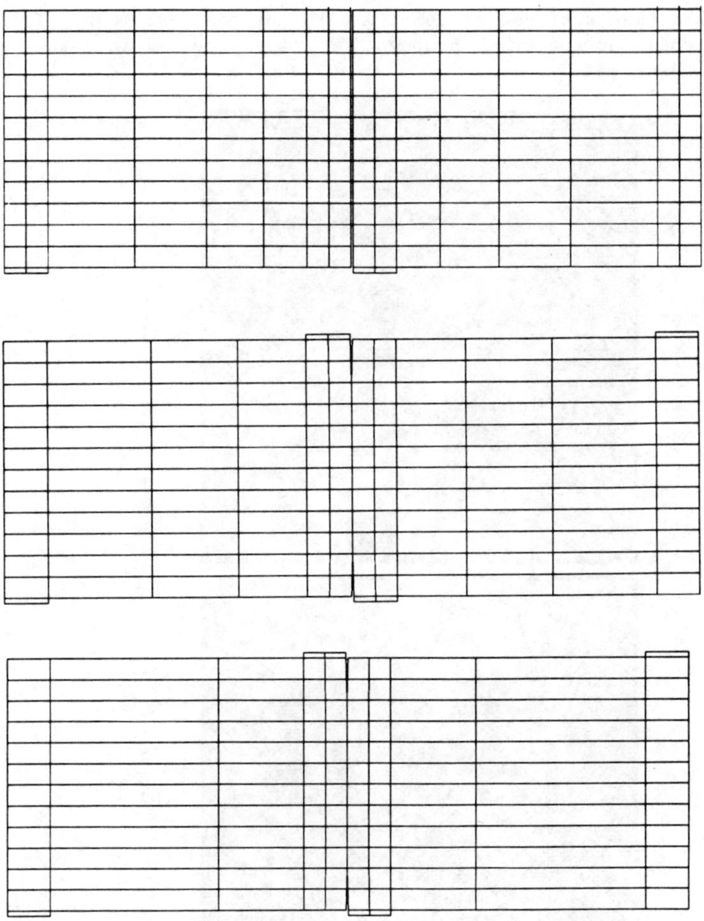

(a)

(b)

(c)

Figure 4. Finite Element Meshes Used for Convergence Studies. (a) Mesh 1. (b) Mesh 2. (c) Mesh 3. Mesh Characteristics Given in Table I.

the direction of this stress is nearly horizontal, that is, nearly perpendicular to a line between the notch tips. Figure 6 shows the same information in a different format. In this figure the principal stresses are plotted as vectors at the element Gauss points. Again, the striking observation shown in Figure 6a is the corridor of sub-horizontal principal tension between the notch tips. This is the first observation which calls into question whether a "shear fracture" occurred in the Bazant and Pfeiffer tests (1985). A quantitative sampling of the principal stresses at the center of the specimen, Point A in Figure 6, is presented in Table I. This table shows that convergence has been achieved at Point A with the coarsest mesh. This mesh will be used for the crack propagation study reported in a subsequent section.

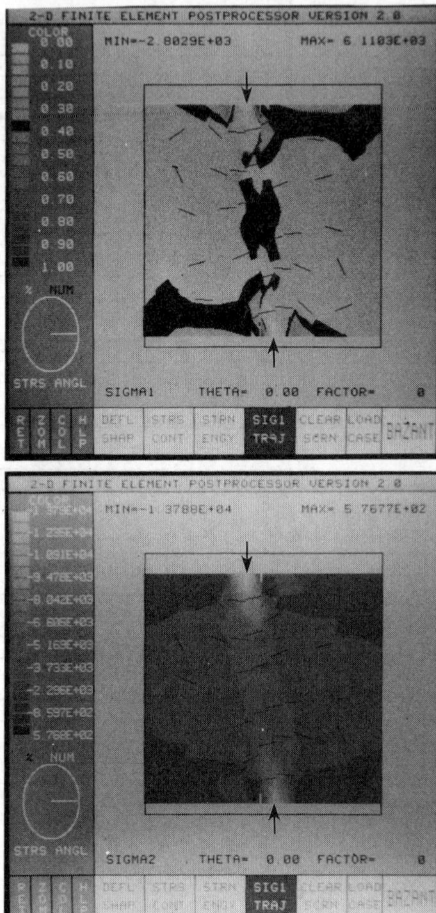

(a)

(b)

Figure 5. (a) Contours of Major Principal Stress in Central Region of the Specimen. Blackened Regions Have High Tensile Stress. (b) Contours of Minor Principal Stress. Vectors are Directions of Major Principal Stress.

Further insight into the stress state in the region between the notch tips can be obtained through Figure 7. This figure shows plots of normal and shear stresses along the line BAC. Line BAC is colinear with one which connects the points of application of the principal loads. It should be noted that,

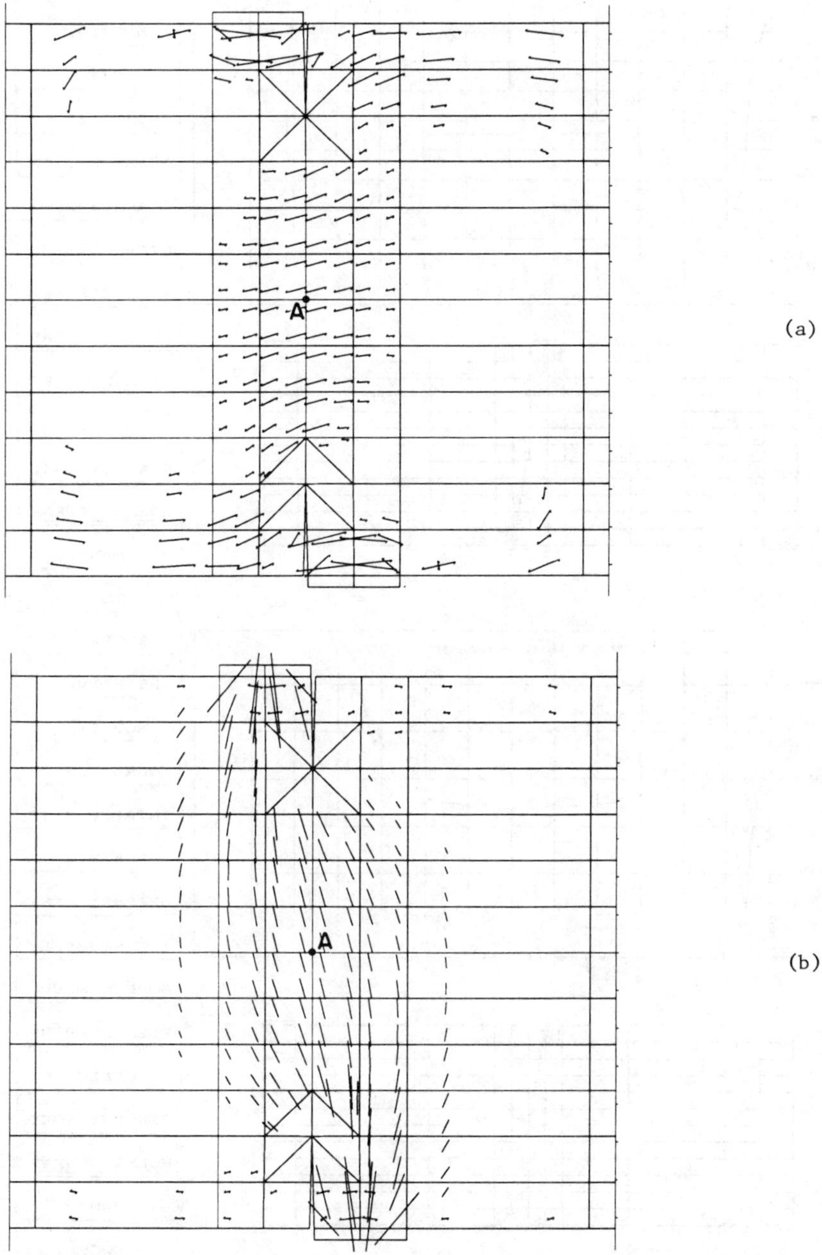

Figure 6. (a) Tensile Principal Stress Vectors in Central Region of the Specimen. (b) Tensile and Compressive Principal Stresses.

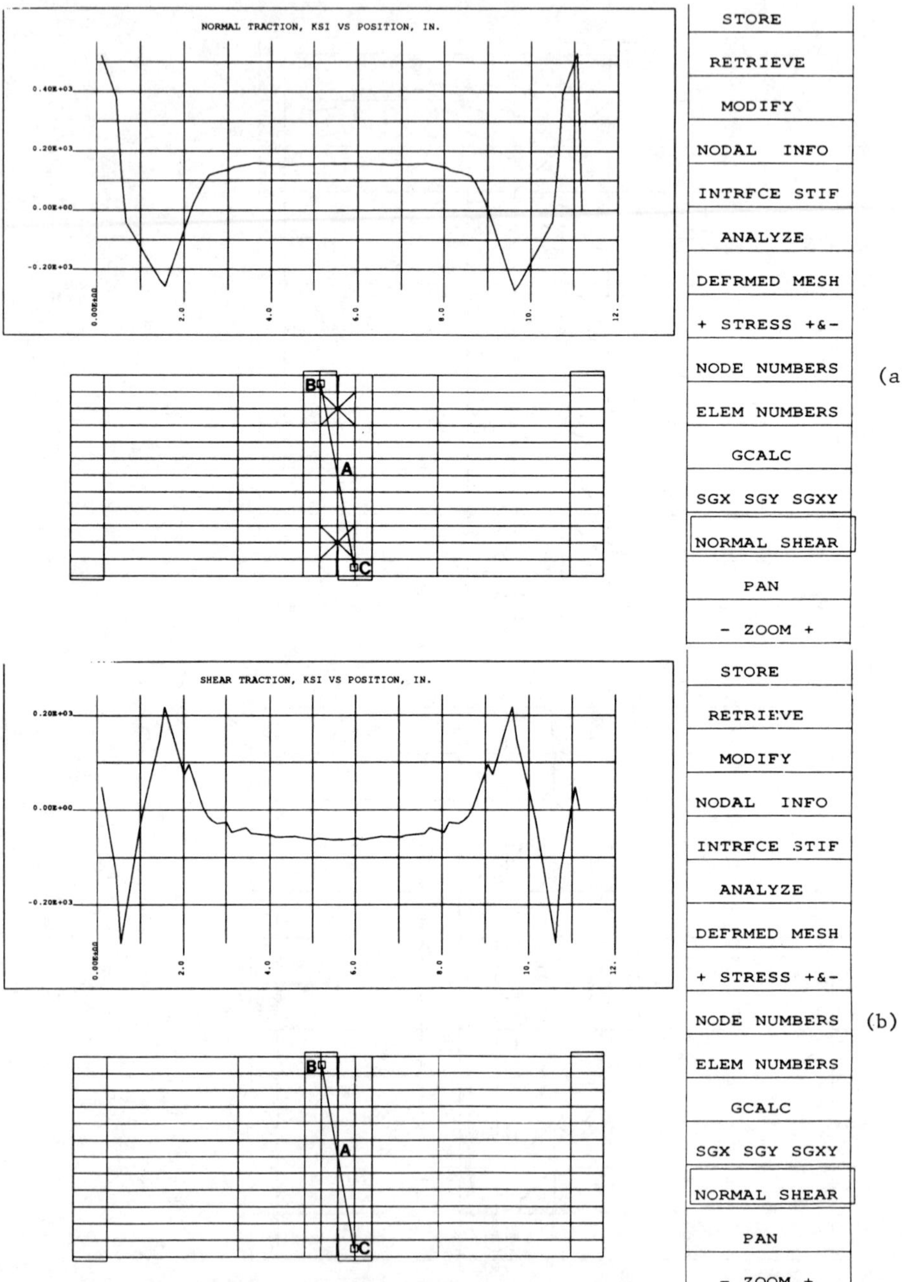

Figure 7. (a) Plot of Stress Normal to the Line BAC. (b) Plot of Shear Stress Along the Plane BAC.

1. The stress normal to Line BAC is very nearly σ_I, and that this stress is fairly uniform over this line.

2. The shear stress along Line BAC does not show the parabolic distribution one expects from elementary beam bending theory. It shows in fact that the minimum shear stress occurs near the center of the beam, exactly the opposite one would expect from beam theory.

These observations serve as a springboard to the discussion of the next section.

Table I

Normalized Principal Stresses and Their Angles at the Specimen Center, Point A, for the Meshes Shown in Figure 5

Mesh	Number of Nodes	Number of Elements	$\frac{\sigma_I db}{P}$	θ_I[a]	$\frac{\sigma_{II} db}{P}$	$\frac{\tau_{max} db}{P}$
1	699	208	0.526	14.7°	-2.10	1.31
2	515	150	0.503	14.7°	-2.11	1.31
3	439	126	0.503	14.7°	-2.12	1.31

[a]Angles measured with respect to the horizontal.

Elasticity Analysis

In this section we will show how two classical problems of the theory of elasticity will at once lend support to the finite element analyses and show that it is unlikely that "shear fracture" was the mode of failure in the Bazant and Pfeiffer tests (1985).

Consider first Figure 1b. This structure was first analyzed using the theory of elasticity by Filon (1902) and is discussed by Timoshenko (1970). The distribution of shearing stresses along the Line nn, the line containing the notches in the Bazant and Pfeiffer (1985) tests, is shown in Figure 8. It may be seen that for small values of the ratio b/c, the distribution does not resemble the parabolic distribution given by beam theory. In fact, this distribution closely resembles that computed for a notched beam, Figure 7b.

The notches in the specimen shown in Figure 3 are, therefore, placed in what would be the region of highest shear stress in the cross-section; their tips are located in a region of <u>decreasing</u> shear stress. This can be seen through the finite element results shown in Figure 9. This figure shows contours of principal shear stress in the region around the notches. The highest principal shears are under the loads, and just at the notch tips, however, the principal shears drop sharply <u>ahead</u> of the notch tips, Figure 7b. The net effect of moving the principal loads closer together is exactly the opposite of what one might intuitively expect: it decreases the peak shear stress in the region between the notch tips. This is the second observation which calls into question whether a "shear fracture" occurred in the Bazant and Pfeiffer tests (1985).

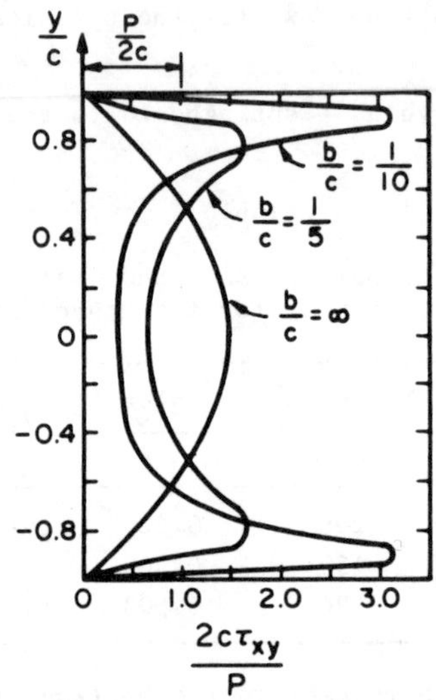

Figure 8. Shear Stress on Plane nn of Figure 1b. From Reference (Timoshenko and Goodier, 1970)

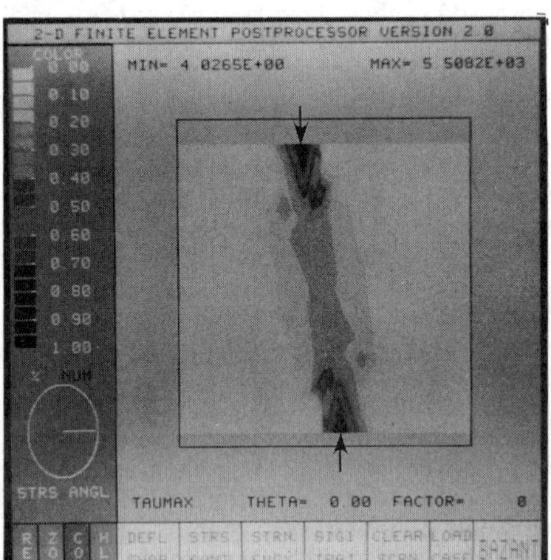

Figure 9. Contours of Principal Shear Stress in the Central Region of the Specimen.

A third observation is suggested by Figure 10. Here a hypothetical circular specimen has been inscribed in the beam. Imagine now that this circular specimen is under diametrically opposed point loads as shown. Such a specimen and load arrangement would comprise the Brazilian (Mellor and Hawkes, 1971) or splitting tensile strength ANSI/ASTM C496-71) test. This hypothetical specimen inscribed as shown in Figure 10 of course differs from a real Brazilian specimen in that it has small tractions distributed along its circumference and the two notches cut into it. It is our contention, however, that these differences have little effect and that the specimens tested by Bazant and Pfeiffer (1985) responded in a manner quite similar to a splitting tensile strength test. In support of this contention consider first a comparison between Figure 7a and Figure 11. Figure 11 shows the distribution of stress normal to the loaded diameter in a Brazilian test. It is obviously quite similar to the distribution computed by finite element analysis for the notched beam. Moreover, the similarity is not just qualitative. The theoretical value of the tensile stress normal to the loaded diameter at the center of the Brazilian test specimen is given by,

$$\sigma = \frac{2P}{\pi dt} \tag{1}$$

where d = diameter of specimen
t = thickness

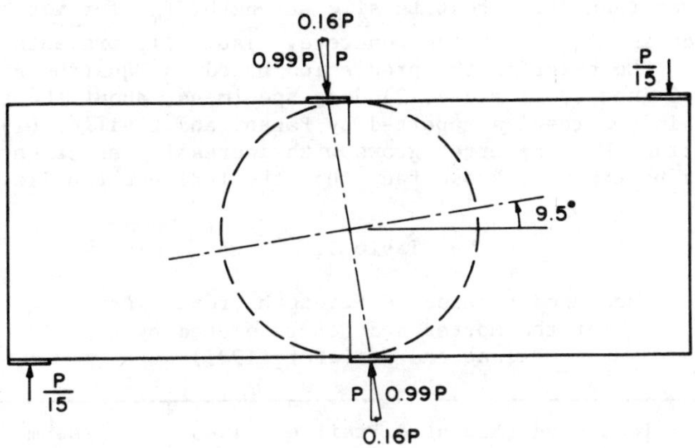

Figure 10. A Hypothetical, Diametrically Loaded Disk Inscribed Within the Beam Specimen.

We will now use Equation (1) to predict the fracture loads, P_c, in the Bazant and Pfeiffer (1985) tests by way of,

$$P_c = \frac{\pi dt}{2} \sigma_T \tag{2}$$

where σ_T = splitting tensile strength

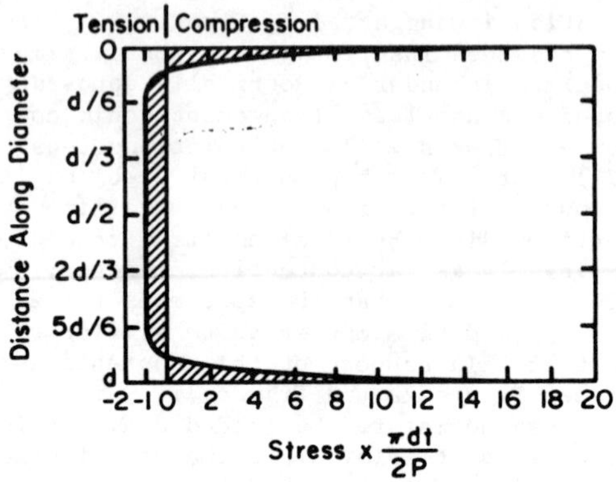

Figure 11. Theoretical Distribution of Stress Normal to the Loaded Diameter in a Brazilian Test from Neville (1981).

To do this we must use a tensile strength value for the materials tested. Bazant and Pfeiffer (1985) tested both mortar and concrete beams. Material properties which they measured or which we assumed are given in Table II. Note that, in accord with well known observations (Neville, 1981), we have assumed a splitting tensile strength, σ_T, which is lower than the direct tensile strength, f'_T, for mortar, and a value which is higher for the concrete. Table III presents our predictions. The error in the prediction based on Equation 2 is, for the 1.5 and 3 inch (38.1 and 76.2) deep specimens, about the same as the scatter in the results reported by Bazant and Pfeiffer (1985). The observation that the error grows with increasing specimen size can be explained by the well known fact that the indirect tensile strength,

Table II

Measured or Assumed Strength Properties for the Mortar and Concrete Used by Bazant and Pfeiffer (1985)

Material	Measured (Bazant & Pfeiffer, 1985)		Assumed
	f'_c psi (MPa)	f'_T psi (MPa)	σ_T psi (MPa)
Mortar	7100 (49.0)	530 (3.66)	503[a] (3.47)
Concrete	5500 (38.0)	410 (2.83)	430[b] (2.97)

[a] Assumed to be 95% of the direct tensile strength.

[b] Assumed to be 105% of the direct tensile strength.

σ_T, is specimen size dependent: as specimen size increases, σ_T decreases. The constant values of σ_T used here are upper bounds on the strengths of the larger specimens. We believe that the results shown in Table III are not coincidental. They are well supported by the analysis results already presented and by more sophisticated analyses to be described in the section on non-linear fracture propagation analyses.

Table III

Comparison of Observed and Predicted Maximum Loads

Type of Test	Specimen Depth d,[a] in (mm)		Mean Maximum Measured Load, P, lb (kN)		Predicted Maximum Load, Equation 2 P_B, lb (kN)		$\frac{P_B - P}{P}$ %
	1.5	(38.1)	1440	(6.41)	1519	(6.76)	5.5
Mode II	3.0	(76.2)	2873	(12.78)	3039	(13.52)	5.8
Concrete	6.0	(152.4)	5490	(24.43)	6078	(27.04)	10.7
	12.0	(304.8)	10000	(44.50)	12158	(54.10)	21.6
	1.5	(38.1)	1730	(7.69)	1778	(7.91)	2.8
Mode II	3.0	(76.2)	3283	(14.61)	3555	(15.81)	8.2
Mortar	6.0	(152.4)	5410	(24.07)	7111	(31.64)	31.4
	12.0	(304.8)	9633	(42.87)	14221	(63.28)	47.6

[a] All specimens had a thickness, t, of 1.5 in (38.1 mm).

The "Shear Fracture" Mechanism

To explain the fractures observed in their tests, Bazant and Pfeiffer (1985) have proposed a "shear fracture" model based on Figure 12. They propose that a shear fracture connecting the notch tips is "... likely to form as a zone of tensile microcracks with a predominantly 45°-inclination which only later connect by shearing." They call the connecting links between the microcracks inclined struts. The final damage mechanism in their model is a compression-shear failure of these struts.

We believe that the analysis results shown in Figures 5a and 6a cast doubt on the microcrack mechanism proposed in (Bazant and Pfeiffer, 1985). It is clear that a continuous corridor of relatively high tensile stress exists between the notch tips. Further, and importantly, any tensile cracking in this region would be at an angle of only about 10 degrees from the vertical, not at about 45 degrees suggested by Figure 12. Figure 8 indicates that elementary beam theory is inadequate to describe the stress field in the central area of this beam: the plane of τ_{max} is not vertical, but is inclined at about 35 degrees to the vertical.

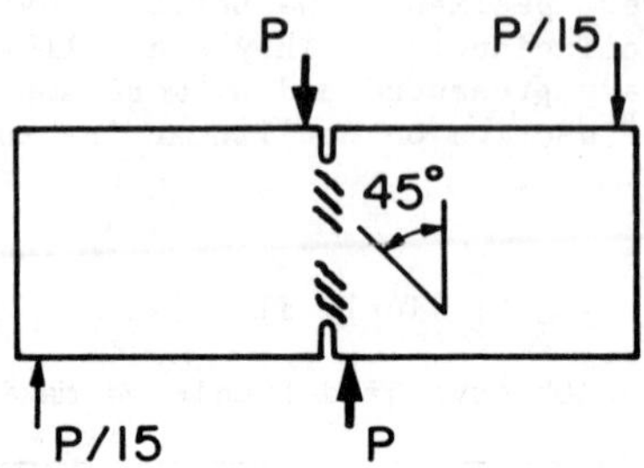

Figure 12. Model for Mode II Fracture Arising From a Band of Mode I Microcracks. From Reference (Bazant and Pfeiffer, 1985).

The Tensile Fracture Mechanism

We believe that the mechanism of fracture in the specimens tested by Bazant and Pfeiffer (1985) is straightforward:

1. Tensile cracks first propagate from the notch tips, just as in the tests of Arrea and Ingraffea (1981), Figure 2. These cracks are quite load-stable, however. It takes a substantial increase in load to keep them propagating.

2. At a load level higher than that which causes cracking from the notch tips, a new tensile crack nucleates near the center of the beam, just as it does in the Brazilian test. This crack would nucleate at an angle of only about 10 degrees from the vertical.

3. The tips of this central crack propagate upwards and downwards, toward the notch tips. As shown by Figure 6a, there is ample tensile stress available to propel this central crack almost up to the notch tips.

4. There will remain a short ligament between the tip of the central crack and the notch tip which will be in high, nearly vertical compression. We believe that the fracture process is completed by crushing (intense microcracking) of this small region.

These steps in our proposed fracture mechanism are described in quantitative and graphical detail in the next section.

Non-Linear Fracture Propagation Analyses

In support of the tensile fracture model described in the previous section, we have performed fracture propagation analyses of a 12 inch (305 mm) deep concrete beam tested in (Bazant and Pfeiffer, 1985). The analyses employed the non-linear, discrete cracking model described in References (Ingraffea and Gerstle, 1984) and (Ingraffea et al., 1984). The three key elements in this model are:

1. A "crack" is a locus of damage localization: the process zone is the crack itself. As the damage accumulates, the crack opens and the capacity for tensile stress transmission across the crack decreases. The constitutive model for the process zone used in the present analyses is shown in Figure 13. It is described by two parameters, the local direct tensile strength, f'_T, and the characteristic crack-opening displacement, δ_o.

2. As the damage process develops, the shear transfer capacity of the "crack" also diminishes. The constitutive model for shear transfer as a function of COD at each point along the "crack" is discussed in References (Arrea and Ingraffea, 1981) and (Ingraffea et al., 1984).

3. The cracking process is geometrically modeled by the finite element analysis. Rather than simulating the cracking by using the smeared crack approach, FEFAP automatically remeshes locally to accommodate a discrete representation of the crack.

Analysis of the mesh shown in Figure 4c indicated that at a load of about P = 6000 lbs (26.7 kN), cracking would initiate at the notch tips as shown in Figure 14. For cracking in this region we have assumed a local tensile strength equal to f'_T, 410 psi (28.3 MPa) and a δ_o = 0.005 inch (.13 mm). This corresponds to a G_{Ic} of about 1 lb/in (0.18 N/mm). The displaced shape shown in Figure 14b is highly magnified. The apparent overlap of elements in this and subsequent figures is not real; rather it is an artifact of the high amplification factors used to display deformed shapes meaningfully. The computed COD at the mouth of the crack (at the tip of the notch) is only 5×10^{-4} inch (1.3×10^{-2} mm), and the crack is less than an inch (25.4 mm) in length. Such a crack would not be visible to the naked eye. It would, however, probably be visible with the fluorescent particle technique used by Arrea and Ingraffea (1981) to delineate cracking as shown in Figure 2.

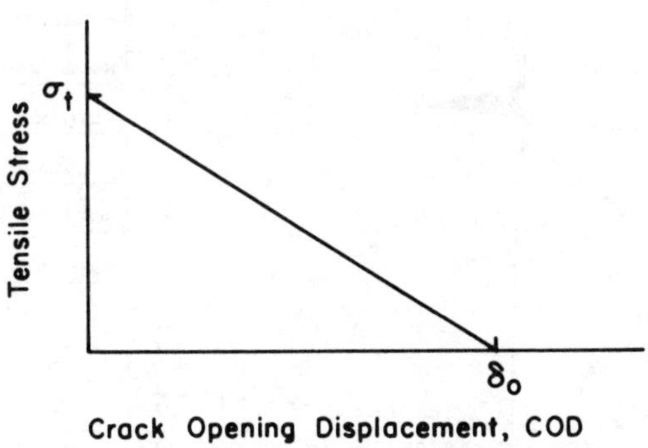

Figure 13. Process Zone Constitutive Model Used for the Non-Linear Fracture Mechanics Analyses. f'_T is Variable, δ_o = 0.005 in (0.13 mm).

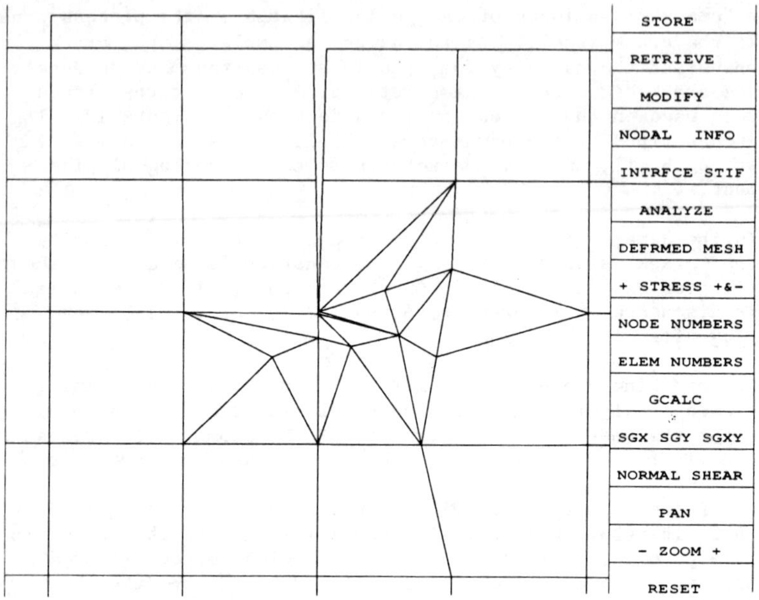

(a)

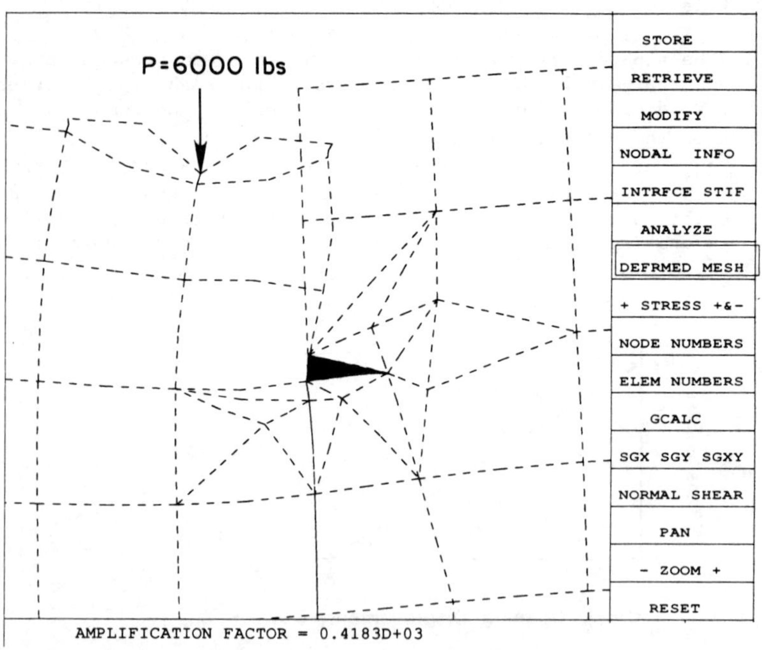

(b)

Figure 14. Region Around the Upper Notch After First Crack Increment. (a) Local Remeshing. (b) Highly Amplified Displaced Shape. Darkened Area is Damage Zone.

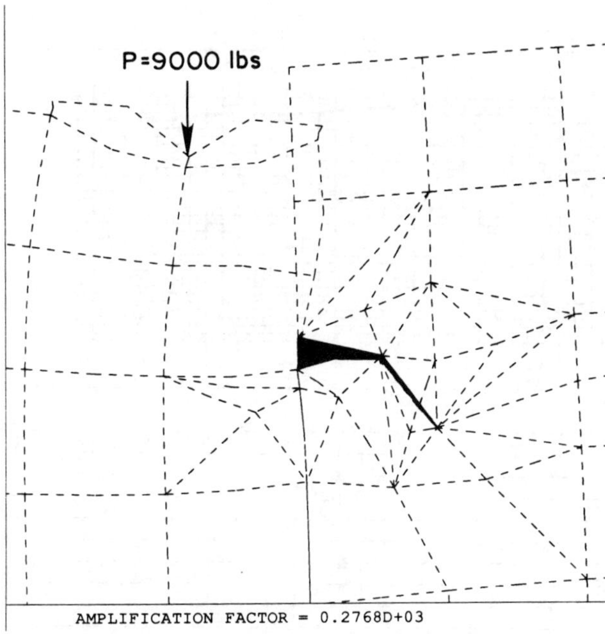

Figure 15. Highly Amplified Displaced Shape Around Upper Notch After Two Increments of Cracking.

At a load level of about P = 9000 lbs (40.05 kN), the cracking process at the notch tips had progressed as shown in Figure 15. The crack-mouth-opening displacement had increased to about 1×10^{-3} inch (2.54×10^{-2} mm) and the crack had begun to turn as indicated. Arc length of the crack was now about 1.4 inches (34 mm).

At a load level of 10000 lbs (44.5 kN), the notch-tip cracks had not lengthened, and the stress state between the notch tips now came into play. Figure 16 shows a plot of stress normal to the line indicated in the figure. This stress is very nearly σ_I and has exceeded 300 psi (2.07 MPa). We believe that this stress level is sufficient to have nucleated a new crack in the central region of the beam. Since the uniaxial tensile strength was reported to be 410 psi, and since σ_{II} at the center of the beam was computed to be about 1200 psi compression (8.28 MPa) at this load level, the local tensile strength would likely be only about 320 psi (2.20 MPa) as indicated by the operation shown in Figure 17.

At this point in the analysis, two slightly different paths were taken. In one case, a potential crack path was defined in the mesh along a line connecting the two notch tips. This was done by inserting a row of interface elements along this line. The load was then increased until this crack "unzipped" as shown in Figure 18. This

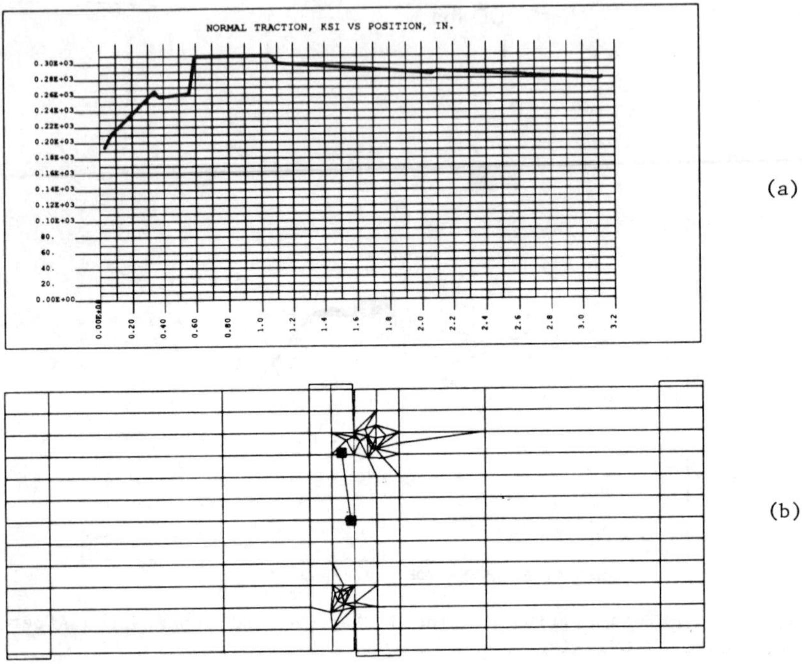

Figure 16. (a) Plot of Stress Normal to the Line Connecting the Blackened Squares in (b).

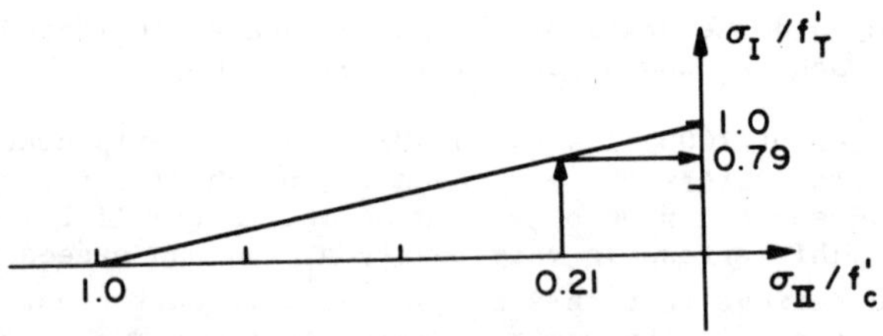

Figure 17. Strength-Interaction Relationship for Concrete in the Tension-Compression Region.

occurred at a load of P = 13000 lbs (57.85 kN). The tips of the central crack have propagated to within one element, about 1/4 inch (6.4 mm), of the notch tips. At this point the principal compressive stress in this element is well beyond f'_c. Although interesting, this analysis path is somewhat unsatisfying in that the central crack trajectory was forced, and, because it was not normal to the principal tension, the predicted failure load was too high.

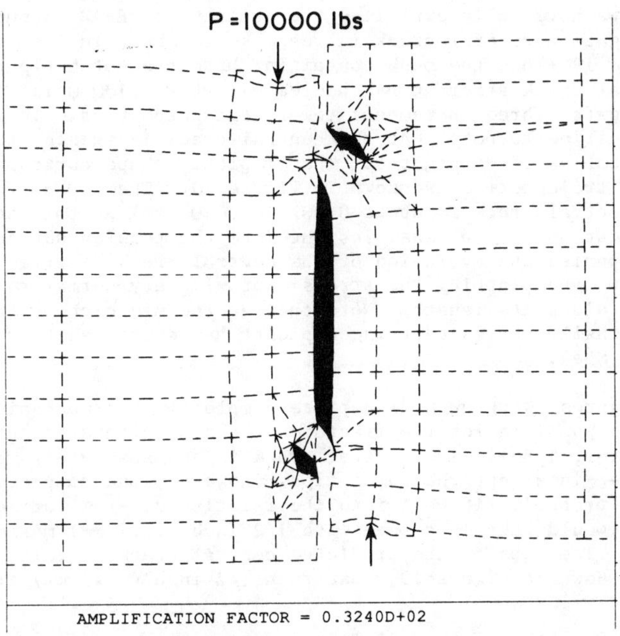

Figure 18. Final Deflected Shape for Analysis With Vertical Central Crack.

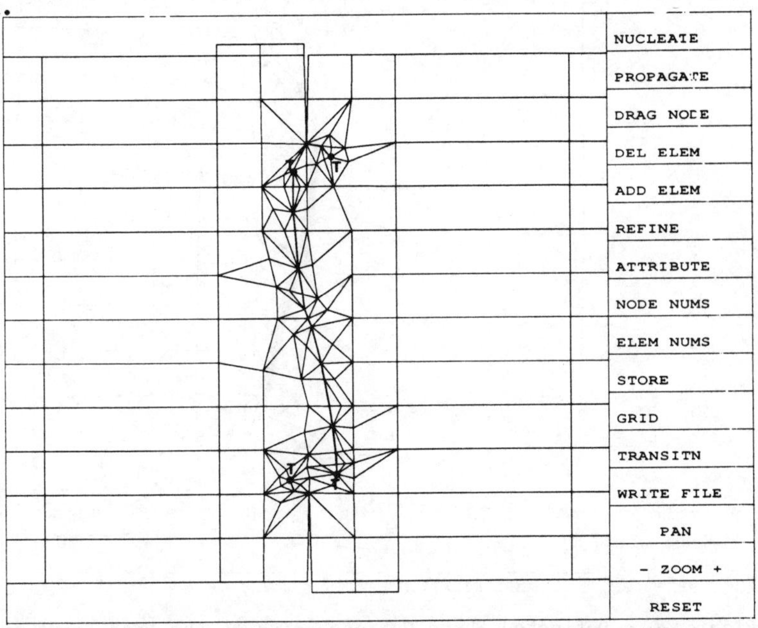

Figure 19. Central Region of the Mesh After 3 Increments of the Central Crack. Crack Tips Indicated by a "T."

We have remedied this problem by taking a second analysis approach. Here we have fully exploited the ability of FEFAP to nucleate and propagate a central crack in response to the principal stresses. Figure 19 shows the mesh containing both the notch tip cracks and the central crack after it has nucleated, at P = 10000 lbs (44.5 kN), and propagated three increments toward the notch tips. It is most interesting to note that the central crack is tending toward the vertical as it propagates. The displaced shape corresponding to these crack trajectories is shown in Figure 20. The COD at the midpoint of the central crack is about 0.002 in (0.05 mm) at this stage of propagation. As a final insight into the tension softening which accompanies the evolution of the central crack we offer Figure 21. Here we are plotting the stress approximately normal to the central crack along its length. Note that at the midpoint of the central crack the tensile stress carrying capacity has decreased to about 180 psi (1.24 MPa).

Figures 19 through 21 capture completely the mechanisms we believe are responsible for the fracture of the specimens tested in (Bazant and Pfeiffer, 1985). The central crack is not absolutely vertical and does not yet intercept the notch tips. However, the computed trajectory would probably fit well into the relative zig-zag corridor that a real crack would take in a beam with 1/2 inch (12.7 mm) maximum aggregate size. The tips of the predicted central crack are, in the configuration shown in Figure 19, less than 1/2 inch (12.7 mm) from the notch tips.

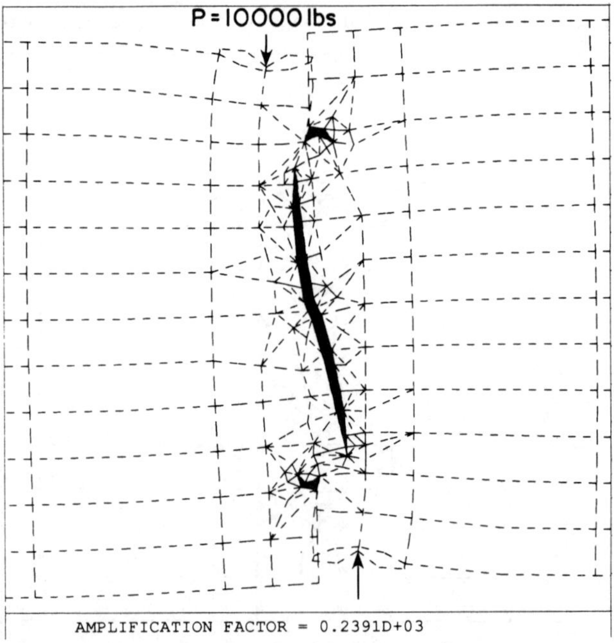

Figure 20. Deflected Shape Corresponding to the Mesh Shown in Figure 19.

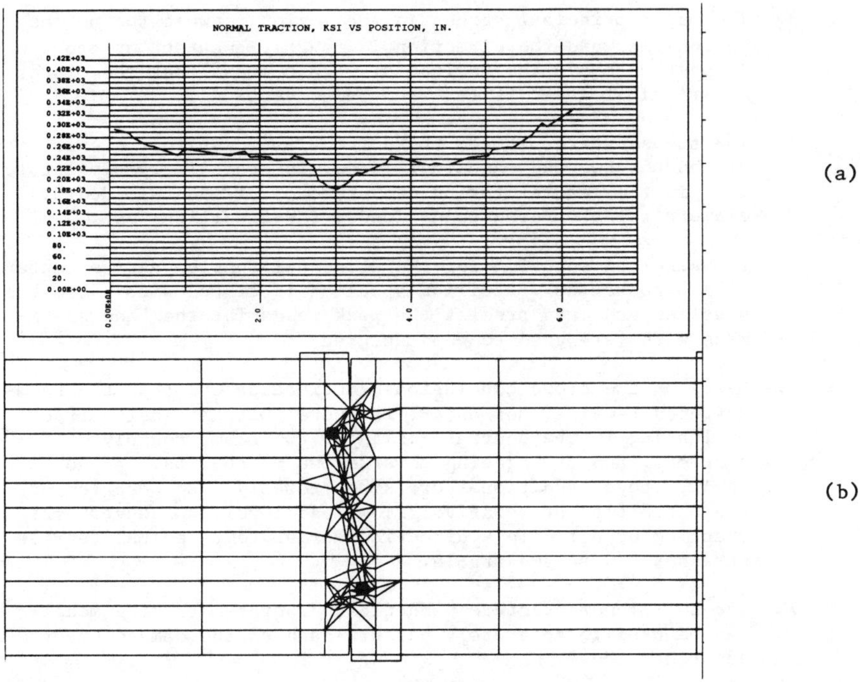

Figure 21. (a) Plot of Stress Normal to the Line Connecting the Blackened Squares in (b). P = 10000 lbs (44.5 kN).

Conclusions

We have attempted to provide an alternative explanation for the behavior observed and modeled by Bazant and Pfeiffer (1985) in their fracture tests of concrete beams. We have employed classical elasticity solutions, linear finite element analysis, and non-linear fracture mechanics analysis to assemble an explanation for, and a computer simulation of, the observed behavior. Based on our analyses, we have reached the following conclusions:

1. Limited tensile cracking from the notch tips, although not reported in (Bazant and Pfeiffer, 1985), is likely to have occurred and should be observable using a crack detection technique more sensitive than the naked eye.

2. By moving the principal loads closer together than in the Arrea and Ingraffea (1981) tests, the intensity of the shear stress in the region between the notch tips is decreased rather than increased. The shear stress distribution, magnitude and trajectory, in this region cannot be predicted using elementary beam theory.

3. The major principal stress in the region between the notches is tensile, and the direction is roughly constant at about 10 degrees above the horizontal. This means that the plane of principal shear is far from being vertical.

4. The stress state between the notch tips, despite the existence of the notches and the minor loads, is startlingly similar to that in the central region of a cube or cylinder loaded by diametrically opposed point loads, the Brazilian Test.

5. By using the solution for the principal tension at the center of a hypothetical Brazilian specimen inscribed within the beam specimen, one can predict the peak loads for the beam specimens with very good to good accuracy.

6. Based on the above conclusions, it is apparent that the beams suffered tensile, not shear, fracture, with a tension crack nucleating in the central region of the beam, roughly vertical, and propagating towards the notch tips. As noted in 1., above, additional cracking probably also occurred at the notch tips at relatively low load levels. However, this cracking occurred in a direction normal to principal tensile stresses and became stable.

7. The non-linear fracture mechanics, discrete cracking model can be used to accurately capture each of the damage events described above.

Acknowledgements

The research reported here was partially supported under NSF Grant 8351914. The analyses were performed using the facilities of the Cornell University Program of Computer Graphics and the Computer-Aided Design Instructional Facility. The authors would like to thank the developers of the analysis program, FEFAP-G, Dr. Walter Gerstle and Mr. Paul Wawrzynek.

References

ANSI/ASTM C496-71, Standard Test Method for Splitting Tensile Strength of Cylindrical Concrete Specimens.

Arrea, M., Ingraffea, A.R., "Mixed-Mode Crack Propagation in Mortar and Concrete," Department of Structural Engineering Report 81-13, Cornell University, 1981.

Bazant, Z.P., Pfeiffer, P.A., "Tests of Shear Fracture and Strain Softening in Concrete," Proc., 2nd Symposium on the Interaction of Non-Nuclear Munitions on Structures, Panama City Beach, Florida, April 15-19, 1985.

Filon, L.N.G., Transactions of the Royal Society (London), Ser. A, Vol. 201, 1902, p. 67.

Ingraffea, A.R., "Mixed-Mode Fracture Initiation in Indiana Limestone and Westerly Granite," Proceedings 22nd U.S. Symposium on Rock Mechanics, Cambridge, MA, 1981, pp. 186-191.

Ingraffea, A.R., Gerstle, W., Gergely, P., Saouma, V., "Fracture Mechanics of Bond in Reinforced Concrete," Journal of the Structural Division, ASCE, Vol. 110, No. 4, April, 1984, pp. 871-890.

Ingraffea, A.R., Gerstle, W., "Non-Linear Fracture Models for Discrete Crack Propagation," Application of Fracture Mechanics to Cementitious Composites, S.P. Shah, editor, Martinus Nijhoff Publishers, 1984, pp. 171-209.

Ingraffea, A.R., Saouma, V., "Numerical Modeling of Discrete Crack Propagation in Reinforced and Plain Concrete," Chapter 4 in Application of Fracture Mechanics to Concrete Structures, G.C. Sih and A. DiTommaso, editors, Martinus Nijhoff Publishers, 1985.

Mellor, M., Hawkes, I., "Measurement of Tensile Strength by Diametral Compression of Discs and Annuli," Engineering Geology, Vol. 5, 1971, pp. 173-225.

Neville, A.M., Properties of Concrete, Third Edition, Pitman Publishing, 1981, pp. 549-552.

Timoshenko, S.P., Goodier, J.N., Theory of Elasticity, Third Edition, McGraw-Hill Book Company, 1970, pp. 57-58.

COMMENT ON INGRAFFEA AND PANTHAKI'S ANALYSIS OF SHEAR FRACTURE TESTS OF CONCRETE

By

Zdeněk P. Bažant[1] and Phillip A. Pfeiffer[2]

In the preceding paper in this volume [1], Ingraffea and Panthaki argued that the mechanism of failure of the shear-loaded specimens tested at Northwestern University [2] was not shear fracture but splitting compression similar to that observed in the Brazilian split cylinder test. Meritorious though certain aspects of their argument may be and are gratefully appreciated by the authors, the essence of their negative criticism nevertheless consists of overstatement and conjecture. Their criticism is an articulate elaboration on the well-known difficulty in devising shear fracture specimens that would fail in a pure two-dimensionally antisymmetric mode. The fact is that at present one must either accept imperfect shear fracture specimens or remain ignorant.

Some aspects of the shear fracture specimen can be discussed in terms of linear elastic fracture mechanics (LEFM), which generally yields an acceptable approximation for the fracture of concrete despite its nonlinearity, provided that the variation of the energy required for crack growth is somehow determined separately. Under the assumption of linearity, the loading used in the tests can be decomposed into symmetric and antisymmetric parts corresponding to fracture Modes I (tensile) and II (in-plane shear); see Fig. 1. Ideally, the antisymmetric loading shown in Fig. 1 should be applied in the

[1]Professor of Civil Engineering and Director, Center for Concrete and Geomaterials, Northwestern University, Evanston, IL., 60201.

[2]Graduate Research Assistant, Northwestern University.

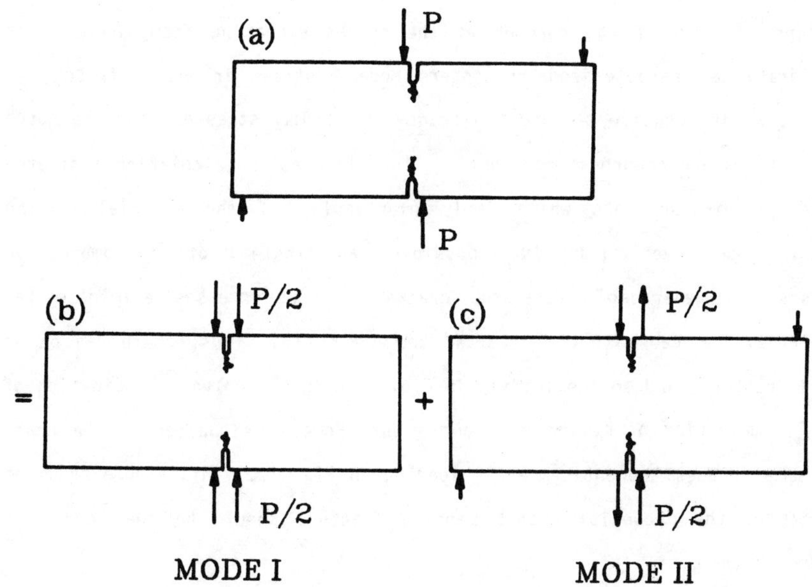

Fig. 1 Decomposition of the Loading Used In Shear Fracture Tests Into Its Symmetric and Antisymmetric Parts.

tests, however, it would be much more difficult to realize, due to the need to apply tensile loads on concrete and to control a proportional increase of a greater number of loads. Nevertheless, the superimposed symmetric loading (Fig. 1) should not alter the basic mode of failure, for the following reasons.

The loading in Fig. 1b, which bears much more similarity to the Brazilian tests than does the actual loading used (Fig. 1a), produces in the ligament section transverse normal stresses which are tensile in the middle and compressive near the notches, as is indicated by the finite element results obtained at Northwestern University (Fig. 2) as well as those of Ingraffea and Panthaki. The symmetric loading produces a symmetric stress singularity at the tips of each of the notches at the cracks emanating from them. This singularity is characterized by nonzero Mode I stress intensity factor K_I. However, K_I is negative because the transverse normal stresses near the notch or crack tips are compressive. Thus, a linearly elastic calculation indicates horizontal displacements which imply overlapping of the material on the cracks. Such overlapping is impossible, and instead of it compressive stresses are generated across the cracks. These stresses eliminate the singularity and cause that $K_I = 0$ at the crack tip. Thus, there can be no Mode I fracture, and so the loading in Fig. 1a cannot involve a combination of modes (combination of K_I and K_{II}) but a pure Mode II situation at the crack tip, same as does the antisymmetric loading in Fig. 1c. This simple argument invalidates the proposition that tensile fracture should be the source of failure in the test.

An exception to the preceding argument is the initial state at which the crack extension from the notch is zero. Since the notch has a finite width, material overlapping does not occur (except if the displacements are too

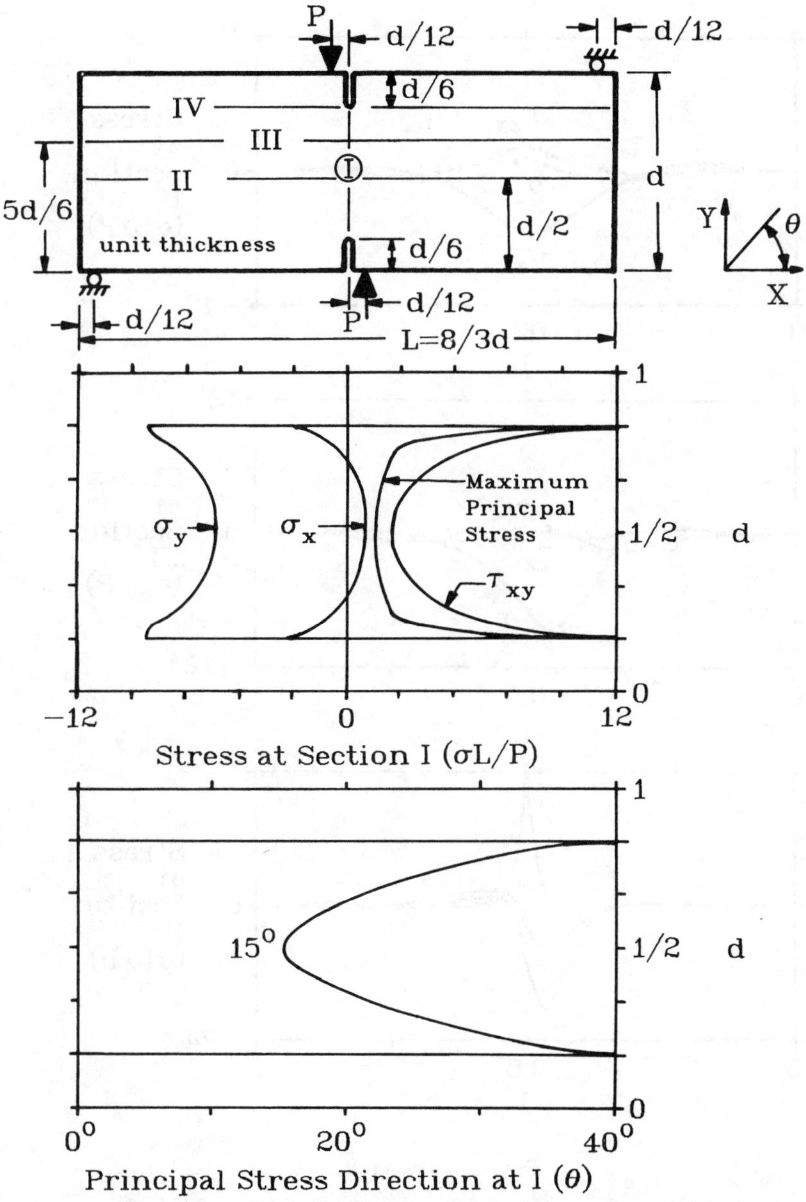

Fig. 2 Results of Linear Elastic Finite Element Analysis for the Distributions of Stresses and Principal Stress Inclinations Along Various Cross Sections (compare to Ingraffea and Panthaki's Fig. 11).

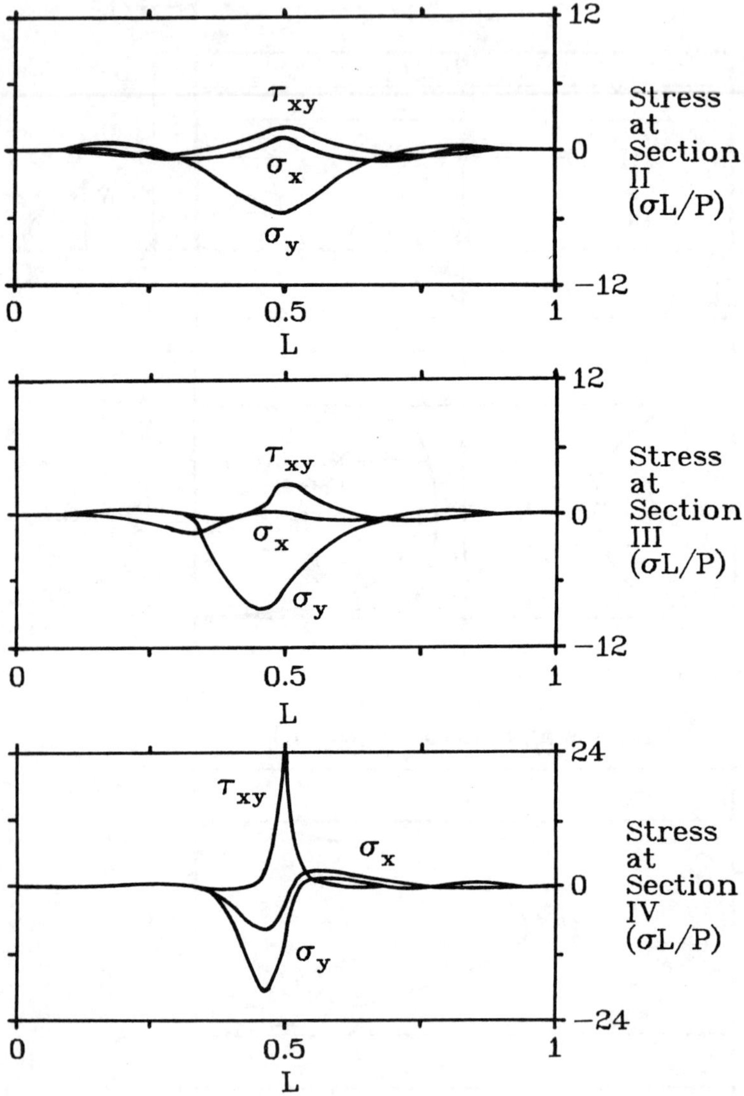

Fig. 2 - (Continued)

large), and so a negative value of K_I is possible for the symmetric loading part (Fig. 1b). However, this fact is irrelevant for the failure of the specimens tested, because the failure, i.e., the maximum load, did not occur right at the beginning of crack propagation but later, after a finite growth of the cracks from the notches.

According to linearly elastic fracture mechanics, there is no possibility for the cracks to begin in the middle of the ligament section rather than at the notches, because there is no stress singularity there. Thus, as long as linear elastic analysis is adequate for the overall response of concrete specimens, tensile fracture can have no significant role in the failure.

This conclusion is also supported by linearly elastic finite element results based on elements larger than aggregate size which are too coarse to represent stress singularities. These results (Fig. 2) indicate that the maximum principal stress, which is tensile near the notch or crack tip and occurs in a strongly inclined direction, is much larger in magnitude than the maximum principal stress (tensile) in the center of the ligament. Thus, the crack cannot be assumed to start at the center of the cross section but must initiate from the notch tips and propagate continuously toward the center. Now, as clearly seen from the tests, the cracks propagate straight along the ligament section rather than in strongly inclined directions from the notches, while the maximum principal tensile stress is strongly inclined. Thus, the cracks which begin at the notches cannot be considered as tensile. They represent shear cracks, although microscopically they may consist of a row of inclined tensile microcracks prior to the formation of the final continuous fracture.

More realistic finite element results can be obtained by nonlinear analysis based on the crack band theory, which has been shown to agree well with many different types of fracture test data; see Fig. 3. Although this analysis indicates a somewhat smaller inclination of the tensile microcracks (assumed to be normal to the maximum principal stress at the peak stress point), the inclination of these microcracks is still significant; 31° as shown in Fig. 3. Thus, the band of microcracks emanating from the notches cannot be considered as tensile fracture but as shear fracture affected by tensile stresses. Furthermore, while the microcrack inclination near the notches is smaller than that from linear elastic analysis, the inclination obtained for the middle of the ligament section is larger (22° in Fig. 3). This indicates that even in the latest stage of failure, after the peak load, the overall fracture is not tensile.

Neither the linear nor the nonlinear finite elenent analysis based on the crack band theory permits the crack to initiate at the middle of the cross section, which is the central assumption in Ingraffea and Panthaki's argument. This is because the present shear fracture test is fundamentally different from the Brazilian test in that there exist large tensile stresses near the notches. Recourse to the nonlinear finite element results is strongly supported by the fact that the ratio of the calculated fracture energy for this type of test to that for the Mode I fracture is about the same as observed experimentally, i.e., approximately 30:1. A large value of the fracture energy for shear failure is obtained from the crack band theory because shear failure requires not only the creation of inclined tensile cracks, but also compression failure of the material between the cracks due to compressive stresses parallel to the cracks. The energy of this compression failure, which is absent in the mechanism of Mode I fracture, is far larger than the energy dissipated by tensile cracking.

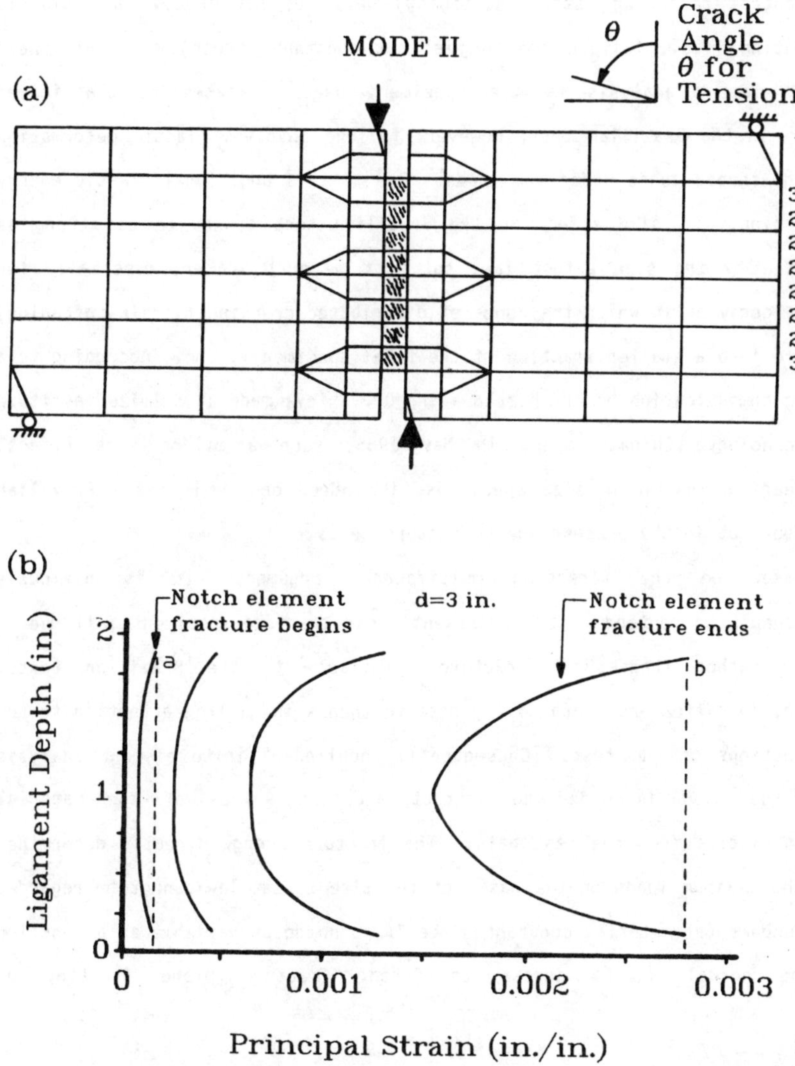

Fig. 3 Crack Inclinations Obtained by Nonlinear Finite Element Analysis According to the Crack Band Theory and Principal Strain Distribution Along the Ligament at Various Loads, (a) Strain corresponding to the peak stress, b) Strain at which the stress is reduced to zero; compare to Ingraffea and Panthaki's Fig. 11).

Another argument for dissimilarity from the Brazilian test is provided by the size effect, analyzed in the paper [2]. It was found that the maximum loads for geometrically similar specimens of different sizes agree quite well with the size effect law. This law can apply only for failures due to distributed fracturing (strain-softening) which do not involve any plastic dissipation (i.e., failure due to yield at constant stress). The nonlinear finite element analysis of the specimens used indicates no plasticized zones. In the Brazilian test, however, failure involves plastic deformation in wedge-shaped zones under the loads. Due to this additional plastic energy dissipation, the size effect in the Brazilian test cannot be as strong as indicated by the size effect law, but must be much weaker, especially for larger specimens in which the zones of distributed cracking (strain-softening) localize into a smaller fraction of the total specimen volume. According to a private communication by Dr. Hazekawa and Dr. Shioya made at Shimizu Institute of Technology, Chiba, Japan, in May 1985, such a milder size effect, disappearing for larger size specimens, is indeed observed in the Brazilian test, but not in the present shear fracture tests.

Based on the foregoing multifaceted argument, it is obviously questionable to contend that the present shear fracture specimens fail due to tensile rather than shear fracture, similarly to the Brazilian tests. However, Ingraffea and Panthaki [1] deserve thanks in calling attention to the imperfections of the test. Consequently, nonlinear finite element analysis should be used in order to extract accurate values of the material characteristics from the test data. The fracture energy directly determined from the maximum loads on the basis of the size effect law cannot be regarded as a fundamental material constant since it is no doubt variable as a function of the normal stress across the shear fracture plane (testing of longitudinally prestressed specimens, designed to show this dependence, is in fact in progress at Northwestern University). It is probably easier to carry out a sophisticated nonlinear finite element analysis of an imperfect shear fracture test (Fig. 1a) than to implement the perfect antisymmetric loading (Fig. 1c).

References

1. Ingraffea, A. R., and Panthaki, N. J., "Analysis of Shear Fracture Tests of Concrete Beams", Paper in this Volume; also: Preprints, U. S. - Japan Seminar on Finite Element Analysis of Reinforced Concrete Structures, Tokyo, May 21-24, 1985, pp. 71-91.

2. Bazant, Z. P., and Pfeiffer, P. A., "Tests of Shear Fracture and Strain Softening in Concrete", Proc., Second Symposium on the Interaction of Nonnuclear Munitions with Structures, Panama City Beach, Florida, April 15-19, 1985, ed. by P. Thompson, Eglin Air Force Base.

CONCRETE CRACK PROPAGATION - STUDY OF MODEL PARAMETERS

David Darwin,* M.ASCE

The effects of key parameters on smeared cracking representations for the finite element analysis of concrete members are discussed. Both reinforced and plain concrete members are studied. Nonlinearity is limited to cracking of the concrete and yielding of the reinforcement. Influences of grid refinement, load step size, concrete tensile strength, and element type are considered for reinforced flexural members. The effects of the shape of the descending branch of the tensile stress-strain curve, fracture energy, grid refinement, load step size, width of the process zone, and constraints imposed on crack angles are studied for plain concrete fracture specimens. For the reinforced members, load-deflection curves exhibit convergence with grid refinement. Below a certain magnitude, load increment size does not affect the stiffness or the response in stable regions of the load-deflection curves. Concrete tensile strength has a negligible influence on the members studied. For the fracture specimens, fracture energy, tensile strength, and shape of the descending branch interact to determine the stiffness and general behavior of the specimen. The width of the process zone has little influence on response. The importance of proper crack orientation is demonstrated.

Introduction

Work has been underway at the University of Kansas since 1980 to study the parameters that affect the finite element analysis of reinforced concrete. The emphasis of the work has been on the cracking response of concrete. To date, the research has consisted of two separate investigations.

The first investigation concerns the effects of grid size on the response of flexural members in which cracking is governed by a limiting tensile stress criterion. In addition to grid refinement, the type of element, the type of member, the load increment size, and the tensile strength of the concrete were considered.

The second investigation concerns the parameters that control the behavior of plain concrete fracture specimens modeled using a strain softening representation in tension. In this work, the effects of the

*Prof. of Civil Engrg. and Dir. of the Structural Engineering and Materials Laboratories, Univ. of Kansas, Lawrence, KS 66045.

shape of the descending branch of the tensile stress-strain curve, fracture energy, grid refinement, load step size, width of the process zone, and constraints imposed on crack angles were investigated.

This paper presents an overview of this work and summarizes the key findings. The details of the studies are presented in a number of reports and papers (6-8,10-12).

Stress Controlled Smeared Cracking

General.--The first portion of the study (7,8) looked at smeared cracking models that use a limiting tensile stress criterion to establish crack formation (Fig. 1a). Bazant and Cedolin (1-3) had expressed a concern that solutions using this technique would not converge with grid refinement. They felt that the calculated stress in advance of the "crack tip" would increase without bound with reduced element size. They concluded that the cracking criteria used for reinforced concrete members must be a function of the element size if the results were to be objective.

Smeared cracking models with a limiting tensile stress criterion have been used for many years. This study was designed to evaluate the impact of grid size on typical reinforced concrete members. The study concentrated on flexurally reinforced members, with different span to depth ratios. Beam response varied from flexure dominated to shear dominated.

The material models used in the study limited the nonlinear behavior to yielding/strain hardening in the reinforcement (Fig. 1b) and cracking in the concrete (Fig. 1a). Following crack formation, the following constitutive relations were used to represent the concrete.

$$\begin{Bmatrix} d\sigma_1 \\ d\sigma_2 \\ d\tau_{12} \end{Bmatrix} = \begin{bmatrix} 0 & 0 & 0 \\ 0 & E & 0 \\ 0 & 0 & \beta G \end{bmatrix} \begin{Bmatrix} d\varepsilon_1 \\ d\varepsilon_2 \\ d\gamma_{12} \end{Bmatrix} \qquad (1)$$

in which E = modulus of elasticity; ν = Poisson's ratio; $d\sigma_1$, $d\sigma_2$, $d\tau_{12}$ = incremental stresses; and $d\varepsilon_1$, $d\varepsilon_2$, $d\gamma_{12}$ = incremental strains; subscripts 1 and 2 refer to the material axes (direction 1 is normal to the first crack); G = shear modulus = $E/2(1 + \nu)$; and β = shear retention factor ($0<\beta<1$).

A single element type, the simple constant strain bar element, was used to represent the reinforcement.

Two elements were used to model the concrete, the linear (4-node) quadrilateral isoparametric element and the quadratic (8-node) quadrilateral isoparametric element. 2 x 2 integration was used for the 4-node element, while both 2 x 2 and 3 x 3 integration were used for the 8-node element. The effect of the element representation on response was studied.

The finite element solutions were obtained using the Newton-Raphson technique. The model was loaded incrementally. The unbalanced forces due to nonlinear behavior were reapplied in successive iterations until convergence was attained.

Three simply supported beams were studied with span to depth ratios of 12:1, 5:1, and 2:1. The beams had a single layer of reinforcement at the bottom edge. The steel had a Young's modulus of 200 GPa, a strain hardening modulus of 10 GPa, and a yield stress of 414 MPa. The reinforcement ratio was 1.5%. The concrete had a Young's modulus of 24.8 GPa, a tensile strength of 2.76 MPa, Poisson's ratio of 0.2, and a shear retention factor β of 0.4.

Findings.--The results of the study were aimed at answering four questions.

1) Do load-deflection curves converge with grid refinement? If yes, in what manner, and what governs the type of convergence?
2) Are numerical or response problems introduced when the tensile stress is immediately reduced to zero upon crack formation?
3) What is the impact of load increment size on the response, i.e., how path dependent is the response?
4) What are the effects of concrete tensile strength on response?

The answer to these questions can be demonstrated with results for two of the three beams studied, the slender beam (span to depth = 12:1) and the deep beam (span to depth = 2:1) (Fig. 2).

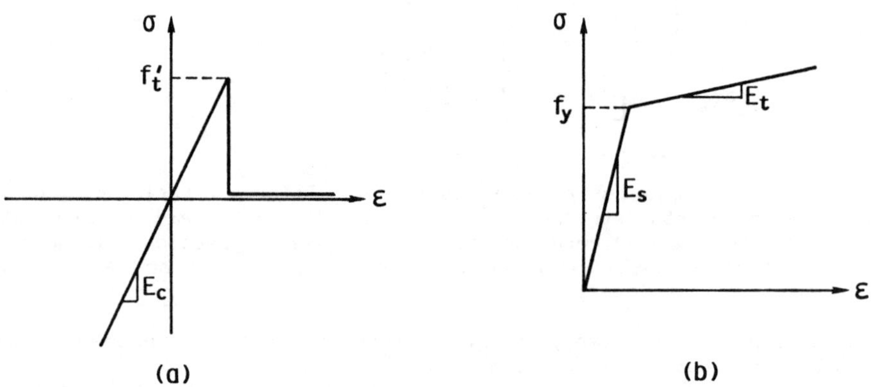

Fig.1 Material Representation: (a)Concrete; (b)Reinforcement

Fig. 3 and 4 show the load-deflection curves for the slender beams under distributed load, with the linear and the quadratic (reduced integration) elements, respectively. The model with quadratic elements with full integration gave virtually identical results with the linear elements and, therefore, is not shown. Crack patterns for the beams are shown in Fig. 5 and 6.

The load-deflection curves for the linear element converge, producing successively more flexible results with increasing grid refinement (as the grid changes from one element to five elements through the depth), as expected for most finite element solutions. The quadratic elements with reduced integration show progressively stiffer specimen response. This unusual trend can be traced to the development of zero energy deformation modes in cracked quadratic elements with reduced integration (7). The load-deflection responses for both element types rapidly converge to a common solution.

The crack orientations for the linear element are quite erratic for the coarse grids due to the limited ability of the element to model shear stress. As expected, the quadratic element gives a good showing even for one element through the depth.

The finer grids also exhibit regions without cracks, surrounded by cracked elements. This is due to the "stress relief" provided by the adjacent cracked elements, or Gauss points in the case of the quadratic elements. The coarse grids are unable to provide the stress relief and, therefore, exhibit no crack localization.

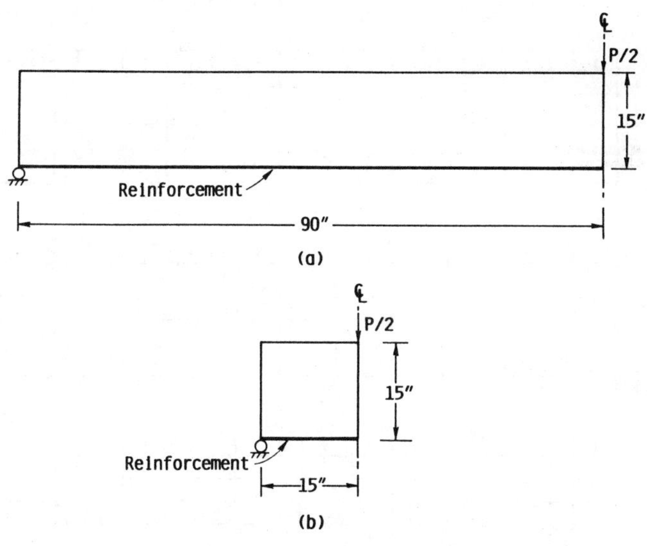

Fig.2 (a)Slender Beam; (b)Deep Beam (1 in. = 25.4.mm)

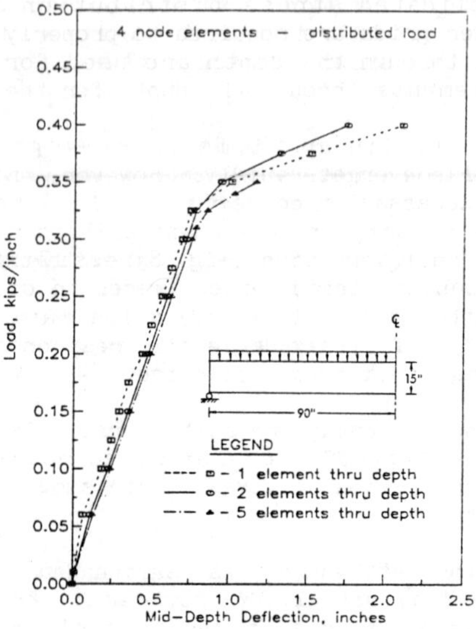

Fig.3 Slender Beam, Linear Elements (1 kip = 4.45 kN, 1 in. = 25.4 mm)

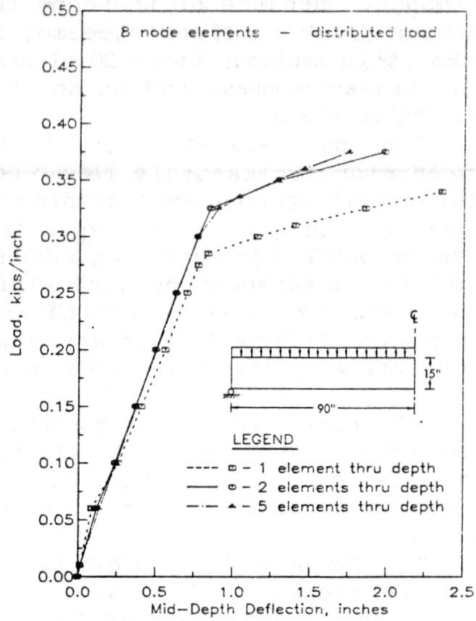

Fig.4 Slender Beam, Quadratic Elements (Reduced Integration)

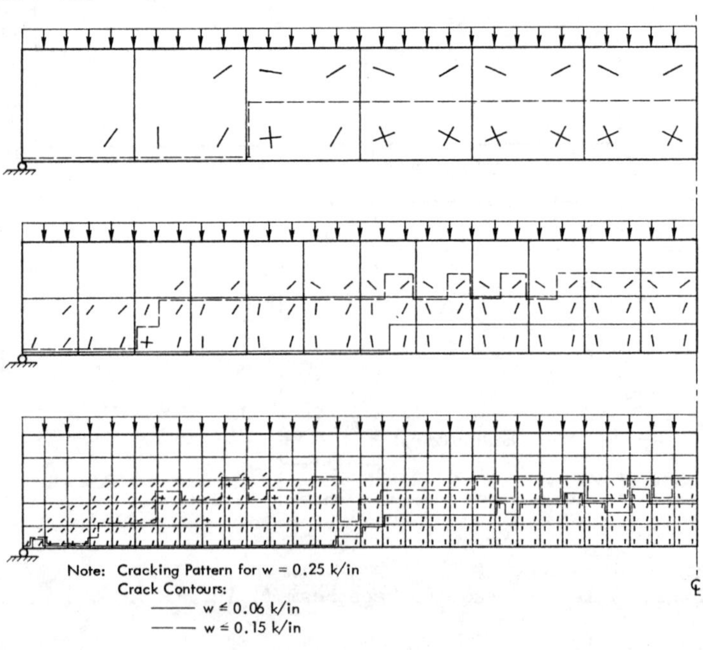

Fig.5 Slender Beam, Linear Elements (1 kip = 4.45 kN)

Typical load-deflection results for the deep beam are illustrated in Fig. 7, 8 and 9. In this case, the concentrated load is placed at midspan. Because of the more complicated stress distribution, compared to the slender beams, finer grids are required to properly model this member. Up to 20 elements through the depth are used for the linear element and up to 10 elements through the depth for the quadratic element.

As seen in Fig. 7, the linear model converges rapidly. However, as discussed below, the solution is not necessarily correct.

The quadratic element with reduced integration (Fig. 8) exhibits oscillating convergence. Part of the problem is the formation of horizontal cracks at mid-depth over the support (Fig. 10) which cause an upward deflection at that location, thus increasing the apparent deflection. The model with two elements through the depth does not have these cracks and thus appears much stiffer. The model with 10 elements through the depth has horizontal cracks, but does not exhibit an upward displacement over the support. For this particular example and element, there is clearly a level grid refinement needed to obtain satisfactory results. These results cast doubt on the ability of the quadratic element with reduced integration to properly represent cracked elements. The problem, again, is due to the unrealistic softness of the elements, due to the generation of spurious zero energy modes with the formation of cracks within the elements (7).

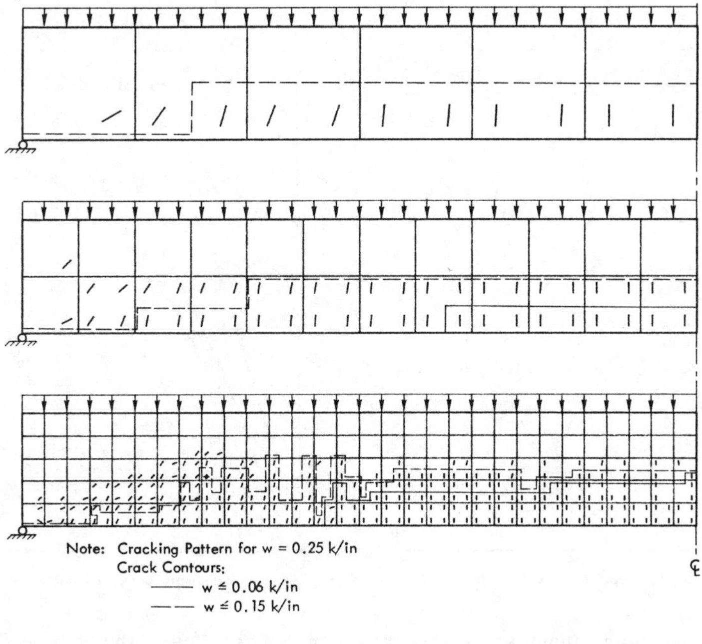

Fig.6 Slender Beam, Quadratic Elements (Reduced Integration)

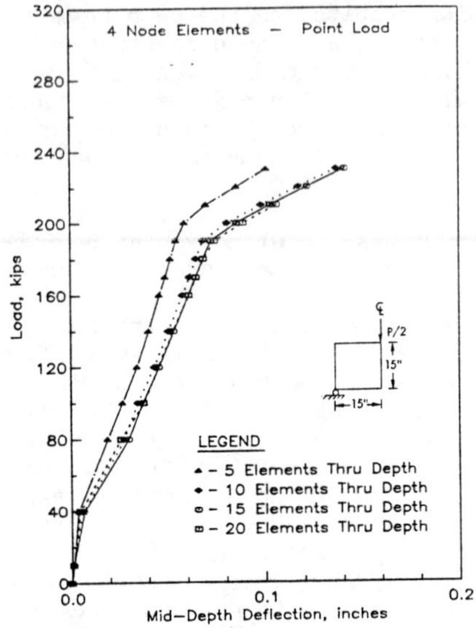

Fig.7 Deep Beam, Linear Elements (1 kip = 4.45 kN, 1 in. = 25.4 mm)

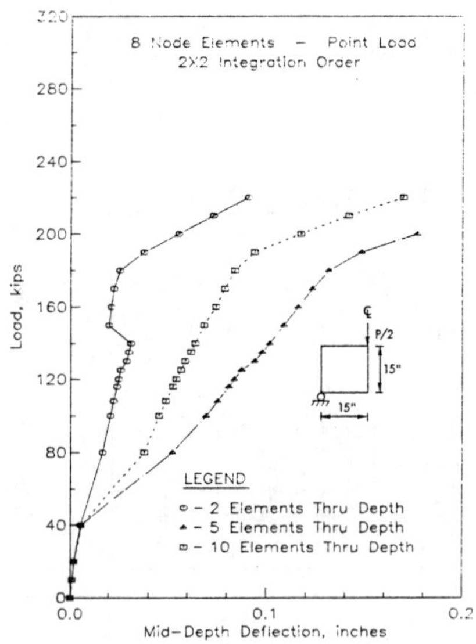

Fig.8 Deep Beam, Quadratic Elements (Reduced Integration)

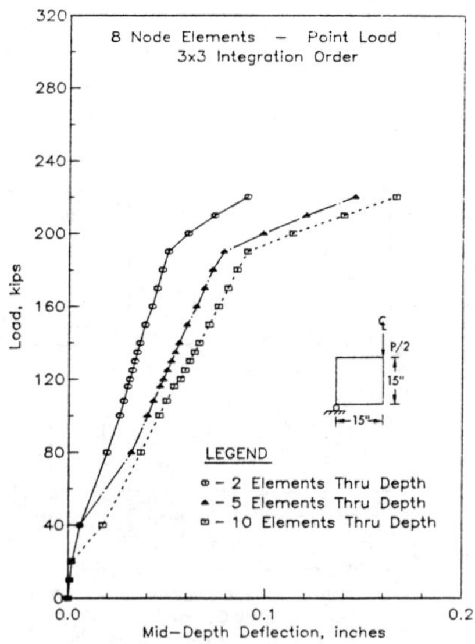

Fig.9 Deep Beam, Quadratic Elements (Full Integration)

The results for the model with the fully (3 x 3) integrated quadratic element (Fig. 9) show increased flexibility with grid refinement.

With increased grid refinement, the results for the linear and fully integrated quadrilateral should match. However, as shown in Fig. 11, this is not always true. The difference here turns out to be due, not to the element type, but to the sensitivity of the elements, with the given load step size, to the transition of beam behavior from that of a flexural member, Fig. 12a, to that of a tied arch, Fig. 12b, (shown here for a beam with 15 elements through the depth).

The problem is that the load steps are too large for the 4-node element to properly represent the transition.

As illustrated in Fig. 13 for the model with 15 elements through the depth, the correct global behavior can be represented if the size of the load steps is reduced. Fig. 14 compares the results for models with 15 and 20 elements through the depth. These figures illustrate two important facts. First, in cases where the strength of the structure is not governed by the fracture of concrete and the load increment size is below a certain level, the smeared cracking representation with a limiting tensile strength criterion is satisfactory. Second, some aspects of the solution, such as the transition from beam to tied arch behavior, are sensitive to both load step size and grid refinement. It is precisely these aspects of behavior that will require a fracture mechanics analysis for proper modeling (1-3).

The need for more refined procedures depends on the nature of the structure and its mode of failure. For most cases, the procedures shown here would be satisfactory. However, in structures for which there is not an alternative equilibrium configuration, the transition load may constitute the "failure load." Unreinforced structures and certain structures with very low reinforcement ratios are included in this category. Fracture mechanics based analyses may then be necessary to accurately predict the transition loads. The cracking

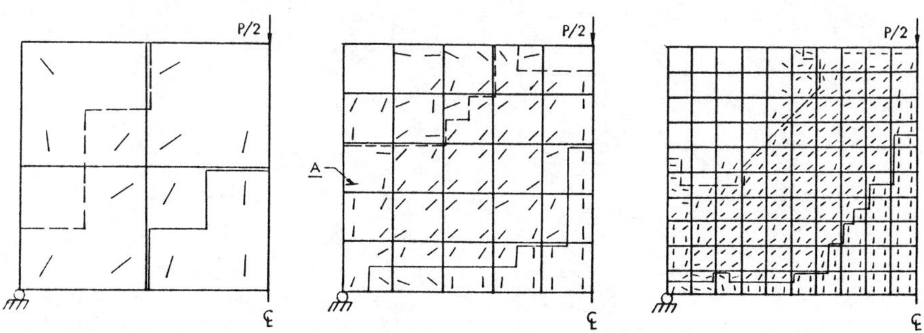

Note: Cracking Pattern for P = 160k
Crack Contours:
——— P ≤ 40k
— — — P ≤ 80k

Fig.10 Deep Beam, Quadratic Elements (Reduced Integration) (1 kip = 4.45 kN)

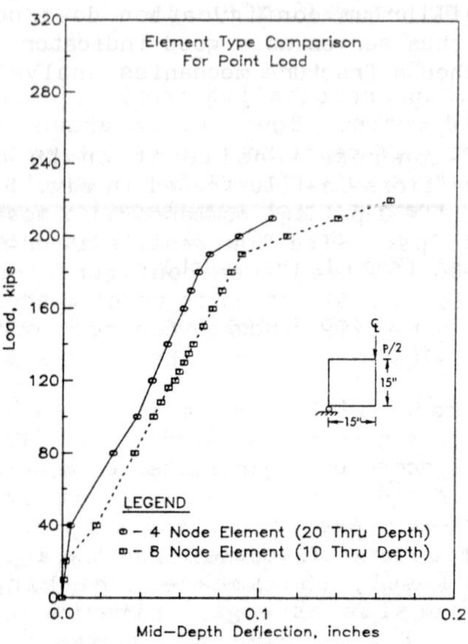

Fig.11 Deep Beam, Linear Elements and Quadratic Elements (Full Integration)

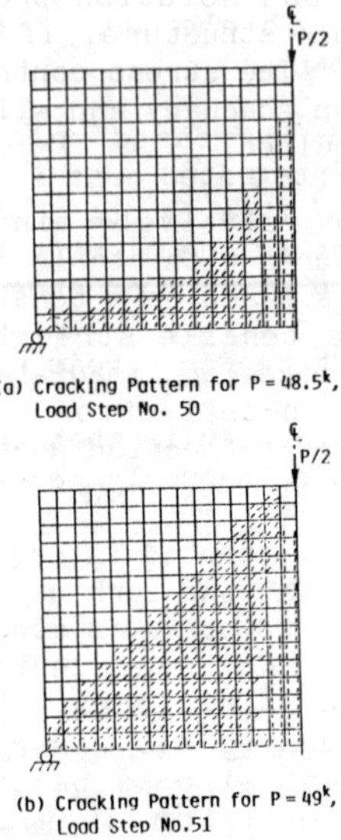

Fig.12 Cracking Patterns for Load Increment Size Effects (1 kip = 4.45 kN).

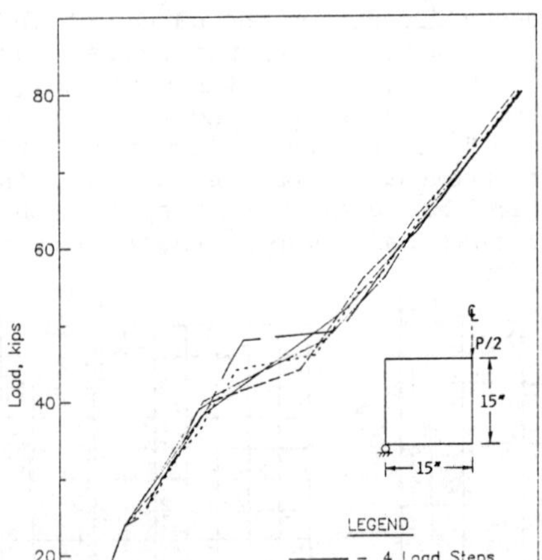

Fig.13 Effect of Load Increment Size on Load-Deflection Curves (1 kip = 4.45 kN, 1 in. = 25.4 mm)

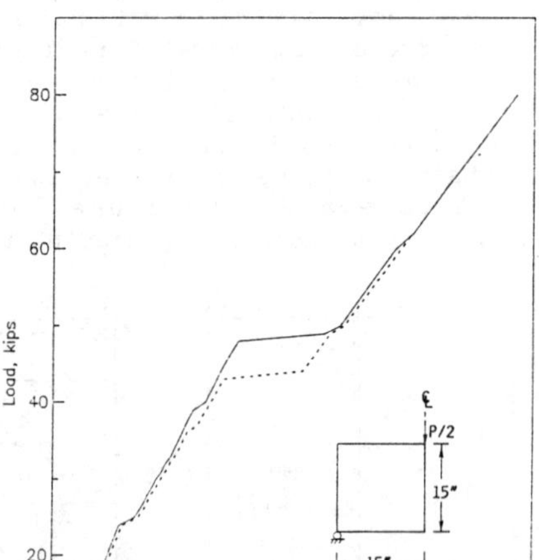

Fig.14 Effect of Load Increment Size and Grid Refinement on Load-Deflection Curves

model and solution procedures employed in this study will predict an unstable structure, if another equilibrium configuration does not exist. The stress controlled model thus serves as a good indicator of its own adequacy and will indicate when a fracture mechanics analysis is required.

The relatively minor effect of concrete tensile strength on structures with stable crack configurations is illustrated in Fig. 15. Results based on a tensile strength of 400 psi are compared to a model with a tensile strength of only 10 psi. The low tensile strength model makes an immediate transition to the tied arch configuration. The load-deflection curve for the low tensile strength model matches the load-deflection curve for the high tensile strength concrete once the higher strength model undergoes its transition. The results are identical thereafter.

Key Conclusions.--The key findings of this study can be summarized as follows.

Load-deflection curves for reinforced concrete members of the types considered exhibit general convergence towards a common solution with increased grid refinement. Flexure controlled members converge more rapidly than shear controlled members. The type of convergence, whether from the stiff or the flexible side, depends on the geometry, loading, element type and numerical integration order (for the quadratic element).

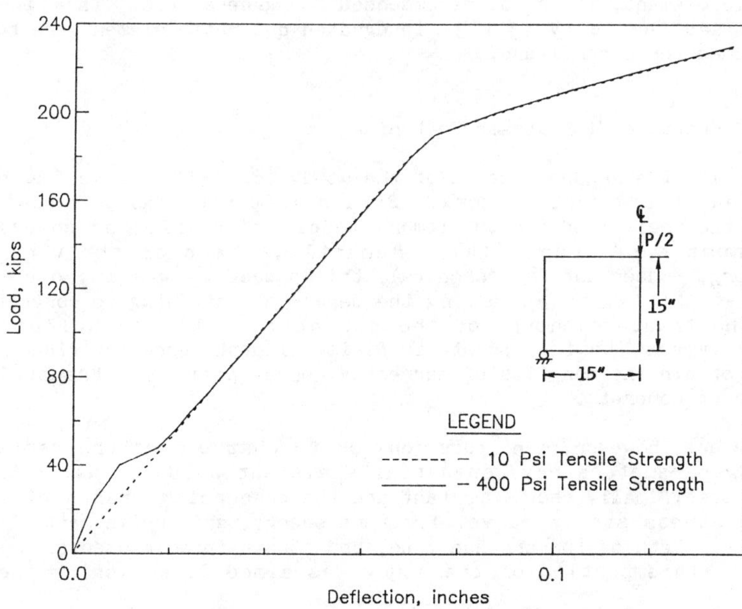

Fig.15 Effect of Concrete Tensile Strength on Load-Deflection Curves (1 kip = 4.45 kN, 1 in. = 25.4 mm)

Detailed analyses for a flexure reinforced deep beam show that below a certain magnitude, the load increment size does not affect the macroscopic stiffness or strength predictions. For a specific grid refinement, the load increment size does influence the load at which the transition in structural behavior from a beam to a tied arch occurs. The analysis may not detect the existence of such a transition if large load increments are used.

Concrete tensile strength has only a minor effect on the overall load-deflection behavior for the deep beam. Tensile strength controls the initial cracking and the load at which the stable cracked configuration is attained.

The stress controlled-smeared crack scheme is suitable for macroscopic strength and deformation analyses of members in which the limit strength is not solely governed by cracking. Unreinforced and very lightly reinforced members may require an analysis using fracture mechanics procedures. The stress controlled-smeared crack representation adequately models the general softening that results from cracking in common reinforced concrete members. Such members invariably achieve stable equilibrium configurations following extensive crack formation when load is transferred to the reinforcement. The failure of the solution procedure adopted in the study to reach a stable cracked configuration would strongly indicate the need for a fracture mechanics investigation.

Because of the response variations with the reduced integrated quadratic element, it is not recommended for general use. The linear element and the fully (3 x 3) integrated quadratic element are both recommended for general use.

Concrete Fracture With Strain Softening

General.--The second portion of the study (6,10-12) was a logical extension of the earlier work. Strain softening has been used to improve the realism of finite element models of reinforced concrete for a number of years (14). Recently, based on the work of Hillerborg, Modeer and Petersson (9), this procedure was improved by Bazant and Oh (4) by adjusting the degree of softening to correctly match the fracture energy of the concrete. This technique is relatively easy to incorporate in finite element representations and thereby obtain the benefits of correctly representing the fracturing behavior of concrete.

There are, however, many questions as to what effect the various modeling parameters have on a finite element solution using this model. Specifically, how important are the descending branch of the tensile stress-strain curve, fracture energy, grid refinement, load step size, width of the process zone, and constraints imposed on crack angles? This portion of the study was aimed at answering these questions.

To concentrate on the cracking response of concrete, a plain concrete fracture specimen was investigated (Fig. 16).

The concrete material model extended Bazant and Oh's smeared crack compliance formulation to include a provision for unloading. The model represents concrete as a microcracking material (Fig. 17a). The material between the microcracks remains linear and elastic. In contrast, the "material" within the microcracks is assumed to be in a uniaxial state of stress, with a strain that is dependent only on the stress transferred normal to the cracks. It is useful to think of the microcracks as being lumped together, with the cracked material consisting of two linear elastic blocks connected by ligaments having a limited stress transferring capability (Fig. 17b).

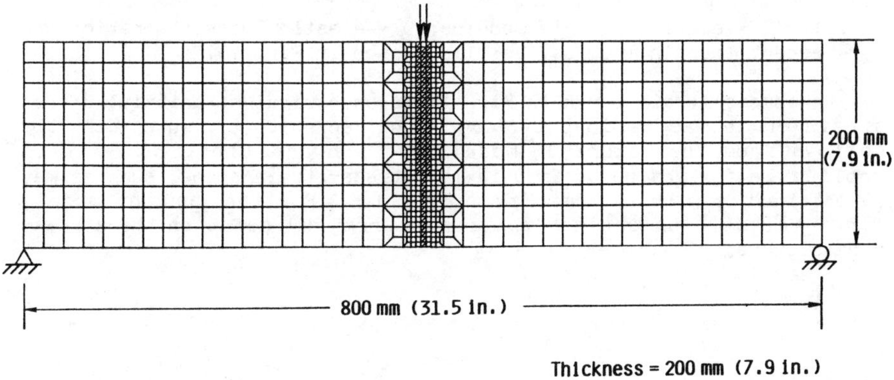

Fig.16 Finite Element Model of Notched Beam

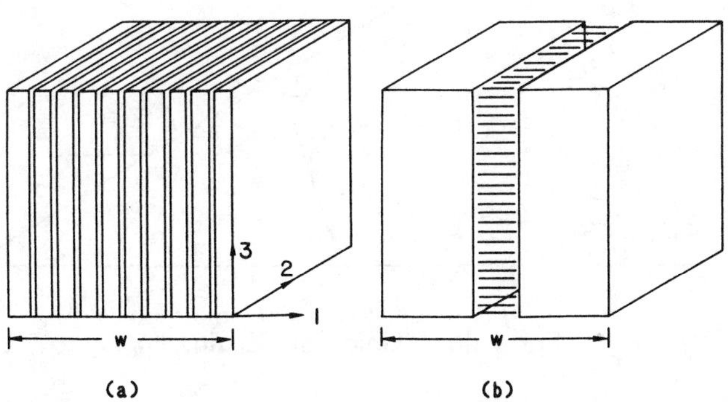

Fig.17a Microcracked Zone

Fig.17b Lumped Approximation of Microcracked Zone

Under monotonic loading, the material follows a stress-strain curve like the one shown in Fig. 18. Considering only principal stresses, the strains, ε, are related to the applied stresses, σ, in the general three-dimensional case as follows (4).

$$\begin{Bmatrix} \varepsilon_1 \\ \varepsilon_2 \\ \varepsilon_3 \end{Bmatrix} = \begin{bmatrix} E_s^{-1} & \nu E^{-1} & -\nu E^{-1} \\ & E^{-1} & -\nu E^{-1} \\ \text{sym} & & E^{-1} \end{bmatrix} \begin{Bmatrix} \sigma_1 \\ \sigma_2 \\ \sigma_3 \end{Bmatrix} \tag{2}$$

in which E = elastic Young's modulus; ν = elastic Poisson's ratio; and E_s = secant Young's modulus, as illustrated in Fig. 18.

The stress-strain curve in Fig. 18 is an equivalent uniaxial stress strain curve, as defined by Darwin and Pecknold (5), which directly follows from the Bazant-Oh model (4). In addition, Eq. (2) is applicable for any point in uniaxial stress-strain space, such as may be obtained upon unloading (Fig. 18) (11). For the current model, linear elastic unloading is assumed from any point on the stress-strain curve.

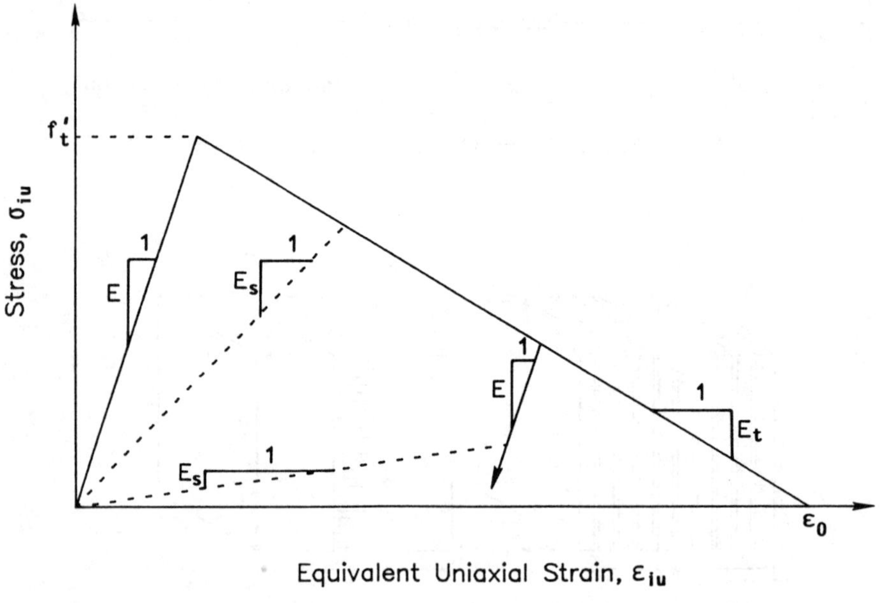

Fig.18 Equivalent Uniaxial Stress-Strain Curve for Tension Softening Material with Unloading

Eq. (2) is specialized to the plane stress case and expanded to include shear, τ and γ, as follows (11).

$$\begin{Bmatrix}\sigma_1\\ \sigma_2\\ \tau_{12}\end{Bmatrix} = \frac{1}{(1-\nu^2E_1E_2/E^2)}\begin{bmatrix}E_1 & \nu E_1E_2E^{-1} & 0\\ & E_2 & 0\\ \text{sym} & & (1-\nu^2E_1E_2/E_2)\beta G\end{bmatrix}\begin{Bmatrix}\varepsilon_1\\ \varepsilon_2\\ \gamma_{12}\end{Bmatrix} \quad (3)$$

in which E_1, E_2 = secant moduli; β = shear retention factor; and G = $0.5\ E/(1+\nu)$. In this study β was taken equal to zero.

As illustrated in Fig. 18, concrete is treated as a linear elastic material in tension, until the peak stress, f_t', is exceeded. The stress then drops with increasing equivalent uniaxial strain, reaching zero at strain ε_o. The fracture energy, G_f, is equal to the product of the area under the equivalent uniaxial stress-strain curve and the width of the fracture zone, w. In finite element analysis, w equals the weighted width of the finite element, corrected to account for the angle of the crack with respect to the element orientation (4,11).

Four different descending branch shapes were considered, as shown in Fig. 19.

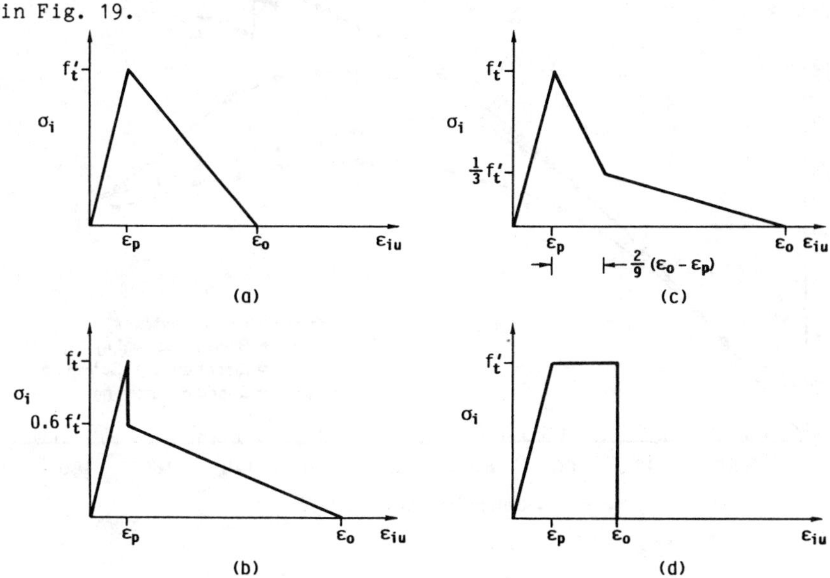

Fig.19 Concrete Models in Tension: (a) Linear Softening; (b) Discontinuous Softening; (c) Bilinear Softening; (d) Dugdale Softening

A notched bend specimen, originally analyzed by Petersson (13), was selected for study (Fig. 16). The beam is 200 mm wide, 200 mm deep, and 800 mm long, with a notch length of 80 mm . The fracture energy was assumed to be 100 N/m , the tensile strength 4 MPa, Young's modulus 40 GPa and Poisson's ratio 0.2. Additional analyses were performed using fracture energies of 50 N/m and 200 N/m. The hatched area represents the nonlinear part of the model.

The models were loaded by imposing displacements at the nodes indicated in Fig. 16.

Findings.--The effect of the shape of the descending branch of the tensile stress-strain curve is illustrated in Fig. 20. At low loads, the load-deflection curves are initially the same. At increasing displacements, the solutions become successively stiffer going from the discontinuous model to the bilinear model, the linear model, and finally the Dugdale model. The more rapid the dropoff in stress, the lower member capacity and the softer the response. A comparison of process zone ($\varepsilon > \varepsilon_p$, see Fig. 19) depth shows that the differences in model stiffness can be explained largely by the progress of the process zone through the specimen (Fig. 21). At every displacement above 60 μm, the stiffer the response, the shorter the process zone.

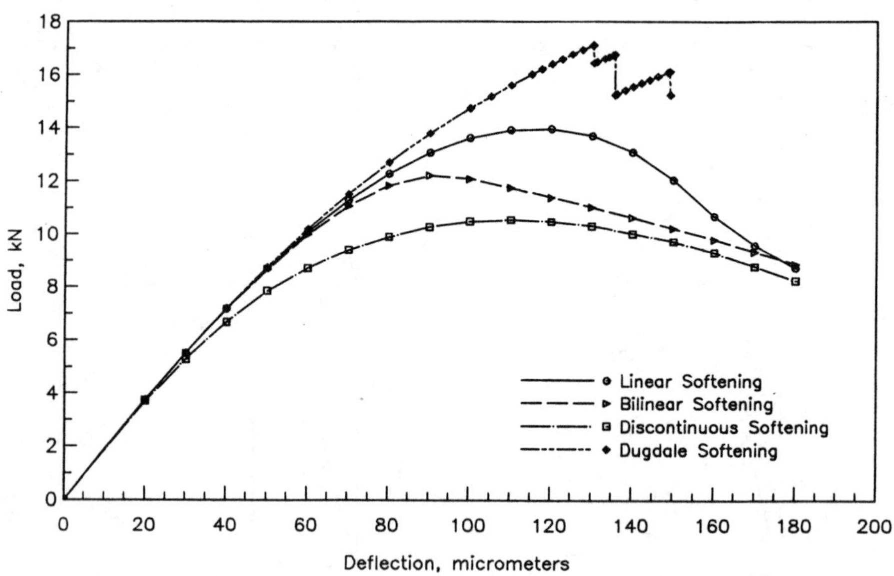

Fig.20 Effect of Concrete Tensile Response on Load-Deflection Curves

Fig. 21 illustrates the cracking patterns obtained at a deflection of 150 μm. In this case, a three element wide nonlinear zone is permitted. Elements outside the center band crack in each case. In most cases, the side elements unload immediately, and the crack localizes into the center column of elements. However, in the case of the Dugdale model, which does not provide any immediate stress relief, only isolated points unload before the terminal strain, ε_0, is exceeded near the base of the process zone.

On the other hand, the discontinuous model provides for immediate relief the instant ε_p is exceeded. The process zone is initially one-element wide and only broadens slightly at higher displacements. In contrast for the other models, the process zone encompasses the entire nonlinear zone.

The cracking patterns illustrate other common facets: The formation of primary cracks which are oblique instead of vertical, followed by the formation of secondary transverse cracks somewhat later. The oblique cracks are due to small shear stresses at the Gauss points, which are located just off center. The transverse cracks appear to help the model represent a single vertical crack.

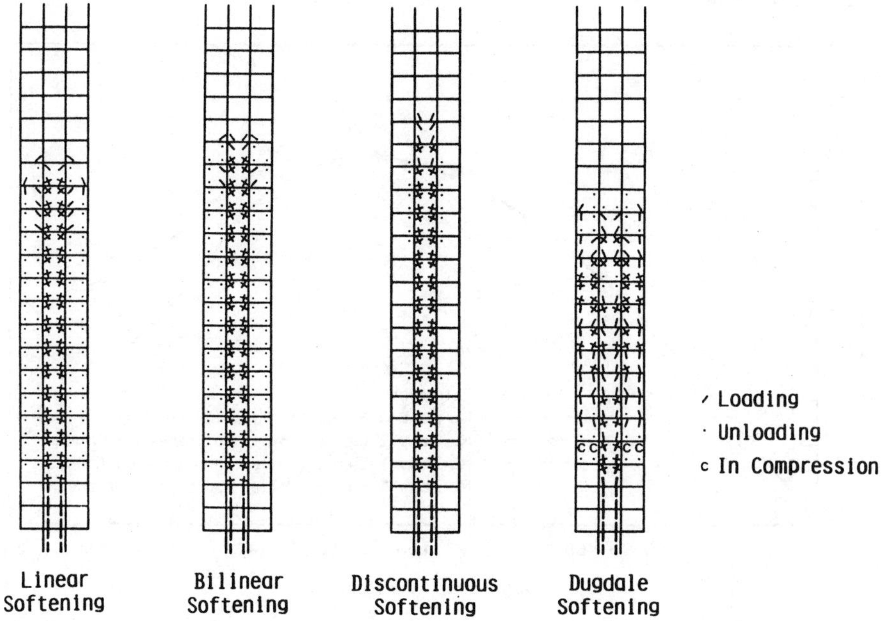

Fig.21 Crack Patterns for Beam at Deflection, δ = 150 μm

Fig. 22 compares the results obtained with the smeared model to the results obtained with Petersson's equivalent discrete model (13). The strain softening model with a linear descending branch is used. Fig. 22 also shows the effects of crack angle, process zone width and restriction on transverse cracking.

The curve marked "A" represents Petersson's results, as well as the smeared crack model with cracks constrained to form vertically. Little difference is apparent. If the cracks are free to form at the orientation of the principal tensile stress, curves "B" and "C" are obtained for the three-element wide and one-element wide nonlinear zones, respectively. These curves illustrate that while the width of the process zone has little effect on specimen response (largely because of crack localization), the crack angle is extremely important. The importance of the crack angle is further emphasized, since the process zone lengths for curves "A", "B" and "C" differ only slightly. The clear lesson is that the inability to correctly model crack angle may increase the predicted stiffness and strength of a structure.

Transverse cracks form in cases "A", "B" and "C". The transverse cracks continue to open for the beams with the oblique cracks ("B" and "C"), but unload after formation for the beam with the constrained vertical cracks ("A").

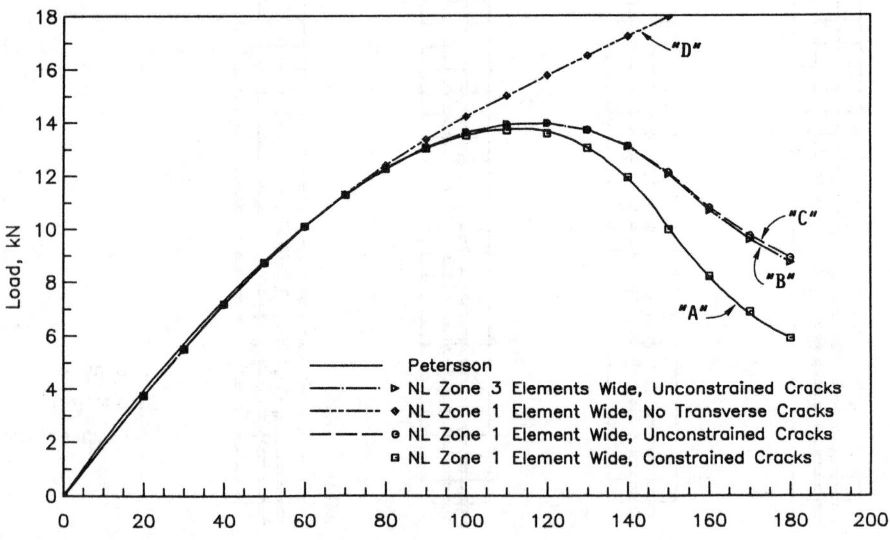

Fig.22 Effect of Crack Angle Constraint and Width of Nonlinear Zone on Load-Deflection Curves of Beam with Linear Softening

Curve "D" illustrates the role of the transverse cracks when the primary cracks form at an oblique angle. The transverse cracks help to represent the reduced stiffness obtained with a true vertical crack. In this case, transverse cracks are prevented from forming, and an unrealistically stiff response is obtained.

The effect of fracture energy on beam response is illustrated in Fig. 23 for the discontinuous model. As might be expected, fracture energy primarily increases specimen toughness, rather than strength. A doubling of the fracture energy from 50 to 100 N/m increases the maximum load by only 14.9%. Doubling it again to 200 N/m increases the peak load by 11.4%. This observation suggests that only a reasonable estimate of the fracture energy of concrete is required for analysis. This observation may also explain the insensitivity of many of the early finite element models to the particular cracking criterion selected.

The study also considered the effects of load step size and grid refinement on specimen response and found the model to be objective with respect to both.

Key Conclusions.--The results of this study lead to the following findings.

The post-peak tensile behavior of the concrete controls the load-deflection response of the structure. As the slope of the descending

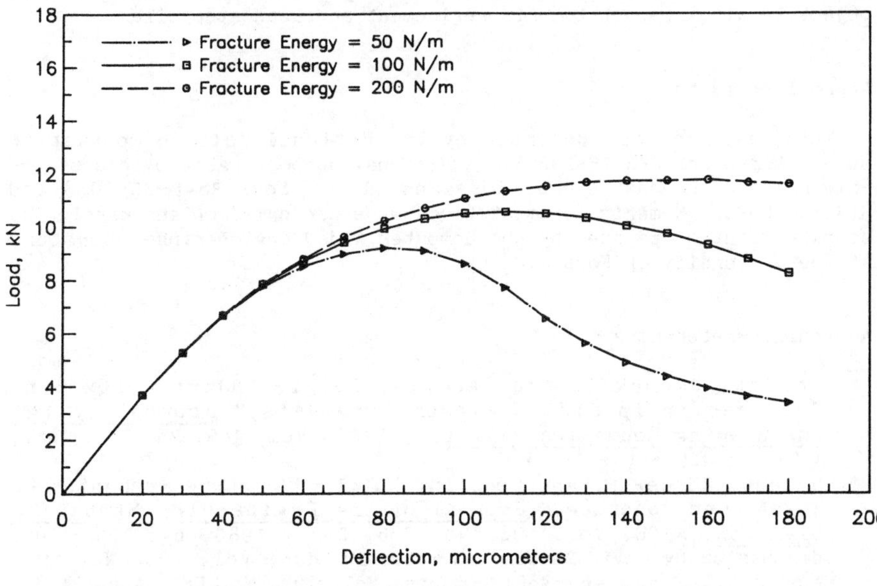

Fig.23 Effect of Fracture Energy on Load-Deflection Curves of Beam with Discontinuous Softening

branch of the tensile stress-strain curve becomes less severe, the process zone extends more slowly and the load capacity of the specimen increases.

Increasing the fracture energy primarily toughens, not strengthens the specimen. The load capacity of the beam is insensitive to small changes in fracture energy. This observation may explain why many models (14) are insensitive to the precise choice of cracking criterion, as controlled by tensile strength, degree of strain softening, or element size.

The width of the nonlinear zone has a negligible effect on the behavior of the fracture specimen. If the constitutive of model permits unloading and the tension softening model is capable of relieving stresses adjacent to an opening crack, the crack localizes into a region one element wide. Minimal energy is absorbed in forming the microcracks that subsequently unload.

Both primary and secondary cracks may open in an attempt to simulate a single more nearly vertical crack.

Crack angles in a structure must be properly represented if reliable predictions of structural response are to be made. Misoriented cracks may cause the stiffness and ductility of a structure to be overestimated. In turn, predictions of load capacity and failure mode may be in error.

The Bazant-Oh (4) formulation with provisions for unloading is objective with respect to grid refinement and load step size.

Acknowledgements

This research was supported by the National Science Foundation under NSF Grant PFR 79-24696. Additional support was provided by the University of Kansas General Research Allocations 3842-X0-0038 and 3131-X0-0038. Numerical computations were performed on the Harris 500 computer system operated by the Computer Aided Engineering Laboratory at the University of Kansas.

Appendix.--References

1. Bazant, Zdenek P. and Cedolin, Luigi, "Blunt Crack Band Propagation in Finite Element Analysis," Journal of the Engineering Mechanics Division, ASCE, Vol. 105, No. EM2, Apr., 1979, pp. 297-315
2. Bazant, Zdenek P. and Cedolin, Luigi, "Fracture Mechanics of Reinforced Concrete," Journal of the Engineering Mechanics Division, ASCE, Vol. 106, No. EM6, Dec., 1980, pp. 1287-1306; discussion by David Darwin and Robert Dodds, Vol. 108, No. EM2, Apr., 1982, pp. 464-466; closure, Vol. 108, No. EM2, Apr., 1982, pp. 466-471.

3. Bazant, Zdenek P. and Cedolin, Luigi, "Finite Element Modeling of Crack Band Propagation," Journal of Structural Engineering, ASCE, Vol. 109, No. 1, Jan., 1983, pp. 69-92.
4. Bazant, Zdenek P. and Oh, Byung H., "Crack Band Theory for Fracture of Concrete," Materiaux et Constructions, Vol. 16, No. 93, May-June 1983, pp. 155-177.
5. Darwin, David and Pecknold, David A., "Analysis of RC Shear Panels Under Cyclic Loading," Journal of the Structural Engineering Division, ASCE, Vol. 102, No. ST2, Feb. 1976, pp. 355-369.
6. Darwin, David; Leibengood, Linda D. and Dodds, Robert H., "Structural Aspects of Tension Softening in Concrete," presented at the Fifth Engineering Mechanics Division Specialty Conference, ASCE, University of Wyoming, Laramie, August 1-3, 1984, Proceedings, Vol. 2, pp. 1389-1392.
7. Dodds, Robert H.; Darwin, David; Smith, Jerry L. and Leibengood, Linda D., "Grid Size Effects with Smeared Cracking in Finite Element Analysis of Reinforced Concrete," SM Report No. 6, University of Kansas Center for Research, Inc., Lawrence, Kansas, Aug. 1982, 118 pp.
8. Dodds, Robert H.; Darwin, David and Leibengood, Linda D., "Stress Controlled Smeared Cracking in R/C Beams," Journal of Structural Engineering, ASCE, Vol. 110, No. ST9, September 1984, pp. 1959-1976.
9. Hillerborg, Arne; Modeer, Matz and Petersson, Per-Erik, "Analysis of Crack Formation and Crack Growth in Concrete by Means of Fracture Mechanics and Finite Elements," Cement and Concrete Research, Vol. 6, No. 6, Nov. 1976, pp. 773-782.
10. Leibengood, Linda D.; Darwin, David and Dodds, Robert H., "Tension Stiffening and Compression Softening in Concrete," presented at the Fourth Engineering Mecanics Division Specialty Conference, ASCE, Purdue University, West Lafayette, Indiana, May 23-25, 1983, Proceedings, Vol. II, pp. 1043-1046.
11. Leibengood, Linda D.; Darwin, David and Dodds, Robert H., "Finite Analysis of Concrete Fracture Specimens," SM Report No. 11, University of Kansas Center for Research, Inc., Lawrence, Kansas, May 1984, 120 pp.
12. Leibengood, Linda D.; Darwin, David and Dodds, Robert H., "Parameters Affecting FE Analysis of Concrete Structures," submitted for publication to the Journal of Structural Engineering, ASCE.
13. Petersson, Per-Erik, "Crack Growth and Development of Fracture Zones in Plain Concrete and Similar Materials," Report TVBM-1006, Division of Buiding Materials, University of Lund, Sweden, 1981, 174 pp.
14. Task Committee on Finite Element Analysis of Reinforced Concrete Structures, State-of-the-Art Report on Finite Element Analysis of Reinforced Concrete, American Society of Civil Engineers, New York, N.Y., 1982, 545 pp.

ANALYSIS OF RC WALLS BY PLASTIC-FRACTURING THEORY

By Hiroshi Yamaguchi,[1] M.AIJ and Setsuro Nomura,[2] M.AIJ

ABSTRACT: The purpose of this study was to verify the application to reinforced concrete members, of the analytical model based on plastic-fracturing theory proposed by Bažant et al.(3). Inelastic analysis of reinforced concrete shear walls (4 specimens) subjected to monotonic loading and repeated reversal loading in plane stress, were carried out using the finite element method. The smeared cracking model for crack representation was used and two-directional cracking was provided for repeated reversal loading. Concrete was assumed to be orthotropic for post-cracking behavior and the tension stiffening effect was considered. The Bauschinger effect was taken into account for the constitutive relations of reinforcing steel. For finite element representation,an isoparametric quadrilateral element was used for concrete;a smeared model,which has axial stiffness only,was used for reinforcing steel, and perfect bond was assumed.

1. INTRODUCTION

For inelastic analysis of reinforced concrete members using the finite element method, it is necessary to model the material nonlinearities of concrete, steel and the interaction between them. The constitutive relations of concrete are the most basic problem. Several constitutive models for these relations have been proposed so far, but all these models have certain inherent advantages and disadvantages. Recently, an available constitutive model based on plastic-fracturing theory was proposed by Bažant et al. The model allows a more realistic and comprehensive representation of various aspects of inelastic behavior of concrete. However, the application of the model to the numerical analysis of reinforced concrete members has not been reported ,so that limits regarding its applicability remain to be determined. So,to verify the application of the analytical model based on plastic-fracturing theory, the following analyses were carried out: the analyses of shear walls subjected to monotonic loading which were selected by JCI as analysis verification models, and the analysis of a shear wall subjected to repeated reversal loading tested by the authors.

2. REVIEW OF PLASTIC-FRACTURING THEORY

In the plastic-fracturing theory, it is assumed that the inelastic behavior of concrete consists of both plastic and fracturing deformations, and the incremental plasticity is combined with fracturing theory(Fig.1). Plastic deformation which results from plastic strain (slip) $d\epsilon^{pl}$ is irreversible. This deformation does not cause the

[1] Research Engineer at Fourth Research Center of Technical Research and Development Institute,Japan Defense Agency, Fuchinobe 2-9-54, Sagamihara-City, Kanagawa, Japan

[2] Professor of Science University of Tokyo,Department of Architecture. Higashikameyama 2641,Yamazaki,Noda-City,Chiba,Japan

decrease of elastic moduli. In contrast to plastic deformation, fracturing deformation which results from microcracking is reversible. This is accompanied by the decrease of elastic moduli and represents the strain softening (i.e. the decrease of stress at increasing strain) due to fracturing stress decrement $d\sigma^{fr}$. Plastic strain increments are prescribed by the plastic loading surface and the flow rule. Fracturing stress decrements are prescribed by the fracturing loading surface in the same manner as plastic strain increments. For the purpose of representing inelastic behavior under cyclic loading, the jump-kinematic hardening is introduced, where the center of the current loading surface is allowed to jump instantaneously into the current stress and strain states when loading reverses into unloading. The constitutive relation given by the theory yields

$$\{ d\sigma \} = [D_c]\{ d\epsilon \} \qquad \text{-----------(1)}$$

$$[D_c] = [D^{el}] - [D^{pl}] - [D^{fr}]$$

in which $[D^{el}]$ is the elastic stiffnes matrix,and $[D^{pl}]$ and $[D^{fr}]$ express the decrease of stiffness due to plastic strain increments and fracturing stress decrements,respectively. The stiffness matrix $[D_c]$ is not symmetric and does not generally have an orthotropic form. The typical behavior of plain concrete modeled by this theory is shown in Fig.2.

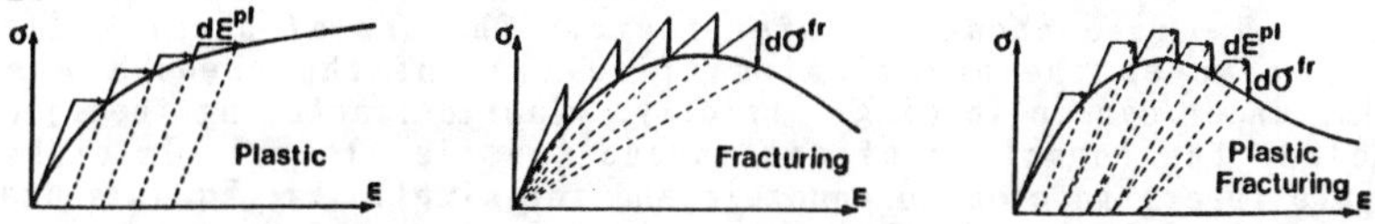

Fig.1 Plastic,Fracuring and Plastic-Fracturing Deformations

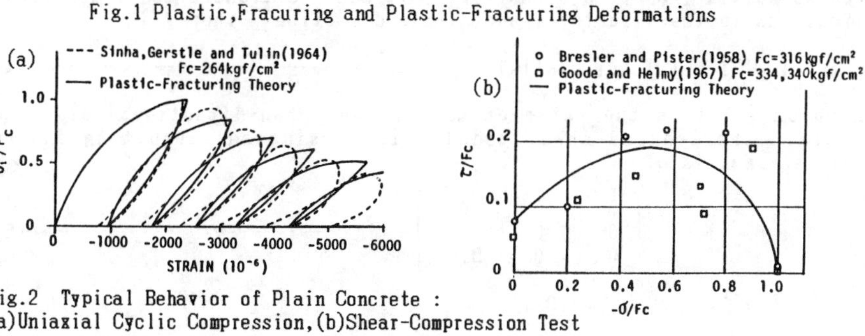

Fig.2 Typical Behavior of Plain Concrete :
(a)Uniaxial Cyclic Compression,(b)Shear-Compression Test

3. ANALYTICAL MODEL

The constitutive relations of plain concrete in plane stress are predicted by the theory with several material parameters,which are given as functions of the strength of concrete, proposed by Bažant et al., as follows

$$\begin{Bmatrix} d\sigma_{11} \\ d\sigma_{22} \\ d\sigma_{12} \end{Bmatrix} = \begin{bmatrix} D_{11} & D_{12} & D_{13} \\ D_{21} & D_{22} & D_{23} \\ D_{31} & D_{32} & D_{33} \end{bmatrix} \begin{Bmatrix} d\epsilon_{11} \\ d\epsilon_{22} \\ d\epsilon_{12} \end{Bmatrix} \qquad \text{------------------------(2)}$$

$$\{ d\sigma \} = [D_c] \{ d\epsilon \}$$

When the maximum principal tensile stress σ_1 exceeds a certain value given by the biaxial strength envelope curve(5), cracks are assumed to develop in the plane normal to the σ_1 direction. For the analysis of a specimen subjected to cyclic loading, two-directional cracking is provided. The second crack need not be normal to the first crack ,but it is assumed that the angle between them should be larger than 30 degrees. This restriction on the angle of two-directional cracking influences considerably the opening and the closing of cracks due to cyclic loading. An open crack is assumed to close when the strain normal to the crack becomes both less than the strain (ϵ_{cr}) at which the crack has opened and less than zero. A closed crack is assumed to open when the stress normal to the previous crack becomes tension. A smeared cracking model is used as a crack representation . Thus, the plastic-fracturing theory is applied to cracked concrete which is assumed to be orthotropic with respect to the crack axis. Noting that $D_{13} = D_{23} = D_{31} = D_{32} = 0$ so as to obtain zero normal stress on the crack planes, the stiffness of cracked concrete yields

$$[D_{cr}] = \begin{bmatrix} 0 & 0 & 0 \\ 0 & D_{22}' & 0 \\ 0 & 0 & D_{33}' \end{bmatrix} \qquad \text{------------------------(3)}$$

$$D_{22}' = D_{22} - D_{12} \cdot D_{21} / D_{11} \qquad D_{33}' = \beta_s \cdot D_{33}$$

in which β_s is a shear transfer factor. The use of shear modulus, $\beta_s \cdot D_{33}$ removes the numerical difficulties of the shear transfer mechanism under cyclic loading. Strictly, plastic-fracturing theory does not include the formation of continuous tensile cracks. In order to apply this theory to cracked concrete and for simplicity, Eq.3 is used. Because matrix $[D_{cr}]$ is orthotropic, it must be transformed from crack directions to the structural coordinate directions. Then

$$[D_c] = [T]^T [D_{cr}][T] \qquad \text{------------------------(4)}$$

in which [T] is the rotation matrix. When two-directional cracking occurs, the reduced shear modulus is considered, resulting in the stiffness matrix of

$$[D_{cr}] = \begin{bmatrix} 0 & 0 & 0 \\ 0 & 0 & 0 \\ 0 & 0 & D_{33}'' \end{bmatrix} \qquad \text{------------------------(5)}$$

$$D_{33}'' = \beta_t \cdot D_{33}'$$

in which β_t is a reduction factor taken as $\beta_t = 0.5$.

For the purpose of taking the Bauschinger effect into account in the constitutive relations of reinforcing steel, the model proposed by Agrawal et al.(Fig.3)(1) is used in this paper. This model was used by Bažant et al.(2) and Tanigawa et al.(6),as well. By defining the tangent modulus E_s as shown in Fig.3, the stiffness matrix of reinforcing steel which is represented by a smeared model with axial stiffness yields

$$[D_s] = \begin{bmatrix} p_{11} \cdot E_{S11} & 0 & 0 \\ 0 & p_{22} \cdot E_{S22} & 0 \\ 0 & 0 & 0 \end{bmatrix} \qquad \text{------------------------(6)}$$

in which, p_{11}, p_{22} are the percentages of reinforcing steel within the element . The stiffness matrix [D] for reinforced concrete becomes

$$[D] = [D_c] + [D_s] \qquad (7)$$

To improve the expression for cracking , the tension stiffening effect is considered and represented as follows under the name of so-called discontinuous linear unloading model(Fig.4). When the stress in concrete reaches its tensile strength, the first crack forms and the concrete stress drops not to zero but only some part(γ_t) , because the concrete between cracks still carries some tensile stress, and according to the strain increase,the tensile stress drops to zero ,and the strain at this stage is named ϵ_t . For finite element representation,an isoparametric quadrilateral element is used for concrete. As for the bond behavior, perfect bond is assumed here for simplicity,though more realistic bond behavior should be considered for further accuracy.

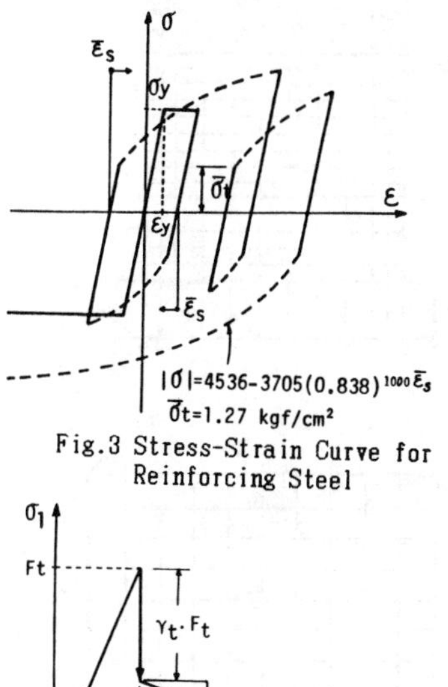

Fig.3 Stress-Strain Curve for Reinforcing Steel

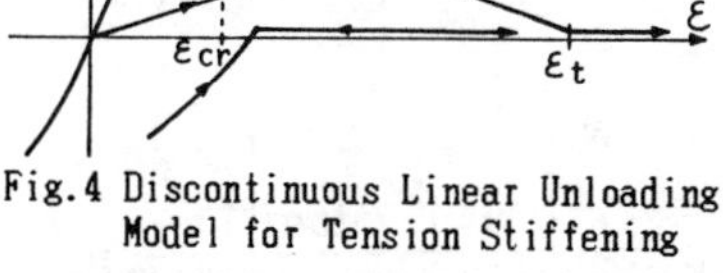

Fig.4 Discontinuous Linear Unloading Model for Tension Stiffening

Table.1 Strengths of Concrete

Specimen No.	Compressive Strength F_c(kgf/cm²)	Tensile Strength F_t(kgf/cm²)
#1	297.0	23.6
#3	233.0	21.3
#4	170.0	16.3
#5	316.0	29.3

Table.2 Strengths of Reinforcing Steel

Specimen No.	Size	σ_y	σ_u	Note
#1	D6	3990	5170	Hoop
	D6	3620	5560	Wall Reinf.
	D10	3530	5130	Stir.,Beam
	D13	3680	5500	Column
#3	6φ	3740	4180	Wall Reinf. Hoop,Stir.
	D13	3240	4700	Beam,Column
#4	6φ	2850	4000*	Beam,Column
	13φ	2850	4000*	Wall Reinf. Hoop,Stir.
#5	5φ	3500*	5531	Wall Reinf. Hoop,Stir.
	D10	4190	6107	Column
	D13	4323	6413	Beam

σ_y=Yield Strength(kgf/cm²)
σ_u=Tensile Strength(kgf/cm²)
* =Default value

4. SPECIMENS FOR VERIFICATION OF ANALYTICAL MODEL

The specimens for verification of the analytical model are the shear walls(#1,#3,#4) which are some of the specimens selected by JCI for verification of analytical models (4) subjected to monotonic increasing loading, and the shear wall (#5) subjected to repeated reversal loading. The specimen #1 is a single story and single span shear wall and the

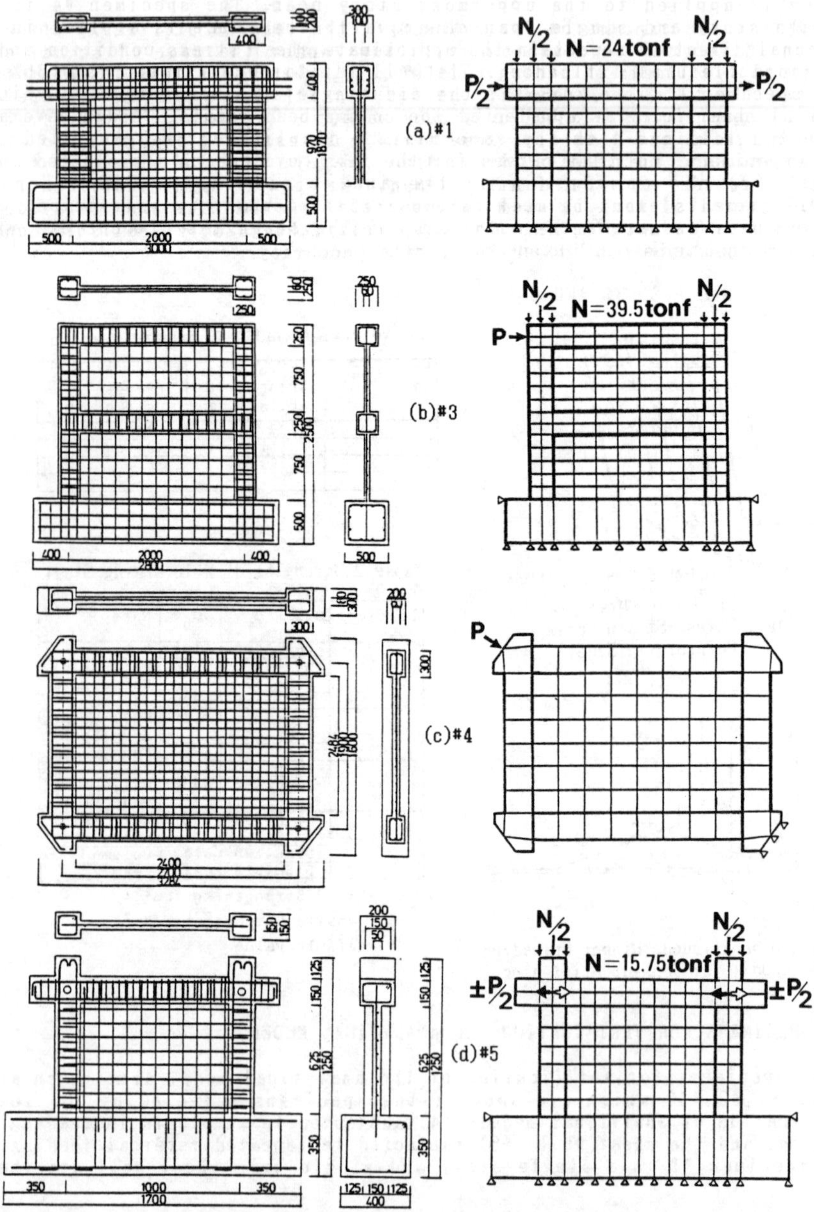

Fig.5 Dimensions, Reinforcements,Loading Methods and
Finite Element Representations for Specimens (unit;mm)

wall reinforcement ratio is 0.85%. The lateral shear force is applied at the one end of the beam as a push and at the other end as a pull at the same time.The specimen #3 is a two story and single span cantilever type shear wall and the wall reinforcement ratio is 0.21%. The lateral shear force is applied to the uppermost story beam. The specimen #4 is a single story and single span shear wall. The force is applied in a diagonal direction to obtain the approximate shear stress condition and to avoid flexural influence . The wall reinforcement ratio is 0.35% . The specimen #5 is a single story and single span shear wall in which lateral shear force is applied at the column-beam joints. Though several repeated loadings with the same maximum displacement were applied in the experiment, only the curves for the first cycle each are plotted in Fig.7,(d) and compared with the analytical ones. Dimensions, reinforcements, loading methods and finite element representations of these specimens are shown in Fig.5 and the strengths of concrete and reinforcing steel are listed in Table 1 and 2.

5. ANALYTICAL RESULTS AND CONSIDERATIONS

The shear transfer factor β_s and the parameters γ_t , ϵ_t of the tension stiffening effect are decided by a parametric study.The examples of load-displacement curves under the influence of these parameters for specimen #1 are shown in Fig.6 . The value of maximum capacity is influenced considerably by the shear transfer factor. As the results of the parametric study,these parameters are decided as follows

$$\beta_s = 0.25 \ , \ \gamma_t = 0.5 \ , \ \epsilon_t = 20 \cdot \epsilon_{cr} \qquad \text{------------(8)}$$

The comparison of analytical results using these parameters with experimental ones are listed in Table 3,and the load-displacement curves are shown in Fig.7. The analytical initial rigidities are slightly larger than experimental ones, but as shown in Table 3, the values of shear force at flexural cracking, shear cracking or at yielding of reinforcing steel give good agreement with experimental ones. The maximum analytical shear forces plotted in Fig.7 or Table 3 are the ones which are calculated within the possible range of convergence, since the values of tangential stiffness modulus are very small near the maximum shear forces and it is difficult to get sufficient convergence. So the values of displacement at this stage are not accurate. This is observed especially in specimen #3 which has large plastic deformation after yielding of main reinforcment. In the analysis of the specimen #5 subjected to repeated reversal loading, each hysteresis loop area is smaller than the experimental one. The area of hysteresis loops is influenced considerably by the shear transfer mechanism and bond slip, but it is not sufficient to evaluate these factors in this analytical model, and the modeling of these factors are the subject of a future study. The crack patterns are shown in Fig.8. Computed cracks tend to spread more widely over the specimen compared with the experimental ones, but their inclinations in the wall almost coincide with those observed in the experiment. Especially the crack pattern at the adjacent frame of specimen #4 satisfactorily corresponds with the experimental one (Fig.8,(e),(f)). The two-directional cracks occur slightly at the column of specimen #5, and the opening and the closing of cracks agree well under the repeated reversal loading (Fig.8,(i) and (j)). Fig.9,(a) and (b) show the stress of concrete at the base of the column in the compression region and the strain up to cracking of the specimen #1. The principal compressive stress reaches the maximum value at P=120 tonf and

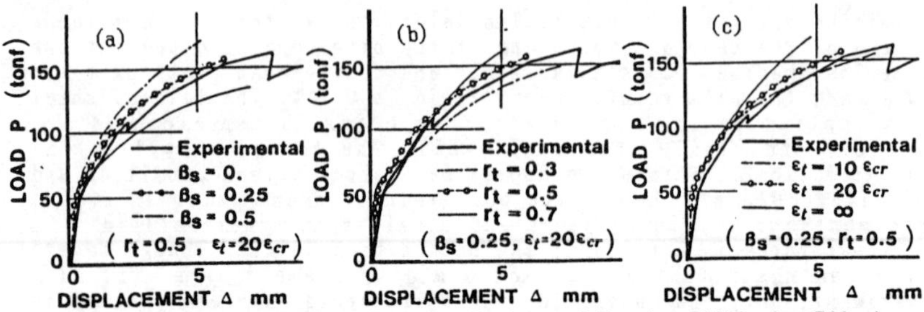

Fig.6 Influences of Shear Transfer Factor and Tension Stiffening Effect
(a) βs , (b) γt ,(c) ϵt

Table.3 Compasison of Analytical Results with Experimental Results(unit;tonf)

Specimen No.		#1	#3	#4	#5
Shear Force at Flexural Cracking	Experimental	30.0	20.0		6.0
	Analytical	30.0	20.0		10.0
Shear Force at shear Cracking	Experimental	30.	36.0	41.0	6.0
	Analytical	40.0	30.0	35.0	10.0
Shear Force at Yielding Reinf.	Experimental	125.0	62.0		26.8
	Analytical	125.0	58.5		23.0
Maximum Shear Force	Experimental	161.0	68.1	92.6	
	Analytical	152.0	62.5	90.0	

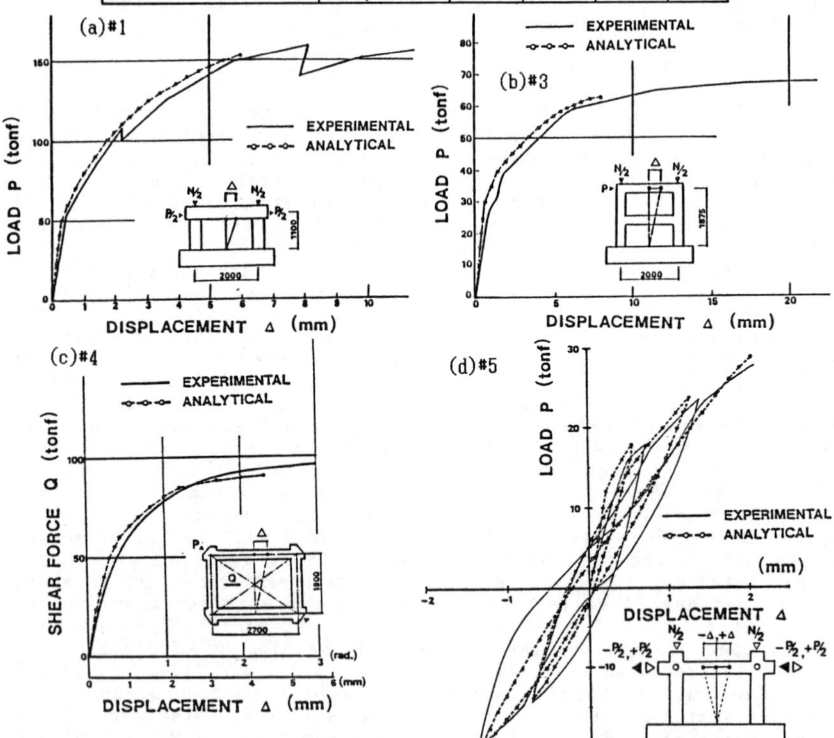

Fig.7 Load-Displacement Relations

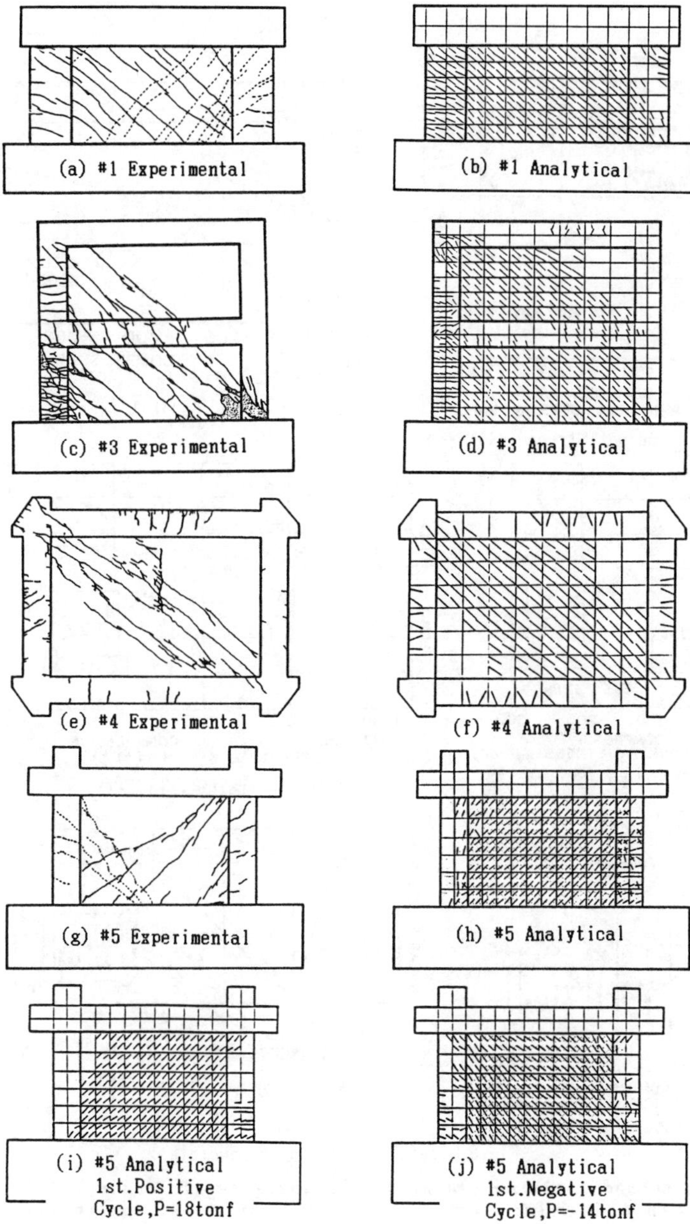

Fig.8 Crack Patterns

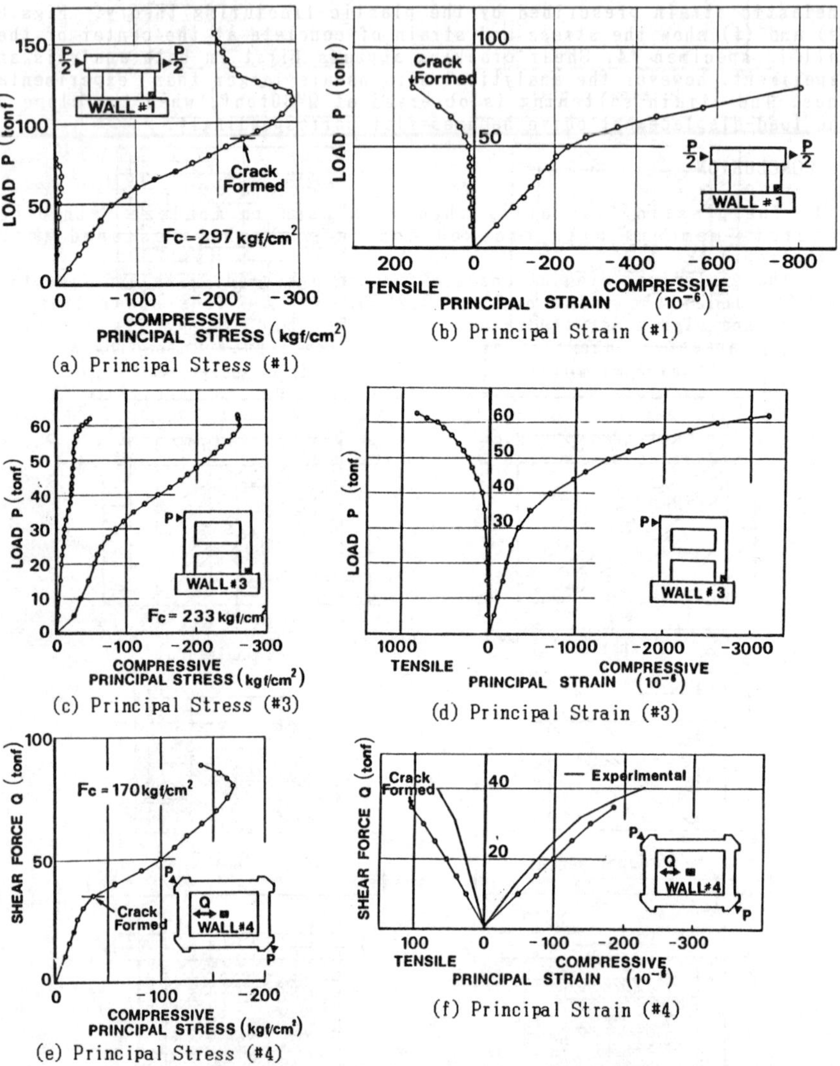

(a) Principal Stress (#1)

(b) Principal Strain (#1)

(c) Principal Stress (#3)

(d) Principal Strain (#3)

(e) Principal Stress (#4)

(f) Principal Strain (#4)

Fig.9 Stress and Strain in Concrete

decreases subsequently according to the increase of loading. This shows that strain softening occurs at this point , although the strain softening and crushing at this point are not observed in experiment. The strain increases rapidly when cracks spread over the wall. Flexural cracking occurs at the point when the principal tensile strain reaches 180 microstrains. Concerning specimen #3,the analytical stress and strain of concrete at the same position are shown in Fig.9,(c),(d). In the analysis,cracks do not occur at this point, and the stress and strain states under biaxial compression may be evaluated. Though there are no experimental data ,the present analytical model seems to represent a more realistic biaxial compressive stress condition. The principal tensile strain reaches 900 microstrains,which is caused by the inelastic strain prescribed by the plastic-fracturing theory. Figs.9, (e) and (f) show the stress and strain of concrete at the center of the wall in specimen #4. Shear cracking appears first in both analysis and experiment, however the analytical strains are larger than experimental ones. The strain softening is observed at Q=80tonf, when the slope of the load-displacement curve becomes flat correspondingly.

6. CONCLUSION

1. The plastic-fracturing theory is used to analyze reinforced concrete members with cracked concrete which is assumed to be orthotropic.

2. The analytical predictions are in good agreement with experimental results under the value of β_s = 0.25 where β_s is a shear transfer factor, and also under the values of $\gamma_t = 0.5, \epsilon_t = 20.0 \cdot \epsilon_{cr}$ which are the parameters used in the discontinuous linear unloading model of tension stiffening effect.

3. In the analysis of repeated reversal loading, the analytical model used in this paper gives smaller areas of hysteresis loops than the experimental ones, but the peak displacements of each cycle given by the analysis almost coincide with the experimental ones.

APPENDIX.-REFERENCES

(1) Agrawal,G.L.,Tulin,L.G., and Gerstle,K.H.,''Response of Doubly Reinforced Concrete Beams to Cyclic Loading,'' ACI Journal,Vol.62, No.7,July 1965,pp.823-836.

(2) Bažant,Z.P., and Bhat,P.D.,''Prediction of Hysteresis of Reinforced Concrete Members,'' Journal of the Structural Division,ASCE,Vol.103, No.ST1,Jan. 1977,pp. 153-167.

(3) Bažant,Z.P.,and Kim,S.S.,''Plastic-Fracturing Theory for Concrete,'' Journal of the Engineering Mechanics Division,ASCE,Vol.105,No.EM3, June 1979,pp.407-428.

(4) JCI, ''Collected Experimental Data of Specimens for Verification of Analytical Models,'' Proceedings of JCI 2nd Colloquium on Shear Analysis of RC Structures,JCI-C6,Oct. 1983, pp.43-54.

(5) Kupfer,H.,and Gerstle,K.,''Behavior of Concrete Under Biaxial Stresses,''Journal of the Engineering Mechanics Division,ASCE, Vol.99,No.EM4,Aug.1973,pp.852-866.

(6) Tanigawa,Y.,Kosaka,Y.,and Yamada,K.,''Inelastic Analysis of Reinforced Concrete Based on Endochronic Theory,''Transaction of AIJ,No.326,Apr.1983,pp.78-90.

A FINITE ELEMENT MODEL FOR CRACKED CONCRETE

Hiromichi YOSHIKAWA*, Tada-aki TANABE**

SYNOPSIS

The crack strain method is developed in order to represent the nonlinear behavior of cracked reinforced concrete. In this paper, the mathematical definition of the crack strain and the constitutive equations of the crack strain for two typical cracking modes are discussed. Then the stiffness equation for the finite element incorporating the crack strain is presented, which is formulated by the principle of virtual work. Finally, the numerical examinations for simple cases under proportional loading are demonstrated and compared with experimental results and existing major theories.

It is shown that the use of crack strain concept and its constitutive equation enables us to successfully treat the typical behavior due to cracks such as tension stiffening and aggregate interlock.

1. INTRODUCTION

The formation of tensile cracks in concrete cause physical discontinuities, whose behavior are quite localized and different from that seen in continuum mechanics. Two approaches have been developed to cope with this in finite element analyses; the smeared crack model and the discrete cracking model [8].

Instead of these approaches, following Tanabe's original studies ([1], [2]), we introduce the concept of crack strains which are mathematically defined as the derivative of discontinuous displacements due to cracks. Then the observed total strain is assumed to be decomposed into elastic strain characterized by Hooke's law and the crack strain viewed as an inelastic strain. According to the first study by Tanabe, et al. [1], it is found that the proposed crack strain method contributes to a reasonable formulation for the analysis of a cracked body and can represent various nonlinearities such as tension stiffening, aggregate interlock and crack dilatancy through the crack strains. A finite element formulation incorporating the crack strains is then developed by means of the principle of virtual work.

* Research engineer at the Technical Research Institute of Hazama-Gumi, Ltd. Yono-City, Saitama-Pref., 338 Japan.

** Professor in the Civil Engineering Department, Nagoya University. Furo-cho, Chikusa-ku, Nagoya-City, 464 Japan.

In this paper, the definition of crack strain and its implementation into a finite element analysis are presented and a few numerical examples are presented together with the results calculated using other major theories.

2. MECHANISM OF CRACK SURFACES

It is generally considered that stress crossing a crack is almost continuous, while displacements jump at a crack. Shown in Fig. 1 (b), σ_n^c and τ_{nt}^c are the normal and tangential stresses on the crack surfaces, respectively, and W_n is the relative normal displacement (crack width) and w_t is the relative tangential displacement (slip) between opposite surfaces of a crack. (Subscripts n and t refer to the normal and tangential directions of cracks.) The relation between these four quantities should be determined based on mechanisms of cracked surfaces and a constitutive model relating the four quantities is necessary for a finite element analysis.

When the analytical treatment for cracked reinforced concrete in the plane stress state is investigated, a mechanism on the cracked surfaces is classified into two typical modes. One is a simple state in which a crack continues to open, yet no slip occurs, where $\sigma_n^c \geqq 0$ (tension), $\tau_{nt}^c = W_t = 0$. The other is a rather complicated mode with tangential slip, W_t, bringing about aggregate interlock on the crack with surfaces, producing not only a shear stress, τ_{nt}^c, but also a normal compressive stress, σ_n^c (<0), for which Bažant, et al. ([4], [7]) built up the refined models. We call the first model the separation mode (S-mode) and the second mode the frictional contact mode (F-mode). These are identified as,

$$\text{Separation mode} \qquad \sigma_n^c \geqq 0,\ \tau_{nt}^c = W_t = 0$$
$$\text{Frictional contact mode} \qquad \sigma_n^c < 0,\ |\tau_{nt}^c| > 0,\ |w_t| > 0 \tag{1}$$

The S-mode is observed in a uniaxially tensioned member, in a reinforced concrete panel under pure shear and in reinforced concrete subject to pure bending, undergoing proportional loading. Releasing the stored energy of elastic tensile strain is of essential importance in this mode.

When representing cracked concrete in the F-mode, one has to take into account aggregate interlock and crack dilatancy. It is pointed out by Bažant and Tsubaki [10] that crack dilatancy causes additional tensile stress in the reinforcement crossing the crack. This may be in accordance with the fact that the shear load capacity that is semi-analytically obtained by Yoshikawa, Umehara and Tanabe is smaller than that calculated from the equilibrium condition alone [3].

Instead of treating the discontinuous displacement field directly, we try to make use of the relation of applied stresses and crack strains that is mathematically equivalent to the discontinuous displacement field. This is described subsequently.

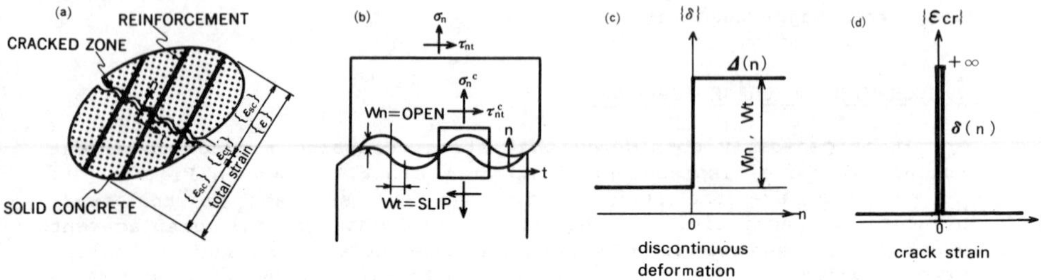

Fig. 1 (a) Cracked Reinforced Concrete, (b) Discontinuities and Stresses of Concrete on the Crack Surfaces, (c) Discontinuous Deformation Defined by Step Function $\boldsymbol{\Delta}(n)$, and (d) Crack Strain Defined by Dirac's Delta Function $\boldsymbol{\delta}(n)$.

3. DEFINITION OF CRACK STRAIN

The total deformation of cracked concrete is considered to be a combined product of discontinuities in cracked zones and elastic/plastic deformations between them. Therefore, the total strain of concrete containing cracks, $\{\varepsilon\}$, is assumed to be expressed by the sum of the two values as follows.

$$\{\varepsilon\} = \{\varepsilon_{sc}\} + \{\varepsilon_{cr}\} \tag{2}$$

in which $\{\varepsilon_{sc}\}$= the elastic strain in the solid concrete and $\{\varepsilon_{cr}\}$ = the crack strain that is regarded as the inelastic strain converted from discontinuous deformation (Fig. 1(a)). In general, the infinitesimal strain, ε_{ij}, is described as the derivative of displacement, u_i, such that

$$\varepsilon_{ij} = \frac{1}{2}(u_{i,j} + u_{j,i}) \tag{3}$$

in tensorial form. Then the crack strains in the two-dimensional field may be defined in a similar manner as,

$$\{\overline{\varepsilon}_{cr}\} = \{\varepsilon_n{}^{cr} \;\; \varepsilon_t{}^{cr} \;\; \gamma_{nt}{}^{cr}\}^t = \{\frac{\partial \delta_n}{\partial n} \;\; \frac{\partial \delta_t}{\partial t} \;\; \frac{\partial \delta_n}{\partial t} + \frac{\partial \delta_t}{\partial n}\}^t \tag{4}$$

In Eq. (4), δ_n and δ_t denote the distributions of discontinuous displacements in the n-t coordinate system and are expressed as

$$\delta_n = W_n \boldsymbol{\Delta}(n), \; \delta_t = W_t \boldsymbol{\Delta}(n) \tag{5}$$

(See Figs. 1(c), (d)). Here, $\boldsymbol{\Delta}(x)$ represent a step function, i.e., $\boldsymbol{\Delta}(x) = 0$ for $x < 0^-$, $\boldsymbol{\Delta}(x) = 1$ for $x > 0^+$. Substituting Eq. (5) into Eq. (4) leads to the following expression for crack strain,

$$\{\overline{\varepsilon}_{cr}\} = \{W_n \;\; 0 \;\; W_t\}^t \boldsymbol{\delta}(n) \tag{6}$$

where $\delta(x)$ is Dirac's delta function; $\delta(x) = 0$ for $|x| > 0$ and $\delta(x) = +\infty$ for $x = 0$, which is related to the step function such that,

$$\frac{d}{dx}(\Delta(x)) = \delta(x), \int_{-\infty}^{+\infty} \delta(x)dx = \int_{-0}^{+0} \delta(x)dx = \Delta(x>0) = 1 \tag{7}$$

Eqs. (4) through (6) show that the crack strain can be mathematically defined by means of the general form of Eq. (3).

From a macroscopic point of view, Eq. (6) can be written as the averaged quantities over the control area S_c, namely,

$$\{\varepsilon_{cr}\} = \frac{1}{S_c}\iint\{\overline{\varepsilon}_{cr}\}dS_c = \frac{L_c}{S_c}\int_{\ell_c}\begin{Bmatrix} W_n \\ 0 \\ W_t \end{Bmatrix}\delta(n)dn = \frac{1}{\ell_c}\begin{Bmatrix} W_n \\ 0 \\ W_t \end{Bmatrix} \tag{8}$$

especially when cracks are densely distributed. In Eq. (8), L_c is a crack length and l_c is a characteristic length of the control area, being calculated from the definition of $l_c = S_c/L_c$ (Glemberg and Samuelsson [11]). The value of l_c may be interpreted as the average crack spacing provided that the cracks are regularly distributed. The above result indicates that the so-obtained averaged crack strains coincide with those which are given from the discontinuous deformations divided by the crack spacing.

The above discussion and formulation are reduced to relations of the crack strain and the total strain, the crack strain and the stress of the uncracked part of concrete, or the crack strain and the strain of the reinforcement across a crack. We assume here the following relation between the crack strains and the stresses of the solid concrete in the incremental form for computational convenience, because the inelastic deformations are supposed to be accompanied by the changes of the applied stresses.

$$\begin{Bmatrix} d\varepsilon_n^{cr} \\ d\varepsilon_t^{cr} \end{Bmatrix} = \begin{bmatrix} S_{nn} & 0 \\ 0 & S_{tt} \end{bmatrix}\begin{Bmatrix} d\sigma_n^c \\ d\sigma_t^c \end{Bmatrix} \tag{9}$$

$$\begin{Bmatrix} d\varepsilon_n^{cr} \\ d\gamma_{nt}^{cr} \end{Bmatrix} = \begin{bmatrix} f_{nn} & f_{nt} \\ f_{tn} & f_{tt} \end{bmatrix}\begin{Bmatrix} d\sigma_n^c / E_c \\ d\tau_{nt}^c / G_c \end{Bmatrix} \tag{10}$$

(Here, E_c and G_c are Young's modulus and shear modulus of the solid concrete, respectively.) Eq. (9) is prepared for S-mode and Eq. (10) for F-mode. In Eq. (9), S_{nn} represents the tension stiffening effect immediately after the crack initiation and S_{tt} the degrading compressive stiffness parallel to the crack direction, which is pointed out by Vecchio and Collins [13], whereas no shear component is considered. As for the crack strains in F-mode, f_{nn} and f_{tt} in Eq. (10) govern the normal and the shear stiffness on the crack surface, respectively and the off-diagonal terms, f_{nt}, f_{tn}, represent the coupling effect between the normal and the shear components. The value of f_{tt} can especially reflect the shear stiffness due to aggregate interlock, and increases as the opposite crack surfaces are less closely in contact. (No crack slip occurs when $f_{tt} = 0$.)

All the terms in [S] of Eq. (9) and [f] of Eq. (10) need to be determined from analytical modeling as well as experimental results. The authors are now developing the present method with this objective.

4. STIFFNESS EQUATION FOR A FINITE ELEMENT INCORPORATING CRACK STRAINS

Since the stresses of reinforcement $\{\sigma_s\}$ respond to the total strains $\{\varepsilon\}$ and the crack strains do not contribute to the stresses of concrete $\{\sigma_c\}$, the constitutive equations for these two materials can be expressed as,

$$\begin{aligned}\{\sigma_c\} &= [D_c]\{\varepsilon_{sc}\} = [D_c](\{\varepsilon\} - \{\varepsilon_{cr}\}) \\ \{\sigma_s\} &= [D_s]\{\varepsilon\}\end{aligned} \tag{11}$$

Here, $[D_c]$ and $[D_s]$ are the corresponding stiffness matrices for the reinforcement and the intact concrete, respectively. (Note that $\{\sigma_s\}$ represents stresses smeared out over an element.)

We now use the principle of virtual work to construct a stiffness equation for a finite element. Consider a finite element under a set of nodal forces $\{F\}$ that are associated with nodal displacements $\{u\}$. Let the virtual displacement be $\{u^*\}$ and the corresponding virtual strains be $\{\varepsilon^*\}$. Equating the work done externally to that done internally gives,

$$\{u^*\}^t\{F\} = \int_V \{\varepsilon^*\}^t\{\sigma\}dV \tag{12}$$

Making use of the strain-displacement relation such as $\{\varepsilon\} = [B]\{u\}$ and the stress-strain relations in Eqs. (11), Eq. (12) becomes the following stiffness equation.

$$\{F\} = \int_V [B]^t([D_s] + [D_c])[B]dV\{u\} - \int_V [B]^t[D_c]\{\varepsilon_{cr}\}dV \tag{13}$$

Moreover, if crack strains are expressed by the total strains such that,

$$\{\varepsilon_{cr}\} = [\xi]\{\varepsilon\} = [\xi][B]\{u\} \tag{14}$$

, then this results in,

$$\begin{aligned}\{F\} &= \int_V [B]^t[D_c][B]dV\{u\} - \int_V [B]^t[D_c][\xi][B]\{u\}dV \\ &= \int_V [B]^t[D_c]([I_0] - [\xi])[B]dV\{u\} = \int_V [B]^t[D_{cr}][B]dV\{u\}\end{aligned} \tag{15-1}$$

$$[D_{cr}] = [D_c]([I_0] - [\xi]),\ [I_0] = \text{Unit matrix} \tag{15-2}$$

As an example of Eq. (14), the following crack strains are possible,

$$\{\varepsilon_{cr}\} = \begin{Bmatrix} \varepsilon_n^{cr} \\ \varepsilon_t^{cr} \\ \gamma_{nt}^{cr} \end{Bmatrix} = \begin{bmatrix} 1 & \nu_c & 0 \\ 0 & 0 & 0 \\ 0 & 0 & \beta \end{bmatrix} \begin{Bmatrix} \varepsilon_n \\ \varepsilon_t \\ \gamma_{nt} \end{Bmatrix} \tag{16}$$

assuming that the elastic strain perpendicular to the crack surfaces is released to be zero and some portion of the shear stress is transmitted across the cracks. By substituting Eq. (16) to Eq. (15-2), $[D_{cr}]$ becomes,

$$[D_{cr}] = \begin{bmatrix} 0 & 0 & 0 \\ 0 & E_c & 0 \\ 0 & 0 & (1-\beta)G_c \end{bmatrix} \tag{17}$$

It is found that the so-obtained stiffness matrix is identical to the so-called smeared cracking model [9], which has long served a number of investigators, partly because of its simplicity.

In nonlinear finite element computations, Eq. (13) is treated as the initial strain problem where equivalent nodal loads are converted from the inelastic strains, while Eq. (15) corresponds to the variable stiffness method in which the stiffness $[D_{cr}]$ is incrementally updated according to the crack strains. It is expected that both of these finite element techniques can be easily incorporated into the existing finite element codes with the incrementally linearized formulation.

5. NUMERICAL EXAMPLES AND COMPARISONS WITH OTHER MODELS

In the present chapter, parametric studies are shown for three cases: a reinforced concrete member in uniaxial tension, an orthogonally reinforced concrete panel in pure shear, and a direct shear model. The results are compared to values calculated by other theories. A single four-noded isoparametric finite element is employed and loading is carried out proportionally and monotonically. The stiffness equation (13), modified in the incremental form, is used for the nonlinear finite element computations.

① A Reinforced Concrete Member in Uniaxial Tension (S-mode)

The first model of the numerical calculations is a reinforced concrete uniaxial member subjected to a single tension force, which is in the separation mode. We apply Eq. (9), whose first diagonal term, S_{nn}, is the only one considered here and defined as,

$$d\varepsilon_n^{cr} = S_{nn} d\sigma_n^{c}, \quad S_{nn} = m\varepsilon_n^{sc} \tag{18}$$

in which ε_n^{sc} is the elastic tensile strain of the solid concrete normal to the cracks and m, named as a releasing factor, controls the rate of the released elastic tensile strain. Although the releasing factor, m, is supposed to be influenced by the level of the applied stress, the bond characteristics and the properties of concrete, we assume it to be simply constant. The calculations are then carried out with the varying material parameters, m and p, for each example, and results are shown in Figs. 2. (p is the reinforcement ratio.)

The relationships between the applied stress σ_n and the total strain ε_n are drawn for different reinforcement ratios p in Fig. 2 (a) and for the changing releasing factor m in Fig. 2 (b). Those curves are interpreted as the deformational response of a composite material which is composed of the reinforcement and the cracked concrete. Fig. 2 (b) exhibits that the curves of the total strain starts to deviate from the purely elastic line, $E_c + pE_s$, immediately after the crack initiation, and then approaches the stiffness of the reinforcement alone, pE_s, and that the rate of the degrading stiffness of the cracked member is well controlled by means of the releasing factor m.

Fig. 2 (c) shows the relationships between the elastic tensile stress and the total strain of the concrete for the different reinforcement ratios, where the strain-softening, a decline of stress at increasing strain, is observed. In the figure, the values calculated from the formula,

$$\frac{1}{E^*} = \frac{1-\lambda}{pE_s} + \frac{\lambda}{E_c+pE_s}, \tag{19}$$

(λ: Contribution ratio of concrete to stiffness, $0 \leqq \lambda \leqq 1$)

proposed by Adham and Isenberg [6] and the tension stiffening method are also shown. The tension stiffening method, originally proposed by Scanlon [8], is a traditional technique in finite element analyses to express the declining stress of the cracked concrete. Herein, we use the following equation for a descending branch of cracked concrete as tension stiffening.

$$\sigma_n^c = f_t \eta^{(\varepsilon-\varepsilon_o)/0.002}, \quad (0 \leqq \eta \leqq 1) \tag{20}$$

in which f_t is the tensile strength of the concrete, ε_o is the corresponding strain, and η is a constant, which may be influenced by the chosen element size. Fig. 3 compares the crack strain method and the tension stiffening method.

As can be seen in Fig. 2 (c), the curves by the crack strain method in the strain-softening region descend more quickly with the increase of the amount of reinforcement, and Adham and Isenberg's model provides the same tendency. On the other hand, the values calculated by the tension stiffening method has no relation with the reinforcement ratio because the stress-strain relation is prescribed independent of the material properties and the reinforcement ratio.

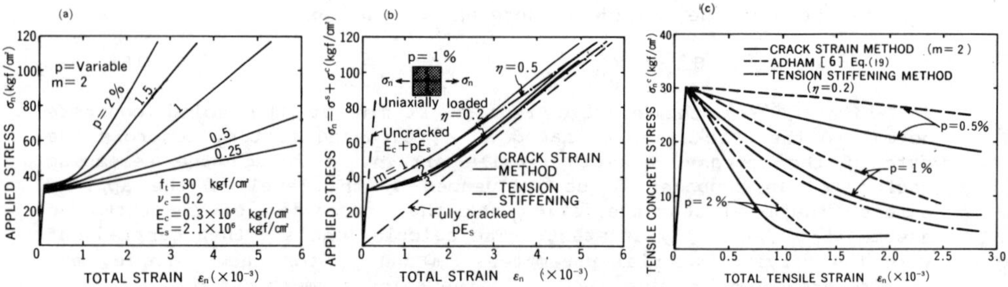

Fig. 2 Numerical Calculations for a Uniaxially Tensioned Member, (a), (b) the Total Strain the Applied Stress Relation, (c) the Total Strain∿the Concrete Stress Relation.

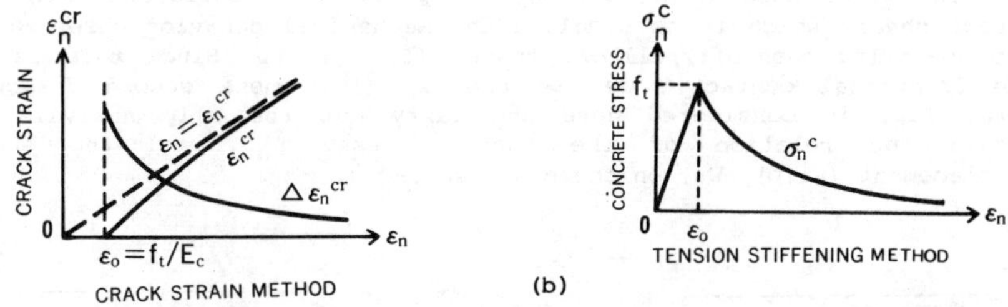

Fig. 3 A Comparison of Two Methods for the Gradual Stress-Declining

There has been discussion that the strain-softening is not a property pertaining to material but stems from a structural characteristic [12], and may depend on the rigidity of a testing apparatus. The numerical results in Fig. 2 (c) show that the strain-softening appears to be influenced by the reinforcement ratio. These may lend support to this discussion.

② A Reinforced Concrete Panel in Pure Shear (S-mode)

The second example is an orthogonally reinforced concrete panel subjected to pure shear stress, which is again in the separation mode and Eq. (18) is applied. The shear strain responses in Fig. 4 (a) exhibit the gradual transition from the uncracked state, G_C, toward the fully cracked state, $(0.5pE_SE_C)/(2pE_S+E_C)$, depending on the releasing factor m. Fig. 4 (b) shows the stresses of concrete, σ_n^C, σ_t^C, and the stresses of reinforcement, σ_x^s, σ_y^s.

This behavior, shown in Fig. 4, is similar to the previous case and other cases deforming in the S-mode.

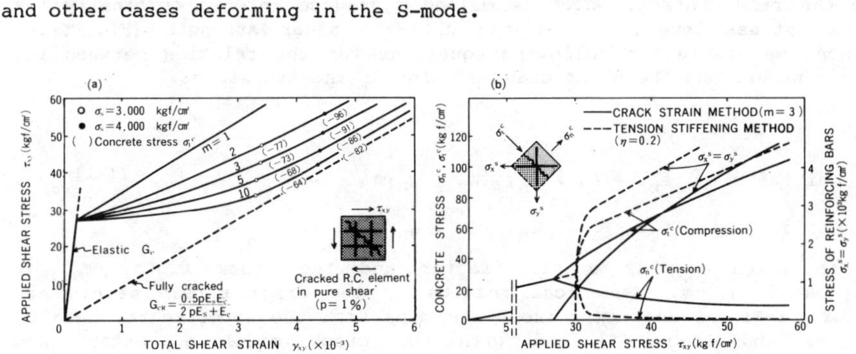

Fig. 4 Numerical Calculations for a Reinforced Concrete Panel in Pure Shear

③ A Precracked Concrete Element in Direct Shear (F-mode)

The final case is a horizontally precracked concrete element in direct shear, which is to simulate the mechanical behavior observed in the so-called push-off/pull-off tests ([5], [15]). Since this is in the frictional contact mode, we use Eq. (10) whose second diagonal term, f_{tt}, is considered here and carry out the slip analysis to obtain the relation of the shear stress, τ_{nt}, and the shear displacement (slip), W_t, on the crack surfaces.

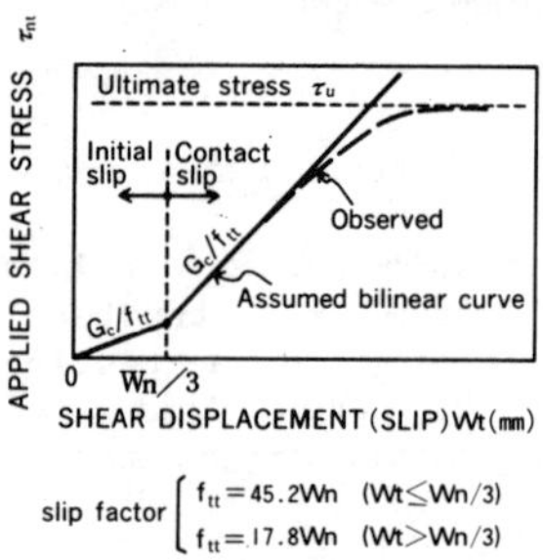

Fig. 5 Simplified Model for Shear Slip ($W_t \sim \tau_{nt}$ Relation)

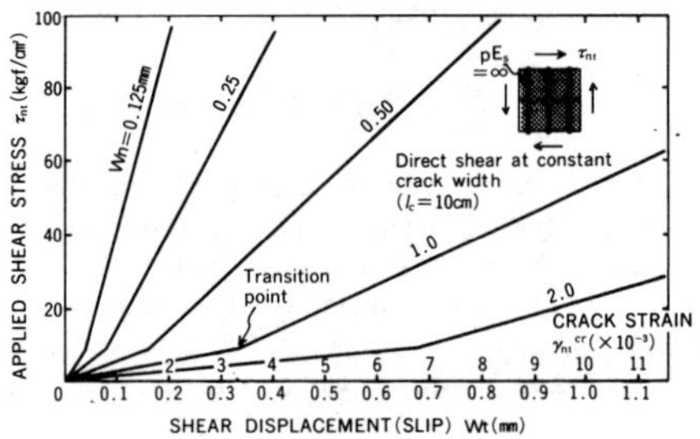

Fig. 6 Numerical Calculations for a Direct Shear at Constant Crack Width (Crack Strain Method)

According to experimental observations, at the initial shear load free slippage occurs on the crack plane that is not in close contact; the further application of the shear force makes the cracks stiffer due to the firm contact, which is called aggregate interlock; finally the shear stress levels off at the ultimate shear strength (Fig. 5). Hence, we assume the following equations for the relation between the shear stress and the shear crack strain for the two stages.

$$d\gamma_{nt}^{cr} = f_{tt} d\tau_{nt}^{c}/G_c$$

$$\text{Initial slip} \quad f_{tt} = 45.2\, W_n, \ (W_t \leqq W_n/3) \tag{21}$$

$$\text{Contact slip} \quad f_{tt} = 17.8\, W_n, \ (W_t > W_n/3)$$

in which f_{tt} is a slip factor and the crack width, W_n, is expressed in mm. Numerical results by the crack strain method are shown in Fig. 6 and comparisons are made with the experimental data by Paulay and Loeber [5] and with the crack model by Bažant and Gambarova [4] in Figs. 7 (a) and (b), respectively. The calculated values are in relatively good agreement with the experiments and Bažant's analytical model. Since Eq. (21) is assumed to be merely bilinear, the example cannot follow the realistic behavior in the final stage represented by Bažant's model. It is also possible to express the increase of the crack opening, W_n, accompanied by the shear displacement, W_t, (crack dilatancy), once the off-diagonal terms in [f] are considered.

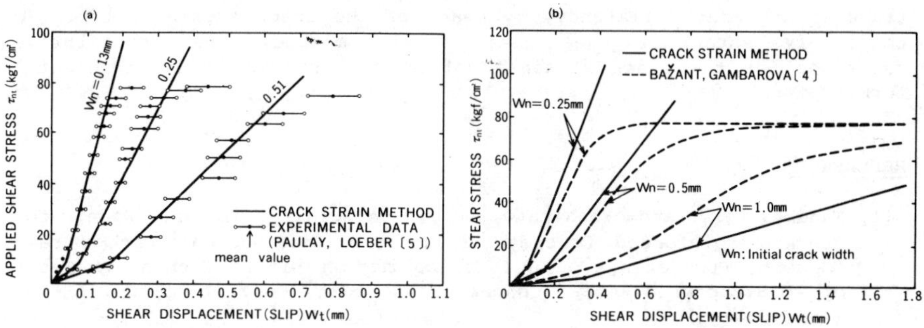

Fig. 7 Comparisons of Crack Strain Method with, (a) Paulay and Loeber's Test Data [5], and (b) Bažant's Analytical Model [4].

Another feature of the crack strain method is that the crack strain can be related to the empirical formulae for the interface shear stiffness per unit area, K_{IST}, proposed by some researchers [8]. Since K_{IST} is defined as $K_{IST} \equiv \tau_{nt}{}^{c}/W_t$, one can obtain the following equation,

$$K_{IST} \equiv \frac{\tau_{nt}{}^{c}}{W_t} = \frac{\tau_{nt}{}^{c}}{\ell_c \gamma_{nt}{}^{cr}} = \frac{G_c}{\ell_c f_{tt}} \tag{22}$$

For example, assuming that $G_c = 0.125\ 10^6$ kgf/cm^2 and $l_c = 10$ cm, the above equation results in,

$$K_{IST} = 28 \sim 80/W_n, \ (W_n \text{ in cm}) \tag{23}$$

which resembles the experimental formula, $K_{IST} = 36/W_n$, proposed by Aoyagi and Yamada [14].

From the aforementioned numerical examinations, it can be said that the crack strain method provides a technique to simulate most of the essential behavior of the cracked reinforced concrete and that the typical nonlinearities can be represented by means of only two terms, S_{nn} and f_{tt}. Further investigations are necessary to fully establish all the terms in Eqs. (9) and (10), these are, in other words, the "fourth" constitutive equations to describe the stress field for the cracked material.

6. CONCLUSION

To make a more accurate prediction of the mechanical behavior of cracked reinforced concrete, fracture mechanics and the concept of a discontinuous field should be introduced. However, these methods may require a great deal of effort and high computational costs if one applies them to the analysis of large-scale structures such as nuclear power facilities. One advantage of the authors' proposed procedure is that the crack strain method can readily be implemented into ordinary finite element codes, requiring no additional isolated interelements like bond links. Moreover, the proposed method can systematically represent essential behavior, such as tension stiffening, shear transfer and crack dilatancy, by means of the crack strain vector. To completely establish the crack strain method, the constitutive relations for the crack strains must be constructed in an incremental form.

REFERENCES

[1] Tanabe, T., and Yoshikawa, H.,: A Finite Element Model for Cracked Reinforced Concrete Structures Introducing Crack Strain Concept, Proceedings of JCI Colloquium on Finite Element Analysis of RC Structures, Japan Concrete Institute, Dec., 1984, pp. 65-72. (in Japanese)

[2] Tanabe, T., Iwata, M., and Yoshikawa, H.,: Introduction of Crack Strain to the Analysis of a Discontinuous Body, 5th International Conference on Numerical Methods in Geomechanics, Nagoya, April, 1985.

[3] Yoshikawa, H., Umehara, H., and Tanabe, T.,: Comprehensive Evaluation of Major Theories on the Ultimate Strength of Reinforced Concrete Panels Subjected to In-Plane Shear Forces And the Proposed Semi-Analytical Method for Estimation of Ultimate Strength, Concrete Library of JSCE No. 4, Dec., 1984, pp. 261-281.

[4] Bažant, Z.P., and Gambarova, P.,: Rough Cracks in Reinforced Concrete, Journal of the Structural Division, ASCE, Vol. 106, No. ST4, Proc. Paper 15330, April, 1980, pp. 819-842.

[5] Paulay, T., and Loeber, P.J.,: Shear Transfer by Aggregate Interlock, SP42-1, American Concrete Institute, 1974, pp. 1-15.

[6] Adham, S., Bhaumik, A., and Isenberg, J.,: Reinforced Concrete Constitutive Relations, AFWL-TR-74-72, Feb., 1975.

[7] Bažant, Z.P., and Tsubaki, T.,: Slip-Dilatancy Model for Cracked Reinforced Concrete, Journal of the Structural Division, ASCE, Vol. 106, No. ST 9, Proc. Paper 15704, Sept., 1980, pp. 1947-1966.

[8] The Task Committee on Finite Element Analysis of Reinforced Concrete Structures of the Structural Division Committee on Concrete and Masonry Structures: State-of-the-Art Report on Finite Element Analysis of Reinforced Concrete, ASCE, 1981.

[9] Chen, W.F.,: Plasticity in Reinforced Concrete, McGraw-Hill, 1982.

[10] Bažant, Z.P., and Tsubaki, T.,: Concrete Reinforcing Nets; Optimum Slip-Free Limit Design, Journal of the Structural Division, ASCE, Vol. 105, No. ST1, Proc. paper 14344, Feb., 1979, pp. 327-346.

[11] Glemberg, R., and Samuelsson, A.,: A General Constitutive Model for Concrete Structures, Nordic Concrete Research, pp. 29-61.

[12] Kawamoto, T., and Ichikawa, Y.,: Mechanics of Soft Rocks and Numerical Analysis, Journal of the Society of Materials Science Japan, Vol. 31, No. 347, Aug., pp. 770-779. (in Japanese)

[13] Vecchio, F., and Collins, M.P.,: Stress-Strain Characteristics of Reinforced Concrete in Pure Shear, Final Report of IABSE Colloquium on Advanced Mechanics of Reinforced Concrete, Delft, June, 1981, pp. 211-255.

[14] Aoyagi, Y., and Yamada, K.,: Strength and Deformation Characteristics of Reinforced Concrete Shell Elements Subjected to In-Plane Forces, Proceedings of the Japan Society of Civil Engineers, No. 331, March, 1983, pp 167-180. (in Japanese)

[15] Mattock, A.H., and Hawkins, N.M.,: Shear Transfer in Reinforced Concrete - Recent Research, Journal of Prestressed Concrete Institute, Vol. 17, No. 2, March-April, 1972, pp. 55-75.

EXPERIMENTAL AND CONSTITUTIVE ASPECTS OF CONCRETE FAILURE

By Kaspar Willam[1], M.ASCE, Bryan Hurlbut[2], and Stein Sture[3], M.ASCE

ABSTRACT

The failure analyses of concrete structures are plagued (i) by the constitutive difficulties in describing the transition from brittle to ductile fracture of concrete under increasing confinement, and (ii) by the numerical intricacies related to structural response predictions of contained material instabilities due to strain-softening and/or due to lack of normality in conjunction with non-associated plastic flow concepts.

This paper presents novel experimental observations of post-critical response phenomena when cylindrical concrete specimens are loaded under displacement control in the form of direct tension and triaxial compression experiments. The investigation focuses on the problem whether or not "proper" material properties can be extracted from such experiments which characterize not only stiffness and strength behavior but also the ductility in the post-peak response regime. In this context a fracture energy based strain-softening plasticity formulation will be developed in terms of the fundamental Mode I type cracking which accounts for the transition from brittle to ductile fracture under increasing triaxial confinement.

INTRODUCTION

The failure behavior of cementitious materials such as mortar and concrete can be broadly classified into (i) tensile cleavage in the form of Mode I type cracking, and (ii) decohesive faulting in the form of Mode II type frictional slip. The tensile fracture mode typically leads to a sudden discontinuity at peak strength and a very brittle appearance of the load-deformation response when the direct tension experiment is considered. Recent experiments on notched concrete specimens by Reinhardt [15], Gopalaratnam and Shah [6] as well as the direct tension experiments by Hurlbut [9] on unnotched specimens all indicate that little hardening precedes the onset of cleavage failure which results in

---

[1] Professor of Civil Engineering, Campus Box 428, University of Colorado, Boulder, CO 80309, U.S.A.

[2] Staff Consultant, Applied Mechanics Incorporated, 12600 West Colfax Ave., Suite C360, Lakewood, CO 80215, U.S.A.

[3] Associate Professor of Civil Engineering, Campus Box 428, University of Colorado, Boulder, CO 80309, U.S.A.

the separation of the test specimen into two competent parts. In contrast, shear fracture exhibits far more ductile behavior, whereby hardening in the pre-peak regime is followed by a gradual softening regime during the process of progressive decohesion down to the residual strength level which is characterized by granular friction only. The triaxial compression experiments on concrete by van Mier [13] and Hurlbut [9] illustrate clearly the increase of ductility with increasing confining pressure. In fact the splitting tension mode of failure in uniaxial compression transforms gradually into a faulting mode of failure with increasing confinement and ultimately diffuses into distributed damage throughout the entire specimen. As a result under sufficiently high confinement there is a third response regime which does not exhibit any softening whatsoever and which is governed by compactive hardening due to pore collapse and related inelastic processes of the composite microstructure of concrete.

Since the failure of plain and reinforced concrete structures depends to a large extent on the ductility and thus the governing post-peak response characteristics of the concrete the main thrust of this investigation is directed towards the softening behavior. We will focus on recent experimental evidence obtained at the University of Colorado within a series of displacement controlled post-critical experiments on low strength concrete specimens featuring shear dilatancy, hardening as well as softening in tension and shear. There have been various attempts to correlate these global phenomena to different processes in the microstructure such as debonding between the cement matrix and the aggregate, progressive microcracking within the cement matrix and ultimately fracture of the aggregate. Quite noteworthy is the recent work by Ortiz [14] who used the mixture theory of interacting continua to model these evolutionary processes within each phase of the composite. This development is closely related to the continuous damage formulation of Mazars [12] who expressed Mode I type cracking and Mode II type shear fracture in terms of maximum tensile strain conditions.

In the following we will concentrate on the fundamental issues of strain-softening, the transition from brittle to ductile fracture, and the decrease of shear dilatancy with increasing confinement. Special attention will be paid to the extraction of fracture properties from the post-critical response regime which are independent of a particular test geometry and the given set of boundary conditions. This task is complicated by the fact that the post-critical response behavior is accompanied by severe strain localization and actual discontinuities within the test specimen. For this purpose we will expand the composite fracture formulation which has previously been developed by the authors [19] in order to incorporate the fracture energy release rate concept within the strain-softening formulation of an equivalent elastic plastic continuum.

POST-PEAK EXPERIMENTS

In conjunction with a substantial research effort on finite elements and localized failure in cementitious materials two series of experiments were recently completed which involved stable post-critical loading of cylindrical concrete specimens in direct tension and triaxial compression, (NX size samples with D = 2.125 in. (54 mm) diameter and H = 4.25 in. (108 mm) height). The concrete mix was based on a water-cement ratio of W/C = 0.833, a mix grading of C:S:G = 1:3.16:3.19 and a maximum aggregate size of 3/8 in. (9.5 mm) which in turn gave a uniaxial compressive strength of f'_c = 3.2 ksi (22.1 MPa) at 28 days. The concrete was cast into 1 x 3 x 3 ft. (0.3 x 1.0 x 1.0 m) slabs which were kept in a fog room until testing except for the period for preparation of the specimens. The cylindrical concrete samples were cored from the slab and milled with a diamond grinding wheel to the prescribed height with a tolerance of $\Delta H = \pm 0.02$ in. (± 0.5 mm). The exposed voids at the cylindrical surface of each specimen were filled with putty in order to prevent intrusion of the elastomeric membrane into the concrete voids in the case of high confining pressures.

Two series of post-peak experiments were carried out with the aid of a MTS loading system consisting of a standard 110 kip (489 kN) loading frame with servo control and function generator units. These were comprised of a series of direct tension and a series of triaxial compression experiments which will be described below.

Direct Tension Experiment

In order to clarify our understanding of the tensile fracture mechanisms the following objectives were pursued with the aid of direct tension experiments:

(i) Uniqueness of the load-displacement response in the post-peak regime,
(ii) Progressive damage and stiffness degradation during loading and unloading cycles, and
(iii) Size effects related to different height specimens.

A special technique was developed by Hurlbut [9] to stabilize the post-critical response behavior of the cylindrical tension specimens. The basic difficulty of maintaining stable post-peak control was resolved by incorporating two axial Linear Virtual Displacement Transducers (LVDT) which were attached to the loading platen as illustrated in Fig. 1. In contrast to the recent tension tests by Reinhardt [15] and Shah [6] unnotched specimens were utilized in order to avoid any bias towards fracture mechanics by introducing a well defined macrofect in the sample to start with. Therefore the actual location of tensile cracking was undetermined and the stroke of the loading actuator had to be controlled by the two axial LVDT's in order to monitor the post-peak response behavior. In return a premature fracture mechanics interpretation of the tensile cracking process was avoided and the question of a volume versus surface dominated failure process could be explored.

It is important to keep in mind that the tensile test specimen was placed into the loading frame 12 hrs before actual testing. The cylindrical sample was attached to the specially prepared solid loading platens by means of a thin layer of a high strength structural epoxy cured during the 12-hr. period under a slight compressive load of P_z = -100 lbs (-445N). Clearly the epoxy layer provided some lateral restraint and was therefore responsible for a non-uniform stress state near the end section. In fact, the stress concentration at the corner of the specimen-platen interface led in some cases to premature bond failure or delamination.

Fig. 2 depicts the axial load-extension response of a typical direct tension experiment conducted at a displacement rate of 1.67×10^{-6} in/sec. (4.24×10^{-8} m/sec). We observe a nearly linear ascending branch up to the peak load. At this point the axial tension suddenly decreased without noticeable formation of a crack. Upon close inspection a single transverse hairline crack appeared around the periphery of the sample at an overall axial displacement of 1.5×10^{-3} in. (0.04 mm) when the residual tensile strength diminished to near zero load level. At that time the specimen had separated into two competent parts which remained entirely intact. Upon continuous loading the increasing displacements reflected simply further crack opening accompanied by rigid body movement of the two separated portions of the specimen.

Considering the reproducibility of the post-critical response branch under monotonic loading the question arises if the residual tensile strength can be recovered during loading-unloading cycles and if the associated stiffness is affected by the degrading strength. Loading-unloading experiments were performed in order to explore the residual tensile strength and the associated tangent stiffness. Fig. 3 depicts the resulting load-displacement response when the stroke is reversed to zero displacement and subsequently increased again. The plot illustrates clearly that tensile bonds still remain intact in the softening branch and that the residual tensile strength is fully recovered during reloading. In fact the post-peak load-displacement response curve approaches that of the monotonic experiment very much along the line of strain-softening elastic-plastic behavior with substantial permanent deformations. We observe a slight decrease of elastic stiffness in the first cycle and a substantial reduction when the tensile crack has fully developed very much along the results on notched tension specimens reported recently by Reinhardt [15]. There is some elastic-inelastic coupling, however, it is not nearly so pronounced as to fully support the progressive fracturing theory of Dougill [5] or the continuous damage theory of Mazars [12], Janson and Hult [10], Krajinovic and Fonseka [11] which resort to secant formulations and thus fully reversible deformations upon unloading.

In order to probe the fracture mechanics hypothesis of a single macrodefect an additional monotonic tension test was carried out with a specimen of half the height and the same cross-section. In the case of a constant fracture energy release rate the area under the P-u diagram (actually the nominal axial stress-extension plot) should remain constant, assuming that the tensile strength is not affected by the two different volumes and their different probability of failure. On the other hand if the concept of a continuous distribution of microdefects were to be adopted along the line of the smeared cracking approach, then the area under the P-u diagram, i.e., the strain energy dissipation during the transverse crack process should be reduced by half. Fig. 4

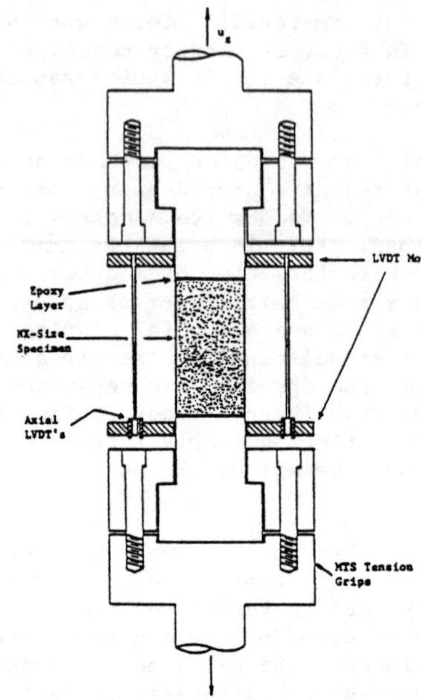

Figure 1 Direct Tension Apparatus

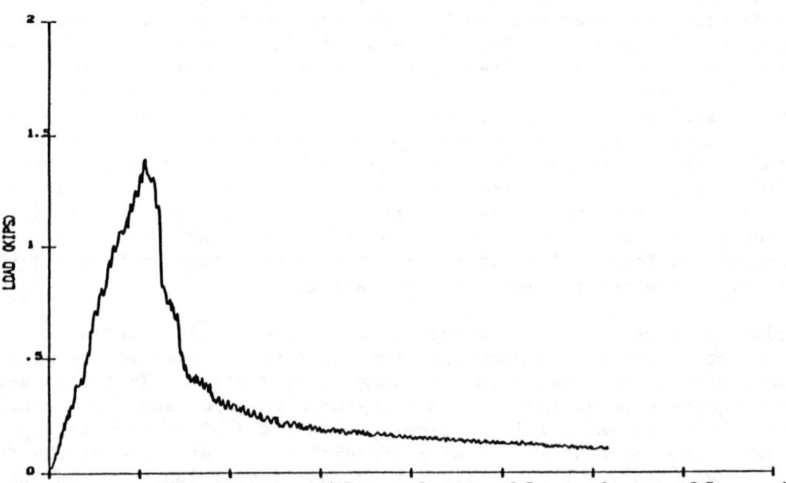

Figure 2 Direct Tension Test of Full Size NX-Specimen

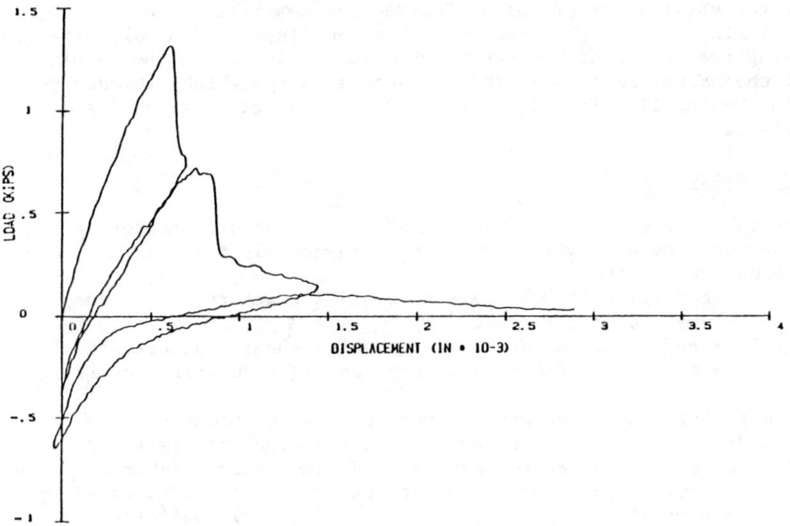

Figure 3 Cyclic Tension Test of Full Size NX-Specimen

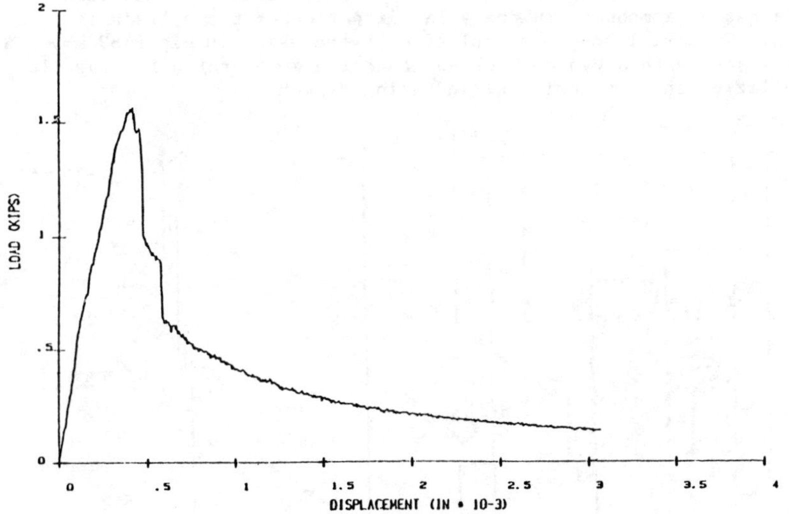

Figure 4 Monotonic Tension Test of Half Size NX-Specimen

illustrates the results of the half-size specimen test. The plot clearly indicates that the area under the P-u diagram is hardly affected by halving the height of the tension specimen. Therefore the results support the postulate of a constant fracture energy which ultimately controls the localization of distributed microdefects into a single macrodefect.

Triaxial Compression Experiments

In order to clarify our understanding of shear fracture the following objectives were pursued with the aid of axisymmetric triaxial compression experiments:

(i) Transition from brittle to ductile fracture at various levels of confinement,

(ii) Shear dilatancy during progressive shear fracture, and

(iii) Stiffness degradation during loading - unloading cycles.

Conventional triaxial compression experiments were conducted in a Hoek cell device [8] which had been modified at the University of Colorado, Boulder, in order to measure axial and lateral deformations of cylindrical NX-size specimens. The Hoek cell which is illustrated in Fig. 5 has a capacity of up to σ_r = 10 ksi (68.9 MPa) confining pressure. It is instrumented for measuring lateral deformations at the mid-height with the aid of six strain gaged cantilevers spaced at 60° intervals around the circumference. The cell pressure is provided by hydraulic fluid and is transmitted to the specimen by a flexible polyurethane membrane of 0.05 in. (1.3 mm) thickness. The confining pressure has no component in the axial direction of the cylindrical specimen. The axial load was applied with the same 110 kip (489 kN) MTS loading system with servo control and function generator units capable of stabilizing the post-critical softening branch.

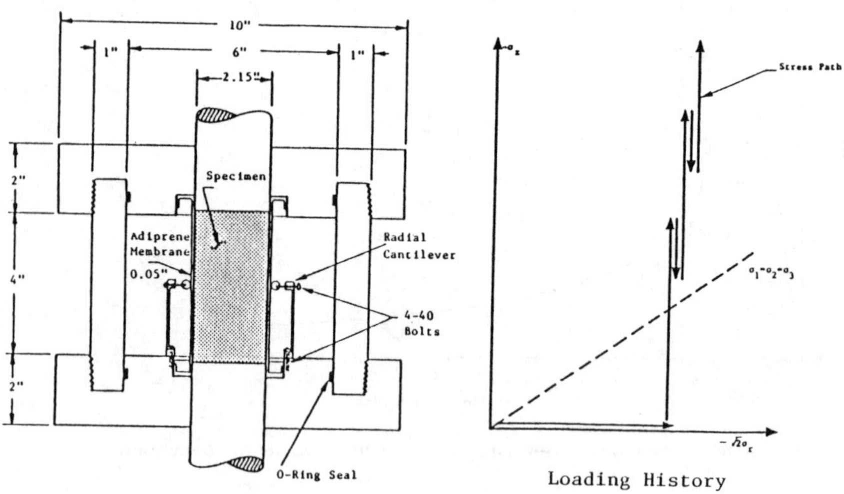

Figure 5 Triaxial Compression Apparatus

All the triaxial compression tests were conducted by pressurizing first the cell to the prescribed confining pressure which was held constant while the axial load was applied under stroke control at a displacement rate of 1.67 x 10^{-4} in/sec (4.24 x 10^{-6} m/sec).

Fig. 6 depicts the axial load-displacement behavior at four different confining levels σ_r = 100, 500, 1000 and 2000 psi (0.69, 3.45, 6.89 and 13.79 MPa). Again experimental results are plotted in the form of global load-displacement results rather than in the form of stress and strain measures since localization of fracture takes place in the softening regime which requires a proper homogenization concept. The four plots clearly demonstrate the increase of ductility with increasing confinement which suppresses the softening tendencies. The partial unloading cycles indicate very little hysteretic energy dissipation. The linear elastic unloading-reloading behavior hardly exhibits any degradation of stiffness. This suggests that the elastic-plastic coupling remains negligible and conjectures little damage accumulation of the elastic material texture not only in the pre- but also in the post-peak regime. The resulting failure modes showed a distinct transition from splitting tension under low confinement, over shear faulting under intermediate confinement, to diffuse damage at high confinement when no softening was observed altogether.

The dilatation characteristics are shown in Fig. 7 in the form of lateral versus axial displacement response curves. Again the lateral expansion diminishes rapidly with increasing confinement pressure. This indicates that volumetric dilatation accompanies strain-softening and that the transition from brittle to ductile fracture is related to the transition from volume dilation to volume compaction.

If the axial and radial displacements in Fig. 7 were converted into nominal strains their ratio would remain fairly constant in the softening regime and would decrease rapidly with increasing confinement. It was therefore felt that the inelastic dilatancy ratio $\alpha = \dot{\varepsilon}_r^f / \dot{\varepsilon}_z^f$ could provide a proper fracture characteristics which would, however, depend on the confining stress. The post-peak response at low confinement is accompanied by large inelastic volume dilation which decreases rapidly under increasing confining stress and results in inelastic compaction rather than dilation.

The triaxial strength results are summarized in Fig. 8 with a plot of the test data in the Rendulic plane where the data are shown on the compressive meridian. The diagram depicts the maximum strength envelope in conjunction with the residual strength data. The residual strength diminishes to zero at zero confinement and approaches rapidly the maximum strength envelope with increasing confinement at the transition point from brittle to ductile fracture when softening is suppressed altogether. Since the entire test program was focused on the softening behavior under low confinement no attempt was made to extract the hardening behavior which had been explored in a separate research effort by Stankowski and Gerstle [17].

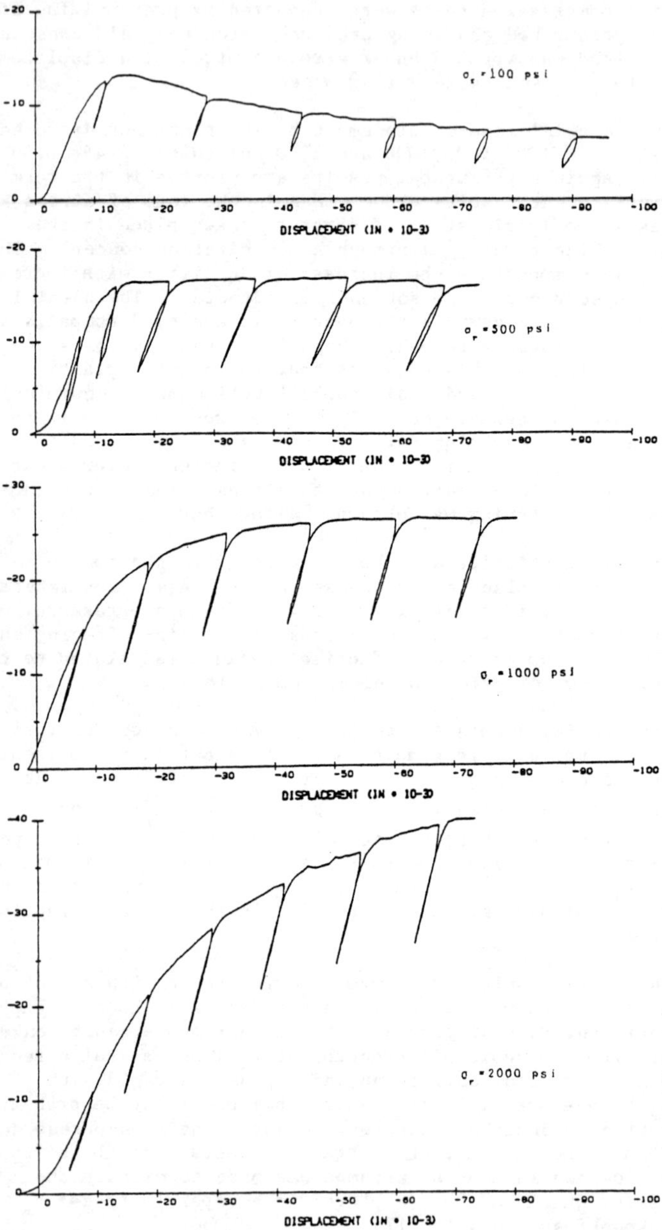

Figure 6 Conventional Triaxial Compression Tests with Partial Unloading

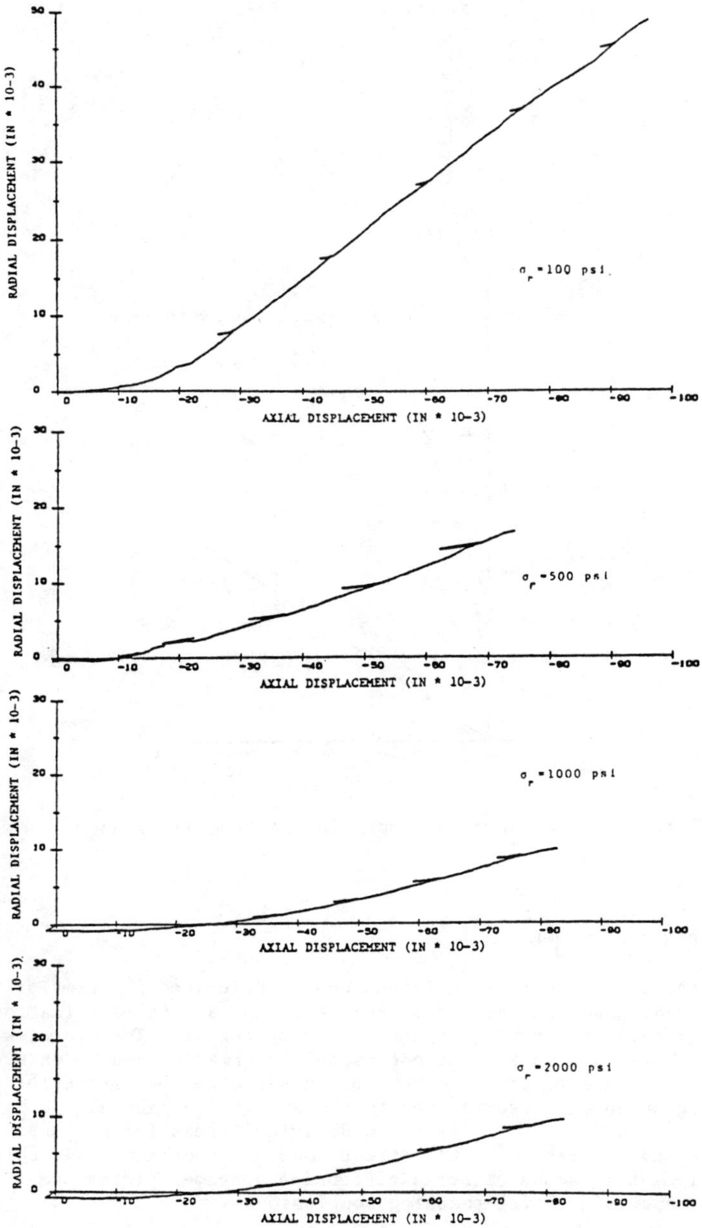

Figure 7 Dilatation Response of Conventional Triaxial Compression Tests with Partial Unloading

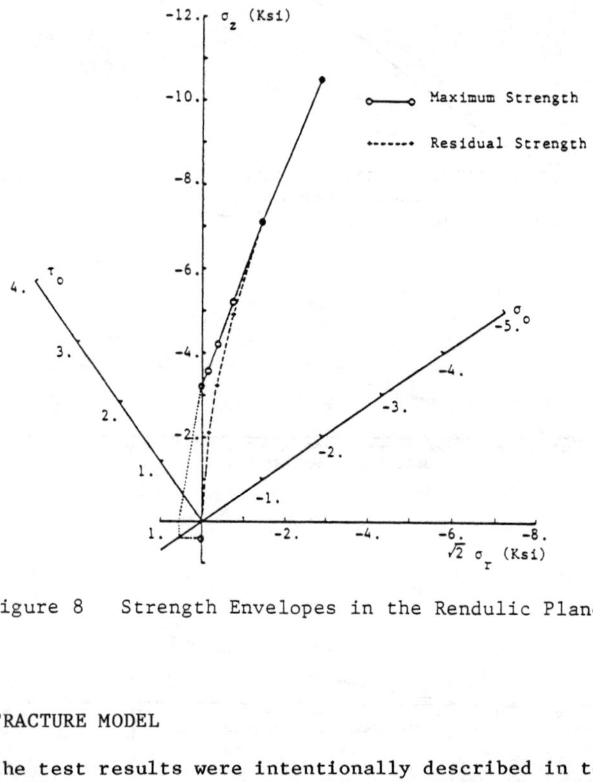

Figure 8 Strength Envelopes in the Rendulic Plane

COMPOSITE FRACTURE MODEL

Up to now the test results were intentionally described in terms of overall load-displacement diagrams because of the strain localization during progressive fracturing in the softening regime. Therefore, we have to ask ourselves whether or not useful information can be extracted from the softening data which is not only generic to the particular test specimen but which can also be used in the analysis of actual structures. In order to clarify the underlying issues let us consider the "simple" configuration of the direct tension experiment. In this case the elementary study of an axial tension specimen yields some surprising results with far-reaching conclusions.

Consider a uniaxial tension specimen of height H and cross-section A with the uniaxial tensile strength σ_t. The incremental stiffness relation follows the elementary axial force-deformation relation

$$\dot{\sigma}_z = \frac{\dot{P}}{A} = \frac{E_T}{H} \dot{u}_z \tag{1}$$

where E_T denotes the tangent modulus, $\dot{\sigma}_z$ the nominal stress rate and $\dot{u}_z$ the axial rate of extension. If we assume a simple bilinear material law with $E_T = E$ for hardening and $E_T = -E$ for softening, uniform degradation of strength without localization will result in the bilinear response depicted in the left plot of Fig. 9 for monotonically increasing extension. Moreover, the total strain energy dissipation for tensile fracture is

$$U_f = \frac{\sigma_t^2}{E} AH \tag{2}$$

which is thus proportional to the height H of the uniaxial tension specimen. This implies that in the limit the energy dissipation would diminish to zero as the height of the specimen were reduced to zero, a fact which has been rightly criticized by Bazant et al. at several occasions [1,2,3]. On the other hand, if we would adopt the notion of a constant fracture energy release rate for the formation of a single crack, then the energy expended for generating a crack of unit area is

$$G_f^I = \frac{dU_f}{dA} = \frac{\sigma_t^2}{E} H \tag{3}$$

which should remain constant irrespective of the size of the specimen. This could be simply accomplished by adjusting the softening slope according to the size of the softening zone, i.e. $E_T = -E/3$ for half the specimen height, in order to maintain a constant fracture energy release rate as indicated in the right plot of Fig. 9.

In order to probe this notion of constant versus variable fracture energies involved, let us compare the two direct tension experiments illustrated in Figs. 2 and 4. Converting the axial load into nominal axial stress, the area under the σ_t-u_f diagram can be readily evaluated, and yields the following average values for the full and the half size NX- specimens:

$$\left. G_f^I \right|_1 = 0.322 \text{ lbf/in} = 56 \text{ N/m} \qquad \text{for} \quad H = 4.25 \text{ in} = 108 \text{ mm}$$

$$\left. G_f^I \right|_{1/2} = 0.354 \text{ lbf/in} = 62 \text{ N/m} \qquad \text{for} \quad H = 2.125 \text{ in} = 54 \text{ mm} \tag{4}$$

The fracture energy release rate slightly increases rather than decreases with decreasing specimen height. Therefore, the direct tension experiments fully support the fracture energy release rate concept. This is not surprising since the specimen ultimately fails in the form of a single localized macrodefect and not in the form of continuous debonding of microdefects which are uniformly distributed over the entire length of the specimen.

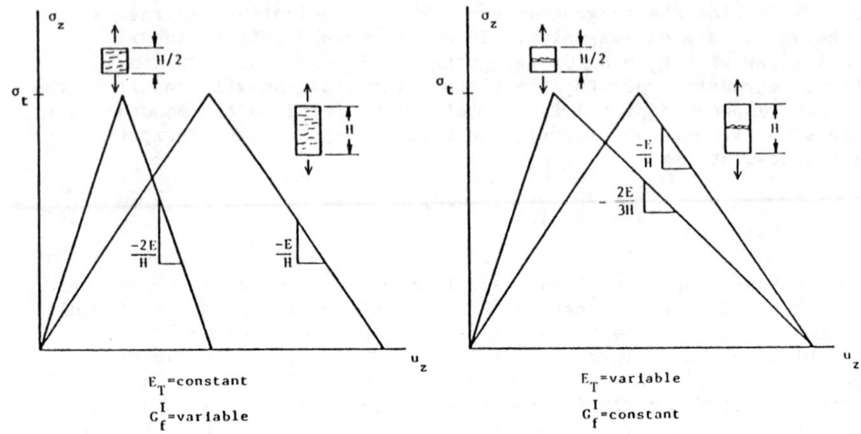

Figure 9 Fracture Energy for Different Specimen Heights

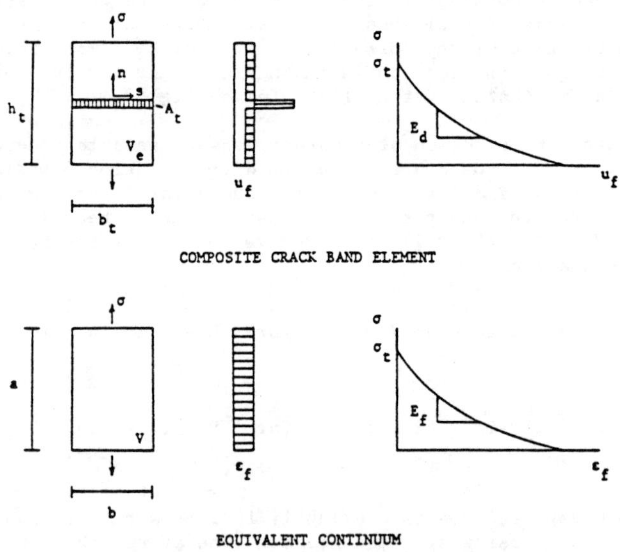

Figure 10 Composite Fracture Model for Tensile Cracking

In the following discussion we will adopt this rationale to develop a strain-softening plasticity model which accounts for the localization of fracture and which monitors constant fracture energy release rates in tension and shear. Conceptually, this approach corresponds to the modification of the softening modulus according to the geometry very much along the line of the elementary example above.

Fracture Based Strain-Softening Model for Tension

The composite fracture formulation is in essence an attempt to incorporate constant fracture energy release rate in the strain-softening formulation of an equivalent elastic-plastic continuum.

For simplicity let us consider the elementary volume V = abt which is subjected to the tensile state of stress depicted in Fig. 10. The maximum stress criterion results in the localized crack band $A_t = b_t t$ which is oriented normal to the direction of major principal stress "n". Incipient cracking is controlled by the initially isotropic fracture condition

$$F(\boldsymbol{\sigma},k) = \sigma_n - \sigma_t = 0 \qquad \text{where } \sigma_n = \sigma_1 \text{ with } \sigma_1 \geq \sigma_2 \geq \sigma_3 \qquad (5)$$

$$\text{and} \quad \sigma_t = \sigma_t(k)$$

For the first active crack the normal tensile stress coincides with the major principal stress $\sigma_n = \sigma_1$. The tensile strength σ_t describes the degradation of bond stresses across the crack band as a function of the tensile crack opening u_f in analogy to the fictitious crack band model of Hillerborg, Modeer and Peterson [7].

Restricting our attention to Mode I type cracking the fracture potential $Q(\boldsymbol{\sigma})$ coincides with the fracture condition $F(\boldsymbol{\sigma},k)$ along the line of an associated flow rule and degenerates in the case of primary crack initiation to the unit vector in three space. The associated fracture rule

$$\dot{\boldsymbol{\varepsilon}}_f = \dot{\lambda}\,\mathbf{m} \quad \text{where } \mathbf{m} = \frac{\partial Q}{\partial \boldsymbol{\sigma}} = \begin{bmatrix} \frac{\partial Q}{\partial \sigma_1} \\ \frac{\partial Q}{\partial \sigma_2} \\ \frac{\partial Q}{\partial \sigma_3} \end{bmatrix} = \begin{bmatrix} 1 \\ 0 \\ 0 \end{bmatrix} \qquad (6)$$

defines direction and magnitude of the tensile strain-softening response in accordance to the flow rule of traditional plasticity theory.
In view of the strain-space setting of the computational algorithm, the consistency condition

$$\dot{F} = \frac{\partial F}{\partial \sigma}^{t} \mathbb{E}(\dot{\varepsilon} - \dot{\lambda} m) - \left(\frac{\partial \sigma}{\partial k} \frac{\partial k}{\partial \varepsilon_f}\right)^{t} \dot{\lambda}\, m = 0 \tag{7}$$

leads to the tensile fracture parameter $\dot{\lambda}$,

$$\text{where} \quad \dot{\lambda} = \frac{m^t \mathbb{E}\, \dot{\varepsilon}}{E_f + m^t \mathbb{E}\, m} \quad \text{for } m^t \mathbb{E} \dot{\varepsilon} > 0 \quad \text{Progressive Cracking}$$

$$\text{and} \quad \dot{\lambda} = 0 \quad \text{for } m^t \mathbb{E} \dot{\varepsilon} \leq 0 \quad \text{Elastic Behavior} \tag{8}$$

where the additive decomposition of strain rate $\dot{\varepsilon} = \dot{\varepsilon}_e + \dot{\varepsilon}_f$ has been inferred and where $\mathbb{E}$ denotes the elastic material law in three space. Let us now consider the hardening or rather the softening modulus E_f which provides the key to incorporate the fracture energy concept within the framework of strain- softening plasticity. Formally, the softening modulus E_f in Eq. 8 is defined by

$$E_f = \left(\frac{\partial \sigma_t}{\partial k} \frac{\partial k}{\partial \varepsilon_f}\right)^{t} m \tag{9}$$

In order to control the fracture energy release rate via the strain-softening modulus E_f we will pursue the following approach: Consider the invariance of the softening branch for different specimen heights in the σ_t-u_f space and the inherent definition of fracture energy in terms of

$$G_f^I = \int_0^{u_r} \sigma_t \, du_f \tag{10}$$

Then it is only natural to monitor softening directly in terms of the fracture modulus E_d which denotes the slope of σ_t-u_f in Figure 10. The mapping between the crack opening displacement du_f and the equivalent tensile fracture strain $dk = d\varepsilon_f$ leads to the following definition of the fracture energy based strain-softening modulus

$$E_f = \left(\frac{\partial \sigma_t}{\partial u_f} \frac{\partial u_f}{\partial \varepsilon_f} \frac{\partial \varepsilon_f}{\partial \boldsymbol{\varepsilon}_f}\right)^{t} m \quad \text{where} \quad \dot{\varepsilon}_f = \langle \dot{\boldsymbol{\varepsilon}}_f^{t}\, \dot{\boldsymbol{\varepsilon}}_f \rangle_+^{1/2} = \sqrt{\langle \dot{\varepsilon}_1 \rangle_f^2 + \langle \dot{\varepsilon}_2 \rangle_f^2 + \langle \dot{\varepsilon}_3 \rangle_f^2} \tag{11}$$

In Eq. 11 the MacCayley brackets extract the tensile components of the fracture strain rate $\dot{\varepsilon}_f$. This provides the key for the homogenization of the localized crack discontinuity over the elementary volume of size $V = abt$. In fact, if we denote by h_t the height of the elementary volume normal to the crack band $A_t = b_t t$, then the projection along the normal vector "m" yields the geometric "gage length" h_t for the evaluation of the softening modulus for cracking in direction "1"

$$E_f = E_d \, h_t \qquad \text{where } E_d = \frac{\partial \sigma_t}{\partial u_f} \; ; \; h_t = \frac{\partial u_f}{\partial \varepsilon_f} \; ; \quad \left(\frac{\partial \varepsilon_f}{\partial \boldsymbol{\varepsilon}_f}\right)^{t} m = 1 \tag{12}$$

As a consequence of this definition of the strain-softening modulus E_f the resulting tangential material law

$$\boxed{\dot{\sigma} = \mathbb{E}_T \dot{\varepsilon}} \qquad \text{where } \mathbb{E}_T = \mathbb{E} - \frac{\mathbb{E}\, \mathfrak{n}\, \mathfrak{n}^t \mathbb{E}}{E_f + \mathfrak{n}^t \mathbb{E}\, \mathfrak{n}} \tag{13}$$

will depend on the mesh size through the geometry of the elementary volume similar to non-local continuum theories.

The composite fracture theory is based on the equivalence of the fracture energy in the crackband A_t in the composite element with that of the equivalent continuum element V. This energy equivalence was previously utilized by the authors [19] in order to express the surface dominated fracture process in terms of an equivalent volume dominated failure phenomenon

$$\int_{A_t} \sigma_t du_f dA = \int_V \sigma^t d\varepsilon_f dV \tag{14}$$

where the individual terms are defined in Fig. 10. Assuming linear softening to simplify the algebra the tensile energy release for the formation of a single crack band ΔA_t within the composite element is simply

$$\Delta U_t^{CE} = -G_f^I \, \Delta A_t = \frac{1}{2} \frac{\sigma_t^2}{E_d} b_t t \qquad \text{where} \quad E_d = \frac{d\sigma_t}{du_f} = -\frac{1}{2} \frac{\sigma_t^2}{G_f^I} \tag{15}$$

The energy release characteristic of the equivalent continuum element is defined by the energy dissipated during softening up to rupture

$$\Delta U_t^{EQ} = \frac{1}{2} \frac{\sigma_t^2}{E_f} V = \frac{1}{2} \frac{\sigma_t^2}{E_f} abt \qquad \text{where} \quad E_f = \frac{d\sigma_t}{d\varepsilon_f} \tag{16}$$

From the equivalence of the energy release during the formation of the crack band, $\Delta U_t^{CE} = \Delta U_t^{EQ}$, the governing strain-softening modulus E_f in the homogenized continuum can be expressed in terms of the fracture modulus E_d or the equivalent fracture energy release rate G_f^I

$$E_f = E_d \frac{V}{A_t} = -\frac{\sigma_t^2}{2G_f^I} \frac{V}{A_t} \tag{17}$$

In the case of uniaxial tension in the axial direction the geometric properties of the crack band are readily expressed in terms of those of the elementary volume, i.e. $A_t = b_t t$ and $h_t = a$, thus

$$E_f = E_d h_t \quad \text{where } h_t = \frac{V}{A_t} \tag{18}$$

We observe that this expression is identical with the definition of the strain-softening modulus E_f developed previously in Eq. 12. This result clearly demonstrates that strain-softening in the equivalent continuum element depends on the geometry of the elementary volume and thus exhibits a pronounced size effect. In other words, the fracture modulus E_f has to be adjusted to the mesh size of the particular finite element idealization if the fracture energy release rate G_f^I should remain constant. As a result the fracture mechanics concept leads to a non-local format of the governing softening properties which is fundamentally different from the local (geometry independent) format of strain-softening plasticity and progressive fracture theories.

In conclusion let us consider the particular concrete tested in the direct tension experiment. For linear softening the fracture modulus is readily evaluated from the nominal uniaxial tensile strength σ_t = 400 psi (2.76 MPa) and the fracture energy release rate for Mode I type cracking, G_f^I = 0.322 lb/in. (56 N/m).

$$E_d = \frac{d\sigma_t}{du_f} = -\frac{\sigma_t^2}{2G_f^I} = -248.45 \ \frac{\text{kip}}{\text{in}^3} = -67.44 \ \frac{\text{N}}{\text{mm}^3} \tag{19}$$

The characteristic crack length which defines the range of debonding within an elastic perfectly brittle comparison solid is in this case

$$\ell_t = \frac{G_f^I \, E}{\sigma_t^2} = 5.63 \text{ in} = 143 \text{ mm} \qquad \text{where } E = 2800 \text{ ksi} = 19305 \text{ MPa.} \tag{20}$$

Since the characteristic length is larger than the specimen height, linear elastic fracture mechanics does hardly apply. Along the same vein, previous suggestions by Bazant et al [1] of three times the aggregate size for the width of the crack band are also open to question.

The numerical stability of strain-softening computations imposes definite restrictions on the localization of fracture within a given finite element mesh which were recently discussed by the authors [20]. We recall that the softening formulation infers unstable material behavior with a corresponding energy release. As long as these material instabilities are localized and contained within the overall structure, stability in the large can be monitored by displacement control since the structural stiffness remains positive definite. On the other hand, since stability in the small is violated by the loss of positive definiteness of the tangent material law $\mathbb{E}_T$ in Eq. 13, we can examine the limiting condition when $\mathbb{E}_T \to -\infty$, which corresponds to perfectly brittle behavior. In the case of tensile softening this takes place when the denominator in the fracture correction of the elastic material stiffness vanishes, i.e. when the following condition holds

$$E_f + \mathbb{n}^t \, \mathbb{E} \, \mathbb{n} = 0 \tag{21}$$

Let us consider for definiteness the case of plane strain and the normality vector n in Eq. 6. Then the limiting condition for the softening modulus reduces to

$$- E_f \leq E \frac{1-\nu}{(1+\nu)(1-2\nu)} \qquad \text{where } E_f = E_d\, h_t \tag{22}$$

In other words the mesh size "h" must be such that the following constitutive constraint is satisfied

$$- h \leq \frac{E}{E_d} \frac{1-\nu}{(1+\nu)(1-2\nu)} \tag{23}$$

For the concrete properties above and $\nu = 0.2$ this constraint results in the maximum height of the elementary volume

$$h \leq \frac{2800}{248.45} \frac{0.8}{1.2 * 0.6} = 12.52 \text{ in} = 318.06 \text{ mm} \tag{24}$$

for which stable post-critical strain control may be maintained. We observe that this mesh size constraint is quite restrictive if large scale massive concrete structures such as dams or pressure vessels are considered. It is a direct consequence of the fracture energy basis of the equivalent strain-softening model which increases the ductility with decreasing mesh size and, vice versa, increases the brittleness with increasing mesh sizes.

Strain-Softening Regime and Transition of Brittle-Ductile Failure

In the following let us probe whether the concept of a composite fracture model for Mode I type tensile cracking could be extended to the Mode II type shear fracture in order to capture the pronounced softening response in low confined compression. In the previous approach Day [4] and the authors [19] utilized the Mohr-Coulomb condition as point of departure augmented by an independent tension cut-off. In that case tensile debonding was controlled by the fracture energy release rate G_f^I for Mode I type cracking similar to the strain-softening formulation for crack bands delineated above. On the other hand, decohesion due to frictional slip was monitored by the fracture energy release rate G_f^{II} for Mode II type fracture governing the formation of shear bands. Because of the tenfold increase of uniaxial compressive versus tensile strength, the value of G_f^{II} was approximately 50 times larger than G_f^I when faulting along a single shear plane was considered along the line of the Mohr-Coulomb failure hypothesis. At this stage it was noted that mixed mode shear fracture is actually initiated by tensile splitting and the formation of a single shear band is thus the result of coalescing crack feathers along an inclined slip plane. Therefore, the postulate of an independent faulting mechanism is really the consequence of the limited resolution of current tools to trace each individual tensile crack preceding the formation of a single shear band.

Since tensile cracking and shear faulting are clearly interrelated we resort here to a continuous failure condition which comprises both tensile as well as shear fracture within a single criterion according to Hoek and Brown [8]. This empirical criterion was originally developed to characterize the triaxial strength of rock specimens which were failed in the conventional triaxial compression experiment. The three parameter failure model neglects the influence of the intermediate principal stress and can be viewed as a combination of the Mohr-Coulomb criterion with the tension cut-off condition very much along the line of the parabolic failure envelope originally proposed in 1933 by Leon for concrete [16].

$$F(\sigma, k) = \sigma_1 - \sigma_3 - (c\sigma_c^2 - m\sigma_c\sigma_1)^{1/2} = 0 \tag{25}$$

σ_c denotes the uniaxial compressive strength and m and c define the frictional and cohesive strength material parameters to be calibrated from the uniaxial tension experiment and a third condition when "cohesive" damage is considered with $c<1$. Neglecting hardening behavior before the peak strength is reached the uniaxial compression experiment yields with $\sigma_1 = 0$ and $c = 1$

$$-\sigma_3 = (c\,\sigma_c^2)^{1/2} \qquad \text{where } \sigma_c = f'_c = 3.2 \text{ ksi} = 22.06 \text{ MPa} \tag{26}$$

Thus under uniaxial compression $\sigma_3 = -3.2$ ksi as long as intact cohesive behavior prevails with $c = 1$. The uniaxial tension experiment on the other hand leads to calibration of m via solution of the quadratic equation Eq. 25 with $\sigma_1 = \sigma_t$ and $\sigma_3 = 0$

$$m = \frac{c\sigma_c^2 - \sigma_t^2}{\sigma_c\sigma_t} \qquad \text{where } \sigma_t = f'_t = 0.4 \text{ ksi} = 2.76 \text{ MPa} \tag{27}$$

This condition yields $m = 7.9$ for the frictional strength parameter at peak strength. Fig. 11 illustrates the fit of the failure envelope with the experimental strength values in terms of the Mohr diagram depicting the normalized shear strength versus effective normal stress, $\tau_n - \sigma_n$.

Let us now examine whether the triaxial strength envelope could be used to describe softening behavior in tension and shear, whereby both the uniaxial tensile strength and the uniaxial compressive strength diminish to zero in the post-peak regime. Fig. 12 delimits the strain-softening regime in terms of the maximum strength and the residual strength envelopes which coalesce at the transition point P of brittle-ductile failure. The direction of inelastic fracture strain rates has been superimposed on a separate diagram in order to illustrate normality and feasible formats of the corresponding plastic potential.

If we rewrite the failure condition Eq. 25 then strain-hardening/softening can be accommodated by the variation of the two parameters m and c.

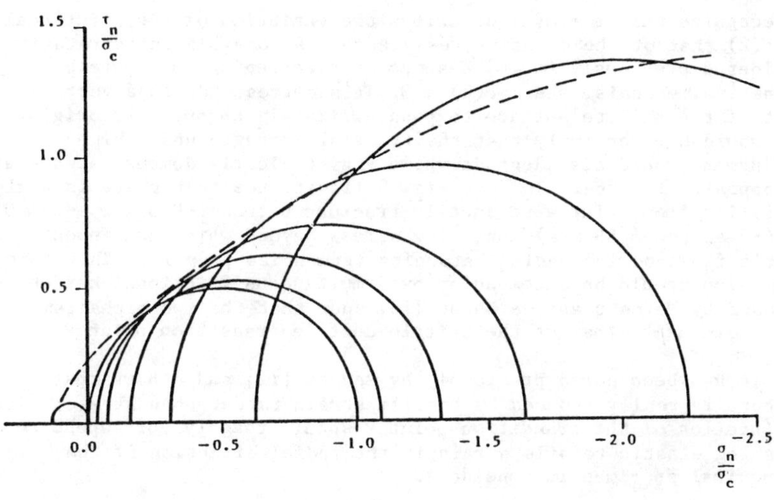

Figure 11 Mohr's Diagram of Triaxial Strength Data

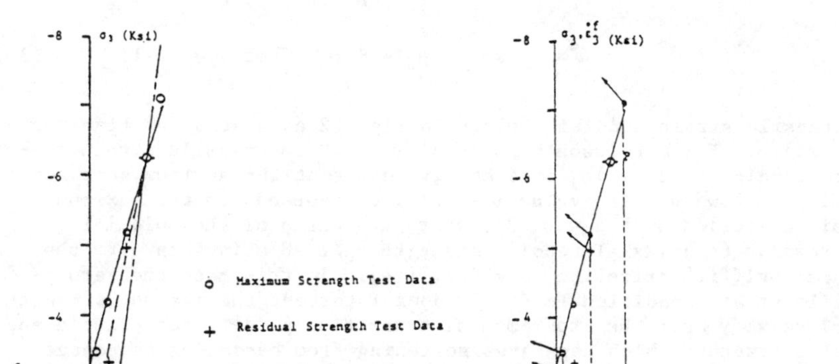

Figure 12 Brittle-Ductile Transiton Point and Inelastic Dilatancy

$$F(\sigma,k) = \tau - (c - m\sigma)^{1/2} = 0 \qquad \text{where } \tau = \left|\sigma_1-\sigma_3\right|/\sigma_c \text{ and } \sigma = \sigma_1/\sigma_c \tag{28}$$

$$\text{with} \quad m_o = 7.9 \text{ and } c_o = 1$$

We recognize that m = m(k) describes the variation of the frictional and c = c(k) that of the cohesive resistance. A possible interpretation considered previously in [4] assumes that softening is a purely decohesive mechanism where c(k) → 0. This corresponds to a vertical shift of the failure surface to pass ultimately through the origin. This would however imply that the triaxial strength under high confinement would also degrade while Fig. 6 clearly demonstrates that the opposite is true. In fact Fig. 8 illustrates that there is a clear transition from brittle to ductile fracture between $-0.5 \leq \sigma_r \leq -1.0$ ksi (-3.45 to -6.89 MPa) confining stress beyond which continuous plastic flow and increasing hardening tendencies prevail. Therefore, decohesion should be accompanied by simultaneous frictional hardening as proposed by Vermeer and de Borst [18] such that the two mechanisms compensate each other at the brittle-ductile transition point P.

It has been noted previously by Mazars [12] and others that fracture is really induced by tensile strain rather than stress. Thus the location of the transition point P should roughly correspond to a vanishing elastic tensile strain in the radial direction if the cylindrical specimen is considered

$$\varepsilon_r^e = \frac{1}{E}\left(\sigma_r - \nu\left(\sigma_\theta + \sigma_z\right)\right) \tag{29}$$

This yields the following condition for $\varepsilon_r^e = 0$ and $\sigma_r = \sigma_\theta$:

$$\sigma_r = \frac{\nu}{1-\nu}\,\sigma_z \quad \text{and} \quad \sigma_3 = 8\,\sigma_1 \quad \text{for } \nu = 0.111 \tag{30}$$

The tensile strain condition plots in Fig. 12 as a straight line through the origin. For a Poisson's ratio of $\nu = 0.2$ the tensile strain cut-off corresponding to $\sigma_3 = 4\sigma_1$ does hardly intersect the maximum strength envelope. However, the value $\nu = 0.111$ corresponds to the maximum strain criterion with $\sigma_3=-\sigma_c+8\sigma_1$ where the ratio of the uniaxial compressive to uniaxial tensile strength $\sigma_c/\sigma_t=8$ coincides with the biaxial Griffith criterion, see e.g. [16]. In this case the zero tensile strain condition in Eq. 30 does intersect the maximum strength envelope very near the experimental transition point P for brittle and ductile fracture which separates softening from hardening response regimes. Therefore we could regard the tensile cap beyond the residual strength envelope as a transient strength regime which ultimately collapses to the residual strength level upon continuous tensile straining. In fact the trace of the residual strength envelope is captured fairly well by the following modification of Eq. 28 for purely frictional resistance

$$F_r(\sigma,k) = \tau - (-m_r\sigma)^{1/2} = 0 \qquad \text{where } m_r=11.96 \text{ and } c_r=0 \tag{31}$$

From the directions of the inelastic strain rates in Fig. 12 we observe that the trace of the maximum strength envelope furnishes an acceptable format for the inelastic potential $Q(\sigma)$ to be used as fracture rule for strain-softening under mixed mode fracture processes.

Fracture Based Strain-Softening Model for Mixed Mode Fracture

Considering the strain softening formulation for Mode I type fracture in the previous section, let us examine the pertinent generalization to mixed mode fracture along the line of the combined softening model for tension and shear.

Starting from the maximum strength condition Eq. 28 we recognize that decohesion is accompanied by frictional hardening. During failure the fracture rule defines the direction and magnitude of inelastic fracture strains in the homogenized continuum.

$$\dot{\boldsymbol{\varepsilon}}_f = \dot{\lambda}_f \, \mathbf{m} \qquad \text{where} \qquad \mathbf{m} = \frac{\partial Q}{\partial \boldsymbol{\sigma}} \tag{32}$$

From Fig. 12 we recognize that the direction of inelastic strain rate is fairly well defined by normality to maximum strength envelope. Therefore, the fracture potential is in this case described by Eq. 28 with $m = m_o$ and $c = c_o$

$$Q(\boldsymbol{\sigma}) = \tau - (c_o - m_o \sigma)^{1/2} = 0 \qquad \text{where } m_o = 7.9 \quad \text{and} \quad c_o = 1 \tag{33}$$

The gradient of the fracture potential defines the direction and thus the inelastic dilatation in principal stress space

$$\mathbf{m} = \frac{\partial Q}{\partial \boldsymbol{\sigma}} = \begin{bmatrix} \dfrac{\partial Q}{\partial \sigma_1} \\ \dfrac{\partial Q}{\partial \sigma_2} \\ \dfrac{\partial Q}{\partial \sigma_3} \end{bmatrix} = \frac{1}{\sigma_c} \begin{bmatrix} 1 + \dfrac{1}{2} \dfrac{m_o}{(c^* - m_o \sigma)^{1/2}} \\ 0 \\ -1 \end{bmatrix} \tag{34}$$

Similar to the Mohr-Coulomb condition there is no inelastic component in the direction of intermediate principal stress. We also note that the fracture component in the minor direction does not depend on the frictional resistance m because the normal stress σ in Eq. 28 is defined only in terms of the major principal stress σ_1 and not in terms of the average value of the major and minor normal stress used in the Mohr-Coulomb condition. For computational reasons the fictitious cohesion $c^* = \tau^2 + m_o \sigma$ has been introduced in order to accommodate "elastic" trial stress states beyond the failure condition which are restricted to major principal stress values $\sigma_1 < c_o\sigma_c/m_o$.

The fracture parameter $\dot{\lambda}_f$ determines the magnitude of inelastic straining during mixed mode fracture. With the aid of the consistency condition $\dot{F} = 0$ the fracture parameter $\dot{\lambda}_f$ reads in the case of a non-associated flow rule where $F \neq Q$

$$\dot{\lambda}_f = \frac{\mathbf{m}^t \mathbb{E}\, \dot{\boldsymbol{\varepsilon}}}{E_f + \mathbf{m}^t \mathbb{E}\, \mathbf{n}} \qquad \text{for } \mathbf{m}^t \mathbb{E}\, \dot{\boldsymbol{\varepsilon}} > 0$$

$$\dot{\lambda}_f = 0 \qquad \text{for } \mathbf{m}^t \mathbb{E}\, \dot{\boldsymbol{\varepsilon}} \leq 0 \tag{35}$$

In this case the vectors $\mathbf{m}$ and $\mathbf{n}$ denote the gradients of the current loading surface $F(\boldsymbol{\sigma},k)$ and the fracture potential $Q(\boldsymbol{\sigma})$, the latter of which was previously developed in terms of principal stresses. In analogy to Eq. 34 we find with $c^* = \tau^2 + m\sigma$

$$\mathbf{m} = \frac{\partial F}{\partial \boldsymbol{\sigma}} = \begin{bmatrix} \dfrac{\partial F}{\partial \sigma_1} \\[2mm] \dfrac{\partial F}{\partial \sigma_2} \\[2mm] \dfrac{\partial F}{\partial \sigma_3} \end{bmatrix} = \frac{1}{\sigma_c} \begin{bmatrix} 1 + \dfrac{1}{2} \dfrac{m}{(c^* - m\sigma)^{1/2}} \\[2mm] 0 \\[2mm] -1 \end{bmatrix} \tag{36}$$

The principal issue of the non-associated softening formulation revolves around the proper definition of E_f. Formally we obtain from the consistency condition $\dot{F} = 0$ the following expression for the instantaneous strain-softening modulus

$$E_f = -\left(\frac{\partial F}{\partial k} \frac{\partial k}{\partial \boldsymbol{\varepsilon}_f}\right)^t \frac{\partial Q}{\partial \boldsymbol{\sigma}} \qquad \text{where } k = k(\varepsilon_f) \tag{37}$$

We recall that decohesion is accompanied by frictional hardening both of which are mobilized by tensile straining during compressive splitting in the brittle failure regime. Therefore it is natural to adopt again the definition of equivalent tensile fracture strain $\dot{\varepsilon}_f$ in Eq. 11 which played the role of the kinematic damage metric of the homogenized continuum subjected to tensile cracking.

$$dk = d\varepsilon_f \quad \text{where} \quad \dot{\varepsilon}_f = \langle \dot{\boldsymbol{\varepsilon}}_f^t\, \dot{\boldsymbol{\varepsilon}}_f \rangle_+^{1/2} = \sqrt{\langle \dot{\varepsilon}_1 \rangle_f^2 + \langle \dot{\varepsilon}_2 \rangle_f^2 + \langle \dot{\varepsilon}_3 \rangle_f^2} \tag{38}$$

We note that the MacCayley brackets only extract the positive, i.e. the tensile fracture strain rates from the inelastic strain rate tensor $\dot{\boldsymbol{\varepsilon}}_f$.

In case of compressive splitting the chain rule of differentiation leads to the following definition of the strain-softening modulus in terms of the tensile fracture behavior

$$E_f = - \frac{\partial F}{\partial c} \frac{\partial c}{\partial \sigma_t} \frac{\partial \sigma_t}{\partial u_f} \frac{\partial u_f}{\partial \varepsilon_f} \left(\frac{\partial \varepsilon_f}{\partial \varepsilon_f}\right)^t \frac{\partial Q}{\partial \sigma} \tag{39}$$

Recollecting that tensile splitting in compression as well as shear faulting is preceded by distributed microcracking and that the actual crack surface is in this case a multiple of that in the direct tension experiment, we resort to the fundamental definition of Mode I type cracking. For the current concrete this fracture property is characterized by $G_f^I = 0.322$ lbf/in. (56 N/m) which corresponds to the area under the tensile softening law in Fig. 13.

The individual terms in Eq. 39 are defined as follows

$$\frac{\partial F}{\partial c} = \frac{-1}{2(c - m\sigma)^{1/2}} \qquad \text{with } dm = -4.06\ dc \tag{40}$$

The interdependence between the cohesive strength parameter and the tensile strength results in our case to

$$\frac{\partial c}{\partial \sigma_t} = 2.5 \qquad \text{where } \sigma_t = 0.4 \text{ ksi} = 2.76 \text{ MPa} \tag{41}$$

The relationship between the degrading tensile strength σ_t and the crack opening displacement u_f was previously characterized by the fracture modulus of the uniaxial tension experiment, see Fig. 10.

$$E_d = \frac{d\sigma_t}{du_f} = \frac{d^2 G_f^I}{du_f^2} \tag{42}$$

An exponential fit of the normalized test data is shown in Fig. 13 which leads to the following decaying softening function

$$E_d = -\sigma_t \left(\frac{5}{u_r}\right) e^{-5 \frac{u_f}{u_r}} \tag{43}$$

where σ_t denotes the uniaxial tensile strength and u_r the crack opening displacement at rupture when the interfacial bond stresses have diminished to zero (in our case u_r = 0.002 in. = 0.05 mm, from Fig. 2).

It remains to relate the crack opening displacement u_f to the tensile fracture strain ε_f of the equivalent continuum. In Fig. 7 we have seen that the lateral extension is over ten times larger than the axial extension in the direct tension experiment shown in Fig. 2. This

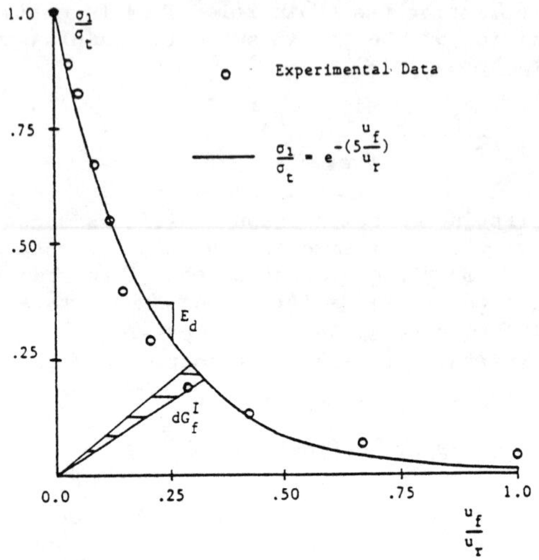

Figure 13 Experimental Softening Model

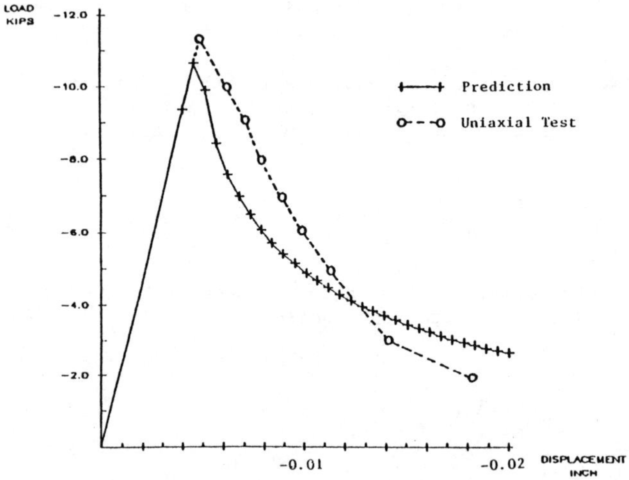

Figure 14 Softening Response in Uniaxial Compression

should be of no surprise since the splitting compression mode of failure is far more distributed which is in turn responsible for the eight to tenfold increase of the uniaxial compressive over the tensile strength values.

A fracture energy argument similar to that developed for tensile cracking leads the following relationship between the crack opening displacement and the equivalent tensile fracture strain in compression

$$\frac{du_f}{d\varepsilon_f} = \frac{V}{A_c} \simeq h_c \tag{44}$$

where the volume fraction A_c/V represents the microcrack density in compression. We observe that Eq. 44 plays the fundamental role of the geometric "gage length" and has the simple physical interpretation of the crack spacing h_c in the case of splitting compression. From the global observations it may be evaluated approximately in terms of the ratio between the fracture energy release rate in tension G_f^I and that in shear G_f^{II}. Thereby it is understood that the second quantity by itself is already the result of homogenizing distributed tensile cracks into a single shear band with the surface area A_s which has the same order of magnitude as the crack surface in direction tension, i.e. $A_t \simeq A_s$. From the equivalence of fracture energy

$$G_f^I \; \Delta A_c = G_f^{II} \; \Delta A_t \tag{45}$$

we can find the unknown fracture surface under splitting compression from

$$\Delta A_c = \frac{G_f^{II}}{G_f^I} \Delta A_t \tag{46}$$

This yields in the case of the uniaxial compression test with G_f^{II} = 27.2 lbf/in (4763 N/m) the following measure of longitudinal crack spacing

$$h_c \simeq \frac{V}{A_c} = \frac{G_f^I}{G_f^{II}} \frac{V}{A_t} = 0.05 \text{ in} = 1.28 \text{ mm} \tag{47}$$

which should be compared with the characteristic length ℓ_t = 5.63 in. (143 mm) in the case of Mode I type cracking. The ratio G_f^{II}/G_f^I represents the magnification of tensile crack surface when compared with that of direct tension test. This ratio increases with increasing confining pressure in order to simulate the diffuse damage mechanism around and beyond the brittle-ductile transition point. Using the experimental data described earlier, a first order approximation yields the following expression for the fracture energy ratio as a function of the minor principal stress σ_3

$$\frac{G_f^{II}}{G_f^{I}} = 1.0 - 84 \left(\frac{\langle\sigma_3\rangle}{f'_c}\right)^5 \tag{48}$$

This ratio reduces to one in the case of triaxial tension when $\langle\sigma_3\rangle = 0$. It increases rapidly with increasing confinement $\sigma_3 < 0$ and reduces the strain-softening modulus to $E_f \rightarrow 0$ as $h_c \rightarrow 0$ which corresponds to perfectly plastic behavior when $\sigma_3 \ll 0$.

In conclusion let us summarize the tangential material law for frictional hardening and non-associated decohesion

$$\boxed{\dot{\boldsymbol{\sigma}} = \mathbb{E}_T\, \dot{\boldsymbol{\varepsilon}}} \quad \text{where} \quad \mathbb{E}_T = \mathbb{E} - \frac{\mathbb{E}\, \mathbf{n}\, \mathbf{m}^t\, \mathbb{E}}{E_f + \mathbf{m}^t \mathbb{E}\, \mathbf{n}} \tag{49}$$

$$\text{and} \quad E_f = \frac{2.5}{2(c - m\sigma)^{1/2}}\, E_d\, \frac{V}{A_c} \left(\frac{\partial \varepsilon_f}{\partial \boldsymbol{\varepsilon}_f}\right)^t \frac{\partial Q}{\partial \boldsymbol{\sigma}}$$

The strain-softening formulation above is based on the fracture energy release rate G_f^I and the concomitant exponential decay function for E_d using energy equivalence concepts to homogenize the underlying failure phenomena. Fig. 14 illustrates the performance of the fracture energy based strain-softening plasticity model when the post-peak response of the uniaxial compression test is considered. In this case only the exponential softening law for tensile cracking in Fig. 13 was utilized, in order to predict the softening response under compression.

CONCLUDING REMARKS

A strain-softening elastic-plastic formulation has been presented for the description of the fracture mechanisms in concrete which range from brittle Mode I type cracking in tension to ductile mixed mode fracture in compression. The composite fracture theory is based on the fracture energy content of the homogenized continuums which is subjected to equivalent strain softening mechanisms in tension and shear.

We conclude that the combined mechanism of cohesive softening and frictional hardening covers the entire failure spectrum of concrete which is monitored solely by the fracture modulus E_d of the underlying Mode I type cracking. We note that an overall constitutive formulation for triaxial concrete behavior should include a hardening cap in compression in addition to the tensile strain-softening mechanism for brittle failure in tension and shear.

ACKNOWLEDGEMENT

This investigation was supported by the U.S. Air Force Office of Scientific Research Contract AFOSR 82-0273 with the University of Colorado, Boulder under liaison of Lt. Col. Lawrence Hokanson, which is gratefully acknowledged.

APPENDIX I - REFERENCES

[1] Bažant, Z.P. and Cedolin, L., "Blunt Crack Band Propagation in Finite Element Analysis," Trans. American Society of Civil Engineers, EMD Vol. 105, No. EM2 (1979), pp. 297-315.

[2] Bažant, Z.P. and Belytschko, T.B., "Continuum Theory for Strain-Softening," Trans. American Society of Civil Engineers, EMD Vol. 110, No. 12 (1984), pp. 1666-1692.

[3] Belytschko, T. and Bažant, Z., "Strain-Softening Materials and Finite Element Solutions," ASME/WAM'84, New Orleans, Symposium on Constitutive Equations, Macro and Computational Aspects, ASME Vol. G00274, (K. Willam, ed.), New York (1984), pp. 253-277.

[4] Day, S.E., "A Softening Plasticity Model for Concrete," M.S. Thesis, CEAE Department, University of Colorado, Boulder, Colorado, 1985.

[5] Dougill, J.W., "On Stable Progressively Fracturing Solids," J. Appl. Math. Physics, ZAMP, Vol. 27 (1976), pp. 423-436.

[6] Gopalaratnam, V.S. and Shah, S.P., "Softening Response of Concrete in Direct Tension," Technological Institute, Report, Northwestern University, Evanston, 1984.

[7] Hillerborg, A., Modéer, H. and Petersson, P.E., "Analysis of Crack Formation and Crack Growth in Concrete by Means of Fracture Mechanics and Finite Elements," Cement and Concrete Research, Vol. 6 (1976), pp. 773-781.

[8] Hoek, E. and Brown, E.T., "Underground Excavation in Rock," The Institution of Mining and Metallurgy, London, 1980.

[9] Hurlbut, B., "Experimental and Computational Investigation of Strain-Softening in Concrete," M.S. Thesis, CEAE Department, University of Colorado, Boulder, Colorado, 1985.

[10] Janson, J. and Hult, J., "Fracture Mechanics and Damage Mechanics, a Combined Approach," J. de Mechanique Appliquee, Vol. 1 (1977), pp. 69-84.

[11] Krajinovic, D. and Fonseka, G.U., "The Continuous Damage Theory of Brittle Materials, - Part I: General Theory," ASME J. Appl. Mech. Vol. 48 (1981), pp. 809-815.

[12] Mazars, J., "Application de la Mecanique de l'Endommagement au Comportment Non Lineaire et a la Rupture du Beton de Structure," Ph.D. Dissertation, Université Pierre et Marie Curie, Paris, 1984.

[13] van Mier, J.G.M., "Strain-Softening of Concrete under Multiaxial Loading Conditions," Ph.D. Dissertation, University of Technology, Eindhoven, 1984.

[14] Ortiz, M., "A Constitutive Theory for the Inelastic Behavior of Concrete," Division of Engineering Research Report, Brown University, Providence, Rhode Island, 1984.

[15] Reinhardt, H.W., "Fracture Mechanics of an Elastic Softening Material like Concrete," Heron Publ. Vol. 29, No. 2, Delft, 1984.

[16] Romano, M., "On Leon's Criterion", Meccanica (1969), pp. 48-66.

[17] Stankowski, T. and Gerstle, K.H., "Simple Formulation of Concrete Behavior Under Multiaxial Load Histories," J. ACI. Vol. 82, No. 2, 1985, pp. 213 - 221.

[18] Vermeer, P.A. and De Borst, R., "Non-Associated Plasticity for Soils, Concrete and Rock," Heron Publ. Vol. 29, No. 3, Delft, 1984.

[19] Willam, K.J., Bićanić, N., and Sture, S., "Constitutive and Computational Aspects of Strain-Softening and Localization in Solids," ASME/WAM '84 Meeting, New Orleans, Symposium on Constitutive Equations, Macro and Computational Aspects, ASME Vol. G00274, (K. Willam ed.), New York (1984), pp. 233-252.

[20] Willam, K., Pramono, E. and Sture, S., "Stability and Uniqueness of Strain-Softening Computations", Europe-U.S. Symposium on Finite Element Methods for Nonlinear Problems, NIT Trondheim, Norway, Aug. 12-16, 1985. Structures Research Series 8503, University of Colorado, Boulder, Colorado, 1985.

SIZE EFFECT ON SHEAR STRENGTH OF REINFORCED CONCRETE BEAM

Toshiyuki Shioya* and Hirotaka Kawasaki**

SYNOPSIS

Nonlinear finite element analysis was performed with the objective of investigating the relevance of the size effect on actual shear strengths of reinforced concrete beams. The size effect on concrete strength and the shear transfer across crack surfaces are considered in this analysis. The analytical results are compared in detail with experimental results regarding deflection, crack propagation, and strains in reinforcing bars and concrete. The analytical results show good agreement with the experimental results. Thus, in estimating shear strengths of large reinforced concrete structures, it is evident that the effect of size on concrete strength and shear transfer must be considered.

1. INTRODUCTION

Reinforced concrete structures in recent years have been gradually increasing in size as a result of advances made in materials and improvements in design and construction techniques. One of the problems accompanying increases in size is the evaluation of shear strength. It has been clarified that the shear strength of a reinforced concrete beam without shear reinforcement gradually decreases as the effective depth of the beam increases, as revealed in experimental studies of Kani [1], Taylor [2], and Iguro, Shioya, Nojiri and Akiyama [3]. Generally called the "size effect," this has been taken up in specifications such as the CEB-FIP "Model Code for Concrete Structures" [4], the Japan Society of Civil Engineers "Limit State Design of Concrete Structures (Proposal)" [5], and others. Meanwhile, analytical studies of the size effect on the shear strength of reinforced concrete beams have been performed by Blaauwendraad [6], Bazant and Kim [7], and others. On the other hand, examples of analysis are still scarce, and the present situation shows that comparison studies with experimental results also are inadequate. This report is based on a comparison study with experimental results of an analysis made by a nonlinear finite element method, with re-examinations especially made of material properties used in the analysis.

---

* Research Engineer, Institute of Technology, Shimizu Construction Co.,Ltd., No.4-17, Etchujima, 3-chome, Koto-ku, Tokyo, Japan.
** Structural Engineer, Civil Engineering Development Department, Shimizu Construction Co.,Ltd., No.13-16, 3-chome, Minato-ku, Tokyo, Japan

2. OUTLINE OF THE EXPERIMENTAL STUDY BY IGURO, SHIOYA, NOJIRI AND AKIYAMA [3]

The shear strength of large reinforced concrete beams under a uniformly distributed load was investigated. The compressive strength of concrete was approximately 23.5 MPa (3410 psi), and the reinforcement ratio of a cross section where shear failure occurred was p = 0.4%. The size ratio of widths and reinforcing bar diameters was matched, and the effective depths were varied at the six levels of d = 10, 20, 60, 100, 200 and 300 cm (3.9, 7.9, 23.7, 34.9, 78.7, 118 inch). The ratio of loading span "ℓ" to "d" was constant at $\ell/d = 12$, roughly corresponding to $a/d = 3$ in case of a concentrated load. The cross sections of the beams and the method of testing are shown in Fig. 1. According to the results of the experiments, the size effect remains even at d > 100 cm (39.4 inch), which is outside the scope of previous experiments, as shown in Fig. 2, and the size effect at d > 100 cm (39.4 inch) is inversely proportional to the fourth root of "d". Regarding the cracking pattern, the principal cracks at shear failure are at a spacing of d/2 irrespective of the size of "d", and are roughly of the same number. Thus when "d" is large, the crack width is large compared with the case of "d" being small. And since the aggregate interlock is reduced when the crack width becomes large, it is thought that the shear strength becomes small when "d" is large. Also based on the fact that the virtual flexural tensile strength calculated by the load at which reinforcing bar strain gauges detected cracks becomes small when the height "H" of the specimen becomes large, it was reported as being conceivable that the size effect on the flexural tensile strength itself is one of the factors causing the size effect on shear strength.

Based on the foregoing, the causes of size effect mainly may be trimmed down to the following two points:

1) Reduction in aggregate interlock between crack surfaces
2) The size effect on concrete strength -- mainly tensile strength

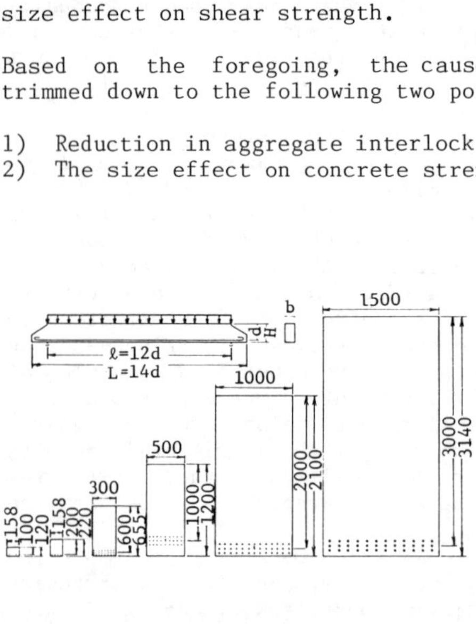

Fig. 1 Iguro et al. -- Test specimens

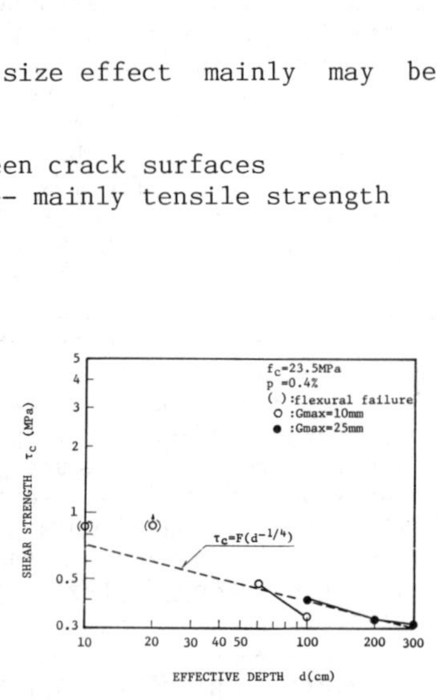

Fig. 2 Iguro et al. -- Test results

3. STUDY CONCERNING CONSTANTS USED IN ANALYSES

3.1 Concrete Strength

Up to now, the size effect of concrete strength itself apparently has not been seriously considered in nonlinear finite element analysis. Therefore, experimental data concerning the size effect on material strength will be briefly summarized.

(1) Compressive Strength

A typical example showing the influence of size on a cylindrical specimen with a height two times its diameter is shown in Fig. 3 [8]. This figure shows the various diameter-cylinder strength ratio, with the diameter of the specimen taken on the abscissa, and the compressive strength of a ϕ15x30-cm (ϕ5.9x11.8-inch) cylindrical specimen on the ordinate for reference. Although there is some amount of scatter depending on the water-cement ratio, the compressive strength is decreased inversely proportional to the tenth root of the diameter as shown in Fig. 3, and it appears there is a lower limit at around a diameter of 60 cm (23.7 inch). In the past, the fact that the compressive strength of concrete in a member has been taken to be about 85 percent of cylinder strength is thought to be the result of considering such a size effect on compressive strength. According to this reference material, as shown in Fig. 4 and 5, the influence of specimen dimensions on the modulus of elasticity and Poisson's ratio is very weak.

Further, with regard to the compressive strength of cracked concrete, Collins and Vecchio [9], based on experiments concerning RC panels, showed that -- depending on the instance -- compressive strength of a RC panel is decreased by about 60 percent from that of the cylinder strength. One reason for this is thought to be that the reduction in compressive strength due to the high slenderness ratio of the concrete struts between cracks contains a kind of size effect, and it is thought necessary to consider this depending on the objective of analysis.

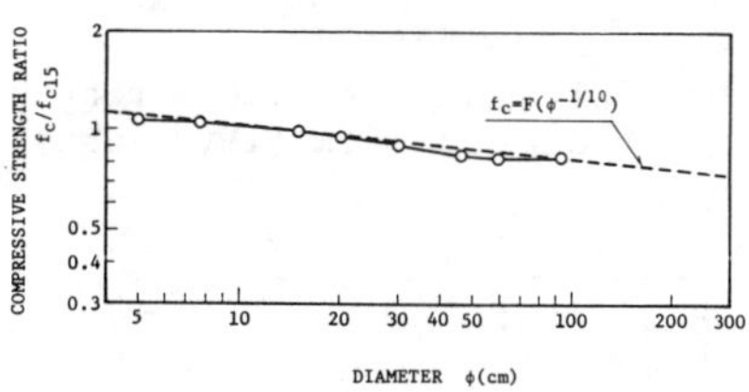

Fig. 3 Blanks -- Compressive strength (where f_{c15}:compressive strength of ϕ15x30-cm (ϕ5.9x11.8-inch))

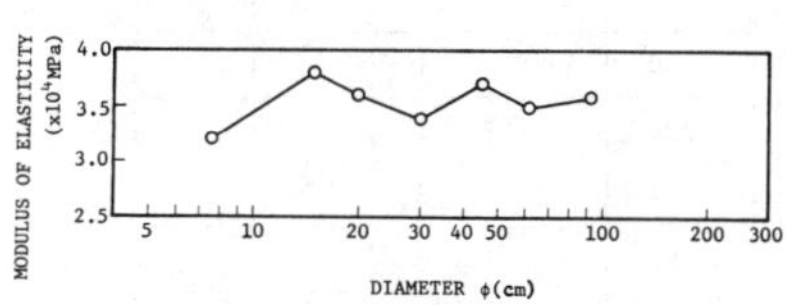

Fig. 4 Blanks -- Modulus of elasticity

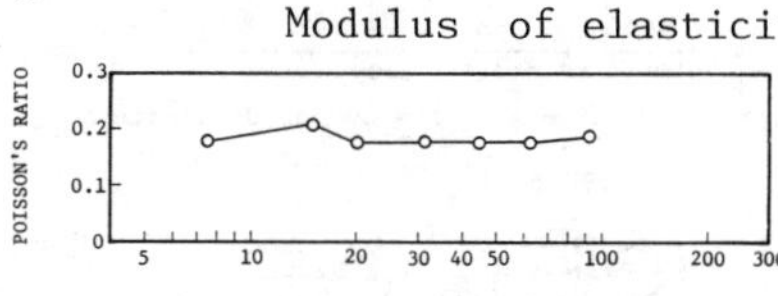

Fig. 5 Blanks -- Poisson's ratio

(2) Tensile Strength

There are few cases of experiments concerning the size effect on the tensile strength of concrete. This is thought to result from the fact that, in the design of reinforced concrete structures, calculations were made without expecting anything of tensile strength, and it had been judged not to be especially important. In past studies, specimen sizes were up to about ϕ = 30 cm (11.8 inch) and dimensions were extremely small. Therefore, the results of experiments recently conducted at the Institute of Technology of the authors' company are described in the following.

Splitting tensile strength tests were performed by using the specimens shown in Table 1, with diameter and length as parameters. The compressive strength of a ϕ10x20-cm (ϕ3.9x7.9-inch) cylindrical specimen was 30.7 MPa (4450 psi) (Gmax = 25 mm (1 inch)). Figure 6 shows tensile strength f_t in specimens with constant length L = 50 cm (19.7 inch) and, diameter "ϕ" varied at 10 cm (3.9 inch)$\leqq \phi \leqq$300 cm (118 inch). The size effect is seen up to ϕ < 50 cm (19.7 inch) with tensile strength approximately 60 percent of that of control test pieces (ϕ 10x20-cm (ϕ3.9x7.9-inch)). The size effect was not seen at ϕ > 50 cm (19.7 inch). Figure 7 shows tensile strength when, with specimens of diameter of ϕ = 100 cm (39.4 inch), length "L" was varied at 10 cm (3.9 inch) $\leqq$ L $\leqq$ 100 cm (39.4 inch). Differences in tensile strength due to length "L" were not recognized. Consequently, although past studies discussed the size effect with the specimen volume as the parameter, according to these experiments, it was thought it would be more rational for the area of the uniformly distributed stress field to be adopted as the parameter. From this viewpoint, it appears there is a difference in the size effect between tensile and compressive strength. Izumi and Mihashi [10] have explained this difference by the fracture theory based on the stochastic fracture theory.

Table 1 Specimens

Diameter ϕ (cm)	Length L(cm) 100	50	20	10
300		1		
200		1		
100	3	3	3	3
50		3		
20		3		
10		3		

Note: 1 cm = 0.394 inch

Numerals in Table show the number of specimen

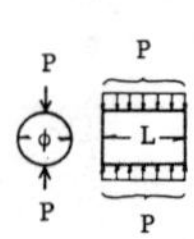

$$f_t = \frac{2P}{\pi \phi L}$$

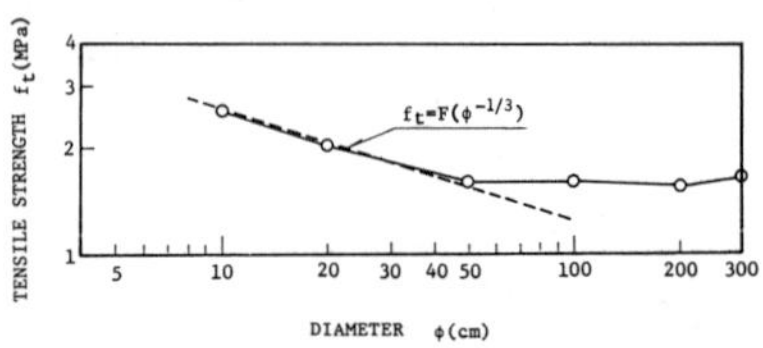

Fig. 6 Splitting tensile strength (length L=50cm (19.7 inch))

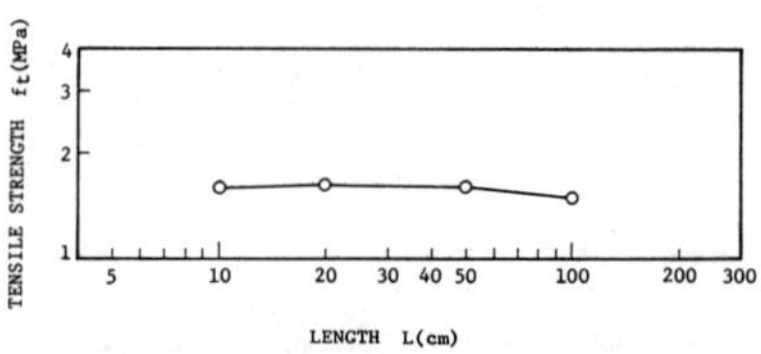

Fig. 7 Splitting tensile strength (diameter ϕ=100cm (39.4 inch))

(3) Flexural Tensile Strength [3], [11], [12]

Flexural tensile strength also decreases in inverse proportion to the cubic to fourth roots of the height "H" of the specimen when the specimen size becomes larger (Fig. 8). In the past, tensile strength and flexural strength were explained mainly according to the tensile plasticity of concrete [13], but it appears that there is a size effect on the tensile strength of concrete, and in the flexural tensile strength test, the zone of tensile stress within the specimen is small (Fig. 9). It is thought that the size effect on tensile strength itself also is a cause of flexural tensile strength becoming larger than tensile strength. Whereas according to Figs. 6, 8, and 9, the size effect on tensile strength is not recognizable at ϕ = 50 cm (19.7 inch), the experimental results reveal that the size effect on flexural strength is recognizable up to around H = 150 cm (59.1 inch). This can be comprehended conceptually.

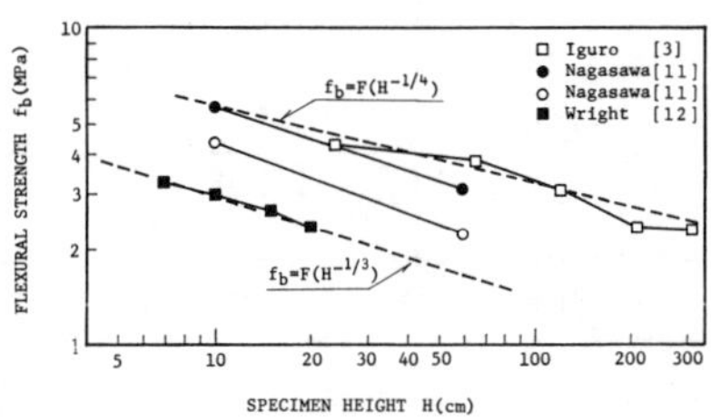

Fig. 8 Flexural tensile strength

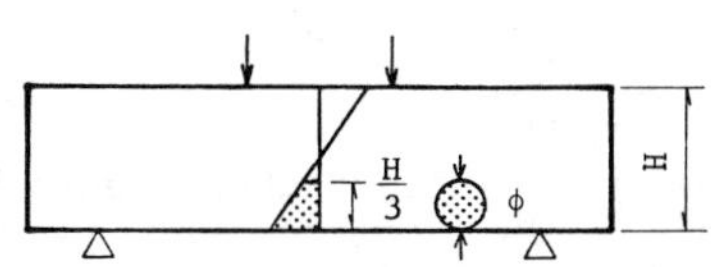

Fig. 9 Flexural and splitting tensile strength

(4) Handling of Size Effect on Concrete Strength in Analysis

In considering the above-mentioned size effect on concrete strength, the failure surface is modified as appropriate and used in this analysis. With regard to the criterion of crack occurrence, the following equation, a modification of the equation in JSCE "Limit State Design of Concrete Structures (Proposal)" [5] is used

$$f_{ts} = 0.27 f_c^{2/3} (\phi / 15)^{-1/4}, \qquad (1)$$

provided that ϕ = 50 cm (19.7 inch) is used when ϕ > 50 cm (19.7 inch). Also, the plastic range was hypothesized as shown in Fig. 10, so that the flexural tensile strength for H = 15 cm (5.9 inch) would be 1.76 times the splitting tensile strength for ϕ = 15 cm (5.9 inch) (according to the equation in JSCE "Limit State Design of Concrete Structure (Proposal)" [5]).

Regarding the failure criterion on the tension side, it was hypothesized that the failure mode of the reinforced concrete beam without shear reinforcement would be that in

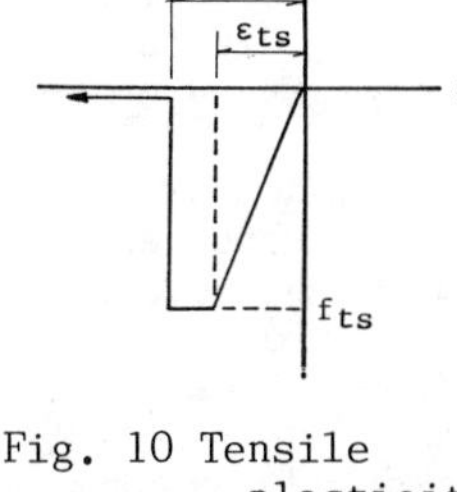

Fig. 10 Tensile plasticity

the form of concrete teeth, as shown in Fig. 11; f_{ts} was selected to give $\phi = H/3 = d/6$, since it denoted flexural failure of the concrete tooth portion of $H = d/2$.
The following equation was used as the uniaxial compressive failure criterion:

$$f_{cs} = f_c(\phi/15)^{-1/10}, \quad (2)$$

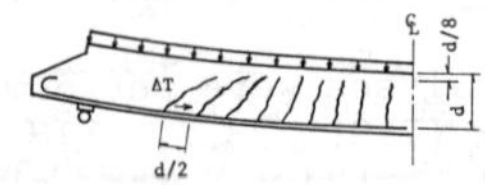

Fig. 11 Failure mode

where, f_{cs} was selected on the assumption that the compressive zone of the beams was $\phi = d/8$. Shown in Fig. 12 are the failure surfaces determined by this method for specimens of d = 20 cm (7.9 inch) and d = 300 cm (118 inch). It is assumed that there are no size effects on the modulus of elasticity and Poisson's ratio.

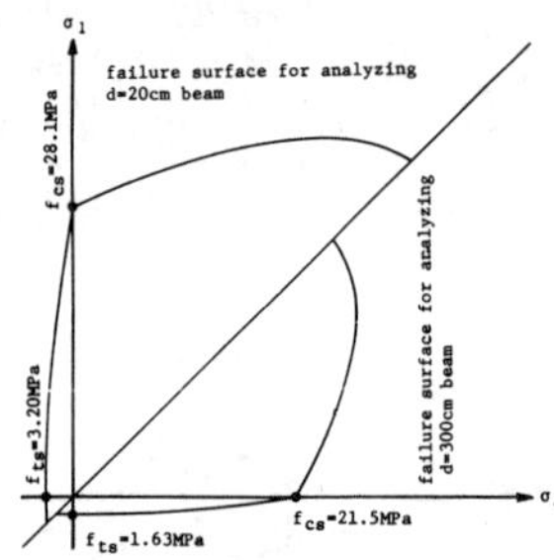

Fig. 12 Failure surface (f_c=23.5MPa (3410psi))

3.2 Shear Transfer Across Crack Surfaces (Interlocking Action of Aggregate Between Crack Surfaces)

According to direct shear tests by Aoyagi and Yamada [14] of cross sections having perpendicularly crossing reinforcing bars, as shown in Fig. 13, after cracks have been formed in concrete, the transfer of shear force at the crack surfaces is dependent on crack width. Figure 13 shows that shear force is transferred by aggregate interlock until crack width becomes about 0.5 mm (0.02 inch). It is even stated that 50 to 70 percent of the shear forces transferred after cracking are due to aggregate interlock [15] [16].

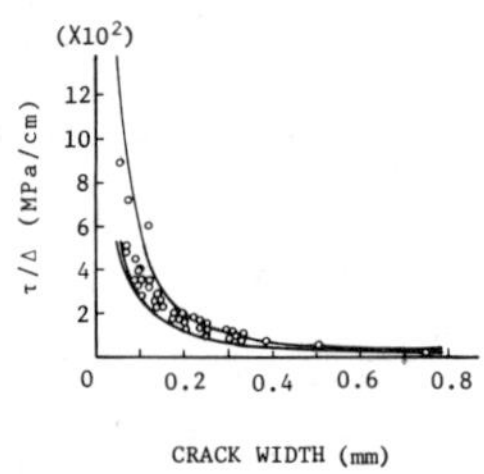

Fig. 13 Shear transfer by aggregate interlock

A number of methods of modeling the transfer of shear forces at crack surfaces have been proposed in accordance with the methods of handling the cracks. In the method of separating nodes, it is expressed by considering the relation of shear force and slippage with regard to rigidity of the connecting element between crack nodes. In the method where all elements are made orthogonal and anisotropic, shear rigidity after cracking is multiplied by α $(1 > \alpha > 0)$ to reduce the rigidity upon which it is considered.

In this analysis, cracks in the vicinity of the cross section of shear failure were taken to be discrete cracks, and with the experimental results of Aoyagi and Yamada [14] as reference, shear rigidities at crack surfaces were made to be dependent on crack width. Since it was presumed that flexural cracks at the middle of specimens would not affect the results, the analysis was performed with these as smeared cracks and assuming $\alpha = 0.3$.

4. ANALYTICAL MODEL

The analysis program is a visco-elastoplastic analysis program introduced as Program B in a previous paper [17], and will not be described herein.

Analyses were performed regarding the specimens of d = 20, 60 and 300 cm (7.9, 23.7, 118 inch) among the experiments conducted by Iguro et al., [3]. The element idealization mesh is shown in Fig. 14. Constants used in the analysis are listed in Table 2. Reinforcing bars were considered to be bilinear bar elements. As for bonding, although there was a close relation with shear strength, many points still need to be clarified, and a conventionally perfect bond was assumed for the analysis.

Table 2 Constants for analysis

	Item	d (cm)		
		300	60	20
Concrete	Modulus of Elasticity (1/3secant)	20600MPa	20600MPa	20600MPa
	Poisson's Ratio (1/3secant)	0.167	0.167	0.167
	Ultimate Compressive Strain	-1800μ	-1800μ	-1800μ
	Compressive Strength	21.5MPa	25.2MPa	28.1MPa
	Tensile Strength	1.63MPa	2.43MPa	3.20MPa
	Ultimate Tensile Strain	119μ	177μ	233μ
Joint	Tensile Strength	1.63MPa	2.43MPa	3.20MPa
	Ultimate Tensile Strain	119μ	177μ	233μ
Re-Bar	Modulus of Elasticity	206000MPa	206000MPa	206000MPa
	Yield Point	363MPa	363MPa	363MPa

Note: 1 MPa = 145 psi

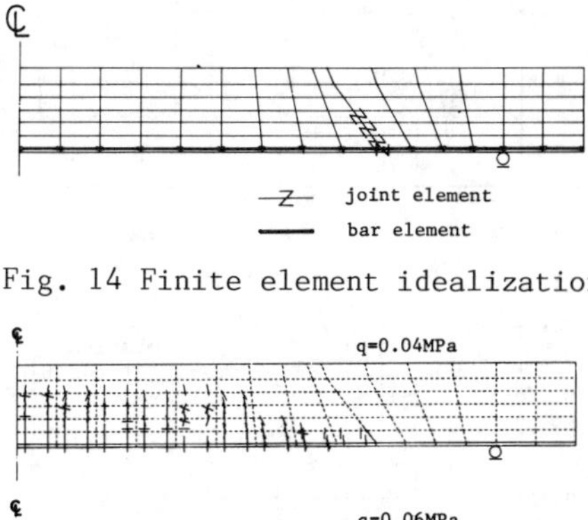

Fig. 14 Finite element idealization

5. ANALYSIS RESULTS

An example of comparisons between results of the analysis and experiments regarding the condition of crack development is given in Fig. 15. Comparisons of analytical and experimental results concerning load-deflection curves are shown in Figs. 16 to 18. As examples of load-reinforcing bar strains, the strains of reinforcing bars in cross sections 1.5 d inside from supporting points in specimens of d = 60 cm (23.7 inch) and d = 300 cm (118 inch) are shown Fig. 19 and 20. The value of RC calculations in the figures were obtained by the conventional reinforced concrete calculation formula.

From these figures, it may be stated that analytical and experimental results correspond well with each other.

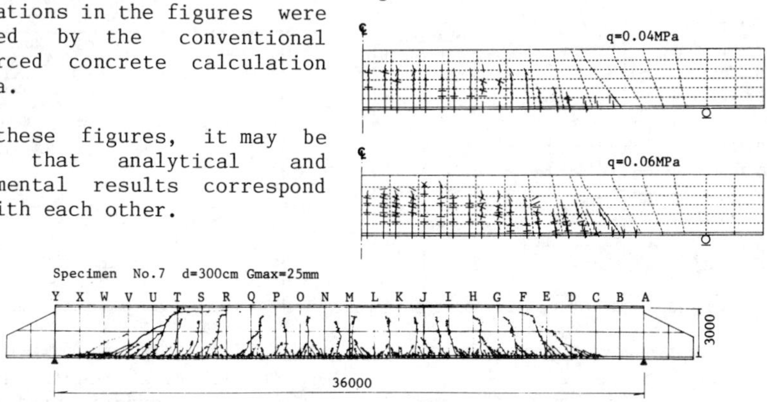

Fig. 15 Cracking Patterns

6. SHEAR STRENGTH

Maximum loads judged to occur when the solutions diverged, are shown in Fig. 21 as analysis values.

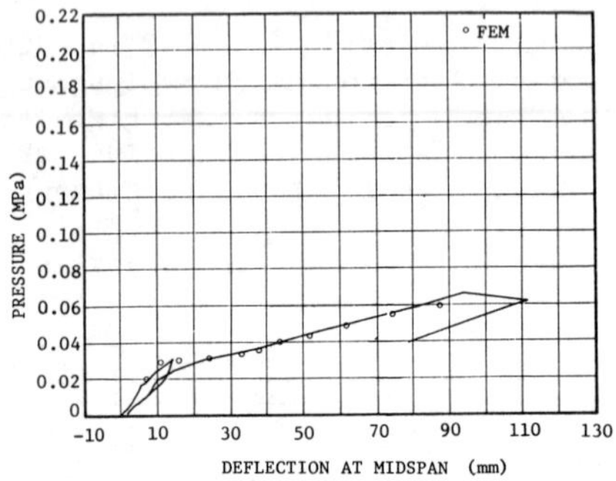

Fig. 16 Load-deflection (d=300cm (118 inch))

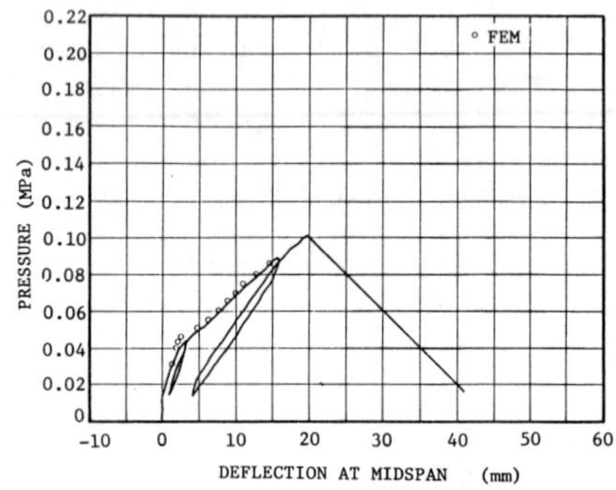

Fig. 17 Load-deflection (d=60cm (23.7 inch))

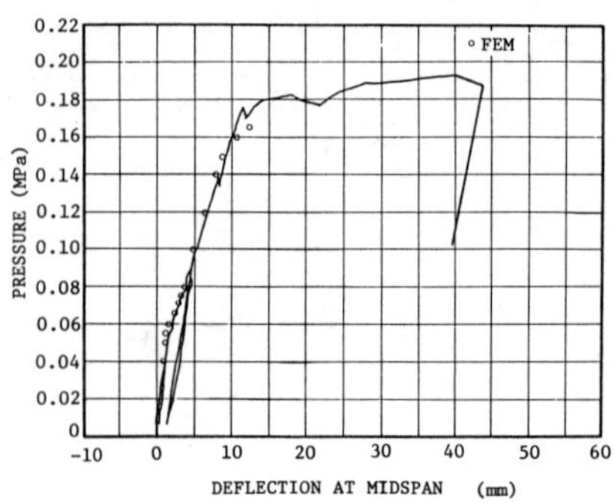

Fig. 18 Load-deflection (d=20cm (7.9 inch))

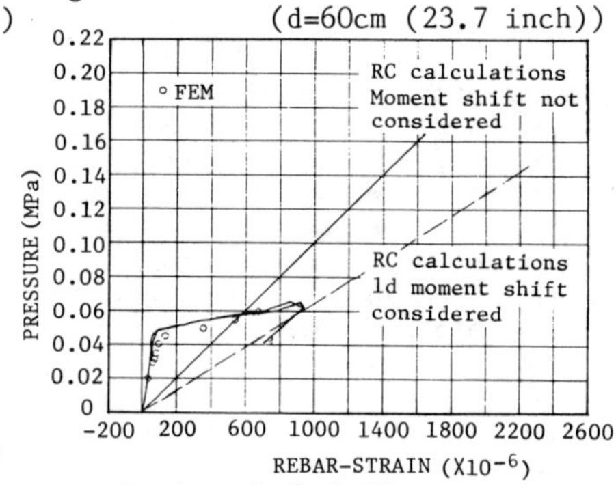

Specimen No.7 d=300cm Gmax=25mm

Fig. 19 Load-rebar strain (d=300cm (118 inch))

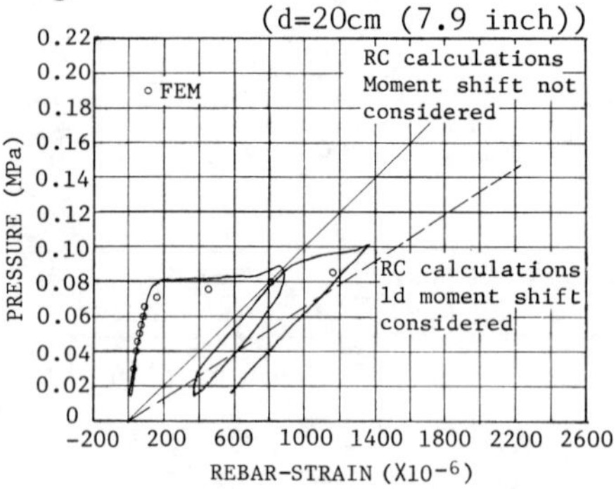

Specimen No.3 d=60cm Gmax=10cm

Fig. 20 Load-rebar strain (d=60cm (23.7 inch))

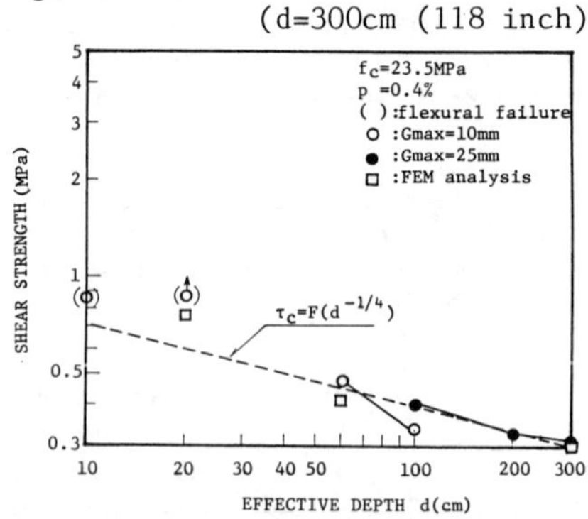

Fig. 21 Analytical results and experimental results of Iguro et al.

7. CONCLUSIONS

This report contains outlines of an experimental study concerning the size effect of reinforced concrete beams without shear reinforcement, especially reexaminations of material properties used in the analysis, with analysis conducted by the nonlinear finite element method and comparisons made with experimental results. The shear strength of a reinforced concrete beam without shear reinforcement, with the maximum size of aggregate being the same, decreases in inverse proportion to roughly the fourth root of the effective depth of the beam. The reasons for this may be considered to be the size effect on concrete strength, and a reduction in interlocking action of aggregate between crack surfaces. Upon study, therefore, regarding material properties, a rough summary may be made as follows:

1) The rate of strength decline due to size effect on compressive strength is small, with a decrease in inverse proportion to the tenth root of specimen diameter ϕ ; the lower limit value is reached at ϕ = 60 cm (23.7 inch). Effects due to specimen size are not evident in the modulus of elasticity and Poisson's ratio.

2) The rate of strength decline due to the size effect on splitting tensile strength is high, with a decrease in inverse proportion to the cubic root of specimen diameter ϕ ; the lower limit value is reached at ϕ = 50 cm (19.7 inch). Further, a difference in strength according to differences in specimen length "L" is not evident, and although past studies often have discussed the size effect with the specimen volume as the parameter, it was deemed more rational to adopt the area of uniform-stress field as the parameter.

3) Flexural tensile strength also decreases in inverse proportion to the cubic to the fourth root of specimen height.

4) The transfer of shear force at the crack surfaces is dependent on crack width.

On attempting a nonlinear finite element analysis by giving consideration to the above-mentioned conditions, there was rough correspondence between analytical and experimental results. Therefore, when estimating the behavior of a large structure based on the results of small-scale experiments at the laboratory level, especially where member strength is determined by failure of the concrete, it is deemed necessary to evaluate the size effect on concrete strength as described in this report. Regarding causes and methods of quantitatively evaluating the size effect on concrete strength, although these already have been taken up on a research basis, it is considered necessary that further examinations be made.

REFERENCES

[1] Kani, G. N. J.: "How Safe Are Our Large Reinforced Concrete Beams?", Journal of ACI, 1967, pp. 128- 141.
[2] Taylor, H. P. J.: "Shear Strength of Large Beams," Proc. of ASCE,

Nov. 1972, pp. 2473- 2490.
[3] Iguro, M., Shioya, T., Nojiri, Y., and Akiyama, H.: "Experimental Studies on Shear Strength of Large Reinforced Concrete Beams under Uniformly Distributed Load," Proc. of JSCE, No. 348/V-1, Aug. 1984, pp. 175- 184 (in Japanese). Also, Concrete Library International, JSCE, No. 5, Aug. 1985, pp. 137-154 (in English).
[4] CEB-FIP: Model Code for Concrete Structures, 1978.
[5] JSCE: Limit State Design of Concrete Structures (Proposal), Concrete Library, No. 52, Dec. 1983 (in Japanese).
[6] Blaauwendraad, J.: "The Use of the Rough Crack Model of Walraven and the Fictitious Crack Model of Hillerborg in F.E. Analysis," IUTAM, 1983.
[7] Bazant, Z. P. and Kim, J. K.: "Size Effect in Shear Failure of Longitudinally Reinforced Beam," Journal of ACI, Sept.-Oct. 1984, pp. 456- 468.
[8] Blanks, R. F. and McNamara, C. C.: "Mass Concrete Test in Large Cylinders," Journal of ACI, Jan. - Feb. 1935, pp. 280- 303.
[9] Collins, M. P. and Vecchio, F.: "The Response of Reinforced Concrete to In-Plane Shear and Normal Stresses," University of Toronto, ISBN 0-7727-702-8, Mar. 1982.
[10] Izumi, H. and Mihashi, H.: "A Stochastic Theory for Concrete Fracture (Part III)," Proc. of AIJ., No. 310, Dec. 1981, pp. 1- 9 (in Japanese).
[11] Nagasawa, Y., Uchida, Y. and Okada, T.: "Bending Behavior of the Lapped Splice for High Amplitude-Low Cyclic Loading," Proc. of JCI, 6th Conference, June 1984, 377- 380 (in Japanese).
[12] Wright, P. J. F.: "The effect of the method of test on the flexural strength of concrete," Magazine of Concrete Research, No. 11, Oct. 1952.
[13] Okada, K. and Muguruma, H.: Concrete Engineering Handbook, Asakura Shoten, Nov. 1981 (in Japanese).
[14] Aoyagi, Y. and Yamada, K.: "Strength and Deformational Characteristics of Reinforced Concrete Shell Elements Subjected to In-Plane Forces," Proc. of JSCE. No. 331, Mar. 1983, pp. 167-180 (in Japanese).
[15] Fenwick, R. C. and Paulay, T.: "Mechanisms of Shear Resistance of Concrete Beams," Journal of Structural Division, ASCE, Vol. 94, ST 10, Oct. 1968, pp. 2235-2350.
[16] Taylor, H. P. J.: "Investigation of the Forces Carried in Reinforced Concrete Beams in Shear by Interlock of Aggregate," Cement and Concrete Association, London, TRA 42, 447, 1979, 22 pp.
[17] Shioya, T., Kawasaki, H. and Kunita, J.: "Nonlinear Finite Element Analysis of Circular Reinforced Concrete Slabs," Proc. of JCI 2nd colloquium on Shear Analysis of RC Structures, Oct. 1983, pp. 183-190 (in Japanese).

SHEAR TRANSFER ACROSS INTERFACES IN BOUNDARY ELEMENT ANALYSIS OF CONCRETE STRUCTURES

by Tatsuya Tsubaki [1]

ABSTRACT

A method to deal with the shear transfer across an interface of a concrete structure is outlined for general-purpose numerical analysis methods, in particular, the boundary element method (BEM). Comparing results with those obtained by the finite element method (FEM) for a couple of examples, it is found that the boundary element method and the combination solution method of FEM and BEM are more effective and efficient for the analysis involving shear transfer across interfaces, in general, than a straight FEM analysis.

INTRODUCTION

It is well known that the shear stresses transferred across interfaces in concrete such as crack surfaces and bond surfaces have a significant influence on the behavior of concrete structures, and many investigations on this subject have been carried out. The mechanism of the shear transfer across interfaces has gradually become clear by many experimental studies. Based on those experimental studies, various analytical models of shear transfer have been developed, such as a macroscopic model to characterize the average behavior of concrete structures and a microscopic model taking into account the geometry of interfaces (1,2,4,5,6,8,13). The final objective of these models seems to be to determine a constitutive equation which can be implemented in a general-purpose structural analysis program.

In the following, the characteristics of various shear transfer models are briefly reviewed, and a method to deal with the shear transfer across interfaces in numerical analysis procedures such as the finite element method (FEM), the boundary element method (BEM), and the combination solution method of FEM and BEM is summarized. Then, the applicability of these analysis methods will be illustrated in the numerical examples for the analysis of a concrete plate subject to in-plane forces, a massive concrete structure with external constraints subject to thermal loading, and the bond between concrete and a steel plate.

---

[1] Associate Professor, Dept. of Civil Engineering, Yokohama National University, Yokohama, Japan.

ANALYSIS OF SHEAR TRANSFER

The shear transfer mechanisms to be considered herein are as follows:

(1) Aggregate interlock and friction between crack surfaces (cf.Fig.1);
(2) Dowel action of reinforcing bars across crack surfaces (cf.Fig.2);
(3) Bond between concrete and steel reinforcement.

In the following, a method to express these shear transfer mechanisms across interfaces will be considered.

The general shear transfer conditions across an interface can be expressed as follows, using the local coordinate system of the interface (cf. Figs.3,4)(11).

$$f_n^1 - f_n^2 = 0 \qquad (f_n^1 < 0 \,,\ f_n^2 < 0) \qquad (1)$$

$$u_n^1 + u_n^2 = 0 \qquad (2)$$

$$f_t^1 - f_t^2 = 0 \qquad (3)$$

$$u_t^1 + u_t^2 = 0 \qquad (| f_t^k | < -\mu f_n^k \,,\ k = 1, 2) \qquad (4)$$

$$f_t^k = \pm \mu f_n^k \qquad (k = 1, 2) \qquad (5)$$

where f, u, and μ stand for the traction, the displacement, and the friction coefficient, respectively. Subscripts n and t stand for the components in the normal and tangential directions of the interface, respectively, and superscripts 1,2 indicate the domain to which the contact point of the interface belongs. Eqs.(1) and (2) respectively stand for the continuity conditions of the normal tractions and the displacements of the interface. Eq.(3) is the continuity condition of the tangential tractions of the interface. Eq.(4) is the condition of no slip (adhesion), and Eq.(5) is that of frictional slipping which follows the Coulomb friction law. Also, $f_n > 0$ implies that the interface condition changes to the state of separation. In such a case, the two contact points on the interface are treated separately, so that the continuity and friction conditions may not be given.

Finite Element Method

A procedure to deal with the above-mentioned shear transfer across an interface by the finite element method is shown in the following. In the finite element method, the friction relationship Eq.(5) can be expressed by approximating the surface tractions of contact points of the interface by the nodal forces of each element. The stiffness equation for the interface in the local coordinate system is expressed as follows.

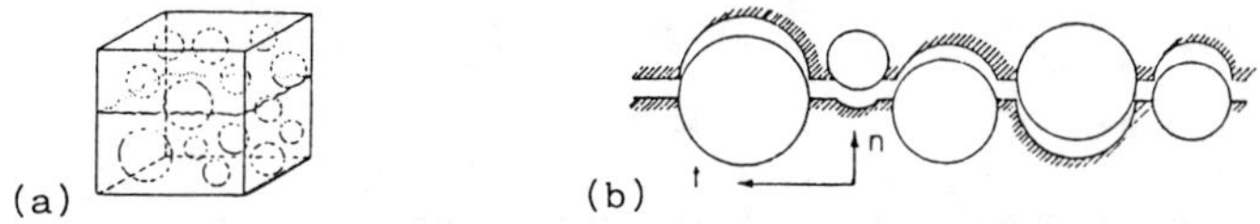

Fig.1 Schematic Figure of Crack Surface :
(a) Cracked Concrete; (b) Crack plane
(From Ref.13)

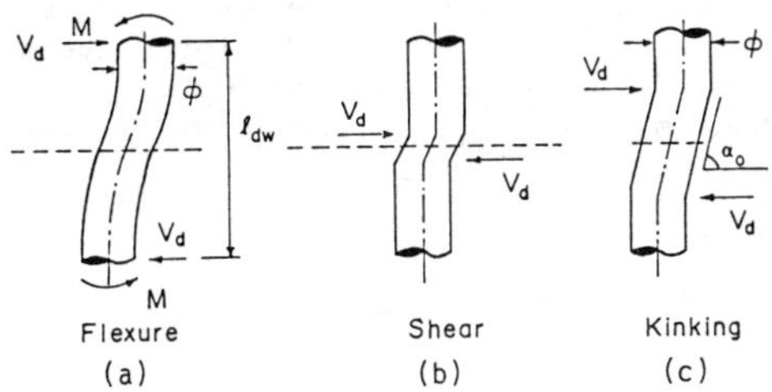

Fig.2 Types of Dowel Action

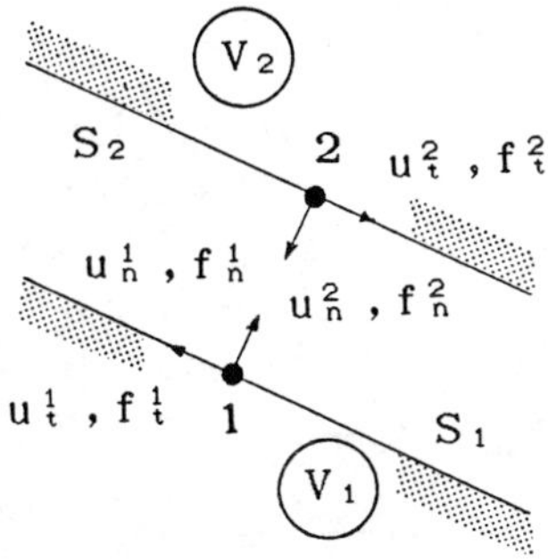

Fig.3 Schematic Figure of Interface

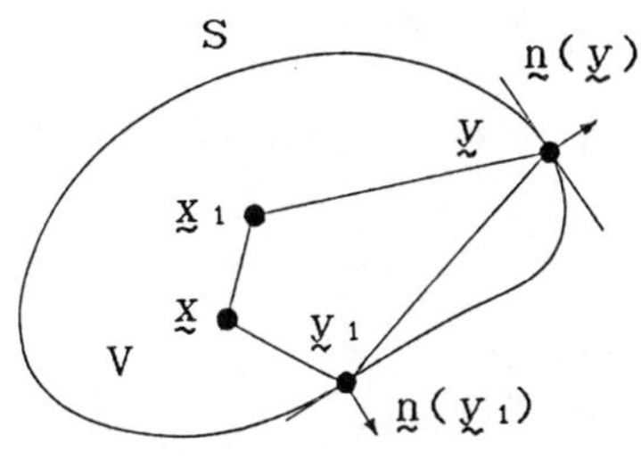

Fig.4 Symbols of Domain and Boundary

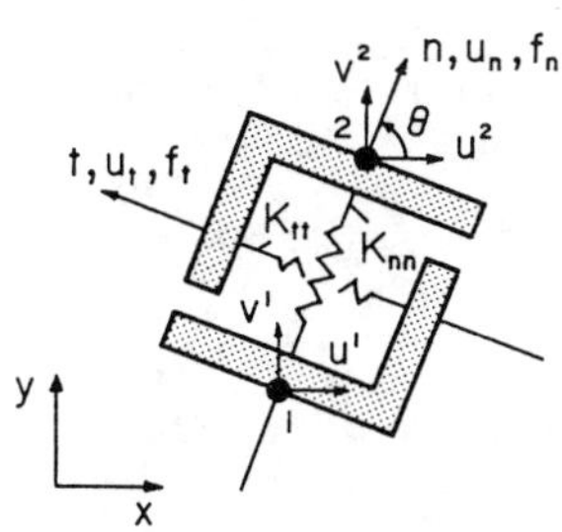

Fig.5 Linkage Element

$$k_{\ell m} u_m = f_\ell \qquad (\ell, m = 1, 2, \cdots, n) \tag{6}$$

where all the variables are defined in the local coordinate system, and $k_{\ell m}$, u_m, and f_ℓ are elements of the stiffness matrix, displacements, and nodal forces, respectively. The letter n stands for the number of degrees of freedom. The indices used twice follow the summation convention. Eq.(6) can be used for the separation condition of an interface, since two independent nodes are used for the points in contact on the interface. In case that the interface is in the adhesion condition, both of the displacements u_n and u_t of the interface are continuous, and they must satisfy Eqs.(2) and (4). In general, Eq.(6) is modified for $u_i = u_j$ as follows.

$$\overline{k}_{\ell m} u_m = \overline{f}_\ell \qquad (\ell, m = 1, 2, \cdots, n) \tag{7}$$

where

$$\overline{k}_{im} = k_{im} + k_{jm} \quad (m \neq i, m \neq j) \quad ; \quad \overline{k}_{mi} = k_{mi} + k_{mj} \quad (m \neq i, m \neq j) \quad ;$$

$$\overline{k}_{ii} = k_{ii} + k_{ij} + k_{ji} + k_{jj} \quad ; \quad \overline{k}_{jm} = 0 \quad (m \neq j) \quad ; \quad \overline{k}_{mj} = 0 \quad (m \neq j) \quad ;$$

$$\overline{k}_{jj} = 1 \quad ; \quad \overline{f}_i = f_i + f_j \quad ; \quad \overline{f}_j = 0 \tag{8}$$

In Eq.(7), $\overline{k}_{\ell m}$ is a symmetric matrix when $k_{\ell m}$ is symmetric.

If the interface is in the frictional slipping condition, Eq.(6) must be changed so that a linearly dependent relationship for the nodal forces, Eq.(5), as well as the displacement continuity condition, may be satisfied. In general, Eq.(6) is arranged for $f_i = \mu f_j$ as follows.

$$\overline{\overline{k}}_{\ell m} u_m = \overline{\overline{f}}_\ell \qquad (\ell, m = 1, 2, \cdots, n) \tag{9}$$

where

$$\overline{\overline{k}}_{im} = k_{im} - \mu k_{jm} \quad ; \quad \overline{\overline{k}}_{jm} = k_{jm} \quad ; \quad \overline{\overline{f}}_i = f_i - \mu f_j \quad ; \quad \overline{\overline{f}}_j = f_j \tag{10}$$

As is obvious from Eq.(10), $\overline{\overline{k}}_{\ell m}$ in Eq.(9) is, in general, a non-symmetric singular matrix.

Besides the above method in which the stiffness equation is directly modified, it is also possible to use the generalized linkage element(cf. Fig.5). The contact points on an interface are expressed by the nodal points. Then the relationship between the nodal forces f_n, f_t and the relative displacements u_n, u_t of the nodes is expressed as follows.

$$\begin{Bmatrix} f_n \\ f_t \end{Bmatrix} = \begin{bmatrix} k_{nn} & k_{nt} \\ k_{tn} & k_{tt} \end{bmatrix} \begin{Bmatrix} u_n \\ u_t \end{Bmatrix} \tag{11}$$

$$\begin{Bmatrix} u_n \\ u_t \end{Bmatrix} = \begin{bmatrix} -c & -s & c & s \\ s & -c & -s & c \end{bmatrix} \begin{Bmatrix} u^1 \\ v^1 \\ u^2 \\ v^2 \end{Bmatrix} \quad ; \quad c = \cos\theta \; ; \; s = \sin\theta \tag{12}$$

Setting $k_{tn} = \mu k_{nn}$, $k_{nt} = k_{tn} = 0$ in Eq.(11), the friction relationship of Eq.(5) is obtained. The displacement continuity in the normal direction, Eq.(2), is approximately obtained by setting $k_{nn} \gg 1$. The adhesion condition, $|f_t| < -\mu f_n$ $(f_n < 0)$, is approximated by $k_{nt} = k_{tn} = 0$, $k_{nn} \gg 1$, $k_{tt} \gg 1$, and the separation condition is obtained by $k_{nn} = k_{nt} = k_{tn} = k_{tt} \ll 1$.

Boundary Element Method

The relationship between the displacement u_i and the traction f_i on the boundary surface S is expressed in the local coordinate system of the interface as follows (cf. Fig.4). The following integral equation has a general form which can be used for a thermal stress analysis(3,9). The indices i,k, etc., stand for the components in the x_i and y_i directions, and follow the summation convention.

$$C_{ki}(\underset{\sim}{y})\, u_i(\underset{\sim}{y}) + \int_S f^*_{ki}(\underset{\sim}{y}, \underset{\sim}{y_1})\, u_i(\underset{\sim}{y_1})\, dS(\underset{\sim}{y_1}) = \int_S u^*_{ki}(\underset{\sim}{y}, \underset{\sim}{y_1})\, f_i(\underset{\sim}{y_1})\, dS(\underset{\sim}{y_1}) + b_k(\underset{\sim}{y}) \tag{13}$$

where u_i, f_i, and b_K stand for the displacement, the traction, and the body force term due to, e.g., temperature θ, respectively. u^*_{Ki} is the fundamental solution of a two-dimensional isotropic elastic body (plane strain), f^*_{Ki} is the traction component corresponding to u^*_{Ki}, and $C_{Ki}(\underset{\sim}{y})=0.5\,\delta_{Ki}$. Discretizing the integral equation (13) with a suitable interpolation function matrix $\underset{\sim}{\Phi}$ for $\underset{\sim}{u}$ and $\underset{\sim}{f}$, the quantities for node i of element j can be expressed as follows. Note that N stands for the number of elements and point y corresponds to node i of element j.

$$\underset{\sim}{C}{}^j_i\, \underset{\sim}{u}{}^j_i + \sum_{k=1}^{N} \underset{\sim}{H}{}^{jk}_i\, \underset{\sim}{u}^k = \sum_{k=1}^{N} \underset{\sim}{G}{}^{jk}_i\, \underset{\sim}{f}^k + \underset{\sim}{b}{}^j_i \tag{14}$$

$$\underset{\sim}{H}{}^{jk}_i = \int_{S_k} \underset{\sim}{f}{}^{*j}_i\, \underset{\sim}{\Phi}{}^T_u\, dS \quad ; \quad \underset{\sim}{G}{}^{jk}_i = \int_{S_k} \underset{\sim}{u}{}^{*j}_i\, \underset{\sim}{\Phi}{}^T_f\, dS \tag{15}$$

where u^k and f^k stand for the vectors of nodal quantities of element k. Summing up each quantity for element j and using a matrix expression, Eq.(14) becomes the following global equations.

$$H\,U = G\,F + B \tag{16}$$

where H consists of C_i^j and H_i^{jk}. Also, U, F, and B are vectors obtained by assembling u^k, f^k, and b^k, respectively. C_i^j contributes to the diagonal terms of H, and is determined by the conditions for a rigid body motion.

In case the domain V is divided into small parts by interfaces in a concrete structure, Eq.(16) needs to be modified using the method of subdivision. In the following, a case in which the domain is divided into two regions as shown in Fig.3 is considered. As for the conditions of an interface, considered are the separation, adhesion, and frictional slipping conditions, depending on the shear transfer condition across the interface. Then, Eq.(16) becomes as follows for the case of separation.

$$\begin{bmatrix} H_1^k & H_2^k \end{bmatrix} \begin{Bmatrix} U_1^k \\ U_2^k \end{Bmatrix} = \begin{bmatrix} G_1^k & G_2^k \end{bmatrix} \begin{Bmatrix} F_1^k \\ F_2^k \end{Bmatrix} + B^k \qquad (\, k = 1,\, 2 \,) \tag{17}$$

where the upper and lower indices stand for the domain to which the element belongs and the domain in contact, respectively. When the shear transfer at the interface is perfect, i.e., for the case of adhesion, Eq.(16) is changed as:

$$\begin{bmatrix} H_1^1 & H_2^1 & -G_2^1 & O \\ O & -H_1^2 & -G_1^2 & H_2^2 \end{bmatrix} \begin{Bmatrix} U_1^1 \\ U_2^1 = -U_1^2 \\ F_2^1 = F_1^2 \\ U_2^2 \end{Bmatrix} = \begin{bmatrix} G_1^1 & O \\ O & G_2^2 \end{bmatrix} \begin{Bmatrix} F_1^1 \\ F_2^2 \end{Bmatrix} + \begin{Bmatrix} B^1 \\ B^2 \end{Bmatrix} \tag{18}$$

If the shear transfer at the interface is accompanied by frictional slipping, the following equation must be solved.

$$\begin{bmatrix} H_1^1 & (H_2^1)_t & (H_2^1)_n & -\{(G_2^1)_n \mp \mu (G_2^1)_t\} & O & O \\ O & O & -(H_1^2)_n & -\{(G_1^2)_n \mp \mu (G_1^2)_t\} & (H_1^2)_t & H_2^2 \end{bmatrix} \begin{Bmatrix} U_1^1 \\ (U_2^1)_t \\ (U_2^1)_n = -(U_1^2)_n \\ (F_2^1)_t = (F_1^2)_t \\ (U_1^2)_t \\ U_2^2 \end{Bmatrix} = \begin{bmatrix} G_1^1 & O \\ O & G_2^2 \end{bmatrix} \begin{Bmatrix} F_1^1 \\ F_2^2 \end{Bmatrix} + \begin{Bmatrix} B^1 \\ B^2 \end{Bmatrix} \tag{19}$$

Combination Solution Method

Since concrete has nonlinear material properties, the boundary element method which is most suitable for a linear system may not be applied to the analysis of concrete structures in general. In the analysis of highly nonlinear behavior of concrete structures, however, it is reported that the so-called combination solution method is effective (12). This method uses finite elements for a nonlinear region and boundary elements for the region in which the nonlinearity may be neglected. The boundary element region is transformed into the finite element region by using a distribution matrix which converts surface tractions into nodal forces. This method is effective and attractive in the analysis of concrete structures because it has the advantages of both FEM and BEM. In particular, it is useful because various types of finite elements can be combined with boundary elements. For the analysis involving shear transfer across an interface, the nonlinear material properties at the interface can be expressed by linkage finite elements with nonlinear stiffness coefficients.

APPLICATIONS OF SHEAR TRANSFER MODELS

The above-mentioned analysis procedures for FEM, BEM, and the combination solution method will be illustrated in the following by three numerical examples.

Concrete Plate Subject to In-Plane Forces

Consider a square concrete plate with the side length of a and a straight interface passing the centroid with an angle of θ (cf. Fig.6(a)) (11). Forced displacements Δ_x and Δ_y are applied uniformly to simulate a simple shear condition under compression (cf. Fig.6(b)). The values of the parameters are selected as $\Delta_y = \Delta_x/50$, $\theta = 0$, $\pi/16$, $\pi/8$ (radian). The friction coefficients used in the analysis are: $\mu = \infty$ for adhesion and $\mu = 1$ for frictional slipping. The concrete is assumed to be elastic. Under these conditions, the shear deformation of the plate with frictional slipping and the traction distribution of both the interface and the external boundary surfaces are investigated by BEM and FEM.

The element discretization is shown in Figs.6(c),(d). The distribution of the normal traction f_n is shown in Fig.7. Separation is observed at the ends of the interface for $\theta = \pi/16$ and $\theta = \pi/8$. Also, f_n increases as θ becomes large, and takes a large value at the center of the interface. Similarly, the distribution of the tangential traction f_t of the interface is shown in Fig.8. It is observed that the solutions of BEM and FEM are in good agreement.

Thermal Stresses of Massive Concrete under External Constraints

Consider a slab which consists of two kinds of concrete, as shown in Fig.9(a) (9). The boundary conditions are shown in Fig.9(a). The boundary element discretization is shown in Fig.9(b).

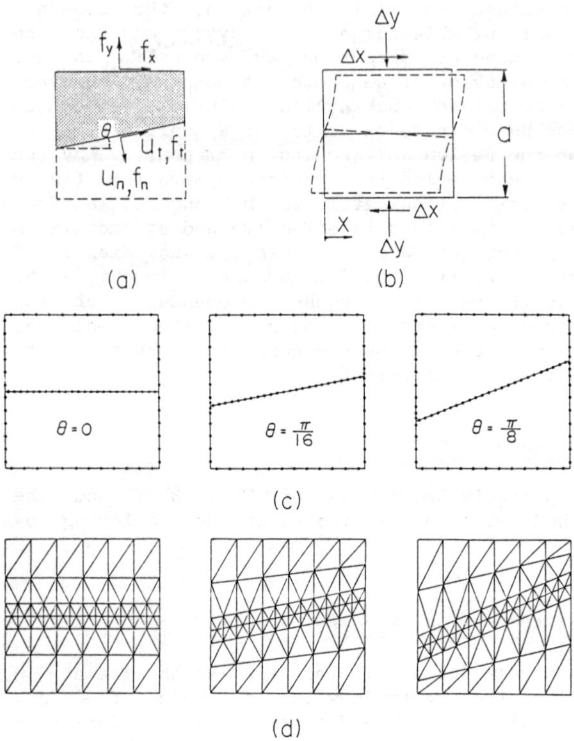

Fig.6

Concrete Plate Subject to In-Plane Forces : (a,b) Loading Conditions; (c) Boundary Element Discretization; (d) Finite Element Discretization (From Ref.11)

Linear elements with interpolation functions such that the displacements and tractions vary linearly within an element are used. The element discretization is finer at the corners of the domain so that the numerical error would be small. Since a stationary state is considered, the body force term due to temperature is expressed by a surface integral, not by a volume integral, as in a general case. Therefore, it is not necessary to divide the domain into small regions to evaluate a volume integral numerically, by which the calculation and the element discretization become simple. The finite element discretization by constant strain triangular elements is also shown in Fig.9(c).

The results of the case in which the temperature difference between the upper and lower surfaces is 10°C are shown in Fig.10. Due to the deformation of the concrete slab and the distribution of the traction along the interface, it is possible to have separation or frictional slipping if the bond of the interface is not perfect. Also, it is found that the boundary element method, using 25% less nodes and 40% less elements, yields about the same result as the finite element method. For the case of a periodic temperature change, see Ref.8.

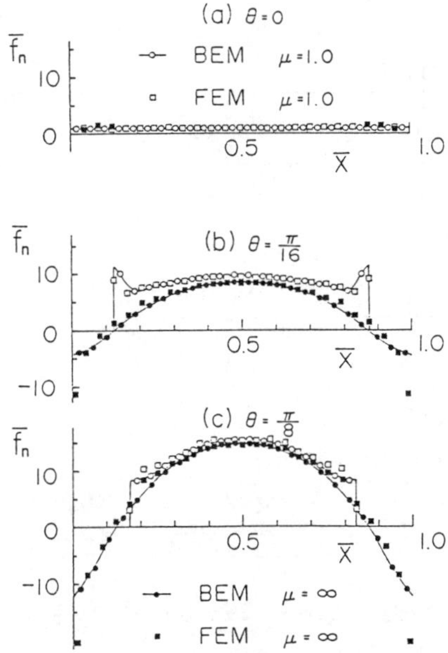

Fig.7 Normal Traction at Interface (From Ref.11)

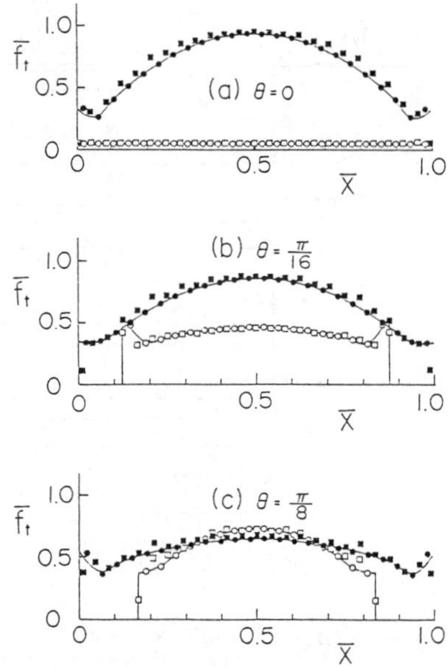

Fig.8 Tangential Traction at Interface (From Ref.11)

Bond Analysis of a Concrete Member

Consider a push-off test of a steel plate embedded in a concrete block (cf. Fig.11) (12). The relationship between shear (bond) stress and shear slip is illustrated in Fig.12. In this example, the nonlinearity involved in bond-slip is confined in the vicinity of the steel plate. Therefore, the concrete which remains linear is discretized by boundary elements and the nonlinear behavior of the interface is expressed by linkage finite elements. The element discretization is shown in Fig.13 and the relationship between the load and the displacement (slip) of the steel plate is shown in Fig.14. It is observed that the combination solution method needs about 40% less nodes (degrees of freedom) than FEM. For a bond analysis by FEM, see Ref.5, for example.

CONCLUSIONS

An analysis procedure to deal with shear transfer across interfaces in concrete structures is investigated. The boundary element method (BEM) and the combination solution method are found to be more effective and efficient than the finite element method (FEM) because BEM needs less nodes than FEM, i.e., the total number of degrees of freedom is less. This tendency seems to become clearer for a three-dimensional

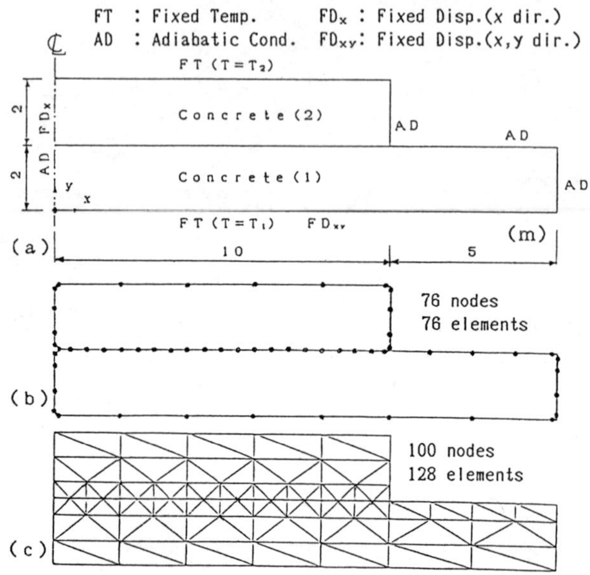

Fig.9 Boundary Conditions and Element Discretization :
(a) Boundary Conditions;
(b) Boundary Element Discretization;
(c) Finite Element Discretization
(From Ref.9) (1m=3.28ft)

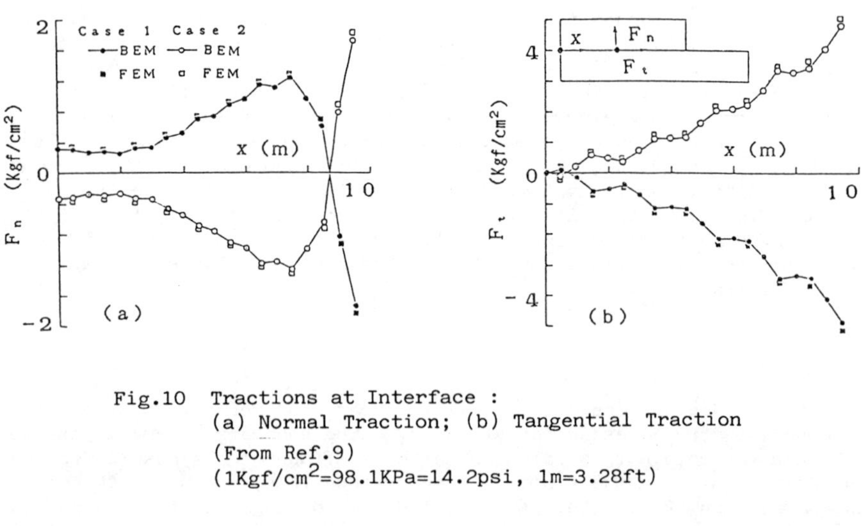

Fig.10 Tractions at Interface :
(a) Normal Traction; (b) Tangential Traction
(From Ref.9)
($1Kgf/cm^2$=98.1KPa=14.2psi, 1m=3.28ft)

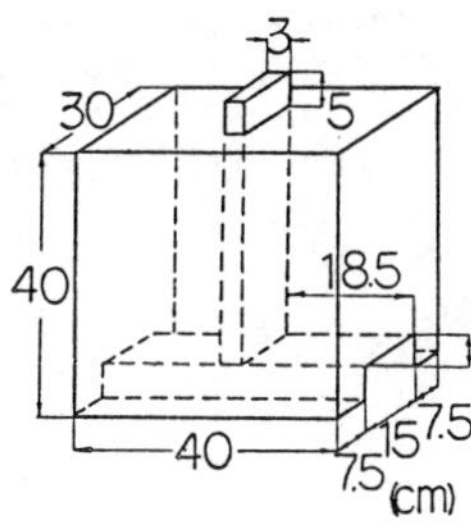

Fig.11 Size of Specimen
(From Ref.12)
(1cm=0.394in)

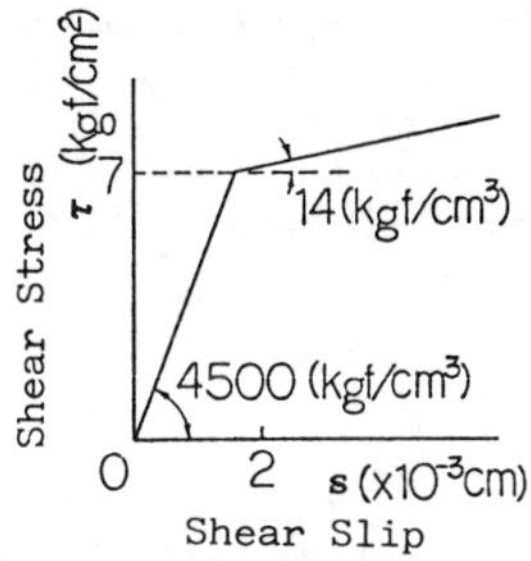

Fig.12 Material Property of Interface
(From Ref.12)
(1Kgf/cm^2=98.1KPa=14.2psi,
1cm=0.394in)

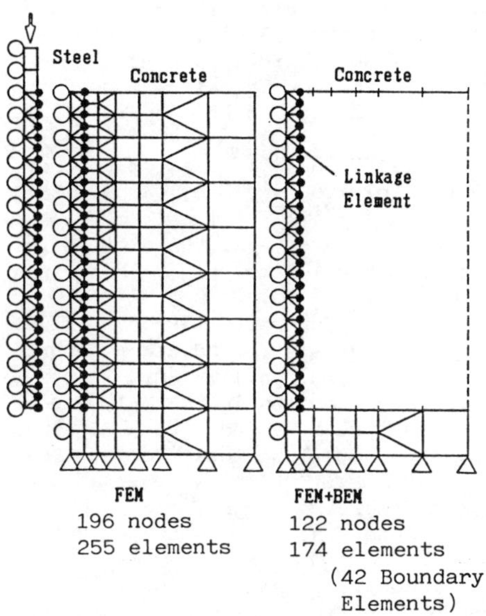

Fig.13 Element Discretization and Boundary Conditions
(From Ref.12)

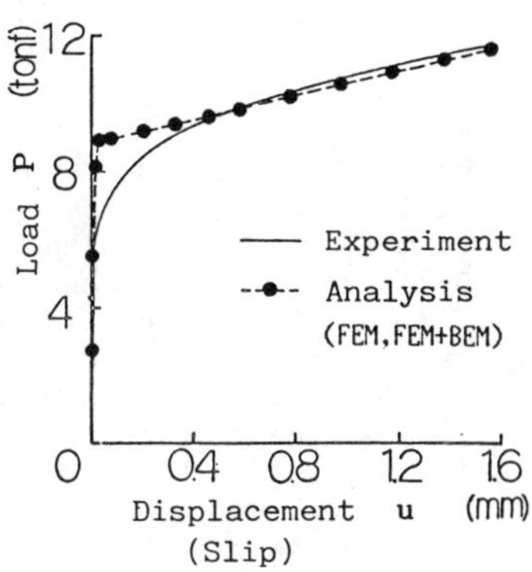

Fig.14 Load-Displacement Relationship
(From Ref.12)
(1tonf=9.81KN=2.21Klbf,
1mm=0.0394in)

problem. Also, the combination solution method seems to be useful because it has the advantages of both FEM and BEM. Finally, it would be of interest to investigate the computational efficiency of BEM and the combination solution method for other structural problems with various constitutive models of the interface shear transfer.

REFERENCES

1. Aoyagi, Y., "Analysis Approach to the Design of Orthogonally Reinforced Concrete Shell Elements Subjected to In-Plane Shear," Proc. of JCI Colloquium on Shear Analysis of RC Structures, JCI-C4E, JCI, Sep., 1983, pp.111-130.
2. Bazant, Z.P., and Gambarova, P.G., "Crack Shear in Concrete : Crack Band Microplane Model," J. of Structural Engineering, ASCE, Vol.110, No.9, pp.2015-2035, 1984.
3. Brebbia, C.A., "The Boundary Element Method for Engineers," Pentech Press, London, 1978.
4. Buyukozturk, O., Connor, J.J., and Leombruni, P., "Research on Modelling Shear Transfer in Reinforced Concrete Nuclear Structures," Nuclear Engineering and Design, Vol.59, No.1, 1980, pp.67-83.
5. Fardis, M.N., and Buyukozturk, O., "Shear Transfer Model for Reinforced Concrete," Proc. of ASCE, Vol.105, No.EM2, April, 1979, pp.255-275.
6. Fardis, M.N., and Buyukozturk, O., "Shear Stiffness of Concrete by Finite Elements," Proc. of ASCE, Vol.106, No.ST6, June, 1980, pp.1311-1327.
7. Reinhardt, H.W., Blaauwendraad, J., and Vos, E.,"Prediction of Bond between Steel and Concrete by Numerical Analysis," Materials and Structures, Vol.17, No.100, 1984, pp.311-320.
8. Tsubaki, T., "Shear Stiffness of Cracked Concrete," Proc. of JCI Colloquium on Shear Analysis of RC Structures, Introductory Report, JCI-C4E, Japan Concrete Institute, Sep., 1983, pp.131-140.
9. Tsubaki, T., "Application of Boundary Elements in Thermal Stress Analysis of Massive Concrete Structures Subject to External Constraints," Trans. of JCI, Japan Concrete Institute, Vol.6, 1984, pp.667-674.
10. Tsubaki, T., "Thermal Stresses in Concrete Subject to Periodic Temperature Change," Trans. of JCI, Japan Concrete Institute, Vol.6, 1984, pp.347-354.
11. Tsubaki, T., Yamanobe, S., and Ikeda, S., "Analysis of Concrete Plates Subject to In-Plane Forces by the Finite Element Method and the Boundary Element Method," Trans. of JCI, Japan Concrete Institute, Vol.6, 1984, pp.493-502.
12. Tsubaki, T., Hashimoto, T., Ikeda, S., and Yamanobe, S., "Analysis of Bond in a Concrete Member Using the Boundary Element Method," Trans. of JCI, Japan Concrete Institute, Vol.7, 1985.
13. Walraven, J.C., "Fundamental Analysis of Aggregate Interlock," Proc. of ASCE, Vol.107, No.ST11, November, 1981, pp.2245-2270.

DISCRETE LIMIT ANALYSIS OF R/C SHEAR WALLS

BY MASATOSHI UEDA[1] AND TADAHIKO KAWAI[2]

ABSTRACT: Based on experimental evidence of solids under the ultimate states of loading, the second author has proposed a discrete limit analysis in 1976. The models consist of rigid bodies and two types of connecting springs, one of which resists the dilatational deformation, while the other one resists the shear deformation. The theoretical basis of the models is briefly described first. Then application of these models is proposed to analyze concrete structures for which the cracking effect may play a vital role in their structural behavior. Verification studies were conducted on the analysis of the shearing type walls.

INTRODUCTION

Failure mechanisms such as crack initiation, shear slip, partial compressive collapse or tension failure of reinforcement of reinforced concrete structures are very complicated and extremely difficult to analyze by the finite element method. Today, the finite element method is considered a promising analysis tool, but to achieve adequate accuracy of the FEM solutions for nonlinear problems there does exist substantial difficulties, including formulations which satisfy the compatibility conditions between elements.

To overcome such drawbacks of the nonlinear finite element method when applied to reinforced concrete structures, the second author has proposed a family of new discrete models called " Rigid Bodies-Spring Models" (RBSM), in 1976.

1: M. Ueda is a chief research engineer at Takenaka Komuten Technical Research Laboratory, Tokyo, Japan. He will submit his doctoral thesis to the Waseda University in the near future. His research interests are related to the nonlinear analysis of reinforced concrete shell structures such as nuclear power plants, using the finite element and other discrete numerical methods. He is a member of the JCI Research Committee and AIJ Technical Committee.

2: T. Kawai is a professor in the Institute of Industrial Science at the University of Tokyo, Tokyo, Japan. He received his Ph. D. of Engineering Degree in 1957 from Lehigh University. He is a leader of the FEM developments in Japan, and his recent research interests are related to a discrete method of failure load analysis of solids and structures. He is a committee member of the founding council of the International Association for Computational Mechanics, and the associate editor of several international journals on computational mechanics.

In this paper, reinforced concrete wall panels are analyzed under a combination of axial and shear forces using RBSM models. In particular the propagation of fracture and the ultimate strength were obtained by load-increment method. Furthermore these results were compared with experiments.

Rigid Body-Spring Model Inplane Triangular Element

Fig. 1 shows the two dimensional rigid body-spring model. The details of these models are given in many other papers published by the authors [3, 4, 7] and therefore only an outline of the theoretical formulation is mentioned here. It is well known that the essential feature of plastic deformation is the slip or separation movement, and at the limiting state, solids or structures may move like a set of rigid bodies. They are assumed to be in equilibrium with external loads and reaction forces of the spring system which is distributed over the contact surface of two adjacent bodies. Therefore, a new inplane triangular RBSM element is introduced by defining three spring constants kd, ks and kr. Using this model, it is very easy to express the behavior of cracks, slip and compressive collapse in the plastic range.

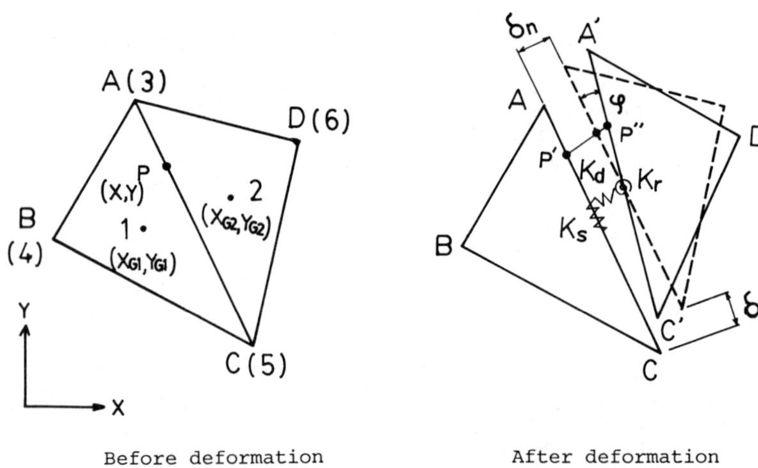

Fig. 1 RBSM in-plane triangular element

There exist two approaches proposed for discrete modeling of reinforced concrete structures. As shown in Fig. 2, one approach is the adoption of composite elements consisting of concrete elements and reinforcing bar elements, while the other is the adoption of sandwich plates composed of orthotropic materials.

The latter method may be imperative for analysis of reinforced shear walls, since such idealization is well established in modern design practice of heavyly reinforced concrete structures.

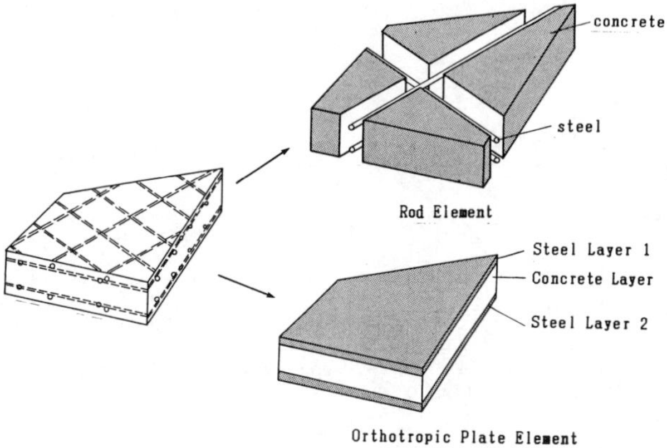

Fig. 2 Modeling of a reinforced concrete element

CHARACTERISTICS OF CONCRETE

The stress-strain relations of concrete in uniaxial compression are approximated by a trilinear curve as shown in Fig. 3. The stresses at the first and second yielding level are Fc/2 and 0.95 Fc, respectively. Tension-stiffening effect is also taken into consideration. Shear stiffness reduction ratio of Cedolin/Dei Poli [1] is adpted, as shown in Fig. 4.

The failure surface of concrete loaded in its triangular plane is divided into 7 regions as shown in Fig. 5 according to the state of stress. Stages 2 and 10 are decided according to the normal strain ε_n. After the stress state reaches the stage of 5, shear slip occurs and shear strain between the element boundary is assumed to reach the constant value γ_u, and then spring value of boundary is replaced zero at the stage of shear fracture.

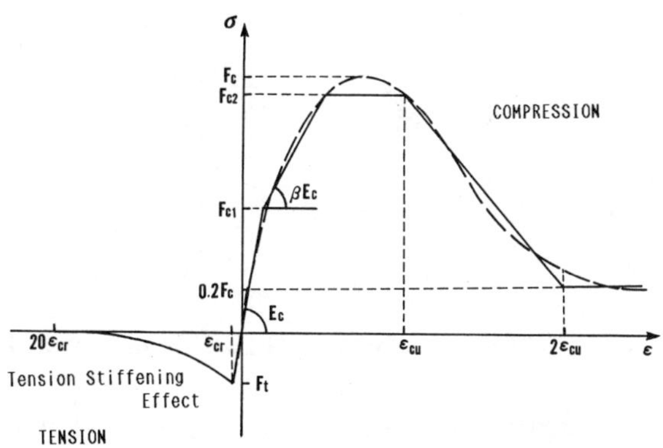

Fig. 3 Stress-strain relation of the concrete material

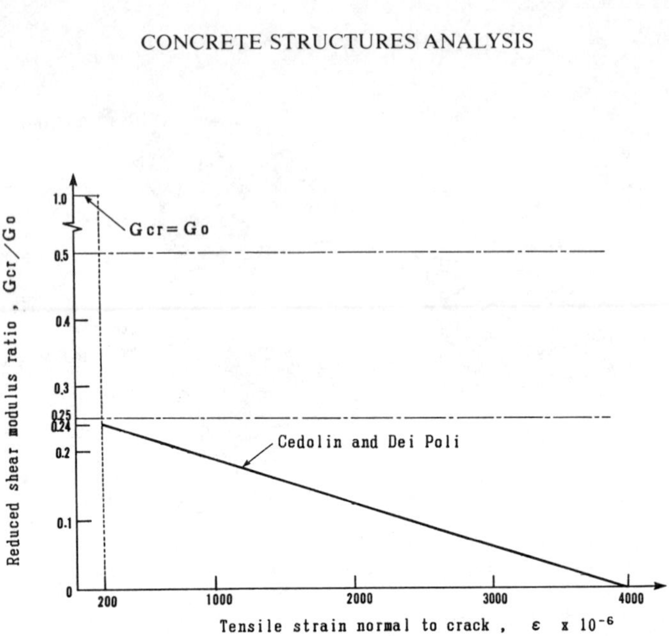

Fig. 4 Reduced shear modulus ratio G_{cr}/G_o

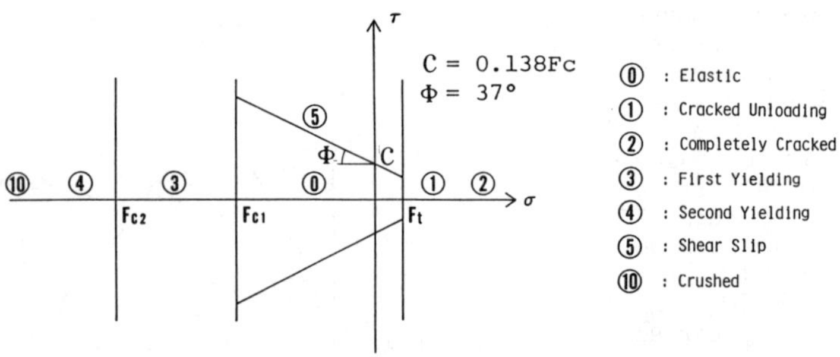

Fig. 5 Adopted yield surface of RBSM concrete model

Comparison Between Analytical And Experimental Results

To evaluate a new proposed discrete limit analysis method, both an analytical and experimental investigation were conducted, using shear wall structures as a family of specimens for a parametric study. A typical specimen and the reinforcing bar arrangement used are shown in Fig. 6. The experimental loading consisted of a constant vertical compressive force (with σ_0=20kg/cm^2), and a varying horizontal force. Ultimated failure was assumed to occur when a mixed compressive-shear fracture pattern occurs at a compression side of the base of the specimen at the yield stage under flexure loading.

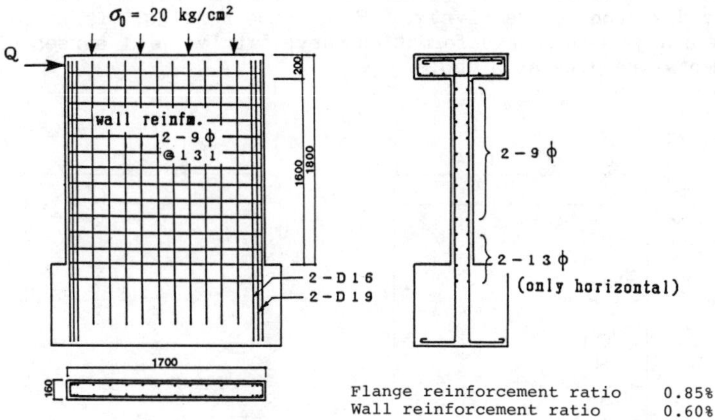

Concrete
Fc = 210 kg/cm²
Ft = 19 kg/cm²
Ec = 2.1×10^5 kg/cm²

Reinforcing bar
Wall bar fy = 4150 kg/cm²
Flange bar fy = 3850 kg/cm²
Es = 2.1×10^6 kg/cm²

Fig. 6 Shear wall specimen and reinforcing bar arrangements (A-20)

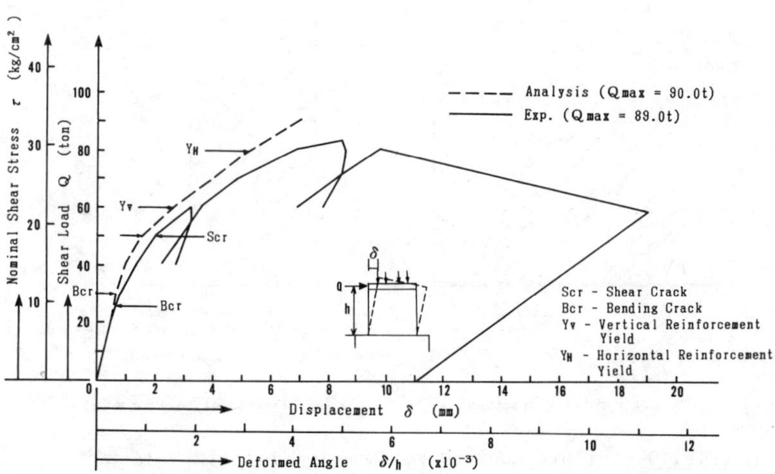

Fig. 7 Load-displacement curve (A-20)

A comparison between experimental and analytical load-deformation curves is depicted in Fig. 7. The analytical and experimental loads at which bending cracks occur are 25 ton and 30 ton, respectively. The loads at which shear cracks occur are 50 ton, and the maximum loads are 89 ton and 90 ton, respectively. From these results, it is concluded that the analytical load-deformation curve fairly well agrees with the experimental results as a whole.

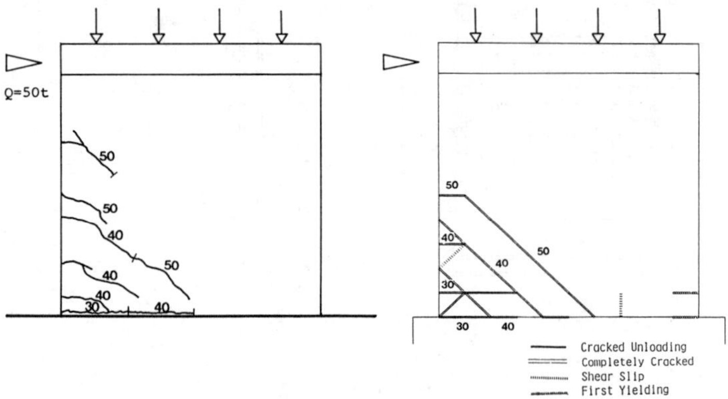

Crack pattern observed   Analytical crack pattern

Fig. 8 Crack pattern at Q = 50 ton (A-20)

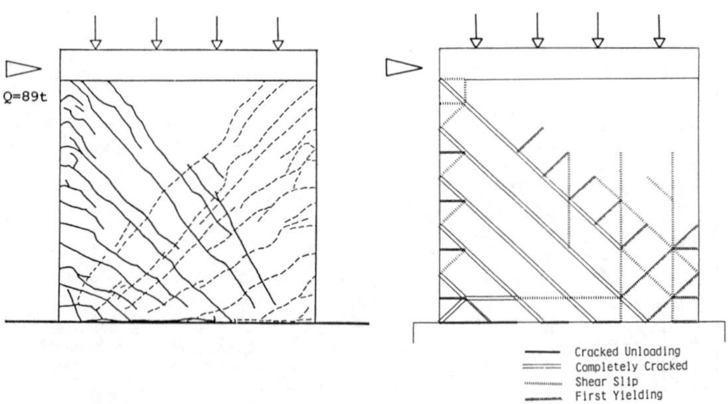

Crack pattern observed   Analytical failure mode

Fig. 9 Failure mode at maximum loading stage (A-20)

On the other hand, with respect to stiffness, the analytical results overestimate the experimental ones. It should be remembered that the experiment was conducted with varying horizontal loadings. Fig. 8 shows the analytical and experimental crack pattern at the stage of the horizontal applied force of 50 ton, and Fig. 9 shows the crack patterns at the stage of maximum load. In the analysis the final failure occurs at the base on the compression side of the concrete, and the deformation mode is shown in Fig. 10. This result indicates that the failure mode is a combination of flexure and shear fracture.

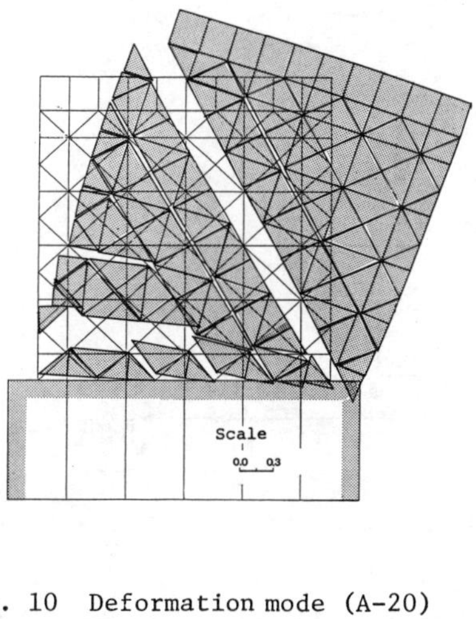

Fig. 10 Deformation mode (A-20)

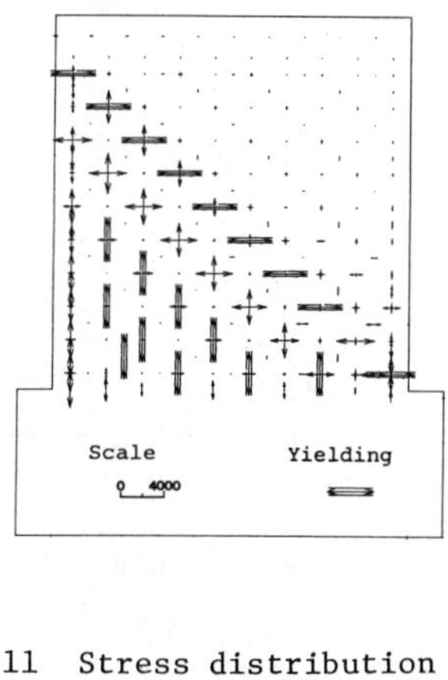

Fig. 11 Stress distribution of the reinforcements (A-20)

The analytical crack initiation load, the cracking state and also the final failure mode agree fairly well with the experimental results. Although strains in the reinforcement were not measured in the experiment, the analytical stress distribution of the reinforcement is shown in Fig. 11.

As far as the behavior of the reinforcement in a shear failure is concerned, the results show a clear inclination of the reinforcement to yield in the horizontal direction but not in the vertical direction.

Parametric Study

As a parametric study, four examples were investigated analytically, combining the cases of a vertical force and no vertical force with flange reinforcing ratios equal to or twice that of the original specimen. The results are shown in Table 1. Case A-20 is the original example. The resulting load-deformation curves are shown in Fig. 12. The maximum failure loads obtained by the analysis, and the flexural and shear strengths calculated by a general formula are also shown in Table-1. The analysis predicted for case A-20 a concrete crushing failure at a load of 90.0 ton , a failure due to yielding of the reinforcement in case A-00 at a load of 77.5 ton , and for case B-20 and B-00 shear failures at loads of 97.5 and 85.0 ton , respectively. Both the failure mode and the ultimate strength predicted by the general formulas agreed fairly well with the results of the analysis.

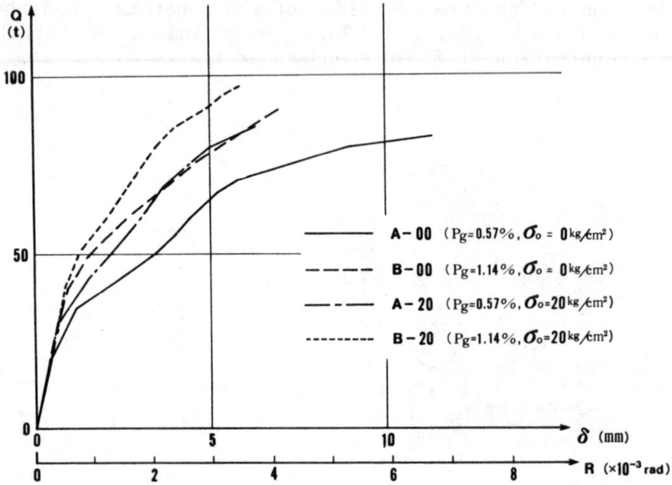

Fig. 12 Load-displacement curve of each cases

Table 1 Results due to analysis and general formula of each cases

Model	Parametric factor		Failure loads by analysis using RBSM					Failure loads calculated general formula		
								(1)	(2)	(3)
	Flange reinforcing ratio (%)	Axial stress σ_o (kg/cm²)	Q_{max} (t)	Final failure	$\frac{c\tau_s}{\bar{\tau}}$	$\frac{c\sigma_n}{0.95F_c}$	$\frac{c\sigma_D}{0.95F_c}$	Q_{bu} (t)	Q_{su} (t)	Q_{su} (t)
A-20	0.57	20.0	90.0	Comp. failure	2.05	0.83	0.72	97.0	83.1	91.0
A-00	0.57	0.0	77.5	Ten. yield	2.90	0.71	0.60	72.4	78.6	91.0
B-20	1.14	20.0	97.5	Shear failure	2.02	0.69	0.69	150.4	91.8	91.0
B-00	1.14	0.0	85.0	Shear failure	2.62	0.54	0.57	125.8	87.3	91.0

(1) Equation of flexural failure mode [2]

$$Q_{bu}=M_u/h=(0.9\,a_t\,\sigma_y\,D+0.4\,a_w\,\sigma_{wy}\,D)/h$$

(2) Equation of shear failure mode [2]

$$Q_{su}=\left(\frac{0.068P_{te}^{0.23}(F_c+180)}{\sqrt{M/QD+0.12}}+2.7\sqrt{P_s\,\sigma_y}+0.1\sigma_0\right)b_e\cdot j$$

(3) Equation of sliding shear failure mode [6]

$$Q_{su}=\frac{1}{1.5}(2.56\sqrt{F_c}+3590P_{Smin})\,t\ell$$

The flexural failure mode of case A-00 and the shear failure mode of case B-20 are shown in Figs. 13 and 14, respectively. From these deformation patterns the final failure modes can be clearly distinguished. Case A-00 exhibits a "perfectly" flexural deformation at failure. The failure modes for cases A-20 and B-20 are very similar, overall. In case A-20, wide cracks appeared almost along the entire region of the base on the tension side. On the other hand, in case B-20, diagonal crack and first yield occured initially in the wall, and finally, shear failure occured at the base of the compression side, after opening of a large diagonal crack.

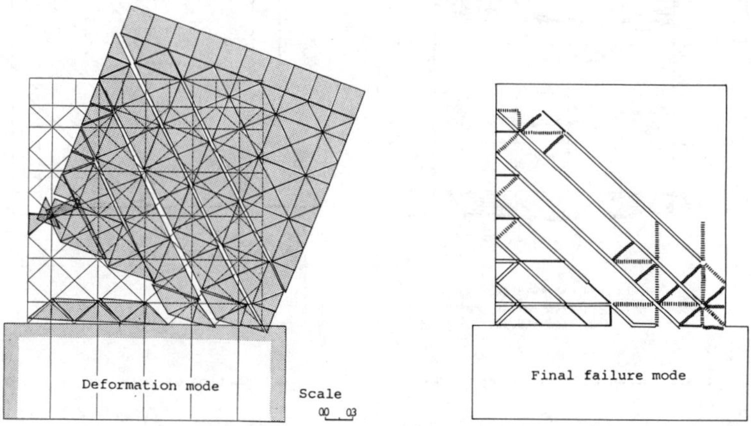

Fig. 13 Deformation and failure mode of A-00 type

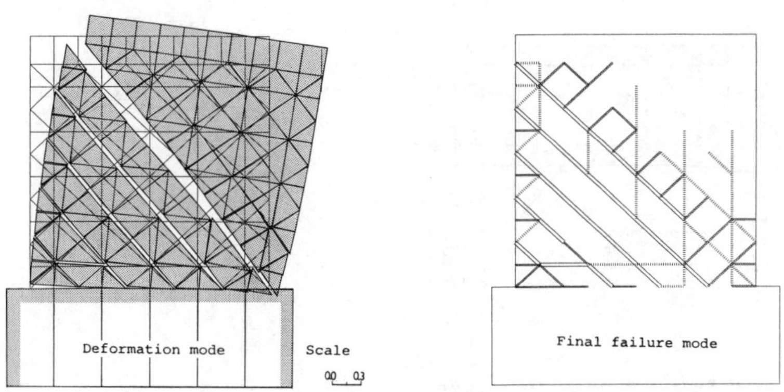

Fig. 14 Deformation and failure mode of B-20 type

The analytical results such as the distribution of shear stresses, $_{c}\tau_{s}$, normal axial stresses, $_{c}\sigma_{n}$, and axial stresses in the diagonal direction, $_{c}\sigma_{d}$, for case A-20 are shown in Fig. 15. These values are measured at the center, upper and lower part of the walls. These results have been normalized by the average shear stress τ and the second yield

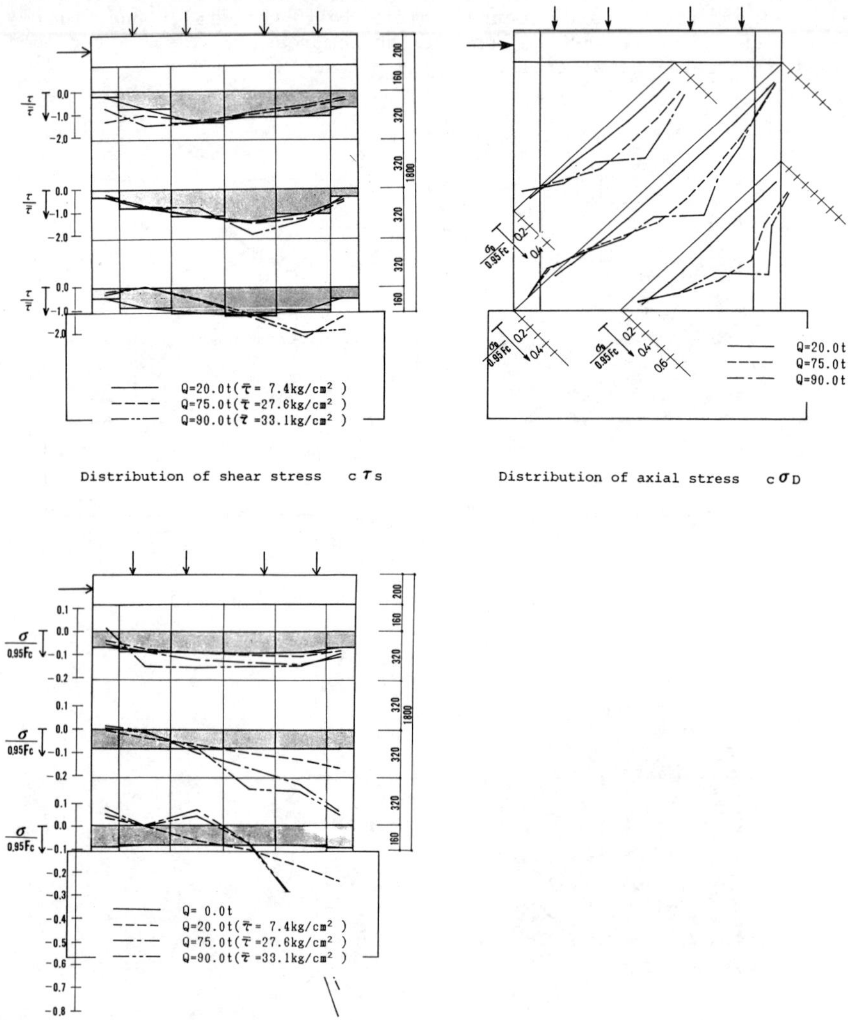

Distribution of shear stress $_{c}\tau_{s}$

Distribution of axial stress $_{c}\sigma_{D}$

Distribution of axial stress $_{c}\sigma_{n}$

Fig. 15 Distribution of shear stress, $_{c}\tau_{s}$, and axial stresses, $_{c}\sigma_{n}$, $_{c}\sigma_{D}$ of A-20 type compression stress 0.95 Fc, respectively. It is clear from these figures that the maximum values of axial and shear stresses in the stress distributions move to the compression zone of the wall in the plastic stage.

Conclusions

Computer simulation results based on discrete limit analysis using the RBSM model proposed by the second author agree fairly well with the experimental results in the case of shear walls. From these results it was concluded that slip or separation behavior of reinforced concrete structures can be expressed clearly, and furthermore final failure states or ultimate strength can be predicted using a fairly coarse discretization.

As a numerical analysis method for reinforced concrete structures, the authors can say that this discrete limit analysis using RBSM models is an intermediate model between the macro-models used in limit analysis and the micro-models used in the finite element method. Also, they expect that this method will be developed further to become an effective tool in the field of analysis of reinforced concrete structures.

Appendix.-References

1. Cedolin, L., and Dei Poli, S. " Finite Element Non-Linear Plane Stress Analysis of Reinforced Concrete, " *Studi e Rendiconti, Construzioni in Cement Armato*, Vol. 13, Politecnico di Milano, pp. 3-33, 1976.
2. Hirosawa, M., " Criterion on the Evaluation of Seismic Safety of Existing Reinforced Concrete Buildings," *Proceedings of the Second Seminar on Repair and Retrofit of Structures*, Michigan, pp. 49-108, May, 1981.
3. Kawai, T., " New Discrete Structural Models and Generalization of the Method of Limit Analysis, " *Finite Elements in Nonlinear Mechanics*, Vol. 2, Norwegian Institute of Technology, Trondheim, pp. 885-906, 1977.
4. Kawai, T., " New Discrete Models and Their Application to Seismic Response Analysis of Structures, " *Nuclear Engineering and Design*, Vol. 48, pp. 207-229, 1978.
5. Toi, Y., " Studies on the Analysis of Plate and Shell Structures Using the Simplified Elements, " *Doctoral Thesis*, University of Tokyo, Dec., 1978.
6. Tomii, M., and Esaki, F., " Predicting Method for Shear Failure Modes of Reinforced Concrete Framed Shear Walls ", *Transactions of the Japan Concrete Institute*, Vol. 4, 1982.
7. Ueda, M., Takeuchi, N., Higuchi, H., and Kawai, H., " A Discrete Limit Analysis of Reinforced Concrete Structures, " *International Conference on Computer Aided Analysis and Design of Concrete Structures*, Split, Yugoslavia, pp. 1369-1384, Sept., 1984.

ANALYSIS OF SHEAR WALL BASED ON COLLINS PANEL TEST

by

Norio Inoue[1], Norihide Koshika[2] and Norio Suzuki[2]

ABSTRACT

This paper presents a finite element analysis method using the Collins' Compression Field Theory for the tension-compression range of cracked concrete which was obtained by Collins and Vecchio's reinforced concrete panel tests. This method was applied to the analysis of shear walls with variable reinforcement ratios and shear span ratios. Comparisons between the analytical and experimental results demonstrate that the sequence of events leading to failure and the shear capacity were captured well and it can be said that this method will become a practical one.

1. INTRODUCTION

Generally, analytical models for reinforced concrete (abbreviated R/C hereafter) structures are either of a macroscopic or microscopic nature. The truss analogy is an example of a macroscopic model. Microscopic models such as those based on the finite element method can be classified into two types, in accordance with the analytical objects as follows: models for the analysis of members like beams, columns, beam-column joints and etc., which utilize relatively precise microscopic descriptions of concrete, steel, bond and etc., and "semimicroscopic" models for the description of entire R/C elements (6).

Regarding the constitutive equations of concrete under bi-axial stresses, mostly the compression-compression region has been discussed in the past, but it has now been recognized that the tension-compression region is important in order to simulate the nonlinear behavior of shear walls subjected to shear forces. Therefore, recently some studies of this region have been carried out energetically (8). Especially, Collins and Vecchio carried out some experiments of R/C panels subjected to pure shear loading or complex stresses and presented constitutive equations for reinforced concrete in the tension-compression region (3). In their reports they indicate that after the cracking both the strength and the stiffness of concrete in the compression region decrease with increasing strain. Thus, recently some analytical studies of these

---

1 Senior Research Engr., Muto Institute of Structural Mechanics, Kajima Corporation, Tokyo, Japan

2 Research Engr., Muto Institute of Structural Mechanics, Kajima Corporation, Tokyo, Japan

characteristics have been carried out (10).

In this paper shear walls with variable reinforcement ratios and shear span ratios are analyzed with a finite element program based on Collins' theory, and comparisons between the analytical and experimental results are made to demonstrate its applicability and to point out the problems to be studied in the future. This approach aims at the development of a semimicroscopic FEM model because the authors believe that choosing elements of a size comparable to entire structural members such as R/C beams or panels will be most appropriate for an efficient analysis method for entire structures.

2. FINITE ELEMENT MODEL

Reinforcing steel is modelled by equivalent plate elements as shown in Fig. 1, and the characteristics of concrete are represented by reinforced concrete elements minus the reinforcement. The nonlinear characteristics are assumed under plane stresses for each component, and equilibrium equations for nonlinear analysis are expressed with total load and secant modulus. The reason for adopting the secant modulus is that the Collins equation is defined for monotonic loading and based on the assumption that one stress state corresponds to only one strain state and that softening after the peak stress is also represented. Our analytical procedure was chosen because it incorporates these basic assumptions.

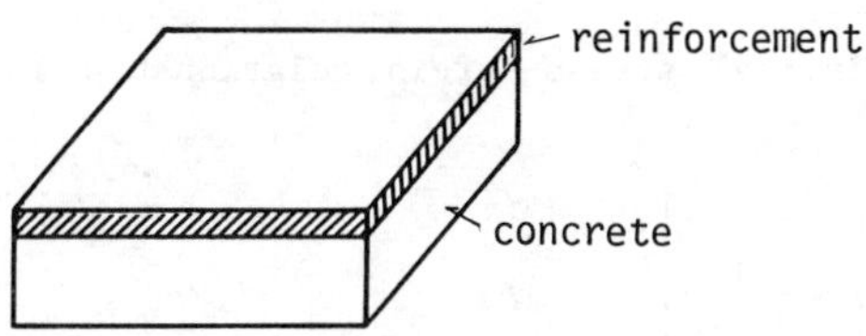

Fig.1 Model of R/C element

2.1 Stress-Strain Relationship

i) <u>Concrete</u>

Concrete is modelled by isoparametric elements with four nodes, and its constitutive equations under biaxial stresses are defined in each stress range as follows:

a. For compression-compression, compression-tension and tension-tension before cracking

This range is simulated using Darwin and Pecknold's orthogonally anisotropic model based on equivalent uniaxial strain, and the biaxial strength envelope is defined with reference to the studies of biaxial states (4) (7), Fig. 2. The stress-strain relationship up to the maximum compressive stress is defined by Saenz's formula, and in the softening region following that the relationship is defined as shown in Fig. 3 using Collins' formula.

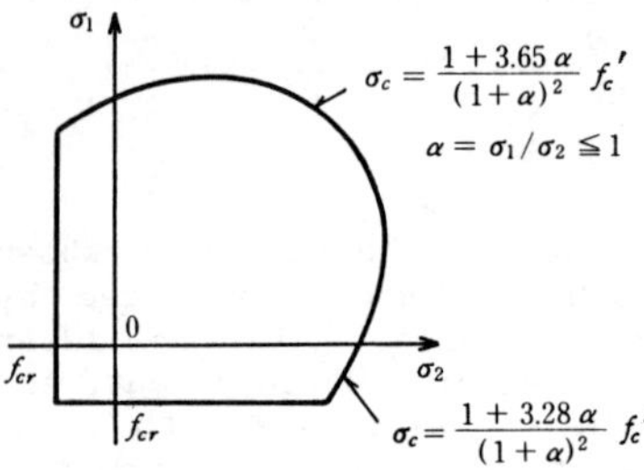

Fig.2 Biaxial strength envelope

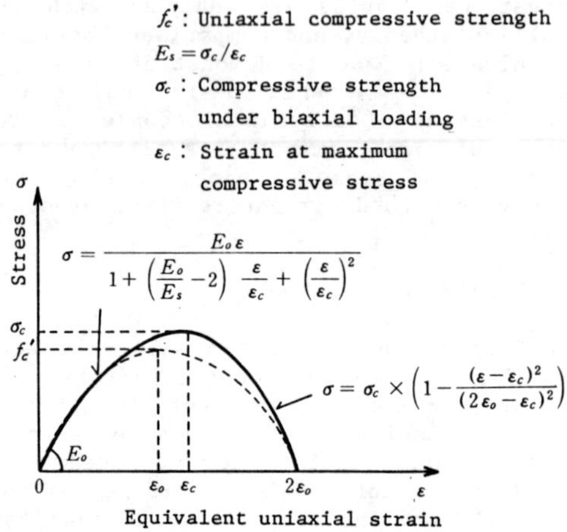

Fig.3 Equivalent uniaxial stress-strain curve of concrete for compression

The principal stress-strain relationship is written as follows (4):

$$\begin{Bmatrix} \sigma_1 \\ \sigma_2 \\ \tau \end{Bmatrix} = \frac{1}{1-\nu^2} \begin{bmatrix} E_1 & \nu\sqrt{E_1 E_2} & 0 \\ & E_2 & 0 \\ \text{sym.} & & \frac{1}{4}(E_1+E_2-2\nu\sqrt{E_1 E_2}) \end{bmatrix} \begin{Bmatrix} \varepsilon_1 \\ \varepsilon_2 \\ \gamma \end{Bmatrix} \quad \cdots\cdots(1)$$

b. For tension-compression after cracking

The Collins theory is introduced in this analysis for this range. Collins derived the principal stress-strain relationship from experimental results of R/C panels. It is written as:

$$\sigma_d = \begin{cases} f_c' \times \left[2 \times \left(\frac{\varepsilon_d}{\varepsilon_o}\right) - \beta\left(\frac{\varepsilon_d}{\varepsilon_o}\right)^2\right] & \varepsilon_d \leqq \varepsilon_p \quad \cdots\cdots(2) \\ f_p \times \left[1 - \frac{(\varepsilon_d-\varepsilon_p)^2}{(2\varepsilon_o-\varepsilon_p)^2}\right] & \varepsilon_p < \varepsilon_d \leqq 2\varepsilon_o \quad \cdots\cdots(3) \end{cases}$$

$$\beta = 0.85 - 0.27\ \varepsilon_{dt}/\varepsilon_d \quad \cdots\cdots(4)$$

$$\varepsilon_p = \varepsilon_o/\beta\ ,\ f_p = f_c'/\beta \quad \cdots\cdots(5)$$

$$\sigma_{dt} = f_{cr} \frac{1}{1+\left(\frac{-\varepsilon_{dt}}{0.005}\right)^{0.5}} \qquad \varepsilon_{dt} < \varepsilon_{cr} \qquad \cdots\cdots(6)$$

σ_d, σ_{dt} : Compressive and tensile principal stress
ε_d, ε_{dt} : Compressive and tensile principal strain
f_c', f_{cr} : Uniaxial compressive strength and tensile strength
ε_o, ε_{cr} : Strain at maximum compressive and tensile stress

The above equations are illustrated in Fig. 4. It is known that the increase in the strain ratio between tension and compression range reduces both the compressive strength of concrete and its strain at the peak stress from Eq.(4), (5) and that Eq.(6) shows the tension stiffening effect. These formulae are used in this analysis with the following modification. The coefficient β in Eq.(4) is originally expressed by the broken line in Fig. 5, but the little modification as indicated by the solid line in Fig. 5 may be necessary, because the upper limit $1/\beta_{ko}$ is dictated by the continuity condition into the tension-compression range before cracking, and the lower limit for $1/\beta$ prevents the compressive strength from being reduced to zero.

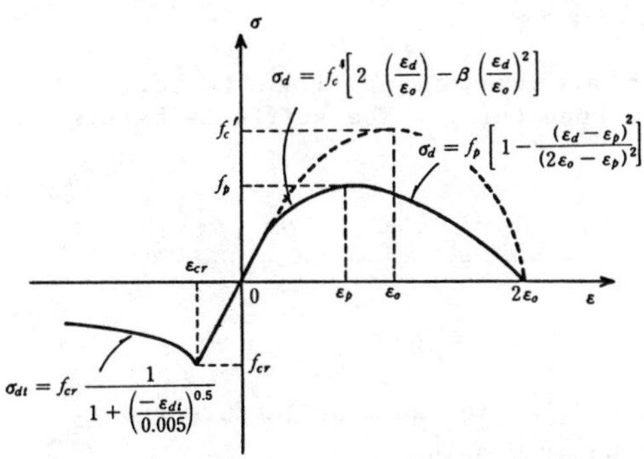

Fig.4 Stress-strain curve for compression and tension after cracking

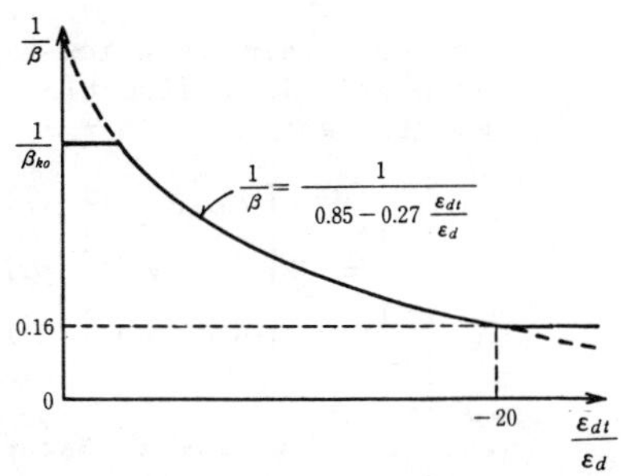

Fig.5 Relationships between the coefficient β and the principal strain ratio

Using these equations the stiffness matrix for cracked concrete [Dc] is defined for the principal axes of concrete as shown in Eq.(7). This description is based on the fundamental assumption of the Collins' Compression Field Theory that "the direction that is subjected to the largest average compressive stress will coincide with the direction that is subjected to the largest average compressive strain".

$$\begin{Bmatrix} \sigma_d \\ \sigma_{dt} \\ \tau \end{Bmatrix} = \begin{bmatrix} E_d & 0 & 0 \\ & E_{dt} & 0 \\ \text{sym.} & & G \end{bmatrix} \begin{Bmatrix} \varepsilon_d \\ \varepsilon_{dt} \\ \gamma \end{Bmatrix} \qquad \cdots\cdots(7)$$

where

E_d : Secant modulus evaluated from Eq.(2), (3)
E_{dt} : Secant modulus evaluated from Eq.(6)

Here G is a kind of shear rigidity which is added to the equation in principal directions which are obtained from the principal strains at the last iteration step, but it does not represent a true aggregate interlock relation. When a R/C panel under uniform stresses is analyzed, any non-zero value of G gives the same result. That is so because, after convergence, G has no influence on the equation (7) in the principal directions. But if the stress condition is not uniform, G might affect the results. In this study G was defined by the formula $36/\varepsilon_{dt}$ (kg/cm^2), based on Aoyagi's study (1).

c. For tension-tension after cracking

Stress-strain relationships are assumed as shown in Eq.(6) for each principal direction independently. The stiffness matrix is written as:

$$\begin{Bmatrix} \sigma_{t1} \\ \sigma_{t2} \\ \tau \end{Bmatrix} = \begin{bmatrix} E_{t1} & 0 & 0 \\ & E_{t2} & 0 \\ \text{sym.} & & G_t \end{bmatrix} \begin{Bmatrix} \varepsilon_{t1} \\ \varepsilon_{t2} \\ \gamma \end{Bmatrix} \qquad \cdots\cdots(8)$$

where

E_{t1}, E_{t2} : Secant modulus in two tensile directions
G_t : Secant shear modulus

ii) Reinforcement

Reinforcement is modelled as an anisotropic plate element with an equivalent axial stiffness, and a bi-linear stress-strain relationship as shown in Fig. 6 is assumed. The displacements of reinforcement are assumed to be compatible with those of the concrete at the nodal points.

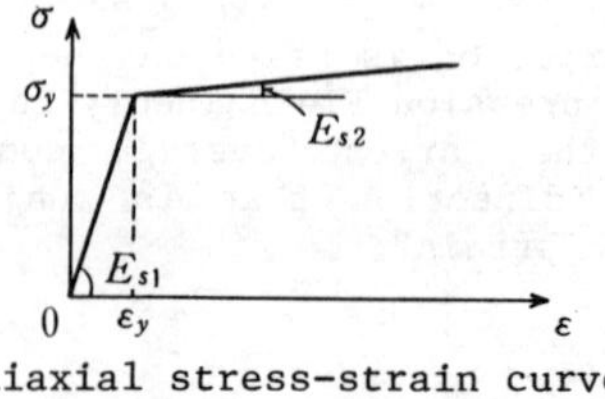

Fig.6 Uniaxial stress-strain curve of steel

2.2 Analytical Method

The equilibrium equations with the secant modulus are set up, and the nodal point displacement results are accepted when the iteration converges within a given percent of the previous cycle, for each load step.

3. ANALYSIS OF R/C SHEAR WALLS

Four R/C shear wall test specimens were analyzed by the finite element method using the above mentioned theory. Below, analytical results are compared with the test results and the propriety of the theory is discussed.

3.1 Description of the Tests

Experimental parametric studies of the R/C shear walls were carried out by Chiba et al. (2). In the tests, three parameters were varied, that is, the axial force, reinforcement ratio and shear span ratio. But in this analysis, the axial forces were kept constant, and only the reinforcement ratio and the shear span ratio were varied. Table 1 shows the selected parameter values and the identification numbers of the specimens that were analyzed and Fig. 7 shows the test specimens. In the web panel deformed bars of 6 mm diameter were used. The spacing of bars was 133 mm, 67 mm and 40 mm for 0.6%, 1.2% and 2.0% respectively.

Table 1 Specimen parameters

M/QD \ Pw	0.6%	1.2%	2.0%	σo
0.6	No.2	No.1	No.5	20 kg/cm^2
0.8		No.9		20 "

Pw : Reinforcement ratio
M/Qd : Shear span ratio
σo : Axial stress

($1 kg/cm^2 = 9.81 x 10^{-2}$ MPa)

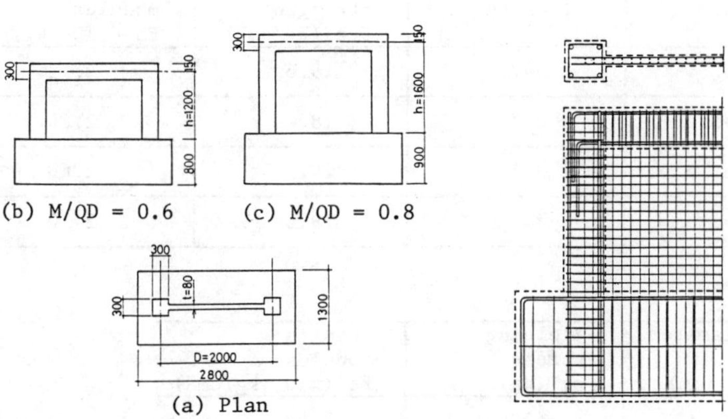

Fig.7 Geometry of test specimens (2)

Fig. 8 shows the loading apparatus. The shear force was introduced by steel beams connected to the R/C beam of the specimen with both pushing and pulling forces. Uniform axial stress conditions were created by applying a vertical force to the rubber mat atop the specimen.

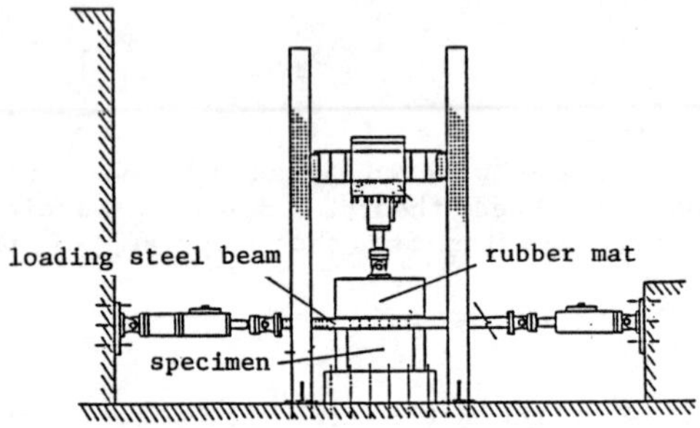

Fig.8 Loading apparatus (2)

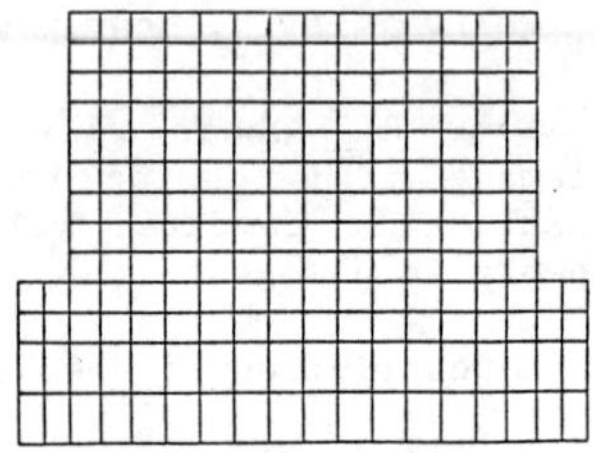

Fig.9 Finite element idealization of the specimen

3.2 Analytical Model

Fig. 9 shows the mesh layout of the FEM model of the specimen with a shear span ratio of 0.6. The reinforcement of the columns and beams was modelled as plate elements with cross sections similar to those of the specimens. The steel loading beams used in the test were modelled as steel beams parallel to the R/C beam of the specimen and were connected thereto by rigid spring elements. The rubber mat was modelled by elastic isoparametric elements, connected to the R/C beam of the specimen by rigid link elements.

Table 2 Material properties

Concrete

($1kg/cm^2 = 9.81x10^{-2}$ MPa)

Specimen No.	Compressive strength fc' (kg/cm^2)	Tensile strength fcr (kg/cm^2)	Young's modulus Eo ($x10^5$ kg/cm^2)
No. 1	347	19.6	2.73
No. 2	297	18.2	2.85
No. 5	353	19.8	2.66
No. 9	341	19.5	3.10

Steel

Diameter (mm)	Yielding stress σy (kg/cm^2)	Young's modulus Es ($x10^6$ kg/cm^2)
6	4200	1.97

Table 2 shows the material properties used in the analysis. Here the tensile strength of concrete was calculated by the formula $0.33\sqrt{fc'}$ (fc': MPa), where fc' was cylinder strength. This formula was used for representing diagonal cracking by Collins, when analyzing R/C panels.

Eq.(6) is used to express the tension stiffening effect. The analysis using Eq.(6) shall be reffered to as "Analysis 1". "Analysis 2" will be explained later.

3.3 Results of Analyses

i) Process of the failure

The observed failure mechanisms were the same in all the tests. These consisted of the following sequence of events: shear cracking, flexural cracking, yielding of the column tensile reinforcement, crushing of the concrete at the bottom of the wall and column in compression, yielding of the tensile reinforcement in the middle of

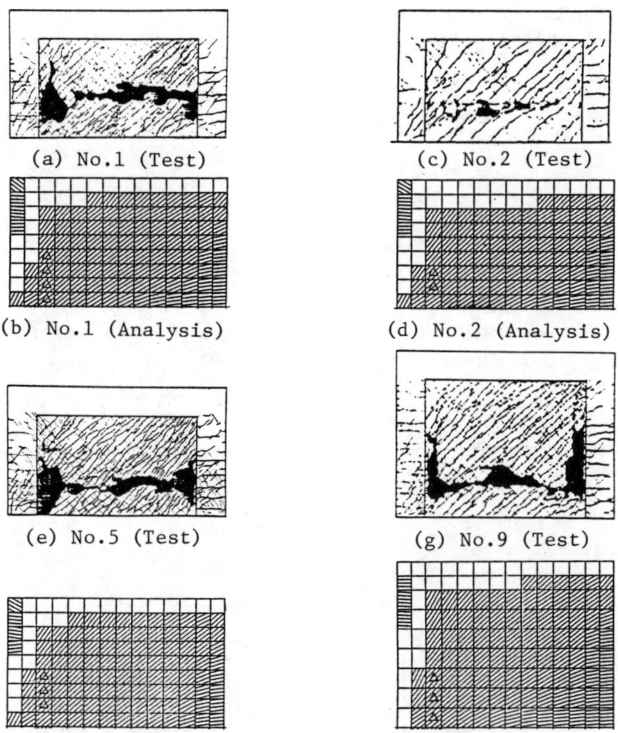

(a) No.1 (Test) (b) No.1 (Analysis) (c) No.2 (Test) (d) No.2 (Analysis) (e) No.5 (Test) (f) No.5 (Analysis) (g) No.9 (Test) (h) No.9 (Analysis)

Fig.10 Crack patterns

△ : Crushing of concrete

the wall, and finally, horizontal shear slip in the wall, leading to destruction of the specimens.

In the analyses, the sequence of events was similar to that observed in the tests: shear cracking, flexural cracking and yielding of the tensile reinforcement of the column and the wall. The theory did not consider the shear slip failure mode. Therefore, the calculated failure modes were different from those observed in the tests. In the analyses, the maximum load was reached when the concrete elements in the corner of the wall were crushed in compression.

Fig. 10 shows the crack patterns of the test specimens observed after the tests and those obtained with the analytical model at the maximum load. The directions of the cracks obtained from the analyses agree well with those observed in the tests.

ii) Force-deformation relationships

Fig. 11 shows the relationships between the shear force and the deflection angle. This analysis did not include rotational deformation due to pullout of the vertical reinforcement from the base. Furthermore, the surrounding members were modelled by equivalent plane elements and therefore the gradual spread of

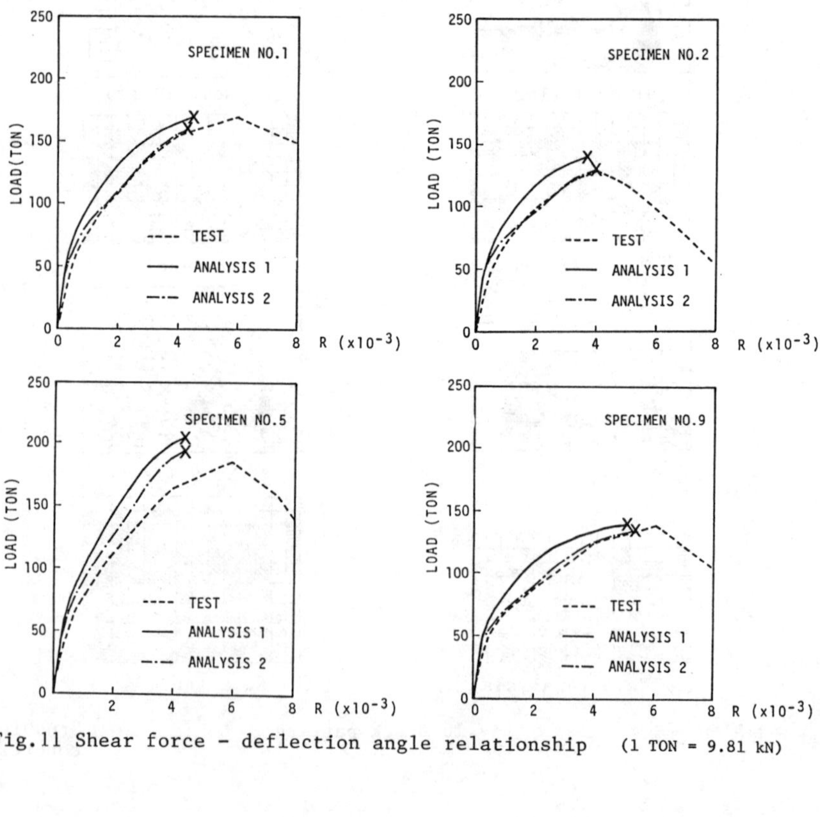

Fig.11 Shear force - deflection angle relationship (1 TON = 9.81 kN)

stresses in the thickness of the member was not represented. This modelling of boundary conditions to a web panel has some influence on deformation but has less influence on shear capacity because the capacity is dominantly subject to the failure mechanism of a panel in a stress field. For these reasons, comparisons between the analytical and experimental results are not performed here. The above mentioned factors should be studied later.

iii) Relationships between the maximum shear stress and the paremeters

Fig. 12 shows the relationship between the maximum shear stress and the wall reinforcement volume. Results of "Analysis 1" captures the tendency of the maximum shear stress to increase with reinforcement volume. The calculated maximum shear stress of the specimen with the reinforcement ratio of 1.2% is similar to the experimental one. For the other two specimens, the calculated values are a little larger than the experimental ones.

Fig. 13 shows the relationship between the maximum shear stress and the shear span ratio. The calculated maximum stresses agree with the experimental ones in the specimens with the shear span ratios of 0.6 and 0.8.

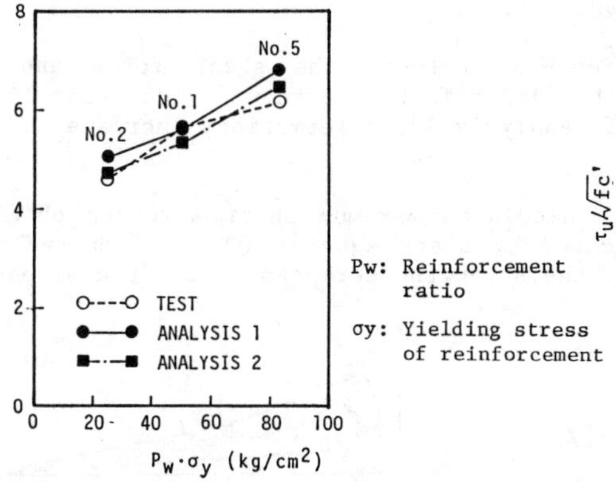

Fig.12 Maximum shear stress-reinforcement volume relationship
($1kg/cm^2 = 9.81x10^{-2}$ MPa)

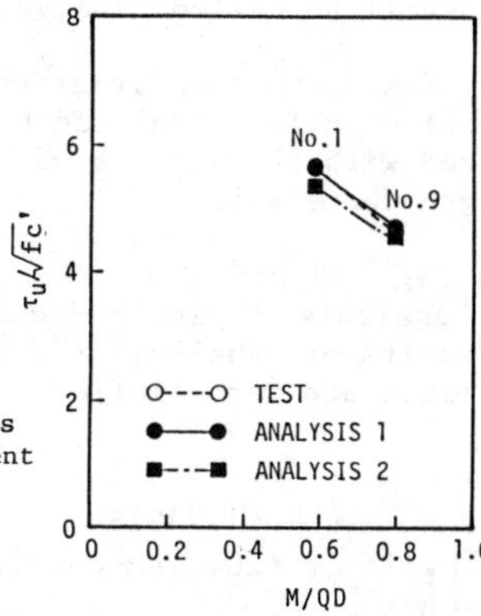

Fig.13 Maximum shear stress-shear span ratio relationship

3.4 Study of the Influence of Tension Stiffening Model

The FEM computer program used herein was based on the theory that the compressive stress-strain relationship of concrete varies in correspondence with the ratio of the tensile principal strain to compressive principal strain (See Eq.(2), (4), (5)). So, if other tension stiffening models were adopted, the calculated tensile strains would be different and then the compressive stress-strain relationship would also change. This means that the results of analysis may be influenced largely by the assumption of the tension stiffening.

Fig. 14 shows the tension stiffening models used by several researchers, where much difference is noticed among them (9). Gilbert and Warner proposed a model like Fig. 15 in which the tension stiffening was changed in accordance with the distance from the reinforcement (5). This idea is based on the fact that the stress-strain relationship of the concrete at a distance from the reinforcement is rather similar to that of plain concrete. Morita and Fujii also pointed out that the distribution of concrete stress cannot be considered uniform in the segment between cracks (9). These points suggest that the tension stiffening model should be changed in accordance with the volume, size and spacing of reinforcement. Furthermore the problem remains that there may be differences between the observed tension stiffening by one dimensional pullout tests and two dimensional panel tests because the relationship between the direction of cracks and that of reinforcement is different. Therefore, it is desirable that the tension stiffening for shear walls be evaluated by panel tests as Collins did. In such cases, scale effects might be important too.

But given the present knowledge, it is difficult to develop a tension stiffening model which considers all the influence of the reinforcing conditions. So in this study the authors assumed hypothetically another tension stiffening model shown in Fig. 16 only to study the influence of the assumption in the analysis of shear walls. The analysis using this model shall be called "Analysis 2".

In Fig. 11, the relationships between the shear force and the deflection angle calculated by "Analysis 2" are shown by dash-dot lines. Compared with the results of "Analysis 1", deformations increase and the strength decreases.

In Fig. 12 and Fig. 13, calculated maximum shear stresses obtained from "Analysis 2" are indicated by black squares (■). Compared with the results of "Analysis 1", the strength decreases. But, the slopes of both lines are very similar.

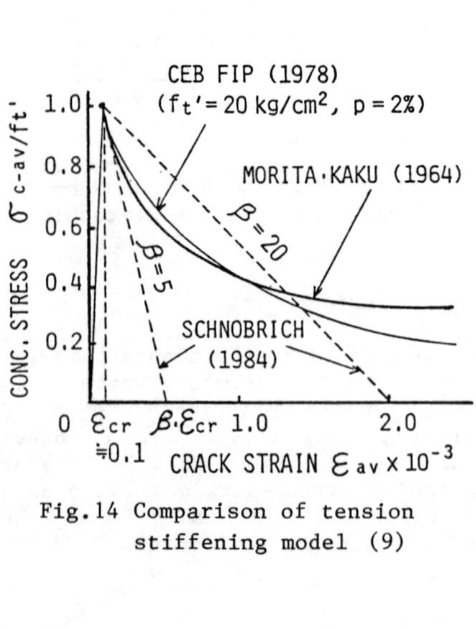

Fig.14 Comparison of tension stiffening model (9)

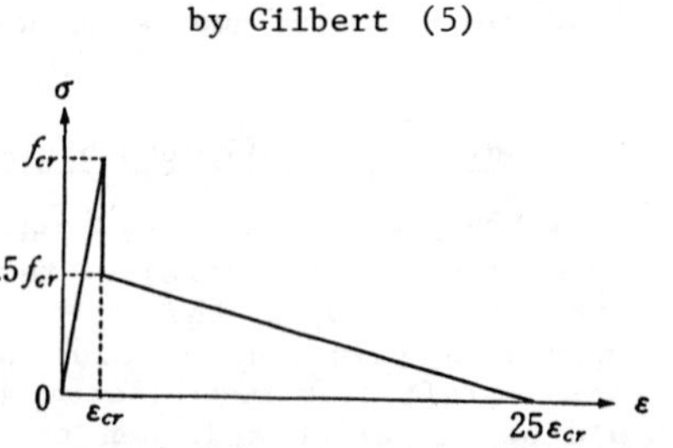

Fig.15 Tension stiffening model by Gilbert (5)

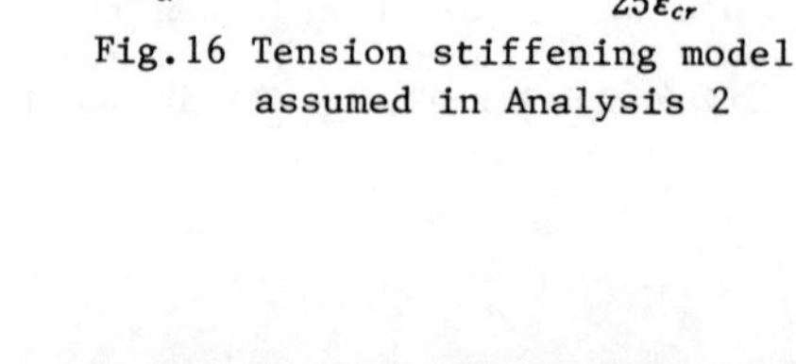

Fig.16 Tension stiffening model assumed in Analysis 2

4. CONCLUSION

A computer program for FEM analysis of R/C shear walls was developed based on the constitutive equation proposed by Collins. The accuracy of the program was checked by comparing the analytical results of R/C shear walls with experiments. As a result the sequence of events leading to failure and the shear capacity were captured well, and it can be said that this program will become a practical one.

Furthermore it was found that the tension stiffening model has a large influence on the analytical results. Therefore, it is desirable that some model be developed, which can adequately represent tension stiffening with consideration for the influnence of the reinforcing conditions.

ACKNOWLEDGEMENT

The authors wish to acknowledge the general guidance of Dr. T. Sugano and valuable suggestions of Mr. T. Miyashita, Mr. H. Morikawa and Mr. Y. Hayami, Kajima Corporation.

REFERENCES

1. Aoyagi, Y., "Analytical Approach to the Design of Orthogonally Reinforced Concrete Shell Elements Subjected to In-plane Shear," Proc. of JCI Colloquium on Shear Analysis of RC Structures, Sep. 1983, pp. 111-130
2. Chiba, O. et al, "Study on the Restoring Force Characteristics of Building (Part 8), (Part 21)," Proc. Annual Conv., AIJ, Sep. 1983, pp. 1509-1510, and Oct. 1984, pp. 2375-2376.
3. Collins, M.P. and Vecchio, F., "The Response of Reinforced Concrete to Inplane Shear and Normal Stresses," ISBN Pub. No. 82-03, Univ. of Toronto, Apr. 1982.
4. Darwin, D. and Pecknold, D.A., "Nonlinear Biaxial Stress-Strain Law for Concrete," ASCE, Vol. 103, EM2, April 1977, pp. 229-241.
5. Gilbert, R.I. and Warner, R.F., "Tension Stiffening in Reinforced Concrete Slab," ASCE, ST12, Dec. 1978, pp. 1885-1900
6. Kano, Y., Minami, K. and Takiguchi, K., "Micro and Macro Models in the Analysis of R/C Member," Proc. of JCI Colloquium on Finite Element Analysis of RC Structures, Dec. 1984, pp. 119-122.
7. Kupfer, H.B. and Gerstle, K.H., "Behavior of Concrete under Biaxial Stresses," ASCE, Vol. 99, EM4, Aug. 1973, pp. 852-866.
8. Maekawa, K., "Constitutive Law of Concrete Based on the Elasto-Plastic and Fracture Theory," Proc. of JCI 2nd Colloquium on Shear Analysis of RC Structures, Oct. 1983, pp. 1-8.
9. Morita, S. and Fujii, S., "Bond Models Used in Finite Element Analyses," Proc. of JCI Colloquium on Finite Element Analysis of RC Structures, Dec. 1984, pp. 35-42.
10. Ueda, M., Seya, H. and Kei, T., "Nonlinear Analysis of RC Cylinder and Box Type Wall Subjected to Shear Force," Proc. of JCI 2nd Colloquium on Shear Analysis of RC Structures, Oct. 1983, pp. 163-172.

NONLINEAR ANALYSIS OF SHEAR WALLS WITH NUMEROUS SMALL OPENINGS

By Kentarou Sotomura* and Yasuyuki Murazumi*

ABSTRACT: Analytical results of shear walls with numerous small openings based on the finite element method are discussed as compared with test results. The test results of six specimens were chosen for this study from a total of fifty-two test results as the first step. Analyses by two methods were conducted and their results were compared with the test results concerning the load-deflection relation, initial cracking load and crack patterns. This paper concludes which method should be applicable to the further study.

1. Introduction

Shear walls in nuclear power plants have several characteristics such as complicated configurations, large thicknesses, significant amounts of web-reinforcement and so on, compared with ordinary buildings.

Besides, numerous openings are required for the purpose of penetration of ducts and pipes. It is generally believed that such openings adversely affect the strength of the shear wall.

There are many studies on the strength of shear walls with a single large opening; however, little information is available on the behavior of shear walls with numerous small openings. In order to establish a rational design method for this kind of walls, we have carried out a series of experiments and analyses, and finally proposed a design method that gives a fairly accurate shear strength of the wall if the crack direction at the failure is correctly assumed.

In the actual design, however, there are many complicated patterns of openings so that it is difficult for the designer to imagine the correct location of the crack at the failure from the limited information obtained by experiments with such shear walls.

In order to find a useful rule for the development of the crack, FEM approaches were adopted.

The analytical results were compared with past experiments.

---

*K.Sotomura is a Manager of the design section in the nuclear facilities engineering department, Taisei Corporation, Tokyo, Japan. He has graduated from Waseda University in 1962. He is engaged in the structural design of nuclear power plant since 1970. He is a member of JCI and AIJ.

*Y.Murazumi is a senior structural engineer in the nuclear facilities engineering department, Taisei Corporation, Tokyo, Japan. He received his Master of Engineering from Kobe University in 1973. He is engaged in the structural design of nuclear power plants since 1978. He is a member of AIJ.

2. Test and Analyses

Fifty-four specimens were tested. The parameters were the opening pattern, the size of openings, the amount of reinforcement, the wall dimension and the diagonal reinforcement between openings. FEM analyses were conducted on six of these specimens shown in Figs. 1 and 2. Each specimen has a 100mm thick shear wall and 48mm diameter circular openings except specimen M3 which has no opening. Cyclic lateral loads were applied at both ends of the upper beam with two hydraulic jacks. Table 1 shows material properties of specimens.

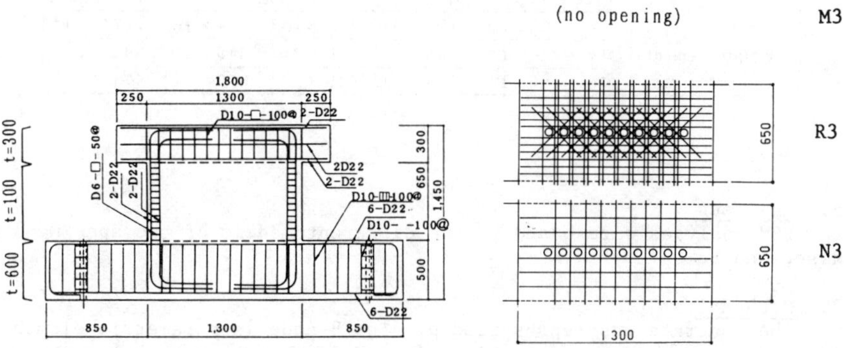

Fig. 1 Details of Specimens M3, N3, R3

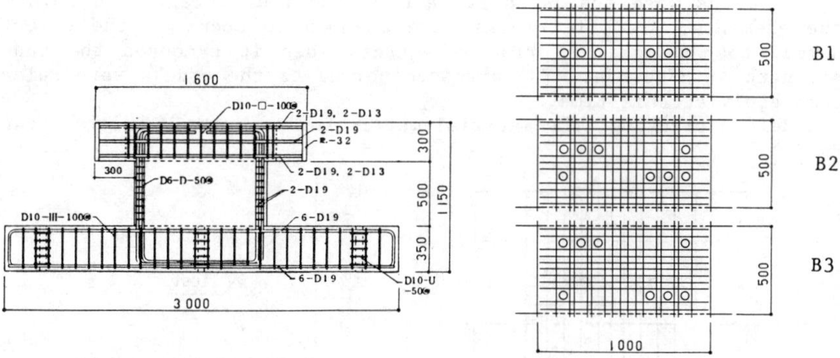

Fig. 2 Details of Specimens B1, B2, B3

Table 1 Properties of Specimens

Specimen			M 3	N 3	R 3	B1, B2, B3
Concrete	Ec	Pa	203×10^8	218×10^8	233×10^8	235×10^8
	σc	Pa	293×10^5	282×10^5	355×10^5	367×10^5
	σt	Pa	283×10^4	275×10^4	284×10^4	294×10^4
Vertical Reinforcement	No. & Size		4−D22	4−D22	4−D22	4−D19
	Es	Pa	198×10^9	196×10^9	198×10^9	188×10^9
	σy	Pa	355×10^6	361×10^6	355×10^6	385×10^6
	σu	Pa	556×10^6	555×10^6	556×10^6	590×10^6
Web Reinforcement	Size & Percent.		D6, 1.28%	D6, 0.64%	D6, 1.28%	D6, 1.28%
	Es	Pa	200×10^9	196×10^9	200×10^9	177×10^9
	σy	Pa	349×10^6	370×10^6	349×10^6	392×10^6
	σu	Pa	570×10^6	540×10^6	370×10^6	576×10^6

3. Analyses

FEM analyses were based on two different models of the specimens as discribed below:

3.1 Method 1

The concrete was represented by 4 or 8-node isoparametric elements. The reinforcing bars were modelled by 2 or 3-node truss elements with equivalent cross sections. Specimens M3, N3, B1, B2 and B3 were analyzed by this method. A typical model (B1) is shown in Fig.3.

The uniaxial stress-strain relationship for concrete was represented by a linear function in tension and a function of third order in compression (see Fig.4).

The assumed biaxial compressive failure envelope for concrete was polygonally approximated (see Fig.5).

Stresses were evaluated at each of the four integration points of the elements. Tensile cracks were assumed to occur in the direction normal to the maximum principal stress when it exceeded the tensile strength of concrete. The stresses normal to the cracks were released completely at this time.

Before cracking, the material matrix for concrete, $[D_c]$ is shown as Eq.(1).

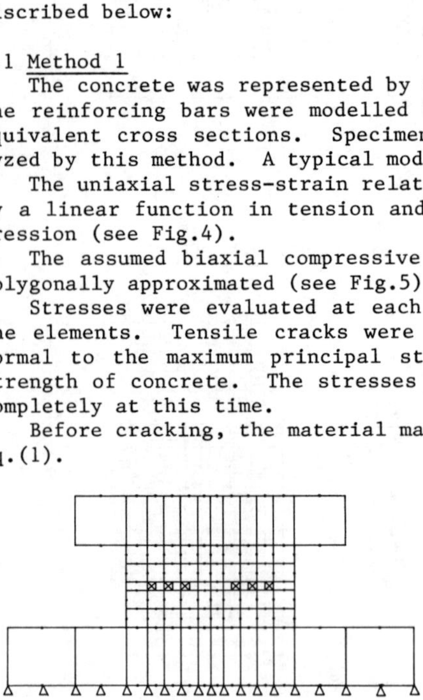

Fig.3 Analytical model by method 1

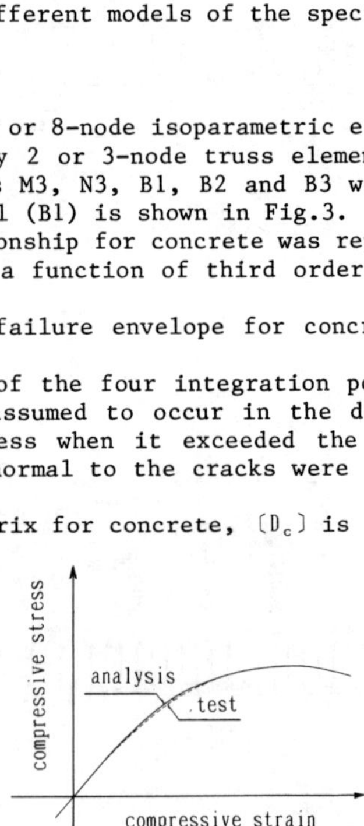

Fig.4 Stress-strain relationship of concrete

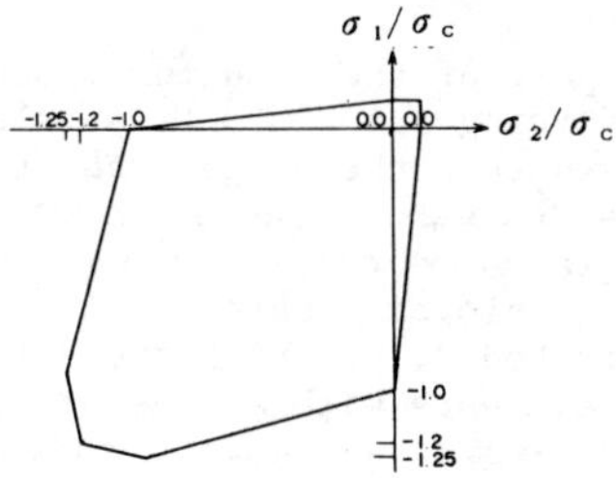

Fig.5 Failure criterion of concrete

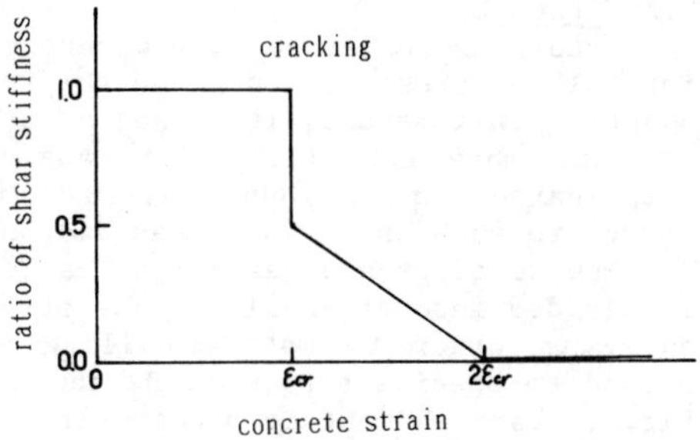

Fig.6 Reduction of shear modulus after cracking

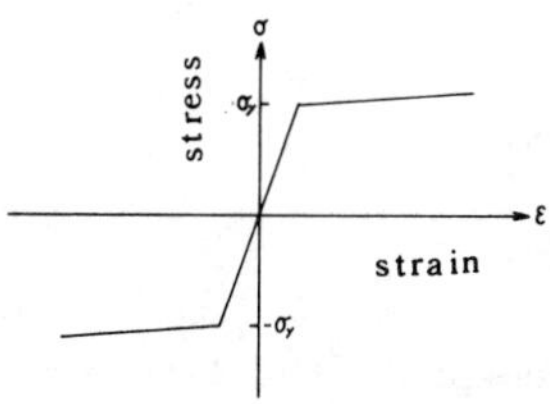

Fig.7 Stress-strain relationship of reinforcement

$$[D_c] = \frac{E_c}{1-\nu^2}\begin{bmatrix} 1 & \nu & 0 \\ \nu & 1 & 0 \\ 0 & 0 & (1-\nu)/2 \end{bmatrix} \quad \text{Eq.(1)}$$

After cracking, in a new local element coordinate system based on the crack face, concrete material matrix $[\overline{D_c}]$ can be written as Eq.(2).

$$[\overline{D_c}] = \begin{bmatrix} 0 & 0 & 0 \\ 0 & E_c & 0 \\ 0 & 0 & \eta G \end{bmatrix} \quad \text{Eq.(2)}$$

In the equations above, Ec denotes Young's modulus of concrete, ν Poison's ratio and G the initial shear modulus of concrete. η means the ratio of the concrete shear modulus after cracking to that before cracking. As shown in Fig.7, η decreases in proportion to the increase of the concrete strain normal to the crack face.

For the reinforcing bars only their axial stiffness is considered. The stress-strain relation is assumed to be elasto-plastic, both in compression and tension (see Fig.7).

Calculations were conducted for one-way loading with a 0.5 ton increment at each step.

3.2 Method 2

According to method 1, the concrete part of the structure is divided into plane stress elements and the reinforcing bars into truss elements. Adopting this method, it cannot be avoided that the larger the structure is, the more complicated the analysis becomes because of the more complicated input. And moreover, it is very difficult to model a structure such as R3 which has diagonal reinforcing bars.

Now we adopt another method called method 2, by which the structure is divided into triangular plane stress elements with a constant strain only, in order to make modelling more easy. We have applied this method to specimens R3, B1, B2 and B3. A typical model B1 is shown in Fig.8. Each element is a composite of one concrete layer and one steel layer with plane stress distributions.

Orthotropic characteristic is considered in the evaluation of the steel layer stiffnesses. The material matrix $[D_s]$ for rectangular and diagonal reinforcing bars is shown as Eq.(3) in the local coordinate system of the element.

$$[D_s] = \sum_{i=1}^{n} \alpha_i [R_i]^T [D_i] [R_i] \qquad \text{Eq.(3)}$$

in which the symbols are defined as follows:

$$[D_i] = \begin{bmatrix} E_s & 0 & 0 \\ 0 & 0 & 0 \\ 0 & 0 & 0 \end{bmatrix}$$

= material matrix for the reinforcement in the i-th direction;

α_i = equivalent thickness of the reinforcement in the i-th direction;

$[R_i]$ = transformation matrix

n = number of reinforcing directions

= 2, for the case with rectangular bars only

= 4, if diagonal bars exist;

E_s = Young's modulus of reinforcing bars.

By superimposing the concrete material matrix $[D_c]$ of Eqs.(1) and (2) onto the reinforcement material matrix $[D_s]$, we obtain the total stiffness matrix for each element.

The properties are the same as in method 1.

Calculations were conducted, as in method 1, for one-way loading with a 0.5 ton increment for each step.

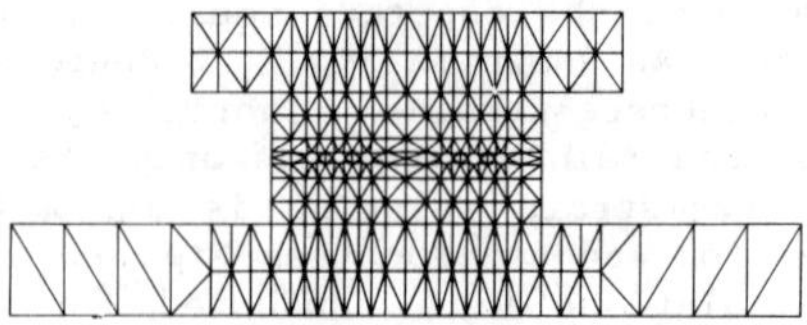

Fig.8 Analytical model by method 2

4. Results of Analyses

4.1 Load-Displacement Relationships

Lateral load-displacement relationships are shown in Fig.9.

In general, method 1 seems to give closer agreement with the experiments than method 2.

As for the ultimate load, however, both analytical methods have the tendency to underestimate the shear capacity of the wall (see Table 2). This is probably so because these analytical methods underestimate the compressive capacity of concrete at the columns near the wall edges compared with the well-confined core concrete in the experiments.

Actually analyses predicted concrete crushing at the bottom of the compression edge, which was not observed in the experiments.

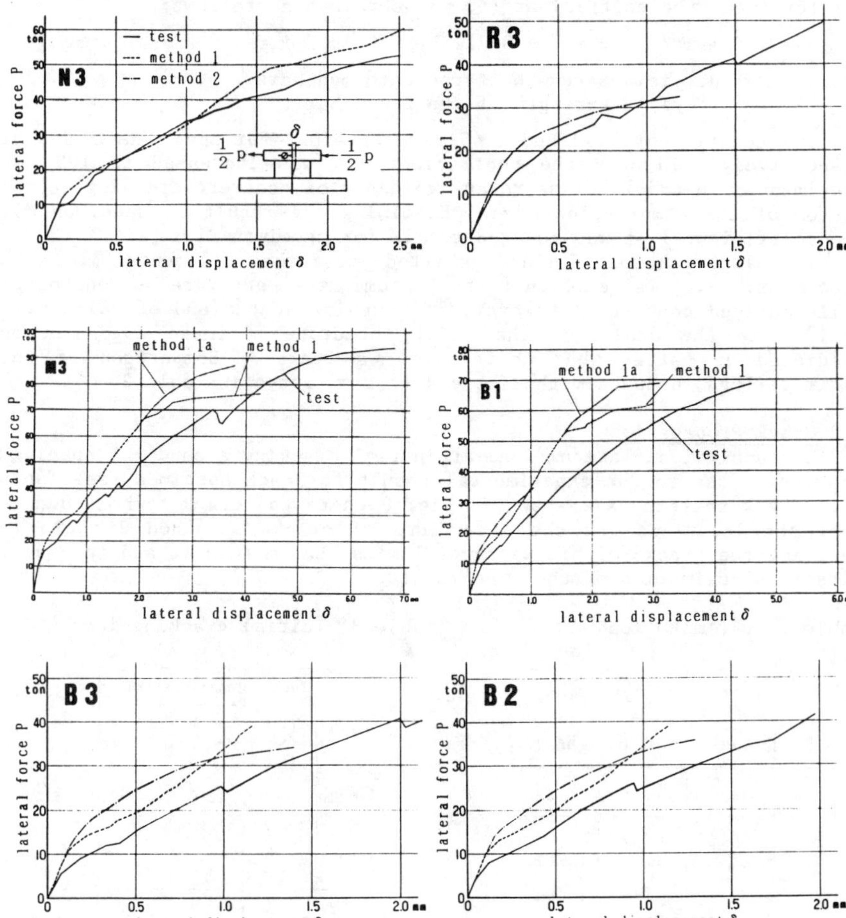

Fig.9 Load-deflection relationships

In order to take into account these circumstances, another analytical method called method 1a was applied to specimens M3 and B1 by which the compressive strength of column core concrete was assumed for specimen B1 to be 1.2 times and for specimen M3 1.5 times as such as the strength of concrete in walls and boundary beams.

Yamada gives the well-confined concrete strength (σ max.) by the following equation.

$$\sigma_{max} = (1 + 0.212\ P_h)\ \sigma_c \qquad \text{Eq.(4)}$$

Where, P_h denotes the percentage of hoop reinforcement.

Although the mild round bars were used for the hoop reinforcement in his study, we adopted the deformed bars.

Taking into account the difference of yield strength of hoop reinforcement the coefficient (ß) was obtained as follows:

$$\beta = \sigma_{max} / \sigma_c = 1 + 0.212\ P_h \times \sigma_d / \sigma_r \qquad \text{Eq.(5)}$$

Hence σ_d: yield strength of the deformed bars.
σ_r: yield strength of the round bars.

The coefficient 1.40 and 1.45 were obtained for specimen M3 and B1 respectively. Firstly the coefficient 1.4 was increased to 1.5 for specimen M3 in order to assure an avoidance of concrete crushing at the bottom of the compression edge. Examining the result of specimen M3, the coefficient 1.45 was reduced to 1.2 for specimen B1.

The location of failure shifted to other elements in both specimens. It is difficult to assume the accurate strength of well-confined concrete. However, if crushing at the end of walls were avoided in the analysis, the fairly accurate ultimate load can be predicted no matter what coefficient was taken. Comparison of the ultimate loads between methods 1 and 1a are listed in Table 2.

4.2 Crack Propagation

The analytical lateral load at initial cracking around the openings was very close to the experimental result for each specimen (see Table 3). Crack patterns are shown in Fig.10 where the crack symbols denote just the locations and the directions of cracks. Method 2 seems to restrict the cracks within narrower limits than method 1, and to give a closer approximation of the experiments.

Table 2 Ultimate load

	Test	Analysis(ton)	
	(ton)	Md. 1	Md. 1a
M 3	94.0	76.2	87.4
B 1	68.7	62.1	68.8

Table 3 Initial cracking load

	Test	Analysis(ton)	
	(ton)	Md.1	Md.2
M 3	17.5	12.2	
N 3	15.0	12.0	
R 3	5.0		11.0
B 1	9.9	9.9	10.0
B 2	8.0	9.9	10.0
B 3	8.9	11.0	10.0

5. Conclusions

It is difficult to describe analytically the cracking and failure response of shear walls with many openings.

This paper has shown that method 1 is useful to estimate the ultimate strength of such shear walls, and method 2 is useful to predict the crack propagations.

We will make some further studies to predict much more exactly the crack direction at the failure by method 2.

References

(1) Sotomura K., Yoshizaki S., et al, "Experimental study of a shear wall with numerous small openings." Transactions of 6th SMIRT 1981, J4/13.

(2) Murazumi Y., et al, "Application of ADINA to Nonlinear Analysis of Reinforced concrete Shear Walls with opening", Computer & structure, Vol.13, No.506, 1981.

(3) Sotomura K., Yoshizaki S., et al, "Ultimate shear strength of shear wall in nuclear power plant." Transactions of 8th SMIRT 1985, H4/4.

(4) Yamada M., Kikuchi T., "Effect of hoop reinforcement to inelastic deformation capability of concrete." Transactions of Kinki division of AIJ 1963 (in Japanese).

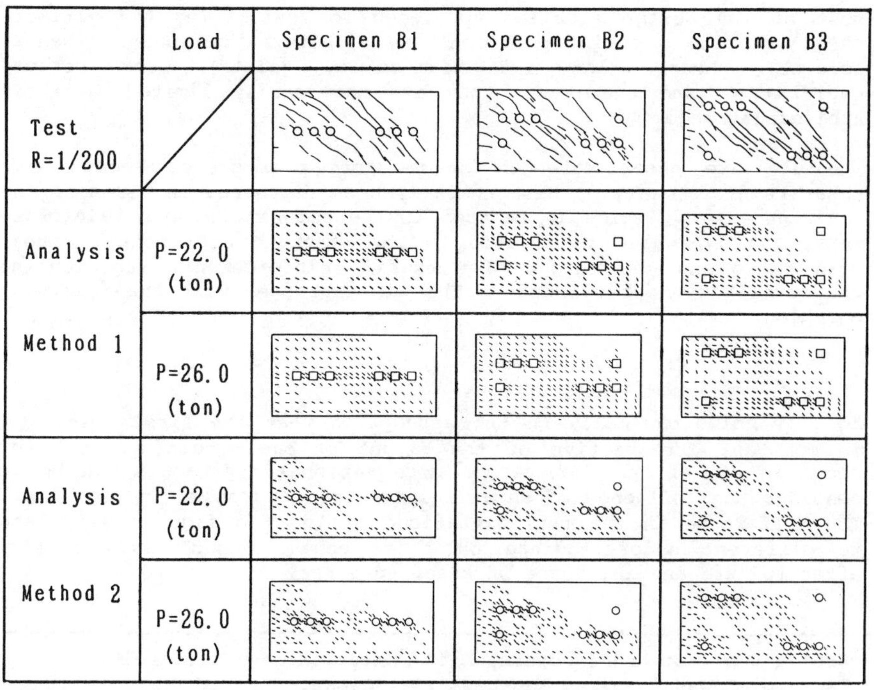

Fig.10 Crack propagation

SHEAR MECHANISM OF REINFORCED CONCRETE MEMBERS

By Yoshikazu KANOH,[1] Koichi MINAMI,[2] Katsuki TAKIGUCHI,[3] and Ichiroh SHIRAISHI[4]

ABSTRACT: A macromodel is proposed to predict analytically the ultimate shear resistance of reinforced concrete members with shear reinforcement. The shear resistance is assumed to consist of two types of mechanism; beam mechanism and arch mechanism, that explain the contributions of shear reinforcement and concrete, respectively. The applicability of the theory is confirmed through the comparison with test results and FEM numerical computations. The analyses show that the so-called "contribution of concrete" to shear strength is not constant but decreases with increase of the amount of shear reinforcement.

1. INTRODUCTION

Most of the design equations provided for estimating the ultimate shear strength of reinforced concrete structural components (beams, columns, shear walls and beam-column joints) were derived empirically, and therefore, they are necessarily limited in their applicable range.

The ultimate shear strength of reinforced concrete members with shear reinforcement, Q, has generally been expressed in terms of the resistances, Q_s, which is related to the amount of shear reinforcement, and Q_c, which is related to the strength of concrete, shear span ratio and the amount of flexural reinforcement, representing the contribution of concrete. This is expressed with the following equation.

$$Q = Q_c + Q_s \qquad (1)$$

Eq. (1) holds basically on the assumption that the first term, Q_c, is constant irrespective of the amount of shear reinforcement, as shown in Fig. 1(a). However, a more rational approach would be to consider the influence of shear reinforcement as shown in Fig. 1(b). The difference in the representations of the two figures will lead to different understandings on shear behavior such as ductility after failure and influence of shear span ratio.

---

[1]Prof., Dept. of Architecture, Meiji Univ., Higashimita 1-1-1, Tama-ku, Kawasaki-city, Kanagawa 214, Japan.

[2]Assoc. Prof., Dept. of Architecture, Osaka Institute of Technology.

[3]Assoc. Prof., Dept. of Architecture, Tokyo Institute of Technology.

[4]Research Engr., Kumagai Gumi Co., Ltd.

For verifying the authors' consideration, a macromodel of the shear resisting mechanism of reinforced concrete members and an analytical procedure to estimate the ultimate shear strength using this model are introduced. The model consists of the two types of mechanisms, the beam mechanism and the arch mechanism. The analytical procedure, based on the theory of plasticity, has been confirmed to be valid for estimating the shear strength of columns [5] [6]. In this paper, the procedure is extended so that the contribution of shear reinforcement can be estimated separately, and the influence of the amount of shear reinforcement on the contributions of the two mechanisms can be examined respectively. The basic procedure is explained. Its applicability is demonstrated by comparing the results obtained from this theoretical procedure with JCI's selected column test results [10]. Furthermore, the theoretical treatment of the function of shear reinforcement is verified by finite element analysis.

2. BEAM MECHANISM AND ARCH MECHANISM

The fundamental characteristics of shear resisting mechanism of reinforced concrete members may be represented as

$$Q = dM/dx = d(Tj)/dx = j(dT/dX) + T(dj/dx) = {}_bQ + {}_aQ \quad (2)$$

in which, Q is the shear force, M is the resisting bending moment, T is the tensile force in the flexural reinforcement, and j is the internal lever arm. The first term, j(dT/dX), represents the shear resistance associated with the change in the internal tensile force with bond action of flexural reinforcement, and this type of the shear resistance mechanism is defined as "BEAM MECHANISM". The second term, T(dj/dX) represents the shear resistance associated with the change in the internal lever arm while the tensile force is constant, and this mechanism is defined as "ARCH MECHANISM" [8]. The development of the shear failure theories have been concentrated on either the beam or the arch mechanism [1], [7] .

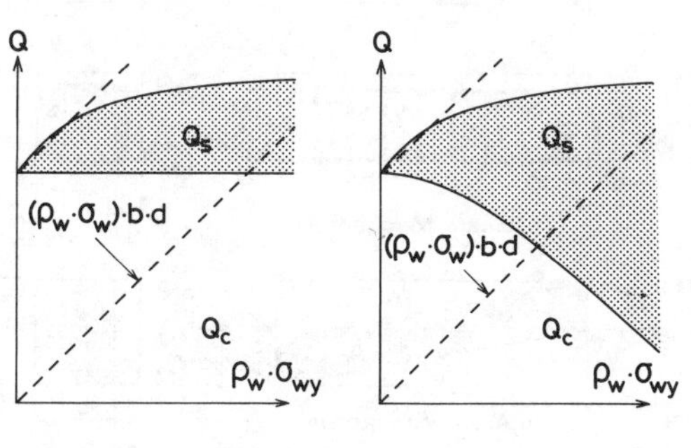

Fig. 1. Contribution of shear reinforcements to shear strength

3. MACROSCOPIC MODEL OF SHEAR STRENGTH OF COLUMNS

In this section, a rational formulation of shear strength is derived using a simple model for the columns subjected to shear with a point of contraflexure at the mid-height, as an illustration.

3.1 Application of Strength Superposition Concept

To develop a unified shear failure theory of reinforced concrete columns with shear reinforcement under combined compression, bending and shear, MINAMI and WAKABAYASHI proposed in Refs. [5] and [6] that the shear resistance of reinforced concrete members can be modeled as the sum of the resistances of the two shear resisting mechanisms; that is, the beam mechanism and the arch mechanism, as illustrated in Fig. 2. The beam mechanism consists of the flexural and the shear reinforcements and the concrete diagonal struts with the width of ${}_bb$ and inclination θ to the column axis. The arch mechanism consists of an unreinforced concrete diagonal strut with the width of ${}_ab$ (= $b - {}_bb$). Suppose that the strength combination (M, N, Q) satisfying the statically admissible stress field is determined for an individual resisting mechanism, the total resistance of the reinforced concrete column is given using the strength superposition concept, as shown in Eq. (3), because the superposition of the two statically admissible stress fields should constitute also a statically admissible state of the column as a whole.

$$M = {}_aM + {}_bM$$
$$N = {}_aN + {}_bN \qquad (3)$$
$$Q = {}_aQ + {}_bQ$$

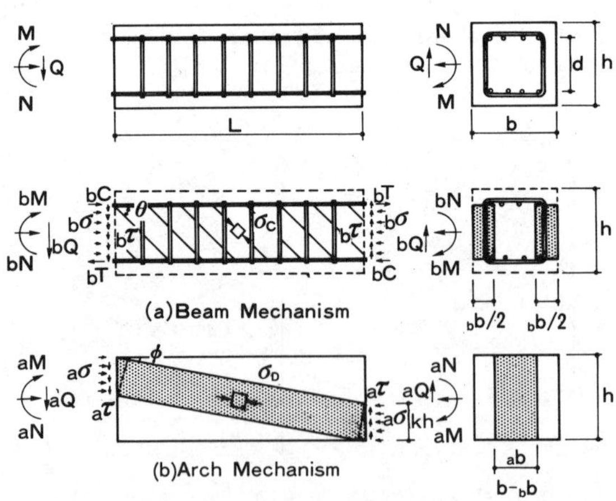

Fig. 2. Resistance mechanism of reinforced concrete columns under combined compression, bending and shear

If the inflection point of the column is assumed to be located in the mid-span of the column:

$$M/Q = {}_aM/{}_aQ = {}_bM/{}_bQ = L/2 \tag{4}$$

where L is the length of column.

3.2 Interaction Equations (${}_bn$ - ${}_bq$ equations) of Beam Mechanism

As shown in Fig. 2(a), a homogeneous compressive field with a uniform compressive stress σ_c is assumed at a constant angle θ to the column axis. The web width ${}_bb$ in the beam mechanism is so defined as to exhibit the full compressive strength in the homogeneous diagonal struts when the flexural reinforcing bars or the shear reinforcement yield. Consideration of the equilibrium equations,

$$\begin{aligned}
{}_bN &= {}_bQ \cot\theta + {}_bC - {}_bT \\
{}_bM &= ({}_bC + {}_bT)\, d/2 \\
{}_bQ &= 2\,{}_bM/L \\
{}_bQ &= {}_bb \cdot \sigma_c \cdot d \cdot \sin\theta \cdot \cos\theta \\
{}_bb \cdot \sigma_c \cdot \sin^2\theta &= b \cdot \rho_w \cdot \sigma_w
\end{aligned} \tag{5}$$

will lead us to the following relationships.

$$\begin{aligned}
-{}_bN + (\eta/d_1 + \cot\theta)\, {}_bQ &= 2\,{}_bT \\
{}_bN + (\eta/d_1 - \cot\theta)\, {}_bQ &= 2\,{}_bC \\
{}_bQ &= b \cdot d \cdot \rho_w \cdot \sigma_w \cdot \cot\theta
\end{aligned} \tag{6}$$

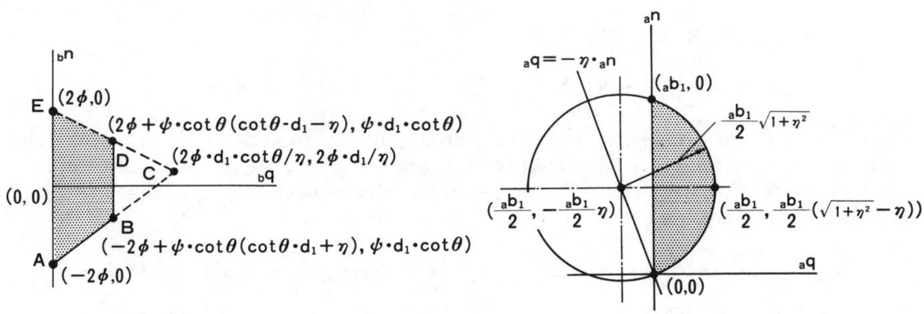

(a) Beam Mechanism (b) Arch Mechanism

Fig. 3. Interaction relationships between compression and shear

in which ${}_bT$ and ${}_bC$ are taken to be positive under tension and compression respectively. The condition that the maximum shear resistance is attained for the beam mechanism gives $\Theta = \pi/4$, $\sigma_c \to F_c$ and $\sigma_w \to \sigma_{wy}$. Hence, the interaction of the non-dimensionalized axial and shear forces for the beam mechanism is shown in Fig. 3(a), where the interaction curve comprises three regions. The interaction equations of those regions are given as follows.

$$
\begin{aligned}
&(a) \quad -2\,\Phi \leqq {}_bn < -2\,\Phi + \Psi\,(d_1 + \eta) \\
&\qquad {}_bq = ({}_bn + 2\,\Phi)\,/\,(1 + \eta/d_1) \\
&(b) \quad -2\,\Phi + \Psi\,(d_1 + \eta) \leqq {}_bn < 2\,\Phi + \Psi\,(d_1 - \eta) \\
&\qquad {}_bq = \Psi \cdot d_1 \\
&(c) \quad 2\,\Phi + \Psi\,(d_1 - \eta) \leqq {}_bn \leqq 2\,\Phi \\
&\qquad {}_bq = ({}_bn - 2\,\Phi)\,/\,(1 - \eta/d_1)
\end{aligned}
\tag{7}
$$

where ${}_bn = {}_bN/b \cdot h \cdot F_c$, ${}_bq = {}_bQ/b \cdot h \cdot F_c$, $\Phi = \rho_t \cdot \sigma_y/F_c$,

$\Psi = \rho_w \cdot \sigma_{wy}/Fc$ and $\eta = L/h$.

3.3 Interaction Equations (${}_an$ - ${}_aq$ equations) of Arch Mechanism

As shown in Fig. 2(b), the arch mechanism is assumed to consist of a diagonal concrete strut having a width ${}_ab$, inclining by ϕ, with uniform compressive stress σ_D. From the consideration of equilibrium for uniform distribution of stress ${}_a\sigma$ and ${}_a\tau$ over the compression region at both ends of the unreinforced concrete strut, the following equations are given:

$$
\begin{aligned}
{}_aQ &= \sigma_D \cdot k \cdot h \cdot {}_ab \cdot \sin\phi \cdot \cos\phi \\
{}_aN &= \sigma_D \cdot k \cdot h \cdot {}_ab \cdot \cos^2\phi \\
{}_aM &= {}_aN\,(1 - k)\,h/2 \\
\tan\phi &= (1 - k)\,h/L
\end{aligned}
\tag{8}
$$

where k is the non-dimensional depth of compressive region at the end of the member. Eliminating ϕ and k from Eq. (8), the interaction equation of the concrete with the width of ${}_ab_1$ $(= 1 - {}_bb_1)$ can be expressed as,

$$({}_aq + {}_ab_1 \cdot \eta/2)^2 + ({}_an - {}_ab_1/2)^2 = ({}_ab_1\sqrt{1 + \eta^2}/2)^2 \tag{9}$$

The interaction of the nondimensionalized axial and shear force is shown in Fig. 3(b), where the interaction curve comprises a circular arc.

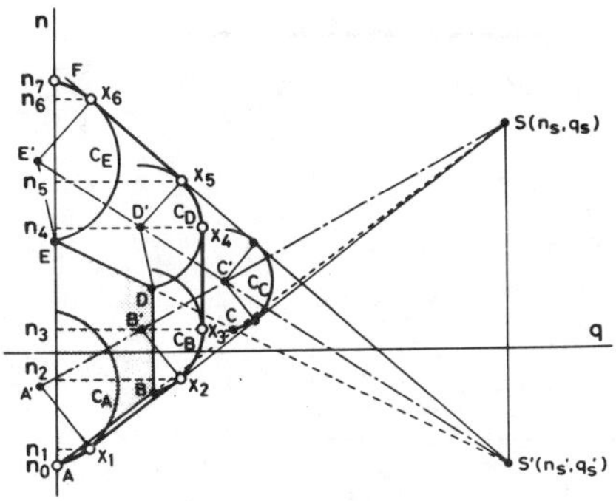

Fig. 4. Envelope of the curves obtained by the extended additive strength concept

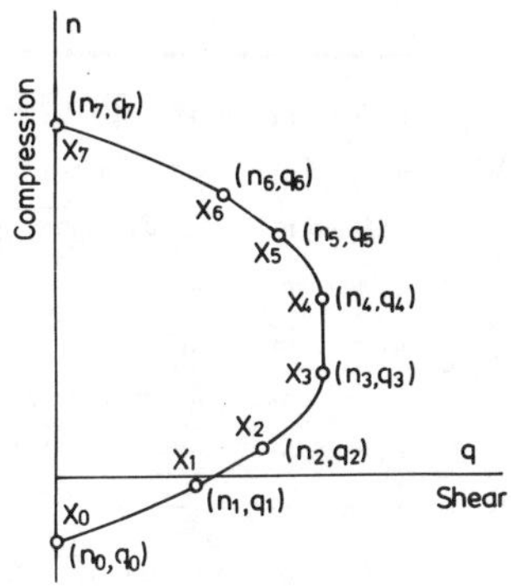

Fig. 5. Schematic example of n - q interaction curves

3.4 Interaction Equations (n - q equations) for Reinforced Concrete Columns

The non-dimensionalized axial and shear force interaction equations for the individual resisting mechanisms are superimposed in order to form the interaction equations for reinforced concrete columns. This concept of strength superposition is illustrated in Fig. 4, where the trapezoid ABDE represents the ${}_bn$ - ${}_bq$ interaction curve for the beam mechanism, and the circles C_A, C_B, C_C, C_D and C_E represent the ${}_an$ - ${}_aq$ interaction cirves for the arch mechanism associated with the change in effective concrete width ${}_ab_1$. For example, the strength of the beam mechanism represented by point B should be added to that of the arch mechanism given by the curve C_B. Similarly, the strength at point D should be added to the curve C_D, and the point E to C_E. Therefore, the n - q interaction curve for a reinforced concrete column is given by the envelope of the curves obtained by the above-mentioned technique. An example of the interaction curve obtained in this manner is shown in Fig. 5. The interaction curve can be expressed by a series of interaction equations, as listed in Table 1 and 2.

The interaction relationships give the lower bound solutions of the n - q combinations at the ultimate stress states, if, for simplicity, the concrete is assumed to undergo large plastic deformations while maintaining a stress equal or nearly equal to its maximum compressive strength, F'_c, while also the steel undergoes plastic deformations.

3.5 m-n-q Interaction Surfaces for Reinforced Concrete Columns

The m-n-q failure interaction surfaces for reinforced concrete columns are illustrated in Fig. 6 with Φ and Ψ as variables. The surfaces are computed by n-q equations in Table 1 with the condition

Table 1. n - q interaction equations

$n_0 \leqq n \leqq n_1$	$q=\{\sqrt{4(n+2\Phi)(1-n-2\Phi)+\eta^2}-\eta\}/2$
$n_1 \leqq n \leqq n_2$	$q=\lambda_1(n-n_1)+q_1$
$n_2 \leqq n \leqq n_3$	$q=\delta\{\sqrt{4na(1-na)+\eta^2}-\eta\}+\Psi\cdot d_1$
$n_3 \leqq n \leqq n_4$	$q=\delta(\sqrt{1+\eta^2}-\eta)+\Psi\cdot d_1$
$n_4 \leqq n \leqq n_5$	$q=\delta\{\sqrt{4nb(1-nb)+\eta^2}-\eta\}+\Psi\cdot d_1$
$n_5 \leqq n \leqq n_6$	$q=\lambda_2(n-n_6)+q_6$
$n_6 \leqq n \leqq n_7$	$q=\{\sqrt{4(n-2\Phi)(1-n+2\Phi)+\eta^2}-\eta\}/2$

$$\lambda_1=[\alpha\cdot\omega_1-\{\alpha\cdot\gamma+\sqrt{(\alpha\cdot\gamma)^2-\omega_1\cdot\rho_1}\}\gamma]/[\beta_1\cdot\omega_1-\{\beta_1\cdot\gamma-\sqrt{(\beta_1\cdot\gamma)^2-\omega_1\cdot\rho}\}\gamma]$$

$$\lambda_2=[\alpha\cdot\omega_2-\{\alpha\cdot\gamma+\sqrt{(\alpha\cdot\gamma)^2-\omega_2\cdot\rho_2}\}\gamma]/[\beta_2\cdot\omega_2-\{\beta_2\cdot\gamma+\sqrt{(\beta_2\cdot\gamma)^2-\omega_2\cdot\rho}\}\gamma]$$

$\alpha=(d_1+\eta)/2$ $\quad\beta_1=(d_1+\eta-1)/2$ $\quad\beta_2=(d_1-\eta-1)/2$ $\quad\gamma=\sqrt{1+\eta^2}/2$

$\delta=(1-2\Psi)/2$ $\quad\rho=\gamma^2-\alpha^2$ $\quad\rho_1=\gamma^2-\beta_1^2$ $\quad\rho_2=\gamma^2-\beta_2^2$

$\omega_1=\alpha^2+\beta_1^2$ $\quad\omega_2=\alpha^2+\beta_2^2$

$na=\{n+2\Phi-\Psi(d_1+\eta)\}/2\delta$

$nb=\{n-2\Phi-\Psi(d_1+\eta)\}/2\delta$

Table 2. Specific values for n - q interaction equations

$X_0: n_0=-2\Phi$	$q_0=0$
$X_1: n_1=\{\beta_1\cdot\gamma-\sqrt{(\beta_1\cdot\gamma)^2-\omega_1\cdot\rho}\}\gamma/\omega_1+(1-4\Phi)/2$	$q_1=\{\alpha\cdot\gamma+\sqrt{(\alpha\cdot\gamma)^2-\omega_1\cdot\rho_1}\}\gamma/\omega_1-\eta/2$
$X_2: n_2=2\delta(n_1+2\Phi)-2\Phi+\Psi(d_1+\eta)$	$q_2=2\delta\cdot q_1+\Psi\cdot d_1$
$X_3: n_3=\delta-2\Phi+\Psi(d_1+\eta)$	$q_3=\delta(\sqrt{1+\eta^2}-\eta)+\Psi\cdot d_1$
$X_4: n_4=\delta+2\Phi+\Psi(d_1-\eta)$	$q_4=\delta(\sqrt{1+\eta^2}-\eta)+\Psi\cdot d_1$
$X_5: n_5=2\delta(n_6-2\Phi)+2\Phi+\Psi(d_1-\eta)$	$q_5=2\delta\cdot q_6+\Psi\cdot d_1$
$X_6: n_6=\{\beta_2\cdot\gamma+\sqrt{(\beta_2\cdot\gamma)^2-\omega_2\cdot\rho}\}\gamma/\omega_2+(1+4\Phi)/2$	$q_6=\{\alpha\cdot\gamma+\sqrt{(\alpha\cdot\gamma)^2-\omega_2\cdot\rho_2}\}\gamma/\omega_2-\eta/2$
$X_7: n_7=1+2\Phi$	$q_7=0$

$m = q\cdot\eta/2$ derived from Eq. 4. The interaction surfaces which enable one to assess the effect of axial force and bending moment upon the shear strength of columns have a concave shape in the region where shear failure is predominant over flexural failure, while the surfaces are convex for columns with much shear reinforcement and larger values of η. Also, it is indicated that the influence of axial force upon the shear strength is less in columns with less shear reinforcement and smaller values of η.

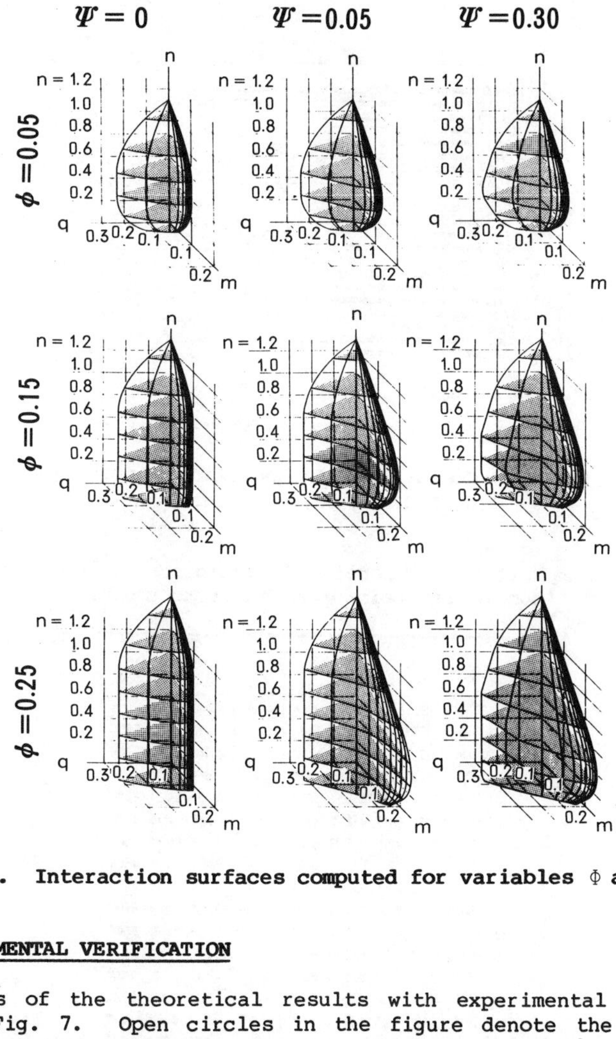

Fig. 6. Interaction surfaces computed for variables Φ and Ψ

4. EXPERIMENTAL VERIFICATION

Comparisons of the theoretical results with experimental ones are shown in Fig. 7. Open circles in the figure denote the measured strengths of JCI's selected column specimens in Ref.[10]. Open and solid column marks denote the shear resisted by the arch mechanism and the beam mechanism, respectively. The ratio of the measured to the theoretical strength scatters between 0.98 and 1.34 (except for specimen EN12 with simple supports) with an average value of 1.12, indicating a close agreement between them. Comparison of predictions with test results are listed in Table 3.

5. EFFECTS OF SHEAR REINFORCEMENT

One of the characteristics of the above mentioned shear strength theory considering the beam and arch mechanism is that the shear

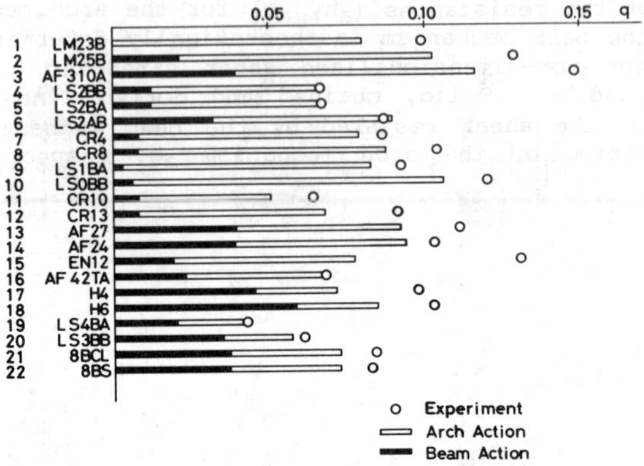

Fig. 7. Comparison of theoretical prediction with JCI's selected test results

Table 3. Comparison of theoretical prediction with test results for JCI's selected reinforced concrete columns

Specimen	Φ	Ψ	d_1	η	n	Measured Strength $q_{meas.}$	Theoretical Strength q	$_aq$	$_bq$
1 LM23B	0.052	0.055	0.72	2	0.107	0.099	0.0800	0.0800	0
2 LM25B	0.098	0.148	0.72	2	0.107	0.129	0.1027	0.0811	0.0215
3 AF310A	0.180	0.172	0.72	2	. 0	0.148	0.1165	0.0767	0.0398
4 LS2BB	0.067	0.030	0.72	3	0.136	0.066	0.0654	0.0527	0.0216
5 LS2BA	0.067	0.058	0.72	3	0.136	0.067	0.0654	0.0527	0.0126
6 LS2AB	0.067	0.113	0.72	3	0.272	0.087	0.0898	0.0576	0.0321
7 CR4	0.115	0.012	0.72	3	0.137	0.086	0.0789	0.0702	0.0086
8 CR8	0.115	0.012	0.72	3	0.261	0.104	0.0872	0.0786	0.0086
9 LS1BA	0.114	0.154	0.72	3	0.136	0.092	0.0828	0.0496	0.0332
10 LS0BB	0.177	0.191	0.72	3	0.136	0.120	0.1061	0.0454	0.0607
11 CR10	0.115	0.012	0.72	4	0.137	0.064	0.0608	0.0522	0.0086
12 CR13	0.115	0.012	0.72	4	0.261	0.091	0.0680	0.0593	0.0086
13 AF27	0.111	0.124	0.72	4	0.580	0.111	0.0921	0.0522	0.0399
14 AF24	0.111	0.236	0.72	4	0.436	0.103	0.0940	0.0540	0.0399
15 EN12	0.256	0.024	0.80	4	0.571	0.131	0.0777	0.0585	0.0192
16 AF42TA	0.257	0.033	0.72	4	−0.110	0.068	0.0687	0.0450	0.0237
17 H4	0.210	0.080	0.73	5	0.100	0.098	0.0718	0.0260	0.0457
18 H6	0.210	0.080	0.73	5	0.200	0.103	0.0843	0.0259	0.0584
19 LS4BA	0.114	0.040	0.72	6	0.136	0.043	0.0439	0.0231	0.0207
20 LS3BB	0.177	0.050	0.72	6	0.136	0.061	0.0572	0.0222	0.0350
21 8BCL	0.158	0.107	0.72	4	0.109	0.084	0.0729	0.0351	0.0378
22 8BS	0.158	0.107	0.72	4	0.109	0.083	0.0729	0.0351	0.0378

strength of columns with shear reinforcement is related to the amount of shear and flexural reinforcements (Φ and Ψ), length-to-depth ratio of columns (η) and the magnitude of axial force (n). Illustrated in Fig. 8 is the strength superposition concept, where $_aI$ and $_bI$ represents the resistance vectors of the arch mechanism and the beam mechanism, respectively. When the axial force is given, a set of the resistances ($_an$, $_aq$) for the arch mechanism, and ($_bn$, $_bq$) for the beam mechanism is theoretically determined. Figure 9 shows how the non-dimensionalized shear strengths q, $_aq$ and $_bq$ varys with Ψ and n. Solid, chained and dotted lines denote the total shear, q, the shear resisted by the beam mechanism, $_bq$, and the shear resisted by the arch mechanism $_aq$, respectively. The

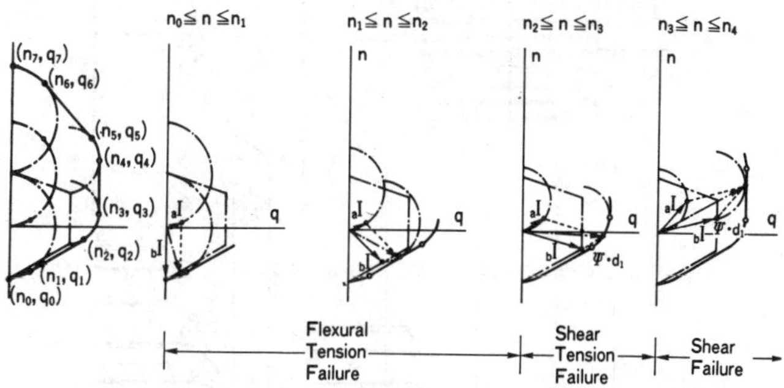

Arch	$_an$	$n-{}_bn$	$n-{}_bn$	$n-{}_bn$	δ
	$_aq$	$(\sqrt{4(n+2\phi)(1-n-2\phi)+n^2}-n)/2$ $(=q)$	$q-{}_bq$	$\delta(\sqrt{4n_a(1-n_a)+n^2}-n)$	$\delta(\sqrt{1+n^2}-n)$
Beam	$_bn$	-2ϕ	$(1+n/d_1)_bq-2\phi$	$-2\phi+\Psi(d_1+n)$	$n-{}_an$
	$_bq$	0	$\dfrac{\lambda_1(n-n_1)+q_1+n/2+(n+2\phi-1/2)/\lambda_1}{1+n/d_1+(1+(n-1)/d_1)/\lambda_1}$	$\Psi\cdot d_1$	$\Psi\cdot d_1$

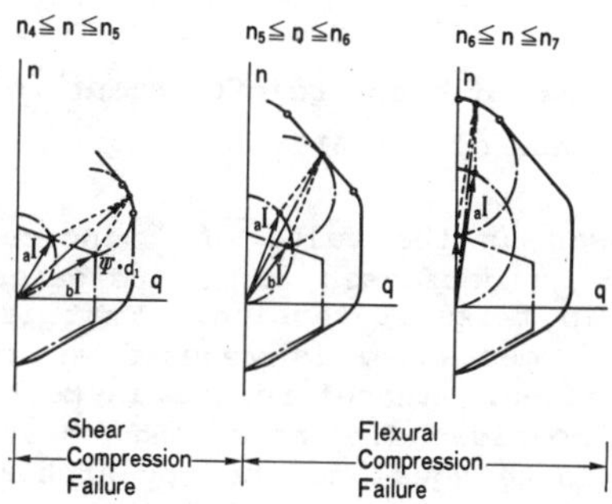

Arch	$_an$	$n-{}_bn$	$n-{}_bn$	$n-{}_bn$
	$_aq$	$\delta(\sqrt{4n_b(1-n_b)+n^2}-n)$	$q-{}_bq$	$(\sqrt{4(n-2\phi)(1-n+2\phi)+n^2}-n)/2$ $(=q)$
Beam	$_bn$	$2\phi+\Psi(d_1-n)$	$(1-n/d_1)bq+2\phi$	2ϕ
	$_bq$	$\Psi\cdot d_1$	$\dfrac{\lambda_2(n-n_6)+q_6+n/2+(n-2\phi-1/2)/\lambda_2}{1+n/d_1+(1-(n+1)/d_1)/\lambda_2}$	0

Fig. 8. Load-carrying capacity for arch mechanism and beam mechanism

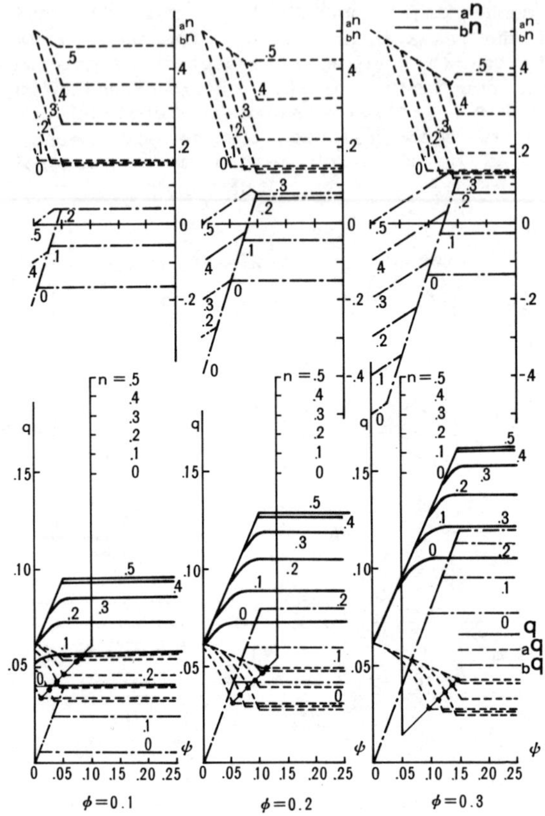

Fig. 9. Contribution of shear reinforcement to shear strength
(for the case of η = 4)

value of $_bq$ increases as the value of Ψ increases. On the other hand, the value of $_aq$ decreases as Ψ increases until the longitudinal bars yield in flexural tension. That is, the load-carrying capacity of the arch mechanism is reduced as the capacity of the beam mechanism increases. Therefore, as is mentioned in chapter 2, it is rational to consider the remaining shear resistance by the arch mechanism not to be equal to that by arch mechanism when there is no shear reinforcement.

6. NUMERICAL VERIFICATION BY FINITE ELEMENT ANALYSIS

6.1 Contents of the Numerical Experiments

A series of numerical analyses were made to clarify the effect of shear reinforcement using the finite element method [9]. Figure 10 shows the column specimens used to investigate the effects of the following three factors on the shear strength of the column;

1) Amount of flexural reinforcement: ρ_t = 0.96% and 1.92%

2) Amount of shear reinforcement: ρ_w = 0.25%, 0.57% and 1.0%

3) Magnitude of axial forces: σ_o = 0.1F$_c$ and 0.3F$_c$

Applying the concept of beam mechanism, the contribution of web reinforcement to shear strength was estimated from the computed stresses in the shear reinforcement and the direction of the principal compressive stress. The results were compared with the prediction by the macromodel analysis described in section 5.

6.2 Models used in the Finite Element Analysis

The finite element mesh used in the analysis is shown in Fig. 11. Concrete was modelled as isoparametric elements with four nodes. For its constitutive law, the theory proposed by Darwin and Pecknold was adopted [2]. The stress-strain relationship was defined as shown in Fig. 12(a), the biaxial strength envelope is based on the Kupfer's study [4].

Main reinforcement was modelled as truss elements, and web reinforcement as equivalent plate elements. The stress-strain relationship for reinforcement is defined as shown in Fig. 12(b).

The effect of the bond between main reinforcement and concrete was expressed by bond links. Bond stress-slip relationship was modelled as shown in Fig. 12(c).

The dimensions and mechanical properties of specimens are listed in Table 4.

The applicability of the model was examined by comparing analytical and test results. Two such comparisons are illustrated in Fig. 13.

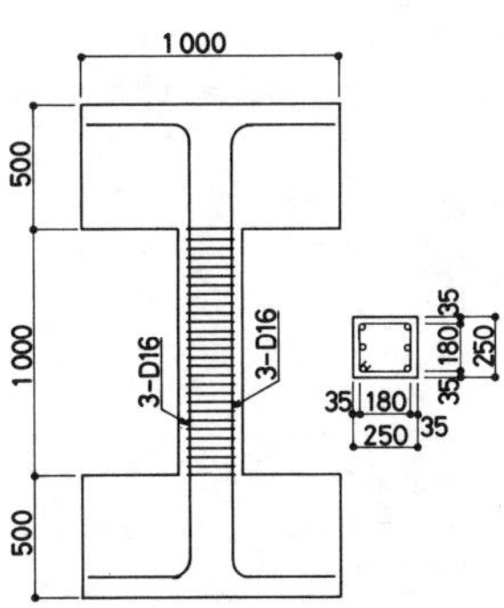

Fig. 10. Column's dimensions

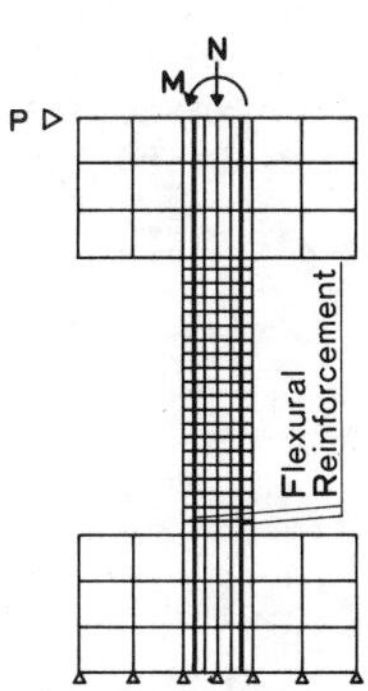

Fig. 11. Finite element mesh for analysis of columns

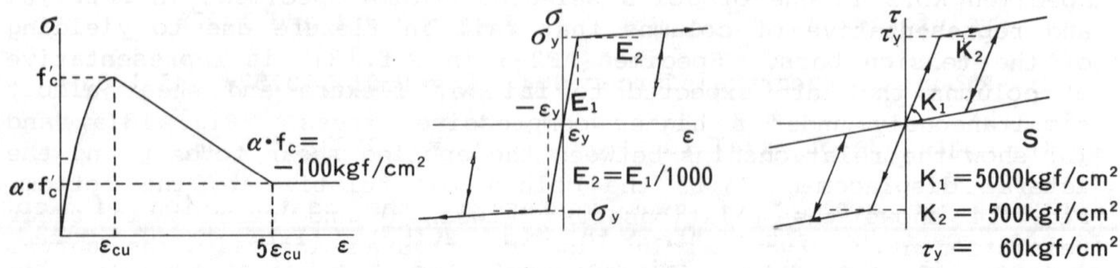

(a) Concrete (b) Steel Bar (c) Bond Action

Fig. 12. Constitutive relation for materials

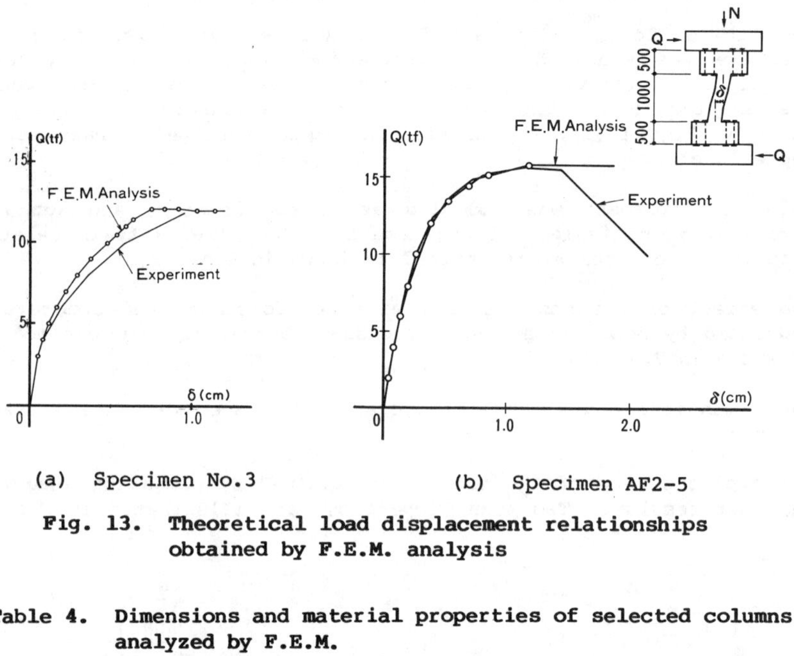

(a) Specimen No.3 (b) Specimen AF2-5

Fig. 13. Theoretical load displacement relationships obtained by F.E.M. analysis

Table 4. Dimensions and material properties of selected columns analyzed by F.E.M.

	NO. 3 (Ref. 3)	AF2-5(Ref. 5)
b × h	25cm×25cm	25cm×25cm
1	100cm	100cm
N	16.4tf	66.3tf
ρ_t	0.96%	0.61%
ρ_w	0.57%	0.75%
F_c	240kgf/cm²	241kgf/cm²
σ_y	3950kgf/cm²	4414kgf/cm²
σ_{wy}	4546kgf/cm²	3804kgf/cm²
$n\ (=N/bhF_c)$	0.109	0.44
$\eta\ (=1/h)$	4	4
$\Phi(=\rho_t\cdot\sigma_y/F_c)$	0.158	0.112
$\Psi(=\rho_w\cdot\sigma_{wy}/F_c)$	0.108	0.118

Specimen No.3 is one of JCI's selected column specimens in Ref.[10] and representative of columns that fail in flexure due to yielding of the tension bars. Specimen AF2-5 in Ref.[3] is representative of columns that are expected to fail in flexure and shear almost simultaneously under a higher compressive stress. Fig. 13(a) and (b) show the relationships between the applied shear force Q and the lateral displacement δ of the column top relative to the bottom. Fig. 14 shows the relationships between Q and strains in shear reinforcement. Open circles in the figures identify the curves obtained by finite element analysis. The analytical and experimental results are in good agreement. Also the failure modes predicted for both specimens were in agreement with those observed in the tests.

6.3 Results of Numerical Computations

Fig. 15 shows the load deflection curves obtained in a series of analyses. These figures show how the maximum load carrying capacity of the specimens under a constant axial stress varies with ρ_w and ρ_t. In the case of ρ_t = 0.96% and $\sigma_o = 0.1F_c$, all the specimens failed in flexure, indicating that the maximum load carrying capacities are constant regardless of the value of ρ_w. In other cases, the maximum load carrying capacity increases as the value of ρ_w increases. Figure 16 shows the principal stress distributions in columns. The figure shows that the stress pattern varies from the one indicating a predominantly diagonal stress flow to the one with a predominant stress flow in the direction of 45 degrees, showing that the failure mechanism gradually shifts from the arch to the beam mechanism with the increase in the value of ρ_w and ρ_t.

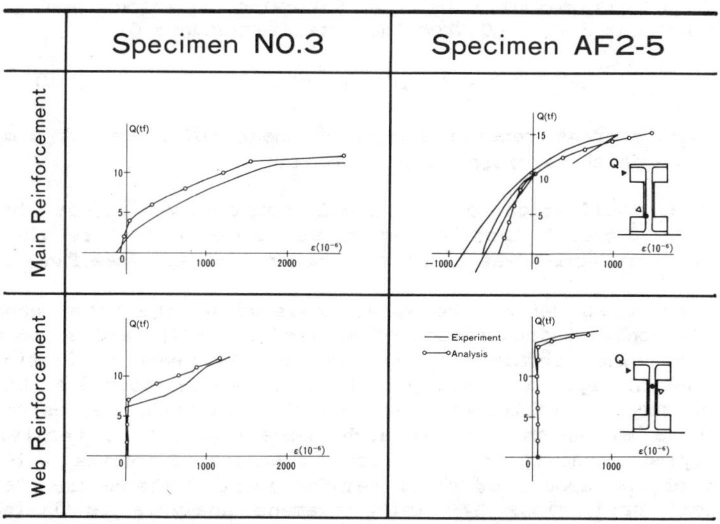

Fig. 14. Strain of shear reinforcement

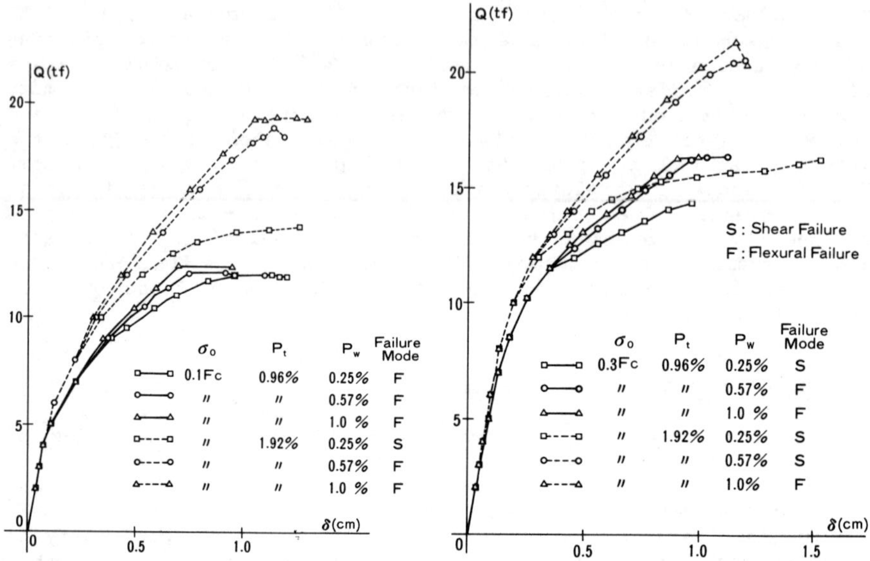

Fig. 15. Summary of parametric numerical analysis

In Fig. 17, the shear resisted by the beam mechanism, ${}_{b}Q$, and the shear resisted by the arch mechanism, ${}_{a}Q$, are plotted as functions of ρ_w. ${}_{b}Q$ was computed by the following equation, and ${}_{a}Q$ was obtained by subtracting ${}_{b}Q$ from the total resistance Q.

$$ {}_{b}Q = b \cdot d \cdot \rho_w \cdot \sigma_{we} \cot\theta \qquad (10)$$

where σ_{we}: average tensile stress of shear reinforcement computed in the flexural tensile region

θ : inclination of principal compressive stress in the flexural tensile region at column end where the beam mechanism was assumed to be predominant (see Fig. 16)

Square open marks denote the shear resisted by the beam mechanism assuming a constant inclination of stress ($\theta = \pi/4$), and solid marks denote the values obtained by the use of the measured inclination illustrated in Fig. 16. From this figure, we can recognize that the amount of shear reinforcement has an effect on both the resistances by the beam mechanism and the arch mechanism. As a result, the load-carrying capacity by the arch mechanism decreases with the increase of the amount of shear reinforcement. The summary of the theoretical predictions by finite element analysis is listed in Table 5.

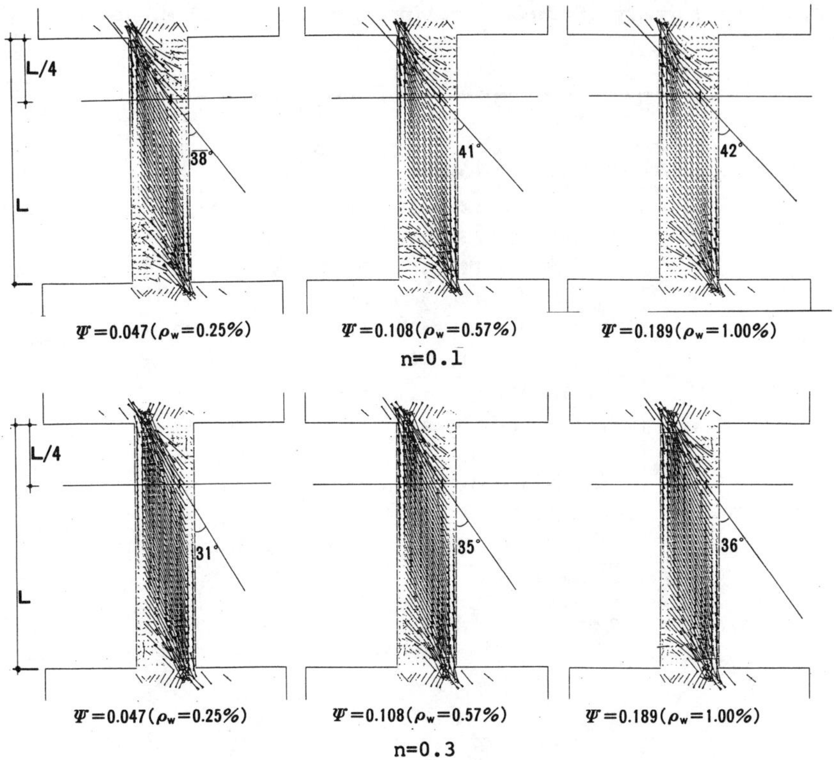

Fig. 16. Principal stress distribution in columns (for the case of Φ = 0.158)

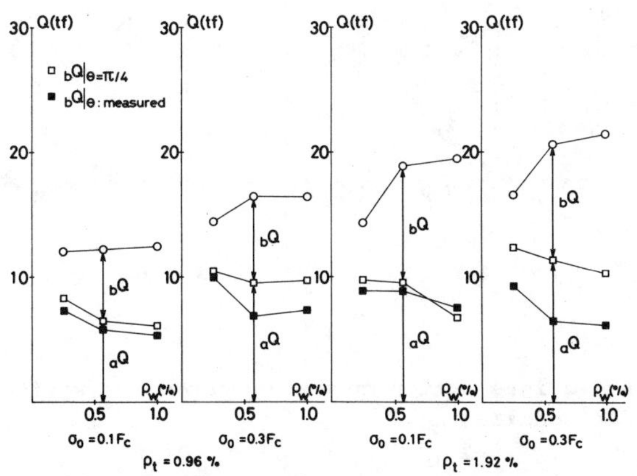

Fig. 17. Theoretical prediction by F.E.M. analysis for specimen No.3

Table 5. Summary of theoretical predictions by F.E.M. analysis

ρ_t (%)	σ_0 (kgf/cm²)	ρ_w (%)	$Q_{max.}$ (tf)	σ_{we} (kgf/cm²)	τ_w (kgf/cm²)	$\theta=\pi/4$ ${}_bQ_1$ (tf)	$\theta=\pi/4$ ${}_aQ_1$ (tf)	θ=measured value θ (degree)	θ=measured value ${}_bQ_2$	θ=measured value ${}_aQ_2$
0.96	$0.1F_c$	0.25	12.0	3683	9.2	3.8	8.2	38	4.8	7.2
		0.57	12.1	2386	13.6	5.6	6.5	41	6.4	5.7
		1.00	12.4	1551	15.5	6.4	6.0	42	7.1	5.3
	$0.3F_c$	0.25	14.4	3785	9.5	3.9	10.5	31	6.5	7.9
		0.57	16.4	2892	16.5	6.8	9.6	35	9.6	6.8
		1.00	16.4	1618	16.2	6.6	9.8	36	9.1	7.3
1.90	$0.1F_c$	0.25	14.3	4398	11.0	4.5	9:8	40	5.4	8.9
		0.57	18.9	4008	22.8	9.4	9.5	43	10.0	8.9
		1.00	19.4	3119	31.2	12.7	6.7	47	11.9	7.5
	$0.3F_c$	0.25	16.6	4137	10.3	4.2	12.4	30	7.3	9.3
		0.57	20.6	3941	22.5	9.2	11.4	33	14.1	6.5
		1.00	21.4	2725	27.3	11.2	10.2	36	15.3	6.1

Notes : b = 25cm, d = 16.4cm

$\tau_w = \rho_w \cdot \sigma_{we}$

${}_bQ_1 = \tau_w \cdot b \cdot d$, ${}_aQ_1 = Q_u - {}_bQ_1$

${}_bQ_2 = \tau_w \cdot b \cdot d \cdot \cot\theta$, ${}_aQ_2 = Q_u - {}_bQ_2$

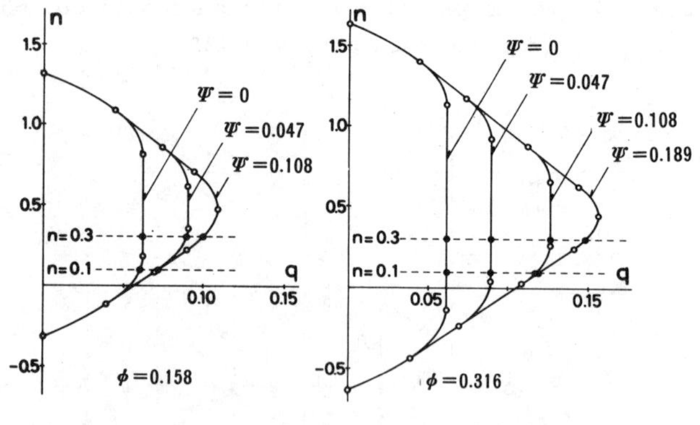

Fig. 18. n - q interaction curves computed for variables Φ and Ψ for specimen No.3

6.4 Comparison of the Computations with the Macromodel Analysis

Non-dimensionalized interaction relationships of axial and shear forces at ultimate, obtained by the macromodel analysis, are shown in Fig. 18 for different combinations of Φ and Ψ. In Fig. 19, the shear resistances estimated by the macromodel analysis are compared with the results of the numerical computations by finite element analysis. Solid, chained and dashed lines in the figure denote the total shear, the shear resisted by the beam mechanism and the shear resisted by the arch mechanism, respectively.

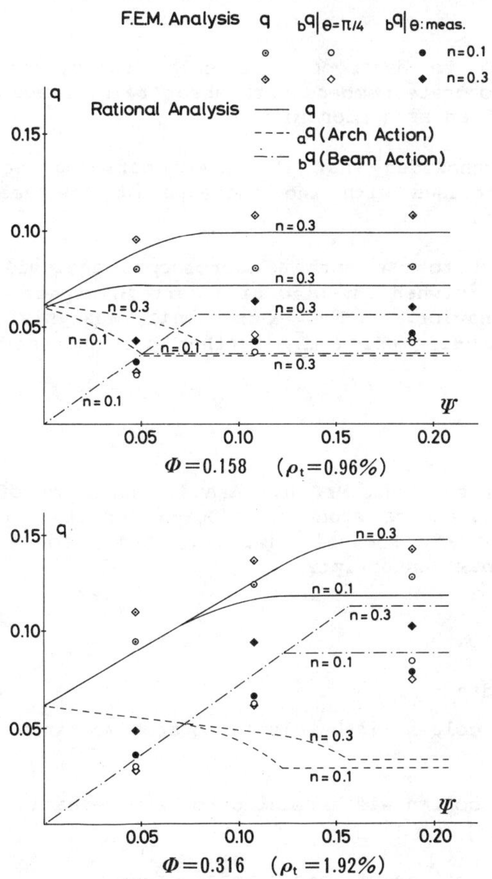

Fig. 19. Comparison between macroscopic analysis and F.E.M. analysis for specimen No.3

The influence of the amount of shear reinforcement on the shear resistance due to the beam mechanism or the total resistance as predicted by the two analysis is almost identical, if we use measured inclinations of compressive stress to evaluate the resistance of the beam mechanism from the computed results of the finite element analysis.

7. CONCLUDING REMARKS

The following conclusions can be drawn from this study.

1) MINAMI's macromodel is useful to represent the shear resistance of reinforced concrete members both qualitatively and quantitatively.

2) It is useful to decompose the shear resistance mechanism of reinforced concrete members with shear reinforcement into a beam mechanism and an arch mechanism.

3) It can be concluded that the shear carrying capacity of the concrete decreases with the increase of the amount of shear reinforcement.

4) It is helpful to use both a macroscopic analysis and a microscopic analysis when investigating certain aspects of reinforced concrete behavior. The macroscopic analysis tool permits parametric studies which may further our understanding of such behavior.

ACKNOWLEDGEMENT

The writers wish to thank Mr. H. TAKAGI, assistant of Meiji Univ., Mr. K. OKAMOTO, graduate student of Osaka Institute of Technology, and Mr. Y. NAKAMURA, Kumagai Gumi Co., Ltd., for helping in the preparation of this manuscript.

APPENDIX - NOTATION

b : Column width

$_ab$: Effective column width related to arch mechanism

$_ab_1$: $_ab/b$

$_bb$: Effective column width related to beam mechanism

$_bb_1$: $_bb/b$

d : Centroid distance of main reinforcement

d_1 : d/h

F_c : Compressive strength of concrete

h : Column depth

L : Column length

M	:	Bending moment in general
m	:	Non-dimensional bending moment (= M/bh^2F_c)
N	:	Axial force in general
n	:	Non-dimensional axial force (= N/bhF_c)
$_an$	:	Non-dimensional axial force related to arch mechanism
$_bn$	:	Non-dimensional axial force related to beam mechanism
Q	:	Shear force in general
q	:	Non-dimensional shear force (= Q/bhF_c)
$_aq$	:	Non-dimensional shear force related to arch mechanism
$_bq$	:	Non-dimensional shear force related to beam mechanism
δ	:	Horizontal displacement of column
η	:	Length-to-depth ratio of column subjected to anti-symmetrical bending and shear (= L/h)
ρ_t	:	Flexural reinforcement ratio
ρ_w	:	Shear reinforcement ratio
σ_y	:	Yield stress of flexural reinforcement
σ_{wy}	:	Yield stress of shear reinforcement
Φ	:	Mechanical degree of flexural reinforcement (= $\rho_t \cdot \sigma_y/F_c$)
Ψ	:	Mechanical degree of shear reinforcement (= $\rho_w \cdot \sigma_{wy}/F_c$)

REFERENCES

[1] Chen, W.F., "Plasticity in reinforced concrete", Chapter 7, Limit Analysis of Perfect Plasticity, McGraw - Hill Book Company, 1982, pp.295-346.

[2] Darwin, D. and Pecknold, D.A., "Nonlinear Biaxial Stress-Strain Law for Concrete", ASCE, Vol.103, EM2, April 1977, pp.229-241.

[3] HIGASHI, Y. and OKUBO, M., "Study on prevention of shear failure of reinforced concrete short columns." (Part 44, Test results of AF2 Series), Summaries of Technical Papers of Annual Meeting, AIJ, Oct., 1977, pp.1499-1500 (in Japanese).

[4] Kupfer, H.B. and Gerstile, K.H., "Behavior of Concrete under Biaxial Stresses", ASCE, Vol.99, EM4, Aug. 1973, pp.852-866.

[5] MINAMI, K. and WAKABAYASHI, M., "Rational analysis of shear in reinforced concrete columns", IABSE colloquium Delft, Final report, Vol.34, 1981, pp.603-614.

[6] MINAMI, K., "Limit analysis of shear in reinforced concrete members", Proceedings of JCI colloquium on shear analysis of RC structures, Introductory report, Sept. 1983, pp.5-28.

[7] Nielsen, M.P., "Limit analysis and concrete plasticity", Prentice - Hall, 1983.

[8] Park, R. and Paulay, T., "Reinforced concrete structures", A Wiley - Interscience Publication, John Wiley & Sons., 1975, pp.278-287.

[9] SHIRAISHI, I., MACHIDA, N., NAKAMURA, Y. and KANOH, Y., "Nonlinear analysis of reinforced concrete shear wall", Transactions of JCI 7th Conference, June 1985, pp.417-420 (in Japanese).

[10] YOSHIOKA, K., "Selected reason of JCI 2nd colloquium on shear analysis of RC structures", Oct. 1983, pp.9-20 (in Japanese).

ISOPARAMETRIC CONTACT ELEMENTS FOR ANALYSIS OF REINFORCED CONCRETE STRUCTURES

Gerhard MEHLHORN
Professor Dr.-Ing.
Gesamthochschule Kassel

Manfred KEUSER
Dr.-Ing.
Ingenieurbüro Bung
Heidelberg

ABSTRACT

Bond-slip between reinforcement, concrete and bond elements is briefly described. The one- and two-dimensional isoparametric contact elements developed in a research project and their bond behaviour are explained. This model may be used very successfully in FE-computer programs to analyze reinforced concrete structures.

1. BOND-SLIP

The load-carrying behaviour of reinforced concrete is influenced by the interaction between both of its components, plain concrete and reinforcement bars. The transfer of load between these two materials at the bar-concrete interface is called bond. The bearing mechanism of bond can be conceived as being comprised of three components: chemical adhesion, friction and mechanical interlock between steel and concrete.

Bond stresses in reinforced concrete members result from the change of the bar force along its length. Therefore the effect of bond becomes evident, for example, at end anchorages of straight bars and strands, at anchorages of reinforcement bars by hooks, at lapped splices of reinforcement bars, and in regions near cracks, see Fig. 1.

Gerhard Mehlhorn is professor of Massivbau at the University/Gesamthochschule Kassel, Mönchebergstr. 7, 3500 Kassel, F.R. Germany

Manfred Keuser is design engineer with Ingenieurbüro Bung, Consulting Engineers, Englerstr. 4, 6900 Heidelberg, F.R. Germany

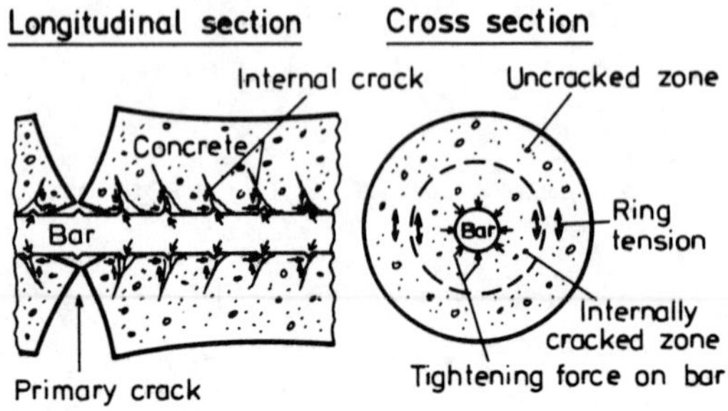

Fig. 1 Reinforced concrete tensile member with crack formation according to Goto [1]

In all these cases, the total force in the reinforcement bar is transmitted from the bar to the concrete or vice versa.

In the simplified analysis of reinforced concrete structures, complete compatibility between concrete and reinforcement is usually assumed, that means, perfect bond is presumed. But this assumption is only valid in those regions where no or only negligible stress transfer between the two components occurs. In regions of high stresses in the contact interfaces, however, for example near cracks, the bond stresses are related to relative displacements between concrete and reinforcement. The assumption of perfect bond in cracked zones would cause infinitely high strains to explain the existence of a finite crack width. In reality, there are different strains in the adjacent regions of the connections between concrete and reinforcement. As a result of this and of the crack propagation, relative displacements, which are usually called b o n d - s l i p, occur between these two components.

2. BOND ELEMENTS

In the Finite Element Method the elements are connected to each other at the nodal points. Therefore adjacent elements have identical displacements at the common nodes. This implies complete compatibility between reinforcement and concrete. If b o n d - s l i p is to be taken into account, special interface elements will have to be used for the connection of the reinforcement elements to the concrete elements.

Considering the modeling of bond-slip, various approaches are known:

The "bond link" element, developed by Ngo/Scordelis [2], connects one node of a concrete element with one node of an adjacent steel element. The link element has no physical dimensions, so that the two connected nodes have the same coordinates. A bond link can be thought to consist of three springs, one parallel and the other two normal to the longitudinal axis of the reinforcement bar.

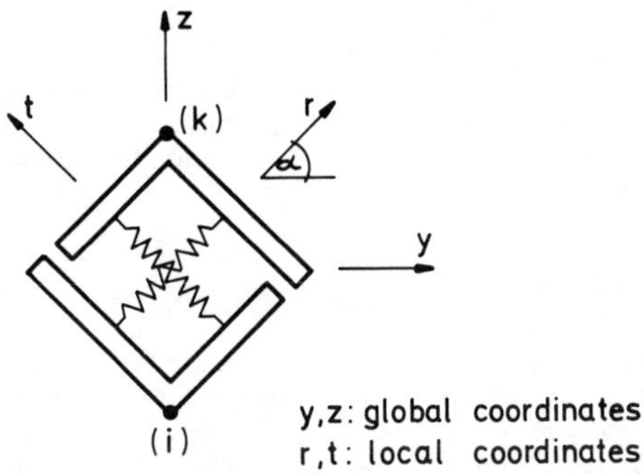

Fig. 2 Bond Link Element according to Ngo/Scordelis [2]

The bond-link element is shown in Fig. 2 for a two-dimensional procedure.

Compared with the bond-link element, the bond zone element (Fig. 3) is significantly different, The most important difference is its finite dimension. In the bond-zone element, developed by de Groot et. al. [3], the contact surface between steel and concrete as well as the concrete in the immediate vicinity of the reinforcement bar are modelled by an adopted material law which considers the special properties of the bond zone.

In the model shown in Fig. 3, the bond stress is conceived as being composed of the sum of the slip resistance and the mechanical interlocking.

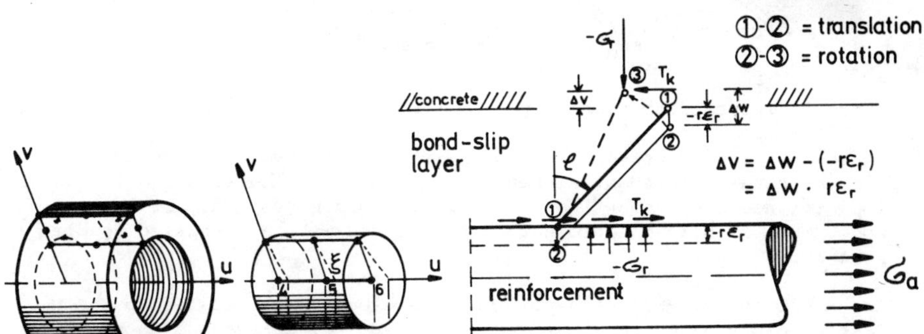

Fig. 3 Bond Zone Element according to [3]

The contact element provides a continuous connection between two elements, based on the assumption of a linear or higher displacement field. Godman et. al. [4] used this element to solve problems of rock mechanics. Hoshino [5] introduced this element to model bond problems in reinforced concrete. He utilized a linear bond-slip relationship. It has been further developed and generalized by Schäfer [6]. In the contributions of Hoshino and Schäfer relative displacements between concrete and reinforcement were considered only in the direction of the reinforcement bar, whereas in a direction normal to the bar complete compatibility was assumed.

3. ISOPARAMETRIC ONE-DIMENSIONAL CONTACT ELEMENT

A modified and more general version of the contact element, which takes into account the normal stresses between concrete and reinforcement, has been developed in a research project supported by the Deutsche Forschungsgemeinschaft [7] [8]. The element is isoparametric and has in the undeformed state no physical dimension in the transverse direction. It has two to four double-nodes (Fig. 4) and uses linear, quadratic or cubic interpolation functions.

It should be noted that the order of the contact element should be compatible with the order of the concrete and reinforcement elements. For example, the connection between a two-node truss element with two nodes of a concrete element is accomplished by a contact element with two double-nodes. If the steel element has three nodes, a contact element with three double-nodes is required to connect it with three nodes of a concrete element.

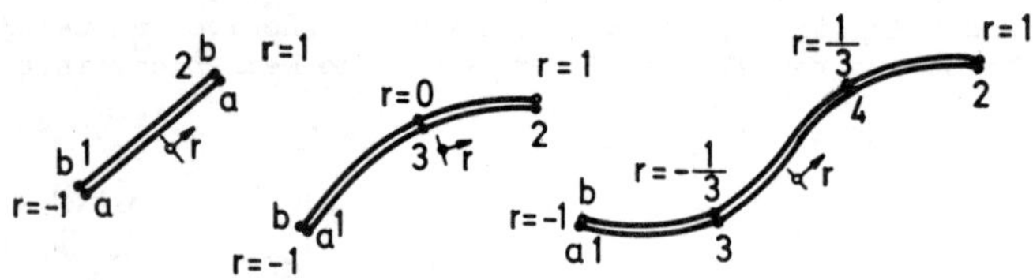

Fig. 4 Two to four-node isoparametric contact elements

In the contact interface the two elements connected by a contact element have independent element nodes; in the unloaded stage two opposite nodes P_i and P_k have identical coordinates. Each of the two nodes with identical coordinates is combined with a so-called double-node ik.

The displacements $\tilde{u}$ at each point of the element are related to the nodal displacements $\tilde{u}_i$ by the shape functions, Eq. 1. Corresponding to this, the local coordinates of any point $\tilde{x}$ are related to the global coordinates of the nodal points $\tilde{x}_i$ in the same way, Eq. 2.

$$u(r) = \sum_{i=1}^{n} N_i(r) \cdot u_i \qquad x(r) = \sum_{i=1}^{n} N_i(r) \cdot x_i$$

$$v(r) = \sum_{i=1}^{n} N_i(r) \cdot v_i \qquad y(r) = \sum_{i=1}^{n} N_i(r) \cdot y_i$$

$$w(r) = \sum_{i=1}^{n} N_i(r) \cdot w_i \qquad z(r) = \sum_{i=1}^{n} N_i(r) \cdot z_i$$

$$\underset{\sim}{u} = \underset{\sim}{N} \cdot \underset{\sim}{u}_i \quad (1) \qquad \underset{\sim}{x} = \underset{\sim}{N} \cdot \underset{\sim}{x}_i \quad (2)$$

The interpolation functions N_i of the two to four nodes adopted from [9] are give in table 1.

Table 1	Interpolation functions of two to four nodes for the one-dimensional elements	
Basic term	Additional term only if node 3 is present	Additional term if nodes 3 and 4 are present
$N_1 = \frac{1 - r}{2}$	$-\frac{1 - r^2}{2}$	$\frac{-9r^3 + r^2 + 9r - 1}{16}$
$N_2 = \frac{1 + r}{2}$	$-\frac{1 - r^2}{2}$	$\frac{9r^3 + r^2 - 9r - 1}{16}$
$N_3 = 1 - r^2$		$\frac{27r^3 + 7r^2 - 27r - 7}{16}$
$N_4 = \frac{-27r^3 - 9r^2 + 27r + 9}{16}$		

As long as no loads are applied to the structure, the two connected points P_i and P_k are in an identical position. In a loaded stage, relative displacements $\underset{\sim}{\Delta}$ between the two connected points will occur, see Fig. 5.

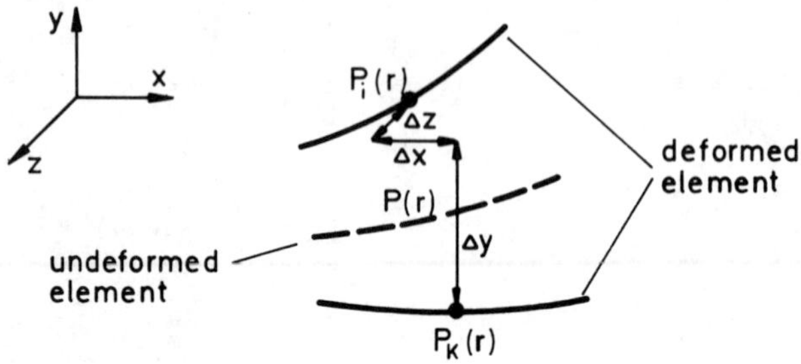

Fig. 5 Relative displacements $\underset{\sim}{\Delta} = [\Delta_x, \Delta_y, \Delta_z]$

The relative displacements $\underset{\sim}{\Delta}$ are given by Eq. 3:

$$\begin{bmatrix} \Delta_x \\ \Delta_y \\ \Delta_z \end{bmatrix} = \begin{bmatrix} -1 & 0 & 0 & 1 & 0 & 0 \\ 0 & -1 & 0 & 0 & 1 & 0 \\ 0 & 0 & -1 & 0 & 0 & 1 \end{bmatrix} \cdot \begin{bmatrix} u_i \\ v_i \\ w_i \\ u_k \\ v_k \\ w_k \end{bmatrix}$$

$$\underset{\sim}{\Delta} = [\underset{\sim}{A}] \cdot \underset{\sim}{u} \tag{3}$$

Substitution of Eq. 1 into Eq. 3 leads to Eq. 4, which relates the relative displacements $\underset{\sim}{\Delta}$ at any point of an element to the nodal displacements $\underset{\sim}{u}_i$.

$$\underset{\sim}{\Delta} = [\underset{\sim}{A}] \cdot \underset{\sim}{N} \cdot \underset{\sim}{u}_i$$

$$\underset{\sim}{\Delta} = [\underset{\sim}{B}] \cdot \underset{\sim}{u}_i \tag{4}$$

with

$$[\underset{\sim}{B}] = [\underset{\sim}{A}] \cdot \underset{\sim}{N}$$

Analogous to Eq. 3, the stresses P at points P_i and P_k are related in Eq. 5 to contact stresses $\underset{\sim}{\tau}$ between two elements, see Fig. 6.

$$\begin{bmatrix} P_{xi} \\ P_{yi} \\ P_{zi} \\ P_{xk} \\ P_{yk} \\ P_{zk} \end{bmatrix} = \begin{bmatrix} -1 & 0 & 0 \\ 0 & -1 & 0 \\ 0 & 0 & -1 \\ 1 & 0 & 0 \\ 0 & 1 & 0 \\ 0 & 0 & 1 \end{bmatrix} \cdot \begin{bmatrix} \tau_x \\ \tau_y \\ \tau_z \end{bmatrix}$$

$$\underset{\sim}{P} = [\underset{\sim}{A}^T] \cdot \underset{\sim}{\tau} \tag{5}$$

$$\begin{bmatrix} \Delta_r \\ \Delta_t \\ \Delta_s \end{bmatrix} = \begin{bmatrix} \underset{\sim}{t}_r \\ \underset{\sim}{n}_t \\ \underset{\sim}{t}_s \end{bmatrix} \cdot \begin{bmatrix} \Delta_x \\ \Delta_y \\ \Delta_s \end{bmatrix}$$

$$\underset{\sim}{\Delta}^L = [\underset{\sim}{T}] \cdot \underset{\sim}{\Delta} \tag{6}$$

$$\begin{bmatrix} \tau_x \\ \tau_y \\ \tau_z \end{bmatrix} = \begin{bmatrix} \underset{\sim}{t}_r^T, \underset{\sim}{n}_t^T, \underset{\sim}{t}_s^T \end{bmatrix} \cdot \begin{bmatrix} \tau_r \\ \sigma_t \\ \tau_s \end{bmatrix}$$

$$\underset{\sim}{\tau} = [\underset{\sim}{T}^T] \cdot \underset{\sim}{\tau}^L \tag{7}$$

The relationship between contact stresses and relative displacements in the contact interface is defined by Eq. 8 in the local coordinate system.

$$\begin{bmatrix} \tau_r \\ \sigma_t \\ \tau_s \end{bmatrix} = \begin{bmatrix} C_{rr} & C_{rt} & C_{rs} \\ C_{tr} & C_{tt} & C_{ts} \\ C_{sr} & C_{st} & C_{ss} \end{bmatrix} \cdot \begin{bmatrix} \Delta_r \\ \Delta_t \\ \Delta_s \end{bmatrix} + \begin{bmatrix} \tau_r^\circ \\ \sigma_t^\circ \\ \tau_s^\circ \end{bmatrix}$$

$$\underset{\sim}{\tau}^L = [\underset{\sim}{C}^L] \cdot \underset{\sim}{\Delta}^L + \underset{\sim}{\tau}^{\circ L} \tag{8}$$

In the special case when the contact behaviour in the three coordinate directions are independent of each other, we obtain the material matrix $[C^{L*}]$ in the form

$$[C^{L*}] = \begin{bmatrix} C_{rr} & 0 & 0 \\ 0 & C_{tt} & 0 \\ 0 & 0 & C_{ss} \end{bmatrix} \tag{8a}$$

C_{rr}, C_{tt} and C_{ss} are the bond-moduli in the r-, t- and s-direction.

To build up the element stiffness matrix and the vector of the nodal forces, the $[\underset{\sim}{C}^L]$-matrix has to be transformed from local to global coordinates, Eq. 9. Then the element stiffness matrix $[\underset{\sim}{k}^m]$ and the vector of element nodal forces $\underset{\sim}{R}^m$ are obtained in global coordinates in Eqs. 10 and 11 and can be assembled into the structure force vector directly without additional effort.

$$[\underset{\sim}{C}] = [\underset{\sim}{T}^T] \cdot [\underset{\sim}{C}^L] \cdot [T] \tag{9}$$

$$[\underset{\sim}{k}^m] = \int_{(A)} [\underset{\sim}{B}^T] \cdot [\underset{\sim}{C}] \cdot [\underset{\sim}{B}] \cdot dA \tag{10}$$

$$\underset{\sim}{R}^m = \int_{(A)} [\underset{\sim}{B}^T] \cdot \underset{\sim}{\tau} \cdot dA \tag{11}$$

and

$$\underset{\sim}{\tau} = [\underset{\sim}{C}] \cdot \underset{\sim}{\Delta} \tag{12}$$

The integrations in the equations above have to be carried out over the contact surface, using the Gauss numerical integration procedure.

The differential dA of the here considered one-dimensional contact element is

$$dA = b \cdot \det J \cdot dr \tag{13}$$

b is the width of the contact surface and det J is the determinant of the Jacobian matrix. For the one-dimensional contact element considered here, det J is equal to the real length L of the element.

Presuming a linear shape function for the contact element and the independence of the contact behaviour in the three coordinate directions of each other, we obtain the relation between element forces and relative displacements, Eq. 14.

$$\begin{bmatrix} R_r^{ik} \\ R_r^{lm} \\ R_t^{ik} \\ R_t^{lm} \\ R_s^{ik} \\ R_s^{lm} \end{bmatrix} = \pi \cdot d \cdot \frac{L}{6} \cdot \begin{bmatrix} 2\cdot G_r & G_r & 0 & 0 & 0 & 0 \\ G_r & 2\cdot G_r & 0 & 0 & 0 & 0 \\ 0 & 0 & 2\cdot G_t & G_t & 0 & 0 \\ 0 & 0 & G_t & 2\cdot G_t & 0 & 0 \\ 0 & 0 & 0 & 0 & 2\cdot G_s & G_s \\ 0 & 0 & 0 & 0 & G_s & 2\cdot G_s \end{bmatrix} \cdot \begin{bmatrix} \Delta_r^{ik} \\ \Delta_r^{lm} \\ \Delta_t^{ik} \\ \Delta_t^{lm} \\ \Delta_s^{ik} \\ \Delta_s^{lm} \end{bmatrix} \tag{14}$$

The bond moduli G_r, G_t and G_s in Eqs. 8a and 14 are obtained experimentally and may be used in linear or nonlinear formulations.

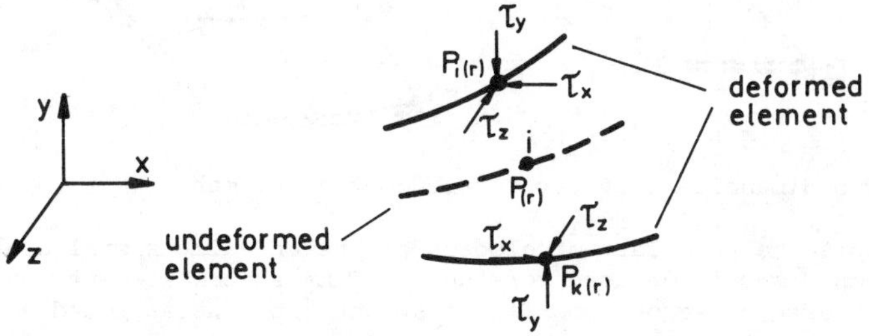

Fig. 6 Stresses in the double-node $P_{ik}(r)$

Generally, the contact behaviour tangential and normal to the contact interface is different, which leads to an anisotropic material matrix $[\underset{\sim}{C}]$. The relative displacements $\underset{\sim}{\Delta}$ must be transformed from global to local coordinates (equation 6) as well as the contact stresses τ are transformed from the local to the global coordinates (equation $\tilde{7}$). The transformation is done by multiplying the respective vector with the transformation matrix [T], which consists of the unit direction vectors parallel and normal to the contact interface at the double-node $P_{ik}(r)$, see Fig. 7.

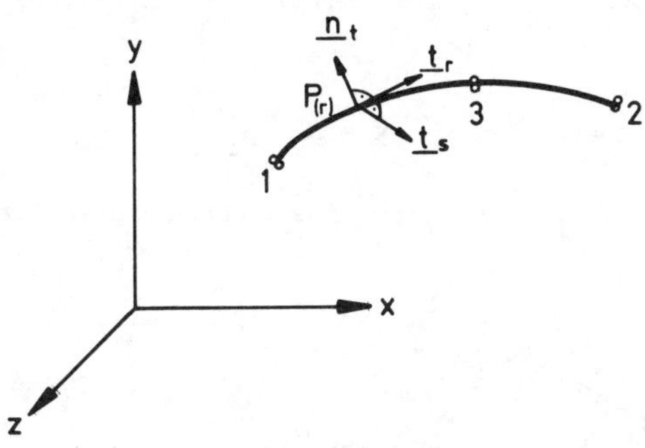

Fig. 7 Normal and tangential vectors at the double-node $P_{ik}(r)$

4. ISOPARAMETRIC TWO-DIMENSIONAL CONTACT ELEMENTS

The isoparametric contact element has been extended (Fig. 8) to be used for the connection of three-dimensional elements as well as to be used in a layered shell element.

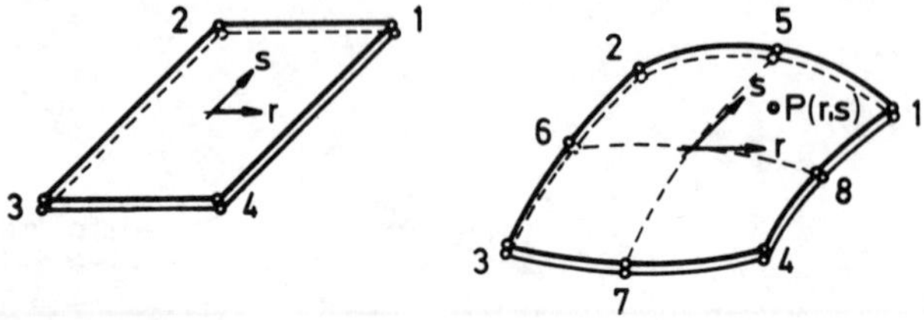

Fig. 8 Two-dimensional isoparametric contact elements (4 to 9 nodes)

In principle the procedure developed for the one-dimensional contact element can directly be used for the two-dimensional element. But we need, of course, a second coordinate s. The coordinates r and s form an orthogonal natural coordinate system (Fig. 9). Again, both, the displacements $\underset{\sim}{u}$ and the geometry are interpolated in the same way, but here in the two-dimensional formulation:

$$u(r, s) = \sum_{i=1}^{n} N_i(r, s) \cdot u_i \qquad x(r, s) = \sum_{i=1}^{n} N_i(r, s) \cdot x_i$$

$$v(r, s) = \sum_{i=1}^{n} N_i(r, s) \cdot v_i \qquad y(r, s) = \sum_{i=1}^{n} N_i(r, s) \cdot y_i$$

$$w(r, s) = \sum_{i=1}^{n} N_i(r, s) \cdot w_i \qquad z(r, s) = \sum_{i=1}^{n} N_i(r, s) \cdot z_i$$

$$\underset{\sim}{u} = \underset{\sim}{N} \cdot \underset{\sim}{u}_i \qquad (15) \qquad \underset{\sim}{x} = \underset{\sim}{N} \cdot \underset{\sim}{x}_i \qquad (16)$$

The interpolation functions N_i of four to nine nodes adopted from [9] are given in table 2.

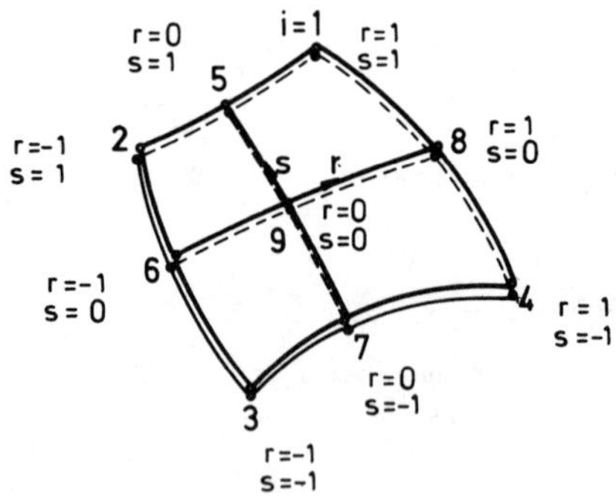

Fig. 9 Numbering of the double-nodes of the isoparametric two-dimensional contact element

Table 2	Interpolation functions of four to nine nodes for the two-dimensional elements				
Basic term	Additional terms for nodes i (five to nine)				
	i = 5	i = 6	i = 7	i = 8	i = 9
$N_1 = \frac{(1+r)\cdot(1+s)}{4}$	$-\frac{N_5}{2}$			$-\frac{N_8}{2}$	$-\frac{N_9}{4}$
$N_2 = \frac{(1-r)\cdot(1+s)}{4}$	$-\frac{N_5}{2}$	$-\frac{N_6}{2}$			$-\frac{N_9}{4}$
$N_3 = \frac{(1-r)\cdot(1-s)}{4}$		$-\frac{N_6}{2}$	$-\frac{N_7}{2}$		$-\frac{N_9}{4}$
$N_4 = \frac{(1+r)\cdot(1-s)}{4}$			$-\frac{N_7}{2}$	$-\frac{N_8}{2}$	$-\frac{N_9}{4}$
$N_5 = \frac{(1-r^2)\cdot(1+s)}{2}$					$-\frac{N_9}{2}$
$N_6 = \frac{(1-r)\cdot(1-s^2)}{2}$					$-\frac{N_9}{2}$
$N_7 = \frac{(1-r^2)\cdot(1-s)}{2}$					$-\frac{N_9}{2}$
$N_8 = \frac{(1+r)\cdot(1-s^2)}{2}$					$-\frac{N_9}{2}$
$N_9 = (1-r^2)\cdot(1-s^2)$					

All equations given in chapter 3 are valid here, too. But the differential dA is now

$$dA = \det J \cdot dr \cdot ds \tag{17}$$

Here, det J ist the determinant of the Jacobian matrix $[\underset{\sim}{J}]$, which is the relationship between the derivatives of the natural coordinates to the local coordinates. Therefore, for the two-dimensional contact element det J is equal to the real surface area of the element.

$$\begin{bmatrix} \frac{\partial}{\partial r} \\ \frac{\partial}{\partial s} \\ \frac{\partial}{\partial t} \end{bmatrix} = \begin{bmatrix} \frac{\partial x}{\partial r} & \frac{\partial y}{\partial r} & \frac{\partial z}{\partial r} \\ \frac{\partial x}{\partial s} & \frac{\partial y}{\partial s} & \frac{\partial z}{\partial s} \\ \frac{\partial x}{\partial t} & \frac{\partial y}{\partial t} & \frac{\partial z}{\partial t} \end{bmatrix} \cdot \begin{bmatrix} \frac{\partial}{\partial x} \\ \frac{\partial}{\partial y} \\ \frac{\partial}{\partial z} \end{bmatrix}$$

$$\frac{\partial}{\partial \underset{\sim}{r}} = [\underset{\sim}{J}] \cdot \frac{\partial}{\partial \underset{\sim}{x}} \qquad (18)$$

5. BOND BEHAVIOUR

The typical bond stress-slip behaviour is shown in a normalized form in Fig. 10.

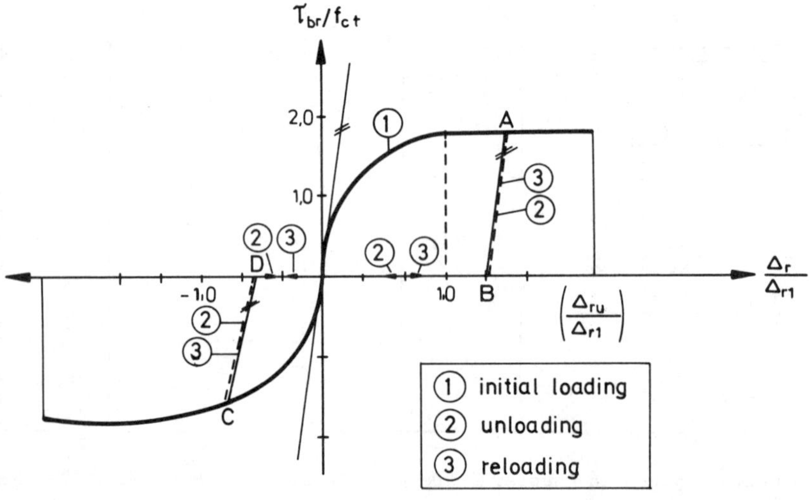

Fig. 10 Bond Stress-Slip Behaviour

The relationship between bond stress and slip is given by Eqs. 19 and 20 which is an extended formula given by Doerr [10]:

$$\frac{\tau_{br}}{f_{ct}} = \begin{cases} \alpha \cdot [5 \cdot (\frac{\Delta_r}{\Delta_{r1}}) - 4.5 \cdot (\frac{\Delta_r}{\Delta_{r1}})^2 + 1.4 \, (\frac{\Delta_r}{\Delta_{r1}})^3] & \text{if } \Delta_r < \Delta_{r1} \\ \alpha \cdot 1.9 & \text{if } \Delta_r \geq \Delta_{r1} \end{cases} \quad (19)$$

$$\alpha = \alpha_1 \cdot \alpha_2 \cdot \alpha_3 \cdot \alpha_4 \quad (20)$$

With

τ_{br}: bond stress in direction r

f_{ct}: concrete tensile strength

Δ_r : relative displacement (slip) in the direction r

Δ_{r1}: limit value for the relative displacement Δ_r. When the relative displacement is greater than Δ_{r1}, the bond stress remains constant until total bond failure occurs. Δ_{r1} depends on the lateral pressure to the axis of the reinforcement bar [10].

Δ_{ru}: ultimate relative displacement

α : factor to consider some of the parameters described below

The factor α_1 considers the size distribution of the aggregates and the consistency of the concrete (influence to α_1 between 0.8 and 1.25) as well as the position of the concrete during casting. Martin/Noakowski [11] found the relation given in Fig. 11. Also Doerr found in principle this influence in his tests.

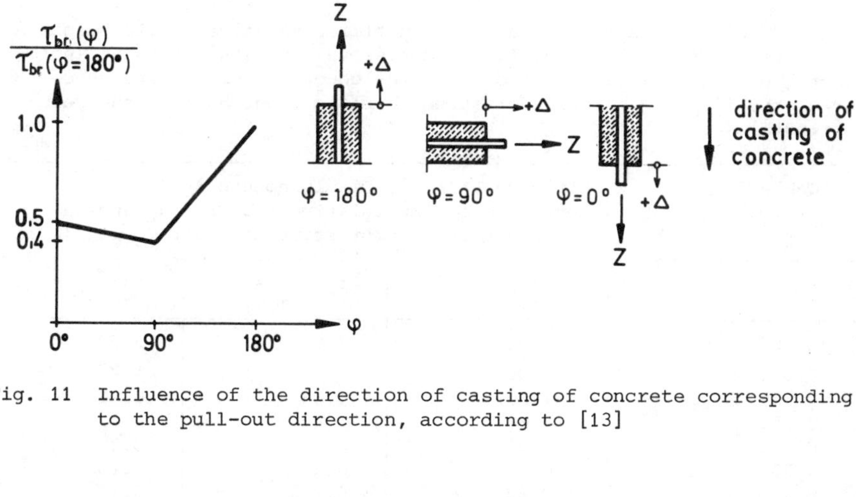

Fig. 11 Influence of the direction of casting of concrete corresponding to the pull-out direction, according to [13]

The factor α_2 considers the geometry of the bar surface, expressed by the relative rib area, which is defined as the quotient of the rib area and the bar surface between two ribs [12]. Because the bond-slip relationship (19) is based on Doerr's tests for ribbed reinforcement of 16 mm diameter, the influence of the geometry of the bar surface is related to the 16 mm diameter bar. For this bar the relative rib area is a_R = 0.065. Martin [13] studied the bond behaviour of reinforced concrete with different geometry of the surface of the tested bar. For relative displacements between 0.01 mm to 0.5 mm he found the bond-stress-slip relationship normalized to the compressive concrete strength. The mean values related to the relative rib area a_R = 0.065 can be represented by the relationship given in Fig. 12.

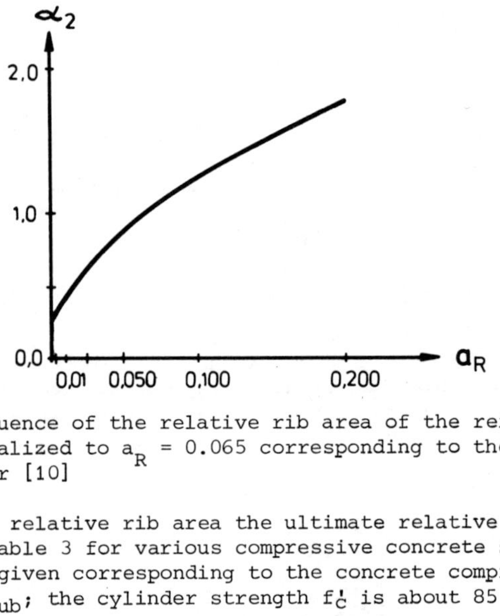

Fig. 12 Influence of the relative rib area of the reinforcement bar normalized to a_R = 0.065 corresponding to the bar tested by Doerr [10]

Related to the relative rib area the ultimate relative displacements Δ_{ru} are given in table 3 for various compressive concrete strengths (note that they are given corresponding to the concrete compressive cube strength, $f'_{c\ cub}$; the cylinder strength f'_c is about 85 % of the cube strength).

Table 3 Ultimate Relative Displacements Δ_{ru} mm depending on the relative rib area a_R and cubic concrete compressive strength $f'_{c\ cub}$ [N/mm²]

a_R	$f'_{c\ cub}$ = 20	$f'_{c\ cub}$ = 40	$f'_{c\ cub}$ = 60
0.005	5.00	3.60	3.00
0.010	3.06	2.16	1.80
0.025	1.50	1.06	0.88
0.050	0.88	0.62	0.52
0.100	0.51	0.38	0.30
0.200	0.30	0.21	0.18
0.400	0.22	0.15	0.13

The factor α_3 considers the transverse pressure or gaping based on the experimental research of Doerr [10] and Eligehausen et. al. [14]. Doerr investigated this influence under rotationally symmetric pressure p leading to the pressure q in the contact surface. With increasing lateral pressure the bond strength increases over-proportional. Eligehausen et. al. studied the influence of lateral pressure in only one direction. These results are summarized in Fig. 13.

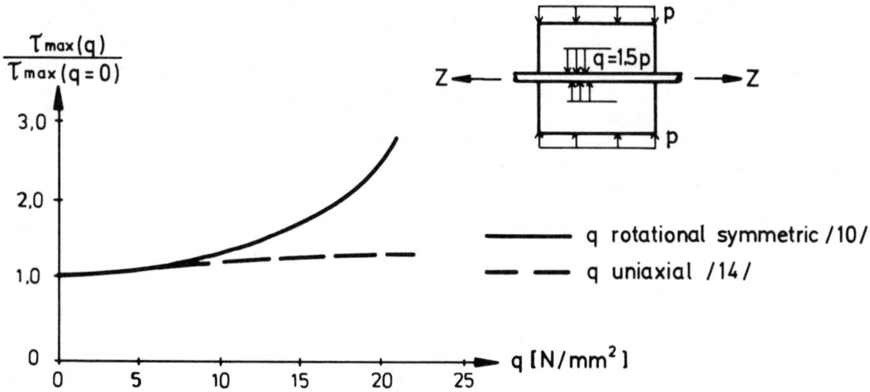

Fig. 13 Increase of bond strength as function of lateral pressure on the contact surface of reinforcement bars

The lateral pressure effects also the value Δ_{r1}. The relationship between the increase of Δ_{r1} as function of the increase of lateral pressure q is approximately linear. So it can be considered by the simple Eq. 21.

$$\Delta_{r1} = 0.06 \text{ mm} + 0.004\ q\ [\text{N/mm}^2] \tag{21}$$

There are no experimental data available to consider the influence of lateral tension on the bond strength. Nevertheless, it is obvious that lateral tension applied to the reinforcement bar will decrease the bond strength. It will be recommended to decrease the factor α_3 depending on the gapsize Δ_t as shown in Fig. 14.

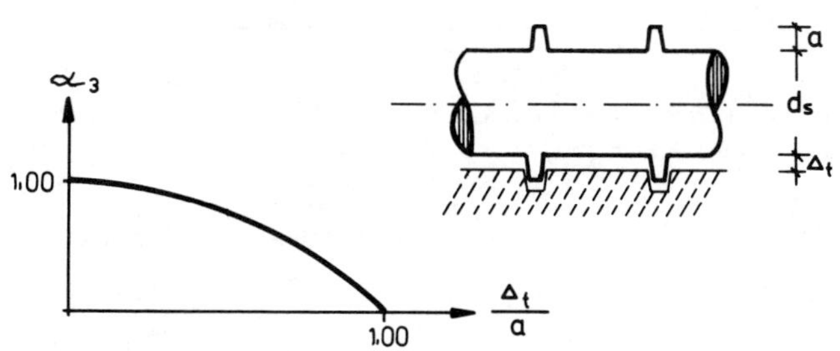

Fig. 14 Decrease of bond strength as function of gap Δ_t between concrete and reinforcement

The factor α_4 considers local damage of concrete in the bond zone caused by the development of secondary cracks. Some authors, especially Nilson [14], have shown that the bond stress-slip behaviour is influenced by local effects. This is due to different local damage of the bond zone. This can be explained by comparison of test specimen used by Nilson and Doerr. Nilson used a test specimen without a special length and without bond between concrete and reinforcement. But Doerr did. To consider these local effects the approximation given in Eqs. 22 can be used. At first, in Eq. 22a, the influence of the strain ε_s in the reinforcement as well as the relation of the yield strength f_{sy} to the ultimate strength f_{su} of the reinforcement will be considered according to Kobarg [16], because the direction of the compression stresses effects these local effects. The direction of the compression stresses is parallel to the secondary cracks. Therefore this effect is considered using the results of Roeder's tests [17]. The results of his tests are summarized in Fig. 15. In a simplified form the influence is given in Eq. 22b. Multiplying these two influence factors leads to Eq. 22 to consider damage effects.

$$\alpha_{41} = 1 - \frac{f_{sy}}{f_{su}} \cdot \frac{\varepsilon_s}{\varepsilon_{sy}} \qquad (22a)$$

$$\alpha_{42} = \sin \phi \qquad (22b)$$

$$\alpha_4 = \alpha_{41} \cdot \alpha_{42} \qquad (22)$$

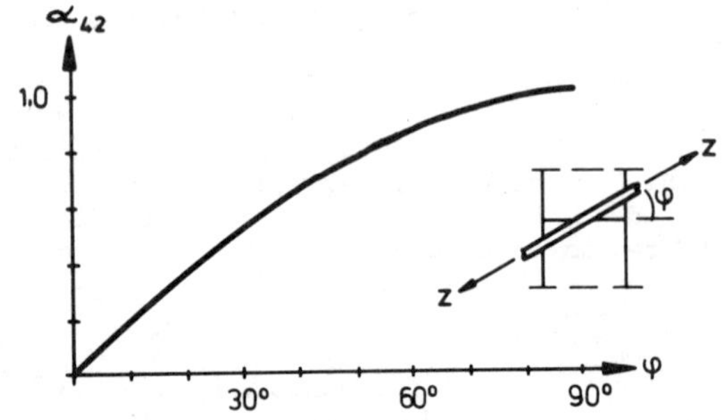

Fig. 15 Decrease of bond strength as function of the angle between the cracks and the reinforcement bar axis

Once all four factors α_1, α_2, α_3, α_4 have been calculated, the factor α can be computed from Eq. 20, and the bond stress-slip relationship can be considered according to Eq. 19.

6. NUMERICAL EXAMPLE

A safe transfer of the tensile forces in reinforcement bars to the concrete is a fundamental requirement for a well designed reinforced concrete structure. In the German reinforced concrete structures code DIN 1045, hooks are required as anchorage of smooth bars. If ribbed bars are used, the anchorage length can be reduced by the use of hooks.

Fundamental experimental and theoretical research on the load carrying behaviour of hook anchorages in concrete blocks has been performed at the Technische Universität München. Test data on pullout tests of hook anchorages with special regard to the different behaviour of smooth and ribbed bars are presented in [20].

To test the quality of both the isoparametric contact element and the bond model, the anchorage of a reinforcement bar with a diameter d_S = 12 mm has been analyzed. The element mesh and the size of the structure are shown in Fig. 16. It should be noted that there is no contact between the reinforcement bar and the concrete up to a distance of 94 mm from the front of the concrete block. The concrete as well as the reinforcement bar are modelled by 8-node quadratic plane stress elements. Consequently, curved contact elements with three double-nodes and a quadratic interpolation function have been used. The 180°-curve can be modelled by only four contact elements at the inside and at the outside respectively. This is an absolute minimum and leads to sufficient results, if no significant nonlinearities in the material like crushing of concrete occur. Otherwise more elements are needed to adequately model the nonlinearities of bond. The results of the FE-analysis of both a smooth and a ribbed bar are shown in fig. 16 in comparison with the test results.

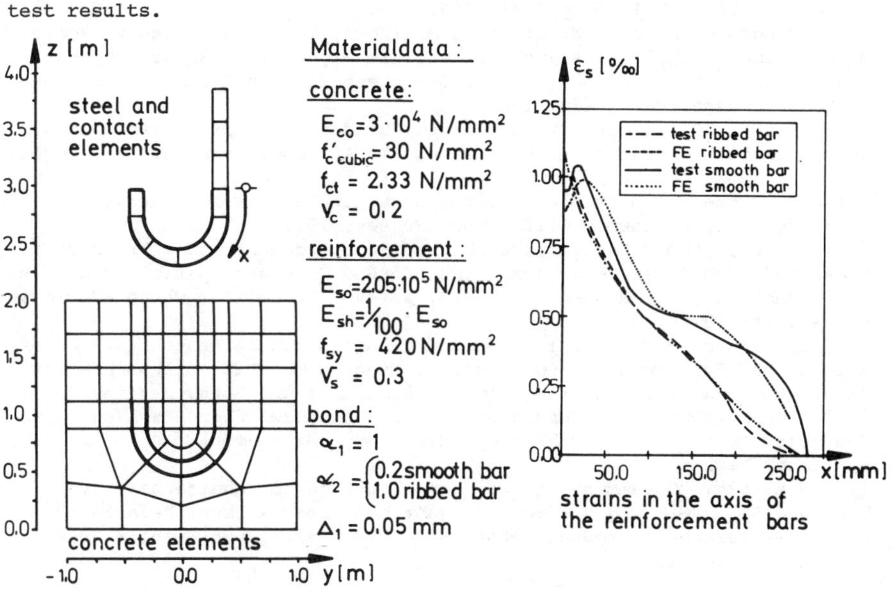

Fig. 16 Reinforcement bar anchorage by a hook

7. FINAL REMARKS

An isoparametric contact element and bond model have been described. This element has been used very successfully in order to analyze reinforced concrete structures. Results are given in the final research report [18] which is in preparation and in the thesis of Keuser [19].

ACKNOWLEDGEMENTS

Support of the Deutsche Forschungsgemeinschaft is gratefully acknowledged.

REFERENCES

[1] Y.Goto, "Cracks formed in concrete around deformed tension bars", ACI-Journal, Vol. 68, April, 1971

[2] D.Ngo, A.C.Scordelis, "Finite element analysis of reinforced concrete beams", ACI-Journal, Vol. 67, March, 1967

[3] A.K.De Groot, G.M.A.Kusters, T.Monnier, "Numerical modelling of bond-slip behaviour", Heron Concrete Mechanics, Vol. 26, No. 1B, 1981

[4] R.Godman, R.Taylor, T.Brekke, "A model for the mechanics of rocks", Soil Mechanics and Foundations, Division 5, 1968, pp. 637-659

[5] M.Hoshino, "Ein Beitrag zur Untersuchung des Spannungszustandes an Arbeitsfugen mit Spanngliedkopplungen von abschnittsweise in Ortbeton hergestellten Spannbetonbrücken", Dissertation, Darmstadt, 1974

[6] H.Schäfer, "A contribution to the solution of contact problems with the aid of bond elements", Computer methods in applied mechanics and engineering 6, 1975, pp. 335-354

[7] D.Dinges, M.Keuser, W.Kolmar, G.Mehlhorn, "Untersuchungen verschiedener Elementsteifigkeitsmatrizen auf ihre Eignung zur Berechnung von Stahlbetonkonstruktionen", Dritter Zwischenbericht zum DFG Forschungsvorhaben, Kassel/Darmstadt, 1983

[8] M.Keuser, G.Mehlhorn, V.Cornelius, "Bond between prestressed steel and concrete - computer analysis using ADINA", Computers and Structures, Vol. 17, 1983, pp. 669-676

[9] K.J. Bathe, "Finite Element Procedures in Engineering Analysis", Prentice-Hall, Englewood Cliffs, New Jersey, 1982

[10] K.Doerr, "Bond Behaviour of Ribbed Reinforcement under Transversal Pressure", Nonlinear Behaviour of Reinforced Concrete Spatial Structures (ed.: Mehlhorn, G., Rühle, H., Zerna, W.) Vol. 1, p. 13, Werner-Verlag, Düsseldorf, 1978

[11] H. Martin, P.Noakowski, "Verbundverhalten von Betonstählen - Untersuchung auf der Grundlage von Ausziehversuchen, Deutscher Ausschuß für Stahlbeton", Heft 319, Verlag von Wilhelm Ernst und Sohn, Berlin 1981

[12] G.Rehm, "Über die Grundlagen des Verbundes zwischen Stahl und Beton", Deutscher Ausschuß für Stahlbeton, Heft 138, Verlag von Wilhelm Ernst und Sohn, Berlin 1961

[13] H.Martin, "Zusammenhang zwischen Oberflächenbeschaffenheit, Verbund und Sprengwirkung von Bewehrungsstäben unter Kurzzeitbelastung", Deutscher Ausschuß für Stahlbeton, Heft 228, Verlag von Wilhelm Ernst & Sohn, Berlin, 1973

[14] R.Eligehausen, E.Popov, V.Bertero, "Local Bond Stress-Slip Relationship of Deformed Bars under Generalized Exhitations", Report No. UCB/EERC-83/23, Earthquake Engineering Center, University of California, Berkeley, 1983

[15] A.H.Nilson, "Bond Stress-Slip Relations in Reinforced Concrete, Report No. 345, Dept. of Civil Enigneering, Cornell University, Ithaca, N.Y., 1983

[16] J.Kobarg, "Verankerung druckbeanspruchter Bewehrungsstäbe unter Querdruck", Institut für Massivbau, Universität Karlsruhe, 1983

[17] F.K. Röder, "Versuchsergebnisse zum Kraftverlauf in schräg zum Riß angeordneten Bewehrungsstäben,", Forschungsbericht Nr. 23 aus dem Institut für Massivbau der TH Darmstadt, 1974

[18] G.Mehlhorn, M.Keuser, W.Kolmar, "Untersuchung verschiedener Elementsteifigkeitsmatrizen auf ihre Eignung zur Berechnung von Stahlbetonkonstruktionen", Abschlußbericht, Teil 1, zum gleichnamigen DFG Forschungsprojekt (Report in preparation)

[19] M.Keuser, "Verbundverhalten für nichtlineare Finite-Element-Berechnungen von Stahlbetonkonstruktionen", Dissertation D17, Darmstadt, 1985, (thesis is to be published in summer 1985)

[20] G.Rehm, H.Martin, H.H.Müller, "Ausziehversuche mit Betonstahlhaken", Materialprüfungsamt für das Bauwesen der TU München, Bericht Nr. 1975, 08. April 1968

BOND-SLIP MODELS IN FINITE ELEMENT ANALYSIS

Shiro MORITA * and Shigeru FUJII **

ABSTRACT

Several problems of bond-slip models used in finite element analysis are discussed. The physical meaning of local slip is defined, and the interface slip is distinguished from the apparent slip, which is caused mainly by the deformation of the surrounding concrete. It is emphasized which one of the slips should be modeled in accordance with the objective of analysis. The local bond versus apparent slip relationship is found not to follow the uniquely defined law, but to vary from location to location depending on the local stress state. Numerical examples are presented to show that the inclined spring model of wedging action is useful to describe the bond-slip mechanism of deformed bars. Further improvement of the bond element is proposed for more general and rational use of the bond-slip model in finite element analysis.

1. INTRODUCTION

The bond transfer of forces across the interface between concrete and steel reinforcement plays an important role in most aspects of reinforced concrete behavior. To take account of bond behavior in analytical studies, several laws to describe the behavior have been developed. The constitutive law between the local bond stress and the local slip at an arbitrary location on the interface has been formulated for general analysis purposes. Though many studies of the bond stress-slip relationship have been reported, considerable uncertainty still exists, because of the complex nature of bond which is affected by many factors.

The finite element most commonly used for modeling bond slip behavior is the linkage element, originally proposed by Ngo and Scordelis(1967). The finite element method, in which the link element is applied to describe the behavior of reinforced concrete, has been widely used. In most cases, the linkage stiffness was determined on the basis of measured bond stress-slip relationships, and a wide range of adopted values could be observed.

From the situation mentioned above, it seems timely to re-examine the effectiveness and the applicability limit of bond slip models in finite element analysis at the present stage. The aim of this paper is to offer a key to answer the following questions:

a) What is the physical meaning of "slip" of bond-slip models ?
b) Is it possible to determine the local bond stress-slip relationship

* Proffessor, Dept. of Architecture, Kyoto Univ., Sakyo-ku, Kyoto, Japan
** Research Associate, Dept. of Architecture, Kyoto Univ., Kyoto, Japan

irrespective of location on the interface ?

c) Is it possible to determine the local bond stress-slip relationship irrespective of existing steel stress at the location ?

d) Is it expected to provide the constitutive bond-slip law which describes the effects of various influencing factors such as end distance, concrete cover, lateral pressure, etc ?

The present study is restricted to bond involving deformed bars.

2. CRITICAL APPRAISAL OF BOND SLIP MODELS

2.1 Definition of slip

The increment of the local slip dSx within the infinite bar element dx at the location x can be obtained as the difference between the bar strain εsx and the concrete strain εcx :

$$\frac{dS_x}{dx} = \varepsilon_{sx} - \varepsilon_{cx} \tag{1}$$

In one-dimensional linear elastic problems, the differentiation of Eq.(1) with respect to x gives:

$$\frac{d^2S_x}{dx^2} = (1 + np)\frac{d\varepsilon_{sx}}{dx} \tag{2}$$

where n and p denote modulus ratio and steel ratio, respectively. The longitudinal force equilibrium for the infinite reinforcement element requires:

$$\tau_x = \frac{E_s\, d_b}{4} \cdot \frac{d\varepsilon_{sx}}{dx} \tag{3}$$

in which τx is the bond stress, Es is the modulus of elasticity of steel and d_b is the bar diameter. From Eq.(2) and Eq.(3), the following equation is obtained:

$$\frac{d^2Sx}{dx^2} = \frac{4\,(1 + np)}{E_s\, d_b}\,\tau_x \tag{4}$$

If the local bond stress τx can be given uniquely as a function of only the local slip Sx, Eq.(4) becomes a second order differential equation with respect to the local slip Sx. On the basis of the simplified theory described above, local bond stress versus local bond slip relationships have been investigated experimentally.

The local slip defined in the preceding theory is not the interface slip, but the slip of the bar relative to the imaginary cross section of concrete over which the transferred stress is distributed uniformly. The slip in this sense is denoted as "apparent slip" hereafter. Fig. 1 shows the often-quoted schematical presentation of stress transfer mechanism originally reported by Goto(1971). Due to the formation of secondary cone-shaped cracks starting at each rib, out-of-plane deformations of the initially plane concrete section become significant. On the other hand, the interface slip, caused mainly by local consolidation or crushing of concrete at the front of the ribs, remains relatively minor.

In experimental investigations the exact measurement of concrete strains at the interface is almost impossible to achieve. Therefore the strains are measured at a certain distance away from the interface. In many studies of bond slip behavior by using prismatic specimens, concrete was assumed as a rigid body and the concrete strain was disregarded in Eq. (1) to derive the local slip. The local slip obtained in such a manner gives an approximate value of the apparent slip and includes concrete-induced slip due to the formation of internal cone-shaped cracks as the dominant part. Nilson (1972) measured the concrete strains along the line a short distance apart from the interface to give the interface slip. Jiang et al.(1984) measured directly the interface slip using a microscopic technique. The different methods used in measuring the local slip led to widely varying test results for bond-slip relationships as shown in Fig. 2. It is noticable to find in Fig. 2 that Nilson's tests gives the stiffer relationships than others.

In finite element analysis, the link element has been used most commonly for modeling bond slip behavior. In determining the constitutive relations of springs in the link element, experimental results such as shown in Fig. 2 were often used. However, it must be noted that in most cases the experimentally obtained local slip is an approximate value of the apparent slip and differs essentially from the interface slip which is basically to be modeled by the link element.

As is apparent from Fig. 1, the real interface slip is negligible in comparison with the apparent one in most practical circumstances in structural members. Nevertheless, the clear distinction has not been made between these two slips, "interface" and "apparent", in modeling bond-slip behavior.

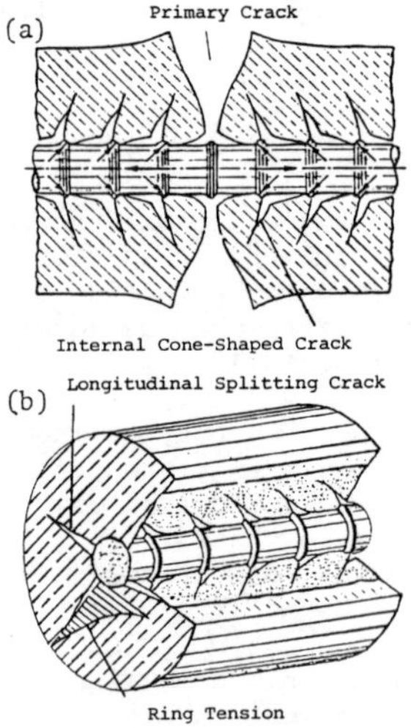

Fig.1 Stress transfer mechanism by bond

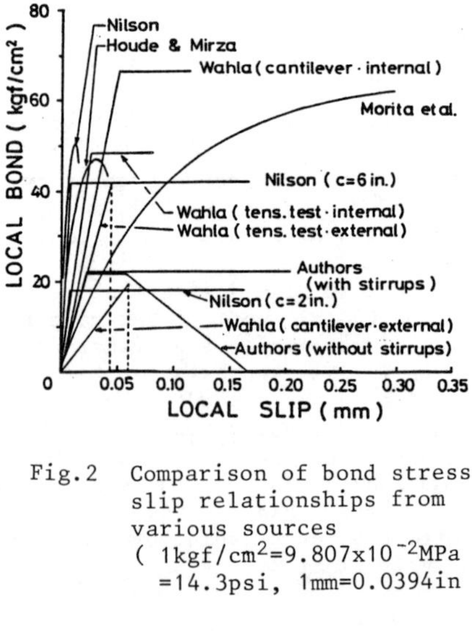

Fig.2 Comparison of bond stress-slip relationships from various sources ($1kgf/cm^2=9.807x10^{-2}MPa$ =14.3psi, 1mm=0.0394in)

2.2 Effect of Bar Size upon Slip Behavior

Because most bond tests have been carried out on only one bar size at a time, existing experimental bond-slip equations could not include bar size as a variable. This leads sometimes to the misunderstanding that bond-slip relations can be used independently of bar diameter. If it is true, Eq.(4) gives different distributions of Sx or τx for different bar diameters. If the apparent slip is proportional to bar diameter at a given bond stress, as is eventually expected, Eq.(4) can be reduced to:

$$\frac{d^2 S_{x_1}}{d{x_1}^2} = \frac{4(1 + np)}{E_s} \tau_{x_1} \qquad (5)$$

where $Sx_1 = Sx/d_b$, $x_1 = x/d_b$ and τx_1 is the local bond stress at the distance x . If τx_1 depends only on Sx_1 , Eq.(5) gives the identical distribution of Sx_1 or τx_1 for different bar diameters.

To verify the above mentioned hypothesis, extensive tensile tests were conducted on the specimens shown in Fig. 3. Deformed bars of 51, 25, 19, 13, 7 and 3mm diameter were used. (Tests for 51mm diameter bars were carried out by Kobe Steel Ltd.) Each specimen had enough concrete cover to simulate the stress state in an infinite half-space volume of concrete and was long enough so that damaged zones at both ends did not overlap. From the measurements of steel strains, the local bond stress x versus Sx/d relationships were reduced neglecting concrete strain, and the following equation was obtained from regression analysis of the test data:

$$\tau x = 888 \cdot (Sx/d_b)^{0.556} \qquad (\mathrm{kgf/cm}^2) \qquad (6)$$

Using Eq.(6), stress distributions were calculated numerically, and the results are plotted in Fig. 4 along with the test data. Fig. 4 indicates that if the distance from the loaded end is normalized with respect to bar diameter the stress distributions are almost identical for bar diameters of 3 to 51mm. This means that the extent of the internal cracking zone is probably proportional to the bar diameter.

In modeling bond versus apparent slip relationships, this nature of slip behavior should be reflected. The stress state, realized in the specimens shown in Fig. 3, is most basic and useful in the sense that, except for the end region of roughly 5 d_b, the local bond slip mechanism is similar at any location, and there exists the unique bond-slip relationship which depends only on material properties and steel geometry. This relationship is available in modeling the potential bond-slip relationship described afterwards in chapter 4. Fracture mechanics approaches should explain this basic behavior relating the size effect of reinforcing bars.

2.3 Radial Force Due to Wedging Action

The other basic aspect of bond behavior is the occurrence of radial forces caused by wedging action of the ribs as shown in Fig. 1. Due to this action, internal longitudinal cracks occur in the vicinity of the bar. With the increase of the slip, splitting of the surrounding

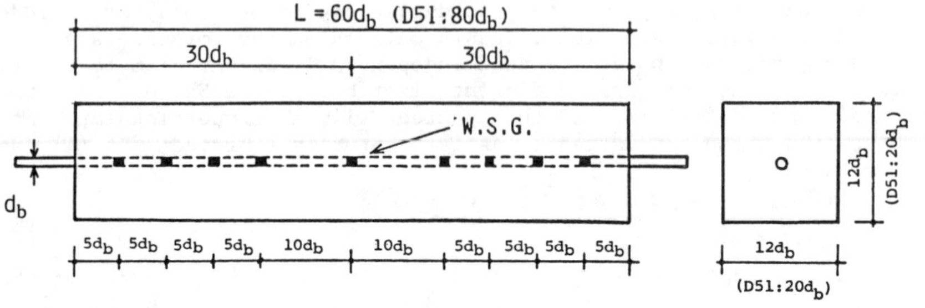

Fig.3 Test specimen to examine bar size effect

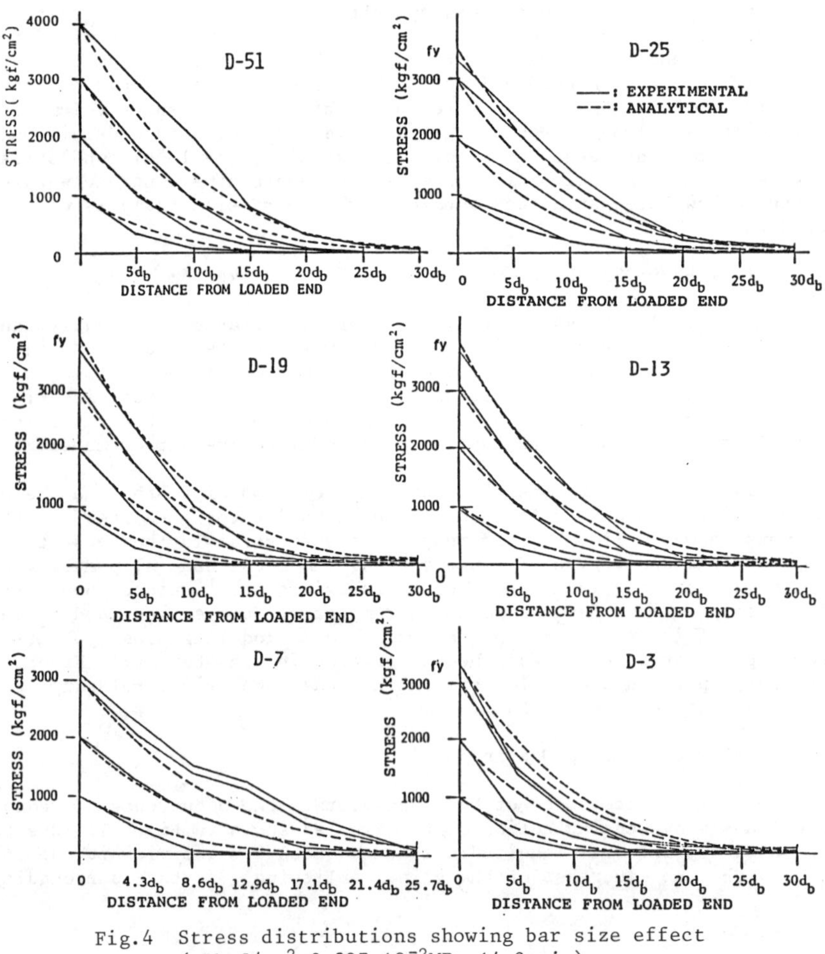

Fig.4 Stress distributions showing bar size effect
($1kgf/cm^2=9.807x10^{-2}MPa=14.2psi$)

concrete develops along the bar axis. The local bond slip behavior is strongly affected by the development of these longitudinal cracks. The peak stress of local bond is often dependent on the resisting capacity against splitting of the entire concrete cover. The provision of transverse reinforcement is effective only to sustain the bond up to a large slip (Morita, Fujii,1982). In order to model bond-slip behavior, the influence of longitudinal splitting of concrete due to radial force should be considered.

2.4 Applicability of Bond-Slip Model

There are at least two levels of approach on which the bond slip model are applied: (a) microscopic analysis of the stress state in the neighborhood of the reinforcement, and (b) macroscopic analysis of the global behavior of a member or a structural assemblage.

In microscopic analysis, the stress transfer mechanism by bond should be treated as the local contact in front of bar ribs, not as the one-dimensional bond slip at the interface. The interface slip is due to crushing of concrete within the limited zone in front of ribs, and can be derived from material properties and the configuration of the bar. It is essential to model appropriately the radial action of bond, as well as shear transfer parallel to the axis of the bar. Stress states in the concrete, especially in the neighborhood of the reinforcement, should be evaluated accurately, and initiation and propagation of internal cracks, both longitudinal splitting and cone-shaped cracks, should be followed analytically.. In this case, fracture mechanics will become a useful tool to follow crack propagation (Ingraffea et al.,1984).

Thus, the local bond stress versus the local apparent slip relationships can be derived as the result of the analysis. Therefore, the bond-apparent slip relationship depends upon the local stress state and in general varies from location to location.

In macroscopic analysis, the bond stress versus the apparent slip relationship may be directly modeled through determining the constitutive law of the one-dimensional link element by making use of experimental data or the results of the microscopic analysis. The stress state of the concrete near the bar is not more realistic, but it is a fictitious one to simplify the behavior. Another practical method to describe the bond effect in a macroscopic level is the tension stiffenning modelling, related to the uniaxial stress-strain behavior of steel or concrete (Gilbert,1978) (Scanlon,1974). Description of the modelling is omitted from the scope of the present study.

Most applications of bond-slip models in finite element analysis have been made at the macroscopic level. However, there always arises the question of how to determine the bond-slip law. In some cases, use was made of the model to explain the behavior of the specimen from which the model was derived. This does not constitute the real verification of the model. There exist too many variables to prescribe the generally applicable law from experimental investigations. The local bond stress versus the apparent local slip relationship, which is useful in the macroscopic approach, should, in the authors' opinion, not be supplied solely from vast test results, but should be the objective of microscopic analysis. Therefore, it is necessary to examine the applicability of the bond-slip model derived from the experimental results at any analytical study.

3. NUMERICAL EXAMPLE

3.1 Models and Material Properties

To realize the concept for microscopic analysis described in the preceding chapter, analytical case studies are presented. Though stress transfer behavior is actually three dimensional, a two dimensional plane stress model is used here for simplicity.

Plate-like reinforced concrete members of 225 mm uniform thickness were analyzed under the restrained end conditions shown in Fig. 5. Each member had three deformed bars of 25 mm nominal diameter (#8) embedded at the central layer for Type A and at the outer layer with thin concrete cover for Type B. Table 1 shows the loading condition applied to the protruding ends of the bars.

Stress transfer by bond at the steel-concrete interface was modeled by the linkage element with no physical dimension which consists of two inclined springs at the angle of ±45° to the bar axis. The linkage element was located at each nodal point to connect steel and concrete elements. Each inclined spring was active only in compression and always gave the linkage force acting at 45°. It was also assumed that the spring inclined at +45° was effective only if the slip at the interface $S>0$ and the other one at -45° was effective if $S<0$. Fig. 6 illustrates the movement of a steel nodal point relative to a concrete nodal point, both of which occupy the same point in the space before loading. The component of the linkage force in the direction of the bar axis gives the bond force, and the normal component gives the radial splitting force. For simplicity, the spring was assumed to behave elastically with the relatively high spring stiffness listed in Table 2. Therefore, the conceptual link element seems to provide a mechanism similar to perfect bonding, but it still gives the splitting or wedging action, which is essential to realize the stress transfer mechanism of deformed bars.

Concrete and reinforcing steel are represented by the eight noded quadrilateral isoparametric elements as shown in Fig. 5. The element stiffness matrix is calculated using Gauss' numerical integration method. The non-linear behavior of concrete is evaluated at each integration point of the element by use of the orthotropic constitutive model proposed by Darwin and Pecknold(1977). Reinforcing steel is conceived as an elastic material. Material properties are listed in Table 2.

The incremental load method was adopted for the nonlinear analysis procedure. At each loading step, concrete cracking and its propagation were checked, and both changes of the element stiffnesses and release of nodal forces by the elements were incorporated.

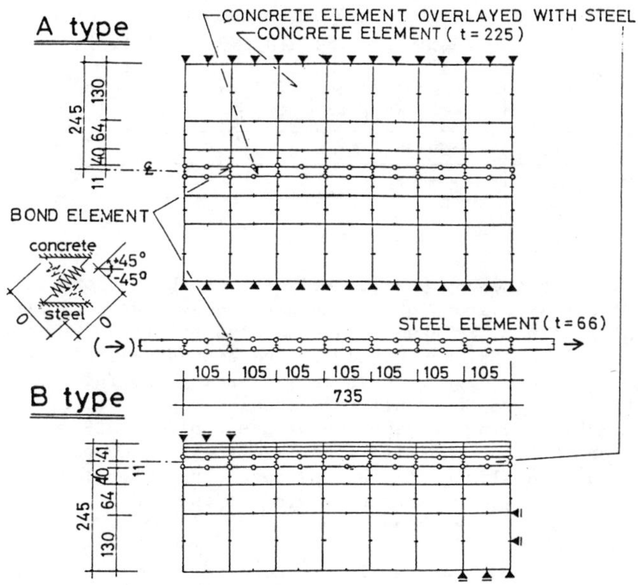

Fig.5 Models for numerical examples (1mm=0.0394in)

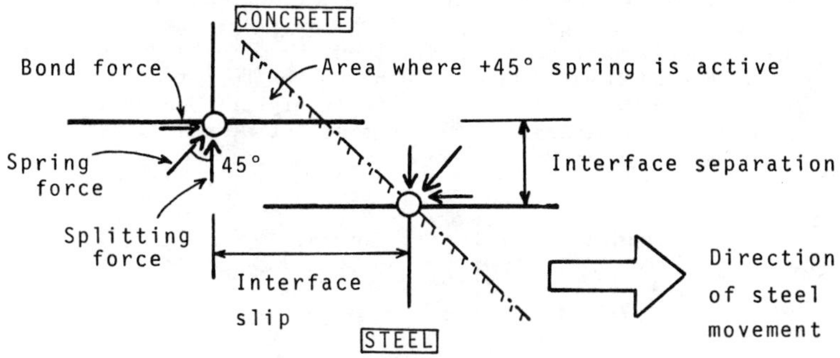

Fig.6 Wedging action modelled by inclined spring element

Table 1 Applied loading

	STEEL	LOADING	SPECIMEN
A-1	3D25	Pull-Only	A-type
A-2	3D25	Push-Only	A-type
A-3	3D25	Push-Pull	A-Type
B-1	3D25	Pull-Only	B-Type

Table 2 Assumed material properties
($1kgf/cm^2=9.807x10^{-2}MPa=14.2psi$)

CONCRETE	E_0= 300000 kgf/cm², f_c' = -240 kgf/cm², f_t' = 18.0 kgf/cm², ε_{cu}= -0.0025, ν= 0.17
STEEL	E_s= 2100000 kgf/cm², ν= 0.34
BOND	K= 10000 kgf/cm³

3.2 Results of Analysis

Slips of bars at the end section relative to the support fixed in the direction of bar axis are plotted against the average bond stress in Fig. 7. Each specimen of Type A gave an almost linearly ascending curve, while Specimen B-1 exhibited limited bond strength caused by splitting of the concrete cover. Push-only loading applied to Specimen A-2 led to a stiffer curve than pull-only loading applied to Specimen A-1. The pull-push loading of equal magnitude applied to Specimen A-3 gave more than one half of the slip of A-1 at the pulled end and more than one half of the slip of A-2 at the pushed end when loaded to a given average bond. This means that loading at one end influences the slip at the other end to a certain extent.

Fig. 8 and Fig. 9 show the distribution of steel stress and bond stress along the bar axis with the crack patterns at several loading stages. The effect of end distance on local bond reduction in the neighborhood of primary cracks is clearly demonstrated at the pull-out end of A-1 or A-3, while at the push-in end of A-2 or A-3, bond stress transfer near the primary crack is much more effective throughout the load application. This finding explains the drift of the zero stress point towards the push-in end of Specimen A-2. The bond stress distribution of B-1 becomes irregular owing to the formation of severe longitudinal and flexural cracks. Relatively higher bond at the free end region is due to the lateral confinement effect of the direct stress provided by the lateral support.

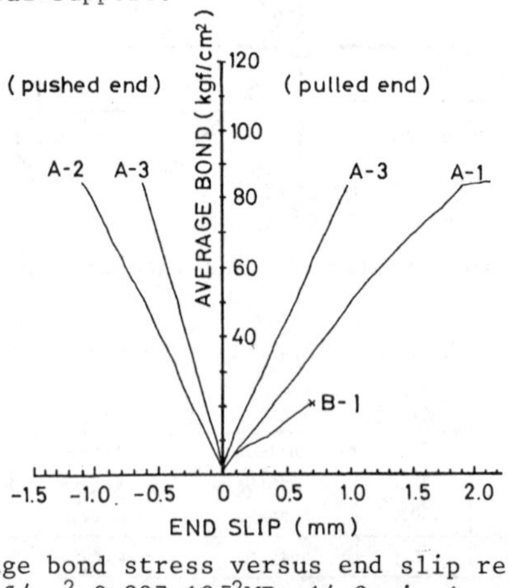

Fig.7 Average bond stress versus end slip relationships
($1kgf/cm^2=9.807x10^{-2}MPa=14.2psi$, 1mm=0.0394in)

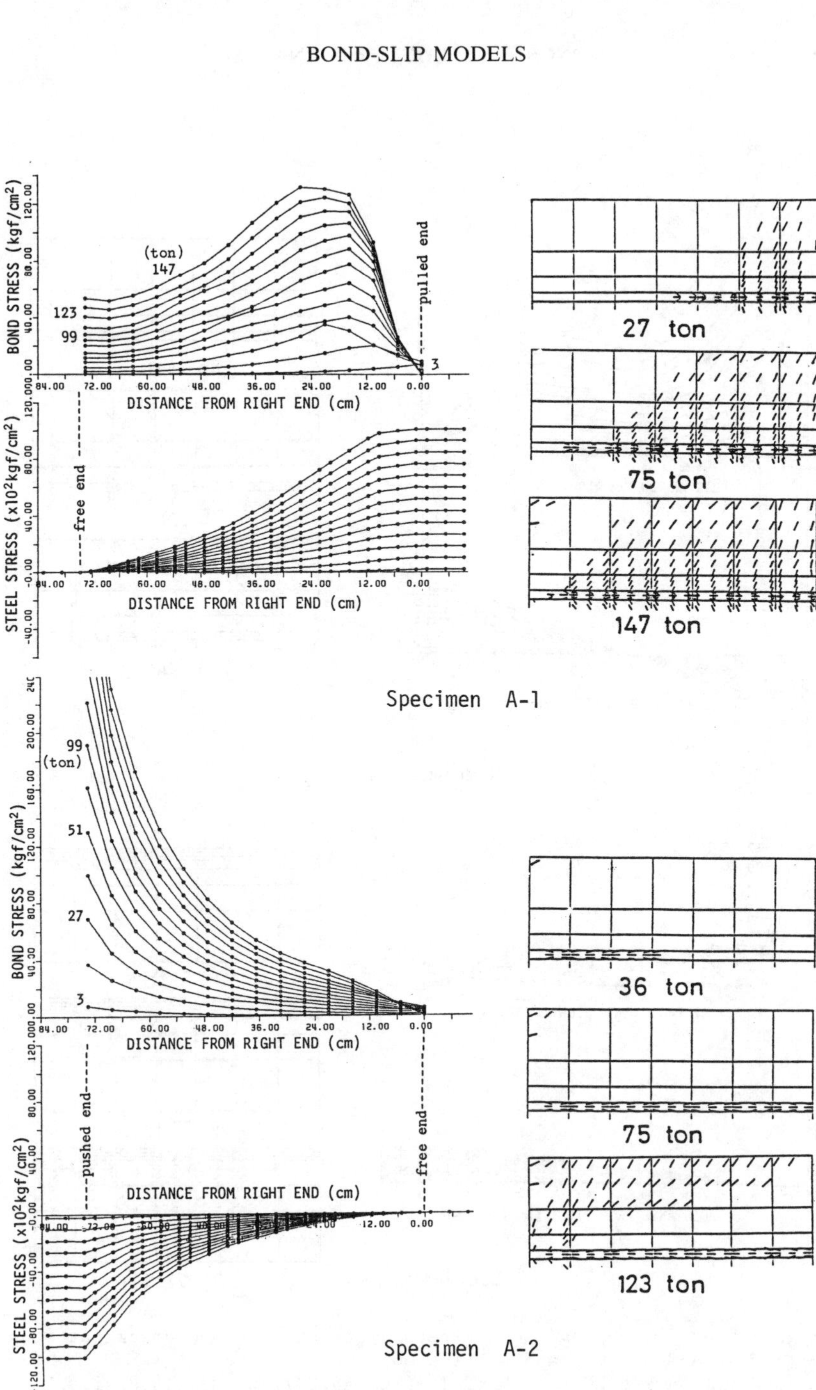

Fig.8 Steel stress and bond stress distributions and crack patterns of A-1 (above) and A-2 (below)
(1ton=9.81kN=2.205kip, $1 kgf/cm^2 = 9.807 \times 10^{-2}$ MPa=14.2psi, 1cm=0.394in)

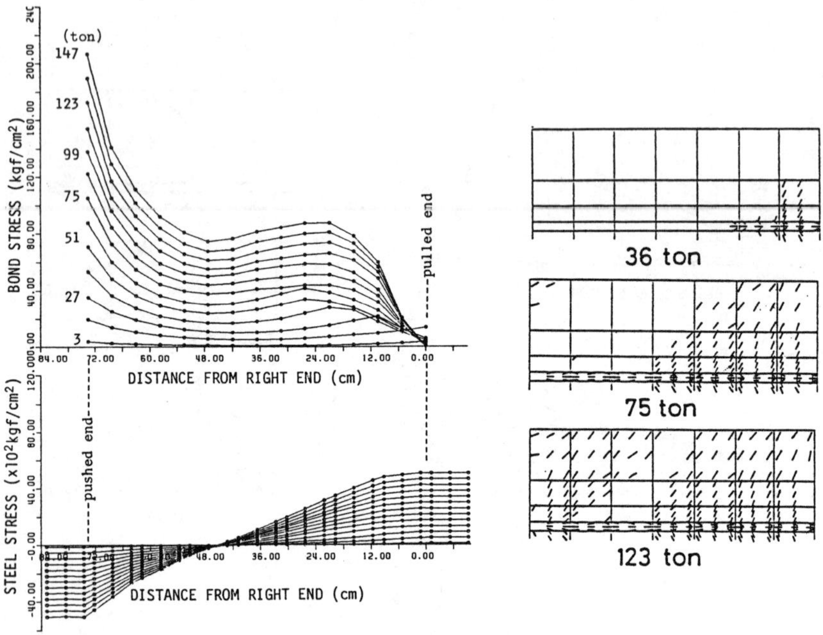

Specimen A-3

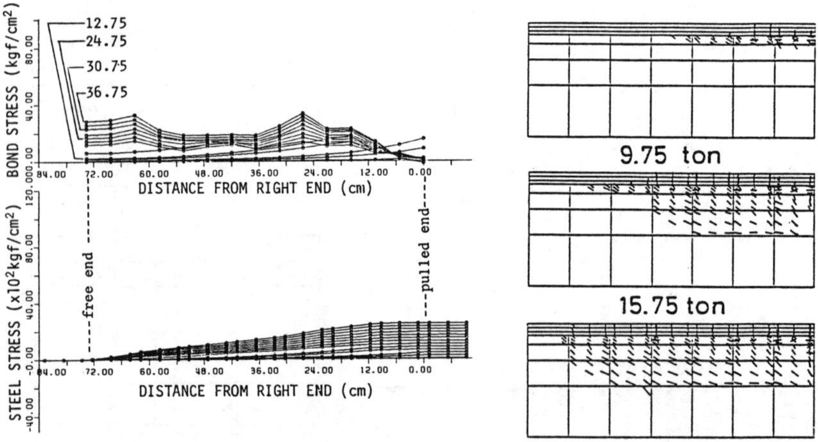

Specimen B-1

Fig.9 Steel stress and bond stress distributions and crack patterns of A-3 (above) and B-1 (below)
(1ton=9.81kN=2.205kip, 1kgf/cm^2=9.807x10^{-2}MPa=14.2psi, 1cm=0.394in)

Fig. 10 shows the local bond stress versus the local apparent slip relationship at every other location of a linkage element for A-1 and B-1. As is often the trend in experimental studies, it is the effect of end distance that the closer the location to the loaded end the more deteriorated the bond-slip behavior when pulled out. Fig. 11 shows also the bond-apparent slip curves: (a) in the neighborhood of the left end, (b) in the middle of the development length, (c) in the neighborhood of the right end. In contrast to the deterioration at the right end region (where bars protrude from the surrounding concrete), the stress transfer at the left end region (where bars enter the concrete) or at the middle region exhibits its potential and behaves similarly, except for B-1, irrespective of the bar stress and the location of the elements.

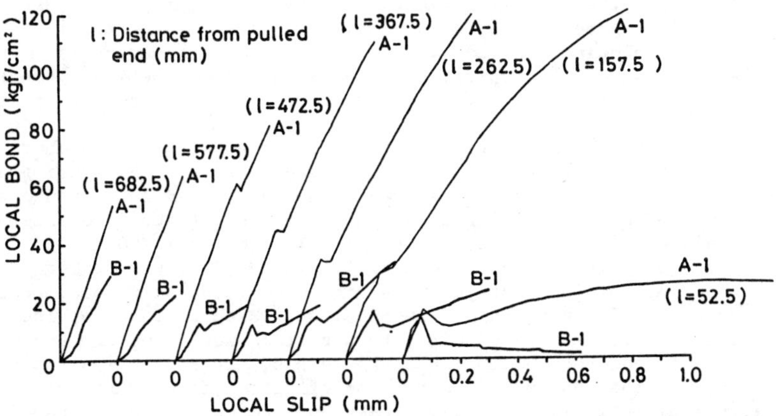

Fig.10 Local bond stress versus local apparent slip relationships of A-1 and B-1 (1kgf/cm^2=9.807x10^{-2}MPa=14.2psi, 1mm=0.0394in)

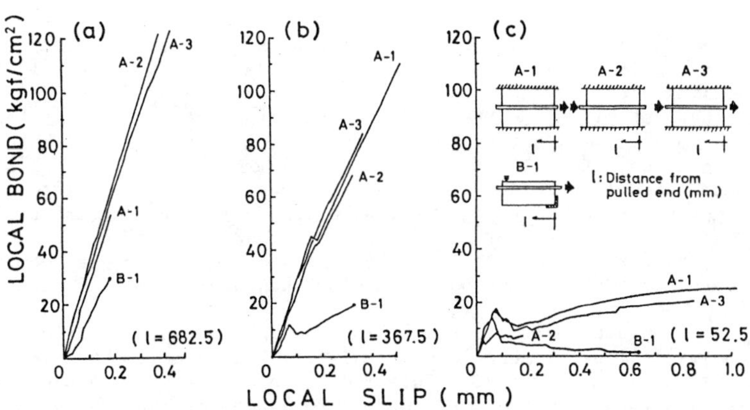

Fig.11 Comparison of local bond stress versus local apparent slip relationships for every models
(1kgf/cm^2=9.807x10^{-2}MPa=14.2psi, 1mm=0.0394in)

The behavior described above from the analysis seems to explain satisfactorily the experimental evidence, at least in a qualitative sense. The inclined spring model introduced in this study can be a useful tool to model the bond mechanism of deformed bars. However, how to determine the stiffness and the angle of the spring still remains open for discussion. Moreover, the applicability of this model seems to depend strongly on the extent of the analytical treatment of the concrete in the neighborhood of the bar.

4. FURTHER IMPROVEMENT OF THE MODEL

4.1 Basic Concept

Although the bond model used in Chapter 3 seems to be useful in investigating generally the bond slip behavior, it requires quite cumbersome calculations to determine the global behavior of even a single member. In the spirit of simplification, some improvements of the model are outlined here, which still remain in the conceptual stage.

It is the authors' primary standpoint that even in a member level analysis the fundamental mechanism of stress transfer by wedging action should be modeled. For simplicity, the concrete deformation due to internal secondary cracking is assumed to occur at the interface and to be involved as a part of the total slip of the bond-slip model. Provision of finer grid size in the vicinity of reinforcement can be eliminated by this assumption. Here, the bond-total slip relationship of the model should be based on the ideal one which will be obtained when the internal cracking is limited to a certain region within the overall geometry of concrete volume. In this sense, this relationship can be denoted as the potential one. The effects of primary cracks and longitudinal entirely-splitting cracks on bond deterioration need not to be modeled, but should be presented as the results of analysis. Based on this viewpoint, the following model is developed.

4.2 Combined Link Element

The linkage element with the inclined spring, used in Chapter 3 is employed again to idealize the wedging action. This element is named "wedge link element" and assumed to behave elastically. In addition to the wedge link element, the ordinary bond link element, which has two springs parallel and normal to the bar axis, is used to model the bond-slip due to internal cracking. This element is named "slip link element". Non-linear and hysteretic behavior of bond-slip can be represented by the constitutive law of the slip link element. Though De Groot et al.(1981) proposed the bond model in which the wedging action was taken into account, the proposed model does not seem to be successful to describe the bond behavior adequately.

The improved model is the "combined link element", which consists of the wedge link element and the slip link element located in series as illustrated in Fig. 12. This model can express the wedging action and bond-slip behavior due to local deformation of surrounding concrete as well.

4.3 Description of the Bond Model

The proposed bond element consists of two linkage elements in series, both of which have no physical dimensions. At the interface between steel and concrete, there exist the radial compressive stress σ_r and the bond stress τ. In the wedge link element, the relationship between these stresses and relative displacements is given by:

$$\begin{Bmatrix} \tau \\ \sigma_r \end{Bmatrix} = k_w \cos^2\phi \begin{bmatrix} \tan^2\phi & \tan\phi \\ \tan\phi & 1 \end{bmatrix} \begin{Bmatrix} S_1 \\ W_1 \end{Bmatrix} \tag{7}$$

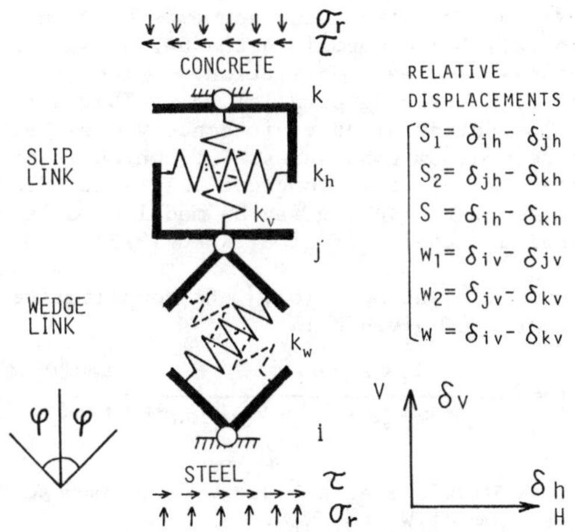

Fig.13 Combined link element

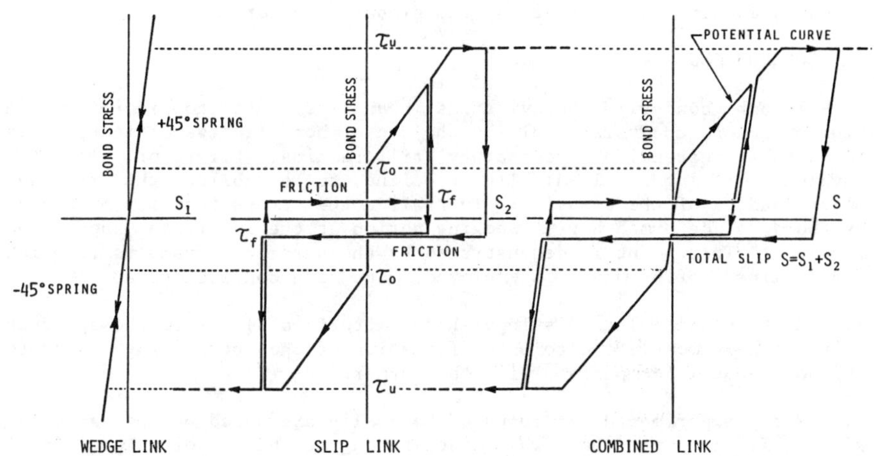

Fig.14 Bond-slip law assumable for combined link element

where S_1 and W_1 are relative displacements between points i and j in the directions parallel and normal to the bar axis, respectively. ϕ is the inclined angle of the spring, and k_w is the spring stiffness in kgf/cm^2 per cm unit. A relatively high value will be given as the spring stiffness, and the angle ϕ is assumed to be nearly 45°.

For the slip link element, the following well-known relationship is available:

$$\begin{Bmatrix} \tau \\ \sigma_r \end{Bmatrix} = \begin{bmatrix} k_h & 0 \\ 0 & k_v \end{bmatrix} \begin{Bmatrix} S_2 \\ W_2 \end{Bmatrix} \tag{8}$$

where S_2 and W_2 are relative displacements between points j and k in the directions parallel and normal to the bar axis. k_h and k_v are the spring stiffnesses in each direction. The assumed deformation characteristics of the spring parallel to the bar axis are illustrated in Fig. 13. The potential curve is conceived as being bi-linear and terminated at the maximum bond stress τ_u, which is deemed to represent the shear capacity of the concrete key extruded between ribs. Hysteretic behavior of the spring may be modeled as shown in Fig. 13 on the basis of previous studies (Morita, Kaku 1973)

Combining Eq. (7) and Eq. (8), the constitutive equation of the combined link element is given by:

$$\begin{Bmatrix} \tau \\ \sigma_r \end{Bmatrix} = \frac{k_h k_v k_w \cos^2\phi}{k_h k_v + k_w k_v \sin^2\phi + k_w k_h \cos^2\phi} \begin{bmatrix} \tan^2\phi & \tan\phi \\ \tan\phi & 1 \end{bmatrix} \begin{Bmatrix} S \\ W \end{Bmatrix} \tag{9}$$

where the relative displacements S and W are composed as the sum of S_1 and S_2, and as the sum of W_1 and W_2 .

This model seems to be useful for the analysis of reinforced concrete members, especially in the case where bond deterioration due to repeated loading is decisive for the global behavior.

5. CONCLUSIONS

Though bond-slip behavior has been taken into consideration in finite element analysis mainly by the use of bond linkage element, it is found that extensive uncertainty still remains. First of all, the physical meaning of local slip is defined to establish the rational understanding of the behavior. To realize the stress transfer mechanism by bond of deformed bars, wedging action at the interface should be modeled. This point is demonstrated by the numerical examples in which the inclined spring model is employed. The main conclusions are:

1. It is essential to distinguish the interface slip from the apparent slip which is mostly due to the deformation of the surrounding concrete and dominates in comparison with the former.

2. Most experimental results of bond-slip are based on the apparent slip. It is important when modeling on the basis of experimental results to distinguish which slip is involved.

3. The inclined spring linkage element is useful in simulating the stress transfer mechanism of deformed bars at the interface.

4. The local bond stress versus the local apparent slip relationship is by no means unique, but is dependent solely on the stress state of the concrete at the location.

5. Applicability or rationality of any interface element is related strongly to the extent of analytical treatment of the concrete, especially in the neighborhood of the bar.

References

1. Darwin,D., and Pecknold,D.A.,"Nonlinear Biaxial Stress-Strain Law for Concrete," ASCE, Journal of the Engineering Mechanics Division, Vol.103, No.EM2, Apr., 1977, pp.229-241.
2. De Groot,A.K., Kusters,G.M.A., and Monnier,Th.,"Numerical Modelling of Bond-Slip Behavior," HERON, the Netherlands, Vol.26, No.1B, 1981.
3. Goto,Y., "Cracks Formed in Concrete Around Deformed Tension Bars," ACI Journal, Vol.68, No.4, April 1971, pp.1491-1505.
4. Gilbert,R.I., and Warner,R,F.,"Tension Stiffenning in Reinforced Concrete Slabs," ASCE, Journal of the Stractural Division, Vol.104, No.ST12, 1978, pp.1885-1899.
5. Ingraffea,A.R., Gerstle,W.H., Gergely,P., and Saouma,V.,"Fracture Mechanics of Bond in Reinforced Concrete," ASCE, Journal of the Stractural Engineering, Vol.110, No.4, Apr., 1984, pp.871-890.
6. Jiang,D.H., Shah,S.P., and Andonian,A.T.,"Study of The Transfer of Tension Forces by Bond," ACI Journal, Vol.81, No.3, May-June 1984, pp.251-259.
7. Morita,S., and Kaku,T.,"Local Bond Stress-Slip Relationship Under Repeated Loading," Preliminary Report, IABSE Symposium, Lisbon, 1973, pp.221-227.
8. Morita,S., and Fujii,S.,"Bond Capacity of Deformed Bars Due to Splitting of Surrounding Concrete," Bond in Concrete, Proc. of International Conference on Bond in Concrete, Paisley, 1982, Applied Science Publishers, pp.331-341
9. Ngo.D., and Scordelis,A.C.,"Finite Element Analysis of Reinforced Concrete Beams," ACI Journal, Vol.64, No.3, Mar., 1967, pp.152-163.
10. Nilson,A.H.,"Internal Measurement of Bond Slip," ACI Journal, Vol.69, No.7, July, 1972, pp.439-441.
11. Scanlon,A., and Murray,D.W., "Time-Dependent Deflections of Reinforced Concrete Slabs," ASCE, Journal of the Structural Division, Vol.100. No.ST9, 1974.

Bond Behavior in the Neighborhood of the Crack

Shizuo Hayashi*, M. AIJ and Seiji Kokusho**, M. AIJ

SYNOPSIS

Experiments were carried out on deformed bars subjected to pull and on reinforced concrete beams subjected to bending moment and shear force. The strains in the longitudinal bar and concrete were measured in detail, and the local bond stress versus the local slip relationships were investigated in the neighborhood of the crack. The maximum bond stress and the bond stiffness were found to be significantly lower near a crack than at some distance from the crack. And no deterioration was observed at distances from the crack of four or more bar diameters. The maximum bond stresses in the longitudinal bars of beams were lower than in the bars subjected to pull-out. One of tested beams was analyzed by the finite element method using a bond and crack link mechanism and considering the bond deterioration near cracks. The analytical results agreed with experimental results very well.

INTRODUCTION

Bond behavior has the same important influence on the mechanical behavior of reinforced concrete members as do the properties of reinforcement and concrete. Since bond behavior is an interaction between reinforcing bars and concrete and is influenced by the stress levels in the bars and conditions of concrete around the bars, it is a complex phenomenon. It is natural to think that bond deterioration is initiated in the neighborhood of cracks. However there do not exist many studies which pay regard to this bond deterioration near cracks.

The authors have carried out pull-out bond tests and experiments on reinforced concrete beams subjected to bending moment and shear force in order to investigate the bond behavior near cracks and to propose a bond deterioration model for the finite element method analysis.

PULL-OUT BOND TEST

Specimens

The pull-out bond tests were carried out in order to investigate the influence of the distance from the loaded end to a bond zone. The specimens shown in Table 1 are divided into four groups identified by the first character in their name (C ~ F).

* Associate Professor, Department of Architecture, Fukui University, 3-9-1 Bunkyo, Fukui City, Fukui, 910, Japan.

** Professor, Research Laboratory of Engineering Materials, Tokyo Institute of Technology, 4259 Nagatsuta, Yokohama City, Kanagawa, 227, Japan.

The bond length of the specimens in groups C and E was 32 mm so that uniform distribution of bond stress can be assumed. Since it was feared that this short bond length and nonsymmetry of the commercially available deformed bars might have some adverse effect on the experimental results, symmetric bars, which were finished by machining, were used. A typical bar specimen of groups C and E is shown in Fig. 1.

Table 1 List of Specimens

Name of Specimens	Position of Bar in Section	L	Bond Length	Remarks
C- I -50	I	5・d	2・d	L : Distance from Loaded End to the Face of the First Lug
C- I -25	I	2.5・d		
C- I -05	I	0.5・d		
C- II -50	II	5・d		Object: Longitudinal Bar of Beam or Column (Thick Cover, Thin Cover, Bar; II, I)
C- II -25	II	2.5・d		
C- II -05	II	0.5・d		
E- III -50	III	5・d		Position of Bar in Section
E- III -25	III	2.5・d		
E- III -05	III	0.5・d		
E- IV -50	IV	5・d		Reinforcement of Slab or Wall (IV, III)
E- IV -25	IV	2.5・d		
E- IV -05	IV	0.5・d		
D- I - (70)	I	——	7・d	
D- II - (70)	II			Minimum Thickness of Cover Concrete = 2.5 x d
F- III - (70)	III			d : Bar Diameter, 16 mm
F- IV - (70)	IV			

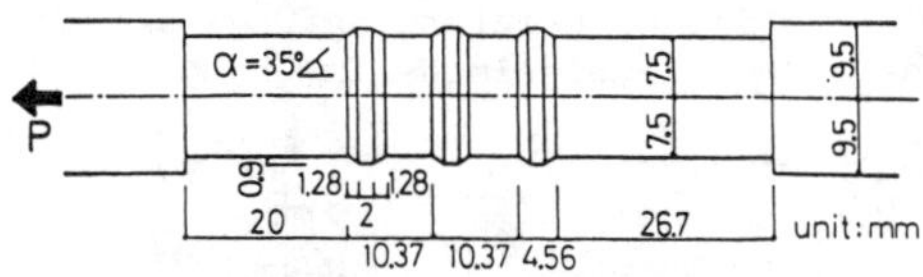

Fig. 1 Bar used in Specimens of Groups C and E

The bond length of the specimens in groups D and F was 112 mm; in these case commercially available deformed bars were used. The strains in the bars were measured at intervals of 30 mm in order to investigate the effects of the distance from the loaded end on the local bond behavior. Grooves were machined into these bars as shown in Fig. 2 in order to measure strains along the bar with a minimum interference with the bond mechanism.

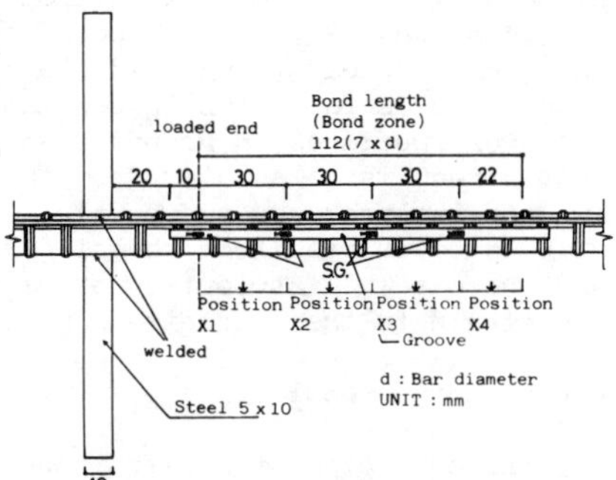

Fig. 2 Bar used in Specimens of Groups D and F

The second character of the Specimen name indicates the position of the tested bar in the section. As shown in the remarks in Table 1, the poaition I and II were selected to simulate longitudinal bars in beams and columns, whereas positions III and IV were selected to simulate reinforcement of walls and slabs.

The third character of the specimen name in groups C and E denotes the distance from the loaded end to the front face of the first lug of the bar.

Examples of arrangement of reinforcement are shown in Fig. 3 and 4. The minimum thickness of concrete cover is 40 mm, which is 2.5 bar diameters.

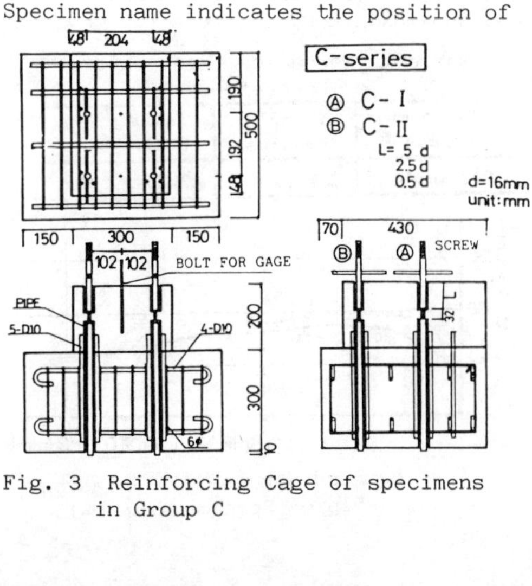

Fig. 3 Reinforcing Cage of specimens in Group C

Mechanical properties of concrete are given in Table 2.

Loading and measuring arrangement

The test set-up is shown in Fig. 5. Load was applied to the bar through the bolt attached to it by means of a screw jack with a capacity of 5 ton and measured by the strain in the bolt. The reaction force was supported by the stub of the specimen so as not to restrain the concrete near the loaded end.

For specimens in groups C and E, bond stress is calculated from the equation $\tau=P/\psi \cdot l_b$ (τ=bond stress in kg/cm², P=load in kg, ψ=distance round of bars in cm and l_b=bond length of 3.2 cm).

For specimens in groups D and F, the local bond stress is calculated from the difference of two strains measured adjacently. And the local slip is calculated from the slippages at both ends and the strain distribution of bars, assuming that the strain in the concrete is zero.

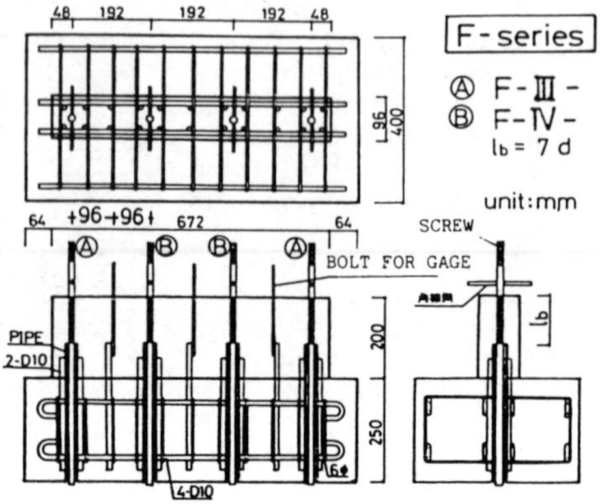

Fig. 4 Reinforcing Cage of Specimens in Group F

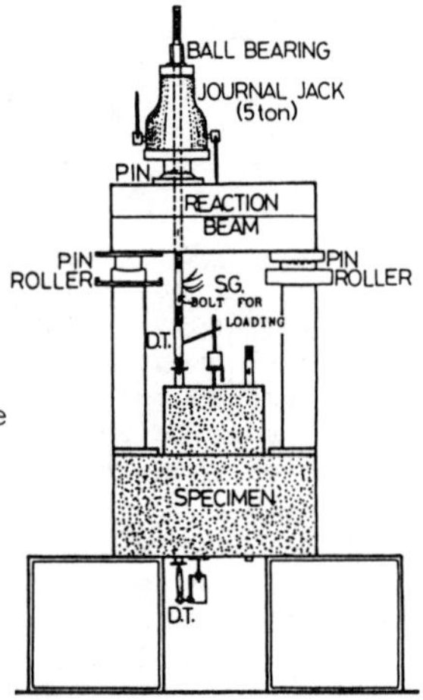

Fig. 5 Testing Set-up

Experimental results

Effects of distance from the loaded end to the bond zone on the relationship between bond stress and slip at the loaded end are shown in Fig. 6 for the three specimens belonging to the group C-I. As the distance from the loaded end is reduced, the maximum bond stress decreases. Figure 7 shows the relationship between the distance from the loaded end and maximum bond stress for the specimens in

Table 2 Mechanical Properties of Concrete

Maximum Strength	Young's Modulus	Tensile Strengtn
342 kg/cm²	224 t/cm²	22.2 kg/cm²

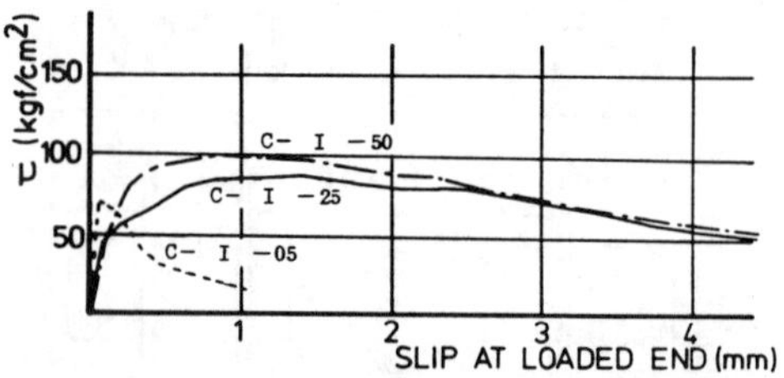

Fig. 6 Bond Stress-Slip Relationship of Specimens in C-I

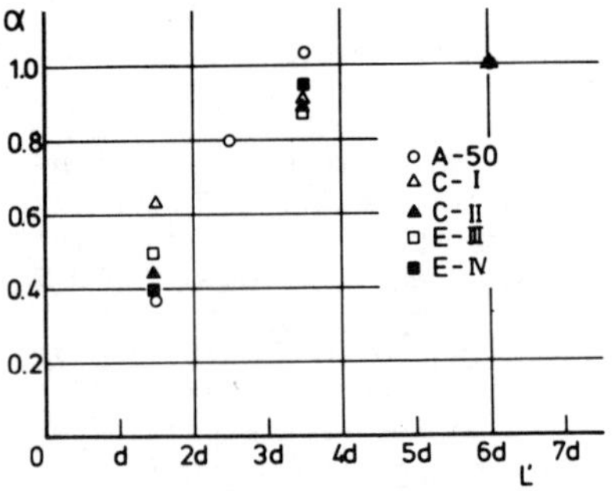

Fig. 7 Relationship between Maximum Bond Stress and Distance from the Loaded End

groups C and E. Specimens in group A-50 have the same testing bar as the specimens in groups C and E and have a concrete cover with a thickness of 5 bar diameters all around the bar as reported previously in [2]. The vertical axis (α) shows maximum bond stress normalized with regard to the maximum bond stress at a distance of 5 bar diameters from the loaded end. The horizontal axis (L') shows the distance from the loaded end to the center of the bond zone, that is, L'=L + 16 mm. When L' is less than 3 bar diameters, α decreases as L' becomes small. When L' is 3.5 bar diameters, α is almost 1.

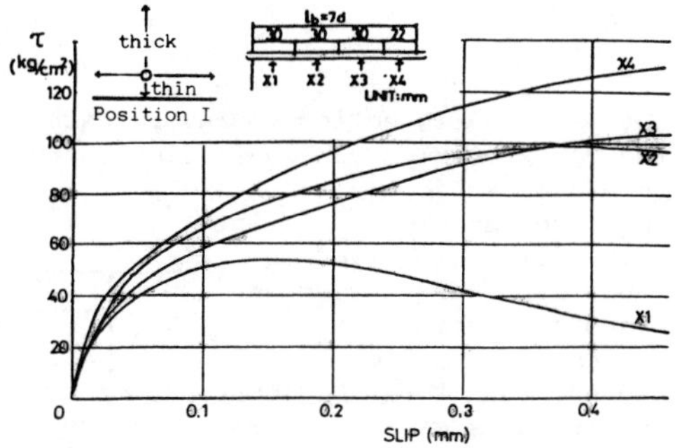

Fig. 8 Local Bond Stress-Slip Relations of D-I-(70)

Figure 8 shows the effects of the distance from the loaded end on the relationship between local bond stress and local slip for the specimen D-I-(70). Figure 9 shows the relationship between the maximum local bond stress and the distance from the loaded end to the measuring point. The vertical axis is now the maximum bond stress normalized with regard to the maximum bond stress at point X4, which is the position 101 mm (6.3 bar diameters) away from the loaded end. Specimens in group B-50 have a concrete cover of 5 bar diameters as reported in [2]. In this case, deterioration of the maximum bond stress occurs when the distance L' is less than 4 bar diameters, too.

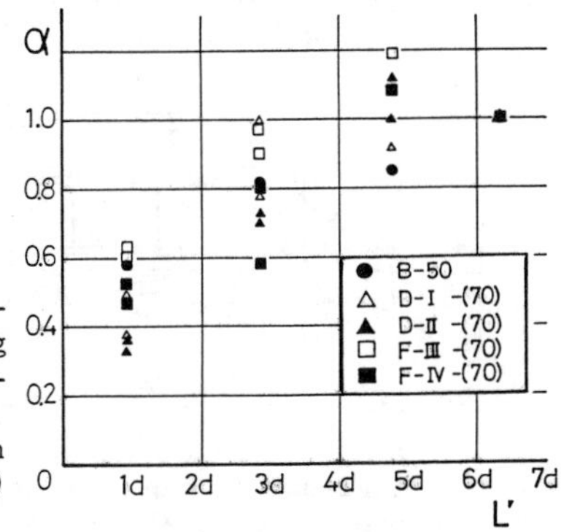

Fig. 9 Relations between Maximum Local Bond Stress and Distance from the Loaded End

The position of the bar in sections I ~ IV have no serious influence on the bond behavior in these experiments.

BOND BEHAVIOR OF BARS IN A BEAM SUBJECTED TO SHEAR AND BENDING MOMENT

Table 3 List of Specimens

No.	Name of Specimens	Longitudinal Bars	Number of Precracks	Intervals of Precracks
1	D16-20	2-D16	2 Precracks	20cm
2	D16-10	2-D16	3 Precracks	10cm
3	D16-0	2-D16	No Precracks	—
4	D19-10	2-D19	3 Precracks	10cm
5	D13-10	3-D13	3 Precracks	10cm

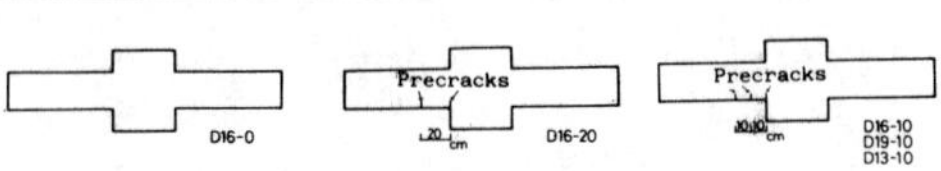

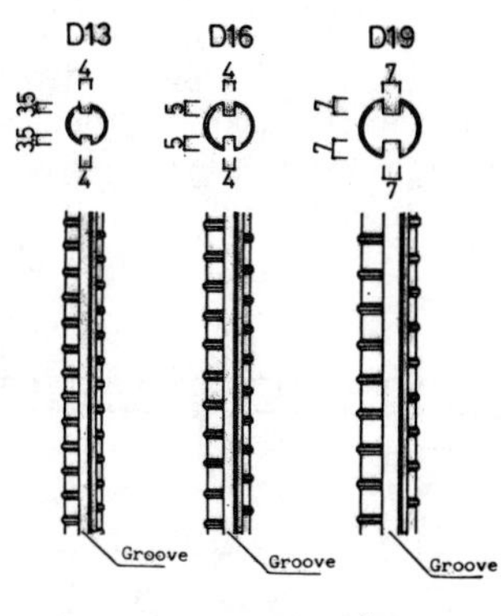

Fig. 10 Details of Longitudinal Bars

Specimens

These experiments were carried out with the intention to investigate the bond behavior of longitudinal bars in a beam subjected to shear and bending moment. A list of specimens is given in Table 3. There are three kinds of longitudinal reinforcement, two D16 bars, two D19 bars and three D13 bars. The products of cross sectional area and yield strength of each kind are almost equal. The details

Table 4 Mechanical Properties of Concrete

Maximum Strength	Young's Modulus	Tensile Strength
218 kg/cm^2	230 t/cm^2	20.1kg/cm^2

Table 5 Mechanical Properties of Reinforcing Bars

Bar	Ungrooved Bar			Area of Grooved Bar
	Yield Strength	Maximum Strength	Elongation	
D13	3.86 t/cm^2	5.68 t/cm^2	25.4 %	1.03 cm^2
D16	3.88	5.69	25.8	1.61
D19	3.70	5.59	28.0	1.92
6 ϕ	3.53	3.96	21.2	—

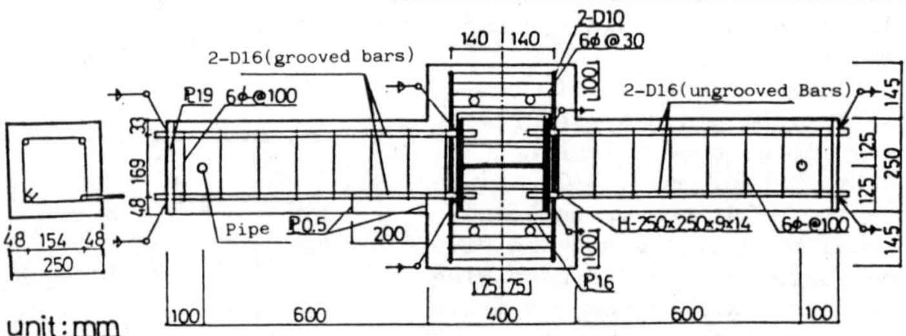

Fig. 11 Reinforcing Cage of Specimen D16-20

of the longitudinal bars are shown in Fig. 10. The mechanical properties and the cross sectional area of the reinforcing bars are listed in Tables 4 and 5. It is necessary to define the position where the bending cracks occurs so that the bond behavior near these cracks can be investigated in detail. Some cracks were introduced prior to loading. The number of prior cracks were two and three, respectively, with 20 cm and 10 cm spacing as shown in Table 3. The reinforcement arrangement is shown in Fig. 11. The left side of the specimen is the test beam and the right side is needed only for loading. In order to make the prior cracks, steel plates of 0.5 mm thickness were set before casting.

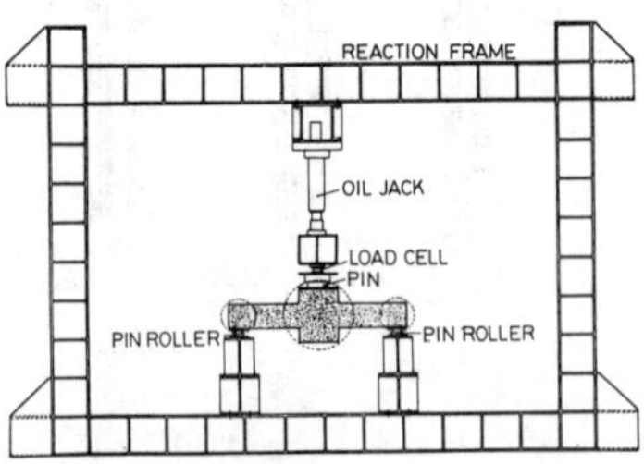

Fig. 12 Loading Set-up

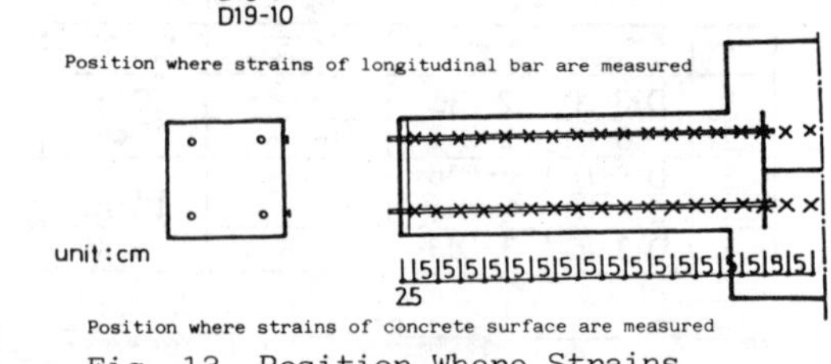

Fig. 13 Position Where Strains were Measured

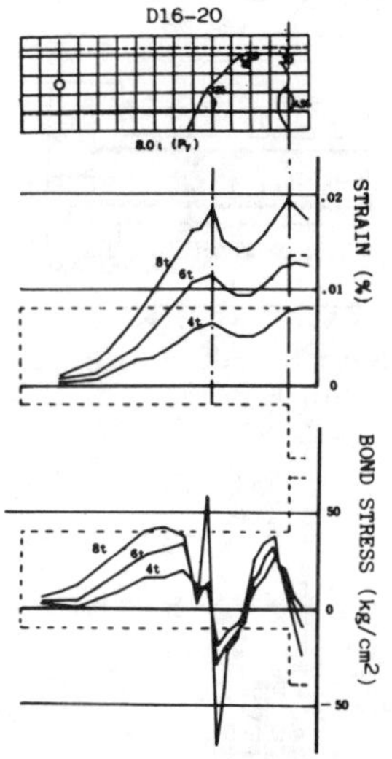

Fig. 14 Crack Pattern, Strain Distribution and Bond Stress Distribution in Specimen D16-20

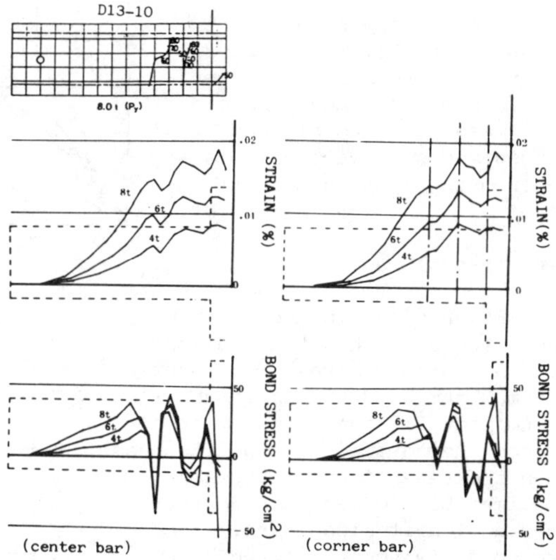

Fig. 15 Crack pattern, Strain Distribution and Bond Stress Distribution in Specimen D13-10

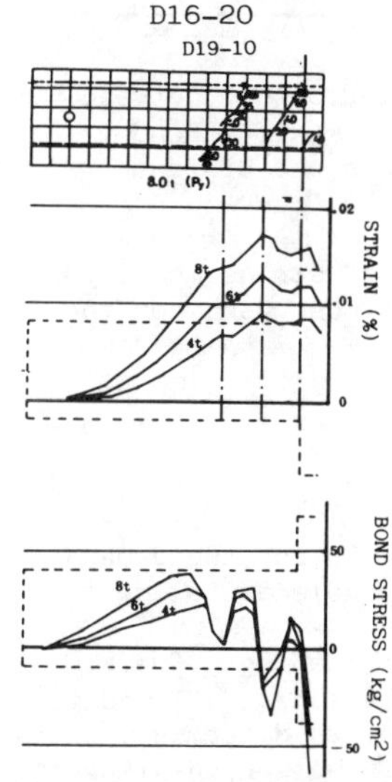

Fig. 16 Crack Pattern, Strain Distribution and Bond Stress Distribution in Specimen D19-10

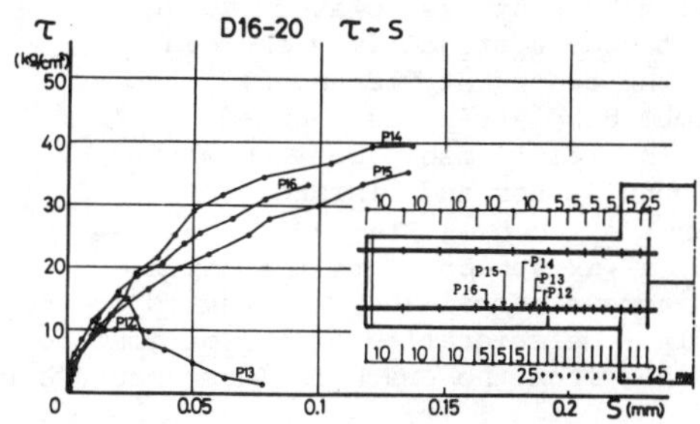

Fig. 17 Local Bond Stress-Slip Relations in Specimen D16-20

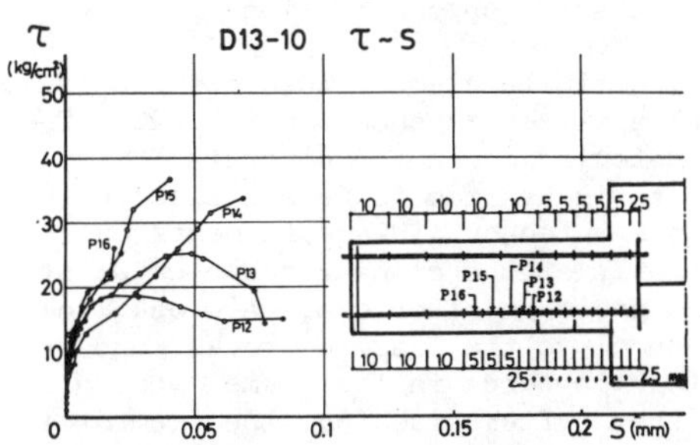

Fig. 18 Local Bond Stress-Slip Relations in Specimen D13-10

Loading and measuring arrangement

The loading arrangement is shown in Fig. 12. Figure 13 shows where strain in the reinforcing bars and concrete were measured. Local bond stress and slip were calculated from the distribution of strain.

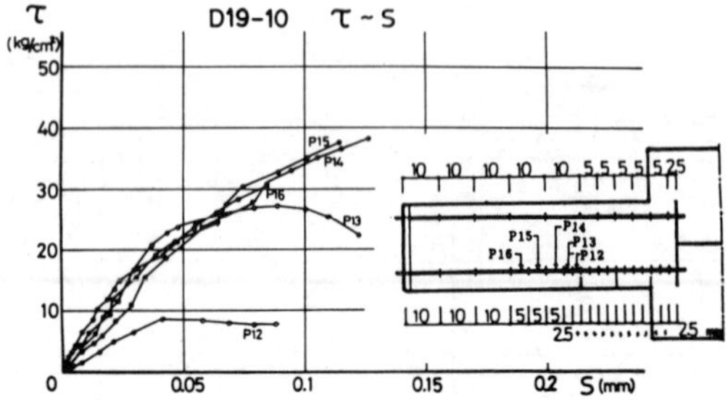

Fig. 19 Local Bond Stress-Slip Relations in Specimen D19-10

Experimental results

Figures 14, 15 and 16 show the crack pattern, strain distribution and bond stress distribution. The maximum bond stress is about 40 kg/cm² for all specimens. The bond stress near cracks does not increase when the shear force increases. Specimen D13-10 has three deformed bars as longitudinal bars. There are no differences between strain distributions center bar and the corner bars. Figures 17, 18 and 19 show the relationship between local bond stress and local slip. The maximum bond stress is about 40 kg/cm² at some distance from the crack. Near the crack, at P12 and P13, the maximum bond stress is reduced.

The relationship between maximum bond stress and distance from cracks is shown in Fig. 20. The vertical axis represents the maximum bond stress normalized with regard to the maximum bond stress measured at P14. Deterioration of the maximum bond stress is observed when the distance from the crack is less than 4 bar diameters.

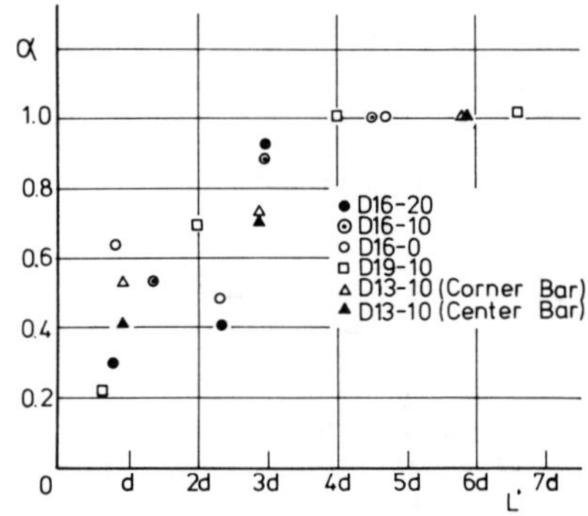

Fig. 20 Relationship between Local Maximum Bond Stress and Distance from a Crack

FINITE ELEMENT METHOD ANALYSES

Finite element apporoximation

The finite element idealization of specimen D16-20 for two-dimensional analyses is shown in Fig. 21. The material models as follows:
(1) material model for concrete
Concrete is assumed to remain elastic throughout the loading. Cracking is represented by linkage elements called crack links and is allowed to occur only at the crack links because it is necessary to determine exactly the position of cracks. A crack link consists of two springs tangent and normal to the assumed cracks, and cracking is assumed to occur when the tensile force in the normal spring reaches the tensile strength of concrete. After cracking, the element forces in two springs are redistributed to the neighbor regions.

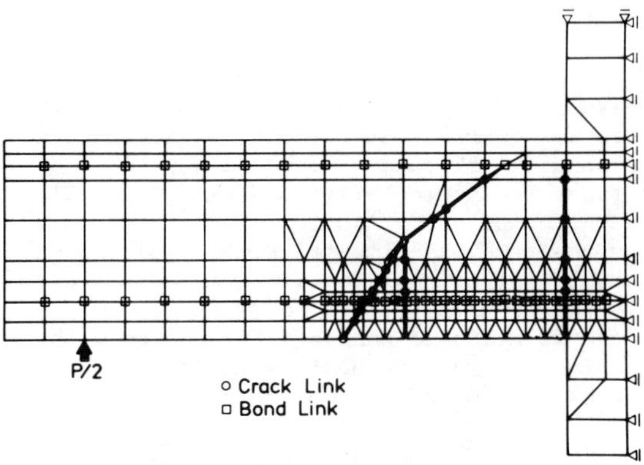

Fig. 21 Finite Element Mesh

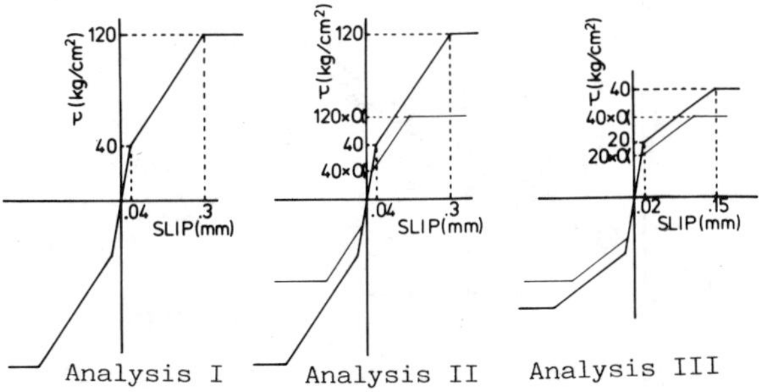

Fig. 22 Material Models for Bond

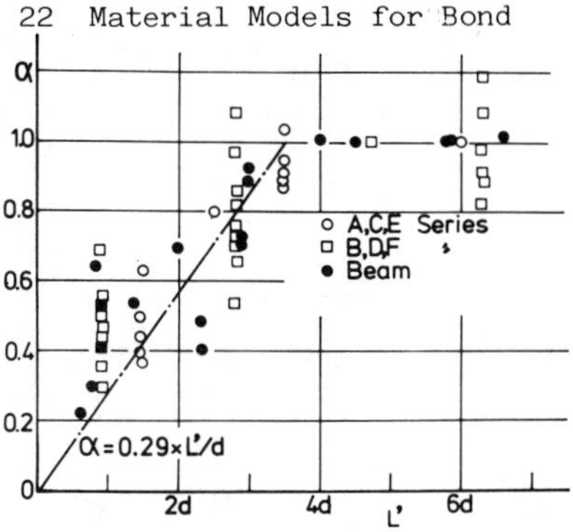

Fig. 23 Coefficient of Local Maximum Bond Stress Deterioration

(2) material model for reinforcement
Reinforcement is represented by line elements and assumed to be elasto-plastic, with the elastic limit defined by the yield strength.

(3) material model for bond
Bond links are used to represent the bond between longitudinal bars and concrete. There are three material models for bond as shown in Fig. 22. The first model, which is used in Analysis I, is derived from the relationship between bond stress and slip at some distance from a crack, as described in Section 2. Therefore, the bond deterioration near the crack is not considered in Analysis I. The second model, used in Analysis II, considers the bond deterioration near cracks. Where the bond stiffness changes, bond stiffnesses are multiplied by α, which is determined from the response at some distance from the crack to the position of the bond link. For the third model, used in Analysis III, the maximum bond stress of the basic bond stress-slip relationship is limited to 40 kg/cm² as shown in fig. 17. The bond deterioration near the crack is taken into consideration in the same manner as in the second model. Though there remain problems to be studied, for example, the effect of concrete strength, in order to determine quantitatively coefficient of bond deterioratin, α, the relationship between α and the distance from the crack proposed here is illustrated in Fig. 23. Since there is not too much difference between the data of figures 7, 9 and 20, these three figures are combined in Fig. 23. The relationship between α and L' is obtained by the least square method as long as L' is less than 4 bar diameters. When L' is larger than that, α is assumed to be equal to 1.

Analytical results

The relationships between shear force and deflection obtained by analyses and experimens are compared in Fig. 24. According to Analysis I, which was carried out without considering the bond deterioration near cracks, the stiffness reduction after cracking is small and the computed deflection at yielding is less than the experimental result. Although the solution of analysis II agree with experimental results better than that of Analysis I, the solution of Analysis III agrees with experimental results best of the three. Figures 25, 26 and 27 compare analytical

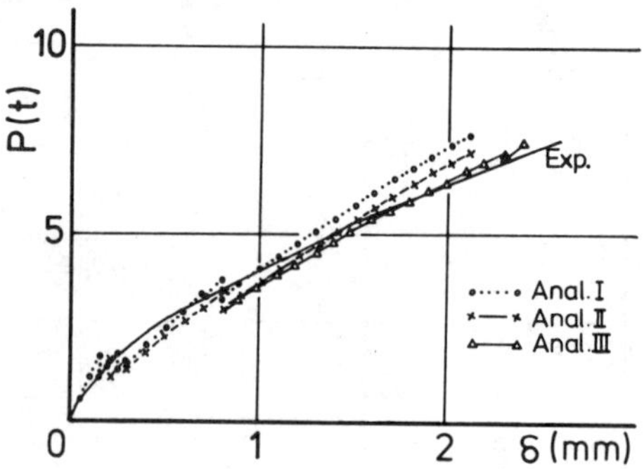

Fig. 24 Comparisons of Shear-Deflection Relationships

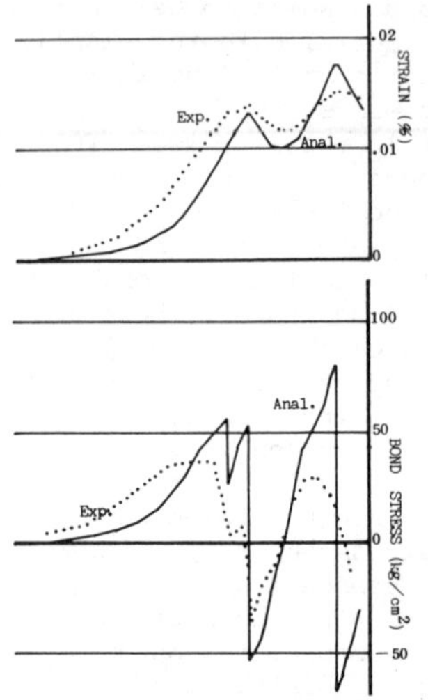

Fig. 25 Comparisons of Strain and Bond Stress Distributions (Analysis I)

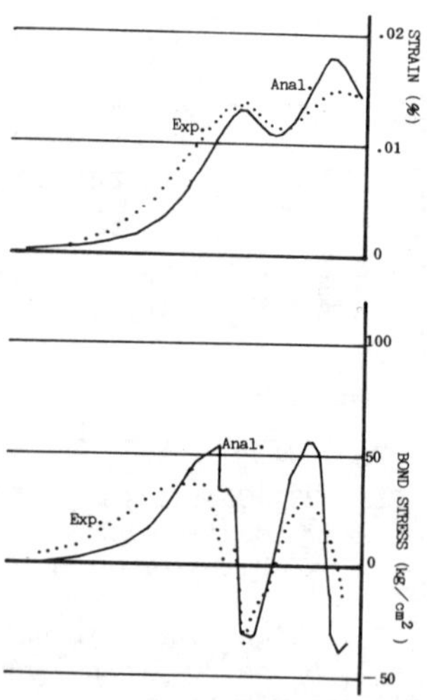

Fig. 26 Comparisons of Strain and Bond Stress Distributions (Analysis II)

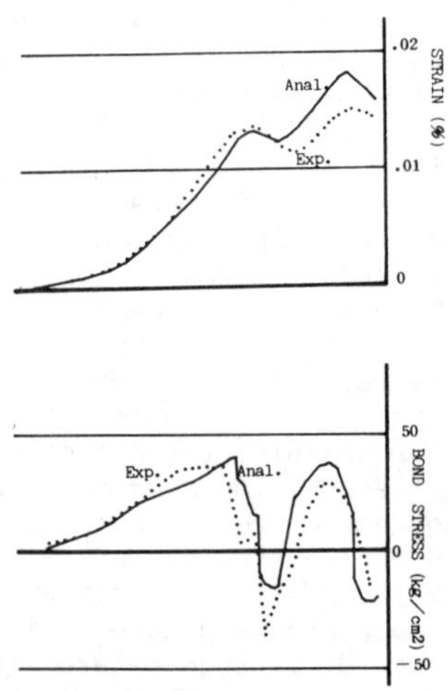

Fig. 27 Comprisons of Strain and Bond Stress Distributions (Analysis III)

and experimental strain and bond stress distributions. The bond stresses obtained by Analyses I and II are larger than the experimental results near the cracks. Since the maximum bond stresses are limited to less than 40 kg/cm^2 in Analysis III , the solutions agree with the experimental results better than the other analyses.

CONCLUSIONS

The conclusions drawn from these experiments and analyses sre as follows:

(1) Near cracks, both bond stiffness and maximum bond stress deteriorate. The local maximum bond stress deteriorates near the crack in proportion to the distance from the crack. When the distance from the crack exceeds 4 bar diameters, no bond deterioration is observed.

(2) The thickness of concrete cover does not effects the bond behavior, as long as the minimum thickness is 2.5 bar diameters.

(3) The results of the analysis which takes the bond deterioration near cracks into consideration agree with experimental results better than those of other analyses, which disregard the bond deterioration.

REFERENCES

[1] S.Hayashi, S.Kokusho and H.Yoshida, "Experiments of Bond Behavior between Deformed Bars and Concrete in the Neighborhood of the Cracks", Transactions of AIJ, No.348, Feb., 1985, pp.86-97

[2] S.Hayashi, S.Kokusho and H.Yoshida, "Fundamental Study of Bond Behavior between Deformed Bras and Concrete", Proc. of AIJ, Sept., 1981, pp.1513-1514

[3] H.Noguchi, "Inelastic Analysis of Reinforced Concrete Beam-Column Joint", Proc. of AIJ, Sept., 1981, pp.1469-1470

[4] G.Plauk and G.Hees, "Finite Element Analysis of Reinforced Concrete Beams with Special Regards to Bond Behavior", IABSE Colloquim, Delft, Netherland, June, 1981, pp.583-598

NONLINEAR SPRING ELEMENT FOR STRAIN-SLIP RELATIONSHIP OF A DEFORMED BAR

By Junichiro Niwa*, Lie-Liung Chou**, Hiroshi Shima***
and Hajime Okamura****

ABSTRACT

Pull out tests of deformed bars embedded in massive concrete were carried out. The strain-slip relationship of a deformed bar can be expressed by a unique equation if the free end slip is negligibly small. This relationship has been formulated in simple form for tension, compression and unloading-reloading cases. An equivalent nonlinear spring element utilizing this strain-slip relationship has been incorporated into a FEM program. Trial calculations to predict the total displacement of a reinforced concrete pier model have been performed to determine the applicability of the spring element.

1. INTRODUCTION

Bond characteristics of a deformed bar embedded in massive concrete are not necessarily the same as those of a deformed bar arranged in normal structural members, such as beams, columns, and walls whose section widths are relatively small and in which cracks may occur frequently. One of the main differences is that free end slip of a bar does not occur in the case of a bar embedded with enough length in massive concrete.

Considering the fact that free end slip is negligibly small, it seems that bond characteristics of deformed bars embedded in massive concrete can not be estimated by existing bond-slip formulae, which are based on experiments with relatively short embedded lengths. In these experiments, the magnitude of free end slip was the same as the slip at the loaded end. Therefore, it is not reasonable to apply present bond-slip models to bars in massive footings.

When free end slip of a bar embedded in massive concrete is zero, boundary conditions for bond-slip relationship become easy. However, it is very rare to obtain experimental data for this type of members. Additional experiments are needed for accumulating the data necessary to formulate a bond-slip model for this case.

---

* Assistant Professor, Department of Civil Engineering, University of Tokyo, Hongo 7-3-1, Bunkyo-ku, Tokyo, Japan

** Chief Engineer, Taipei Railway Underground Project, Taipei, Republic of China

*** Graduate Student, University of Tokyo

**** Professor, University of Tokyo

After investigating the experimental results which have been obtained in this research, it has been found out that slip of a bar is proportional to the strain in the bar. This strain-slip relationship can be incorporated into a finite element analysis program, if this basic relationship is implemented into an equivalent nonlinear spring element. Nonlinear spring elements are easily applied to the analysis of reinforced concrete piers in which the slip of bars from the footing surface should have a large effect on the total displacement of the pier top.

2. PULL OUT TESTS OF DEFORMED BARS EMBEDDED IN MASSIVE CONCRETE

In order to simulate a bar embedded in a footing, experiments were carried out. Experimental variables are the bar diameter, embedded length of the bar and concrete strength. Properties of each specimen are shown in Table 1. Pull out tests were carried out as shown in Fig.1. The bar was arranged vertically in the center of a concrete cylinder of 500 mm diameter. The concrete cylinder diameter was at least 15 times the diameter of the bar. Strains in the bar were measured at intervals of 5 bar diameters from the loaded cylinder surface, together with the slip at the free and loaded ends. The yield point of steel (except for specimen No.12) was intentionally selected to be higher than for normal conditions. The purpose of this was to investigate the bond-slip relationship from near the surface of loaded side deeply to near the free end.

Based on the experimental data, the relationship of slip and strain at a certain point of a bar was examined. The slip was defined as the sum of free end slip and the

Table 1 Properties of specimen

No.	D (mm)	U.L. (D)	E.L. (D)	fc' (MPa)	Remarks
1	19.5	0	50	21.2	Diameter of bars
2	25.4	0	50	21.2	
3	30.7	0	50	21.2	
4	25.4	0	40	27.2	Strength of concrete
5	25.4	0	40	50.0	
6	25.4	10	5	21.8	Embedded length of bars
7	25.4	10	10	21.8	
8	25.4	10	15	21.8	
9	25.4	10	20	21.8	
10	25.4	10	30	21.8	
11	25.4	10	40	21.8	
12	19.5	0	50	19.6	Post yield

D: Diameter of bar, U.L.: Unbonded length
E.L.: Embedded length
fc': Compressive strength of concrete
maximum size of aggregate was 25 mm

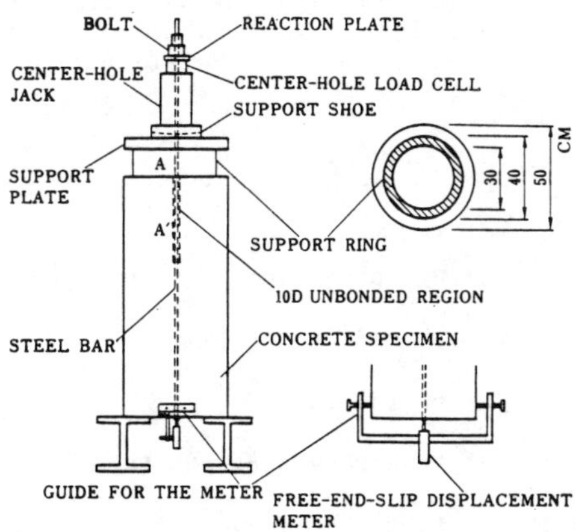

Fig.1 Schematic diagram of testing facility

area under the strain distribution diagram from the free end to a certain point of a bar. In this calculation, the deformation of concrete was neglected.

From the experimental results, it has been found that when free end slip is negligible, the relationship between slip and strain at a certain point of a bar can be said to be constant and independent of the location along the bar[1], except for the bond deterioration zone near the surface of the loaded side of concrete cylinder.

This may be explained precisely as follows. Fig.2 shows the strain distribution curves along the embedded steel at various load steps. As the load increases, the strains in the steel became larger. The shape of the strain distribution curve, however, is similar to the curve for the previous step, only shifted. This is a natural result, considering that stress conditions in the concrete around a bar should be similar from the point at which the bar strain reaches a specific value to the point at which slip and strain of a bar become zero.

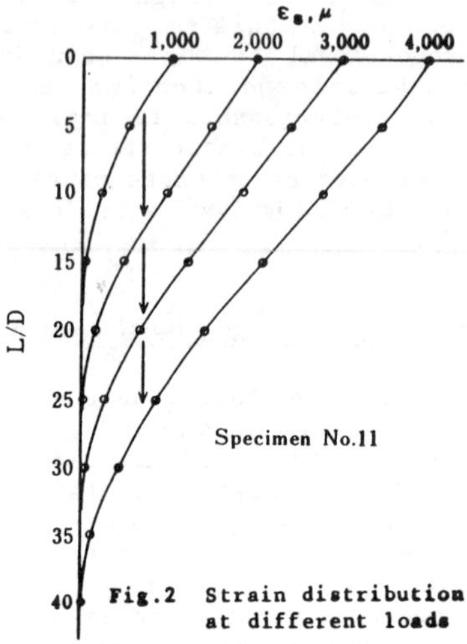

Fig.2 Strain distribution at different loads

Therefore, if a bar is embedded in massive concrete and free end slip does not occur, slip can be represented uniquely by strain of a certain point of a bar.

3. FORMULATION OF STRAIN-SLIP RELATIONSHIP

Based on the experimental data, a strain-slip relationship was formulated. In order to eliminate the effect of bar diameter and concrete strength on the relationship, a non-dimensional slip "s" was defined as follows.

$$s = \frac{1000\ SL}{D} K_{fc} \tag{1}$$

where, SL : slip obtained by the experiment [mm]

D : bar diameter [mm]

$$K_{fc} = (fc'/20)^{2/3} \tag{2}$$

fc': concrete strength (MPa)

It is recognized that the envelope curve of the strain-slip relationship in tension before yielding can be formulated as a function of strain only, independent of bar diameter and concrete strength, as follows.

$$s = 1000\ \varepsilon_s\ (\ 2 + 3500\ \varepsilon_s\ ) \quad (3)$$

where, ε_s : strain of steel, $\varepsilon_s > 0$

This relationship is shown in Fig.3 together with experimental data. It is recognized that eq.(3) agrees well with the experimental data.

If both equilibrium of forces acting on a small length of steel and compatibility of deformations are considered, this strain-slip relationship can be transformed to the bond-strain relationship as follows.

$$\tau = \frac{E_s\ \varepsilon_s\ K_{fc}}{8\ (\ 1 + 3500\ \varepsilon_s\ )} \quad (4)$$

where, τ : bond stress (MPa)

E_s : Young's modulus of steel (MPa)

Fig.4 shows the bond-slip relationship transformed from the strain-slip relationship(eq.(3)).

It was assumed that the same type of strain-slip relationship should hold not only in tension but in compression as well. Using the experimental data for compression steel of reinforced concrete pier model[3], the strain-slip relationship in compression can be formulated as follows.

$$s = 1000\ \varepsilon_s\ (\ 3 - 4500\ \varepsilon_s\ ) \quad (5)$$

where, $\varepsilon_s < 0$

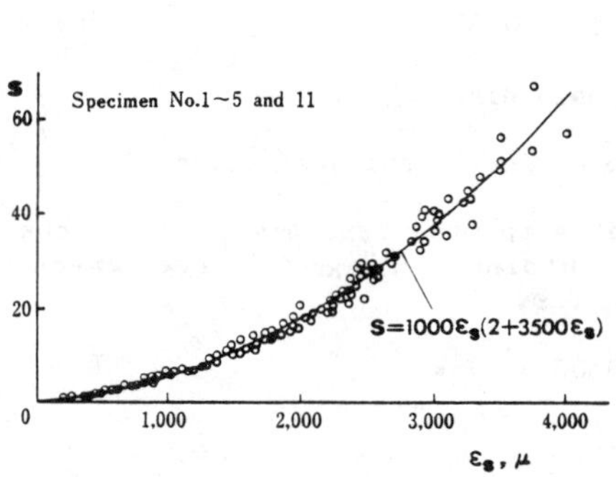

Fig.3 Relationship between steel strain and slip

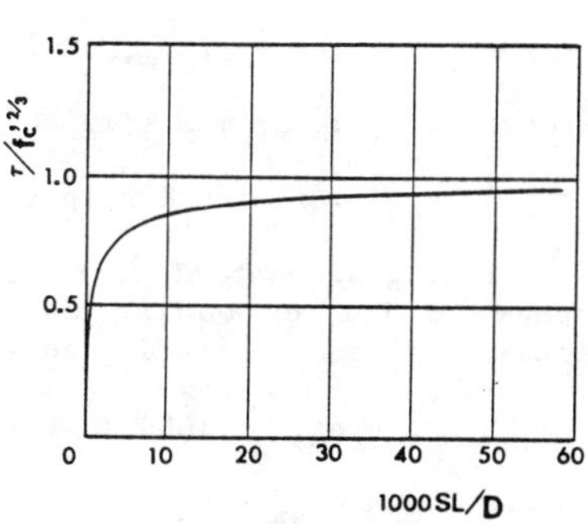

Fig.4 Bond-slip relationship transformed from strain-slip relationship

Slip in eq.(5) is larger than slip in eq.(3) when the absolute value of strain is equal in both cases. It seems that the reason for this is the direction of concrete casting. As a result of bleeding, bleeding water gathers below the ribs of a bar, and the concrete strength in this region decreases in comparison with the region above the ribs. Fig.5 compares the experimental data from the reinforced concrete pier model test[3] with eq.(5). Variations of the experimental data are larger than those for tension in Fig.3. It seems that these variations are due to the same reason why concrete characteristics under the ribs of a bar tend to change.

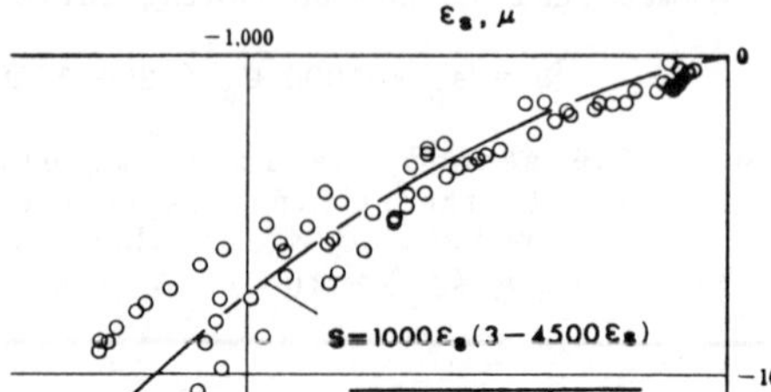

Specimen	D (mm)	fc' (MPa)
P-1	13	18.7
P-2	16	26.5

Fig.5 Strain-slip relationship in compression

Further, considering the application of the strain-slip relationship to analysis of real structural members, the relationship after yielding and for unloading and reloading should be formulated.

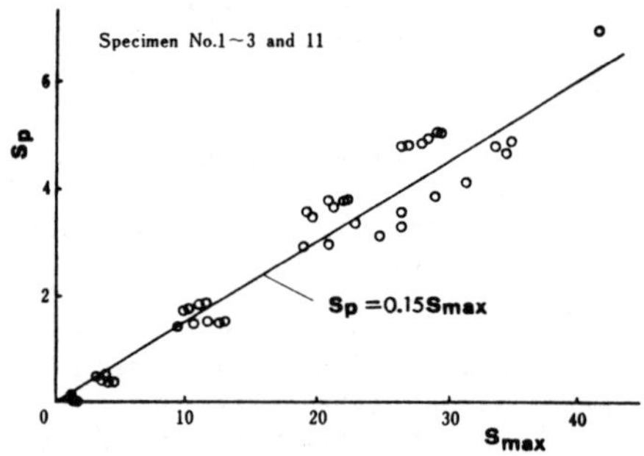

Fig.6 Relationship between maximum slip and residual slip

In order to formulate the strain-slip relationship for unloading, the relationship between the maximum slip before unloading and the residual slip in the perfectly unloaded state was investigated. This relationship as shown in Fig.6 can be expressed by a constant, R.

$$R = 1 - s_p / s_{max} = 1 - 0.15 = 0.85 \tag{6}$$

where, s_{max} : maximum slip before unloading

s_p : residual slip in the perfectly unloaded state

Taking account of this relationship, the strain-slip relationship behavior for unloading can be obtained using a curve which is symmetrical to the loading envelope curve.

$$s = s_{max} - 1000\,\varepsilon'\,(2 + 3500\,\varepsilon')\,R \tag{7}$$

where, $\varepsilon' = \varepsilon_{max} - \varepsilon_s$

ε_{max} : maximum strain before unloading

For reloading, it was recognized that the reloading curve may consist of 3 different branches. The first one is a curve which is symmetrical to the unloading curve with respect to a point (eq.(8)).

$$s = s_p + 1000\,\varepsilon_s\,(\,2 + 3500\,\varepsilon_s\,)\,R \qquad (8)$$

The second one is a curve which is shifted by some amount L parallel to the strain axis from the unloading curve (eq.(9)).

$$s = s_{max} - 1000\,(\,\varepsilon' + L\,)[\,2 + 3500\,(\,\varepsilon' + L\,)]\,R \qquad (9)$$

where, $L = k_1\,\varepsilon_{max}$

$k_1 = 0.1$

The last branch applies for $\varepsilon_s > \varepsilon_j$. It is the curve which is a part of the curve symmetrical to the unloading curve with respect to a point and connects smoothly with eq.(9) at the point $(\,\varepsilon_j\,,\,s_j\,)$ (eq.(10)).

$$s = 1000\,(\,\varepsilon_s - \varepsilon_j\,)\,[\,2 + 3500\,(\,\varepsilon_s - \varepsilon_j + 2\,\varepsilon_i\,)]\,R + s_j \qquad (10)$$

where, $\varepsilon_j = (\,1 - k_2\,)\,\varepsilon_{max}$

$\varepsilon_i = (\,k_1 + k_2\,)\,\varepsilon_{max}$

$s_j = s_{max} - 1000\,\varepsilon_i\,(\,2 + 3500\,\varepsilon_i\,)\,R$

$k_2 = 0.25$

Fig.7 shows the strain-slip relationship of the experimental data and calculated values for both unloading and reloading. It is observed that these equations can predict the experimental data fairly well.

Fig.8 shows the strain-slip relationship after yielding at the loaded end of a bar of specimen No.12. In this specimen, slip increases in proportion to the increase of strain after the point of strain hardening. Mechanical behavior after yielding, however, is influenced significantly by yield point, strain hardening point, stiffness in the

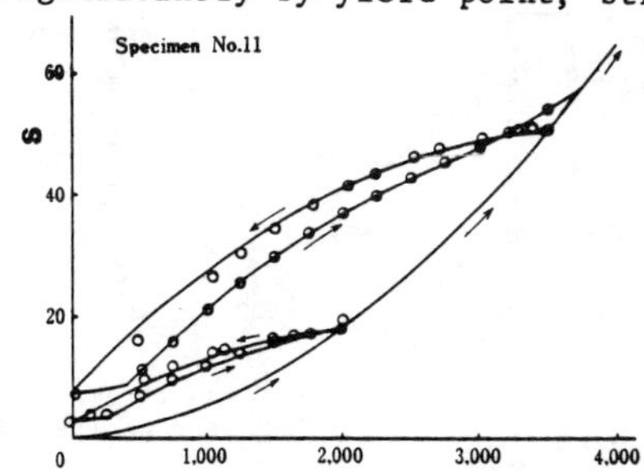

Fig.7 Strain-slip relationship under unloading and reloading

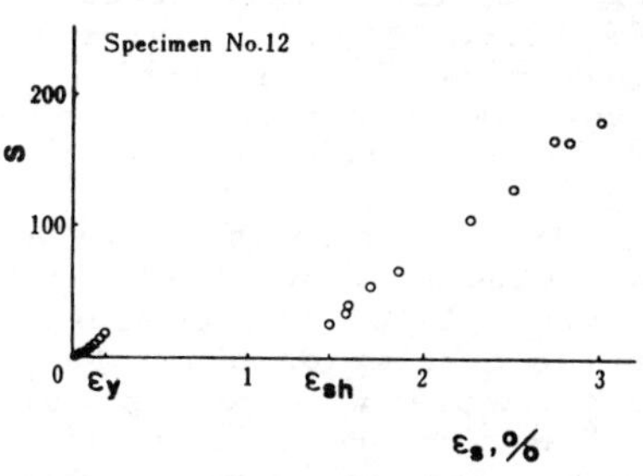

Fig.8 Strain-slip relationship after yielding

strain hardening region and so on. In order to formulate the strain-slip relationship after yielding, it is necessary to accumulate a large number of experimental data. In this report, the strain-slip relationship after yielding was not formulated yet. For the elastic region, however, formulae of the strain-slip relationship can be proposed in simple form and with reasonable accuracy.

4. INCORPORATION OF NONLINEAR SPRING ELEMENT INTO COMPUTER PROGRAM FOR FINITE ELEMENT ANALYSIS

If the relationships of eq.(3) to eq.(10) are used, the strain in a deformed bar can be calculated for a given slip. From this, it follows that the stress can be calculated and finally the tensile force can be obtained by multiplying stresses with the bar area.

Therefore, this sequence of relationships can be converted into a relationship between slip and tensile force of a deformed bar. A nonlinear spring element whose stiffness should be changed according to slip has been incorporated into a computer program for finite element analysis. The flow chart for execution of the calculations with this nonlinear spring element is shown in Fig.9.

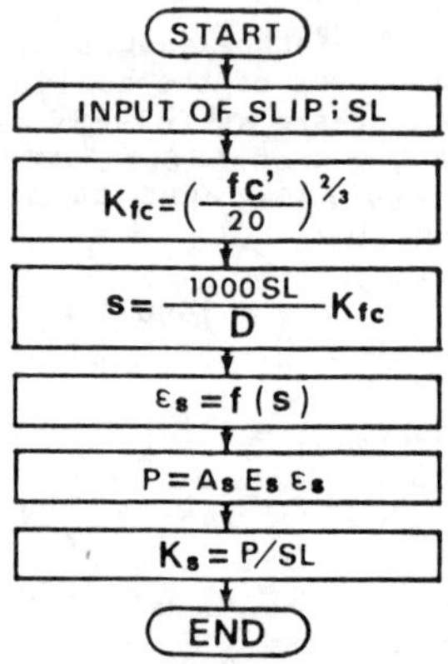

Fig.9 Flow of calculations using the nonlinear spring element

5. FINITE ELEMENT ANALYSIS MAKING USE OF NONLINEAR SPRING ELEMENT

5.1 Benefit of Nonlinear Spring Element

When the load-displacement relationship for a reinforced concrete pier is calculated, the displacement of the pier top usually can vary significantly according to the estimate of bar slip in the footing.

Trial calculations have been made with the nonlinear spring element. Deformed bars embedded in the footing of a reinforced concrete pier were modeled using nonlinear spring elements of zero length. The springs were attached to the boundary between footing and pier. The pier was modeled as a cantilever as shown in Fig.10.

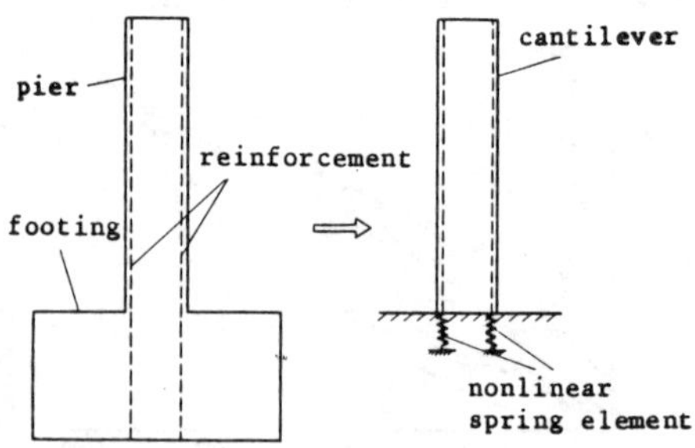

Fig.10 Reduction of RC pier into cantilever and nonlinear spring element

When information about the displacement of a reinforced concrete pier is needed, the massive footing usually has to be analyzed at same time. This new method, however, can omit the analysis of the footing portion and is computationally very effective. If reinforced concrete finite elements are employed, this type of structure can be analyzed easily by only the elements which represent the pier portion and the nonlinear spring elements which represent the footing portion. It may be possible to be analyzed by even a personal computer.

5.2 The Other Assumptions for Finite Element Analysis

The other assumptions beside employment of the nonlinear spring element are as follows.

(1) The constitutive model for concrete is the elasto-plastic fracture model proposed by Maekawa[2].

(2) Tension stiffening of concrete is not considered.

(3) The constitutive model for steel is a perfect bi-linear model. Steel can resist force in the axial direction only.

(4) Steel reinforcement is smeared over an element and is overlapped with concrete.

(5) The element configuration used for analysis is shown in Fig.11.

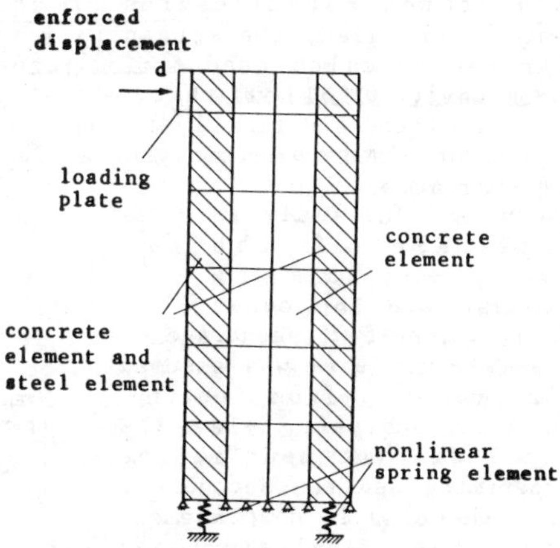

Fig.11 Finite elements configuration

5.3 Specimen

An analysis of Hashimoto's reinforced concrete pier model subjected to cyclic loading was executed. Characteristic of this experiment is the scale of the specimen. The pier was relatively large and the footing was massive. The casting direction of concrete was vertical, the same as for the pull out tests from which the strain-slip relationship was derived.

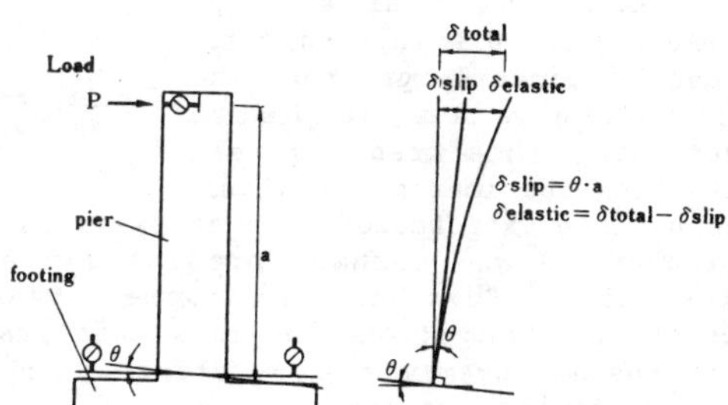

Fig.12 Measurement of displacement and rotation and separation of total displacement

Only the envelope of the cyclic hysteresis curves was extracted in order to compare with the results of analysis. Hashimoto measured the displacement and rotation as shown in Fig.12 and divided the total displacement of the pier top into the elastic displacement in consideration of the occurrence of flexural cracks and the displacement due to the slip of bars from the surface of the footing. The properties of the reinforced concrete pier model are shown in Table 2.

Table 2 Properties of RC pier specimen

ID	a (mm)	h (mm)	b (mm)	fc' (MPa)	reinforcement
P-2	1600	300	250	26.5	D16

Properties of reinforcement

No. and size	Nominal area(mm^2)	Yield point (MPa)	Young's Modulus (MPa)
3 - D16	198.6	395	2.03×10^5

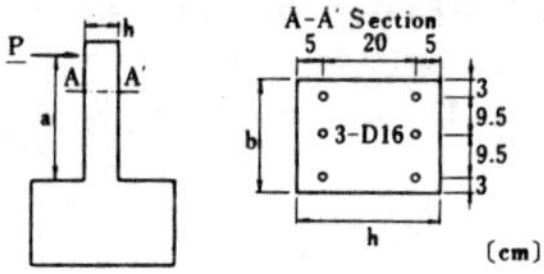

5.4 Results of Analysis

Two types of analysis were conducted. One analysis made use of the nonlinear spring element. In the other analysis, perfect anchorage of the deformed bars was assumed with no slip from footing. The first analysis yielded the total displacement in the experiment and the second one only the elastic displacement component. Analytical and experimental results are shown in Fig.13. It is recognized that the analysis in which perfect anchorage was assumed can not estimate the total displacement. In the analysis with the nonlinear spring elements, the displacement was about 1.5 times larger than in the case with perfect anchorage. This trend agrees with the experimental results.

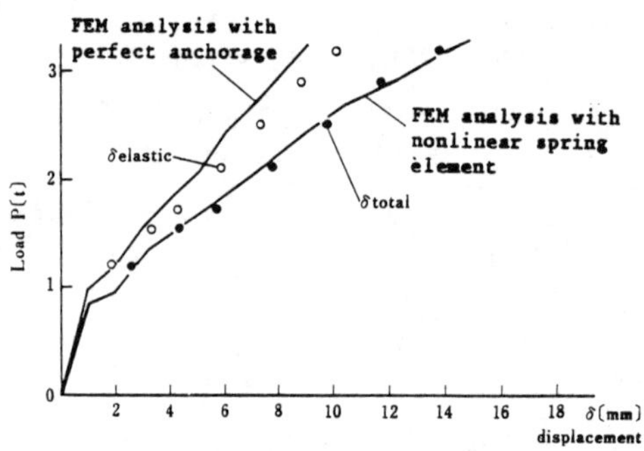

Fig.13 Analytical and experimental load-displacement relationship

Against the expectation, the analysis in which no slip at the surface of the footing was assumed predicted smaller displacements than the experiment. This fact is to some extent due to the inaccuracy of the analytical assumptions. Besides this reason, however, it is suspected that the measurements of rotation of the footing surface might contain some mechanical errors.

If the slip from the footing surface is taken into account and nonlinear spring elements are used, it may be possible to estimate the total displacement of the reinforced concrete pier by analyzing the pier portion only.

6. CONCLUSION

The contents of this report are summarized as follows.
(1) From the results of pull out tests with a deformed bar embedded in massive concrete, it is recognized that the relationship between strain and slip of a bar can be expressed by a unique equation, if the free end slip of the bar is negligibly small.
(2) The strain-slip relationship has been formulated in simple form for tension, compression, unloading and reloading. After yielding, a relationship has not been formulated yet. However, it may be possible to apply a similar method.
(3) This strain-slip relationship can be incorporated into a computer program for finite element analysis very easily, after the relationship has been converted into an equivalent nonlinear spring element.
(4) Trial calculations using the nonlinear spring element have been executed in order to examine its applicability.

7. REFERENCES

[1] L.L.Chou: Analytical Model for Bond of Deformed Bar Embedded in Massive Concrete, Doctor's Thesis of University of Tokyo, March 1984
[2] K.Maekawa, H.Okamura: The Deformational Behavior and Constitutive Equations of Concrete Using Elasto-Plastic and Fracture Model, Journal of the Faculty of Engineering, University of Tokyo,(B),Vol.37,No.2,1983
[3] C.Hashimoto: Division of Displacement of Reinforced Concrete Pier Subjected to Static and Cyclic Horizontal Force, Proc. of Annual Conference of the JSCE, Oct. 1983(in Japanese)

APPLICABILITY OF SMEARED TYPE REINFORCED CONCRETE MODEL BASED UPON ANISOTROPIC CONSTITUTIVE LAW

Nobuaki Shirai*, M.AIJ

ABSTRACT

Elasto-plastic behavior of cracked reinforced concrete is mainly governed by factors such as 1)compressive response of concrete between cracks, 2)bond between concrete and reinforcing bars(tension stiffening) and 3)shear transfer effects due to aggregate interlock between cracked concrete surfaces and dowel action of reinforcing bars crossing cracks. A constitutive law for smeared type reinforced concrete elements may be formulated on the basis of the theory of anisotropy.

However, the applicability of this type of reinforced concrete model to various reinforced concrete members has not been sufficiently studied yet. The purpose of this study is first to clarify the nonlinear behavior of cracked reinforced concrete(the items 1) and 2) in particular), then to present a suitable analytical model, which can be easily incorporated into the anisotropic constitutive law for smeared type reinforced concrete elements for finite element applications, and finally to examine the applicability of the proposed analytical model.

INTRODUCTION

The types of idealization for reinforced concrete elements can be roughly classified into two categories; the discrete idealization, in which concrete and reinforcing bars are seperately discretized into different elements, and the smeared idealization, in which concrete and reinforcing bars are discretized into elements with the same geometrical boundaries and the effects of reinforcing bars are averaged within the pertaining element. Cracks can also be idealized into either the discrete type or the smeared type. This paper shall deal with the reinforced concrete model based upon the smeared idealization.

Constitutive laws for cracked reinforced concrete have been often derived by incorporating the stress-strain relationship of a concrete cylinder, the tension stiffening of concrete and shear transfer effects into the theory of anisotropy. However, it is questionable whether the anisotropic constitutive models proposed in the past work satisfactorily in all cases, because 1)principal stress-strain responses induced in cracked concrete are far different from those in a concrete cylinder according to the previous test results, 2)in spite of the fact that

*Research Associate, Department of Architecture, College of Science & Technology, Nihon University, 1-8, Surugadai, Kanda, Chiyoda-Ku, Tokyo 101.

wall type structures are generally in the combined tensile and compressive stress states, bond and cracking properties of reinforced concrete members under such stress states are not fully understood, and 3)it has been experimentally verified that the aggregate interlock and dowel action are very sensitive to crack widths and spacings, but no reasonable analytical model, which can predict crack widths and spacings correctly, is available.

To cope with the first problem, test results on reinforced concrete panels obtained by Vecchio et al.(1982) are rigorously studied, and consequently the principal stress-strain relationship for cracked concrete is expressed in terms of the principal strain ratio.

To account for the second and third problems, a bond-cracking model(Shirai et al. 1985), which can predict not only the tension stiffening of concrete but also crack widths and spacings, is derived on the basis of the bond theory and the test results on reinforced concrete prisms subjected to transverse pressures and tensile loads conducted by the author et al.(1985). Finally, the applicability of the proposed model is studied through the comparison between test and predicted results of reinforced concrete panels and shear walls.

ANISOTROPIC CONSTITUTIVE LAW FOR CRACKED REINFORCED CONCRETE

It is assumed that the constitutive law for cracked reinforced concrete is expressed in X, Y coordinate system as follows(Shirai et al. 1981).

$$d\{\sigma^{sc}\}_X = \rho^c d\{\sigma^c\}_X + \sum_{i=1}^{N} p_i d\{\sigma^{si}\}_X + \rho^c \sum_{i=1}^{N} d\{\sigma^{bi}\}_X + \rho^c d\{\sigma^a\}_X + \rho^c \sum_{i=1}^{N} d\{\sigma^{di}\}_X$$

$$= [\rho^c [D^c]_X + \sum_{i=1}^{N} p_i [D^{si}]_X + \rho^c \sum_{i=1}^{N} [D^{bi}]_X + \rho^c [D^a]_X + \rho^c \sum_{i=1}^{N} [D^{di}]_X]\ d\{\varepsilon^{sc}\}_X \qquad (1.a)$$

in which,

$$[D^c]_X = E^c_t \begin{pmatrix} C_{cr}^4 & C_{cr}^2 C_{cr}^2 & C_{cr}^3 S_{cr} \\ & S_{cr}^4 & C_{cr} S_{cr}^3 \\ SYM. & & C_{cr}^2 S_{cr}^2 \end{pmatrix} \qquad [D^{bi}]_X = E^{bi}_t \begin{pmatrix} C_i^4 & C_i^2 S_i^2 & C^3 S_i \\ & S_i^4 & C_i S_i^3 \\ SYM. & & C_i^2 S_i^2 \end{pmatrix} \qquad (1.b,c)$$

$$C_{cr} = \cos\theta_{cr}, \quad S_{cr} = \sin\theta_{cr}, \quad C_i = \cos\theta_i, \quad S_i = \sin\theta_i$$

where the subscript i(=1,2,...N) indicates the bar direction; $d\{\sigma^{sc}\}_X$, $d\{\sigma^c\}_X$ and $d\{\sigma^{si}\}_X$ denote, respectively, the stress increments of reinforced concrete, concrete and steel elements; $d\{\sigma^{bi}\}_X$, $d\{\sigma^a\}_X$ and $d\{\sigma^{di}\}_X$ the equivalent concrete stress increments due to tension stiffening, aggreagte interlock and dowel action; $[D^c]_X$ and $[D^{si}]_X$ the material stiffnesses of concrete and reinforcement; $[D^{bi}]_X$, $[D^a]_X$ and $[D^{di}]_X$ the equivalent material stiffnesses due to tension stiffening, aggregate interlock and dowel action; ρ^c the concrete density; p_i the steel ratio; E^c_t the tangent modulus of the uniaxial stress-strain curve of concrete; E^{bi}_t the tension stiffness of concrete due to bond; θ_{cr} the angle between X-axis and the bar axis.

In this study, an analytical model for estimating compressive stress-strain responses of cracked concrete and tension stiffening due to bond between concrete and reinforcement in an arbitrary direction and an approximate formula for predicting crack widths and spacings are developed, and the validity of the proposed model is verified.

STRESS-STRAIN RELATIONSHIP FOR CRACKED CONCRETE ELEMENT

Vecchio et al.(1982) tested orthogonally reinforced concrete panels, which have the size of 89cm x 89cm(35.04in.x35.04in.) and the thickness of 7cm(2.76in.), under shear and normal stresses. Figure 1 shows the normalized compressive principal stress, $|\sigma_2^c/f_c'|$ (positive in tension) -strain, $|\varepsilon_2^c/\varepsilon_c'|$ (positive in tension) relationship, where σ_2^c and ε_2^c are the compressive principal stress and strain of cracked concrete, and f_c' and ε_c' are the compressive strength and corresponding strain of concrete cylinder(positive constants). It is seen that the test values show a wide range of scatters and $\sigma_2^c-\varepsilon_2^c$ responses for cracked concrete are far different from those of concrete cylinder(parabolic equation).

To express these complex behaviors in a unified manner, the $\sigma_2^c-\varepsilon_2^c$ relationship for cracked concrete is idealized into a tri-linear curve

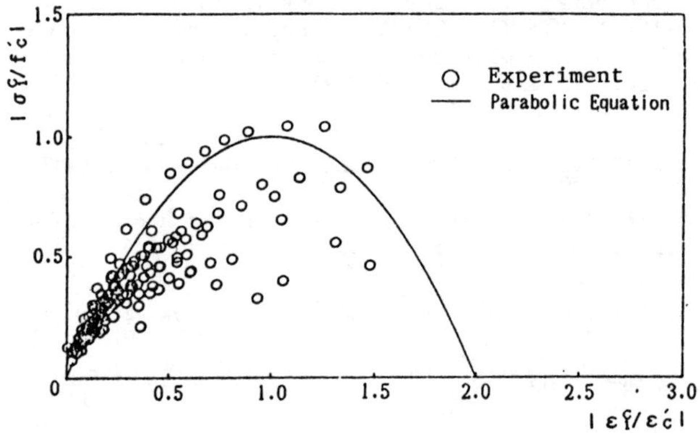

Fig.1 Compressive Principal Stress-Strain Relationship of Cracked Concrete

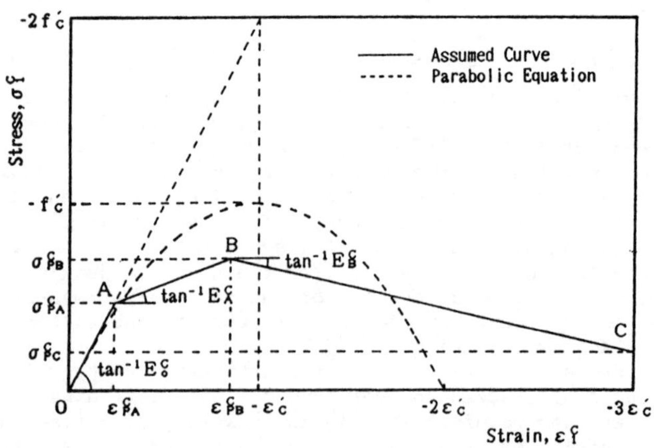

Fig.2 Assumed Principal Stress-Strain Curve of Cracked Concrete

as shown in Fig.2, and the stress-strain relationship of each branch line is expressed in terms of the principal strain ratio, $|\varepsilon_2^c/\varepsilon_1^c|$, where ε_2^c is the principal tensile strain.

(1) Elastic Range; $\overrightarrow{OA}(\varepsilon_{1A}^c, \sigma_{1A}^c), \varepsilon_{1A}^c \leqq \varepsilon_1^c \leqq 0, \sigma_{1A}^c \leqq \sigma_1^c \leqq 0$

$$\sigma_1^c = E_0^c \varepsilon_1^c \tag{2.a}$$

in which, $E_0^c = 2f_c'/\varepsilon_0'$, $\sigma_{1A}^c = -2f_c'/3\beta_m$, $\varepsilon_{1A}^c = -\varepsilon_0'/3\beta_m$

(2) Strain-hardening Range; $\overrightarrow{AB}(\varepsilon_{1B}^c - \varepsilon_{1A}^c, \sigma_{1B}^c - \sigma_{1A}^c), \varepsilon_{1B}^c \leqq \varepsilon_1^c \leqq \varepsilon_{1A}^c, \sigma_{1B}^c \leqq \sigma_1^c \leqq \sigma_{1A}^c$

$$(\sigma_1^c - \sigma_{1A}^c)/(\sigma_{1B}^c - \sigma_{1A}^c) = (\varepsilon_1^c - \varepsilon_{1A}^c)/(\varepsilon_{1B}^c - \varepsilon_{1A}^c) \quad \therefore \sigma_1^c = E_A^c(\varepsilon_1^c - \varepsilon_0') \tag{2.b}$$

in which, $E_A^c = E_0^c/(3\beta_m + 1)$, $\sigma_{1B}^c = -f_c'/\beta_m = \sigma_m^c$, $\varepsilon_{1B}^c = -(\beta_m + 1)\varepsilon_0'/2\beta_m$

(3) Strain-softening Range; $\overrightarrow{BC}(3\varepsilon_0' - \varepsilon_{1B}^c, 0.2f_c' - \sigma_{1B}^c), \varepsilon_{1C}^c \leqq \varepsilon_1^c \leqq \varepsilon_{1B}^c, \sigma_{1C}^c \leqq \sigma_1^c \leqq \sigma_{1B}^c$

$$\sigma_1^c = E_C^c(\varepsilon_1^c + 3\varepsilon_0') - 0.2f_c' \tag{2.c}$$

in which, $E_C^c = -(1 - 0.2\beta_m)E_0^c/(5\beta_m - 1)$, $\sigma_{1C}^c = 0.2f_c'$, $\varepsilon_{1C}^c = 3\varepsilon_0'$

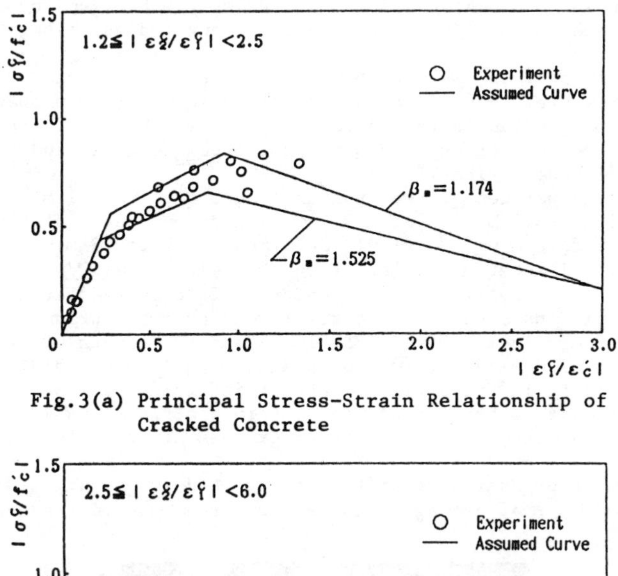

Fig.3(a) Principal Stress-Strain Relationship of Cracked Concrete

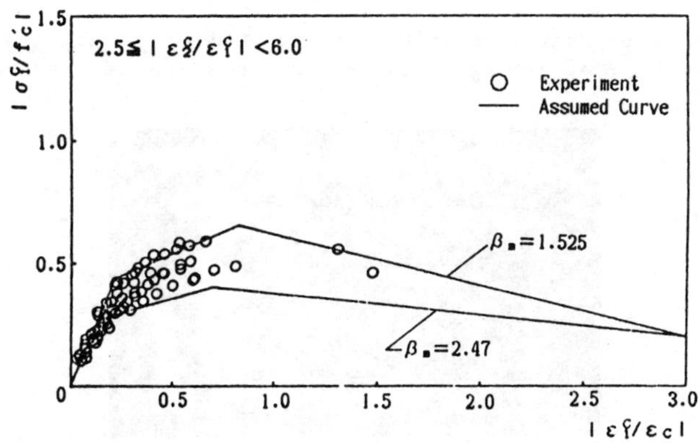

Fig.3(b) Principal Stress-Strain Relationship of Cracked Concrete

where the parameter β_m means the ratio of f_c' to the maximum compressive principal stress of cracked concrete, σ_m^c , and this is approximately written as follows.

$$|\sigma_m^c/f_c'| = 1/\beta_m, \qquad \beta_m = 0.85 + 0.27|\varepsilon_2^c/\varepsilon_1^c| \tag{3}$$

Figures 3(a) and (b) show the comparison between the test values and the values calculated by Eqs.(2.a, b and c).
It is seen that there is a clear correlation between the observed and assumed stress-strain curves.

MODELING OF BOND ACTION BETWEEN CONCRETE AND REINFORCING BARS

Bond Test on Reinforced Concrete Prism under Transverse Pressure

Reinforced concrete prisms were tested under transverse pressures and tensile loads. Figure 4 shows the test setup, and brush type bearing plates were used to minimize frictional resistance between the surfaces of specimens and the bearing plates against longitudinal elongations of specimens.

Two kinds of specimens were prepared, that is, the effects of transverse pressure, type of bar and steel ratio on crack propagation properties were studied in the series I, and the effects of transverse pressure and type of bar on bond stresses, steel stresses and slips were investigated in the series II.

The series I consisted of twenty four concrete prisms with cross sections of 10cm x 10cm(3.94in. x 3.94in.) and the length of 114cm(44.9 in.), each reinforced with a single deformed bar with 13mm(0.512in.) or 19mm(0.748in.) diameter or a plain round bar with 19mm(0.748in.) diameter. Two specimens were tested under different transverse pressures.

The series II consisted of eight reinforced concrete prisms with cross sections of 15cm x 15cm(5.91in. x 5.91in.) and the length of 64cm(25.2in.) , and the bars used were deformed or plain, with 19mm(0.748in.) diameter, as shown in Fig.5. In the series II, the reinforcing bars were grooved along the bar axis on both sides, and strain distributions were measured by strain gauges pasted on the bases of grooves at the interval of 6.25 cm(2.46in.).

The transverse pressures of different magnitudes were applied such that the ratio, σ_N/f_c' , of average compressive stress due to transverse

Fig.4 Loading Setup

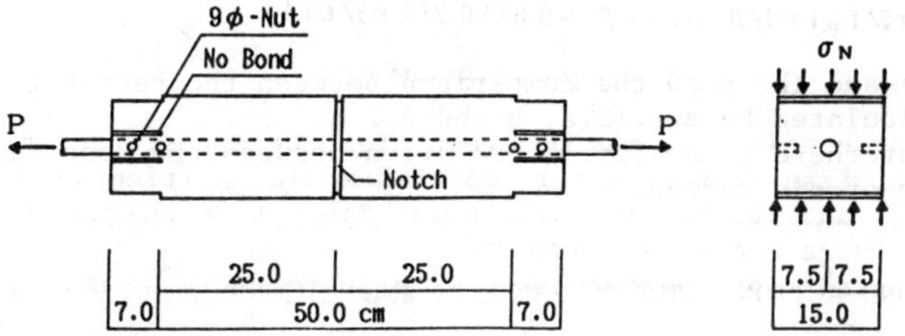

Fig.5 Tensile Bond Specimen(1cm = 0.3937in.)

pressures, σ_N , to compressive strength of concrete, f'_c , was 0, 25, 50 and 75 percent. The specimen identification is, for instance, referred to as D19-25, where D(P) indicates deformed(plain) bars, the next number the bar diameter and the final number indicates the σ_N/f'_c ratio(%).

Figure 6 shows the tensile load P versus average strain ε_{av} curves for specimen D13-25. Figures 7 and 8 show the average crack spacing versus average strain ε_{av} curves for D19 specimens, and the final average crack spacing e_{av} versus transverse pressure σ_N/f'_c curves for D13 specimens, respectively. The analytical values obtained with the bond-cracking model described in the next section are also plotted in Figs.6 and 8. Figure 9 compares the analytical average concrete stress σ^{c}_{av} versus average strain ε_{av} curves of D13 specimens for different transverse pressures. Figure 10 compares the analytical average concrete stress σ^{c}_{av} versus average strain ε_{av} curves for specimens with different steel ratio, p(%). Figure 11 shows the maximum bond stress τ_{max} versus transverse pressure ratio σ_N/f'_c , and the predicted values were calculated by Eq.(8). Figure 12 shows the normalized bond stress $\overline{\tau}(=\tau/\tau_{max})$

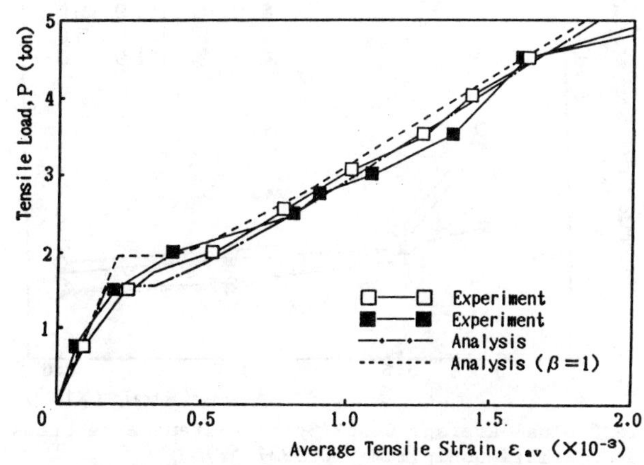

Fig.6 Tensile Load-Average Strain Relationship (1ton = 9.81KN = 2.2Klb)

versus slip $\overline{S}(=S/S_{max})$ relationship for specimen D19-25, where τ and S indicate the bond stress and slip, and τ_{max} and S_{max} mean the maximum values of τ and corresponding values of S. The predicted values shown in Fig.12 were calculated by the following exponential function(Shirai et al.,1985).

$$\overline{\tau} = e\,\ell n\{(e-1)\overline{S}^{\alpha}+1\}/\{(e-1)S^{\alpha}+1\} \tag{4}$$

The following conclusions may be drawn from these test results.
1)The effect of transverse pressure on the tension stiffening of concrete is significant, and the tension stiffening depends on the magnitude of transverse pressure and steel ratio.
2)The average crack spacing decreases with an increase of average strain,

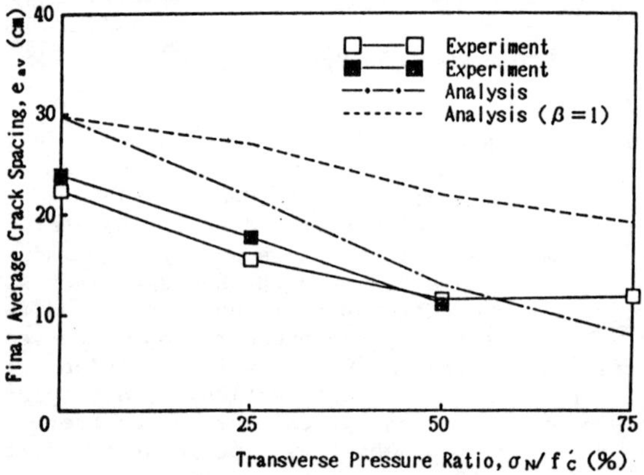

Fig.7 Average Crack Spacing-Average Strain Relationship (1cm = 0.3937in.)

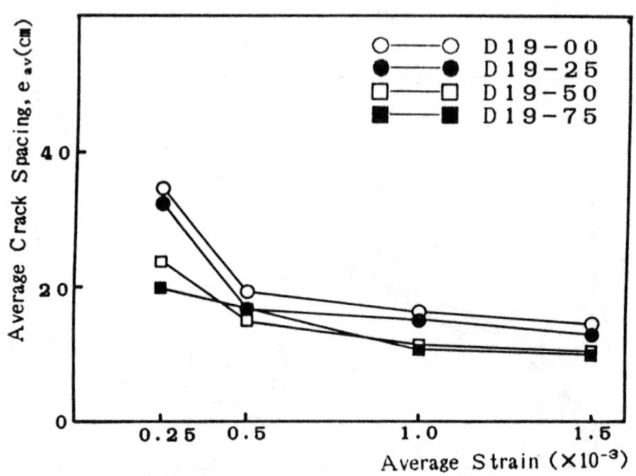

Fig.8 Final Average Crack Spacing-Transverse Pressure Relationship(1cm = 0.3937in.)

and higher the transverse pressure, the smaller the average crack spacing in all cases.
3)The bond stresses increase up to a certain limit with an increase of transverse pressure.
4)The correlation between observed bond stress-slip relations and those predicted by the exponential function of Eq.(4) is relatively good.

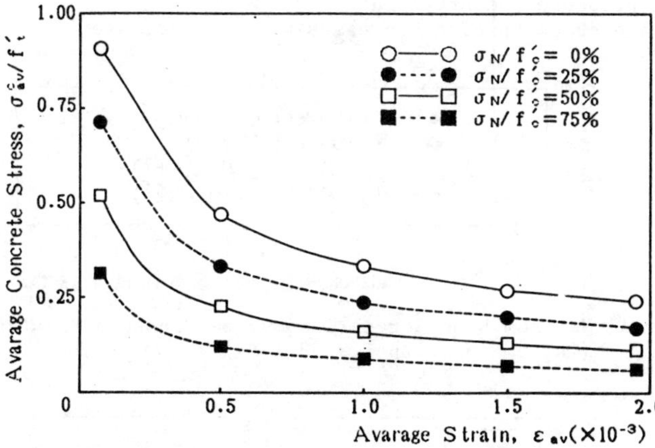

Fig.9 Average Concrete Stress-Average Strain Relationship

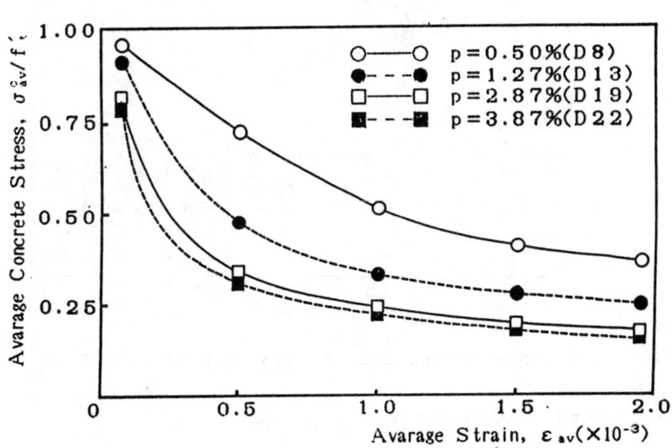

Fig.10 Average Concrete Stress-Average Strain Relationship

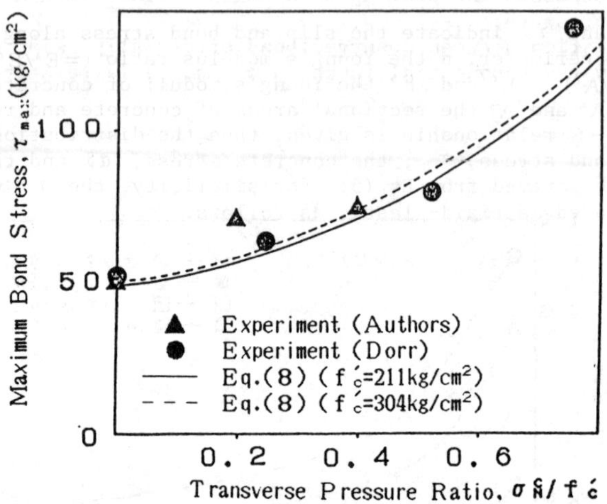

Fig.11 Maximum Bond Stress-Transverse Pressure Relationship ($1kg/cm^2$ = 14.2psi = 98.15Pa)

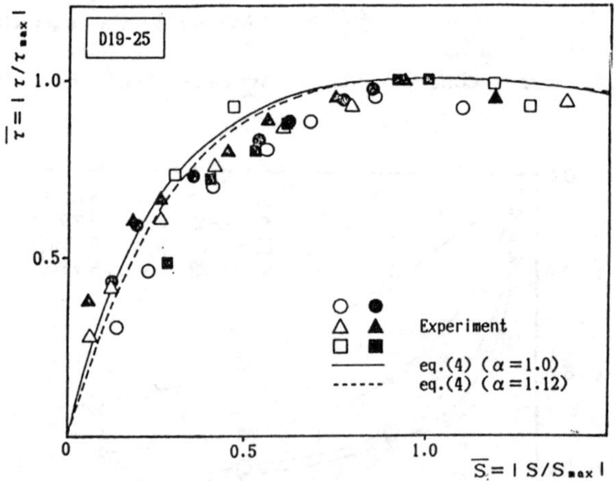

Fig.12 Normalized Bond Stress-Slip Relationship

Derivation of Bond-Cracking Model

Consider cracked concrete segments strengthened by reinforcing bars in an arbitrary direction as shown in Fig.13(a). The reinforced concrete portion between cracks is modeled as an assemblage of tensile reinforced concrete prisms in the direction of the bar axes as shown in Fig.13(b). First, assume that stress distributions of reinforced concrete prisms subjected to tensile forces and transverse pressure may be derived from the following differential equation for bond(see Fig.14).

$$d^2S_x/dX^2=(1+np)\psi\tau_x/E^sA^s \tag{5}$$

where S_x and τ_x indicate the slip and bond stress along the bar axis, ψ the bar perimeter, n the Young's modulus ratio ($=E^s/E^c$), p the steel ratio ($=A^s/A^c$), E^c and E^s the Young's moduli of concrete and reinforcement, A^c and A^s the sectional areas of concrete and reinforcing bar.

If the $\tau-S$ relationship is given, then the distributions of the slip, S_x, the bond stress, τ_x, the concrete stress, σ^c_x and the steel stress, σ^s_x can be derived from Eq.(5). For simplicity, the $\tau-S$ relationship is approximated as rigid-plastic as follows.

$$\tau_x=\tau_y \quad (S_x>0) \tag{6}$$

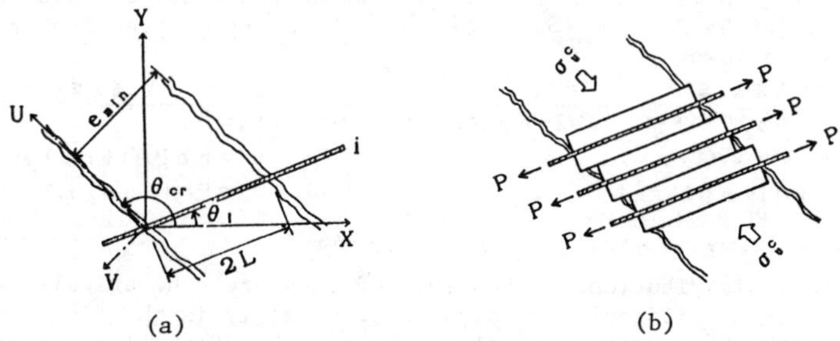

Fig.13 Modeling of Cracked Reinforced Concrete Segment

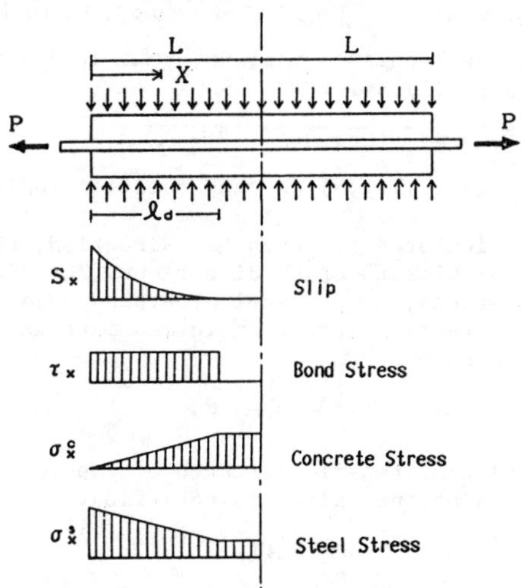

Fig.14 Stress Distributions of Tensile Member

where τ_y is the yield bond stress and may be determined by equating the area under the line expressed by Eq.(4) with α=1 to that by Eq.(6) as follows(Shirai et al.,1985).

$$\tau_y = e\,\tau_{max}/2(e-1) \tag{7}$$

where τ_{max} is the maximum bond stress in which the effect of transverse pressure is considered and may be given for deformed bars as follows (Sato et al.,1982).

$$\tau_{max} = \overline{\tau}_{max}\{1+2(\overline{\sigma}^c_{ra}/f'_c)^2\} - 0.05\overline{\sigma}^c_{ra} \tag{8}$$

in which, $\overline{\sigma}^c_{ra} = (A-1)\sigma_N/2$, $A = \{(1-2\nu^s)G^c + (1-\nu^c)G^s\}/\{(1-2\nu^s)G^c + G^s\}$

where $\overline{\tau}_{max}$ is the maximum bond stress at which σ_N equals to zero, G^c and G^s the elastic shear moduli of concrete and steel, ν^c and ν^s the Poisson's ratios of concrete and steel, and σ_N the compressive concrete stress acting normal to the bar axis. Letting the tensile force at the cracking point be P, then S_x, σ^s_x and σ^c_x can be derived from Eqs.(5) and (6) as follows.

for $0 \leq X \leq \ell_d \leq L$

$$S_x = \gamma x^2/2 - Px/E^sA^s - (\gamma \ell_d^2/2 - P\ell_d/E^sA^s) \tag{9.a}$$

$$\sigma^s_x = P/A^s - E^s\gamma x/(1+np) \tag{9.b}$$

$$\sigma^c_x = pE^s\gamma x/(1+np) \tag{9.c}$$

for $\ell_d \leq x \leq L$

$$S_x = 0 \tag{10.a}$$

$$\sigma^s_x = npP/(1+np)A^s \tag{10.b}$$

$$\sigma^c_x = P/(1+np)A^c \tag{10.c}$$

in which, $\ell_d = P/E^sA^s\gamma$ $\quad \gamma = (1+np)\psi\tau_y/E^sA^s$

The stress distributions of Eqs.(9) and (10) are schematically shown in Fig.14. It is difficult to calculate P directly in the finite element analysis with the smeared type elements, and thus P shall be expressed in terms of the average strain, ε^s_{av} in the direction of the bar axis from Eq.(9.b).

$$\varepsilon^s_{av} = \int_0^{\ell_d}(\sigma^s_x/E^s)\,dx/L, \qquad P = E^sA^s\sqrt{2(1+np)\gamma\varepsilon^s_{av}L/(1+2np)} \tag{11}$$

Tensile concrete stress becomes a maximum, σ^c_{max} at X = L, and σ^c_{max} is given by Eqs.(9.c) and (11) as.

$$\sigma^c_{maxi} = p_iE^{si}\sqrt{2\gamma_i\varepsilon^{si}_{av}L_i/(1+n_ip_i)(1+2n_ip_i)} \tag{12}$$

where the subscript i(=1,2,...,N) of each notation indicates the bar direction.

Although σ^c_{maxi} is calculated for each bar direction, the bar direction is not always perpendicular to crack directions. Therefore, to obtain the maximum tensile stress, $\overline{\sigma}^c_{max}$, acting perpendicular to cracks, σ^c_{maxi} in each reinforcing direction must transformed into the direction parallel to cracks and summed up.

$$\overline{\sigma}^c_{max} = \sum_{i=1}^{N}\sigma^c_{maxi}\sin^2\theta^i_{cr}, \qquad \theta^i_{cr} = \theta_{cr} - \theta_i \tag{13}$$

Letting f'_t be the uniaxial tensile strength of concrete, then a crack occurs when the following inequality is satisfied.

$$\beta f'_t \leq \overline{\sigma}^c_{max}, \qquad \beta = 1.0 - 0.8\,|\,\sigma^c_{u}/f'_c\,| \tag{14}$$

where β is the reduction factor of concrete tensile strength under tension-compression stress states, and σ^c_{u} the concrete compressive stress parallel to cracks.

Assuming that crack spacing becomes a minimum, e_{min}, when the equality of Eq.(14) holds, e_{min} can be derived by substituting $L_i = e_{min}/2|\sin\theta^i_{cr}|$ into Eqs.(12), (13) and (14) as follows.

$$e_{min} = [\beta f'_t / \sum_{i=1}^{N} \{p_i E^{si} \sqrt{\gamma_i \varepsilon^{si}_{av} |\sin^3\theta^i_{cr}| / (1+n_i p_i)(1+2n_i p_i)}\}]^2 \tag{15}$$

The upper bound of crack spacings must be evaluated, that is, the maximum crack spacing, e_{max}, on the basis of such energy criteria as that proposed by Bazant et al.(1983). For simlicity, the average crack spacing, e_{av} and average crack width, W_{av} are calculated by assuming that $e_{max} = k\, e_{min}$.

$$e_{av} = (e_{max} + e_{min})/2 = (1+k)\,e_{min}/2, \qquad W_{av} = \varepsilon_v e_{av} \tag{16.a,b}$$

where ε_v is the average strain normal to cracks and k is assumed to be equal to 2 in this study.

The average concrete stress, σ°_{avi}, in each direction due to the tension stiffening is derived as follows.

$$\sigma^{\circ}_{avi} = \{E^{si} A^{si} \gamma_i \ell_{di}^2 + 2P_i(L_i - \ell_{di})\}/2(1+n_i p_i) A^{\circ}_i L_i \tag{17}$$

The author shall refer to this as the bond-cracking model thereafter.

NUMERICAL EXAMPLES

Analysis of Reinforced Concrete Panels

To verify the validity of the proposed stress-strain relationship for cracked concrete, a finite element analysis was performed on the reinforced concrete panels tested by Vecchio et al.(1982). The specimens

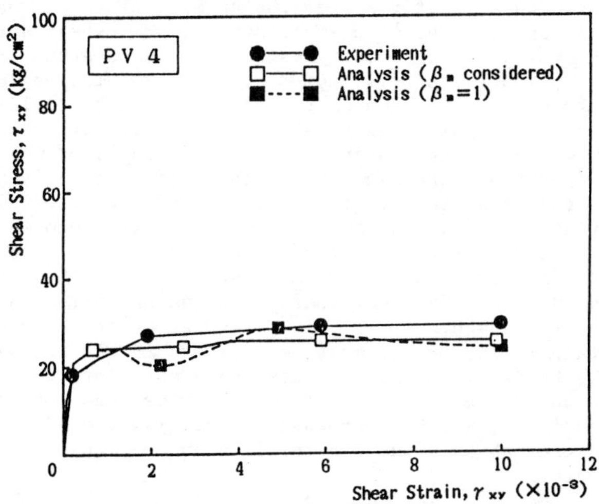

Fig.15(a) Shear Stress-Shear Strain Relationship
(1kg/cm^2 = 14.2psi = 98.15Pa)

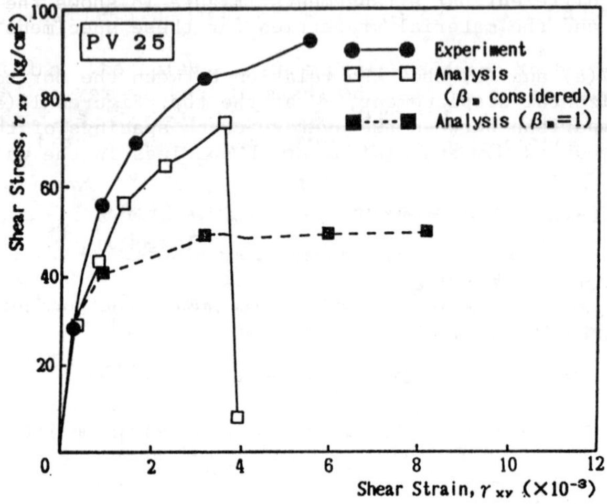

Fig.15(b) Shear Stress-Shear Strain Relationship ($1kg/cm^2$ = 14.2psi = 98.15Pa)

analyzed are PV4 and PV25, and they were tested under pure shear stress, τ_{xy} , and the combination of shear stress, τ_{xy} , and compressive normal stress, $\sigma_x = \sigma_y = -0.69\tau_{xy}$. The observed failure mode of PV4 was the steel yielding, and that of PV25 was the shear failure of concrete.

Figures 15(a) and (b) show the shear stress, τ_{xy} -shear strain, γ_{xy} responses. The values calculated by using the stress-strain relationship of a concrete cylinder, that is, $\beta_m=1$, are also plotted in the figures. In the case of PV4, which failed in the steel yielding, both of the calculated values fairly agree with the test results, and thus β_m has little influence on the shear behavior of reinforced concrete panels. In the case of PV25, on the other hand, the calculated values taking β_m into account tend to agree with the test results. But the values calculated by setting $\beta_m = 1$ give a lower maximum strength than those of the test and indicate failure by steel yielding, which is different from the concrete shear failure observed in the test.

It can be said from this study that the principal stress-strain relationship for cracked concrete may be expressed in terms of the principal strain ratio. But it must be noted that this relation is derived on the basis of the limited test data on reinforced concrete panels tested under specific boundary, geometrical and loading conditions. In addition, it is recommended to clarify the effect of the rotation of principal axes and bar buckling on the shear behavior of cracked concrete.

Analysis of Reinforced Concrete Shear Walls

To verify the validity of the bond-cracking model, a finite element analysis was carried out on reinforced concrete shear wall specimens of one bay and three stories(76W401 and 77W202) subjected to a constant vertical load, N(=13.5 tons =29.76 Klb) and increasing horizontal loads, P(Sato et al.,1980). The two specimens have the same geometrical configuration but different bar arrangements. Figure 16 shows the structural dimensions, and the material properties for these specimens are listed in Table 1.

Figures 17(a) and (b) show the relation between the horizontal load, P and the horizontal displacement, δ at the top. Figures 18(a) and (b) show the comparison between the average crack spacings of the test and the calculated maximum and minimum crack spacings in the wall panel at

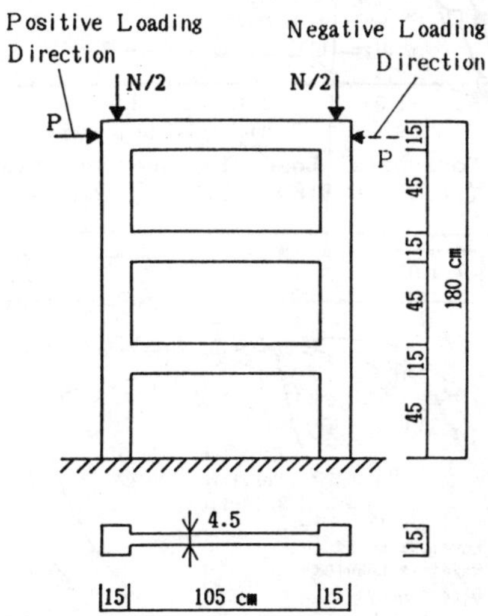

Fig.16 Structural Dimensions of Shear Wall(1cm = 0.3937in.)

Table 1 Material Properties of Shear Wall Specimens
($1kg/cm^2$ = 14.2psi = 98.15Pa)

Specimen	Reinforcement				Concrete		
	Element & Bar Direction		Steel Ratio p (%)	Yield Strength σ_y (kg/cm^2)	Conpressive Strength f'_c (kg/cm^2)	Strain at f'_c ε'_c ($\times 10^{-6}$)	Tensile Strength f'_t (kg/cm^2)
76W401	Beam	X Y	1.27 0.28	3214 4830	226	2240	24.4
	Column	X Y	0.28 4.52	4830 3591			
	Wall	X Y	0.24 0.24	4830 4830			
77W202	Beam	X Y	1.27 0.28	3761 5398	226	2400	27.9
	Column	X Y	0.28 2.53	5398 3761			
	Wall	X Y	0.72 0.72	5398 5398			

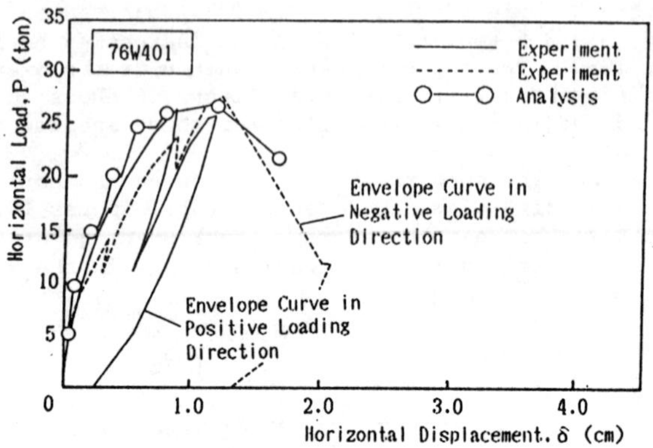

Fig.17(a) Horizontal Load-Displacement Relationship (1ton = 9.81KN = 2.2Klb; 1cm = 0.3937in.)

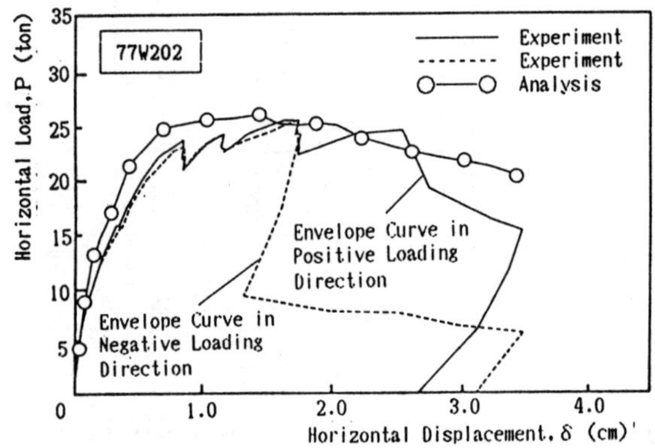

Fig.17(b) Horizontal Load-Displacement Relationship (1ton = 9.81KN = 2.2Klb; 1cm = 0.3937in.)

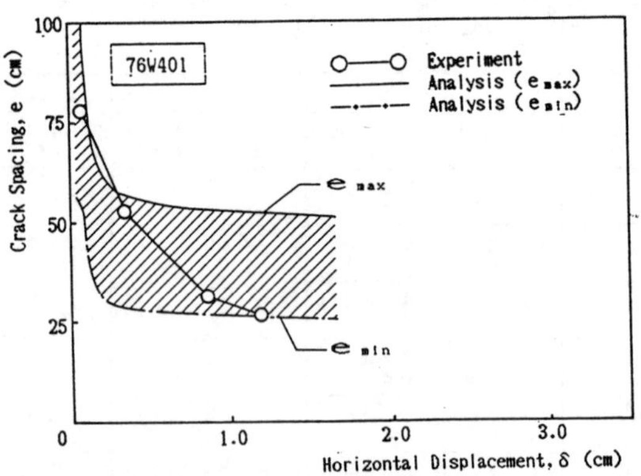

Fig.18(a) Crack Spacing-Horizontal Displacement Relationship (1cm = 0.3937in.)

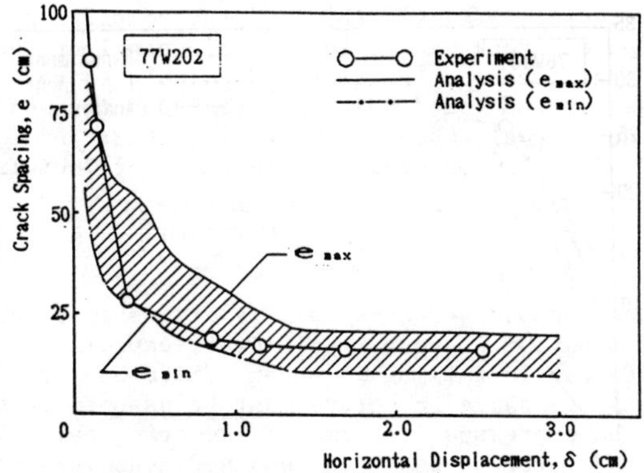

Fig.18(b) Crack Spacing-Horizontal Displacement Relationship (1cm = 0.3937in.)

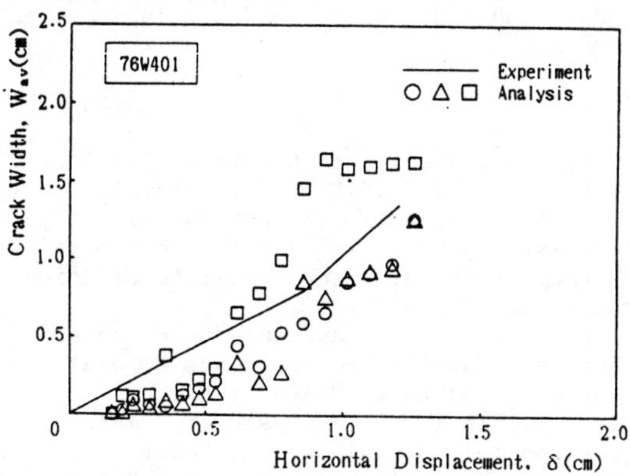

Fig.19(a) Crack Width-Horizontal Displacement Relationship (1cm = 0.3937in.)

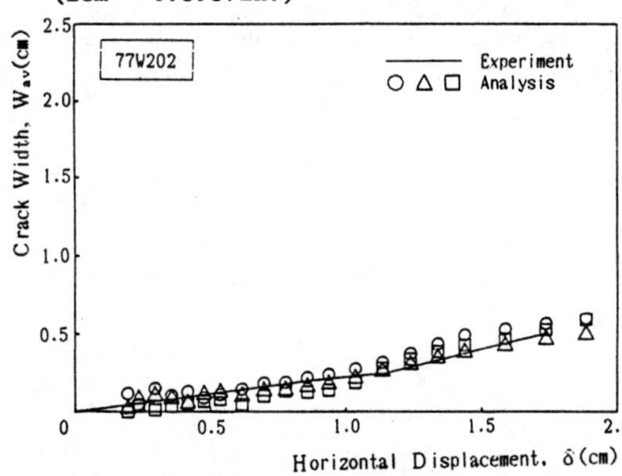

Fig.19(b) Crack Width-Horizontal Displacement Relationship (1cm = 0.3937in.)

the first story. Figures 19(a) and (b) show the comparison between the measured and computed crack widths in the wall panel at the first story. It seems that the agreement of the test and calculated results is fairly good in every case. It was found that the crack spacings and widths of 77W202, whose amount of wall reinforcements was about three times as much as that of 76W401, became smaller than those of 76W401.

CONCLUSIONS

The effect of the assumed stress-strain relationship of cracked concrete on shear behavior is very significant, especially in case of reinforced concrete panels failing in shear. Therefore, the stress-strain relationship of cracked concrete must be assumed carefully, and this relation can be expressed in terms of the principal strain ratio.

The effect of transverse pressure on bond and cracking properties of reinforced concrete members is remarkable, and the proposed bond-cracking model taking the effect of transverse pressure into account can predict the tension stiffening and crack widths and spacings accurately.

REFERENCES

1. Bazant,Z.P. and Oh,B.H.,"Spacing of Cracks in Reinforced Concrete," Proc. of the ASCE, Vol.109, No.ST9, Sept. 1983.
2. Sato,T., Ono,A., Adachi,H., Shirai,N. and Nakanishi,M.,"Experimental Study on Elasto-Plastic Behavior of Reinforced Concrete Shear Wall Structures," Journal of the Research Institute of Science & Technology , Nihon University, No.53, July 1980.
3. Sato,T., Shirai,N., Ono,A. and Adachi,H.,"Cracking and Bond Behaviors of Reinforced Concrete Tensile Members under Lateral Pressure," Trans. of the Japan Concrete Institute, Vol.4, 1982.
4. Shirai,N. and Sato,T.,"Inelastic Analysis of Reinforced Concrete Shear Wall Structures," IABSE Colloquium, Delft, June 1981.
5. Shirai,N. and Sato,T.,"Bond-Cracking Model for Reinforced Concrete," Trans. of the Japan Concrete Institute, Vol.6, 1985.
6. Vecchio,F. and Collins,M.P.,"The Responses of Reinforced Concrete to In-Plane Shear and Normal Stresses," Publication No.82-03, University of Toronto, March 1982.

Thermal Stress Analysis of Massive Concrete

Tada-aki TANABE* Makoto KAWASUMI** Yuji YAMASHITA***

Synopsis

The weak form of the Laplace's diffusion equation together with appropriate boundary conditions were reduced to discrete simultaneous equations for each Fourier term considering the time dependent heat generation rate of concrete. The thermal stresses at each age of concrete due to non-uniform temperature distribution were obtained as well, using Fourier series and considering the variation of Young's modulus of concrete due to cement hydration rate. A creep law for younger age of concrete was developed based on a cement hydration rate for uniaxial load history and developed to 3-dimensional stress field. For the stress analysis after cracking, crack strain is defined with delta function to be treated as initial strain and the relation with the damage tensor of concrete in tension zone is discussed.

1. Introduction

Thermal stress cracking due to the restrained contraction, when elevated temperature caused by cement hydration decreases as a result of heat dissipation, has gained much attention, because crack initiation and crack widths must be controlled to minimize undesirable effects on the durability or permeability of a structure from its very beginning.

The first application of FEM to the thermal stress analysis of a massive concrete structure was carried out in Japan [3] to treat the cracking problem in the base mat concrete of Ikata nuclear power station. In that study, a two dimensional stress analysis for a quadrant of the circular concrete base mat was carried out, and values obtained were compared with the observed values not very successfully. One of the problems raised in that study was the necessity of three dimensional analysis instead of rather crude two dimensional analysis in the case of irregular shaped structures. An other suggestion made in that study was that there may exist slip between the newly placed mass concrete and restraining rock or hardened concrete [12]. If this is the case, the thermal stress analysis of this kind must incorporate such a shear slip mechanism. There have been several attempts to detect real slippage by embedding a displacement sensor between the newly placed concrete and a rock foundation or hardened concrete [5]. However, at the present moment, such slip has not yet been fully confirmed. In view of this problem a large scale thermal stress experiment is now under consideration and is expected to clarify the problem.

* Professor in the Civil Engineering Department, Nagoya University. Fru-cho, Chikusa-ku, Nagoya, 464 Japan

** Senior research engineer at the Civil Engineering Laboratory of Central Research Institute of Electric Power Industry. 1646 Abiko, Abiko-City Chiba-ken, 272-01 Japan

*** Graduate student at the School of Engineering, Nagoya University. Furo-cho, Chikusa-ku, Nagoya, 464 Japan

In this study, theoretical aspects of the problem are investigated. Finite element formulations for 3-D unsteady heat flow and fot 3-D stress analysis with varying Young's modulus and with cracking were studied.

However, the effect of moisture migration and the relevant effects due to it on the cracking are all neglected in this study because the shrinkage effect is relatively small on such a cracking problem that a crack pass through the total width of a large concrete block.

2. Formulation of 3-D Unsteady Heat Flow by Fourier Series

The weak form of the Laplace diffusion equation is expressed as

$$\int_V (\nabla\cdot k\nabla\phi + Q(t) - \rho c\frac{\partial\phi}{\partial t})\delta\phi dv + \int_{s_1} h(\phi - \phi_\infty)\delta\phi ds + \int_{s_2} q\delta\phi ds = 0 \quad \ldots\ldots \quad (2.1)$$

where ϕ denotes temperature, $Q(t)$ denotes heat generation rate due to cement hydration, k denotes the thermal conduction coefficient of concrete, ρ the specific gravity of concrete, c the specific heat of concrete, h the thermal convection coefficient of concrete, and q the heat flux at the boundary. ϕ_∞ denotes the temperature of the outer atmosphere. Then, let us discretize a structure by finite prism elements as shown in Fig. (2.1). The temperature variation along a nodal line of a prism in Fig. (2.2) is expressed by sine series as

$$\phi_i = \theta_{i,o} + \sum_{k=1}^{m}\theta_{i,k}\sin\frac{k\pi z}{l} \quad \ldots\ldots\ldots\ldots\ldots\ldots\ldots\ldots \quad (2.2)$$

It should be noted that the constant term is added to the sine series so that it can express temperature variations at the surfaces of two extreme ends. The temperature variation within the x-y plane is expressed using an appropriate shape function, for example the isoparametric shape function of Eq. (2.3.b). Hence, the temperature at an arbitrary point (x, y, z) is expressed as

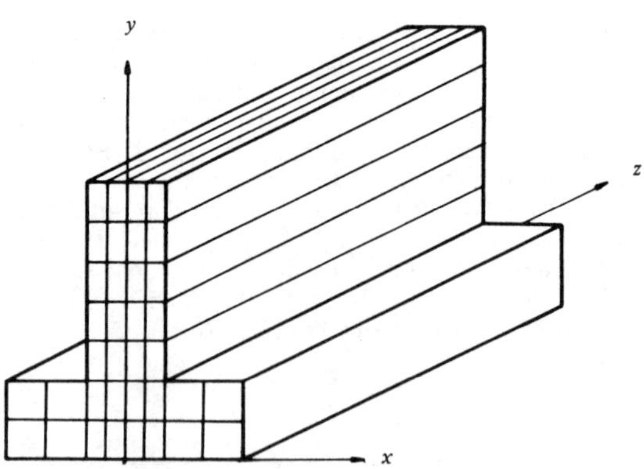

Fig. (2.1) Finite prism discretization of a structure

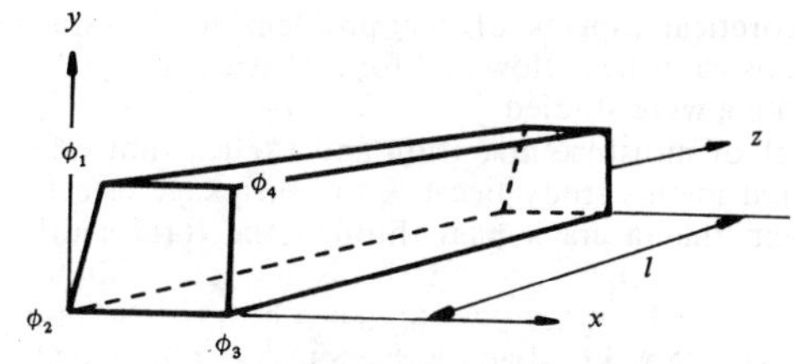

Fig. (2.2) Finite prism element and temperature notation

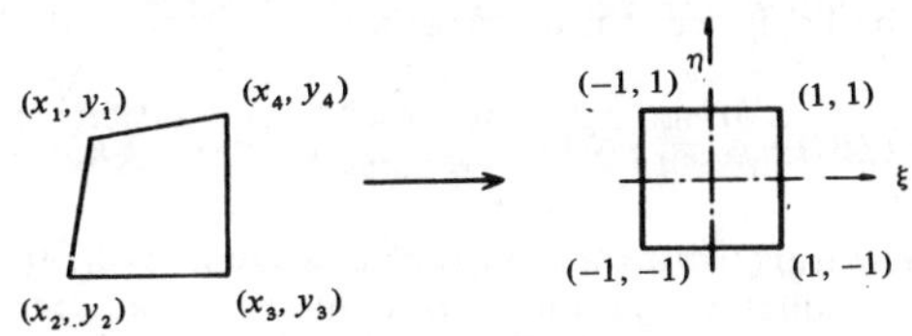

Fig. (2.3) Isoparametric shape function

$$\phi = N_1\phi_1 + N_2\phi_2 + N_3\phi_3 + N_4\phi_4 \quad \ldots\ldots \quad (2.3.a)$$

$$N_i = \frac{1}{4}(1 + \xi_o)(1 + \eta_o) \quad \ldots\ldots \quad (2.3.b)$$

$$\xi_o = \xi\xi_i,\ \eta_o = \eta\eta_i$$

$$x = \sum_{i=1}^{4} N_i x_i \quad \ldots\ldots \quad (2.4.a)$$

$$y = \sum_{i=1}^{4} N_i y_i \quad \ldots\ldots \quad (2.4.b)$$

Substituting Eq. (2.2) into Eq. (2.1), and taking variations in terms of each Fourier term yields

$$[H]_m \{\theta\}_m + [P]_m \left\{\frac{\partial\theta}{\partial t}\right\}_m + \{F\}_m = 0 \quad \ldots\ldots \quad (2.5)$$

where $[H]_m$ denotes the conduction matrix, $[P]_m$ denotes the coefficient matrix for thermal gradient, and $\{F\}_m$ denotes the thermal load vector for the m-th term of the Fourier series. One of the components contributing a load vector is the cement hydration heat rate which is expressed as

$$F_m^i = \int_{-1}^{1}\int_{-1}^{1} \frac{l}{2m\pi} N_i Q(t) \det[J]\, d\xi d\eta \quad \ldots\ldots \quad (2.6)$$

where F_m^i denotes the contribution to the load vector of the m-th term of the series from the i-th nodal line of an element. The length of a prism is expressed as l. In Eq. (1.6), $Q(t)$ should be evaluated at each time increment as the heat generation rate varies with elapsed period.

The finite difference scheme to solve Eq. (2.5) was chosen as

$$\{\theta\}_{m,t} = \{\theta\}_{m,t-\Delta t} + \left(\left\{\frac{\partial\theta}{\partial t}\right\}_{m,t-\Delta t} + \left\{\frac{\partial\theta}{\partial t}\right\}_{m,t}\right)\frac{1}{2}\Delta t \quad \ldots\ldots \quad (2.7)$$

Substituting Eq. (2.7) into Eq. (2.5) yields

$$([H]_m + \frac{\partial}{\partial t}[P]_m)\{\theta\}_{m,t} = [P]_m\left(\frac{2}{\Delta t}\{\theta\}_{m,t-\Delta t} + \left\{\frac{\partial\theta}{\partial t}\right\}_{m,t-\Delta t}\right) + \{F\}_{m,t} \quad \text{...... (2.8)}$$

It should be noted here that the cement hydration heat $Q(t)$ is not only time dependent but also temperature dependent. Hence, rigorously the rate should be determined at each time step according to the state of temperature and the hydration rate. However, the heat generation rate obtained from an adiabatic cement hydration test seems to be a good substitute for it under the arbitrary conditions as it gives reasonably accurate estimate of the temperature distribution within a structure when a structure is massive.

The Heat generation rate of cement under adiabatic condition varies according to the cement type, unit cement content and initial temperature of mixing materials.

For four types of cement which are on the Japanese market, the hydration heat at adiabatic condition was measured at three levels of initial temperature at mixing and shown in Fig. (2.4) to Fig. (2.5) by the JCI Committee on Thermal Stress of Massive Concrete [4, 6].

It is noted that the figures are shown in terms of coefficients of the general expression for adiabatic temperature rise due to cement hydration heat,

$$T(t) = K(1 - e^{-\alpha t}) \quad \text{.......................... (2.9)}$$

where the unit of t is taken as a day. K values are shown in Fig. (2.4.a) to Fig. (2.4.d). In the figures, OPC denotes ordinary portland cement, FC denotes fly-ash cement, BSC denotes blast furnace slag cement and MHPC denotes moderate heat portland cement.

In each figure, the regression line of 20°C mixing temperature are shown together with the correlation coefficients γ. In Fig. (2.5.a) to Fig. (2.5.d), α values are shown with regression lines as well as with the coefficients of correlation. From the figures, it is observed that there exists rather large scatter for the same amount of unit cement content, although it is possible to discern differences between general heat generation rate trends for each type of cement. Hence, it is necessary to make experiments to determine the heat generation rate according to the extent needed for accurate estimates.

The heat conduction coefficient of concrete is much affected by the kind of aggregate used, and experimental values scatter generally. However, its dependency on the hydration rate is considered to be small, as 70 to 80 percent of the concrete volume is occupied by aggregate which is independent of cement hydration.

The specific heat of concrete is also affected by the aggregate used in the concrete, and its dependency on the cement hydration rate is small as well.

The heat convection coefficient between concrete surface and air or between various formwork or heat insulation materials and air are quite important for an accurate temperature estimate. However, we do not yet have ample data so far, owing to the difficulty of execution of experiments. On the other hand, an experiment [11] on heat convection coefficients between cooling pipe and cooling water was carried out by the authors, and the experimental formula proposed was

$$h = 4.75u + 43.0 \ (\mathrm{Kcal/m^2 \cdot hr \cdot {}^\circ C}) \quad \text{.............................. (2.11)}$$

u: velocity of cooling water (cm/sec)

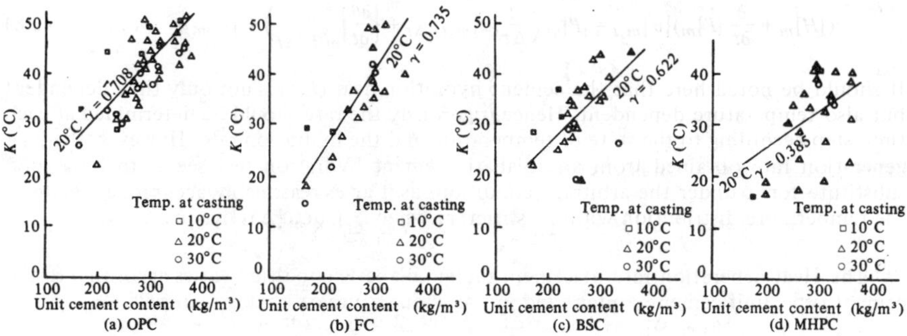

Fig. (2.4) K values of adiabatic temperature rise of concrete

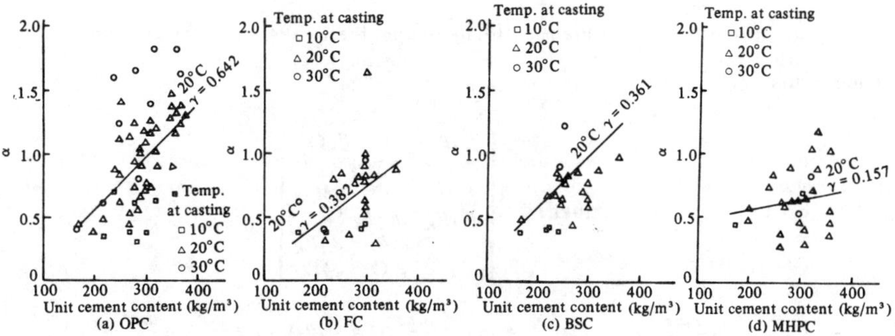

Fig. (2.5) α values of adiabatic temperature rise of concrete

The experimental range for the velocity of cooling water was from 20 cm/sec to 60 cm/sec, using a steel cooling pipe with 2.5 cm diameter which is usually adopted in Japan.

3. Formulation of the Thermal Stress Problem

When the temperature distribution is given, the stress distribution is solved at each time step using similar Fourier series. Displacement increments are expressed at each nodal line as

$$\Delta u_i = \Delta u_{i,o} + \sum_m \Delta u_{i,m} \sin\frac{m\pi z}{l} \qquad \text{(3.1.a)}$$

$$\Delta v_i = \Delta v_{i,o} + \sum_m \Delta v_{i,m} \sin\frac{m\pi z}{l} \qquad \text{(3.1.b)}$$

$$\Delta w_i = \sum_m \Delta w_{i,m} \cos\frac{m\pi z}{l} \qquad \text{(3.1.c)}$$

where Δu_i, Δv_i, Δw_i denote displacement increments in the X, Y, Z directions, repectively. Constant terms $\Delta u_{i,o}$ and $\Delta v_{i,o}$ are added to express the displacements at two end surfaces.

The displacement distribution within the x-y plane was expressed using isoparametric displacement functions, N_i. Then displacement increments at an arbitrary coordinate point (x, y, z) is expressed as

$$\begin{Bmatrix} \Delta u \\ \Delta v \\ \Delta w \end{Bmatrix} = \sum_{i=1}^{4} N_i \begin{Bmatrix} \Delta u_i \\ \Delta v_i \\ \Delta w_i \end{Bmatrix} \qquad (3.2)$$

The stress strain relation is given as

$$\{\Delta\epsilon\} = [B]\{\Delta\delta\} = \sum_{k=0}^{4}[B_k]\{\Delta\delta_k\}_0 + \sum_{m=1}^{r}\sum_{k=1}^{4}[B_k]_m\{\Delta\delta_k\}_m \qquad (3.3)$$

where

$$\{\Delta\delta_k\}_m = \{\Delta u_{1,m}, \Delta v_{1,m}, \Delta w_{1,m}, \Delta u_{2,m}, \Delta v_{2,m}, \Delta w_{2,m}, \Delta u_{3,m}, \Delta v_{3,m}, \Delta w_{3,m}, \Delta u_{4,m}, \Delta v_{4,m}, \Delta w_{4,m}\} \qquad (3.4)$$

and $[B_k]_m$ is expressed as

$$[B_k]_m = \{[B_1]_m, [B_2]_m, [B_3]_m, [B_4]_m\} \qquad (3.5)$$

Each term of $[B_k]_m$ is

$$[B_k]_m = \frac{1}{4}\begin{bmatrix} K_d P_k & O & O \\ O & K_d R_k & O \\ O & O & -M_k \\ K_d R_k & K_d P_k & O \\ O & L_k & J_d R_k \\ L_k & O & J_d P_k \end{bmatrix} \qquad (3.6)$$

where $K_d = 16\sin\frac{m\pi z}{l}/(EA - BD)$, $J_d = 16\cos\frac{m\pi z}{l}/(EA - BD)$, $L_k = \frac{m\pi}{l}\cos\frac{m\pi z}{l}\cdot 4N_k$, $M_k = \frac{m\pi}{l}\sin\frac{m\pi z}{l}\cdot 4N_k$, $P_k = A\frac{\partial N_k}{\partial\xi} - B\frac{\partial N_k}{\partial\eta}$, $R_k = -D\frac{\partial N_k}{\partial\xi} + E\frac{\partial N_k}{\partial\eta}$

and

$$A = (-1+\xi)y_1 + (-1-\xi)y_2 + (1+\xi)y_3 + (1-\xi)y_4 \qquad (3.7.a)$$

$$B = (-1+\eta)y_1 + (1-\eta)y_2 + (1+\eta)y_3 + (-1-\eta)y_4 \qquad (3.7.b)$$

$$D = (-1+\xi)x_1 + (-1-\xi)x_2 + (1+\xi)x_3 + (1-\xi)x_4 \qquad (3.7.c)$$

$$E = (-1+\eta)x_1 + (1-\eta)x_2 + (1+\eta)x_3 + (-1-\eta)x_4 \qquad (3.7.d)$$

The stress and strain relationship considering creep, cracking, and end displacements is given by

$$\{\Delta\sigma\} = [D(t)]\left\{\{\Delta\epsilon\}_t - \{\Delta\epsilon\}_{tem} - \{\Delta\epsilon\}_{crp} - \{\Delta\epsilon\}_{crk} - \{\Delta\epsilon\}_d\right\} \qquad (3.8)$$

where $\{\Delta\epsilon\}_t$ is the total strain increment, $\{\Delta\epsilon\}_{tem}$ the thermal strain increment, $\{\Delta\epsilon\}_{crp}$ the creep strain increment, $\{\Delta\epsilon\}_{crk}$ the crack strain increment which is defined in next section, and $\{\Delta\epsilon\}_d$ is the strain increment due to the displacement at the end surfaces. $D(t)$ is taken as follows

$$[D(t)] = \frac{E_c(t)}{(1+\nu(t))(1-2\nu(t))} \begin{bmatrix} 1-\nu(t) & \nu(t) & \nu(t) & & & \\ \nu(t) & 1-\nu(t) & \nu(t) & & 0 & \\ \nu(t) & \nu(t) & 1-\nu(t) & & & \\ & & & \frac{1-2\nu(t)}{2} & & \\ & 0 & & & \frac{1-2\nu(t)}{2} & \\ & & & & & \frac{1-2\nu(t)}{2} \end{bmatrix} \quad \ldots\ldots (3.9)$$

$E_c(t)$ and $\nu(t)$ are age dependent and should be determined considering the cement hydration rate.

Use of the principle of virtual work and orthogonality condition yield for one prism element and for the m-th Fourier term,

$$\int_V [B]_m^T [D] \cdot [B]_m \{\Delta\delta\}_m dV - \int_V [B]_m^T [D] \{\Delta\epsilon\}_{tem} dV - \int_V [B]_m^T [D] \{\Delta\epsilon\}_{crp} dV - \int_V [B]_m^T [D] \{\Delta\epsilon\}_{crk} dV - \int_V [B]_m^T [D] \{\Delta\epsilon\}_d dV = 0 \quad \ldots\ldots (3.10)$$

When we express the stiffness matrix as

$$[K^I] = \begin{bmatrix} [K_{11}] & [K_{12}] & [K_{13}] & [K_{14}] \\ [K_{21}] & [K_{22}] & [K_{23}] & [K_{24}] \\ [K_{31}] & [K_{32}] & [K_{33}] & [K_{34}] \\ [K_{41}] & [K_{42}] & [K_{43}] & [K_{44}] \end{bmatrix} \quad \ldots\ldots (3.11)$$

then each K_{ij} becomes

$$[K_{ij}] = \int \left[\begin{array}{c|c|c} \begin{array}{l} \frac{l}{2}\left(\frac{1}{\det[J]}\right)^2 \{D_{11}P_iP_j \\ + D_{44}R_iR_j\} \\ + 8\left(\frac{m\pi}{l}\right)^2 l D_{66}N_iN_j \end{array} & \begin{array}{l} \frac{l}{2}\left(\frac{1}{\det[J]}\right)^2 \{D_{12}P_iP_j \\ + D_{44}R_iR_j\} \end{array} & \begin{array}{l} \frac{m\pi}{2}\frac{4}{\det[J]} \{-D_{13}P_iN_j \\ + D_{66}N_iP_j\} \end{array} \\ \hline \text{Sym} & \begin{array}{l} \frac{l}{2}\left(\frac{1}{\det[J]}\right)^2 \{D_{22}R_iR_j \\ + D_{44}P_iP_j\} \\ + 8\left(\frac{m\pi}{l}\right)^2 D_{55}N_iN_j \end{array} & \begin{array}{l} \frac{m\pi}{2}\frac{4}{\det[J]} \{-D_{23}R_iN_j \\ + D_{55}N_iR_j\} \end{array} \\ \hline \text{Sym} & \text{Sym} & \begin{array}{l} 8\left(\frac{m\pi}{l}\right)^2 l D_{33}N_iN_j \\ + \frac{l}{2}\left(\frac{1}{\det[J]}\right)^2 \{D_{55}R_iR_j \\ + D_{66}P_iP_j\} \end{array} \end{array} \right]$$

$$\times \det[J] \times \frac{E_c}{16(1+\nu)(1-2\nu)} d\xi d\eta \quad \ldots\ldots (3.12)$$

Hence the total simultaneous equations are for each Fourier term,

$$[K]_m \{\Delta\delta\}_m - \{F\}_{tem} - \{F\}_{crp} - \{F\}_{crk} - \{F\}_d = 0 \quad \ldots\ldots\ldots\ldots (3.13)$$

The force vector due to thermal strains is given as

$$\{F\}_{tem} = [\{F_1\}, \{F_2\}, \{F_3\}, \{F_4\}] \quad \ldots\ldots\ldots\ldots (3.14)$$

and each $\{F_i\}$ term is expressed as

$$[F_i]_m = \int_{-1}^{1}\int_{-1}^{1} \begin{bmatrix} D_{11}P_i & D_{12}P_i & D_{13}P_i \\ D_{21}R_i & D_{22}R_i & D_{23}R_i \\ \frac{-4m\pi D_{31}}{\det[J]\,l}N_i & \frac{-4m\pi D_{32}}{\det[J]\,l}N_i & \frac{-4m\pi D_{33}}{\det[J]\,l}N_i \end{bmatrix} \frac{\alpha l}{8}$$

$$\times \sum_{i=1}^{4} N_i \{(\theta_{i,m,t} - \theta_{i,m,t-\Delta t}) - \frac{4}{m\pi}(\theta_{i,o,t} - \theta_{i,o,t-\Delta t})\} \times \frac{E_c}{(1+\nu)(1-2\nu)} d\xi d\eta \ldots (3.15)$$

The force vector due to end displacements is not given, and iteration is needed to satisfy the boundary conditions that shear stress and normal stress at end surfaces are zero, however they may be approximated by allowing full free displacement due to temperature variations at the ends.

The steel stress and strain relation is similarly obtained as

$$\Delta\sigma_s = E_s \Delta\epsilon_s = E_s \frac{\partial \Delta w_s}{\partial z} - E_s \Delta\epsilon_{tem,s} \quad \ldots\ldots\ldots\ldots (3.16)$$

and the force displacement relation to be added to the overall equations is obtained as

$$\frac{E_s}{2l}\sum_{i=1}^{4} N_i (m\pi)^2 \Delta w_{i,m} - E_s \alpha_s \sum_{i=1}^{4}\left[\frac{m\pi}{2}(\theta_{i,m,t} - \theta_{i,m,t-\Delta t}) - 2(\theta_{i,o,t} - \theta_{i,o,t-\Delta t})\right] = 0 \quad . \ (3.17)$$

4. Development of Three Dimensional Creep Equations

4.1 Uniaxial Equation

One of the problems of the stress analysis of concrete at a younger age arises from the creep phenomenon of concrete. We need keenly the three dimensional creep equations for concrete of younger age. In the beginning, the creep equation for the uniaxial case based on the hydration-viscosity model is introduced and is developed into three dimensional equations.

While the creep equations of T.C. Powers [9] or Z.P. Bazant [1] based on thermodynamics are applicable only for the recoverable creep and can not let directly be related to the cement-hydration process data, the creep equations by S. Seki [10] based on the hydration-viscosity model is directly derived from the mathematical model of the cement-hydration process.

Considering a one dimensional model of cement paste (as shown in Fig. 4.1), the following assumptions were made to derive a creep equation. The model was originally developed by S. Seki [10].

(1) The shape of a needle of calcium silicate hydrate is approximated by a cylinder. While the number of crystals increases during the process of hydration, the size remains the same as before. This is the result obtained experimentally as well.

(2) An increment of creep strain ($d\epsilon_c$) within a small interval of time is proportional to the stress (σ), and becomes larger as the thickness ($2d$) of the water membrane between crystals increases. On the other hand, the increment becomes smaller when the contact length (l) of crystals, the number of crystals (m) and the viscosity of water (η) increase.
(3) The creep of concrete increases approximately in proportion to the volúme of cement paste in concrete. From these assumptions we have

$$\epsilon(t-t_1)=\int_{t_1}^{t}\frac{1}{E_i(s)}\frac{d\sigma(s)}{ds}ds-\epsilon_0+\left[\epsilon_0^q+\int_{t_1}^{t}\text{sgn}(\sigma)\frac{|\sigma(s)|^q}{\eta_*^q}ds\right]^{1/q}$$
$$+\sum_{i=1}^{m}\frac{1}{\eta_{vi}}\int_{t_1}^{t}\sigma(s)\exp\left[-(t-s)/\lambda_i\right]ds \qquad \text{(4.1)}$$

where η_* is the longitudinal pseudo-viscosity when $q \neq 1$. This η_* can be expressed as follows,

$$\eta_*(t) \equiv \eta\,[V_{HC}/V_{UW}]^{p/q}/(qA_1)^{1/q} \qquad \text{(4.2)}$$

where η is the viscosity of interstitial water and V_{HC} and V_{UW} are, respectively, volume of hydrated cement and of unhydrated water in 1 m^3 of cement paste. Quantitative values of V_{HC} and V_{UW} at a constant temperature (20 ~ 80°C) and at any age can be calculated from the hydration equation with measured hydration parameters [9]. ϵ_0 is the "hidden strain" which expresses the degree of crystal entanglement at the age of loading. q is a parameter which shows the degree of nonlinearity. t and t_1 are the age of concrete and the age at loading. When $q = 1$, η_* coincides with the longitudinal viscosity of concrete.

Neglecting the influence of the initial entanglement of crystals and substituting $\epsilon_0 = 0$ in Eq. (4.1), we obtain a creep expression in which creep curves have the same shapes in tension and compression.

Some examples of identification of creep data of sealed concrete loaded at various ages between 7 days and one year using this creep expression are illustrated in Fig. 4.2. Detailed discussion of the model is found in reference [10].

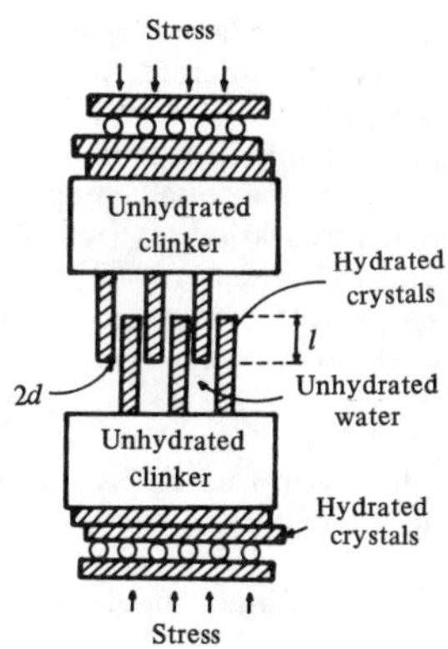

Fig. (4.1) Hydration-viscosity model of creep of concrete

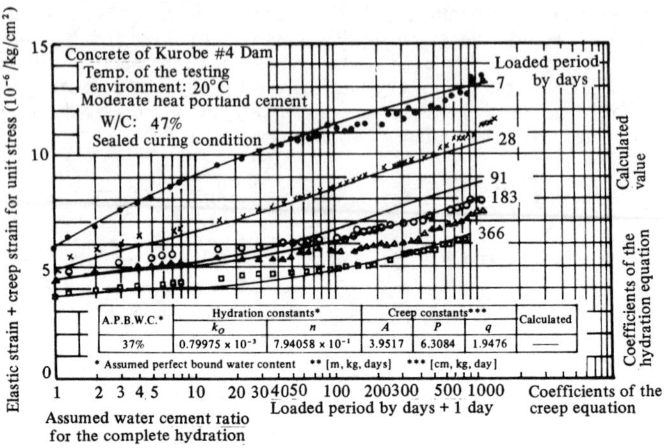

Fig. (4.2) Fitting of Hydration Viscosity Creep Curves of Integral Type [7]

4.2 Three Dimensional Extension of the Creep Equation

In the following, we will extend it to the three dimensional creep formula, taking into account the rate of the hydration of cement and the viscosity of interstitial water.

The accurate extension of the model taking into account three dimensional microstructures is abandoned. Instead, referring to the theory of metal creep, the creep expression is formally extended, taking into account both hydration and viscosity. Thus it includes both cases of time-hardening and strain-hardening as stated in the theory of metal creep in three dimensions. In carying out the extension, it was managed so that the following four conditions are satisfied.

(i) While volume changes due to creep are neglected for metal, creep Poisson's ratio ν_2 $(0 \leqslant \nu_2 \leqslant 1/2)$ must be assumed for concrete, when approximated by linear creep law with respect to creep. The ν_2 is generally different from the elastic Poisson's ratio ν.

(ii) The linear stress creep expression must conform to the conventional creep expression.

(iii) In the case of nonlinear one dimensional creep, the three dimensional expression must reduce to the one dimensional hydration-viscosity creep expression already obtained.

(iv) In the formulation of a uniaxial creep law [7], it was shown that the shape of the tensile creep curve was necessarily different from the compressive one. In the present derivation, the shapes of both creep curves are assumed to be the same for the sake of simplicity.

In the derivation, it should be noted that stress and strain in compression are assumed to be positive in this section and are expressed in tensorial forms.

Odqvist (1934) extended the one dimensional steady creep law to multiaxial stresses for metals as follows [8],

$$\dot{\epsilon}_{ij} = \frac{3k}{2}(3I_2)^{\frac{n-1}{2}} s_{ij} \qquad \cdots\cdots (4.3)$$

where $\dot{\epsilon}_{ij}$ is the strain rate and I_2 is the second invariant of deviatoric strains. Equation (4.3) four uniaxial creep coincides with the following Norton's expression (1929),

$$\dot{\epsilon} = k\,\sigma^n \quad \ldots\ldots\ldots\ldots\ldots\ldots\ldots\ldots \quad (4.4)$$

Then, the following expression has been proposed as a three dimensional expression of the strain hardening creep law,

$$\dot{\epsilon}_{ij} = \frac{3}{2} B\,\sigma_e^{m-1}\,\epsilon_e^{-\mu}\,s_{ij} \quad \ldots\ldots\ldots\ldots\ldots\ldots \quad (4.5)$$

where B, m, and μ are positive constants. σ_e and ϵ_e are, respectively, called effective stress and strain, and defined as follows.

$$\sigma_e = (3I_2)^{1/2} = (\frac{3}{2} s_{ij}\,s_{ij})^{1/2} \quad \ldots\ldots\ldots\ldots\ldots\ldots \quad (4.6)$$

$$\epsilon_e = (\frac{2}{3}\epsilon_{ij}\,\epsilon_{ij})^{1/2} \quad \ldots\ldots\ldots\ldots\ldots\ldots \quad (4.7)$$

Einstein's summation convention is adopted here. We must notice that the coefficient in Eq (4.6) is 3/2 and 2/3 in Eq. (4.7) and that the one dimensional effective stress and strain rate are, respectively, equal to the absolute value of uniaxial stress and longitudinal strain. We must keep in mind that incompressibility has been assumed for the strain.

$$\sigma_e = |\sigma_0|, \qquad \dot{\epsilon}_e = |\dot{\epsilon}| \quad \ldots\ldots\ldots\ldots\ldots\ldots \quad (4.8)$$

The uniaxial expression corresponding to Eq. (4.4) is

$$\dot{\epsilon} \doteq B\,\sigma^m\,\epsilon^{-\mu} \quad \ldots\ldots\ldots\ldots\ldots\ldots\ldots\ldots \quad (4.9)$$

In Eq. (4.5) incompressibility has been assumed as in the other cases of metal creep. On the other hand, Patel Venkatraman and Vafakos (1959) who took into account the compressibility of materials adopted the following s_{ij}' instead of s_{ij} in Eq. (4.5)

$$s_{ij}' = \sigma_{ij} - \alpha\,\sigma_{kk}\,\delta_{ij} \quad \ldots\ldots\ldots\ldots\ldots\ldots \quad (4.10)$$

where $0 < \alpha \leqslant 1/3$.

Here, we develop 3-D creep equations through the analogy. The already derived uniaxial creep equation is [7],

$$\frac{d\epsilon_c}{dt} = \mathrm{sgn}\,(\sigma)\,\frac{|\sigma|^q}{q\,\eta_{*(t)}^q\,(\epsilon_c + \epsilon_0)^{q-1}} \quad \ldots\ldots\ldots\ldots\ldots\ldots \quad (4.11)$$

where ϵ_c is the irrecoverable creep strain. Here, instead of the expression taken in Eq. (4.9), m and μ have the following specially selected values: $m = q$ and $\mu = q - 1$. These restrictions were imposed because concrete creeps approximately in proportion to applied stress. In this expression the influence of temperature and hardening in the process of hydration of cement is expressed by $\eta_*\,(t)$.

The applicability of Eq. (4.11) to the creep data for the specimens loaded increasingly by the step increment, is examined in [7] and the good agreement is shown. However, it is pointed that the creep recovery can not be included in the expression of Eq. (4.9). Hence, Eq. (4.11) is proposed to improve the deficiency, which is the sum of the recoverable and the unrecoverable component of creep strains. However, a study is necessary to include the effect of hardening in the recoverable component of creep strains.

Z.P. Bazant and et al. proposed the different equation [2] to improve the same deficiency starting from the similar equation as Eq. (4.11).

In this paper, the development of Eq. (4.11) to the three dimensional expression is stated.

The 3-D development of the linear component of the recoverable creep strain is already well examined.
Let us explain here why Patel et al. adopted s_{ij}'. Hooke's law states that

$$\epsilon_{ij} = \frac{1}{2G}[\sigma_{ij} - \frac{\nu}{1+\nu}\sigma_{kk}\delta_{ij}], \qquad G = \frac{E}{2(1+\nu)} \quad \text{(4.12)}$$

Therefore, they also adopted

$$s_{ij}' = \sigma_{ij} - \alpha\,\sigma_{kk}\delta_{ij} \quad \text{(4.13)}$$

However, in the case of concrete, we must use ν_2 (creep Poisson's ratio) instead of ν, then, $\alpha = \nu_2/(1+\nu_2)$. As $0 \leqslant \nu_2 \leqslant 1/2$, α is in the range $0 \leqslant \alpha \leqslant 1/3$, as Patel et al. have stated. If incompressibility is assumed, $\alpha = 1/3$. Then s_{ij}' reduces to s_{ij} and Eq. (4.5) is obtained.

In this paper, we extended the uniaxial creep expression which gives the irrecoverable part of creep strain to the following form neglecting the initial resistance,

$$\frac{d}{dt}\epsilon_{ij}^{(c)} = \frac{1}{(1+\nu_2)^{q-1}}\left(\frac{\sigma_e}{2\mu_{*(t)}}\right)^{q-1}\left(\frac{s_{ij}'}{2\mu_{*(t)}}\right)/q\,(\epsilon_e^{(c)'})^{q-1} \quad \text{(4.14)}$$

where $\epsilon_{ij}^{(c)}$ is the irrecoverable part of creep strain, σ_e is effective strain, and

$$\mu_{*(t)} = \frac{\eta_{*(t)}}{2(1+\nu_2)} \quad \text{(4.15)}$$

$$\epsilon_e^{(c)'} = \frac{1}{(1+\nu_2)\sqrt{2}}\,[(\epsilon_{xx}^{(c)} - \epsilon_{yy}^{(c)})^2 + (\epsilon_{yy}^{(c)} - \epsilon_{zz}^{(c)})^2 + (\epsilon_{zz}^{(c)} - \epsilon_{xx}^{(c)})^2 + 6\{(\epsilon_{xy}^{(c)})^2 + (\epsilon_{yz}^{(c)})^2 + (\epsilon_{zx}^{(c)})^2\}]^{1/2} \quad \text{(4.16)}$$

$$= \frac{1}{(1+\nu_2)\sqrt{2}}\,[(\epsilon_1^{(c)} - \epsilon_2^{(c)})^2 + (\epsilon_2^{(c)} - \epsilon_3^{(c)})^2 + (\epsilon_3^{(c)} - \epsilon_1^{(c)})^2]^{1/2}$$

Also,

$$s_{ij}' = \sigma_{ij} - \frac{\nu_2}{1+\nu_2}\sigma_{kk}\delta_{ij} \quad \text{(4.17)}$$

In the above expressions, σ_e is effective stress, $(\epsilon_1^{(c)}, \epsilon_2^{(c)}, \epsilon_3^{(c)})$ are the principal irrecoverable creep strains, μ_* is the pseudo-viscosity of shear. $\epsilon_e^{(c)'}$ will be referred to as "modified effective strain." σ_e and $\epsilon_e^{(c)'}$ are invariant, therefore the development of the theory can be remarkably simplified.

The developed three dimensional creep law is shown to be in conformity with the conventional linear creep equation when $q = 1$, and in conformity with Eq. (4.11) for nonlinear creep at the uniaxial loading case when $q \neq 1$. Thus, conditions (ii) and (iii) are satisfied. Condition (iv) is also satisfied by making $\epsilon_0 = 0$. Condition (i) is automatically satisfied by adopting s_{ij}' instead of s_{ij}. Examination of the developed 3-D creep law for the hydrostatic stress state needs elaboration. However, it has been shown that the equations give the exact solution for a seemingly indefinite form of the equation.

5. Definition of Crack Strain

There are several ways to define the crack strain [13, 14]. We use here the definition that the crack strain is given as the derivative of discontinuity of displacement at a crack. Namely, referring to Fig. (5.1), with the coordinate system taken in such a way that n is as normal to the crack surface and t is the direction of the displacement vector $\boldsymbol{d}$ in the (n, t) plane,

$$\epsilon_{cr,n} = \frac{\partial \Delta w_1}{\partial n} = \Delta w_1 \delta(n) \quad \ldots\ldots\ldots\ldots \quad (5.1.a)$$

$$\gamma_{cr,t} = \frac{\partial \Delta w_2}{\partial n} = \Delta w_2 \delta(n) \quad \ldots\ldots\ldots\ldots \quad (5.1.b)$$

$$\{\epsilon_{cr}\} = [\Delta w_1 \delta(n), 0, 0, \Delta w_2 \delta(n), 0, 0]^T \quad \ldots\ldots\ldots \quad (5.1.c)$$

where $\delta(n)$ is Dirac's delta function.

When the crack orientation is not parallel to the coordinate axis, the same rule of coordinate transformation is applicable and

$$\{\epsilon_{cr}\} = [T]\{\epsilon_{cr}\}_{n,t} \quad \ldots\ldots\ldots\ldots\ldots \quad (5.2)$$

where $[T]$ is the transformation matrix. We call the crack strain defined by Eq. (5.1) crack strain type-1.

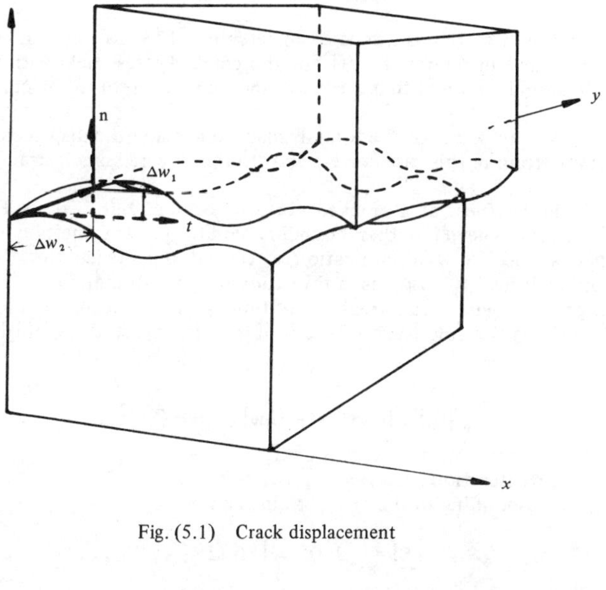

Fig. (5.1) Crack displacement

When crack occurrence is uniform and there exists average crack spacing then the average crack strain is given, with crack spacing l, as

$$\{\epsilon_{cr}\}_{ave} = \frac{1}{l}\int_l \{\epsilon_{cr}\}dx = \frac{1}{l}\{\Delta w\} \quad \text{(5.3)}$$

The crack strain defined by Eq. (5.3) is a smeared crack strain and called crack strain type-2. The concept of crack strain is explained in the following simple derivation as well. When a crack is formed, the stiffness at the cracked portion, ΔV, is reduced and becomes zero with increasing crack width. Then, using FEM formulation, the stiffness reduction is modelled as,

$$\begin{aligned}[K - \Delta K] &= \int_{V-\Delta V} [B]^T [D] [B] dV + \int_{\Delta V} [B]^T ([D] - [\Delta D]) B dV \\ &= \int_V [B]^T [D] [B] dV - \int_{\Delta V} [B]^T [D] [D]^{-1} [\Delta D] [B] dV \end{aligned} \quad \text{(5.4)}$$

Multiplication Eq. (5.4) by displacement yields

$$[\Delta K]\{u\} = \{\Delta F\} = \int_{\Delta V} [B]^T [D] [D]^{-1} [\Delta D] [B] \{u\} dV \quad \text{(5.5)}$$

If we replace $\{\epsilon_{cr}\}$ by

$$\{\epsilon_{cr}\} = [D^{-1}] [\Delta D] [B] \{u\} \quad \text{(5.6)}$$

then,

$$\{\Delta F\} = \int_{\Delta V} B^T \mathrm{D} \{\epsilon_{cr}\} dV \quad \text{(5.7)}$$

Hence, $\{\epsilon_{cr}\}$ is regarded as a crack strain which is treated as initial strain. $[B]\{u\}$ being $\{\epsilon_t\}$, Eq. (5.6) is rewritten as

$$\{\epsilon_{cr}\} = [\mathfrak{D}] \{\epsilon_t\} \quad \text{(5.8)}$$

Evaluation of the $\mathfrak{D}$ matrix is discussed in reference [13] for the cracked stress field with reinforcement and in reference [14] for the cracked stress field with shear transfer and dilatancy. It should be noted that $\mathfrak{D}$ constitutes damage tensor in the cracked zone.

When $\mathfrak{D} \neq \mathrm{I}$, we can solve for the total displacements, and the displacements are continuous. The crack strain in this case corresponds to the smeared crack strain, i.e., type-2.

When $\mathfrak{D} = \mathrm{I}$, the stiffness at a cracked portion becomes zero. To solve for displacements in this case, we need another boundary condition. The method to satisfy the condition depends upon how discretization is carried out. When the displacement is expressed by an analytical function as in this study without discretizing in that direction, we need an expression which can treat discontinuity of displacement. In this case, the crack strain type-1 plays a role to solve the problem. In this study, from Eq. (5.1.c) we assume

$$\{\epsilon_{cr}\} = \{\Delta w\} \delta(x) = \{\Delta w\} \frac{1}{\sqrt{\pi\lambda}} e^{-\frac{x^2}{\lambda}} \quad \text{(5.9)}$$

$\{\Delta w\}$ is determined by iteration such that $\sigma_{at\ crack} = 0$.
The load vector corresponding to the crack strain is given as

$$\{F\} = \int_{\Delta V} B^T D \{\Delta w\} \delta(x) dV \quad \text{(5.10)}$$

and crack width as

$$\{\Delta w\} = \int_l \{\epsilon_{cr}\} dx = \int_l \{\Delta w\} \delta(x) dx \qquad (5.11)$$

For practical application, the shape of the delta function which is represented by a limited number of Fourier terms can not be too steep. Hence, depending upon the shape of the delta function, we need some practical consideration such as the boundary location where the boundary condition is satisfied. We show a numerical example in next section.

6. Numerical Example

The concrete wall is 30 cm wide, 100 cm tall and 1500 cm long and assumed to be constructed on top of a concrete block 120 cm wide, 150 cm tall and 1500 cm long. The cement content is taken as 300 kg/m³, and the steel reinforcement ratio is taken as 0.25 percent of wall sectional area in the longitudinal direction. This is shown in Fig. (6.1).

Accumulated heat generation of concrete is taken as $S(t) = 2.82 \times 10^4 \ (1 - e^{-0.5t})$ Kcal/m³. Heat conduction coefficient and specific heat are taken as 2.0 Kcal/(mhr°C) and 0.286°C/gr, respectively. The heat conduction coefficient of the formwork panel is assumed as h = 0.67 Kcal/(m²hr°C), and when the surface of concrete is exposed to outer air at an age of 2 days, h is taken as 10.0 Kcal/(m²hr°C). The temperature of outer air is assumed to be 18°C to 10°C depending on the concrete age which is shown in Fig. (6.2).

The development of concrete Young's modulus is assumed to be varying from zero to 0.22×10^6 Kgf/cm² (0.22×10^5 Mpa) as shown in Fig. (6.3). Poisson's ratio is assumed to be 0.17 constant, though no creep strain is considered in this case. The kinematic boundary condition is that the vertical displacement of the base surface of the restraining block is restrained.

The temperature computed for the vertical center line at the span center section is shown in Fig. (6.4). The sudden drop of temperature at 2 days is due to the removal of

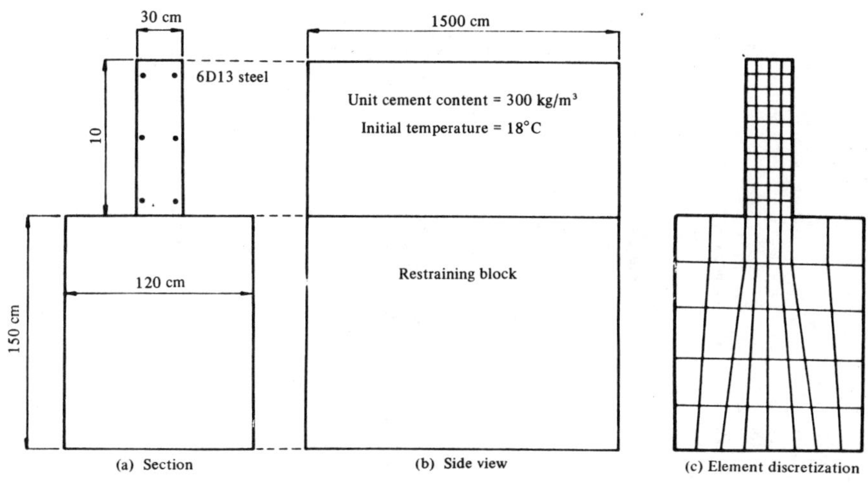

Fig. (6.1) Specimen for numerical example

formwork from the wall. The stabilized temperature distribution is attained within 2 weeks.

Displacements at several time steps are shown in Fig. (6.5). The maximum horizontal expansion is about 0.48 mm and the maximum contraction is about 5.0 mm. Maximum vertical heaving up is about 0.5 mm. It is obvious that newly placed concrete undergoes axial as well as bending deformations, and the effect of the pre-existing block is to restrain those two kinds of deformation.

At the age of 3 days, a crack occurred at the center span, and at the age of 5 days additional cracks appeared on both sides of the center crack, separating the total span into almost 4 quarter spans.

Though the shrinkage effect is neglected in this calculation, its effect on the crack occurrence is considered to be negligibly small in this case as the crack occurs at the age of 3 days and 5 days when the elapsed time after the removal of the formwork is only a day or three.

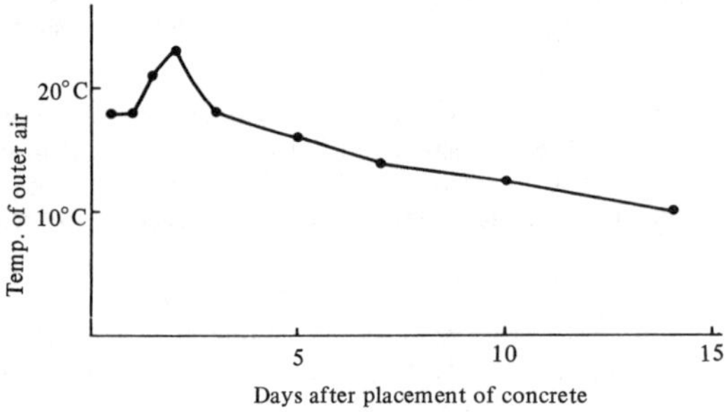

Fig. (6.2) Temperature variation of outer air

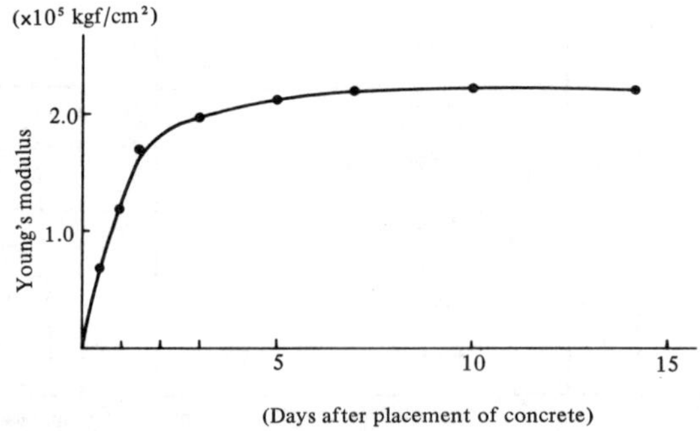

Fig. (6.3) Development of Young's modulus of concrete

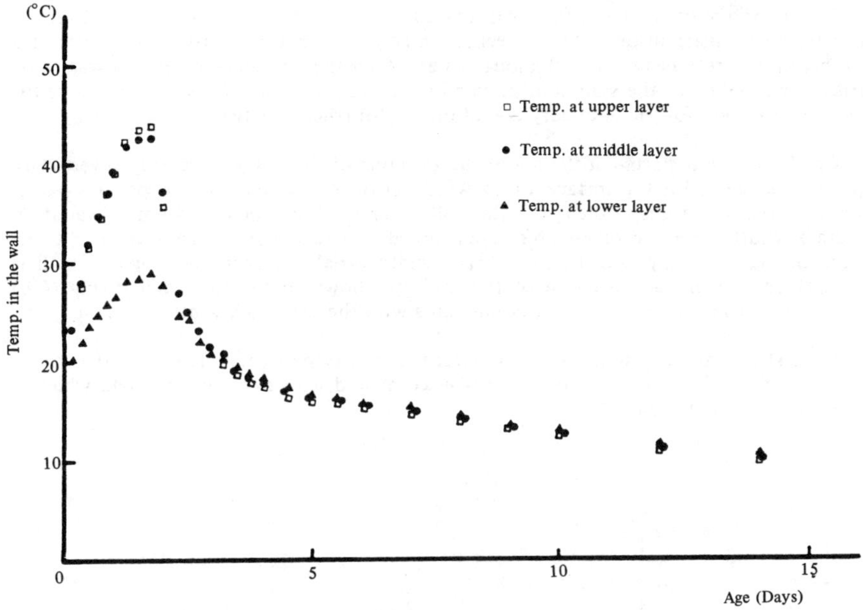

Fig. (6.4) Temp. fluctuation within the wall

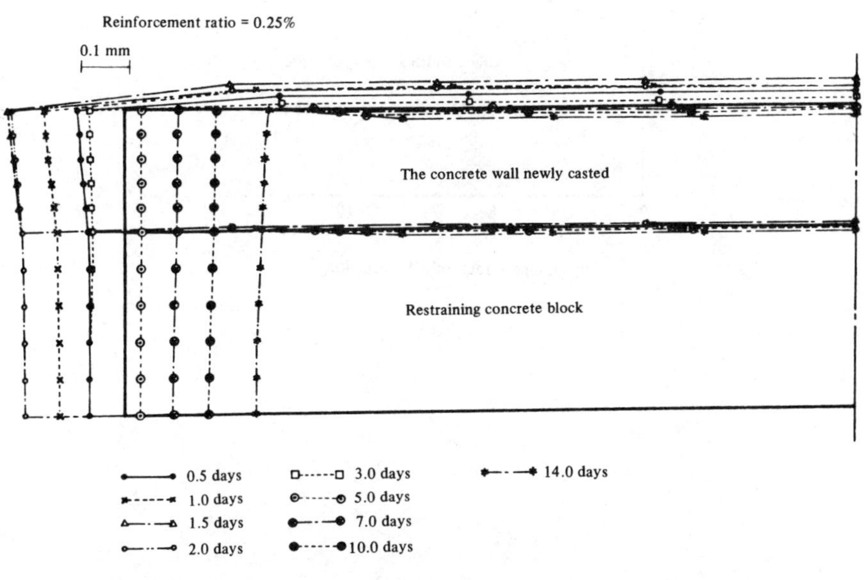

Fig. (6.5) Displacement of the wall at several time steps

Crack initiation was determined by comparing the average stress magnitude in an element section with the tensile strength of concrete at each age. The factor λ which decides the shape of the delta function was taken in this study as $\lambda = 0.9 \times 10^3$ in reference to strain distribution data of steel strain across a crack by axial tension tests of a reinforced concrete beam. And the location at which the boundary condition was to be satisfied was taken at the symmetrical axis of the delta function. However, the value for λ and the location for the boundary condition need further investigation.

Crack initiation started at the middle height layer of the wall and developed vertically up and reached to the top surface. Crack widths at the center and quarter span are shown in Fig. (6.6) as functions of the age of the wall concrete. There seems to be small variation of crack width with age of concrete even though the temperature drops about 5°C between the age of 3 days and 15 days. Crack width variations along the height of wall at the surface and along the width of the wall are shown in Fig. (6.7) and in Fig. (6.8) respectively. These distributions are comparable with the real crack width variation.

Steel strains are shown in Fig. (6.9). After the initiation of concrete cracks, the strains increase rapidly. All these figures are to be compared with experimental data which is now under consideration.

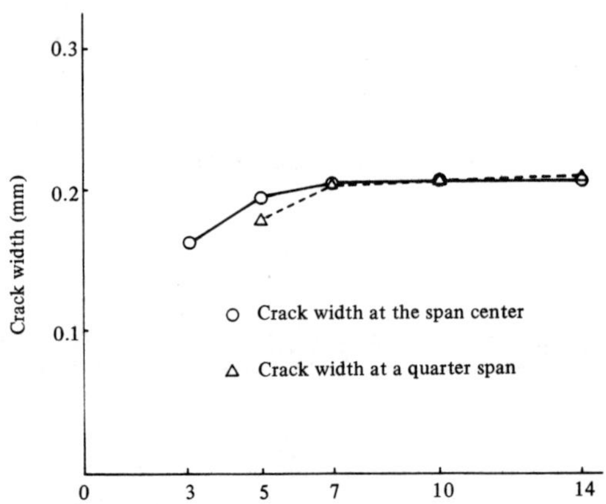

Fig. (6.6) Crack width variation

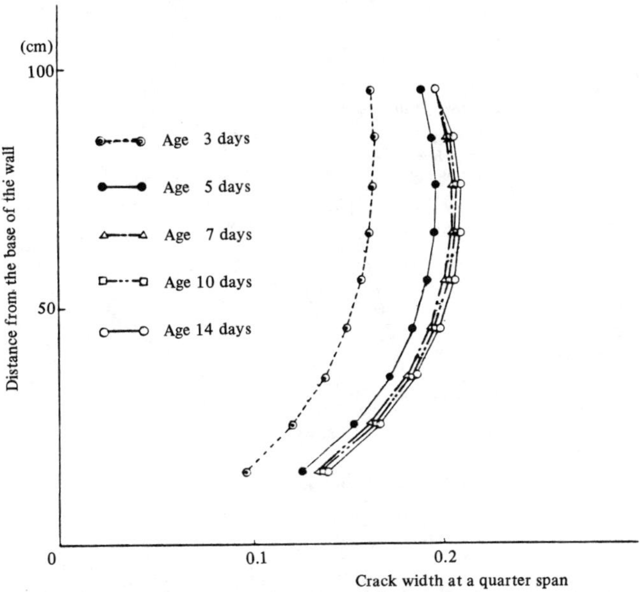

Fig. (6.7) Crack width variation along the height of the wall

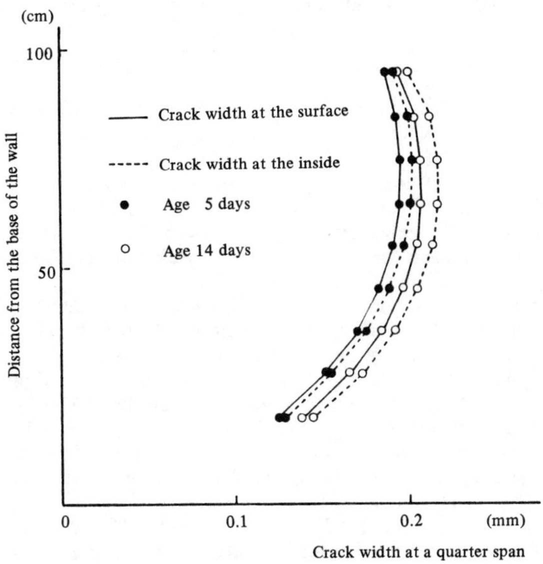

Fig. (6.8) Crack width variation along the width

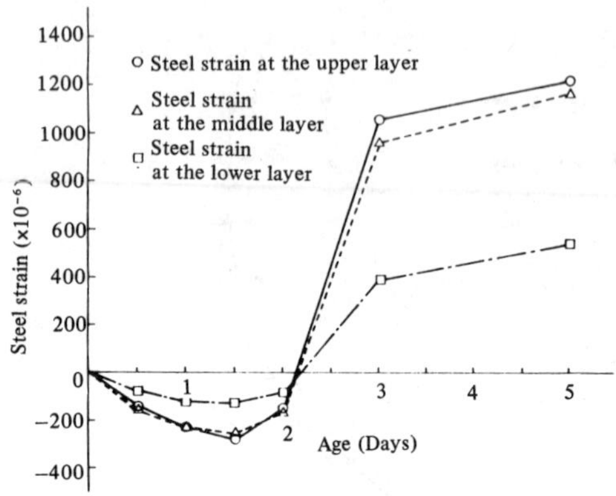

Fig. (6.9) Steel strain fluctuation

7. Conclusion

Formulation for three dimensional thermal stress analysis by the finite prism method was carried out. A creep law for three dimensional stress fields was developed from the rheology model of cement hydration process. Two kinds of crack strain concept were discussed to model the thermal cracking of wall type structures. A numerical example of crack width analysis was shown, which is still to be compared with experimental data.

References

[1] Z.P.Bazant, "Thermodynamics of Interacting Continua with Surfaces and Creep Analysis of Concrete Structures", Nuclear Engineering and Design, Vol.20, No.2, 1972, pp.477–505

[2] Z.P.Bazant, T.Tsubaki and Z.Celep, "Singular History Integral for Creep Rate of Concrete", Journal of Engineering Mechanics, ASCE, June 1983, pp.866–884

[3] A.Haraguchi, T.Tanabe, T.Okazawa and H.Abe, "Thermal Stress Analysis of the Base Mat Concrete at Ikata Nuclear Power Station", Central Research Institute of Electric Power Industry Report, No.74558, June, 1975

[4] T.Hiraga and K.Kurabayashi, "Review of Research in Regard to Adiabatic Temperature Rise of Concrete", Proc. of JCI Colloquium on the Mechanism of Thermal Stress Generation in Massive Concrete Structures, Sept., 1982, pp.1–17

[5] T.Hiraga, K.Kurabayashi, T.Itatani and T.Tanabe, "Thermal Stress in Concrete Structure Cooled by Pipeline", Proc. of JCI 2nd Colloquium on the Mechanism of Thermal Stress Generation in Massive Concrete Structures", March, 1984, pp.35–40

[6] JCI Committee on Thermal Stress in Massive Concrete Structures, "The State of the Art Report on Thermal Stress Evaluation in Massive Concrete Structures, JCI Journal Vol.21, No.8, Aug., 1983, pp.4–16

[7] M.Kawasumi, S.Seki, K.Kasahara and T.Kuriyama, "Creep of Concrete in the Light of Hydration of Cement and Viscosity of Internal Water", Concrete Library of JSCE, No.3, 1984, pp.85–108

[8] F.K.G.Odqvist and J.Hult, "Kriechfestigkeit metallischer Werkstoffe", Springer-Verlag, 1962

[9] T.C.Powers, "Some Observatins on the Interpretation of Creep Data", RILEM Bull. No.33, 1966, pp.381–391

[10] S.Seki and M.Kawasumi, "Creep of Concrete at Elevated Temperatures", Concrete for Nuclear Reactors, ACI Special Publication SP-3-4, Vol.1, 1972, pp.591–638

[11] T.Tanabe, H.Yamakawa and A.Watanabe, "Determination of Convection Coefficient at Cooling Pipe Surface and Analysis of Cooling Effect", Proc. of JSCE, No.344, March, April, 1983, pp.171–179

[12] T.Tanabe, A.Haraguchi and M.Ishikawa, "Restraining Effect of Rock Foundation or a Lower Layer of Concrete on the Thermal Behavior of Newly Placed Massive Concrete", Trans. of JCI, 1983, pp.111–118

[13] T.Tanabe and H.Yoshikawa, "A Finite Element Model for Cracked Reinforced Concrete Structures introducing Crack Strain Concept", Proc. of JCI Colloquium on Finite Element Analysis of RC Structures, December, 1983, pp.65–72

[14] T.Tanabe, M.Iwata and H.Yoshikawa, "Introduction of Crack Strain to the Analysis of a Discontinuous Body", Proc. of 5th Int. Conference on Numerical Methods in Geomechanics, Nagoya, Japan, April, 1984

Behavior Prediction of Confined Concrete Under Cyclic Loadings

Oral Buyukozturk[1] and En-Sheng Chen[2]

A rate-independent constitutive model is proposed for the behavior of concrete in multiaxial cyclic loading. The material composite is assumed to experience a continuous damage process under load histories. The model adopts a damage-dependent bounding surface in stress space to predict the strength and deformation characteristics of the gross material under general loading paths. Reduction in size of the bounding surface as damage accumulates, and the adopted functional dependence of the material moduli on stress and damage permit a realistic modeling of the concrete behavior.

Satisfactory prediction is obtained of the generally nonlinear stress-strain response, degradation in stiffness during load cycles, shear compaction-dilatancy phenomena, and post-failure strain softening behavior. Finite element implementation of the proposed model is feasible and computationally efficient.

INTRODUCTION

Confined concrete in compression is known to manifest higher strength and ductility than concrete in uniaxial compression. In structural members, confinement of concrete may be present through closely spaced transverse reinforcement and from adjoining concrete elements. Therefore in critically stressed areas such as the bases of concrete columns and bridge piers, beam-column joints and hinges, substantial lateral confinement is always provided. Also, behavior of confined concrete under cyclic loads is important in the design for dynamic effects on structures due to earthquakes, winds and waves.

Safety and economic reasons dictate that concrete structures such as nuclear containment pressure vessels and offshore structures be accurately analyzed considering the full range of nonlinear behavior for cyclic loadings. Application of such analyses are often limited by the inadequacy of material models in representing the behavior of multiaxially loaded concrete subjected to load reversals and low-cycle fatigue.

[1]Professor of Civil Engineering, Massachusetts Institute of Technology, Cambridge, MAssachusetts 02139.
[2]Engineer, Brian Watt Associates, Houston, Texas 77032

Numerous efforts have been devoted to the development of analytical models for predicting the short term, rate-independent stress-strain behavior of concrete [1,2,3,6,7,9,10,12,16,17,19,21,22,23,27,28,29,33,34,37,38,41]. However, only a few of these models are capable, with limited success, to describe the complicated multiaxial-cyclic behavior of concrete. In a crude approximation the nonlinear incrementally orthotropic models have been extended to represent cyclic effects [16,17]. However, predictions based on these models are generally limited to the cases where material is stressed proportionally [4].

The development of the plasticity-type models reaches greater generality. Some of the previously proposed models were based on classical plasticity with the Drucker's stability postulate and the associated flow rules [e.g. 6,37]. Models using nonassociated flow rules have also been introduced [e.g. 7]. These models accurately predict inelastic shear compaction-dilatancy behavior but generally, with these predictions, cyclic behavior, such as damage accumulation (fracturing), and degradation in stiffness, have not been well represented.

More physical evidence need to be obtained and incorporated in the development of a realistic material model for concrete. This would include stiffness degradation during progressive damage process, energy dissipation during load cycling, shear compaction-dilatancy phenomena, and post-failure (strain softening) behavior. In viewing these needs some material models involving plasticity theories and damage concepts (fracturing model) have been developed. These include the endochronic model [1,3], plastic-fracturing model [2], elastic progressive fracturing model [18,19,20,29] and bounding surface model of the softening type [22]. Models based on endochronic-fracturing and plastic-fracturing theories are quite powerful, and complex. Elastic-progressive fracturing models are of simpler form. With this model the concept of "fracture surface" in strain space is introduced, which is analogous to the yield surface in hardening plasticity. In some cases many fracture surfaces may be involved which would make the determination of the model parameters difficult and the model computationally infeasible. Moreover, the energy dissipation during load cycles may not be well represented.

The recently developed "Bounding Surface" model is attractive. This model was first developed to characterize hysteretic material behavior of metals [13,14,15,31] and soils [35,36]. Application of this model to concrete has been recently attempted [22]. The model reported in Ref. 22 was shown satisfactory in predicting stress-strain response of concrete under monotonic multiaxial and cyclic uniaxial loads. However, the application of this model to more general loading situations, such as the multiaxial cyclic and nonproportional loadings, is limited.

The model's assumption of a purely plastic behavior further limits its capability. Moreover, this assumption, when applied with the assumption that the direction of the plastic strain increment is along the direction of the deviatoric stress, as proposed in Ref. 22 leads to a singular flexibility matrix. Since the concrete deforms almost linear-elastically at low stress levels, implementation of the elastic strain component in the model is necessary, which also eliminates the problem of matrix singularity.

In this paper a constitutive model for concrete is proposed, which uses the concept of bounding surface further developed to allow for realistic predictions of the behavior of concrete in multiaxial cyclic compressive loadings. The important features of this proposed model are: (1) it adopts a damage concept to characterize progressive stiffness degradation and to capture nonlinear behavior, (2) it incorporates material parameters consistent with the phenomenological behavior of concrete.

In the following sections, fundamental characteristics of concrete behavior are briefly reviewed, and essential elements of the proposed model are described. Finally, simulation analyses are performed and the results are compared with the experiments.

GENERAL DESCRIPTION OF CONCRETE BEHAVIOR

Concrete contains numerous microcracks. The progression of these microcracks with the application of the externally applied loads contributes to the generally observed nonlinear stress-strain behavior and the plastic deformation of concrete. Moreover, cracking, frictional interlocking over the crack interface, and local crushing result in further behavioral complexities, such as dilatancy and compaction during loading and unloading, and stiffening effects during the beginning of unloading cycles. These phenomena which are caused by the existence of microcracks and their propagation under loads are herein termed as "damage" to the concrete . This physical damage process of concrete has to be recognized in constructing a realistic constitutive model [19,20,30,40]. Acoustic emission measurements [40] show that this damage process in concrete under uniaxial compressive loading is continuous, and begins at very low levels of applied strains. The magnitude of the damage increases as strain increases. During subsequent unloading and reloading significant additional damage is observed only when maximum axial strain ever experienced is exceeded. This suggests the use of a strain concept to evaluate the damage in the gross material. It is feasible that in multiaxial loading cases the damage measure be based on a first or higher order strain vector. [25,30].

The stress combinations corresponding to the ultimate strength of concrete under different monotonic loadings are usually represented by the so called "failure surface" defined in the stress space. Various functional forms have been proposed for the failure surface (see e.g. Ref. 11). The failure surface shrinks in size consistently with the falling branch in stress-strain curves, depending on the maximum damage (or maximum strain quantitites) experienced in the material. The tests [8,26,40] indicate that for a given damage level failure stresses for monotonically and cyclically loaded specimens approximately coincide. This suggests the existence of a unique innermost envelope in stress space, corresponding to a certain damage level, which encloses all the possible stress points and shrinks in size as the damage accumulates. This envelope called "bounding surface" in the following context plays an important role in characterizing material properties.

A realistic representation of the stress-strain behavior of concrete may be achieved by decomposing stress and strain into deviatoric and hydrostatic components [9,23,28,29]. Thus, the constitutive behavior can be formulated on the deviatoric plane and along the hydroaxis, i.e., a three dimensional representation is possible in the form of octahedral shear, and volumetric stresses and strains, respectively. Test results [9,38] show that (1) purely increased volumetric stress (strain) does not cause any change in the octahedral shear strain except when the volumetric and octahedral stresses are very high and (2) the volumetric strain is influenced not only by the change in volumetric stress but also by the change in octahedral shear stress and strain. This coupling phenomenon between volumetric and octahedral shear components is generally interpreted as shear compaction and dilatancy effect. For a fixed volumetric stress, Ref. 38 shows that the proportional loading and unloading on the deviatoric plane changes the volumetric strain, in the sense that the deviatoric loading results in both dilatancy and compaction, and unloading and reloading up to a level of the previous maximum deviatoric stress results in only dilatancy.

In the development that follows the stress σ_{ij} $(i,j = 1,2,3)$, and the strain ε_{ij} are normalized with respect to the peak stress, f' and the associated strain in uniaxial loading curve, ε_p, respectively. Positive sign is assigned to compressive stress, and to strain if it represents contraction.

DAMAGE PARAMETER AND BOUNDING SURFACE

Concrete is subjected to continuous damage during the loading process due to microcracking, fracturing, crushing and other interaction between its constituents. Ultrasonic measurements on uniaxial cyclically loaded specimens [40] suggest the use of a strain vector to evaluate the accumulated damage in concrete. An overall assessment of the damage, as postulated in the present work, would be based on the evaluation of the plastic volumetric strain ε_v^p and the plastic octahedral shear strain γ_o. The coupling of these two effects is achieved through a shear compaction-dilatancy factor.

Representation of the damage along hydroaxis is relatively trivial since in this case, the damage can be measured primarily by a scalar quantity and can be taken as a function of either ε_v or I_1. This can be implicitly included in the pure volumetric stress-strain formulation (no volumetric-deviatoric coupling) and explicitly expressed by the degradation of the tangent bulk modulus during loading. On the deviatoric plane the damage accumulation is hereby proposed to be evaluated by the use of a "damage parameter", "K", which is related to γ_o^p. This damage accumulation, as defined on the deviatoric plane is sensitive to the hydrostatic-pressure and is stress path dependent. Under higher confining stress states, the same increase in γ_o^p is expected to produce less damage on the deviatoric plane than that under lower confining stress. Hence, it seems that a realistic modeling of this damage accumulation under complex stress paths can be achieved by defining K in an incremental form as

$$dK = \frac{d\gamma_o^p}{F_1(I_1,\theta)} \tag{1}$$

and

$$K = \int_{\text{loading history}} dK \tag{2}$$

The function F_1 (I_1,θ) is chosen to be proportional to the plastic octahedral shear strain at failure under monotonic loadings corresponding to different I_1 , and θ. The proportionality factor is determined in such a way that K reaches a certain value, say 1, at failure under monotonic loading. Using the test data from Refs. 8,26,27,28, and 32, the following regression forms are proposed:

(1) For deviatoric loading

$$F_1 = \begin{cases} 0.23\ (I_1 + 0.3)^2/F_2 & \text{for } I_1 \leq 3.18 \\ 1.60\ (I_1 - 1.44)/F_2 & \text{for } I_1 > 3.18 \end{cases} \tag{3}$$

and

(2) For deviatoric unloading

$$F_1 = 1.4\frac{F_{1,\max}}{F_2}\left[0.85 - \frac{I_1 + 0.3}{I_{1,\max} + 0.3}\right] \tag{4}$$

where

$$F_2 = (12 + 11\cos 3\theta)^{\frac{1}{6}}$$

$I_{1,max}$ = maximum I_1 before recent unloading

$F_{1,max}$ = maximum F_1 before recent unloading

Definitions of loading and unloading are given at the end of this section.

The bounding surface, as defined, is the innermost locus of stress points which always encloses the current stress point, and is a function of stress invariants and damage parameter (see Fig. 1). This surface is taken to be unique and shared by both monotonic and cyclic loadings. In the present study the bounding surface, F, is proposed to be a function of σ_{ij} (or stress invariants) and K_{max}, the maximum value of K ever experienced by the material:

$$F(\sigma_{ij}, K_{max}) = 0 \tag{5}$$

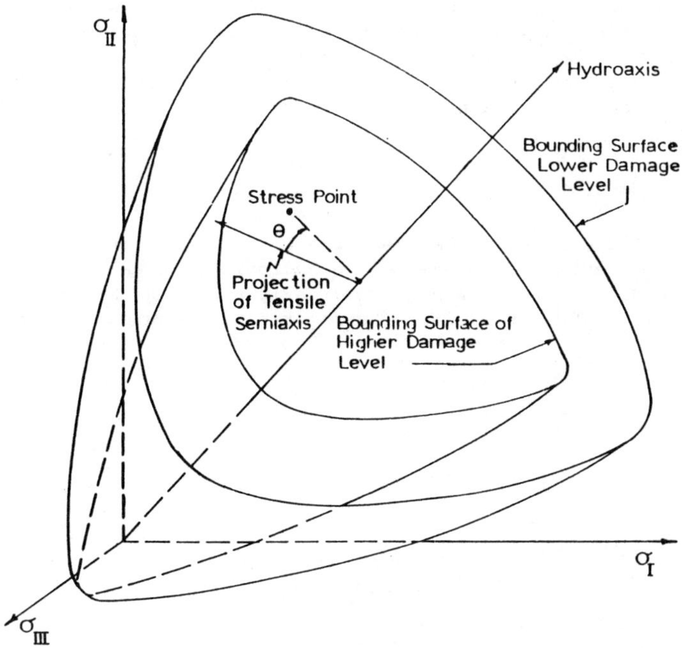

Figure 1. The Bounding Surface

The following function, which is obtained by modifying the failure surface, proposed in Ref. 22 for monotonic loading, to include K_{max}, is chosen for F:

$$F(\sigma_{ij}, K_{max}) = \frac{1.85(\sqrt{J_2} + 0.378\ J_2)(12 + 11 \cos 3\theta)^{\frac{1}{6}}}{I_1 + 0.3} - \frac{40}{K_{max}^2 + 39} \tag{6}$$

As seen, for $K_{max} = 1$ the above equation reduces to the expression of the monotonic failure surface. This arrangement allows adoption of the formulation to different materials with different failure criteria, if desired.

The bounding surface defined in the above described manner will permit to (1) characterize the strength of the material at varying stress/strain states, and (2) evaluate the material moduli by measuring the distance from the present stress point to the bounding surface along a certain direction. This direction for a material with vanishing elastic region, as developed in Ref. 15 and applied in Ref. 22, is along the $d\sigma_{ij}$ direction. For a material with an assumed elastic region, this direction can be determined depending on the kinematic rule adopted, as discussed in Refs. 14 and 15. The present formulation adopts that the distance between the stress point and the bounding surface is measured along the S_{ij} direction. By this definition, the octahedral stress-strain behavior can be characterized. Moreover, since the bounding surface on the deviatoric plane is I_1 dependent, a normalized measure "D" is introduced for this purpose:

$$D = \frac{r}{R} \tag{7}$$

where r is the distance from the projection of the current stress point on the deviatoric plane to the hydroaxis, and R is the distance of the bounding surface from the hydroaxis along the S_{ij} direction (see Fig. 2). Thus, when D=1, the material is assumed to have failed.

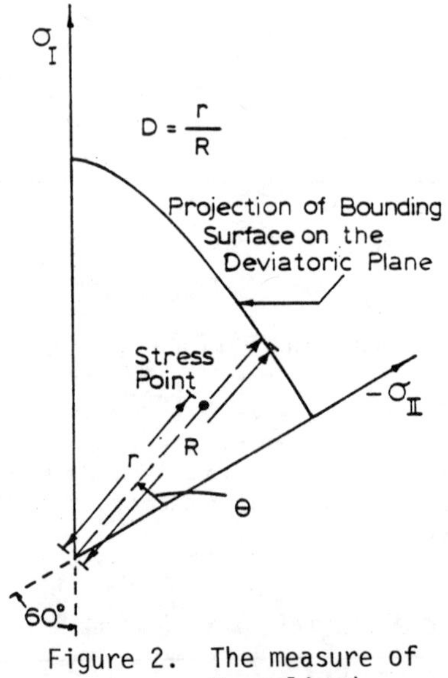

Figure 2. The measure of Normalized Distance D

The loading and unloading along the hydroaxis and on the deviatoric plane are defined as follows:

(1) Hydrostatic loading is defined as the process when $dI_1 \geq 0$, and unloading when $dI_1 < 0$.

(2) Deviatoric loading is defined as the process when $dD \geq 0$, and unloading when $dD < 0$.

Thus, any general stress path can be represented by the combination of these loading and unloading conditions.

CONSTITUTIVE MODEL

It is convenient to decompose strain increment $d\varepsilon_{ij}$ into its deviatoric and volumetric components:

$$d\varepsilon_{ij} = de_{ij} + \delta_{ij} \frac{d\varepsilon_{kk}}{3} \qquad (k = 1,2,3) \tag{8}$$

The deviatoric strain increment can be further decomposed into elastic and plastic components, de_{ij}^{e} and de_{ij}^{P}

$$de_{ij} = de_{ij}^{e} + de_{ij}^{P} \tag{9}$$

The elastic deviatoric strain increment de_{ij}^{e} can be related to the stress increment following Hook's Law

$$de_{ij}^{e} = \frac{1}{H^{e}} \, dS_{ij}$$

where H^{e} is the generalized elastic shear modulus and dS_{ij} is the deviatoric stress increment. The plastic deviatoric strain increment de_{ij}^{P} is assumed to be independent of any volumetric change, and that the projection of de_{ij}^{P} on the deviatoric plane is assumed directly along the projection of the position vector of the stress point. In other words, de_{ij}^{P} is proportional to S_{ij}. This proportionality yields

$$\frac{de_{ij}^{P}}{S_{ij}} = \frac{d\gamma_{o}^{P}}{\tau_{o}} \tag{10}$$

and, assuming incremental linearity, one can write

$$d\gamma_{o}^{P} = \frac{d\tau_{o}}{H^{P}} \tag{11}$$

where the generalized plastic shear modulus H^P, depends on the history of the stress and strain.

The effects of the incremental volumetric stress dI_1 and the incremental octahedral shear stress $d\tau_o$ on the volumetric strain increment $d\varepsilon_{kk}$ are assumed decoupled. The portion of $d\varepsilon_{kk}$ caused by dI_1, $d\varepsilon_{kk,o}$, is calculated as

$$d\varepsilon_{kk,o} = \frac{dI_1}{3K_t} \tag{12}$$

where tangent bulk modulus K_t is assumed to be a function of I_1. The remaining portion of $d\varepsilon_{kk}$, $d\varepsilon_{kk,d}$, is directly associated with the plastic octahedral shear strain increment, $d\gamma_o^P$, by the linear relationship

$$d\varepsilon_{kk,d} = \beta\, d\gamma_o^P \tag{13}$$

where shear compaction-dilatancy factor β is also a function of the stress and strain, and

$$d\varepsilon_{kk} = d\varepsilon_{kk,o} + d\varepsilon_{kk,d} \tag{14}$$

Combining eqs. 8 to 14, and by expressing $d\tau_o = \partial\tau_o/\partial\sigma_{km}\, d\sigma_{km} = S_{km}\, d\sigma_{km}/3\tau_o$ $(k,m = 1,2,3)$, one can obtain the following relationship between $d\varepsilon_{ij}$ and and $d\sigma_{ij}$:

$$d\varepsilon_{ij} = \frac{d\sigma_{ij}}{H^e} + \frac{1}{3H^P\tau_o}\left(\frac{S_{ij}}{\tau_o} + \delta_{ij}\frac{\beta}{3}\right) S_{km}\, d\sigma_{km} + \delta_{ij}\left(\frac{1}{9K_t} - \frac{1}{3H^e}\right) d\sigma_{kk} \qquad (k,m = 1,2,3) \tag{15}$$

DETERMINATION OF THE MODEL PARAMETERS

The parameters of the proposed model are H^e, H^p, K_t, and β which can be determined by fitting the available experimental data. The expressions given below, are based on the data drawn from Refs. 8,9,24,26,27,28,32, and 38, and are representative of the average behavior of the concrete material. The data includes uniaxial and biaxial (monotonic and cyclic), and triaxial (monotonic) loading tests. It is noted that the expressions proposed herein are subject to further calibration as more experimental data become available for general stress conditions.

The results are summarized as follows:

(1) Generalized elastic shear modulus H^e: This is simply taken as the initial shear modulus at the beginning of the loading process

and is determined to be

$$H^{e} = 2.0 \tag{16}$$

This value is used for both deviatoric loading and unloading processes.

(2) Generalized plastic shear modulus H^P: Usually a unique relationship between $d\tau_o$ and $d\gamma_o^p$ is proposed in constitutive models [9,27,28,29]. However, when the material is close to failure, for different stress paths involving large deformations this uniqueness does not hold (see e.g. Ref. 9). Instead, based on this observation, adoption of a unique relationship between dD and dK is hereby proposed:

$$dK = \frac{dD}{H^*}$$

(17)

and

$$H^P = \frac{R}{F_1} H^*$$

For deviatoric loading and unloading H* is determined as follows:

For loading:

$$H^* = \frac{2.4\ (1-D)^{0.65D^2}}{(1 + 0.7\ K_{max}^2)\ A_L} \tag{18}$$

where

$$A_L = \begin{cases} 1.02-0.81\ \dfrac{K_R}{K_{max}} & \text{if } K < K_{max} \\ 1 & \text{if } K = K_{max} \end{cases} \tag{19}$$

and K_R is the associated value of "K" at the beginning of recent loading process.

For unloading:

$$H^* = \frac{2.4}{(1 + 0.7\ K_{max}^2)\ A_u}$$

where

$$A_u = \begin{cases} 0 & \text{if } K < 0.2\ K_{max} \\ \dfrac{K_u - 0.2\ K_{Max}}{0.8\ K_{max}} & \text{if } K \geq 0.2\ K_{max} \end{cases}$$

and K_u is the associated value of "K" at the beginning of recent unloading process.

Note that H^p is negative during the initial stage of the unloading process (see Eqs. 17 and 4) which results from the interlocking behavior during this stage; more physical evidence is needed to justify this behavior.

For the softening branch, H^p can be calculated based on the assumption that, after initial failure, the stress point moves on the failure

surface, and that $dF = \frac{\partial F}{\partial \sigma_{ij}} d\sigma_{ij} + \frac{\partial F}{\partial K_{max}} dK_{max} = 0$. A general

formulation for H^p derived in this way is complex. For the present purpose, an approxi-mation from the simple uniaxial case will suffice. The following form for H^p found from the softening branch of the stress-strain curve is satisfactory:

$$H_p = -0.15\, e^{-0.025(K_{max}-1)^2} \tag{20}$$

This negative value of the plastic shear modulus, which reflects the softening behavior of the concrete after the initial failure, does not cause any instability problem in the cases studied in the next section. For more general and complex stress paths further experimentation should be considered.

(3) Bulk tangent modulus K_t: Volumetric stress-strain relationship of concrete is nonlinear [24,38]. Under compressive hydrostatic loading, the material is first characterized by a decrease in K_t (softening) with increased I_1. Then a stiffening behavior is obtained with increased I_1. For practical purposes the following model, which fits the softening range, is found to be generally acceptable.

$$K_t = \begin{cases} \dfrac{1.2}{1+0.358\, I_1^{1.5}} & \text{, for hydroaxis loading} \\[2ex] 1.2 & \text{, for hydroaxis unloading} \end{cases} \tag{21}$$

(4) Shear compaction-dilatancy factor β: The effect of the deviatoric deformation upon the volumetric strain has been observed. [e.g. 38]. The combined effects of shear compaction and dilatancy on the volumetric strain can be conveniently separated as:

(a) Compaction: This is primarily the result of shear crushing and void reduction by which the material is compacted during shear process. Experimental results [38] indicate that the shear compaction occurs continuously during monotonic deviatoric loading and not during successive unloading and reloading processes until further damage occurs. This suggests that the shear compaction occurs only when $K = K_{max}$.

(b) Dilatancy: shear dilatancy phenomena can be clearly observed during loading, unloading and reloading processes [32,38]. Data drawn from the unloading and the reloading processes is used first to fit a dilatancy factor, β_2, by which $d\varepsilon_{kk,d2} = \beta_2 d\gamma_o$. Combining this result with the monotonic deviatoric loading curve, one can determine a compaction factor β_1, by which $d\varepsilon_{kk,d1} = \beta_1 d\gamma_o^P$, and

$$d\varepsilon_{kk,d} = d\varepsilon_{kk,d1} + d\varepsilon_{kk,d2} = (\beta_1 + \beta_2)\ d\gamma_o^P = \beta\ d\gamma_o^P \qquad (22)$$

Consequently the following functions for β_1 and β_2 are obtained :

$$\beta_1 = \begin{cases} 1.1e^{-30(K_{max}-0.6)^2} & ,\ \text{for } K = K_{max} \\ 0 & ,\ \text{for } K < K_{max} \end{cases} \qquad (23)$$

and

$$\beta_2 = -\ 1.97 \cdot \lambda \cdot e^{-2\lambda^2}$$

where (24)

$$\lambda = D - 0.2\ K_{max}^2$$

The developed model with the defined parameters has been incorporated in a computer program to predict the multiaxial stress-strain response of the concrete under monotonic and cyclic loadings. Note that the flexibility matrix in eq. 15 is asymmetrical due to the fact that the volumetric strains are affected by the deviatoric strain component. Implementation of the model in finite element analysis is computationally efficient.

APPLICATION AND COMPARISON

In this section stress-strain responses predicted by the proposed model are verified for different loading conditions by comparing the predictions with the corresponding experimental results. In these comparisons all stress and strain quantities are normalized with respect to the peak stress, f' and the associated strain, ε_p, respectively, for uniaxial loading. In cases when these values were unknown, they were estimated from the available data in such a way that the initial monotonic stress-strain curve from the test results fit that from the model prediction.

Conventional monotonic loading case is first considered: (1) The biaxial test results obtained by Kupfer, et. al. [32] with $\Delta\sigma_1 = \Delta\sigma$, $\Delta\sigma_2 = \alpha\Delta\sigma$, and $\Delta\sigma_3 = 0$, for the stress ratio $\alpha = 0$, 0.52, and 1 are compared with the predictions in Fig. 3. Good agreement between the test results and the model prediction has been found. (2) Triaxial test results performed by Kotsovos and Newman [27] (see Fig. 4) in which the specimens were first loaded along the hydroaxis with

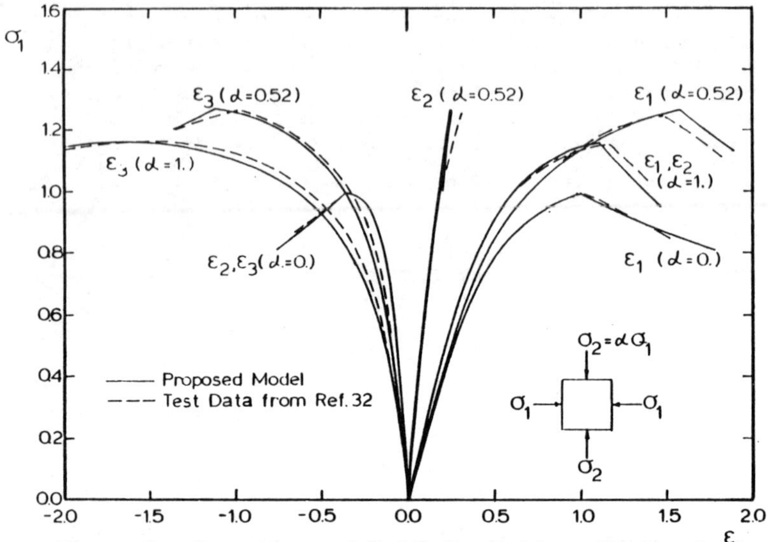

Figure 3. Comparison of Model Prediction with Monotonic Biaxial Compression Test Data

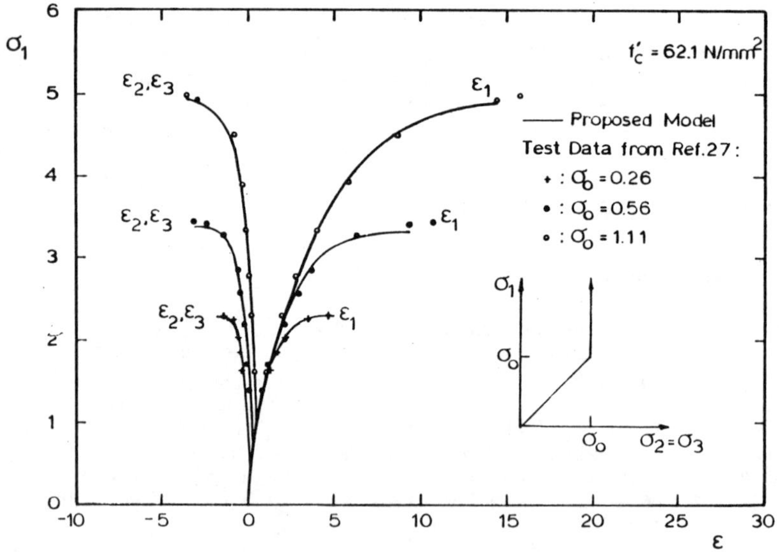

Figure 4. Comparison of Model Prediction with Monotonic Triaxial Compression Test Data (θ = 60°)

$\Delta\sigma_1 = \Delta\sigma_2 = \Delta\sigma_3 = \Delta\sigma_0$ and followed by uniaxial loading $\Delta\sigma_1 = \Delta\sigma$, and $\Delta\sigma_2 = \Delta\sigma_3 = 0$. The model prediction is satisfactory. (3) triaxial results performed by Schickert and Winkler [39] (Fig. 5) in which the specimens were first loaded along hydroaxis with $\Delta\sigma_1 = \Delta\sigma_2 = \Delta\sigma_3 = \Delta\sigma_o$, and then followed by a pure deviatoric loading including: (a) Path 1: $\Delta\sigma_1 = \Delta\sigma$, and $\Delta\sigma_2 = \Delta\sigma_3 = -1/2\Delta\sigma$, (b) Path 2: $\Delta\sigma_1 = \Delta\sigma$, $\Delta\sigma_2 = 0$, and $\Delta\sigma_3 = -\Delta\sigma$, (c) Path 3: $\Delta\sigma_1 = \Delta\sigma_2 = \Delta\sigma$, and $\Delta\sigma_3 = -2\Delta\sigma$. (all $\Delta\sigma > 0$). Generally good agreement between the test results and the prediction is found for the stress-strain behavior.

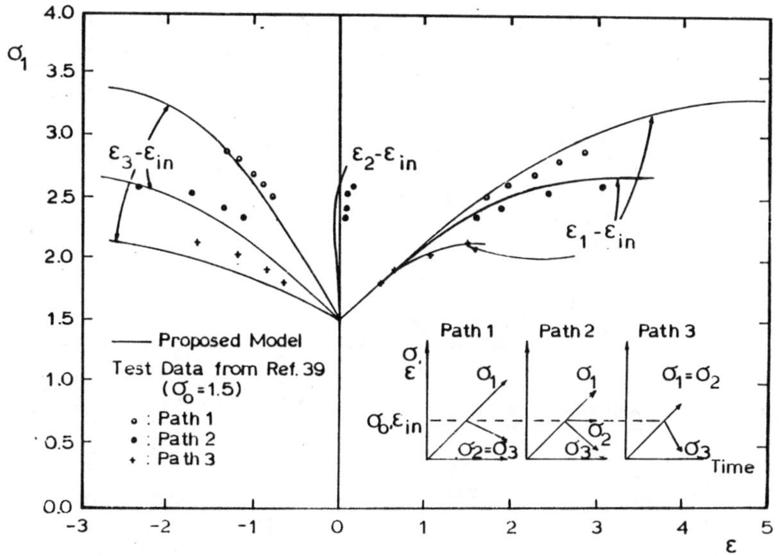

Figure 5. Comparison of Model Prediction with Monotonic Triaxial Compression Test Data (I_1 = Constant)

For cyclic behavior the model predictions are compared with (1) the uniaxial test by Karsan and Jirsa [26] (Fig. 6), (2) the biaxial tests by Beams, et.al [5] (Fig. 7) with constant ratio $\alpha = \sigma_2/\sigma_1 = 1/3$ and (3) the generalized plane strain biaxial test by Buyukozturk and Tseng [8] (Fig. 8) in which a predetermined value of ε_2 was first applied followed by the application of $\Delta\sigma_1 = \Delta\sigma$ with $\Delta\varepsilon_2 = 0$. As seen from Figs. 6, 7 and 8, the model predictions of the behavior of concrete in cyclic loading are satisfactory.

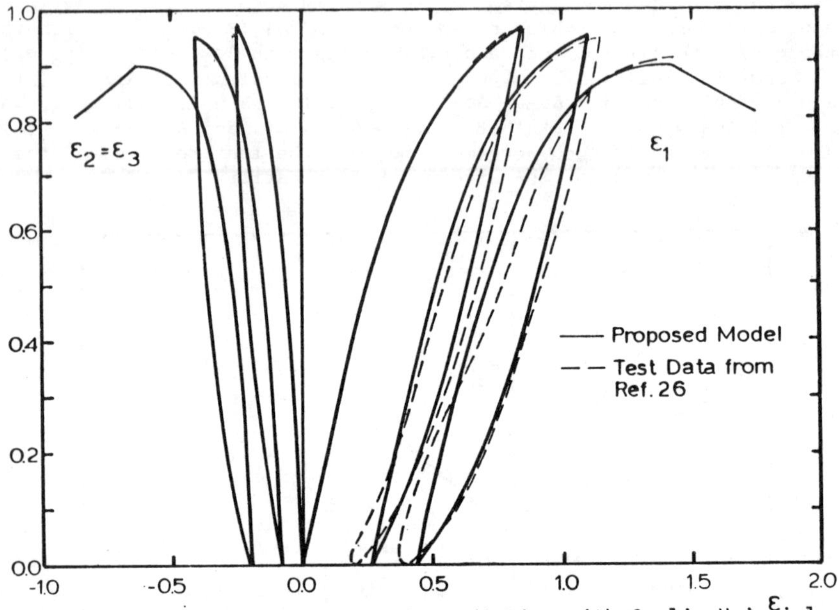

Figure 6. Comparison of Model Prediction with Cyclic Uniaxial Compression Test Data

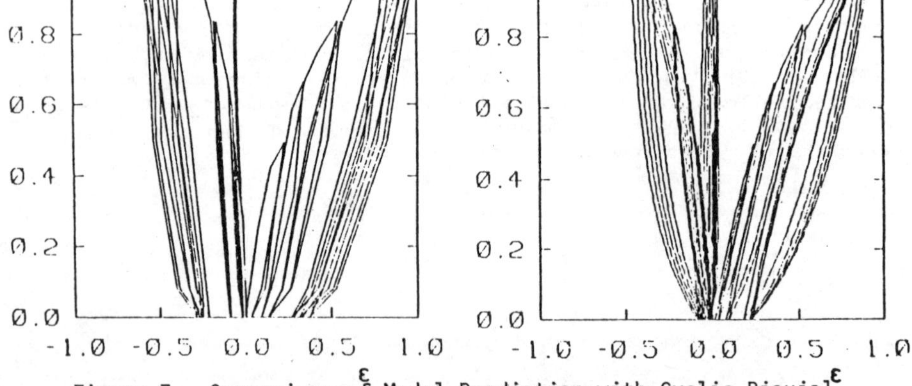

Figure 7. Comparison of Model Prediction with Cyclic Biaxial Compression Test Data: (a) Test Data From Ref. 5
(b) Model Prediction

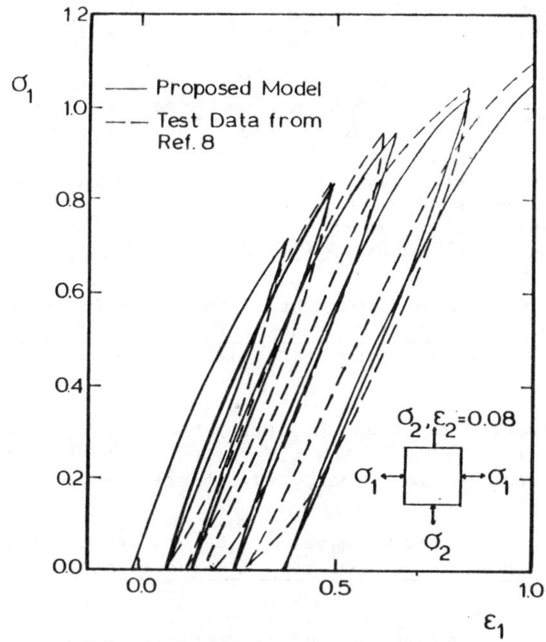

Figure 8. Comparison of Model Prediction with Cyclic Biaxial Compression Test Data (Plain Strain)

CONCLUSION

A rate-independent damage-type bounding surface model is presented for predicting the constitutive behavior of concrete in multiaxial cyclic stress. This model considers that the concrete experiences a continuous damage process during loadings and predicts behavioral characteristics such as stiffness degradation, generally nonlinear stress-strain response, plastic deformation, shear compaction and dilatancy. The reduction in size of the bounding surface with the increased damage level and the adopted functional dependence of the material moduli on stress and damage permit one to realistically model the complex behavior of concrete.

Functional forms of the model parameters are proposed based on the available test results for structural concrete. The model predictions have proven to be satisfactory when compared with the experimental data. Implementation of this model in generalized finite element analysis programs appears promising and is computationally efficient.

APPENDIX I - REFERENCES

1. Bazant,Z.P., and Bhat, P.D., "Endochronic Theory of Inelasticity and Failure of Concrete," Journal of the Engineering Mechanics Division, ASCE, Vol. 102, No. EM4, Proc. Paper 12360, Aug. 1976, pp. 701-721.

2. Bazant, Z.P., and Kim, S.S., "Plastic-Fracturing Theory of Concrete," Journal of the Engineering Mechanics Division, ASCE, Vol. 105, No. EM3, Proc. Paper 14653, June 1979, pp. 407-428.

3. Bazant, Z.P., and Shieh, C.L., "Hysteretic Fracturing Endochronic Theory for Concrete," Journal of the Engineering Mechanics Division, ASCE, Vol. 106, No. EM5, Proc. Paper 15781, Oct. 1980, pp.929-950.

4. Bazant, Z.P., "Comment on Orthotropic Models for Concrete and Geomaterials," Journal of the Engineering Mechanics Division, ASCE Vol. 109, No. EM3, Proc. Paper 18014, June 1983, pp. 849-865.

5. Beams, G.W., Gerstle, K.H., and Ko, H.Y., "Response of Concrete to Cyclic Biaxial Compressive Loads," Prepreint 81-015., ASCE National Convention, Las Vegas, Nev., April 1982.

6. Buyukozturk, O., "Nonlinear Analysis of Reinforced Concrete Structure," Journal of Computers and Structures, Vol. 7, Feb. 1977, pp.149-156.

7. Buyukozturk, O., and Tassoulas, J., "A Constitutive Model for Concrete in Compression," Proceedings of the Third ASCE Annual Engineering Mechanics Division Speciality Conference, Sept. 1979.

8. Buyukozturk, O. and Tseng, T. M. "Concrete in Biaxial Cylic Compression," Journal of the Structural Engineering, ASCE, Vol. 110, No. 3, March 1984, pp. 461-476.

9. Cedolin, L., Crutzen, Y.R.J. and Dei Poli, S., "Triaxial Stress-Strain Relationship for Concrete," Journal of the Engineering Mechanics Division, ASCE, Vol. 103, EM3, Proc. Paper 12969, June 1977, pp.423-439.

10. Chen, A.C.T. and Chen, W.F., "Constitutive Relations for Concrete," Journal of the Engineering Mechanics Division, ASCE, Vol. 101, No. EM4, Proc. Paper 11529, Aug. 1975, pp.465-481.

11. Chen, W.F., "Plasticity in Reinforced Concrete," McGraw-Hill Book Company, 1982

12. Coon, M.D. and Evans, R.J., "Incremental Constitutive Laws and Their Associated Failure Criteria with Application to Plain Concrete," International Journal of Solids and Structures, Vol. 8, 1972, pp. 1169-1183.

13. Dafalias, Y.F. and Popov, E.P., "A Model of Nonlinearly Hardening Materials for Complex Loading," Acta Mechanica, Vol. 21, 1975, pp. 173-192.

14. Dafalias, Y.F. and Popov, E.P., "Plastic Internal Variables Formalism of Cyclic Plasticity," Journal of Applied Mechanics, Transactions of ASME, Dec. 1976, pp. 645-651.

15. Dafalias, Y.F. and Popov, E.P., "Cyclic Loading for Materials with a Vanishing Elastic Region," Nuclear Engineering and Design, Vol. 41, 1977, pp. 293-302.

16. Darwin, D. and Pecknold, D.A., "Inelastic Model for Cyclic Biaxial Loading of Reinforced Concrete," Civil Engineering Studies SRS No. 409, University of Illinois of Urbana-Champaign, Urbana, Ill., July 1974.

17. Darwin, D and Pecknold, D.A., "Analysis of R/C Shear Panels under Cyclic Loading," Journal of the Structural Division, ASCE, Vol. 102, ST2, Proc. Paper 12839, April, 1976, pp.355-369.

18. Dougill, J.W., "On Stable Progressively Fracturing Solids," Zeitschrift fur Angewandte Mathematik und Physik, Vol. 27, 1976. pp. 423-437.

19. Dougill, J.W., Lau, J.C., and Burt N.J., "Towards a Theoretical Model for Progressive Failure and Softening in Rock, Concrete and Similar Materials," Mechanics in Engineering, Proc. ASCE-EMD. Conf. Univ. Waterloo Press, 1977, pp. 335-355.

20. Dougill, J.W., "The Response of Concrete to Short Term Loading," Design of Dams to Resist Earthquake, ICE, London, 1980.

21. Elwi, A.A., and Murray, D.W., "A 3-D Hypoelastic Concrete Constitutive Relationship," Journal of the Engineering Mechanics Division, ASCE, Vol. 105, NO. EM4, Proc. Paper 14734, Aug. 1979, pp. 623-641.

22. Fardis, M.N., Alibe, B., and Tassoulas, J.L., "Monotonic and Cyclic Constitutive Law for Concrete," Journal of the Engineering Mechanics Division, ASCE. Vol. 109 No. EM2 Proc. Paper 17871, April 1983. pp. 516-536.

23. Gerstle, K.H. et al., "Behavior of Concrete under Multiaxial Stress States," Journal of the Engineering Mechanics Division, ASCE, Vol 106, No. EM6, Proc. Paper 15945, Dec. 1980, pp.1383-1403.

24. Green, S.J., and Swanson, S.R., "Static Constitutive Relations for Concrete," Technical Report No. AFWL-TR-72-2, Terra-Tec Inc., Salt Lake City, Utah, April 1973.

25. Kachanov., M. "Continuum Model of Medium with Cracks," Journal of the Engineering Mechanics Division, ASCE, Vol 106, 1980, pp. 1039-1051

26. Karsan, I.D., and Jirsa, J.O., "Behavior of Concrete Under Compressive Loadings," Journal of the Structural Division, ASCE, Vol. 95, No. ST12, Proc. Paper 6935, Dec. 1969, pp. 2543-2563.

27. Kotsovos, M.D., and Newman, J.B., "A Mathematical Description of the Deformational Behaviour of Concrete under Complex Loading," Magazine of Concrete Research, Vol. 31, NO. 107, June 1979, pp. 77-90.

28. Kotsovos, M.D. and Newman, J.B., "Mathematical Description of Deformational Behavior of Concrete under Generalized Stress Beyond Ultimate Strength," ACI Journal, Vol. 77, No.5, Sept.-Oct. 1980, pp. 340-346.

29. Kotsovos, M.D., "Concrete-A Brittle Fracturing Material," Material and Structures, RILEM, 1984.

30. Krajcinovic,D, and Fonseka, G.U. "The Continuous Damage Theory of Brittle Materials: Part 1 -General Theory; Part II -Uniaxial and Plane Response Modes," J. of Applied Mechanics, Transactions of ASCE, Vol. 48, Dec. 1981. pp. 809-824.

31. Krieg, R.D., "A Practical Two-Surface Plasticity Theory," Journal of Applied Mechanics, Transactions of ASME, Sept. 1975, pp. 641-646.

32. Kupfer, H., Hilsdorf, H.K., and Rusch, H., "Behavior of Concrete Under Biaxial Stresses," ACI Journal, Vol. 66, No.8, Aug. 1969, pp. 656-666.

33. Levine, H.S., "A Two-Surface Plastic and Microcracking Model for Plain Concrete" in Nonlinear Numerical Analysis of Reinforced Concrete, edited by Schwer, L.E., ASME, 1982, pp. 27-47.

34. Liu, T.C.Y., Nilson, A.H., and Slate, F.O., "Stress-Strain Response and Fracture of Concrete in Uniaxial and Biaxial Compression," ACI Journal, Proceedings, Vol. 69, No. 5, May 1972, pp. 291-295.

35. Mroz, J., Norris, V.A. and Zienkiewicz, O.C., "An Anisotropic Hardening Model for Soils and its Application to Cyclic Loading," International Journal for Numerical and Analytical Methods in Geomechanics, Vol. 2, 1978, pp.203-221.

36. Mroz, Z., Norris, V.A. and Zienkiewicz, O.C., "Application of an Anisotropic Hardening Model in the Analysis of Elasto-plastic Deformation of Soils," Geotechnique, Vol. 29, No 1, 1979, pp. 1-34.

37. Murray, D.W., Chitnuyanondh, L. Riyub-Agha, K.Y., and Wong, C., "Concrete Plasticity Theory for Biaxial Stress Analysis," Journal of Engineering Mechanics Division, ASCE, Vol. 105, No. EM6, Proc. Paper 15046, Dec. 1979, pp. 989-1006.

38. Scavuzzo, R. et al., "Simple Formulation of Concrete Response to Multiaxial Load Cycles," Proceedings of the International Conference on Constitutive Laws for Engineering Materials, Theory and Application, Jan, 1983, pp. 421-426.

39. Schickert, G., and Winkler, H., "Results of Tests Concerning Strength and Strain of Concrete Subjected to Multiaxial Compressive Stresses," Deutscher Ausschuss fur Stahlbeton, Heft 277, Wilhelm Ernst & Sohn, Berlin 1977.

40. Spooner, D.C., and Dougill, J.W., "A Quantitative Assessment of Damage Sustained in concrete during Compressive Loading," Magazine of Concrete Research, Vol. 27, No. 92, Sept. 1975, pp. 155-160.

41. Willam, K.J., and Warnke, E.P., "Constitutive Model for the Triaxial Behaviour of Concrete," International Association of Bridge and Structural Engineers, Seminar on Concrete Structures Subjected to Triaxial Stresses; Paper III-1, Bergamo, Italy, May 17-19, 1974.

APPENDIX II - NOTATION

The following symbols are used in this paper:

A_L, A_u, λ = model parameters

D = normalized distance

e_{ij}, de_{ij} = deviatoric strain tensor, and its increment

$$e_{ij} = \varepsilon_{ij} - \delta_{ij} \frac{\varepsilon_v}{3}$$

e^e_{ij}, de^e_{ij} = devatoric strain due to elastic response, and its increment

e^p_{ij}, de^p_{ij} = deviatoric strain due to plastic response, and its increment

F = bounding surface

$F_1, F_{1,max}$ = normalization factor, and its maximum value

f'_c = concrete strength in uniaxial compression

H^e	= generalized elastic shear modulus
H^p	= generalized plastic shear modulus
H^*	= generalized modulus between dD and dK
$I_1, dI_1, I_{1,max}$	= first stress invariant, its increment, and its maxium value $I_1 = \sigma_{ii}$ (i=1,2,3)
J_2, J_3	= second and third deviatoric stress invariants $J_2 = \frac{1}{2} S_{ij}S_{ij}$ (i,j = 1,2,3) $J_3 = \frac{1}{3} S_{ij}S_{jk}S_{ki}$ (i,j,k = 1,2,3)
K, dK, K_{max}	= damage parameter, its increment, and its maximum value
K_R, K_u	= associated K value at the beginning of recent loading and unloading process, respectively
K_t	= tangent bulk modulus
R	= distance of bounding surface from the hydroaxis along S_{ij} direction.
r	= distance from the projection of current stress point on the deviatoric plane to the hydroaxis.
S_{ij}, dS_{ij}	= deviatoric stress tensor, and its increment. $S_{ij} = \sigma_{ij} - \delta_{ij}\frac{I_1}{3}$ where δ_{ij} is the Kronecker delta
β	= shear compaction-dilatancy factor
β_1	= shear compaction factor
β_2	= shear dilatancy factor
$\varepsilon_{ij}, d\varepsilon_{ij}$	= strain tensor, and its increment
$\varepsilon_{kk}(\varepsilon_v), d\varepsilon_{kk}, \varepsilon_v^p$	= volumetric strain, its increment, and its plastic component $\varepsilon_v = \varepsilon_{jj}$ (k = 1,2,3)
$d\varepsilon_{kk,o}, d\varepsilon_{kk,d}$	= volumetric strain increment due to isotropic and deviatoric stress/strain increments, respectively.
ε_p	= associated axial strain to f_c' in uniaxial loading

$\gamma_o^p, d\gamma_o^p$ = plastic octahedral shear strain and its increment

$$\gamma_o^p = \sqrt{\frac{1}{3} e_{ij}^P e_{ij}^P} \quad (i,j = 1,2,3)$$

Θ = angle between projections of position vector of principal stress and that of any tensile semiaxis on deviatoric plane

$$\theta = \frac{1}{3} \cos^{-1} \left(\frac{-3\sqrt{3}\ J_3}{2J_2^{3/2}} \right)$$

$\tau_0, d\tau_0$ = octahedral shear stress, and its increment

$$\tau_o = \sqrt{\frac{1}{3} S_{ij} S_{ij}} \quad (i,j = 1,2,3)$$

$\sigma_{ij}, d\sigma_{ij}, \sigma_I, \sigma_{II}, \sigma_{III}$ = stress tension, its increment and its principal values

$\Delta\sigma_i$ = stress increment in i^{th} principal direction

ANALYSIS OF REINFORCED CONCRETE UNDER SHOCK LOADING

Jeremy Isenberg* and Howard S. Levine*

ABSTRACT

Two approaches to finite element modeling of reinforced concrete structures subjected to blast loading are described. The continuum approach, as currently implemented, is best suited to structures which resist loads primarily in the compressive crushing (ductile) regime of concrete behavior. In cases where shear and flexural rotations play a major role, a layered finite element with degrees of freedom tailored to the expected deformational modes appears promising. Concrete constitutive models appropriate to the two different finite element approaches are described, including the three-invariant cap and a time-dependent viscoplastic softening model. Example applications of each are presented.

INTRODUCTION

This paper describes two approaches to finite element modeling of reinforced concrete structures subjected to blast loading. In the continuum approach, the properties of plain concrete, idealized as an elastic, strain-hardening continuum, are superimposed on explicit models of reinforcing bars and liners. This type of model is best suited to structures which resist loads primarily in the confined compressive (ductile) regime of concrete behavior. A second approach described below is specialized to cases where shear and flexural rotations play a significant role. In this case, a layered finite element model with degrees of freedom tailored to the expected deformational mode appears to be a good choice. Preliminary experimentation with a rate-dependent, strain-softening model for the concrete indicates that concentrated localized deformation occurring at shear hinges can be better represented than with ideally plastic models.

Analysis of reinforced concrete structures subjected to blast relies on finite element modeling that is refined on the basis of field testing. For example, correlation with test data shows that analysis of candidate designs relying primarily on ductility achieved by confining the concrete requires explicit representation of bars, liners

*Principals, Weidlinger Associates, 620 Hansen Way, Palo Alto, CA 94304.

and interior plates. Ideally, modeling of the constituents would include inelastic, multi-axial, stress-strain properties of plain concrete including rate and strain-softening effects; inelastic reinforcing bars with dowel action, bond slip and effect of transverse stress on axial strength; and inelastic behavior of liners including wave propagation effects and loss of confinement due to liner buckling. However, only some of these are currently developed and implemented.

In the following, we review the current status of development and application for analysis of protective structures and indicate which developments are planned for the near future.

CONSTITUTIVE MODELS FOR PLAIN CONCRETE

Rate Independent, Strain-Hardening Model

The type of model used predominantly in analysis of protective structures is based on the rate-independent, strain-hardening theory of plasticity. The example which is summarized here is the three-invariant cap model described by Levine (1982) and Levine and Mould (1985). The main features of the model include:

1. An ideally plastic failure surface, a function of three stress invariants.

2. An isotropically strain-hardening cap which introduces volumetric hysteresis and controls dilatancy (i.e. ratio of major to minor principal inelastic strains) for states of stress inside the failure surface.

The failure surface is in the form of a revised Willam and Warnke failure surface, Willam and Warnke (1974), Figs. 1 and 2. Its functional form in terms of stress invariants is given by:

$$f(\sigma_a,\tau_a,\theta) = \frac{\tau_a}{r(\sigma_a,\theta)f_c'} - 1 = 0 \tag{1a}$$

and

$$\sigma_a = \frac{1}{3}(\sigma_1 + \sigma_2 + \sigma_3) = J_1/3 \tag{1b}$$

$$\tau_a = \sqrt{(2/5)}\ \sqrt{J_2'} = \sqrt{2/5}\ \sqrt{(1/2)s_{ij}s_{ij}} \tag{1c}$$

$$\cos 3\theta = (3/5)\ \sqrt{(6/5)}\ \frac{J_3}{\tau_a^3} = \frac{3\sqrt{3}}{2}\ \frac{J_3}{(\sqrt{J_2'})^3} \tag{1d}$$

$$J_3 = (1/3)s_{ij}\, s_{jk}\, s_{ki}\ ;\ s_{ij} = \sigma_{ij} - \sigma_a \tag{1e}$$

where J_1 is the first invariant of the stresses and hence σ_a is the mean pressure; τ_a is related to $\sqrt{J_2}$' the second invariant of the deviatoric stresses; and θ is the Lode angle related to the third invariant of the deviatoric stresses J_3 and also τ_a.

Here $r(\sigma_a,\theta)$ represents the cross-sectional shape in the deviatoric plane and is given by

$$r(\sigma_a,\theta) = \frac{2r_2(r_2^2-r_1^2)\cos\theta + r_2(2r_1-r_2)\sqrt{4(r_2^2-r_1^2)\cos^2\theta+5r_1^2-4r_1r_2}}{4(r_2^2-r_1^2)\cos^2\theta + (r_2-2r_1)^2} \tag{2}$$

The functional forms for r_1 and r_2 used for the model are:

$$r_1(\sigma_a) = a_0 - a_1\, e^{a_2(\sigma_a/f_c')} \tag{3a}$$

$$r_2(\sigma_a) = b_0 - b_1\, e^{b_2(\sigma_a/f_c')} \tag{3b}$$

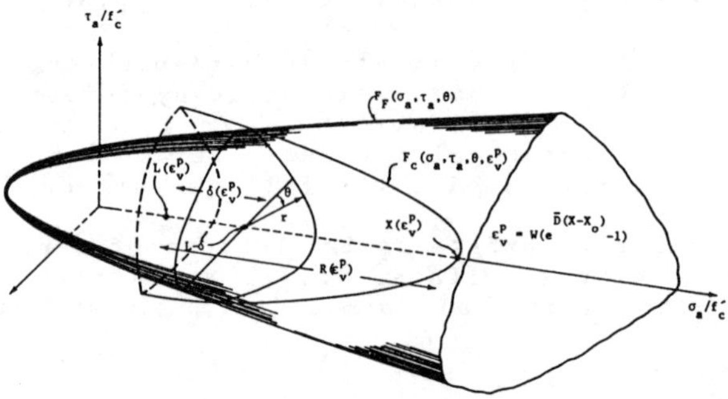

Figure 1. Configuration of cap in relation to failure surface.

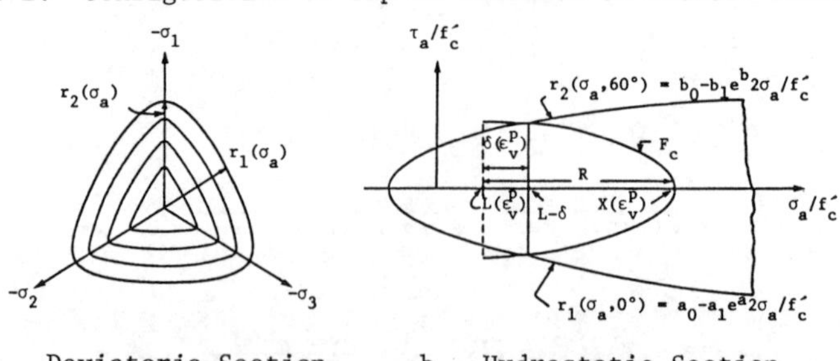

a. Deviatoric Section b. Hydrostatic Section

Figure 2. Cross section of cap and failure surface.

Hence, the meridional failure curves approach asymptotic values at high compression and can never intercept the hydrostat.

The six constants a_0, a_1, a_2, b_0, b_1, and b_2 are fit to the following six conditions:

1. Maximum unconfined compressive strength, f_c'.
2. Maximum uniaxial tensile strength, f_t'.
3. Maximum biaxial compressive strength, f_{cb}'.
4. A point on the triaxial tensile failure surface at high compressive stress (θ = 0 degrees).
5. A point on the triaxial compressive failure surface at high compressive stress (θ = 60 degrees).
6. The triaxial compressive and triaxial tensile curves must intercept the hydrostatic axis at the same value of tensile mean normal stress, i.e.,

$$r_2(\sigma_a^{max}) = r_1(\sigma_a^{max}) = 0.$$

These six conditions are shown in Fig. 3 and how well the failure criterion fits the experimental data is also indicated in the figure.

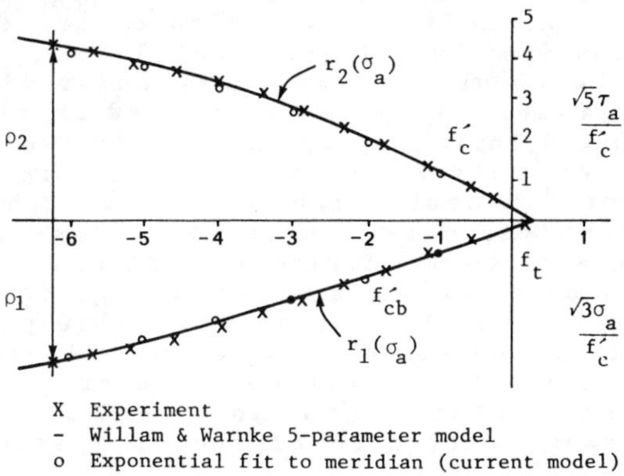

Figure 3. Comparison of fits with triaxial test data.

The functional form of the cap that intercepts the revised Willam and Warnke failure surface in a plane is

$$f_c = \frac{R^2-\delta^2}{r^2(L-\delta,\theta)f_c'^2}\frac{\tau_a^2}{f_c'^2} + \frac{(\sigma_a-L)^2}{f_c'^2} - \frac{(X-L)^2}{f_c'^2} = 0 \tag{4}$$

where

$$X(\varepsilon_v^p) - L(\varepsilon_v^p) = -R(\varepsilon_v^p) \tag{5a}$$

$$\delta = \delta(\varepsilon_v^p) \quad ; \quad R = R(\varepsilon_v^p) \tag{5b}$$

and two forms of hardening rule for the cap have been explored, as follows:

$$\varepsilon_v^p = W \left[e^{\bar{D}(X-X_0)} -1\right]$$

and $$X = X_0 + C_1\varepsilon_v^p + C_2\ell n \cosh(-C_3\varepsilon_v^p) + C_4\ell n \cosh(-C_5 \varepsilon_v^p) \quad (5c)$$

The quantities X, L and R, defined in Figs. 1 and 2, represent the intercept on the hydrostatic axis and maximum value of σ_a for the semi-major axis of the elliptical variation of the cap cross section with regard to σ_a. X_0 defines the initial position of the cap. The quantities W and $\bar{D}$ (or C_1 through C_5) define the hydrostatic hardening and are fit to hydrostatic stress-strain curves. The constitutive laws invoke the conventional flow rules of plasticity theory and the yield and loading criterion.

Usually, the main source of experimental data for choosing values of empirical parameters is static triaxial compression and extension data of the type shown in Figs. 4 through 6. In addition, a series of tests conducted on specimens confined by a light steel jacket, as illustrated in Fig. 7, were used to validate and refine the chosen parameter values. Loads were applied statically to some specimens and dynamically by means of controlled explosions to others. The role of the jacket is to provide a moderate degree of confinement which increases with axial strain. Due to incomplete measurements of stress and strain, it became necessary to use finite element analysis to simulate the tests, including concrete, grooved liner, surrounding soil (where relevant) and the load. This procedure involves assuming a constitutive model for the concrete, simulating response of the specimen, comparing this with measured response, amending the model and repeating the process. An example of a completed simulation is shown in Fig. 8a,b.

The foregoing model treats rate effects only by <u>ad hoc</u> enhancement of strength or stiffness and neglects strain-softening altogether. Recent theoretical advances by Bazant et al. (1984), Belytschko et al. (1984), Sandler (1984), Sandler and Wright (1983), Willam et al. (1984, 2 refs.) have pointed out introducing strain-softening into continuum models results in mesh sensitivity unless special precautions are taken; in dynamic problems, solutions are obtained by introducing rate-dependence into strain-softening constitutive models.

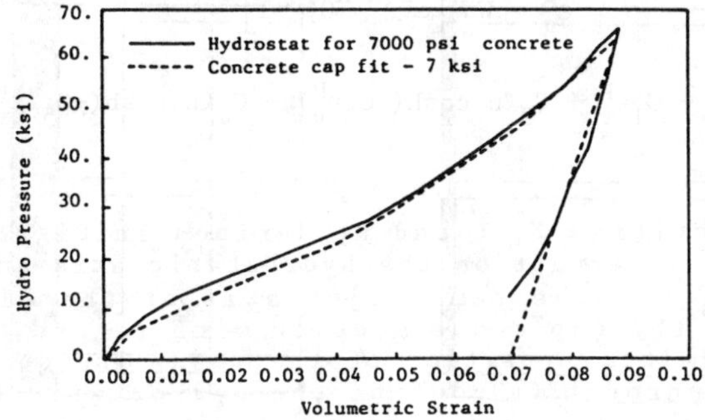

Figure 4. Fit of hydrostat--7 ksi concrete.

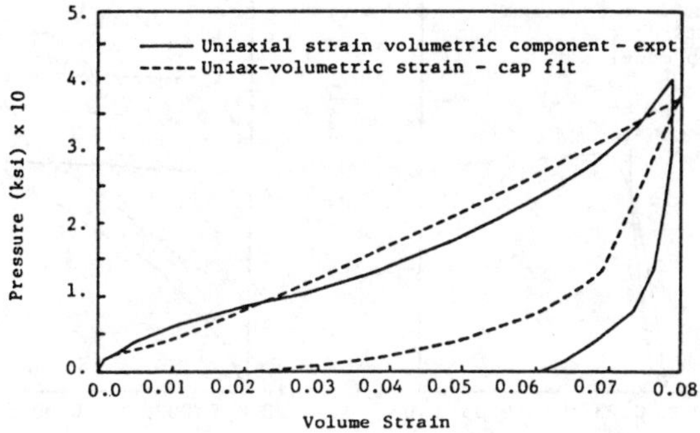

Figure 5. Volumetric component of uniaxial stress-strain path, $f_c{}'$ = 4410 ksi.

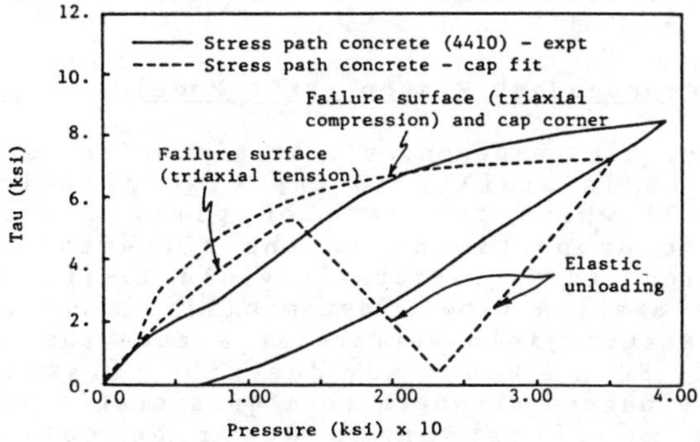

Figure 6. Stress path for uniaxial strain loading of 4410 psi concrete.

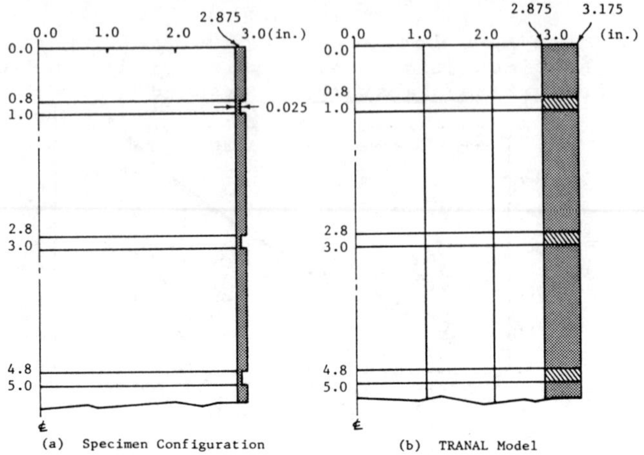

Figure 7. Specimen liner configuration and preliminary TRANAL model.

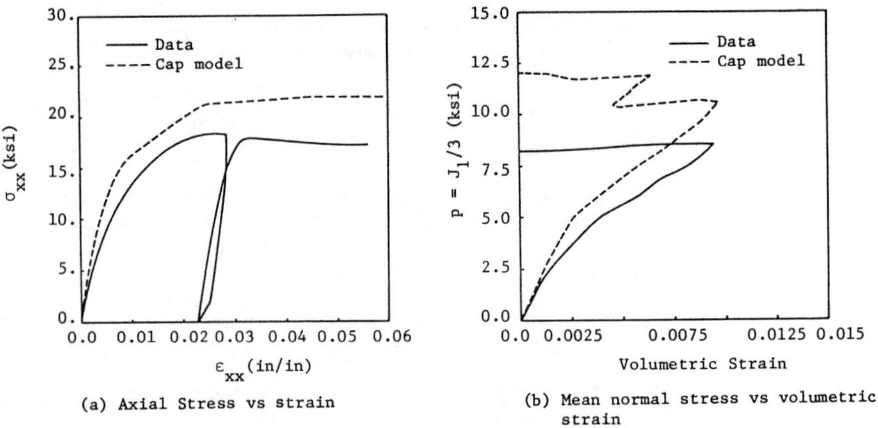

Figure 8. Stress-strain behavior of jacketed specimen subjected to axial load.

Rate- or Time-Dependent Viscoplastic Model

Recently, a rate-dependent, strain-softening model was proposed which is similar to the viscoplastic model of Perzyna (1963) where the rate of plastic straining is assumed to be proportional to the difference between a stress function and a quasistatic yield limit. This leads to stress relaxation from a dynamically induced peak back to the quasistatic yield surface at a rate that depends on strain rate. Such a model may describe relaxation from a dynamically enhanced strength to a quasistatic strength but it requires modification to describe total loss of strength.

The present model described by Isenberg et al. (1984) assumes that the locus of maximum stress states for a concrete specimen defines an initial yield surface, Fig. 9. Within the surface, behavior is assumed to be elastic; this is not a fundamental restriction, however. Upon satisfying the initial yield condition, the yield surface begins to contract at a rate that increases with time spent on the yield surface and with growth of a damage parameter that is an invariant of cumulative plastic strain. A function behaving in this manner is

$$f = \sqrt{J_2'} - B = 0 \qquad (6a)$$

where

$$B = B_{1\Delta} \exp\left(\frac{-Qt^*}{\bar{t}}\right) + B_{10} - B_{1\Delta} \qquad (6b)$$

and B_{10}, $B_{1\Delta}$ = initial, residual effective cohesion, respectively; $Q = g(\varepsilon^p_{ij})$; t^* = total accumulated time on yield surface; and $\bar{t}$ = empirical strain-softening coefficient. Thus, the initial yield surface may contract during some phase of a complicated strain-rate path, may become stationary during an unloading phase and may resume contraction. Underlying the model, which is at the moment speculative, are the ideas that strain-softening is due to development and coalescence of cracks (role of Q in eq. 6b) and that cracking progresses at an inherent rate governed by viscous forces acting on a microscopic level.

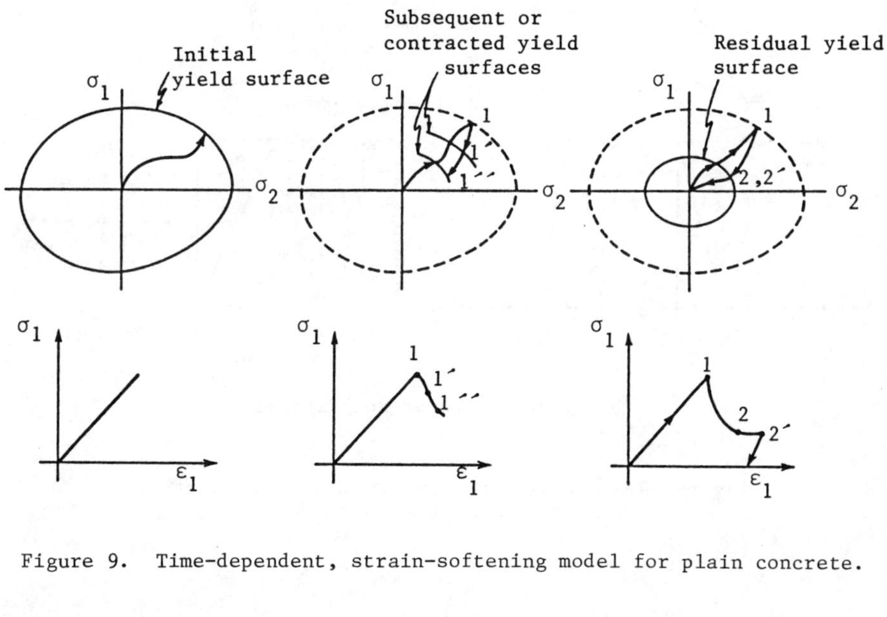

Figure 9. Time-dependent, strain-softening model for plain concrete.

To illustrate the behavior of a specialized version of the model, example stress-strain relations for prescribed strain rates are shown in Fig. 10. The yield surface is specialized to two non-zero stress variables σ_{11} and σ_{12}. The yield surface is

$$f = \sigma_{11}^2 + B_1 \sigma_{11} + B_2 \sigma_{12}^2 + B_3 = 0 \tag{7}$$

which is an ellipse in $\sigma_{11} - \sigma_{12}$ plane.

Figure 10a shows the case of uniaxial compressive strain. The material is elastic until the unconfined compressive strength $f_c' = 5000$ psi is reached. The residual compressive stress is $\sigma_{11} = 500$ psi and it is reached at $\varepsilon_{11} = 0.6$ percent. No special attempt is made here to fit this stress-strain relation to a particular suite of data, which might require adjusting $\bar{t}$, or other adjustments. Figure 10b shows the results of imposing combined shear and compressive strain.

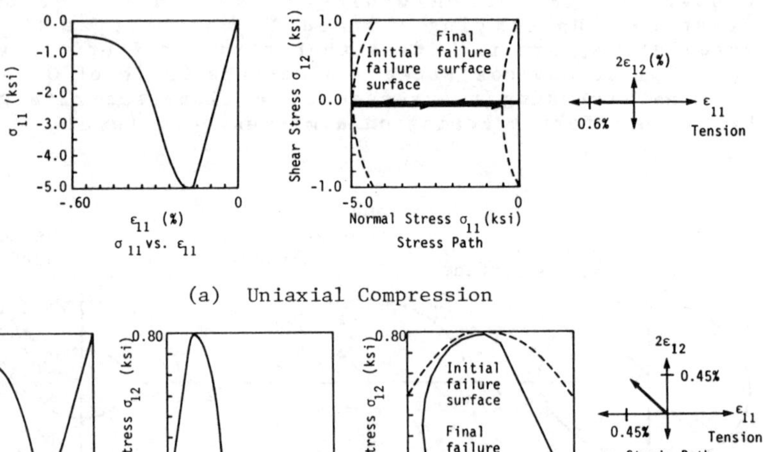

(a) Uniaxial Compression

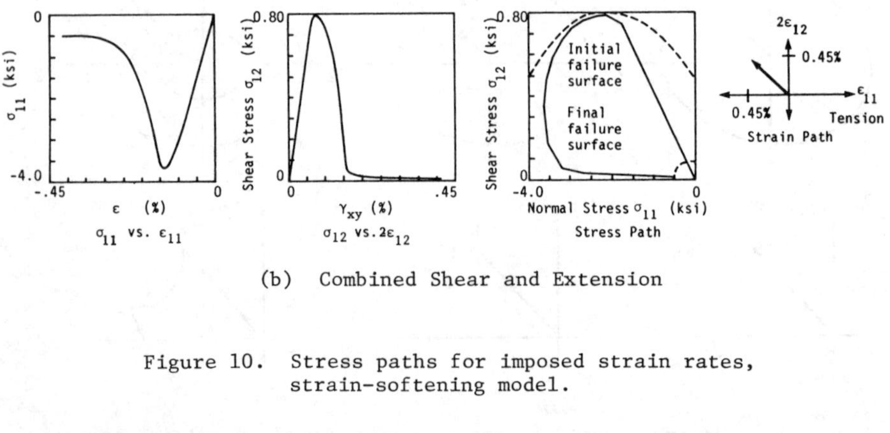

(b) Combined Shear and Extension

Figure 10. Stress paths for imposed strain rates, strain-softening model.

MODELS FOR STEEL ELEMENTS

Steel liners are represented by a four-node plane stress isoparametric quadrilateral with single-point integration. Up to the present, such elements are restricted to small strains. Elastic, ideally plastic constitutive properties are assumed; rate effects are currently taken into account by enhancing the assumed yield strength according to the average strain rate prevailing near peak stress. Displacements of membrane nodes are assumed to be compatible with those of the corresponding nodes of the TRANAL axisymmetric or hexahedral element. Nodal forces for membranes are added to contributions from the adjacent continuum elements.

The major shortcomings of this idealization are first the inability to represent buckling of the liner with resulting loss of concrete confinement, and second the ad hoc treatment of rate effects on strength. The former is considered more important than the latter.

Steel reinforcing bars are represented by means of a massless, one-dimensional element which passes through the centroids of opposing element faces. A set of three such bars embedded in a hexahedral element is shown in Fig. 11a.

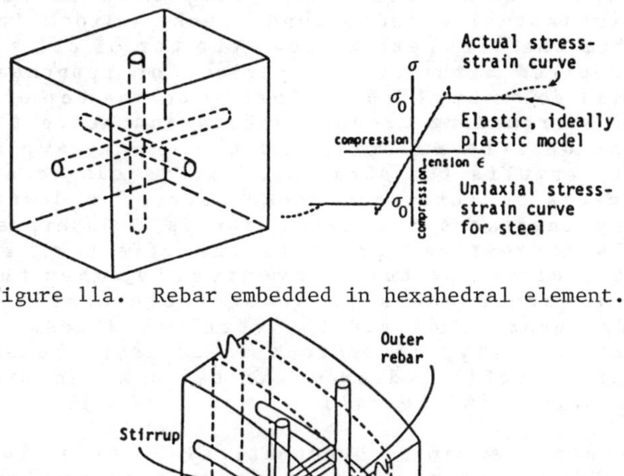

Figure 11a. Rebar embedded in hexahedral element.

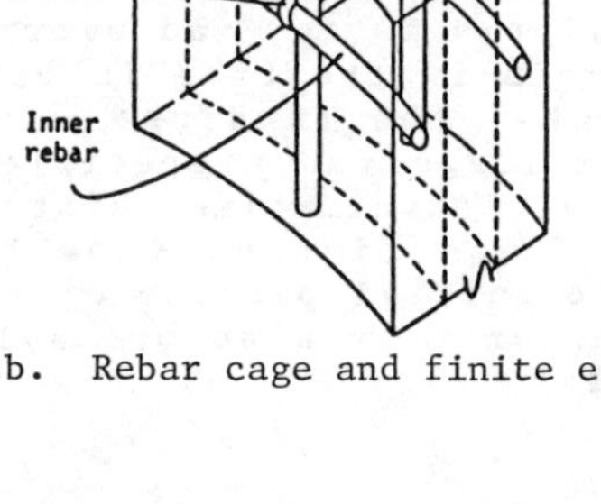

Figure 11b. Rebar cage and finite element model.

This type of model might be used to represent the configuration shown in Fig. 11b, which consists of inner and outer cages of vertical and circumferential reinforcing tied together by radial bars. The strain in each bar is assumed to be equal to the strain in the continuum element evaluated at the continuum element centroid and in the direction of the bar axis. The bar is assumed to be elastic, ideally plastic; the stress is uniquely a function of axial strain and strain history. The forces in the bar are distributed equally to the nodes of the faces through which it passes. The primary goal of modeling reinforcing in this manner is to represent the confining effects of steel bars on plain concrete in cases where the principal directions of strain coincide pretty nearly with the direction of the reinforcing. The procedure is not intended to describe degradation in shear stiffness that accompanies severe cracking; in fact, the present model probably overestimates shear stiffness.

EXAMPLES OF FINITE ELEMENT ANALYSIS OF PROTECTIVE STRUCTURES

Silo-Type Structures

The basic load transfer paths and their influence on structural response are illustrated in Fig. 12 for the case of a simplified elastic silo subjected to airblast and local airblast- induced ground shock. Since the time required for the airblast to load the top of the structure is small compared with natural periods of response, the load is assumed to be applied uniformly to the top of the structure and surrounding ground surface and hence the analytic model can be axisymmetric. As the shock wave in the soil gradually engulfs the structure, it is subjected to gradually increasing interface shear force, or downdrag, which gradually increases the axial stress. Superposed on this gradually increasing stress is the effect of rarefactions from the ends of the tube. Eventually, when the velocity of the structure and soil become more nearly equal, the interface shear loads and the vertical stress in the wall decrease. Finally, if bedrock is nearby, the shock in the soil that is reflected from the bedrock can significantly increase stress in the wall.

An example of a hypothetical design is given in Fig. 13. Figure 14 shows the finite element discretization. The goal of the analysis is to find overpressure levels at which the axial strain in the tube is between 1-2 percent. Figure 15 illustrates the findings in terms of average axial strain versus range for a specified yield. Typical vertical and radial strain-time histories are shown in Figs. 16 and 17. These figures show that most of the damage occurs due to initial passage of the shock through the concrete. Shear drag is also probably a factor in the time histories in Fig. 16.

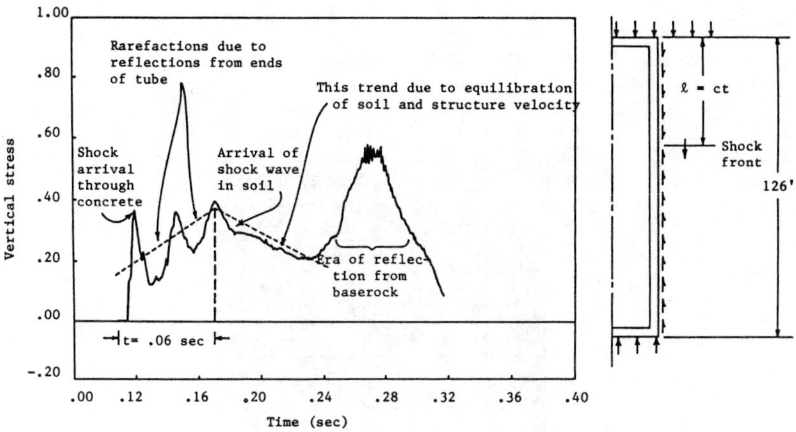

Figure 12. Vertical stress-time history in wall of structure, 80-ft depth (structure assumed to be elastic, compression positive).

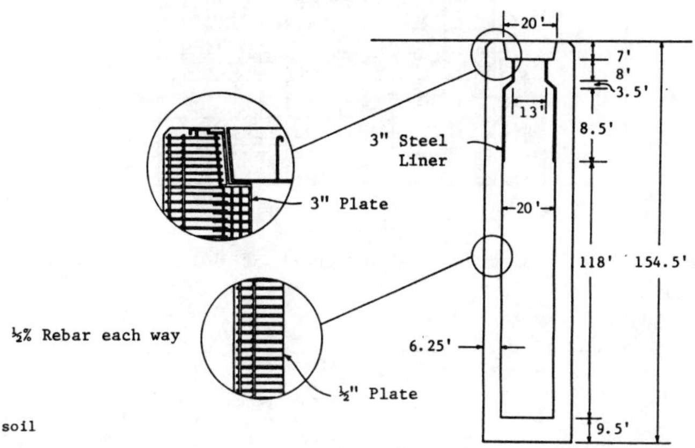

Dry soil

Concrete

f'_c = 9300 psi including dynamic enhancement

Steel

liner (½ inch) f_y = 42 ksi; reinforcing (½% e.w.) f_y = 60 ksi

Figure 13. Vertical shelter.

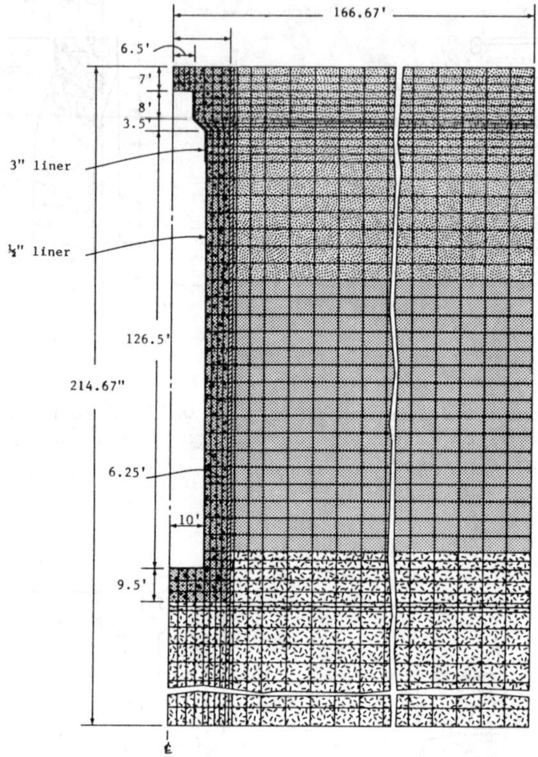

Figure 14. Axisymmetric TRANAL model of vertical shelter.

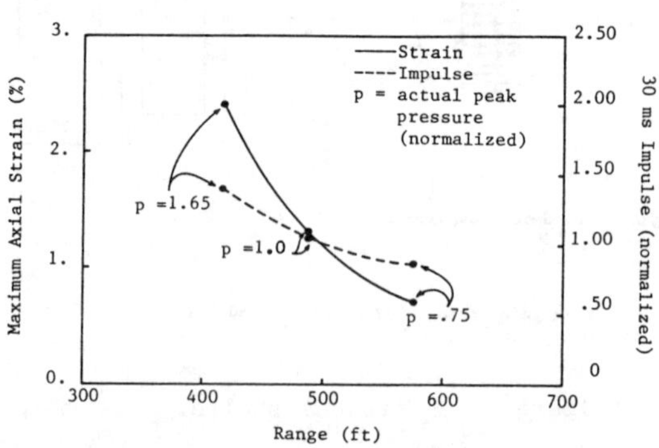

Figure 15. Maximum axial strain and impulse versus range for vertical shelter.

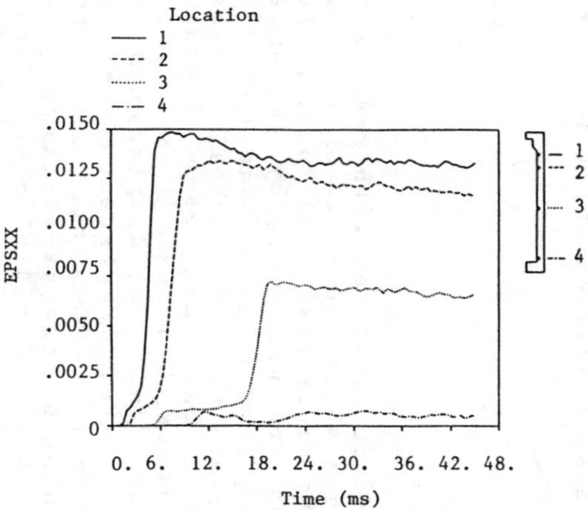

Figure 16. Vertical strain in tube, compression positive.

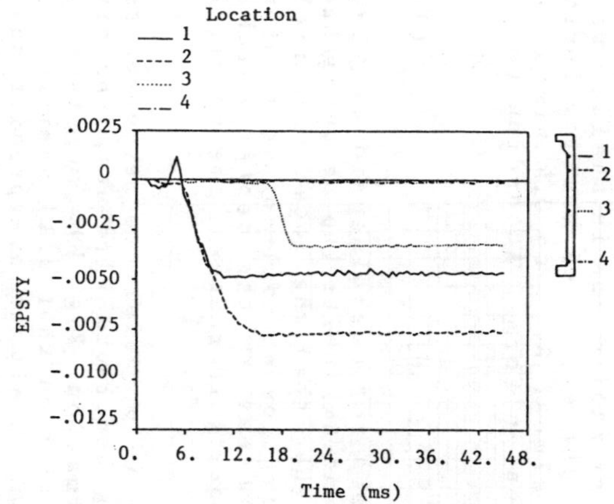

Figure 17. Radial strain in tube, compression positive.

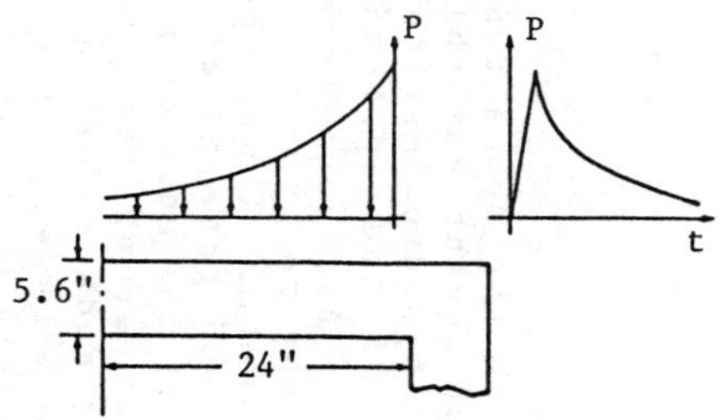

Figure 18. Spatial and time variation of interface pressure on roof slab.

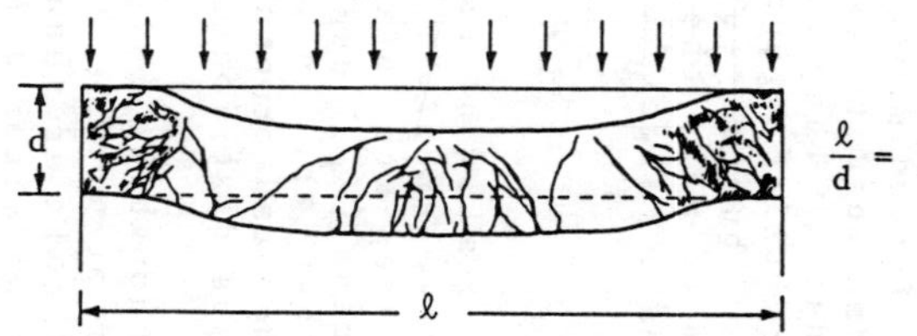

Figure 19. Deformed shape of member exhibiting shear hinge.

Moderately Deep Beams

Reinforced members having span-to-depth ratios of 4 to 10 are designed to resist dynamic pressure loading in one-way bending. They are used as roof slabs in shallow rectangular buried structures. The effect of shallow burial (depth of burial equal to almost 0.15 to 0.4 times the span) is to make the load non-uniform. According to the approximate relationship,

$$\sigma_{int} = \sigma_{ff} + \rho c\,(V_{ff} - V_{str}) \tag{8}$$

which is valid at early times for one-dimensional plane strain wave effects, Wright and Smilowitz (to be published), the stresses are intially higher than the local free field soil stress (σ_{ff}) due to reflection at soil-concrete interface; then the interface stress at the center quickly diminishes as structural velocity increases. The result is the stress distribution shown in Fig. 18.

Although it is possible in principle to represent the behavior of roof slabs using continuum elements, the significant role of shear deformation, Fig. 19, suggests the need for an alternative formulation that will overcome the tendency of current continuum models to overpredict shear stiffness. One such formulation by Ghaboussi (1984) is shown in Figs. 20 and 21, which follows the layered finite element approach suggested by Hand et al. (1973) and Buyukozturk (1977).

The axial and lateral shear strains ε_{xx} and ε_{xy} are related to axial and transverse displacements u and w and shear deformation γ by

$$\varepsilon_{xx} = u_{,x} - y(w_{,x} - \gamma)_{,x} + \frac{1}{2} w^2_{,x} \tag{9a}$$

$$\varepsilon_{xy} = \gamma + (\gamma - w_{,x})u_{,x} \tag{9b}$$

in which the comma denotes differentiation with respect to the subsequent variable. This shear theory is of the same order of approximation as the Timoshenko beam theory. Details of the approximation are presented by Ghaboussi (1984). As shown in Fig. 20, bending rotations θ_1 and θ_2 and shear rotations Γ_1 and Γ_2 are treated as nodal degrees-of-freedom. Thus, the element has 9 degrees-of-freedom: axial and lateral displacements, bending and shear rotations at the ends of the element and axial displacements at the center of the element. Axial displacements at the center node and shear rotations are condensed out at the element level. The remaining 6 degrees-of-freedom are comparable to those of a standard flexural element. Inter-element continuity is enforced for bending rotations only. The element is strictly valid for small to moderate rotations.

For purposes of integration through the depth, the element is subdivided into a number of layers. Within each the properties are assumed to remain uniform (see Fig. 21). Gaussian integration is used along the axis of an element.

The equations of motion may be integrated by either an implicit or explicit algorithm. The present examples use an explicit second order, central difference algorithm to solve the following equations:

$$M\,\ddot{U} = \underline{P} - \underline{R} \qquad (10)$$

where

M = lumped mass (rotational inertias are increased to enhance stability of the solution)

$\underline{\ddot{U}}$ = nodal accelerations

$\underline{P}$ = external forces

$R = \underline{R} = \sum_{i=1}^{NRB} a_i \int_x B^T(y_i)\sigma dx = \int B^T \sigma dV$

and

a_i = cross-sectional area of ith layer

B = nonlinear displacement-dependent, strain-displacement matrix

y_i = depth to i^{th} element

V = volume

σ = stress vector

x = coordinate along length of element

NRB = number of reinforcing bars

Geometric nonlinearity is taken into account in the B-matrix and by adding incremental displacements to the coordinates at the end of each time step (updated Lagrangian approach).

A series of numerical experiments was undertaken based on physical experiments conducted at the U.S. Army Waterways Experiment Station (USAWES). The results of simulations using the rate-independent model are reported by Ghaboussi (1984). To compare the performance of the rate-independent model with that of the rate-dependent one described above, a plane strain simulation of the one-way roof slab was performed using a value of $\bar{t}$ = 0.01 ms for the rate-dependent parameter, and a rate-independent version of the model.

θ_1 and θ_2 = flexural rotation degrees-of-freedom at a nodal point

Γ_1 and Γ_2 = shear rotation degrees-of-freedom at a nodal point

$(U)^T = (U_1 W_1 \theta_1 U_3 W_2 \theta_2 \vdots U_2 \Gamma_1 \Gamma_2)$ the last three degrees-of-freedom to be condensed out

$\phi_1 = \theta_1 + \Gamma_1$

$\phi_2 = \theta_2 + \Gamma_2$ θ_1 and θ_2 = bending rotation degrees-of-freedom

Figure 20. Independent degrees-of-freedom for finite element in RCBEAM.

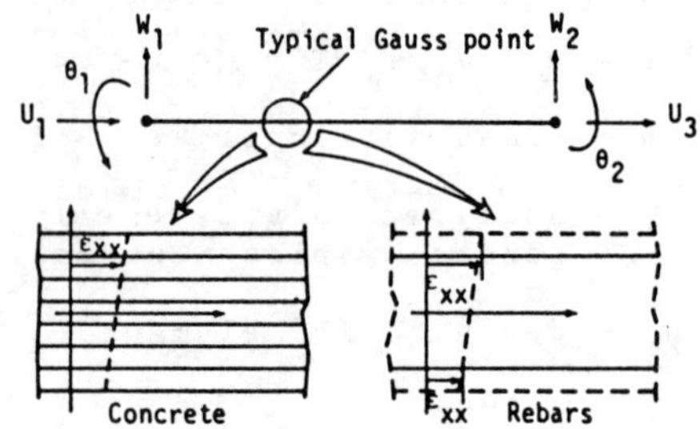

Figure 21. Strain displacement relations for element in RCBEAM.

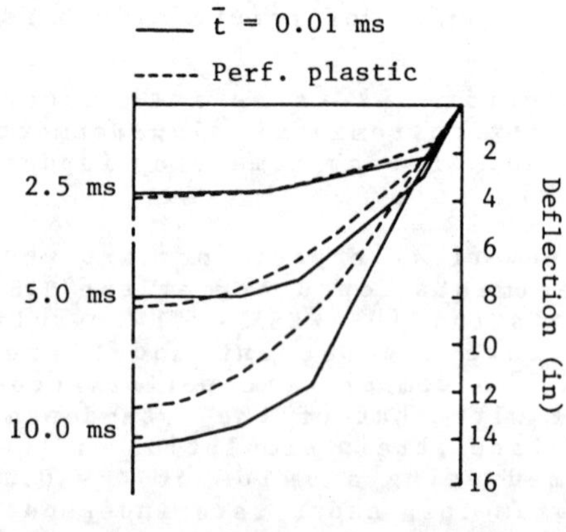

Figure 22. Comparison of calculated deformed shapes using perfectly plastic and strain-softening models.

Figure 22 shows that the softening model quickly forms a sharper hinge near the support, while the perfectly plastic case hinges slightly and flexes more in the midspan region. At later times, the hinges are pulled straight as axial forces develop in the span due to large deformation effects. The softening case tends to represent localized deformation, although this is nearer center span than experimental data suggests, whereas the rate-independent case develops flexural and axial extension modes.

SUMMARY

This paper describes continuum and layered finite element approaches to modeling of protective structures subjected to blast loading. Continuum modeling, with accompanying explicit models of reinforcing bars and liners, is appropriate for such structures as cylinders which are designed to resist loads primarily as internal compressive forces in combination with low order bending. Constitutive modeling for such structures, which rely on achieving ductility through confinement, is of the plasticity theory type. A three-invariant, rate-independent version of the cap model is described above. Since ductility depends partly on correct interaction between reinforcing steel and dilatant concrete, special attention is given to those features controlling dilatancy, including the hardening rule and the shape of the cap. Experimental data on stress paths of interest to cylindrical protective structures, which lie primarily in the triaxial compression octant of stress space, but which involve confining stresses that increase with axial strain, have only recently been developed. The success of such models as the three-invariant, strain-hardening cap in representing the main features of the data is encouraging.

In contrast to continuum modeling for cylindrical structures, the layered approach appears better suited to members where shear deformation plays a major role. A model is described above in which shear rotations are explicitly included with flexural rotations as independent degrees-of-freedom. The layered finite element approach lends itself to defining material properties in terms of the stress-strain properties in each layer from which multi-dimensional moment, thrust and shear interaction characteristics can be developed. Both rate-independent, ideally plastic and time-dependent viscoplastic strain-softening models have been proposed. The latter appears to improve slightly the ability of the model to represent concentrated shear rotation.

APPENDIX A - REFERENCES

Bazant, Z. P., Belytschko, T. B. and Chang, T. P. (1984), "Continuum Theory for Strain-Softening," J. of Eng. Mech., Vol. 110.

Belytschko, T. B. and Bazant, Z. P. (1984), "Strain-Softening Materials and Finite Element Solutions," Constitutive Equations: Macro and Computational Aspects, ASME Winter Annual Meeting, New Orleans, LA.

Buyukozturk, O. (1977), "Non-linear Analysis of Reinforced Concrete Structures," J. of Comp. and Struct., Vol. 7.

Ghaboussi, J., Millavec, W. A. and Isenberg, J. (1984), "R/C Structures Under Impulsive Loading," J. of Struct. Eng., ASCE, Vol. 110, No. 3, pp. 505-522.

Hand, F. R., Pecknold, D. A. and Schnobrich, W. C. (1973), "Nonlinear Layered Analysis of RC Plates and Shells," J. of the Struct. Div., ASCE, Vol. 99, No. ST7, pp. 1491-1505.

Isenberg, J., Richardson, E. and Levine, H. S. (1984), "A Rate-Dependent, Strain-Softening Model for the Dynamic Collapse of Beams in Bending and Shear," Constitutive Equations: Macro and Computational Aspects, ASME Winter Annual Meeting, New Orleans, LA.

Levine, H. S. (1982), "A Two-Surface Plastic and Microcracking Model for Plain Concrete," Nonlinear Numerical Analysis of Reinforced Concrete, Winter Annual Meeting ASME, Phoenix, AZ.

Levine, H. S. and Mould, J. C. (1985), "A Three-Invariant Cap Model for Concrete Soil and Rock: Computational Aspects," ASME/ASCE Summer Mechanics Conference, Albuquerque, NM.

Perzyna, P. (1963), "The Constitutive Equations for Rate Sensitive Plastic Materials," Quarterly Appl. Math., Vol. 20.

Sandler, I. S. (1984), "Strain-Softening for Static and Dynamic Problems," Constitutive Equations: Macro and Computational Aspects, ASME Winter Annual Meeting, New Orleans, LA.

Sandler, I. S. and Wright, J. P. (1984), "Summary of Strain-Softening," Theoretical Foundation for Large Scale Computations of Nonlinear Material Behavior, DARPA-NSF Workshop, S. Nemat-Nasser, ed., Northwestern University, Evanston, IL.

Willam, K. J. and Warnke, E. P. (1974), "Constitutive Model for the Triaxial Behavior of Concrete," IABSE Seminar on Concrete Structure Subjected to Triaxial Stresses, Paper III-1, Instituto Specimentali Modelli e Strutture, Bergamo, Italy.

Willam, K. J., et al. (1984), "Identification of Strain-Softening Properties with Computational Predictions of Localized Fracture," Department of Civil, Environmental and Architectural Engineering, College of Engineering and Applied Science, University of Colorado, Boulder, Structural Research Series 8404, Boulder, CO.

Willam, K. J., Bicanic, U. and Sture, S. (1984), "Constitutive and Computational Aspects of Strain-Softening and Localization in Solids," Constitutive Equations: Macro and Computational Aspects, ASME Winter Annual Meeting, New Orleans, LA.

Wright, J. P. and Smilowitz, R. (to be published), "Uncoupling Approximation for the Dynamic Analysis of Structures Embedded in Hysteretic Media," to appear in J. of Comp. and Struct..

APPENDIX B - NOTATION

σ_a, τ_a	stress invariants related to first invariant (J_1) and second invariant of stress deviator (J_2') tensors, respectively
θ	Lode angle
r_1, r_2	lengths of vectors from a point on hydrostat to yield surface for Lode angles of 0deg. and 60deg., respectively
X, R, L, δ	parameters defining position and size of cap in cap model
ε_v^p	cumulative volumetric plastic strain
Q	damage parameter, a function of accumulated plastic strain
t*	accumulated time on yield surface
$\bar{t}$	relaxation time governing rate of strain-softening
σ_{11}, σ_{12}	direct and shear stress components
σ_{int}	interface normal stress

σ_{ff}, V_{ff}	free-field stress and velocity components normal to interface
V_{str}	velocity of structures in the direction of σ_{int} and at a point where it acts; varies with length along the span
ρ, c	mass density, wave speed of soil
ε_{xx}, ε_{xy}	axial and shear strain components of a one-way slab or beam
u,w	axial and transverse displacements of slab or beam element
x,y	axial and transverse direction coordinates of slab and beam element
γ	shear deformation of beam and slab; varies with length along the span
θ_1, θ_2, Γ_1, Γ_2	bending and shear rotations at ends 1 and 2 of a finite element

Modeling Large Inelastic Cyclic Response of Concrete

Christian Meyer*, M, ASCE and Hugo Delgado-Saavedra**

Abstract

An elastic/viscoplastic material model has been developed to simulate the behavior of plain concrete under arbitrary triaxial stress states and different strain rates. This model has been implemented into a general finite element program to study the behavior of three-dimensional solids under large cyclic loads, together with a general crack monitoring algorithm.

Introduction

There are a number of applications for which the correct analytical modeling of concrete behavior under load is important. Examples are the prediction of failure loads for nuclear reactor containment vessels and certain shell roof structures. In some of these cases sufficiently accurate estimates are possible using common engineering tools and intuition. If the loads are of dynamic nature, and in particular, if they result in large inelastic cyclic response, such engineering estimates are less useful, because the material strength and stiffness undergo a complicated course of degradation as a function of damage accumulation. The prediction of strength and behavior characteristics in such situations has to rely almost exclusively on empirical knowledge based on laboratory tests and field observations. For example, in the area of seismic response of structures a considerable store of such knowledge has been accumulated.

The response of ductile moment resistant concrete frames to strong seismic ground motion is controlled largely by the material behavior within the plastic hinge regions. Only through careful detailing and provision of adequate confinement reinforcement, the concrete is able to undergo without failure the large inelastic strain reversals. Similar behavior is encountered in structural walls. Also the response of protective structures to shock and blast loads is comparable, even though the number of stress reversals is relatively small.

*Associate Professor, **Formerly, Graduate Student, Dept. of Civil Engineering and Engineering Mechanics, Columbia University, New York, N.Y.

It is the objective of this paper to discuss the behavior of concrete in such highly stressed regions and to suggest a model for simulating this behavior, which can be understood only by realizing the three-dimensional nature of the state of stress and strain. A simple standard cylinder subjected to "uniaxial" compression fails in the well-known mode only because of the triaxial stress state that is present in reality. The insight into the material behavior obtained from standard tests is instrumental to the development of mathematical models to simulate more realistic multi-axial stress states.

Test data for two- and three-dimensional stress states are very scarce. Whatever data do exist are generally for loadings that are increased proportionally and monotonically. Since we will present a material model intended to simulate the behavior of plain concrete under arbitrary stress states and different strain rates, the scarcity of test data makes it difficult to calibrate this model. Further verification and evaluation of the model beyond the test data used in this study has to await the availability of data from comprehensive dynamic experiments.

Behavior of Concrete under Cyclic Load

Typical stress-strain curves for concrete under monotonic uniaxial compression are shown in Fig. 1 (1). The nonlinearity starts at stresses somewhere between 0.3 f'_c and $0.5f'_c$, and increases at an accelerating rate with strain, accompanied by gradual propagation of microcracks. After reaching the material strength f'_c, concrete exhibits strain-softening characterized by the appearance of macroscopic cracks. The tensile stress-strain curves are somewhat similar to the compression curves, but with markedly lower strength and less nonlinearity. In many tension tests, concrete behaves more like a linear elastic brittle material, without the softening following cracking.

When concrete is subjected to cyclic compression loading to stresses above approximately $0.6f'_c$, it degrades in both stiffness and strength, Fig. 2 (2). The monotonic stress-strain curve can serve as a reasonable envelope for the peak stresses in the concrete under cyclic loading.

The behavior of concrete under biaxial loading conditions is somewhat different (3). Under equi-biaxial compression, a strength increase of about 16% is observed. The maximum strength of about $1.25f'_c$ is obtained for a stress ratio $\sigma_2/\sigma_1 = 0.52$, Fig. 3. Biaxial compression increases the initial stiffness of the material and the strain at peak stress. In the vicinity of peak stresses, concrete exhibits dilatancy, i.e., an inelastic volume increase, due to the propagation of major cracks. For proportional biaxial loading, the failure of concrete appears to be controlled by a maximum tensile strain criterion. Little is known about the path dependency of the failure surface.

Experimental data on concrete behavior under biaxial cyclic loading conditions are extremely limited. Recently reported test results (4) indicate that, for the load histories employed, strength envelopes appear to be independent of whether the loadings are cyclic or monotonic.

For monotonic triaxial stress states, proportionally applied, the failure surface for plain concrete in principal stress space appears as shown in Fig. 4. It signifies a large increase of strength and ductility as a function of the hydrostatic pressure component, and inelastic dilatancy, i.e., an increase of volume due to large deviatoric strains. The interaction between deviatoric and volumetric behavior is attributed to progressive cracking of the material under increasing deviatoric strains.

Depending on the confinement stress, concrete can act either as a quasi-brittle, or strain-softening, or strain-hardening material. A recent experimental program at the University of Colorado has explored the load-path dependency of the failure surface (5,6), but little is known about the effects of triaxial cyclic loading.

In spite of the dearth of experimental data, a number of statements can be made to summarize a few important features of concrete behavior, which should be borne in mind when attempting to simulate this behavior mathematically.

1. With the onset of cracking, the material becomes clearly anisotropic. Whether a crack is open or closed, the material "remembers" both the existence and the orientation of the crack plane, and the tensile strength normal to the crack plane is zero.
2. The crack not only affects the tensile strength normal to its plane, but also strength in compression and tension in any direction within the plane of the crack. The raggedness of the bounding faces of any concrete element between adjacent cracks, Fig. 5, reduces the effective strength, presumably because of stress concentrations and cross sectional variations.
3. The crack width determines the amount of shear stress that can be transferred across it. Under cyclic loading abrasion of the crack faces quickly reduces the capacity for interface shear transfer, even for nominally closed cracks.

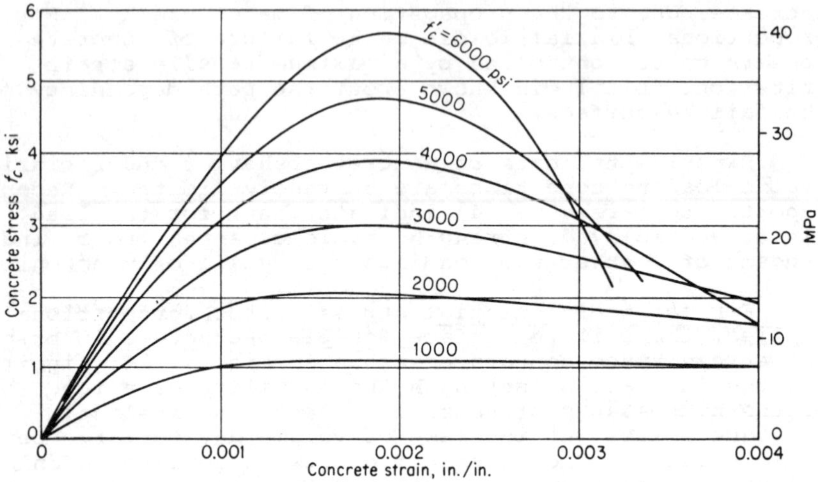

Fig. 1 Monotonic Uniaxial Stress-Strain Behavior of Concrete (Ref. 1)

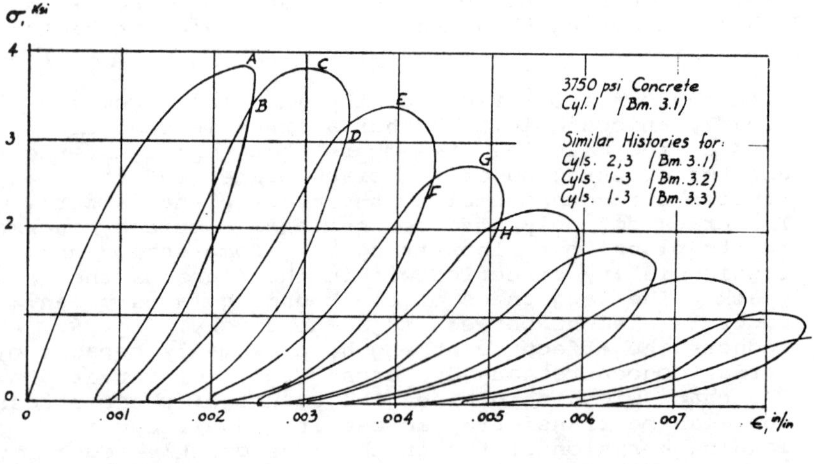

Fig. 2 Cyclic Uniaxial Stress-Strain Behavior of Concrete (Ref. 2)

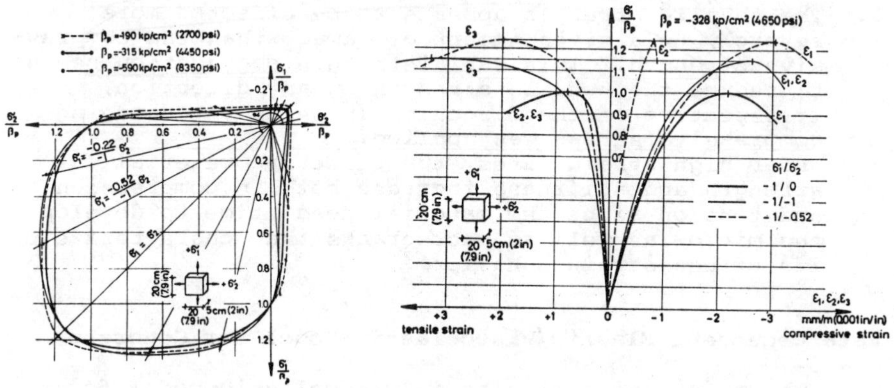

Fig. 3 Biaxial Behavior of Concrete (Ref. 3)

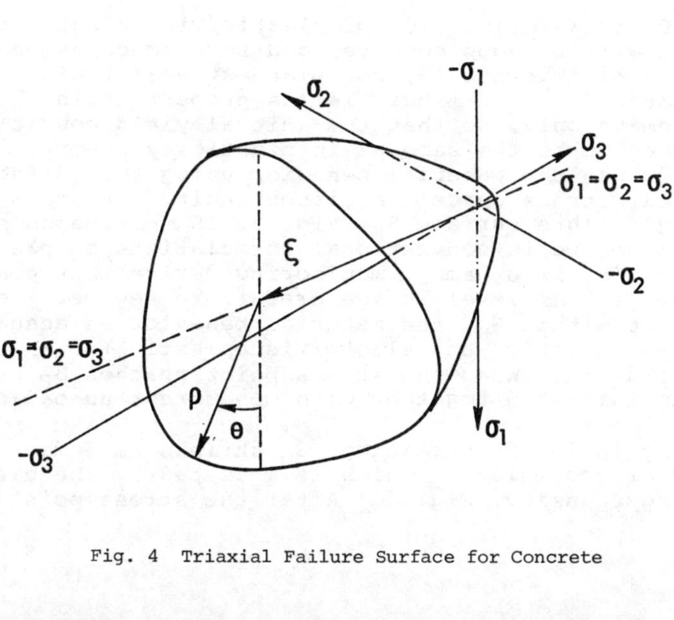

Fig. 4 Triaxial Failure Surface for Concrete

4. Any stress above $0.3f'_c$ to $0.5f'_c$ incurs damage in the material, characterized by increasing levels of microcracking. No matter how this damage is measured, the material "remembers" the amount of damage. This is particularly significant in the case of cyclic loading, as the damage accumulates. This behavior is to some extent comparable to fatigue of metals.
5. Both the material strength and apparent stiffness (e.g. Young's modulus) are functions of the amount of accumulated damage. A single high stress excursion can have the same effects as several excursions to lower stress levels.
6. The tensile strength appears to be affected more severely by a given amount of damage than the compressive strength. This rapid strength degradation seems to reduce the tensile strength in any direction, irrespective of the direction in which the damaging compressive stress was applied.
7. Under high strain rates, the effective material strength and stiffness increase both in compression and tension since the material needs time to develop the micro- as well as macrocracks that characterize its stress-strain behavior.

Rate Dependent Elastic/Viscoelastic Model for Concrete

The rate-dependent elastic/viscoplastic model for concrete under triaxial stress states to be described below should be considered as a first attempt to reproduce most of the characteristics listed above. For further details of this model see Ref. 7.

The Perzyna formulation of elastic/viscoplasticity theory (8) will be adopted here, and some concepts used by Bicanic and Zienkiewicz (9) for plane stress/strain cases are utilized. The model has viscous properties in the plastic domain only, so that the initial yield condition can be taken to be the same as in plasticity theory. The model describes the material behavior using two surfaces in principal stress space: a discontinuity surface S_D and a bounding failure surface S_F, Fig. 6. S_D corresponds to the yield surface in conventional formulations of plasticity theory. S_F is a damage monitoring device that controls the failure stress level in the prefailure regime. For a stress point within S_D, the material behavior is assumed to be linear elastic, and elasto/viscoplastic for a stress point outside S_D. When the stress point reaches S_F, failure is initiated together with concurrent unloading.

During inelastic straining, S_D shrinks as damage of the material accumulates, which is measured by the dissipated energy density, Fig. 7. After the stress point

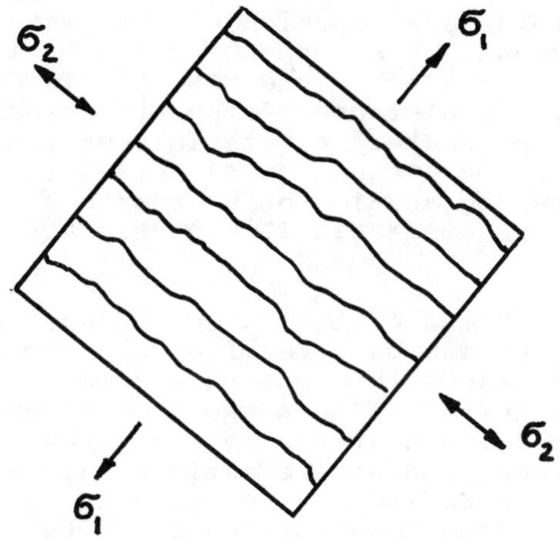

Fig. 5 Concrete Cracked in One Direction

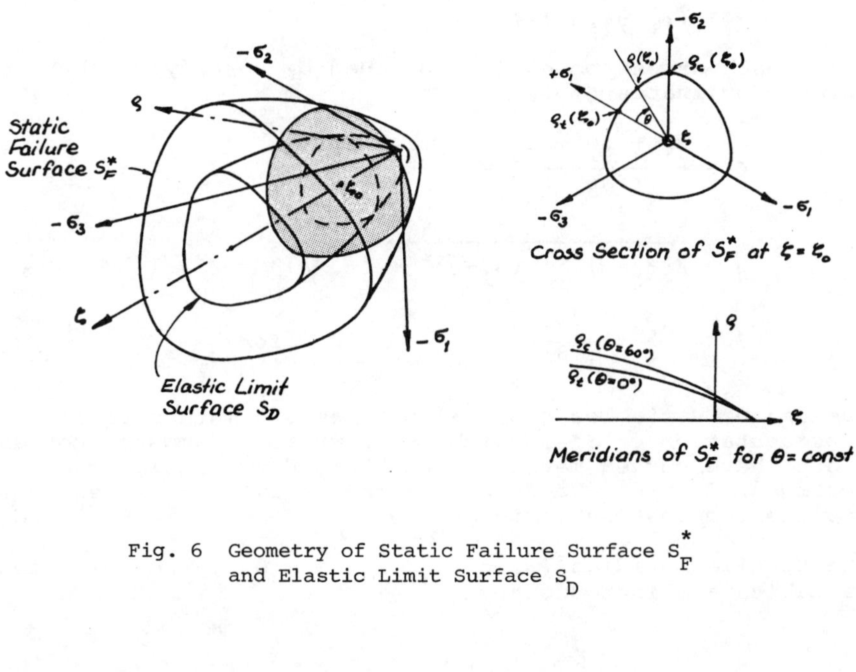

Fig. 6 Geometry of Static Failure Surface S_F^* and Elastic Limit Surface S_D

reaches S_F, S_D shrinks as a function of the post-failure dissipated energy density. According to the basic elastic/viscoplasticity theory, the rate of viscoplastic flow is governed by the distance of the stress point from S_D and the fluidity parameter, a material constant which is considered here to be a function of the elastic strain rate (9). Since the degradation of S_D starts as soon as failure occurs, the viscoplastic flow accelerates in the postfailure regime.

The initial positions of the S_D and S_F surfaces in principal stress space can be determined experimentally. Let S_D^o correspond to the first departure from linear elasticity in slow rate tests (but not too slow to avoid creep effects), and S_F^o to failure stress levels reached with very high loading rates and no inelastic straining. Both S_D^o and S_F^o will be defined in relation to S_F^*, a failure surface corresponding to quasi-static tests, i.e., tests performed to determine such concrete properties as cylinder strength f'_c (with elastic strain rates ~10^{-5}). Henceforth, S_F^* will be referred to as "static failure surface."

Static Failure Surface S_F^*. Any state of stress represented by a point P in the principal stress space can be decomposed into a hydrostatic and a deviatoric part, Fig. 8,

$$\underset{\sim}{\sigma} \equiv (\sigma_1, \sigma_2, \sigma_3) = \underset{\sim}{\zeta} + \underset{\sim}{\rho}$$

i.e., each stress point P is defined by specifying the three coordinates ζ, ρ, θ, where

$$\zeta = \frac{\sigma_1 + \sigma_2 + \sigma_3}{\sqrt{3}} = \sqrt{3}\,\bar{\sigma} = \frac{I_1}{\sqrt{3}}$$

$$\rho = \sqrt{(\sigma_1-\bar{\sigma})^2 + (\sigma_2-\bar{\sigma})^2 + (\sigma_3-\bar{\sigma})^2} = \sqrt{2J_2}$$

$$\cos 3\theta = \frac{3\sqrt{3}}{2}\,\frac{J_3}{J_2^{3/2}} \qquad \text{for } \begin{matrix} 0^\circ \leq \theta \leq 60^\circ \\ \sigma_1 \geq \sigma_2 \geq \sigma_3 \end{matrix}$$

The use of variables ζ, ρ, θ simplifies the mathematical representation of S_F^*, which is known to be smooth, convex, and to have curved meridians and bulged triangular cross sections, Fig. 6. To determine the analytical expressions for the compressive meridian ($\sigma_1 = \sigma_2 > \sigma_3$, $\theta = 60^\circ$) and the tensile meridian ($\sigma_1 = \sigma_2 < \sigma_3$, $\theta = 0^\circ$), the following variables are introduced,

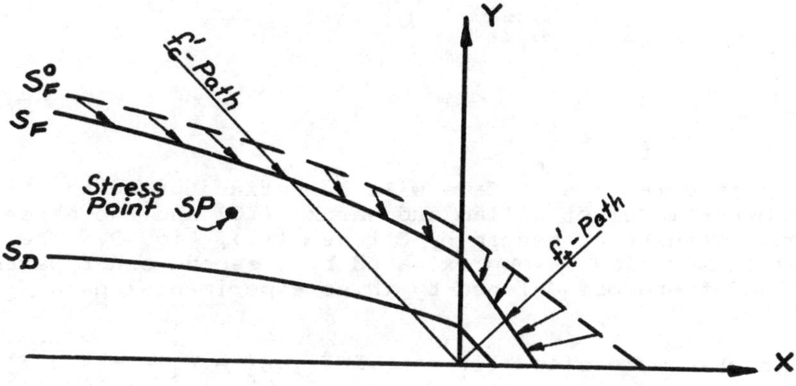

Fig. 7 Degradation of Failure Surface

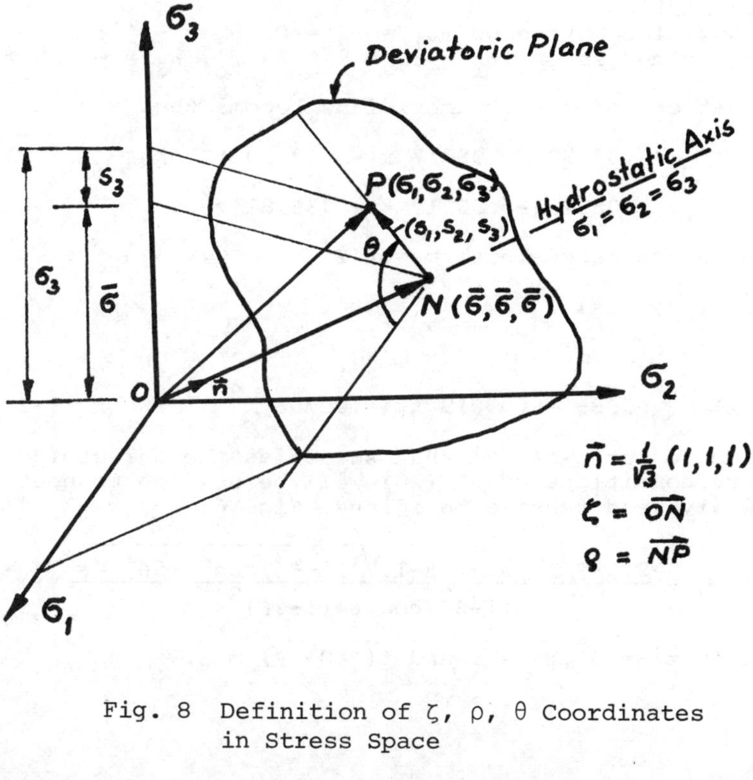

Fig. 8 Definition of ζ, ρ, θ Coordinates in Stress Space

$$x = \frac{\bar{\sigma}}{f_c'} = \frac{\zeta/\sqrt{3}}{f_c'}$$

$$y = \frac{\bar{\tau}}{f_c'} = \frac{\rho/\sqrt{5}}{f_c'}$$

The static failure surface will be defined using the five parameter model of Willam and Warnke (10) and the experimental results presented by Ottosen (11), Fig. 9. The tension meridian is approximated by a second-order parabola and therefore defined by three experimental data points:

1. high compression ($\sigma_1 = -.58f_c'$, $\sigma_2 = \sigma_3 = -4.0\ f_c'$);
2. biaxial compression ($\sigma_1 = 0$, $\sigma_2 = \sigma_3 = -1.16\ f_c'$);
3. uniaxial tension ($\sigma_1 = \eta\ f_c'$, $\sigma_2 = \sigma_3 = 0$). η is defined as the ratio of uniaxial tensile to compressive strength, $\eta = f_t'/f_c'$. According to test data reported by Hatano and Tsutsumi (12), $\eta = .061$ is used here. With the tensile meridian thus defined, its intersection with the x-axis is determined. This point corresponds to a hydrostatic tension state of stress. As the compression meridian also passes through this point, two more data points will define its parabolic description.
4. uniaxial compression ($\sigma_1 = \sigma_2 = 0$, $\sigma_3 = -f_c'$);
5. high compression ($\sigma_1 = \sigma_2 = -1.25\ f_c'$, $\sigma_3 = -6.15\ f_c'$).

The equations of the two meridians become thus

$$Y_c^* = .05988 - .95717\ x - .12410\ x^2$$

$$Y_t^* = .03323 - .53319\ x - .36781\ x^2$$

Y_t^* can be expressed in terms of Y_c^*,

$$Y_t^* = \beta(x)\ Y_c^*$$

where

$$\beta(x) = .55588 - .03019\ x + .00706\ x^2 \qquad x > -2$$

An elliptic curve $f(\theta,\beta)$ that satisfies the threefold symmetry conditions of S_F^* (10) will be used to connect the compressive and tensile meridians, Fig. 6,

$$f(\theta,\beta(x)) = \frac{2(1-\beta^2)\cos\theta+(2\beta-1)\sqrt{4(1-\beta^2)\cos^2\theta+5\beta^2-4\beta}}{4(1-\beta^2)\cos^2\theta+(1-2\beta)^2} \qquad 0^\circ<\theta<60^\circ$$

Note that $f(\theta=60^\circ,\beta) = 1$ and $f(\theta=0^\circ,\beta) = \beta$.

In the tension zone ($x > 0$), the parabolic expression for Y_c^* can be approximated by a straight line that intersects the x-axis at the same point. The static failure surface S_F^* is then completely defined as follows,

$$Y^* = f(\theta, \beta(x))\,[.05988 - .95717\,x - .12410\,x^2] \quad -2 \leq x \leq 0$$

$$= f(\theta, \beta(x))\,[.05988 - .96486\,x] \quad x > 0$$

Discontinuity Surface S_D. In order to simplify the model, a discontinuity surface with circular cross section on the deviatoric plane will be considered, by scaling down an average static failure surface of circular sections,

$$S^*_{F,ave} = \frac{1}{2}\,[S_F^*(\theta=60^o) + S_F^*(\theta=0^o)]$$

Using as the scale factor 0.68, we obtain for $S_D^o(x,y)$,

$$Y = .0317 - .5067\,x - .0547\,x^2 \qquad x \leq 0$$
$$= .0317 - .7502\,x \qquad x > 0$$

According to this surface, nonlinear behavior in uniaxial compression starts at $\sigma = -.16f_c'$. During inelastic straining, the discontinuity surface S_D will be obtained as an expansion of S_D^o in terms of the dissipated energy density W^p

$$W^p = \int_0^{\varepsilon_{kl}} \sigma_{ij}\, d\varepsilon^p_{ij} = \int_0^t \sigma_{ij}\, \dot{\varepsilon}^p_{ij}\, dt$$

This feature allows the model to reproduce hardening of the material.

When the stress point reaches S_F, failure takes place ($W^p = W^*$). In the postfailure regime, S_D shrinks according to the function $e^{-\alpha\chi}$, where χ is the postfailure dissipated energy density, $\chi = W^p - W^*$. If the stress point reaches S_F in the tensile region ($x > 0$), a crack is assumed to form without triggering the failure mechanism, i.e., S_D and S_F remain unchanged.

Failure Surface S_F. The failure surface S_F is assumed to have a shape similar to the static failure surface S_F^*, and has initially the position S_F^o corresponding to stress levels reached at very high monotonic loading rates and no inelastic straining. The scaling coefficient that relates S_F^* and S_F^o is calibrated by using test results of Hatano and Tsutsumi (12). Although the strain rates in that test series range from 10^{-5} to 10 in/in, only the higher rates were used in the calibration process. The model is not expected to reproduce quasi-static tests. The two failure surfaces are related thus by

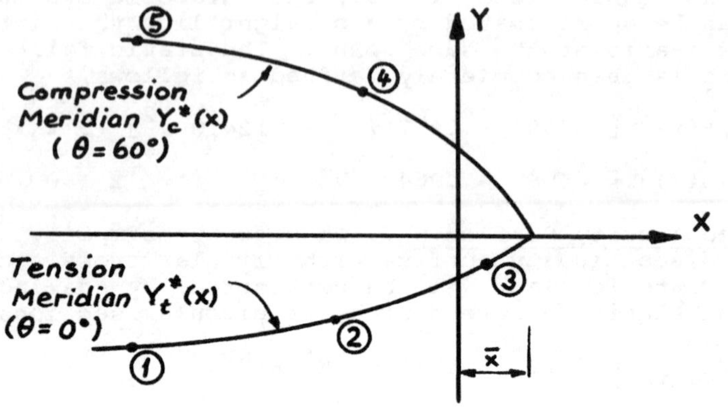

Fig. 9 Experimental Data Points Defining Static Failure Surface

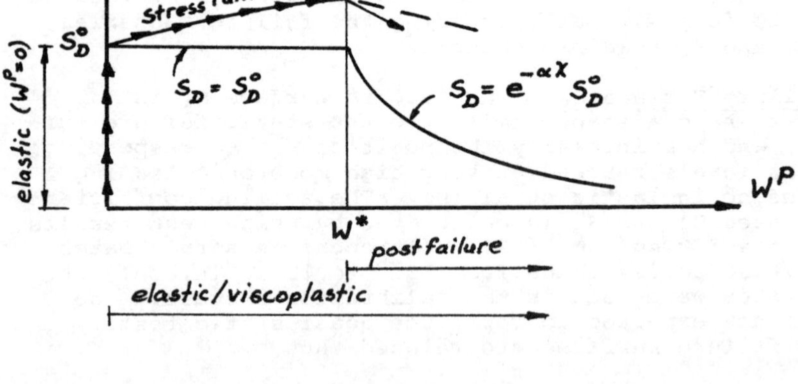

Fig. 10 Behavior of S_D and S_F Surfaces During Viscoplastic Straining

$$S_F^o = f(\theta,\beta)\ [1.51\ S_F^*\ (\theta=60^o)]$$

During viscoplastic straining, S_F^o will be considered to shrink as a function of the viscoplastic dissipated energy density W^p, to model the decrease of concrete strength with accumulation of damage. Since it has been observed that the strength of concrete in tension degrades more rapidly than in compression for the same value of W^p, two different degradation functions will be used, using again results from Ref. 12,

$$k_c = 1 - .0669\ W^p \qquad \text{for } x < 0$$
$$k_t = 1 - 6.724\ W^p \qquad \text{for } x > 0$$

In practical terms, the tensile strength of concrete drops to zero soon after viscoplastic behavior starts, while the compressive strength will have degraded only a little. In the tension zone ($x > 0$), this rapid degradation of S_F will eventually force the degradation of S_D.

The behavior of S_D and S_F as functions of viscoplastic work is illustrated in Fig. 10.

<u>Constitutive Equations</u>. The constitutive equations for this material model are as follows,

$$\dot{\varepsilon}_{ij} = \frac{1}{2G}\dot{s}_{ij} + \frac{1-2\nu}{E}\dot{s}\,\delta_{ij} + \gamma_0 \langle\Phi(F)\rangle \frac{\partial F}{\partial\sigma_{ij}}$$
$$= \frac{1}{2G}\dot{s}_{ij} + \frac{1-2\nu}{E}\dot{s}\,\delta_{ij} + \gamma \langle\Phi(F)\rangle \frac{\partial f}{\partial\sigma_{ij}}$$

where

$F(\sigma_{ij}, \varepsilon^p_{kl}) = \dfrac{f(\sigma_{ij}, \varepsilon^p_{kl})}{k} - 1$ = static yield condition that corresponds to S_D for the model presented here

$k = k(W^p)$ = strain hardening parameter

$\gamma = \gamma_o/k$ = fluidity parameter or viscosity constant

$\dot{s}_{ij} = \dot{\sigma}_{ij} - \dot{s}\,\delta_{ij}$ = deviatoric stress rate

$\dot{s} = \frac{1}{3}\dot{\sigma}_{ii}$ = volumetric stress rate

$\Phi(F)$ = some appropriate function of F

$$\langle\Phi(F)\rangle = \begin{cases} 0 & \text{for } F \leq 0 \\ \Phi(F) & \text{for } F > 0 \end{cases}$$

These constitutive equations indicate that the rate of

increase of inelastic strain is a function of the excess stresses above the static yield criterion (Maxwell-type viscosity law). Elastic components of the strain tensor are considered to be independent of strain rate.

Crack Monitoring Algorithm

The low tensile strength of concrete tempts one to accept highly approximate methods of modeling this material aspect. This intuitive feeling is reinforced by three factors.

1. Due to creep and shrinkage as well as previously applied service loads the concrete can often be considered to be precracked. For example, in a moment resistant frame, dead load and normal occupancy loads are likely to crack the beams at midspan and near the joints, so that in the event of an earthquake the potential plastic regions are already partially precracked.
2. Concrete regions stressed in tension are usually reinforced with steel. That means, once the concrete has cracked, the steel will carry the tension released by the concrete.
3. Experimental data on tensile strength of concrete are scarce and exhibit more statistical scatter than those for other concrete properties.

These observations tend to support the argument that for certain types of structures the tensile strength of concrete and details of crack propagation are not very important. If only the static ultimate strength of certain reinforced concrete members are to be determined, this is certainly the case. But the actual load-deformation behavior, especially under dynamic and cyclic load, is strongly dependent on the accurate representation of cracking behavior.

The crack monitoring algorithm adopted for our model is essentially an extension to three dimensions of the two-dimensional algorithm, outlined by Zienkiewicz et al (13). The "smeared crack" approach is used, with a dual criterion for crack initiation: 1) in terms of strain, a crack is assumed to develop after a principal elastic strain exceeds a critical value; 2) in terms of stress, a crack develops normal to the direction of maximum strain when the stress point reaches the failure surface S_F in the tensile region ($x > 0$). Since the strain criterion for crack initiation is given in terms of elastic strain, rate effects do not enter explicitly, as they do for the criterion in terms of stress.

Once the first crack develops at a point (Gauss

point), a record of its orientation, strain normal to the crack plane, and opening status is kept and constantly updated. Subsequent cracks have to be orthogonal to the first crack. This leaves one degree of freedom for the second crack, and the orientation of a possible third crack at that point is fixed in space.

The formulation of a crack affects the state of stress as follows:

1. Stresses normal to open cracks are reduced to zero.
2. Shear stress transfer across open cracks is reduced linearly down to zero for a certain crack width.
3. Stresses normal to closed cracks are not reduced, whereas shear stresses are reduced by a crack healing factor.

Since the three dimensional crack monitoring algorithm is very complex, only a simplified summary of the computational flow is given in Fig. 11.

FInite Element Implementation

The computational procedure for implementing the concrete model follows the standard elastic/viscoplastic algorithm (14,15), with the addition of:

1. variation of the fluidity parameter as a function of elastic strain rate;
2. change of the discontinuity and failure surface parameters as functions of the dissipated energy density W^p;
3. crack monitoring algorithm.

For the finite element implementation of the model, program DEGDYN, developed by Bicanic (16), has been expanded from two to three dimensions. The program incorporates a rate dependent concrete degradation model combined with a crack monitoring algorithm. It employs a central difference explicit algorithm for integration of the equations of motion, whereas an Euler forward integration scheme is used to discretize the equations controlling viscoplastic straining. Fig. 12 is a simplified flow diagram of the numerical implementation.

A 20-noded isoparametric solid element is used in conjunction with the material model. Several of the subroutines of program NONSAP-C (17) were used for this implementation, such as those for the computation of shape functions and principal strains in three-dimensional space.

CHECK IF PREVIOUSLY CRACKED

A - <u>if not previously cracked</u>
- compute principal elastic strains
- check for crack initiation
- if no crack initiated
  - standard stress evaluation procedure $(\underline{\sigma}_t = \underline{D}\,\underline{\varepsilon}_t^e)$. GO BACK TO MAIN PROGRAM.
- if a crack has been initiated
  - store crack(s) orientation
  - set appropriate flags
  - go to E

B - <u>if previously cracked in one direction</u>
- compute strains in crack direction and find rotation θ about it to new principal directions of strains
- check whether the old crack is open or closed and whether new crack(s) has (have) just opened orthogonal to the direction of the first one
- update appropriate flags
- go to E

C - <u>if previously cracked in two directions</u>
- compute strains in cracks orientation
- check whether cracks are open or closed and if third crack has just opened orthogonal to the first two cracks
- go to E

D - <u>if previously cracked in three directions</u>
- compute strains in cracks orientation
- check whether cracks are open or closed
- update appropriate flags
- go to E

E - <u>Stress evaluation procedure for cracked state</u>
- evaluate stresses in crack(s) orientation
- reduce stress(es) normal to open crack(s) to zero
- reduce shear stress across open cracks depending on crack width and the CROPE parameter
- reduce shear stress across closed cracks depending on the CRHEA parameter
- transform stresses back to global directions

GO BACK TO MAIN PROGRAM

Fig. 11 Cracking Algorithm Computational Flow

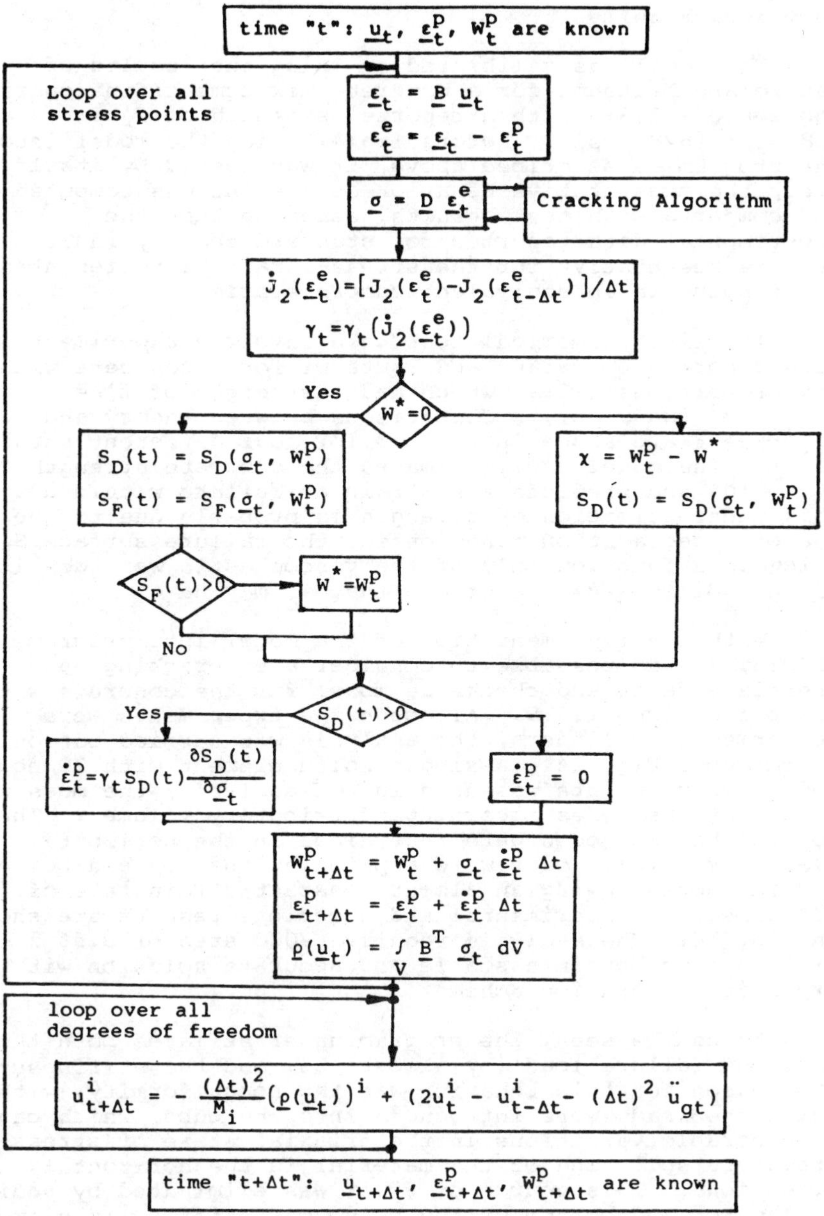

Fig. 12 Computational Flow

Numerical Results

The model was calibrated by using the results of Hatano and Tsutsumi for a concrete mix cement:fine:coarse aggregate = 1:3:5 with a reported strength of f'_c = 4380 t/m^2 (6280 psi). Before implementing the model into the program as described above, it was tested by itself, i.e., the constitutive behavior at a point was computed and compared with test results, assuming that the experimental data reported for standard test cylinders were representative for the stress-strain histories at a Gauss point in an equivalent concrete prism.

The first numerical test case involved experimental data reported by Hatano and Tsutsumi for a concrete with mix proportions 1:2:4, which had a strength of f'_c = 5900 t/m^2 (8460 psi). Comparisons between theory and experiment are shown in Fig. 13 for four different strain rates. The model underestimates the concrete strength by up to 10% and predicts the strain at failure within 5%. This underestimation of strength is probably due to the use of a degradation function for the failure surface S_F, which is a function only of the viscoplastic work density W_p, and disregards the effect that f'_c may have.

With the implementation of the model into program DEGDY3, it was possible to consider also cracking and inertia effects and check the model for the concrete with mix proportions 1:3:5. Although the experiments were performed on cylinders, the analysis was carried out on a 20 cm cube, Fig. 14. A single solid element with 20 nodes and 27 Gauss points was used in the analysis; the mass was lumped at the nodes according to tributary volumes. The top and bottom nodes were restrained in the horizontal plane. Vertical loads were applied at the top and bottom midside nodes, using an almost constant strain rate of .0458 sec^{-1}. Experimental and numerical results are shown in Fig. 14. The analysis required 8000 step of 0.56 x 10^{-5} sec. each to obtain a stable and accurate solution with the explicit integration scheme.

As can be seen, the program underestimates both the recorded failure load (by almost 10%) and the stiffness. The reason for this behavior was the nonuniformity, with which the loads were introduced into the cube, which caused considerable variations in the triaxial state of stress and premature splitting of the material in the horizontal directions. This source of error was eliminated by adding at the top and bottom of the prism one stiff solid element each, Fig. 15, with the sole purpose of enforcing a uniform stress state in the element under consideration. As a result, theory and experiment agree very closely.

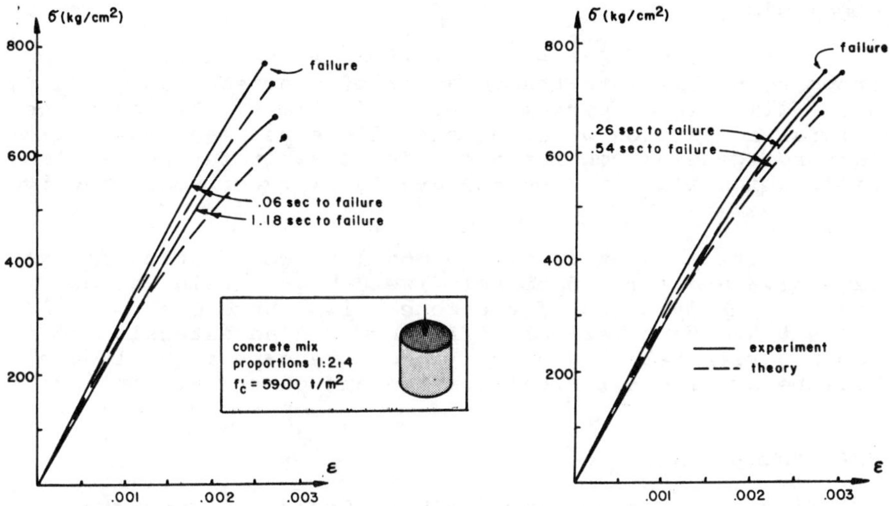

Fig. 13 Model Predictions and Test Results for Concrete Under Uniaxial Compression

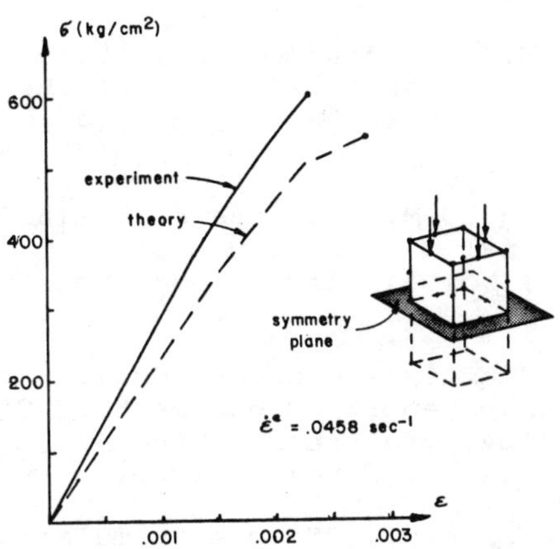

Fig. 14 Theory and Experiment - Single Cube

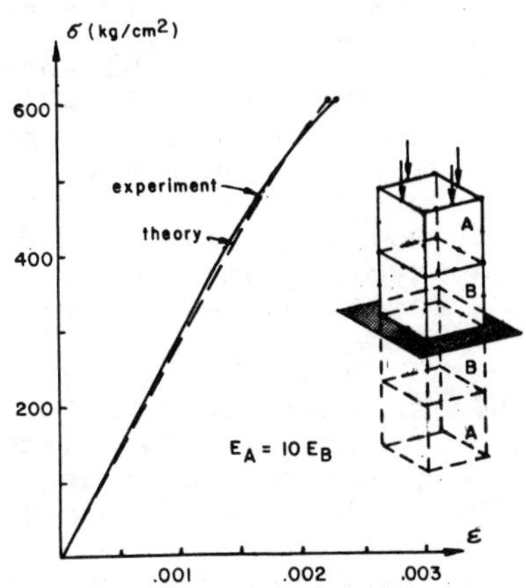

Fig. 15 Theory and Experiment - Two Cubes

Conclusions

A rate dependent elastic/viscoplastic model has been proposed to simulate the behavior of concrete subjected to three-dimensional dynamic states of stress. Although the initial comparison with one specific experiment was satisfactory, general conclusions have to await further evaluation, which also implies the availability of comprehensive test data.

In its present form, the model is computationally very expensive and limited to relatively high strain rates. Further improvements of the model will therefore concentrate on numerical efficiency and an extension to lower strain rates. In that case more experimental data will be available for calibration and numerical testing.

References

1. Winter, G., and Nilson, A.H., "Design of Concrete Structures," 9th Ed., McGraw-Hill, New York, 1979, p. 16.

2. Sinha, E.P., Gerstle, K.H., and Tulin, L.G., "Stress Strain Relations for Concrete Under Cyclic Loading," Journal of the American Concrete Institute, Vol. 61, No. 2, Feb., 1964, pp. 195-211.

3. Kupfer, H.B., Hilsdorf, H.K., and Rüsch, H., "Behavior of Concrete Under Biaxial Stresses," Journal of the American Concrete Institute, Vol. 66, No. 8, Aug., 1969, pp. 656-666.

4. Buyukozturk, O., and Tseng, T.-M., "Concrete in Biaxial Cyclic Compression," Journal of the Structural Division, ASCE, Vol. 110, No. 3, March, 1984, pp. 461-476

5. Stankowski, T., et al., "Tests of Concrete Under Multiaxial Load Histories", Report, Department of Civil, , Environmental and Architectural Engineering, University of Colorado, Boulder, 1983, 102 pp.

6. Stankowski, T. and Gerstle, K.H., "Simple Formulation of Concrete Behavior Under Multiaxial Load Histories," ACI Journal, March-April, 1985, pp. 213-221.

7. Delgado-Saavedra, H., "A Rate Dependent Model for Concrete Under Dynamic Multiaxial Loads," Ph.D. Thesis, Columbia University, New York, May, 1985.

8. Perzyna, P., "Fundamental Problems in Viscoplasticity," Advances in Applied Mechanics, Vol. 9, 1966, pp. 243-377.

9. Bicanic, N., and Zienkiewicz, O.C., "Constitutive Model for Concrete under Dynamic Loading," Earthquake Engineering and Structural Dynamics, Vol. 11, 1983, pp. 689-710.

10. Willam, K.J. and Warnke, E.P., "Constitutive Model for the Triaxial Behavior of Concrete," International Association of Bridge and Structural Engineering, Seminar on Concrete Structures Subjected to Triaxial Stresses, Paper III-1, Bergamo, Italy, May, 1974.

11. Ottosen, N.S., "A Failure Criterion for Concrete," Journal of the Engineering Mechanics Division, ASCE, Vol. 103, No. EM4, Aug., 1977, pp. 527-535.

12. Hatano, T., and Tsutsumi, H., "Dynamic Compressive Deformation and Failure of Concrete Under Earthquake Load," Proc. of the 2nd World Conference on Earthquake Engineering, Vol. III, Tokyo, 1960, pp. 1963-1978.

13. Zienkiewicz, O.C., Fejzo, R., and Bicanic, N., "Experience in Analyzing Plain Concrete Structures Using a Rate Sensitive Model with Crack Monitoring Capabilities," International Conference on Constitutive Laws for Engineering Materials, University of Arizona, Arizona, Jan. 10-14, 1983.

14. Argyris, J.H., Krempl, E., and Willam, K.J., "Constitutive Models and Finite Element Solution of Inelastic Behaviour," Formulations and Computational Algorithms in Finite Element Analysis (Eds. Bathe, K.J., Oden, J.T., and Wunderlich, W.), M.I.T. Press, Cambridge, Mass., 1977.

15. Zienkiewicz, O.C., Owen, D.R.J., and Cormeau, I., "Analysis of Viscoplastic Effects in Pressure Vessels by the Finite Element Method," Nuclear Engineering Design, Vol. 28, 1974, pp. 278-288.

16. Bicanic, N., "Nonlinear Finite Element Transient Response of Concrete Structures," Ph.D. Thesis, C/Ph/50/78, University College of Swansea, Wales.

17. Smith, P.D. and Anderson, C.A., "NONSAP-C: A Nonlinear Stress Analysis Program for Concrete Containments Under Static, Dynamic, and Long-Term Loadings," Los Alamos National Laboratory Report NUREG/CR-0416, Oct. 1978.

ANALYTICAL MODELS FOR CYCLIC LOADING OF RC MEMBERS

by

Hiroshi Noguchi[1], M. ASCE

ABSTRACT: Previous analytical studies on shear and bond deterioration of RC members subjected to reversed cyclic loading are reviewed, and the critical factors which need to be considered in the FEM analysis of RC members under reversed cyclic loading are discussed. First, previous analytical load-deflection predictions are compared with experimental results to review how well the deterioration of the restoring force characteristics can be modelled in the analysis. Modelling methods for the rotation of principal axes and the stress-strain curves of concrete under cyclic stresses, the criterion for opening and closing of cracks, and the bond stress-slip curves under cyclic stresses, are reviewed and discussed in light of the author's analytical models. Finally, the author's analytical examples for shear and bond deterioration of beam-column joints subjected to cyclic loading are presented.

INTRODUCTION

When flexural moment is dominant in properly designed reinforced concrete (RC) members subjected to reversed cyclic loading, simulating seismic loading, the strength is maintained and the deformation capability can remain excellent even after flexural yielding. In this case, the shape of the hysteresis loops are nearly spindle-shaped and the energy absorption capacity is stable. But when shear force is dominant such as in deep beams and short columns, and the bond deterioration of beam bars is advanced in beam-column joints, the shape of the hysteresis loops are contra-S-shaped with lower stiffness at lower load, (i.e. 'pinched effect'), because of crossed shear cracks and bond deterioration. In this case, the energy absorption capacity decreases gradually. Therefore, in design it is desirable to prevent such shear dominated failure. It is important to realize that even in members that yield in flexure and which have high shear reinforcement ratios, the pinched effect is caused by the interaction between flexure and shear during the hysteresis loop after flexural yielding under reversed cyclic loading, and there are cases where shear failures occur after considerable plastic deformations have taken place. The deterioration of the hysteresis characteristics caused by shear and bond deterioration has adverse effects on the behaviour of the whole structure.

---

1. Assoc. Prof. of Struct. Engr., Dept. of Arch. Engr., Chiba Univ., 1 - 33 Yayoi-cho, Chiba, 260 Japan

The above mentioned characteristics of RC structural members under reversed cyclic loading have been confirmed in many experiments. But it seems to be rather difficult to gain a quantitative and systematic understanding of the deterioration micromechanisms of the members from experimental studies only. Because measurements are difficult to take on cracked concrete, the experimental information is usually very limited.

As for analytical studies on members under reversed cyclic loading, for the case of flexure dominating, the material properties have been modelled fairly precisely and the hysteresis behaviour has been simulated rather correctly by the fiber model (Bazant (5), Kosaka and Yamada (17)). Analytical techniques for RC members in which shear dominates or bond deteriorates have been developed since the 1979's keeping up with the developement of nonlinear analytical methods for RC members by FEM. But because the modelling of the material behaviour under reversed cyclic loading is difficult, there are not as many analytical examples as for analyses of monotonic loading. In these analyses, the central aim was to get agreement with the corresponding experimental results on the hysteresis behaviour. It seems to be the present situation that the shear and bond deterioration process of the restoring force characteristics and the decrease of the energy absorption capacity have hardly ever been investigated quantitatively.

PREVIOUS SHEAR ANALYTICAL MODELS FOR CYCLIC LOADING OF RC MEMBERS

The FEM analysis of shear and bond behaviour of RC members subjected to cyclic loading was begun in the early 1970's. Thereafter, the following more than twenty analytical examples or state-of-the-art reports have been published to author's knowledge: Cervenka (7, 8), Muto and Masuda (26), Yuzugullu (52, 53), Imoto and Takeda (12, 13), Darwin (9, 10, 11), Agrawal (1), Shipman and Gerstle (43, 44), Nomura (39), Muto and Inoue (27), Sorensen (46), Sato and Shirai (41, 42), Kokusho (16, 40), Mochizuki (23), Aktan and Hanson (2), Tada and Takeda (48, 49), Meyer (20), Shirai and Sato (45), Kaneko (15), Watanabe (51), Sugano and Inoue (47), ASCE Committee (3), Noguchi and Naganuma (30, 32, 33, 38) and Meyer (21).

In this section of the paper, comparisons of previous representative analytical load-deflection curves with experimental results are reviewed, with the focus on how well the deterioration of the restoring force characteristics can be modelled in the analysis. The analytical models for the constitutive law, the opening and closing of cracks and bond, which are important under cyclic loading in those analyses were reviewed in the author's former report (32).

Muto and Masuda (26) analyzed a cantilever beam subjected to reversed cyclic loading, using the plasticity model based on the Drucker-Prager criterion. The analytical results of load-deflection curves were a little stiffer and could not very well represent the pinch effect just after unloading and lower load, as shown in Fig. 1. But this analysis was the first one and a pioneering effort on the reversed cyclic loading in Japan to author's knowledge.

Imoto and Takeda (12, 13) analyzed a column by the plasticity model based on the Drucker-Prager criterion and considering the downward portion of the stress-strain relationship of concrete after the peak stress. The analytical predictions of load-deflection curves gave reasonable agreement with the experimental results on the whole, as shown in Fig. 2. But the strength increased in the experiment as the number of cyclic loading increased, the rate of stiffness deterioration was almost constant, and the unloading stiffness was larger, as was also pointed out by the investigator.

Darwin and Pecknold (9, 10, 11) analyzed a shear panel tested by Cervenka and Gerstle, using their orthogonally anisotropic model. The analytical predictions of the load-deflection curves gave satisfactory agreement on the whole with the experimental results as commented by the authors and shown in Fig. 3. Also the analytical results for crack pattern, compressive softening of concrete and open cracks gave good agreement with the experimental results. But the discrepancy between the analytical and experimental results is recognized in the load-deflection curves under the intermediate load. As the shape of the hysteresis loop was spindle-shaped and not contra-S-shaped, which usually appears under dominant shear or bond deterioration, the validity of the analytical model for this severe situation does not seem to be clear.

Agrawal, Jaeger and Mufti (1) analyzed a beam tested by Paulay, using the plasticity model based on the von Mises yield criterion. The analytical and experimental load-rotation curves are shown in Fig. 4. As mentioned by the authors, agreement is considered to be good except for the low energy dissipation by the analytical model, and its subsequent stiff behaviour in the second positive loading cycle at higher loads.

Shipman and Gerstle (43, 44) reanalyzed a panel tested by Cervenka, using their updated analytical model in which bond deterioration under load cycles was considered. The analytical residual deflection is only about one-third of the observed residual deflection after the steel has yielded, as shown in Fig. 5. The analytical results of Darwin and Pecknold, who consider the effects of concrete deterioration under load reversals, are shown in Fig. 5. It is interesting that the effects of concrete degradation are responsible for the remaining two-thirds of the observed residual deflection. This report gave us good information for estimating each factor's role.

Nomura, Sato and Ono (39) analyzed short columns tested by Tani and Higashiura, et al., using the bi-linear model based on the Mohr-Coulomb failure criterion. The analytical stiffness just after unloading is softer and the area of the analytical hysteresis loop is much less than the experimental results, as shown in Fig. 6. The authors indicated that this discrepancy might be attributed to the inadequate modelling of bond of reinforcing bars under unloading.

Sorensen (46) compared the analytical results of one of Cervenka's shear panel, using two different material models for concrete under compression: a plasticity model based on the von Mises yield criterion and one based on endochronic inelasticity. The load-deflection curves

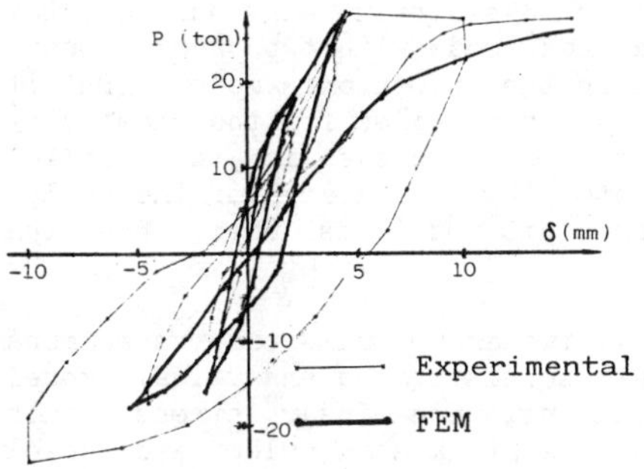

Fig. 1: Load-Deflection Curves for Beam (Muto et al, 1972 (26))

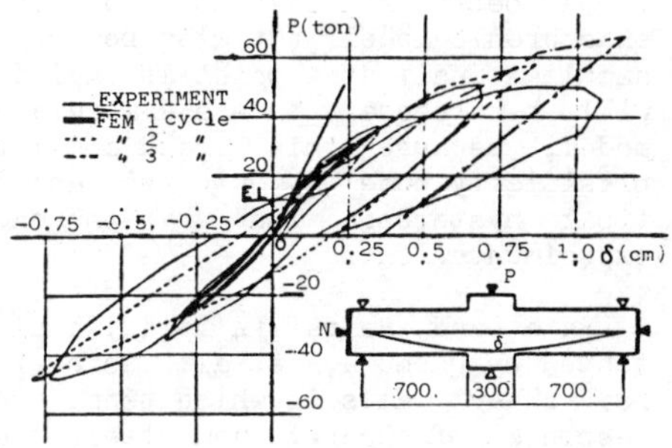

Fig. 2: Load-Deflection Curves for Column (Imoto et al, 1973 (12))

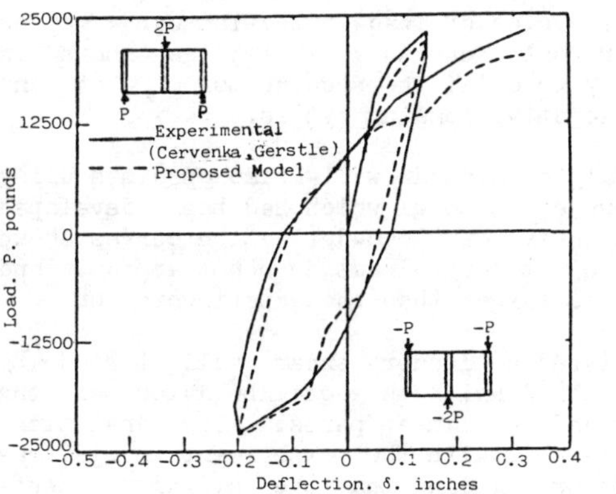

Fig. 3: Load-Deflection Curves for Shear Panel (Darwin et al, 1974 (9))

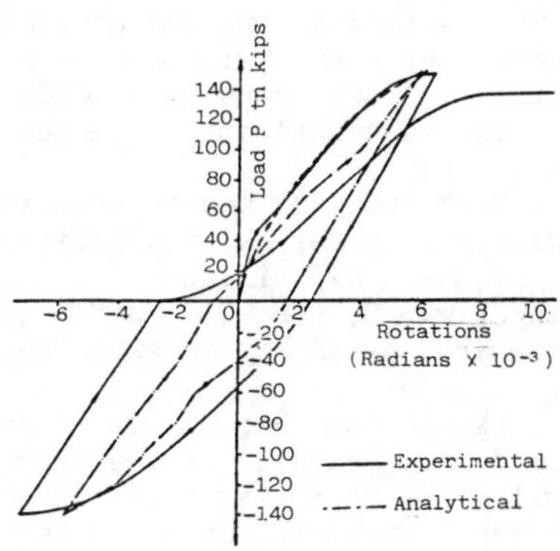

Fig. 4: Load-Rotation Curves for Beam (Agrawal et al, 1976 (1))

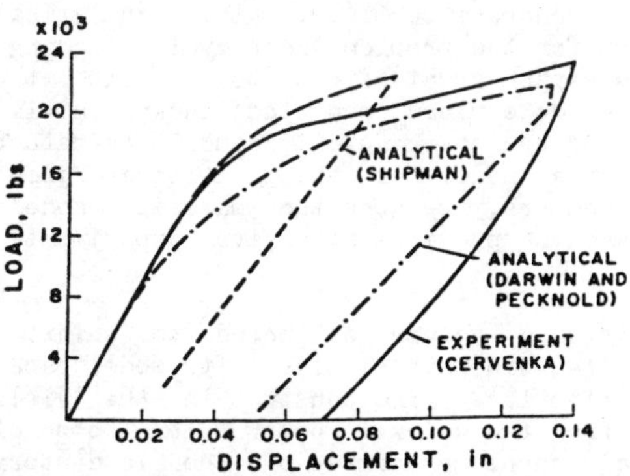

Fig. 5: Load-Displacement Curves for Panel (Shipman et al, 1977 (43))

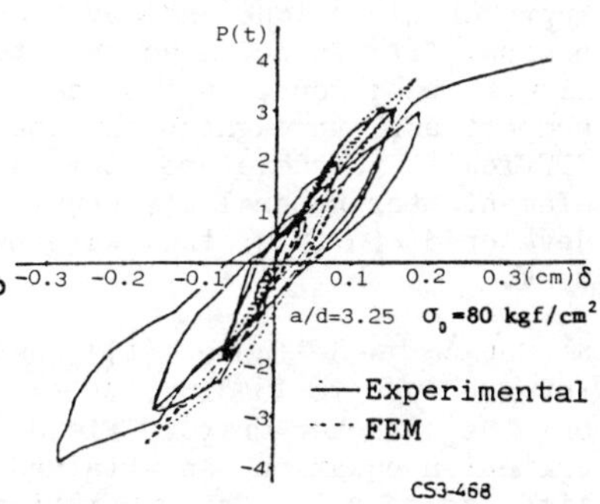

Fig. 6: Load-Displacement Curves for Column (Nomura et al, 1978 (39))

obtained by the endochronic model compare favourably with the experiments, as shown in Fig. 7. The author indicated that this is due to a better representation of cyclic stress-strain behaviour in the endochronic model, but also because a crack mode with two cracks opened simultaneously at a point is included in the endochronic model. But it will be necessary to use a more accurate model also for the plasticity model, because this crack model was considered even in the earlier plasiticity model by Muto (26) and Imoto (12). The area of the analytical hysteresis loop by the endochronic model is less than the experimental one.

Sato and Shirai (41, 42) analyzed two shear wall-frame specimens tested by Umemura, et al. using their sophisticated analytical model for shear walls in which tensile cracking, non-linear stress-strain response of concrete and steel, bond, aggregate interlock and dowel action were included. From studying the bond effect as shown in Fig. 8, the investigators concluded that if bond effects are not considered, the analysis not only underestimates loads for relatively small deflections, but also results in a large discrepancy in cyclic behaviour. The analytical two and one-half cycles of load-deflection curve gives good agreement with the experimental one as a whole, as shown in Fig. 9. But there is a tendency to delay the recovery of strength and to increase deflections in the negative loading cycles.

Mochizuki and Kawabe (23) analyzed a shear wall-frame specimen using Darwin's model (9) and their own crack model which had been developed from basic experiments. In the analytical load-deflection curves shown in Fig. 10, the slip deflection at lower loads does not appear, and the area of the hysteresis loop is larger than the experimental one.

Aktan and Hanson (2) analyzed a slender shear wall tested by Cardenas, et al. by the approach in which the total behaviour of the wall is separated into linear and non-linear parts. The analytical moment-curvature does not agree well with the experimental one, as shown in Fig. 11, but the authors claimed that the overall moment-curvature behaviour was of a reasonable shape, and that the assumption of taking the shear stiffness as a function of the normal stress on the element after the crack has closed led to a good representation of the shear pinching characteristics of concrete sections. This analytical approach is unique and covenient for the problem under cyclic loading, because it is not necessary to expend great effort to describe the linear behaviour, which can be determined from the known initial properties, throughout the loading cycles. But it seems to be rather difficult to develope the material model for the non-linear joint element for general structural members, because the material models developed in this study were based on previous extensive experimental work.

Sugano and Inoue (47) analyzed a column subjected to uniaxial reversed cyclic loading, using three-dimensional plasticity model based on the Drucker-Prager yield criterion. The constant in the yield criterion equation was obtained from the uniaxial compression tests of tied columns. The analytical envelope curve of the load-story displacement relationship gives good agreement with the experimental result as shown in Fig. 12, but the area of the analytical hysteresis

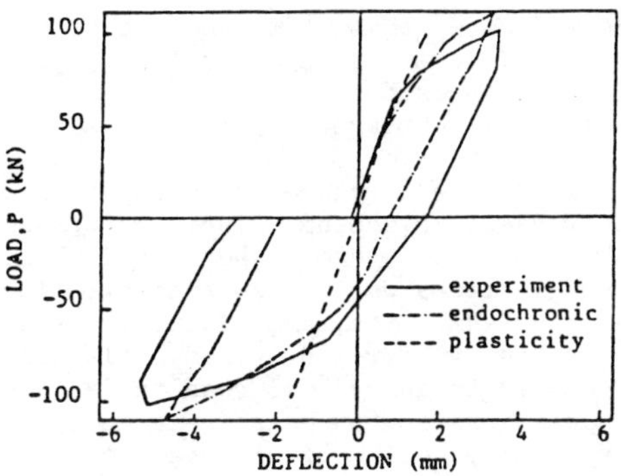

Fig. 7: Load-Deflection Curves for Shear Panel (Sorensen, 1979 (46))

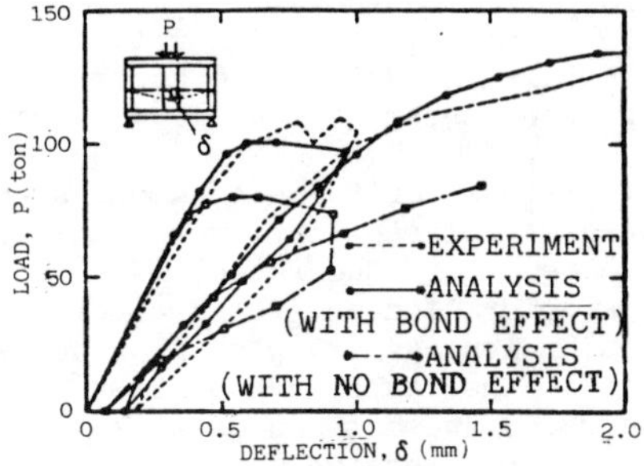

Fig. 8: Load-Deflection Curves for Shear Wall (Sato & Shirai, 1979 (41))

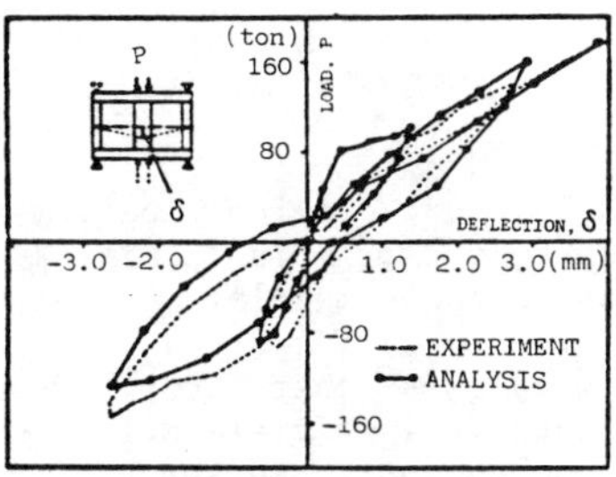

Fig. 9: Load-Deflection Curves for Shear Wall (Sato & Shirai, 1979 (41))

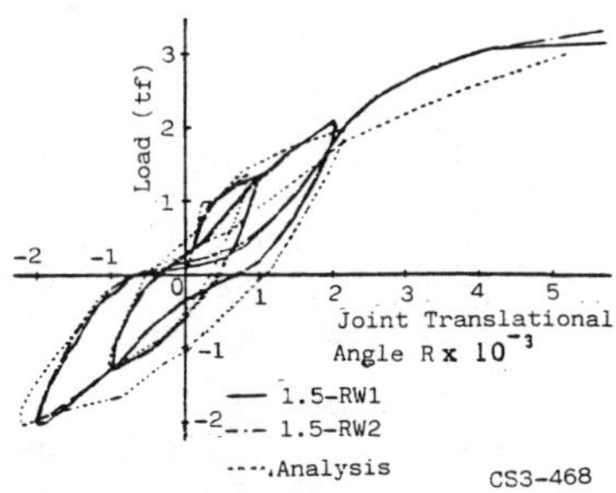

Fig. 10 : Load-Joint Translation Angle Curves for Shear Wall (Mochizuki et al, 1980 (23))

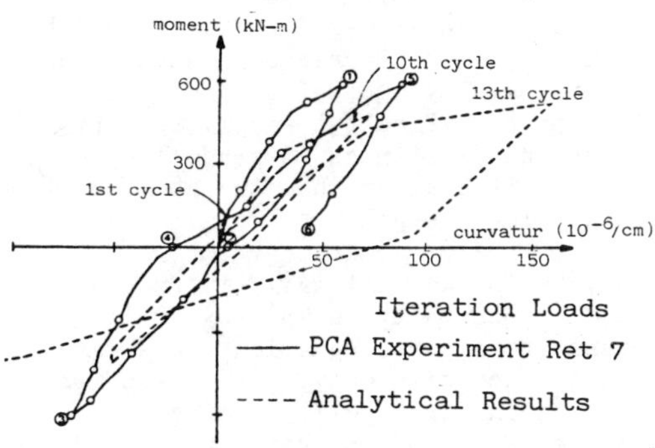

Fig. 11: Moment-Curvature Curves for Shear Wall (Aktan et al, 1980 (2))

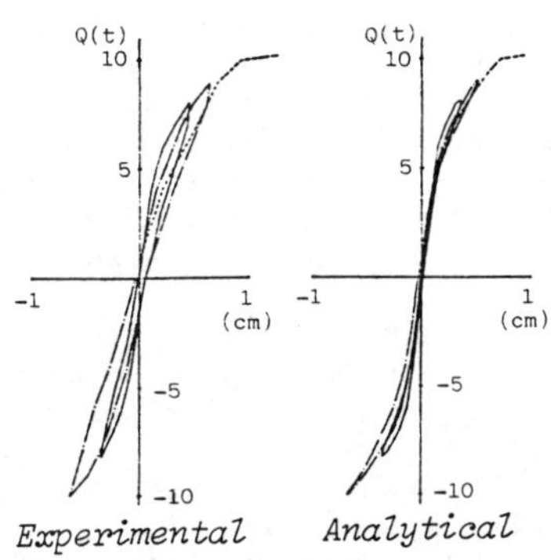

Fig. 12 Load-Story Displacement Curves for Column (Sugano & Inoue, 1982 (47))

loop is very small. The authors indicated that it is necessary to return to various assumptions of each element for investigation. But in this analysis, the accumulation of the residual strain in the ties perpendicular to the loading direction was captured.

In most of these previous studies, agreement between experiments and analytical results for unloading stiffness, pinching effect (contra-S-shaped) of the restoring force characteristics under shear and bond deterioration and the area of hysteresis loops was not sufficient. Too much emphasis was placed on how well the analytical calculations agreed with the experimental results, especially for the load-deflection curves, and not enough attention was paid to analytical results dealing with progressive shear and bond deterioration in each element and energy dissipation capacities under load cycles. It will be necessary in the future to examine efficiently the large amount of information that can be obtained from FEM analysis relevant to these questions.

MODELLING OF CONCRETE

Concrete Models For Cyclic Loading

Previous analytical concrete models for cyclic loading are classified into the following three groups: plasticity models, equivalent uniaxial strain models and endochronic models.

Plasticity models were adopted by many researchers and were used with the von Mises, Mohr-Coulomb and Drucker-Prager yield criteria. It seems that these models have been used commonly because of their simplicity, their direct applicability to steel reinforcement and the efficiency of computation. But it is questionable whether plasticity models are applicable for high stresses in a brittle composite material like concrete. The author has pointed out that the strains in both principal directions obtained by a plasticity model were considerably less than experimental results under high stresses (28). In this regard, refined analytical models in which the increase in strain under high stresses was considered were proposed by Imoto (14) and Miyashita et al. (22). But it seems to be rather difficult to apply the present models to the strain levels when the stiffness degradation under compressive cyclic loading becomes significant or when the compressive hysteresis behaviour affects the tensile strength.

The inelastic endochronic model, which was initially introduced for metals by Valanis, was extended to concrete by Bazant (4). In this model, many characteristic phenomena of concrete behaviour under cyclic loading are considered such as inelastic dilatancy, strain-softening that occurs after peak stress, unloading and cyclic loading. Though this model seems to be well suited for an analysis under cyclic loading, there are yet few remarkable achievements for shear analysis of RC members. This seems to be due to the specification of many constants in the formulation and it will be necessary to simplify this model. It will also be necessary to investigate the validity and applicability of this model in comparison with other models and experimental result as in Sorensen's study (46).

Darwin and Pecknold (9) applied their equivalent uniaxial strain model to a shear panel subjected to reversed cyclic loading, and obtained good agreement with experimental results. This model assumed orthogonal anisotropy in concrete with different tangent moduli in each principal direction. Young's modulus was determined from Saenz's equation by seeking the equivalent uniaxial strain, and the shear modulus was assumed to be independent of direction. In the author's computer program, this model was adopted by modifying the rotation of principal axes and the uniaxial stress-strain curves.

Rotation of Principal Axes in Concrete

The author and his colleagues are attempting to predict the behaviour of various structural members using one single computer program. They analyzed various members such as simple beams (35), beam-column joints (36), panels (34), short columns (31, 35) and shear walls (37) under monotonic loading, and most of the analytical results were in good agreement with the experimental results near the maximum strength. But when they analyzed a beam-column joint in which shear deterioration of the joint concrete and bond deterioration of the main beam reinforcement became significant under reversed cyclic loading, agreement with experiments could not always be obtained with the former analytical model (Noguchi and Naganuma (30)). In this case, it was necessary to establish proper rules for each material model under cyclic stresses and to solve various problems under unloading by the incremental load method.

The author has used Darwin's orthogonally anisotropic model based on the equivalent uniaxial strain (9) in his concrete model. In this model, the change in equivalent uniaxial strain due to rotation of principal axes which is within ± 45 deg. from the original principal axes at the first loading stage is:

$$\Delta\varepsilon_{iu} = \frac{\sigma_{inew} - \sigma_{iold}}{E_i} \qquad \text{..... (1)}$$

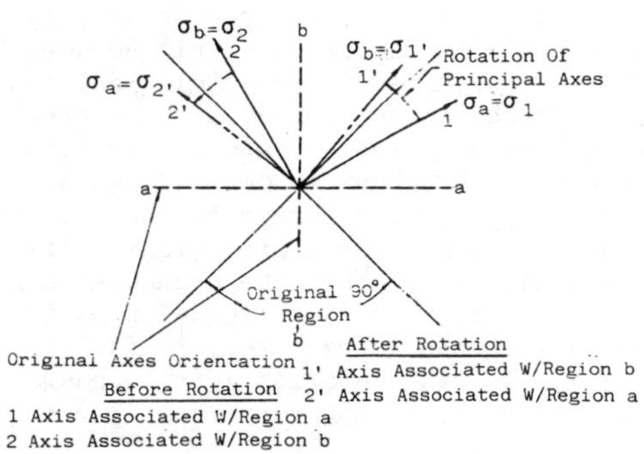

Fig. 13: Rotation of Principal Stress Axes (Darwin, 1974 (9))

in which σ_{iold} and E_i correspond to the ith principal stress axis before the load increment is applied, and σ_{inew} corresponds to the ith axis after the load is applied. The value for $\Delta\varepsilon_{iu}$ obtained in Eq. (1) is added to the previous value of ε_{iu} to give the totals for equivalent uniaxial strain on the newly oriented material axes. Even when the incremental load is smaller, the principal axes sometimes rotate greatly. When the principal axes rotate over ± 45 deg. from the original principal axes, the material axes are so transposed as to coincide with the new principal axes as shown in Fig. 13.

As for the rotation of the principal axes, Bazant (6) indicated that the non-linearity of concrete response is due to the existence of certain defects produced within the microstructure by previous deformations. The rotation of the axes of orthotropy means that we rotate these defects against the material. This implies that defects are assumed to be caused solely by the current stress state. This is certainly not true for concrete materials in general. Maekawa (19) also indicated that the applicabilty of previous plane stress constitutive equations was not proved under the rotation of the principal axes. He carried out biaxial loading tests of concrete with stress hysteresis where the principal axes rotate discontinuously and showed that their elastic-plastic and fracture model is applicable under the rotation of the principal axes. Recently he also carried out biaxial loading tests of concrete with continuous rotation of the principal axes and observed unsymmetrical and anisotropic stiffness behaviour and shear dilatancy with the intoduction of shear stress at higher stress states (18). The author hopes that such a basic experimental investigation will be deepened in the future in order to clarify the limitations of the analytical models by investigating the various assumptions.

Under the present conditions, an analytical example of a beam-column joint which is subjected to an axial force at first and then to reversed cyclic shear forces on the beams is introduced and investigated using the orthogonally anisotropic model based on the equivalent uniaxial strain.

In Darwin's model, the material axes are not transformed if the principal axes rotate within ± 45 deg. from their original position. Therefore, when the principal axes rotate nearly 90 deg. traversing the original position halfway during an incremental load, the material axes are not transformed although the new principal axes are located nearly at the position of the former principal axes. In this case, as the direction of E_i is different by about 90 deg. from that of σ_{inew}, the computational error of $\Delta\varepsilon_{iu}$ becomes larger if there is much difference in stiffness between the two principal directions, like under higher compressive and tensile stresses. When the accumulated rotation of the principal axes is over 45 deg. from the original position, the material axes are transformed even if the current incremental rotation is rather small. In this case, as the rotation gets to nearly 90 deg. in consequence, a similar error as in the former example occurs.

The following method was adopted in order to prevent the discontinuity of the equivalent uniaxial strain due to transformation and to reduce the error when computing the equivalent uniaxial strain with the rotation of principal axes. The new equivalent uniaxial strain is computed from the values which are transformed for new principal axes from the former equivalent uniaxial strain and principal strain and the new principal strain. In this method, the rotation angle from the original position, $\Delta\theta_t = \theta_{new} - \theta_o$ is computed at each loading stage, where θ_{new} is a new principal direction and θ_o is a principal direction at the first incremental loading stage. If $45° < \Delta\theta_T < 135°$, the material axes are transformed and various parameters on each principal axis (tangential stiffness, equivalent uniaxial strain, principal stress and others) are substituted. Subsequently the rotation angle from the former principal axis, $\Delta\theta_I = \theta_{new} - \theta_{old}$ is obtained, provided that $\Delta\theta_I = \theta_{new} - \theta_{old} - 90°$ just after the transformation of the material axes, where θ_{old} is the former principal direction. The components, $\{\varepsilon_{iuold}'\}$ and $\{\varepsilon_{iold}'\}$ in the new principal direction are obtained from transforming the former equivalent uniaxial strain $\{\varepsilon_{iuold}\}$ and principal strains $\{\varepsilon_{iold}\}$ by $\Delta\theta_I$. The incremental principal strains for the new principal axes, $\{\Delta\varepsilon_i\}$ are defined as the difference between the principal strains $\{\varepsilon_i\}$ and $\{\varepsilon_{iold}'\}$. The incremental equivalent uniaxial strain, $\{\Delta\varepsilon_{iu}\}$ is defined from the following equation by $\{\Delta\varepsilon_i\}$ and the equivalent Poisson's ratio, ν_u.

$$\{\Delta\varepsilon_{iu}\} = \frac{1}{1 - \nu_u^2}\begin{bmatrix} 1 & \nu_u \\ \nu_u & 1 \end{bmatrix}\{\Delta\varepsilon_i\} \qquad \text{..... (2)}$$

The new equivalent uniaxial strain is obtained from the following equation:

$$\{\varepsilon_{iu}\} = \{\varepsilon_{iuold}'\} + \{\Delta\varepsilon_{iu}\} \qquad \text{..... (3)}$$

In this method, the equivalent uniaxial strain is obtained from the principal strain and the equivalent Poisson's ratio. This is induced from experiences that when the tangential stiffness, E_i becomes nearly zero and the incremental principal stress, $\Delta\sigma_i$ gets smaller in Darwin's method ($\Delta\varepsilon_{iu} = \Delta\sigma_i / E_i$), the error of $\Delta\varepsilon_{iu}$ becomes larger and compression failure of concrete suddenly occurs with the rapid increase of the equivalent uniaxial strain. The equivalent uniaxial strains are imaginary strains, and do not transform under rotation of principal axes (3). Here, this method is used as a convenient method to minimize errors as much as possible in computation of the equivalent uniaxial strains under rotation of principal axes.

In the analysis of beam-column joints, very small shear forces were applied to both side beams, and the principal axes in the joint were inclined at 45 degree angle from the beam and column axes in order to make the original position of the principal axes just before axial loading on the column. By this procedure, the principal axes at higher loading stages, when concrete is stressed inelastically, form an angle of 45° with the beam and column axes and coincide with the original position. The principal axes are transformed at lower loading stages, when concrete is stressed nearly elastically under unloading or reloading of lower load. The transformation and large rotation of principal axes was prevented near the direction where the defects

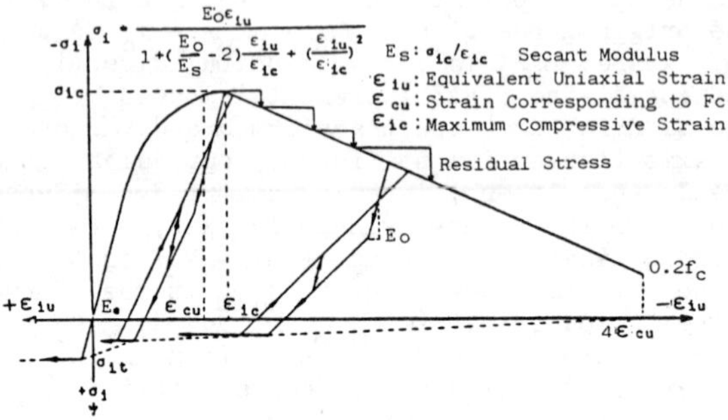

Fig. 14: Equivalent Uniaxial Stress-Strain Curves under Cyclic Loading

appeared in concrete and the analytical load-joint distortion curves and progress of failure gave good agreement with the experimental results. (33, 38)

Stress-Strain Curves Under Cyclic Loading

The equivalent uniaxial stress curves under cyclic loading, as shown in Fig. 14, were used in this analysis, originally based on Darwin's model (9, 11). Concrete is assumed to be linear elastic to be linear elastic to failure in tension. When concrete tensile stress exceeds the tensile strength, the stiffness is reduced to the value of one-hundredths except at concrete nodal points composing the crack-link element. After the concrete has been subjected to a compression load cycle, the tensile strength is reduced by a factor (down to one half) which depends on the maximum compressive stress level reached. This rule is based on Morita's study (25). The post-crushing behaviour of concrete, the strain-softening portion of the stress-strain curve, was represented by the step-by-step releasing method of the residual stresses as shown in Fig. 14.

MODELLING OF CRACKS

Crack Models For Cyclic Loading

Analytical models for cracking are divided into two groups. One is the discrete crack model, in which nodes are separated by setting up cracking surfaces between neighbouring elements, and the other one is the smeared crack model, in which the cracked element is assumed to be orthogonally anisotropic (Noguchi and Inoue (29)).

The smeared crack model has been used in most of the previous studies of reversed cyclic loading. This approach assumes innumerable cracks in an element, so that crack spacing and width are hard to evaluate and the crack pattern depends on the size of an element. But it is easy to handle even for cyclic loading and structures in which the prediction of crack patterns is difficult, because the separation of nodes are not necessary (29). As for a criterion for judging the opening and closing of cracks, which is most important in the analysis under reversed cyclic loading, a number of proposals were made, e.g., compressive force, plastic strain, crack width in which the Poisson's effect is considered, experimental equations for stress and crack width, strains when a crack occurs and others. The plastic strain criterion has been used most commonly. Even though the opening and closing of a crack is important, most of the above criteria are based on simple assumptions or empirical equations which have a limited range of application.

Criterion for Opening of Crossed Cracks

In the author's analytical model, the discrete crack model has been used. The crack-link element, shown in Fig. 15, is inserted between two nodes on both surfaces of a crack along the crack path which was predicted from test results or a previous elastic analysis. When the principal stress at a crack nodal point exceeds the modulus of rupture, the crack is initiated by setting the spring stiffnesses both normal and parallel to the crack surface from the initial large values to zero, and the nodal forces released by the crack are applied to the two crack nodal points. The reasons for using this model are as follows:

1) Research-directed analysis of RC members in which it is not so difficult to predict the crack pattern is chosen as the objective.

2) As the linearly varying strain triangular element with six nodal points is used in the computer program, it is possible to judge the opening and closing of a crack by the stress criterion of nodal points on the surface of the crack.

3) It is possible to represent the shear transfer across a crack by the reduction of spring stiffness.

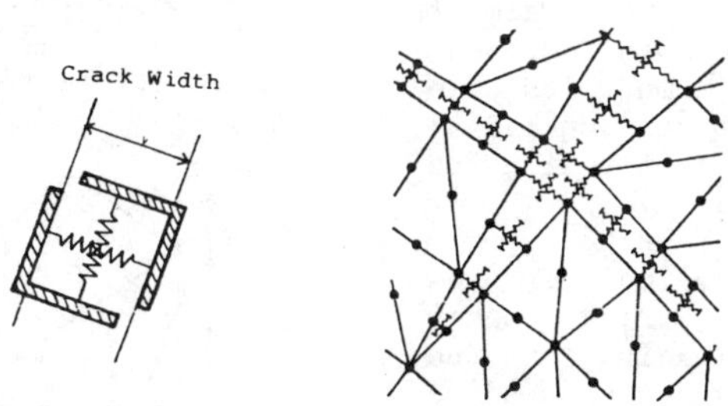

Fig. 15: Crack-Link Element

4) The discrete crack model is suitable to deal with those shear problems, where a few cracks are distinguished near the ultimate strength, or shear sliding failure occurs along a line which intersects the former cracks at an acute angle.

The following two phenomina appeared when problems were analyzed under reversed cyclic loading by this model. At first, some crossed cracks under negative loading opened under positive loading. The second is that the overclosing of a crack occurred under unloading and two concrete elements overlapped. These problems had negative effects on the load-deflection curves.

For the first problem, after some trial and error, it was realized that an actual crack forms not straight but in zigzags and that the crack direction does not always coincide with the principal stress direction. Therefore, a crack was assumed to open when the normal stress at an angle of 30 deg. to the normal line of a crack surface exceeded the tensile strength, even if the principal stress direction deviated more than 30 deg. from the normal line of a crack surface. According to this method, the positive cracks propagated in a similar way as in experiments and almost no negative cracks opened under positive loading in the analysis of beam-column joints under reversed cyclic loading.

Criterion for Opening and Closing of a Crack Considering the Effect of Local Contact

As for the second problem of overclosing of a crack under unloading, which was mentioned in the preceding paragraph, the following model was assumed with reference to the studies of Darwin (9) and Morita (25). When a crack is closing, the value of the spring stiffness of a crack link increases gradually as shown in Fig. 16. The proposed model represents the following effect of local contact between crack surfaces; as the shear force increases the slip parallel to the crack surface (w_h in Fig. 16), the recovery of the stiffness occurs earlier and the local contact effect becomes significant. It was found from the analysis using the model that the overclosing of a crack was prevented almost completely if the load increment was small enough, in which case also the analytical unloading stiffness agreed better with the experimental one.

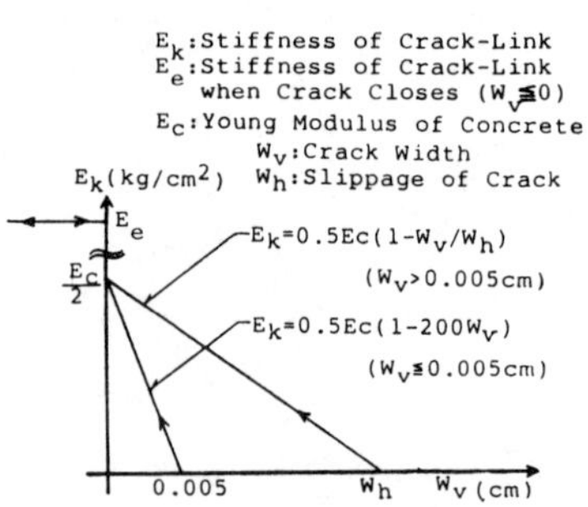

Fig. 16: Stiffness of Crack-Link When Crack is Closing

As the criterion for re-opening of a crack under reversed cyclic loading, at the beginning a crack was assumed to be re-opened when the incremental crack width was positive and the crack width grew larger than that when the crack had closed at the last time. But as a

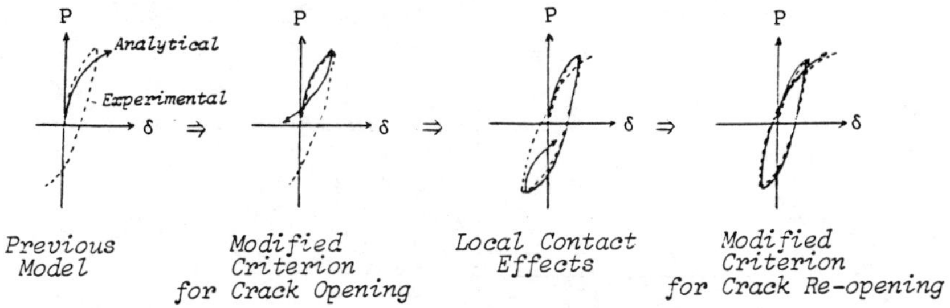

Fig. 17: Development of Load-Deflection Curves According to Modification of Crack Model

result of the modelling of the local contact effect, the problem arose where to judge closing of a crack. From some investigations, it was concluded that it was impossible to judge the re-opening of a crack by the crack width. In this model, the spring stiffness of a crack link increases gradually and the compressive force is accumulated in the crack link when the crack is closing. This compressive force must turn into tension for the re-opening of the crack. When the accumulative spring force turns positive (tensile), the nodal stress on the crack surface is also checked. Only when the stress normal to the crack is tensile, the crack is re-opened. Satisfactory results were obtained by this judging method in the analysis of beam-column joints under reversed cyclic loading (Noguchi and Naganuma (33, 38)). In this analysis, it was recognized that the crack propagation was not unnatural, and good results could be obtained for the simulation of actual cracking behaviour.

The development process of load-deflection curves with the above-mentioned modification of the crack model is shown in Fig. 17.

MODELLING OF BOND

Bond Models For Cyclic Loading

In the analysis of members under reversed cyclic loading, it has been common to assume perfect bond, or to use bond links. Sato and Shirai used a tension-stiffening model for the analysis of shear walls to represent bond (41).

When reinforcing bars are uniformly distributed throughout concrete elements as in a panel or a shear wall, it is assumed that the effect of bond deterioration is relatively small. But in Shipman and Gerstle's analysis of a panel under reversed cyclic loading, which was mentioned in Chapter 2, it was pointed out that bond deterioration can also be important, together with concrete deterioration, even in the case of panels. The influence is more pronounced in situations where

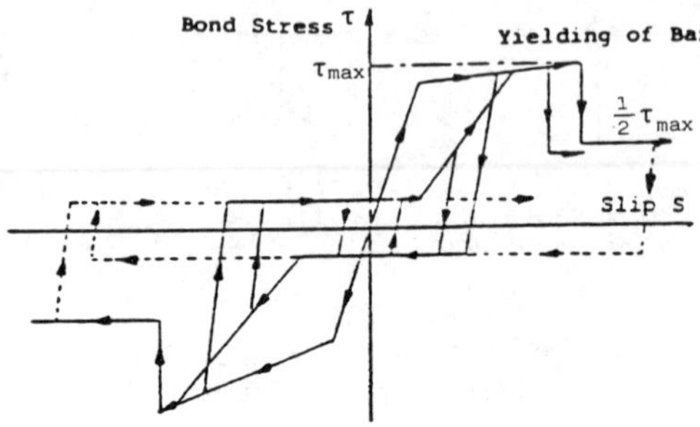

Fig. 18: Bond Stress-Slip Relationships under Reversed Cyclic Loading

the reinforcement is concentrated such as in beams, columns and beam-column joints, and it is expected that the analytical results are influenced significantly by the accuracy of the bond models for these members.

In the author's model, a bond link was used to model the bond mechanism of RC members under reversed cyclic loading. The bond deterioration near a crack, after the yielding of a longitudinal bar and beyond the maximum bond strength, and the changes in bond characteristics with stress conditions of concrete around a longitudinal bar are all considered in defining the bond characteristics for the whole member. The effect of the wedging action of a deformed bar on the behaviour of the whole member is an objective for further investigation.

Bond Stress-Slip Curves Under Cyclic Loading

Slip characteristics parallel to the bar axis in this analytical model were obtained from the bond stress-slip relationships under reversed cyclic loading, as shown in Fig. 18, which were originally proposed by Morita and Kaku (24). A modification of Morita's model is that when the bond stress reaches the maximum bond strength or when a longitudinal bar yields, half of the bond stress is released and the bond stiffness is set to zero, as shown in Fig. 18. When bond failure has occurred in both positive and negative directions, the bond stress-slip curve is assumed to be a parallelogram, as shown in Fig. 18, and as found in the experimental studies of Viwathanatepa (50). The spring stiffness of a bond link normal to the bar axis was assumed to be linear and elastic.

ANALYTICAL EXAMPLES OF SHEAR AND BOND DETERIORATION UNDER REVERSED CYCLIC LOADING

The author's analytical examples of shear and bond deterioration

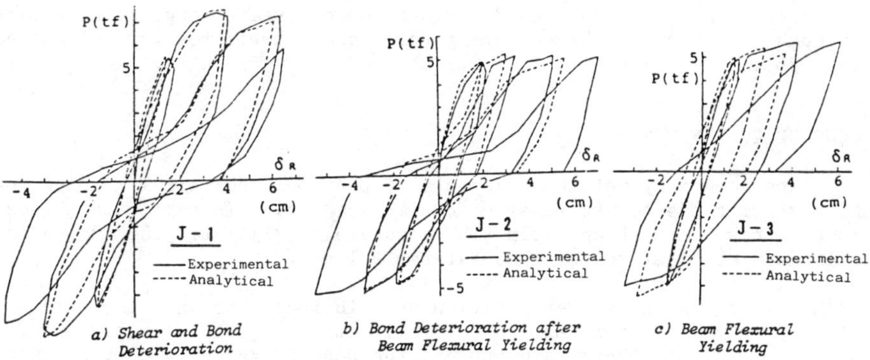

Fig. 19: Load-Story Displacement Relationships for Beam-Column Joints (33, 38)

under reversed cyclic loading, using the concrete, crack and bond models described above will now be described.

Three RC cross-type beam-column joints, J-1, J-2, J-3 were analyzed (33, 38). These correspond to half-scale specimens, tested by Kamimura for J-1 and by Tada for J-2 and J-3. They differed in the amount of reinforcing bars and in material properties. Specimen J-1 was a joint failure type, and J-2 and J-3 were beam yielding types. For specimen J-3, the bond strength and yielding strength of beam longitudinal bars within the joint were set three times and two times, respectively, as large as those in J-2, in order to prevent the pull out of beam longitudinal bars from the joint. Therefore, joint failure and bond deterioration of beam longitudinal bars for J-1, and bond deterioration of beam longitudinal bars for J-2, were considered to be dominant as deterioration factors. It was intended to eliminate these two deterioration factors for J-3.

The comparisons of analytical load-story displacement curves with the experimental results are shown in Fig. 19. For specimens J-1 and J-2, the analytical results gave good agreement with the experimental results for the shape of hysteresis loops. The analytical restoring force characteristics adopted the contra-S-shape with poor energy dissipation capacity, as cracking in the joint and bond deterioration of beam longitudinal bars progressed under the subsequent cyclic loading. This tendency was more pronounced for J-2 in which bond deterioration became significant after yielding of beam longitudinal bars. For specimen J-3, the analytical restoring force characteristics maintained the stable spindle-shape hysteresis loops. In the test of J-3, local stress concentrations occurred in the joint because the anchorage of beam longitudinal bars were insufficient in the joint. As a result, the joint failed during the second negative loading, and the analytical results ceased to correspond to the test results.

In this analysis, comparisons between analytical and experimental results were made also for crack propagation, progress of failure, joint behaviour, bond slip of beam longitudinal bars through the joint. From the analysis of components of the story displacement, the effects of joint failure and bond deterioration of beam longitudinal bars through the joint on the restoring force characteristics were discussed (33, 38).

CONCLUDING REMARKS

There are many debatable aspects of previous analytical models for RC members subjected to reversed cyclic loading. In particular, these models have not been able to accurately simulate the unloading stiffness and the area of the hysteresis loops.

Modelling methods for rotation of principal axes and the stress-strain curves of concrete under cyclic stresses, the criterion for opening and closing of a crack and the bond stress-slip curves under cyclic stresses were reviewed and discussed through explanation of the author's analytical models. These analytical models are especially important in the FEM analysis of shear and bond behaviour of RC members subjected to reversed cyclic loading.

The author's analytical results of beam-column joints that had been tested previously gave a good match with the experimental results, and it became possible to discuss the shear and bond deterioration process and their effects on the restoring force characteristics in more detail.

In the future, it will be important to rationalize the present analytical model by investigating the rotation of principal axes in concrete, the criterion for opening and closing of a crack, the quantification of bond characteristics in structural members and other controversial points in the previous studies by basic experiments and parametric analysis. At the same time, it will be necessary to deepen the understanding of the deterioration process of the restoring force characteristics by analyzing RC members more widely by comparison with experiments. Subsequently, the development of reinforcement details which improve ductility may be investigated and, eventually, design equations for ductility may be proposed as a result of the parametric analysis.

ACKNOWLEDGEMENTS

The author would like to express his appreciation to Mr. Kazuhiro Naganuma, a former graduate student, currently at Technical Research Institute, Ohbayashi-gumi Corporation, who helped to successfully develop the analytical models and analyze beam-column joints. The author also is sincerely grateful to Mrs. Agnes Kaneko of the Department of Civil Engineering, University of Toronto, for typing the manuscript.

REFERENCES

1. Agrawal, A.B., Jaeger, L.G. and Mufti, A.A., "Crack Propagation and Plasticity of RC Shear-Wall under Monotonic and Cyclic Loading," Proc. of the Conf. on FEM in Engrg., Univ. of Adelaide, 1976, pp. 37.1-18.
2. Aktan, H.M. and Hanson, R.D., "Nonlinear Cyclic Analysis of Reinforced Concrete Plane Stress Members," ACI Publication SP-63, 1980, pp. 135-152.
3. ASCE Task Committee, "State-of-the-Art Report on Finite Element Analysis of Reinforced Concrete," ASCE, New York, 1982, pps. 542.
4. Bazant, Z.P. and Bhat, P.D., "Endochronic Theory of Inelasticity and Failure of Concrete," Jour. of EM-Div., Proc. of ASCE, Vol. 102, No. EM4, Aug. 1976, pp. 701-722.
5. Bazant, Z.P. and Shieh, C.L., "Hysteretic Fracturing Endochronic Theory for Concrete," Jour. of EM-Div., Proc. of ASCE, Vol. 106, No. EM5, Oct. 1980, pp. 929-950.
6. Bazant, Z.P., Final Report of IABSE Colloquium, 1981, General Discussion of Session 2, Part 1, pp. 490-491, and Ref. (3): "Chapter 2. Constitutive Relations and Failure Theories," Appendix, pp. 136-137.
7. Cervenka, V. and Gerstle, K.H., "Inelastic Analysis of RC Panels: Theory," Publication, IABSE, Vol. 31-II, 1971, pp. 31-45.
8. Cervenka, V. and Gerstle, K.H., "Inelastic Analysis of Reinforced Concrete Panels: Experimental Verification and Application," Publication, IABSE, Vol. 32-II, 1972, pp. 25-39.
9. Darwin, D. and Pecknold, D.A., "Inelastic Model for Cyclic Biaxial Loading of RC," Univ. of Illinois, UILU-ENG, 74-2018, 1974, pps. 169.
10. Darwin, D. and Pecknold, D.A., "Analysis of RC Shear Panels under Cyclic Loading," Jour. of ST-Div., Proc. of ASCE, Vol. 102, No. ST2, Feb. 1976, pp. 355-369.
11. Darwin, D. and Pecknold, D.A., "Nonlinear Biaxial Stress-Strain Law for Concrete," Jour. of EM-Div., Proc. of ASCE, Vol. 103, No. EM2, Apr. 1977, pp. 229-241.
12. Imoto, K. and Takeda, T., "Nonlinear Analysis of Reinforced Concrete Members by FEM: Part 1," Proc. Annual Conv., AIJ, Oct. 1973, pp. 505-506. (in Japanese)
13. Imoto, K. and Takeda, T., "Nonlinear Analysis of Reinforced Concrete Members by FEM: Part 2 and Part 3," Report, Technical Research Institute, Ohbayashi-gumi, Ltd., No. 8, 1974, pp. 96-105. (in Japanese)
14. Imoto, K., "An Analytical Method for Three-Dimensional Nonlinear Responses of Prestressed Concrete Nuclear Reactor Vessels," Report, Technical Research Institute, Ohbayashi-gumi, Ltd., No. 21, 1980, pp. 1-13. (in Japanese)
15. Kaneko, Y., "Shear Analysis of Reinforced Concrete Members under Cyclic Loadings," Proc., JCI Colloquium on Shear Analysis of RC Structures, Introductory Report, JCI-C4E, June 1982, pp. 39-48.
16. Kokusho, S., Morimoto, T. and Hayashi, S., "Study on Restoring Force Characteristics of RC Members - Differences of Bond Slip in Longitudinal Bars and Restoring Force Characteristics by Three Loading Methods," Proc., Annual Conv., AIJ, Sept. 1980, pp. 1801-1802. (in Japanese)
17. Kosaka, Y., Tanigawa, Y. and Yamada, K., "Inelastic Analysis of

Concrete by Endochronic Theory," Trans. of AIJ, No. 326, 330, April, Aug. 1983, pp. 78-90, pp. 9-23. (in Japanese)

18. Li, Baolu and Maekawa, K., "Deformational Behaviour of Concrete under Principal Stress Rotation (Biaxial Compression-Tension Area)," Proc., JCI Colloquium on Finite Element Analysis of RC Structures, JCI-C8, Dec. 1984, pp. 29-34. (in Japanese)
19. Maekawa, K., "Flow Rule and Plane Stress Constitutive Equations in Concrete," Concrete Journal, Vol. 21, No. 8, Aug. 1983, pp. 103-121. (in Japanese)
20. Meyer, C., "Dynamic Finite Element Analysis of Reinforced Concrete Structures," Introductory Report, Vol. Band 33, IABSE Colloquium, Delft, 1981, pp. 65-84.
21. Meyer, C., "Earthquake Analysis of Structural Walls," Proc., Int. Conf. on Computer Aided Analysis and Design of Concrete Structures, Split, Yugoslavia, Part II, Sept. 1984, pp. 1479-1492.
22. Miyashita, T. and Hayamizu, Y., "Stress-Strain Relationships of Two-Dimensional Concrete Computed from Plasticity Theory," Proc., Annual Conv., AIJ, Oct. 1982, pp. 1669-1670. (in Japanese)
23. Mochizuki, S. and Kawabe, S., "Experiment of Shear Wall Restricted by Frame under Repeated Pure Shear: Part 3: Finite Element Analysis," Proc., Annual Conv., AIJ, Sept. 1980, pp. 1619-1620. (in Japanese)
24. Morita, S. and Kaku, T., "Bond Characteristics between Reinforcing Bars and Concrete under Cyclic Loading," Trans. of AIJ, No. 229, March 1975, pp. 15-24. (in Japanese)
25. Morita, S., Kaku, T. and Sudo, E., "Deformation of Reinforced Concrete Prism under Reversed Cyclic Loading," Proc., JCI Annual Conv., 1979, pp. 329-332. (in Japanese)
26. Muto, K., Tsugawa, T., Miyashita, T. and Masuda, K., "Nonlinear Analysis of Reinforced Concrete Members by FEM (Part 4: Under Repeated Forces)," Proc., Annual Conv., AIJ, Oct. 1972, pp. 685-686. (in Japanese)
27. Muto, K., Sugano, T., Miyashita, T. and Inoue, N., "Three-Dimensional Nonlinear Analysis of Reinforced Concrete Columns by FEM," Proc., Annual Conv., AIJ, Sept. 1978, pp. 1779-1780. (in Japanese)
28. Noguchi, H., "Finite Element Nonlinear Analysis of Reinforced Concrete : Part 1: Stress-Strain Relationships of Concrete under Biaxial Stresses," Trans. of AIJ, No. 252, Feb. 1977, pp. 1-11. (in Japanese)
29. Noguchi, H. and Inoue, N., "Analytical Techniques of Shear in Reinforced Concrete Structures by Finite Element Method," Proc., JCI Colloquium on Shear Analysis of Reinforced Concrete Structures, Introductory Report, JCI-C4E, June 1982, pp. 57-92.
30. Noguchi, H. and Naganuma, K., "Nonlinear Finite Element Analysis of Reinforced Concrete Beam-Column Joints: Part 3: Behaviour under Unloading and Reversed Loading," Proc., Annual Conv., AIJ, Oct. 1982, pp. 1665-1666. (in Japanese)
31. Noguchi, H. and Ihzuka, S., "Nonlinear Analysis of The Effects of Bond Chracteristics on Mechanical Behaviour of RC Short Columns," Proc., Annual Conv., Sept. 1983, pp. 2105-2106. (in Japanese)
32. Noguchi, H. and Naganuma, K., "Shear Analytical Model of Reinforced Concrete Members under Reversed Cyclic Loading," Proc., JCI 2nd Colloquium on Shear Analysis of RC Structures, JCI-C5, Oct. 1983, pp. 97-104. (in Japanese)
33. Noguchi, H. and Naganuma, K., "Nonlinear Analysis of Reinforced

Concrete Beam-Column Joints under Reversed Cyclic Loading," Proc., JCI 2nd Colloquium on Shear Analysis of RC Structures, JCI-C5, Oct. 1983, pp. 139-146. (in Japanese)

34. Noguchi, H. and Naganuma, K., "Nonlinear Parametric Analysis of Reinforced Concrete Panels," Proc., JCI 2nd Colloquium on Shear Analysis of RC Structures, JCI-C5, Oct. 1983, pp. 147-154. (in Japanese)
35. Noguchi, H. and Masuda, A., "Nonlinear Analysis of JCI Selected Specimens of RC Beams and Columns," Proc., Annual Conv., JCI, July 1984, pp. 513-516. (in Japanese)
36. Noguchi, H. and Terasaki, H., "Nonlinear Analysis of JCI Selected Specimens of RC Beam-Column Joints," Proc., Annual Conv., JCI, July 1984, pp. 661-664. (in Japanese)
37. Noguchi, H. and Maruta, M., "Nonlinear Analysis of JCI Selected Specimens of RC Shear Walls," Proc., Annual Conv., JCI, July 1984, pp. 741-744. (in Japanese)
38. Noguchi, H. and Naganuma, K., "Nonlinear Finite Element Analysis of Restoring Force Characteristics of Reinforced Concrete Beam-Column Joints," Proc., Eighth World Conf. on Earthquake Engrg., San Fransisco, July 1984, pp. 543-550.
39. Nomura, S., Sato, K. and Ono, H., "Analysis of Hysteresis of Reinforced Concrete Members under Cyclic Loading by FEM," Proc., Annual Conv., AIJ, Sept. 1978, pp. 1469-1470. (in Japanese)
40. Osanai, T., Yoshida, H., Hayashi, S. and Kokusho, S., "Experimental Study on Bond Slip of Reinforced Concrete Members Subjected to Flexural Moment and Shear Force," Proc., Annual Conv., AIJ, Sept. 1980, pp. 1795-1796. (in Japanese)
41. Sato, T., Shirai, N., Ihzuka, M. and Kato, S., "Hysteretic Characteristics and Inelastic Analysis of Reinforced Concrete Shear Walls: Part 1 and Part 2," Proc., Annual Conv, AIJ, Sept. 1979, pp. 1475-1478. (in Japanese)
42. Sato, T., Shirai, N. and Kato, S., "Hysteretic Characteristics and Inelastic Analysis of Reinforced Concrete Shear Walls: Analytical Bond Model," Proc., Annual Conv, AIJ, Sept. 1980, pp. 1803-1804 . (in Japanese)
43. Shipman J.M. and Gerstle, K.H., "Bond Deterioration in Concrete Panels under Load Cycles," ACI Annual Conv., March 1977.
44. Shipman J.M. and Gerstle, K.H., "Bond Deterioration in Concrete Panels under Load Cycles," ACI Jour., Feb. 1979, pp. 311-325.
45. Shirai, N. and Sato, T., "Analysis of Reinforced Concrete Shear Wall Structures - Applications and Experimental Verifications," Final Report, Vol. Band 34, IABSE Colloquium, Delft, 1981, pp. 687-700.
46. Sorensen, S.I., "Plasticity and Endochronic Inelasticity in Finite Element Analysis of RC," Final Report, Vol. Band 29, IABSE Colloquium, Copenhagen, 1979, pp. 59-66.
47. Sugano, T. and Inoue, N., "Three-Dimensional Analysis of Reinforced Concrete Column under Cyclic Shear Forces: Part 1 and Part 2," Proc., Annual Conv., AIJ, Oct. 1982, pp. 1233-1236. (in Japanese)
48. Tada, T. and Takeda, T., "Analytical Study on Deformation Characteristics of Plastic Hinge at Ends of RC Member and Bond Deterioration Process at Neighbouring Anchorage Zone: Part 1-7," Proc., Kanto Dist. Symp., AIJ, 1980-82. (in Japanese)
49. Tada, T. and Takeda, T., "Analytical Study on Bond Deterioration Process in RC Members," Report, Technical Research Institute,

Ohbayashi-gumi, Ltd., No. 22, 1981, pp. 49-53. (in Japanese)

50. Viwathanatepa, S., Popov, E.P. and Bertero, V.V., "Effects of Genaralized Loading on Bond of Reinforcing Bars Embedded in Confined Concrete Blocks," Report No. UCB/EERC-79/22, Univ. of California, Berkeley, Aug. 1979, pps.293.
51. Watanabe, F., "Bond and Shear in Reinforced Concrete," Proc., JCI Colloquium on Shear Analysis of RC Structures, Introductory Report, JCI-C4E, June 1982, pp. 93-110.
52. Yuzugullu, O. and Schnobrich, W.C., "Finite Element Approach for the Prediction of Inelastic Behavior of Shear Wall Frame Systems," Civil Engineering Studies, SRS No. 386, Univ. of Illinois, May 1972.
53. Yuzugullu, O. and Schnobrich, W.C., "A Numerical Procedure for the Determination of the Behavior of a Shear Wall Frame System," ACI Jour., July 1973, pp. 474-479.

NUMERICAL PROBLEMS IN NON-LINEAR FINITE ELEMENT ANALYSIS OF THE POST-FAILURE BEHAVIOR OF STRUCTURAL SYSTEMS

Koichi Maekawa* Jun Yamazaki** Takeshi Higai***

An investigation is made of the sources of problems and peculiarities encountered during finite element analysis of strain softening structures. Typical problems are first demonstrated by using a reinforced concrete deep beam structure. Second, a one-dimensional three-element example structure is used to demonstrate the problems of solution divergence, oscillations, or non-uniqueness of solution. Effects are discussed of the material models, such as hysteretic as contrasted to non-linear elasticity, load increment step size, tangent stiffness and non-negative stiffness convention.

1. INTRODUCTION

1.1 Problem

There exists the need to evaluate reserve strength, deformation capacity, and the deterioration process of reinforced concrete structures under repetitive or reversal loadings. However, successful results seem to be scarce with respect to reliable and convergent solutions in the strain softening region of the load-displacement relationship. Two possible sources of the difficulty are the ambiguities in representation of the material properties when dealing with large deformations, and the numerical problems such as the divergence of the solution process.

This paper will first report problems encountered in analyses using an existing finite element analysis program which takes into account material non-linearity [1][2][3], and second investigate the process of oscillation and divergence of the solution using a one-dimensional analysis model structure.

1.2 Existing Finite Element Methods of Analysis for Strain Softening

Two aspects of the analytical procedure govern the reality of the solution for structural analyses in the strain softening region of the load-displacement relationship. The one is the treatment of the negative element stiffness when the stiffness at the integration point goes into a strain softening region, and the other is the convergence algorithm.

* Assistant lecturer, Department of Civil Engineering, University of Tokyo, 7-3-1, Hongo, Bunkyo-Ku, Tokyo 113. JAPAN

** Associate Professor, Department of Civil Engineering, Tokyo Metropolitan University, 2-2-1,Fukazawa, Setagaya-Ku, Tokyo 158. JAPAN

*** Associate Professor, Department of Civil Engineering, Yamanashi University, 4-3-11, Takeda, Kofu, Yamanashi 400. JAPAN

In some of the existing analysis programs[5] there is an accepted convention to let the argument of the element stiffness equal to zero (or small enough positive number) if that element is in the strain softening region, and then to assemble the structure stiffness matrix. Next the unbalanced nodal forces are relaxed until a convergence criterion is satisfied. During this process, the decrease in stresses may be considered, or the stresses may be assumed to remain constant. For the former procedure no report that demonstrate convergence is known to the writers. In the latter procedure, the unbalanced forces are relaxed in the next step, which means that the actual constitutive relation has not been used.

A method referred to as the hyper elliptic function method[6] has been effectively used for analyses of geometrically non-linear problems. No report is known to the writers of successful application of this method to the strain softening regime with concrete structures.

Those are the reasons which motivated the investigation of divergence and oscillation of the solution process.

2. EXAMPLE OF DIVERGENCE IN ANALYSES OF DEEP BEAM (CASE STUDY)

The structure to be analysed is shown in Fig.1 along with the finite element mesh [10]. Prescribed displacements are imposed at the node A, and the reaction at that node is computed. The convergence algorithm follows Newton's iteration. For simplification the non-linear elastic material properties are as shown in Fig.2. For this model there is a one-to-one relationship between stress and strain. The tangent stiffness is used to form the structure stiffness matrix.

Fig.3 shows the load-displacement relationship that satisfies equilibrium. The convergence process is also shown for the reaction associated with the prescribed displacements (case A). The equilibrium equation is given by Eq.1. The convergence index is defined by Eq.2 and denoted by ϕ. The convergence criterion is prescribed by setting ϕ equal to or less than 1.0%.

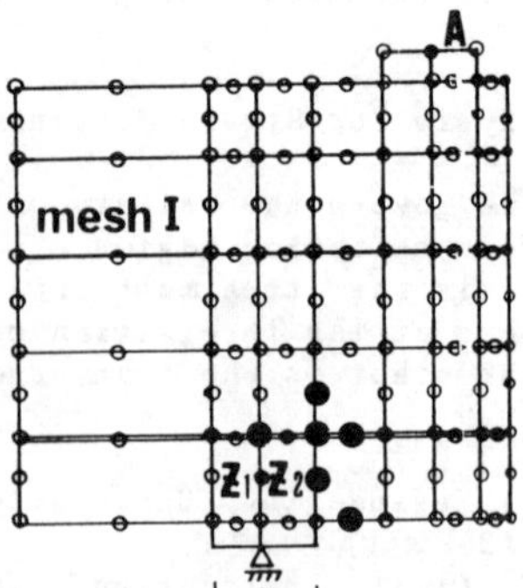

Fig.1 Mesh layout of example structures [10]

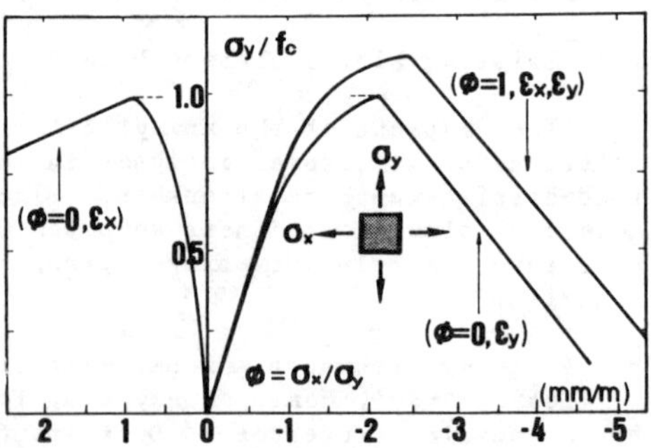

Fig.2 Non-linear elasticity model of stress-strain relationship

$$F_i = F_{eq,i}, \quad F_{eq,j} = \iiint_V B_{ji}\sigma_i dv \tag{1}$$

$$\phi = \sqrt{\sum_i (\Delta F_i)^2}/R_A, \quad \Delta F_i = F_i - F_{eq,i} \tag{2}$$

where, F_i is an external nodal load, $f_{eq,i}$ is equivalent nodal force, ΔF_i is unbalanced nodal force, R_A is reaction at node A, B_{ij} is strain matrix, and i is degree of freedom.

At step number 7, when the stiffness of elements z_1 and z_2 in Fig.1 entered into the strain softening region, the solution diverged. Fig.3 exhibits that the computed reaction at node A diverged under increasing numbers of iterations with some oscillation.

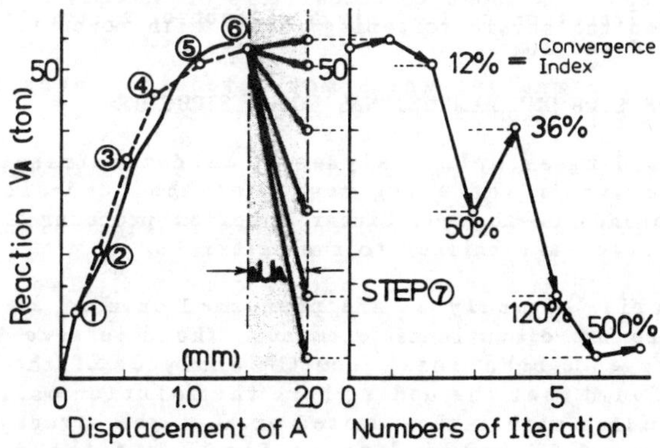

Fig. 3 Displacement vs. reaction and convergence process (case A)

The next analysis was performed as follows. The iteration was terminated after one execution, and the unbalanced nodal forces carried over to the next step (case B, refer to Fig.4). This process resembles the incremental integration[7]. In appearance the results seem to depict reasonable behavior, but in reality error accumulates as the load increment advances. The convergence index ϕ soon exceeded 100%. This can hardly be accepted as a correct solution.

Next, the stiffness of the element which was strained into the strain-softening region was set equal to zero, and the convergence procedure executed with iteration cycle numbers of 1, 5, and 10. The load-displacement relationships for these respective cases are given in Fig.4 (case C). Convergence was slow. The reaction kept decreasing as the number of cycles of iteration was increased. Deviation from equilibrium also became more pronounced.

Next, a path-dependent elasto-plastic-fracture model[8] was used to describe the material properties. This model recognizes negative stiffness. The problem is the same as before with reference to Fig.1. Newton's iteration is again used. The results are shown in Fig.5 (case D). In this case, convergence is attained even in the strain-softening region up to step 9. However, divergence suddenly occurs in the next increment, when the unbalanced nodal forces are relaxed due to cracking occurring in the element surrounding the strain softening element.

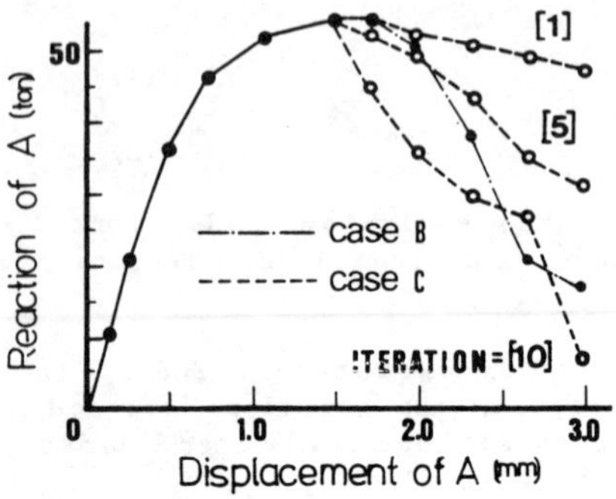

Fig.4 Effect of iteration cycle number 1, 5 and 10

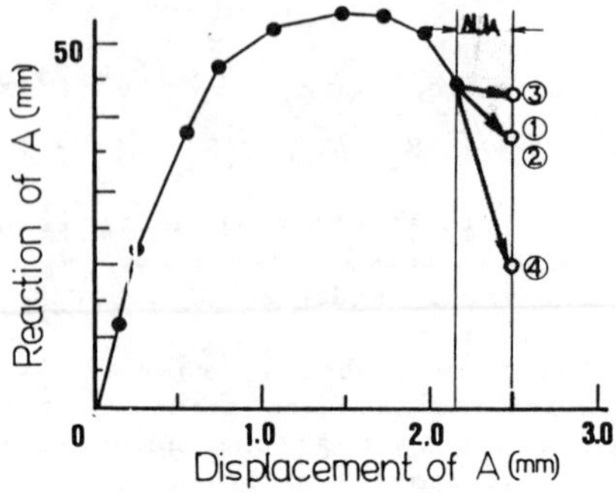

Fig.5 Effect of hysteretic material (case D)

3. EXAMPLE ANALYSES OF ONE-DIMENSIONAL MODEL STRUCTURE

In the preceding example analyses it is demonstrated that the behavior in the strain softening region of the load-displacement relationships depends on the non-linear solution procedure, load step and the constitutive law assigned to the materials.

The next series of analyses are preformed on a model structure composed of three one-dimensional elements, the objective being more detail of convergence behavior. From the analyses of the preceding section it was found that the nodes where the solution oscillates or diverges are confined within the limited area of the structure. Those nodes are indicated by solid dots in Fig.1, and the nodes where solution converges are indicated by open circles for the case A at step 7. The objective of this section is to simulate the behavior of those problematic nodes.

Three elements are connected in a row as shown in Fig.7. This stress-strain relationship is given in Fig.6.

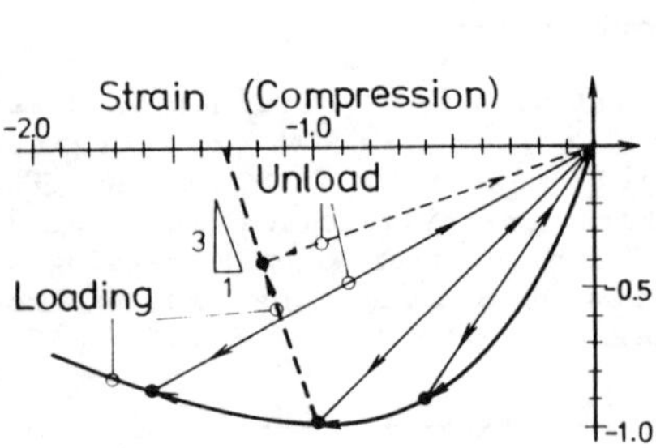

Fig.6 1D inelastic stress-strain relation

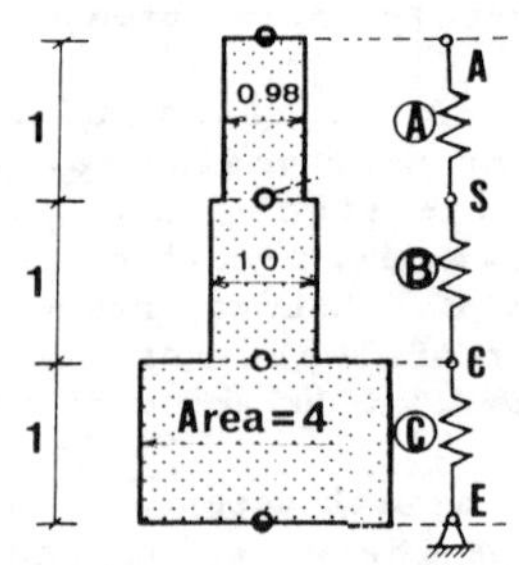

Fig.7 1D model structures

The solid lines show the first model, and the broken lines show the second model. (The second model is to be used in the discussion of section 3.3.) Prescribed displacement is imposed on node A, and the displacement of the internal node S is then computed. The cross sectional area of element A is set equal to 98% of that of element B. The graduated variation in cross sectional area of the three elements is to simulate the stress pattern in the critical reigon of the preceding example deep beam structure, where the stress flow is funneled into a narrow channel.

3.1 Strain Softening Materials Model

Comparison is made of the three different step sizes for the applied displacement increment at node A after the last stable step 9 which is just before attainment of the maximum resisting strength of the structure. Fig.8 shows the load-displacement relationships for the three cases. Fig.9 shows the variation in the strain in the elements with the progress of iteration. The structural stiffness matrix to be used while relaxing the unbalanced nodal forces are tangential stiffnesses even for the strain softening region. (The data with solid dots in Fig.9 are for the case of non-negative stiffness, which is to be discussed in section 3.2.)

Type 1 is for the smaller step size of load increment. Convergence is attained monotonically after two cycles.

Type 2 is for the larger step size of the load increment. Convergence is attained but after a larger number of iterations. For successive load increments also solutions are obtained with two or three cycles of iteration for both types. However, the load-displacement relationships differed grossly (Fig.8).

Type 3 is for an intermediate step size of load increment. Now the solution oscillates as shown in Fig.9 and convergence is not attained. This dependence of the solution on the step size of the load increment does not occur in cases with positive tangent stiffnesses.

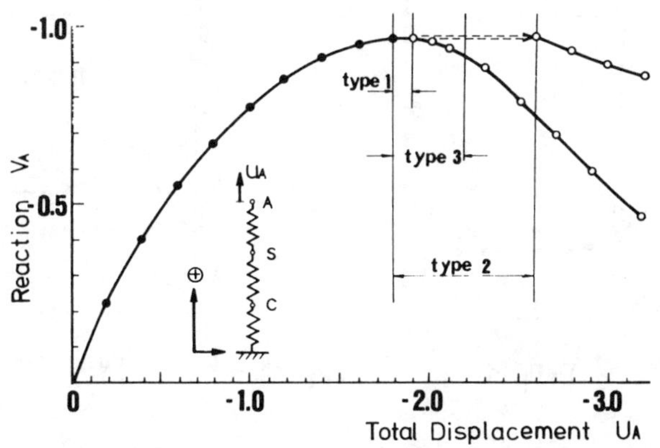

Fig.8 Effect of displacement increment step size (type 1,2,3)

For the type 1 solution element A (weaker element) is in the strain softening region, whereas element B is on the unloading path (from strain hardening region). This situation agrees with the theoretically correct solution. (See Fig.9)

For the type 2 solution both element A and element B are in the strain softening region. Equilibrium is satisfied.

For the type 3 solution it is assumed that the element A (weaker element) is in (the unloading path from the) strain hardening region, whereas element B is in strain softening region.

In the following discussion 'mode 1,2,3' refers to the state of the strains in the two elements. The meaning of 'type 1,2,3' is same as above.

Next, the equilibrium equation, Eq.1, is solved analytically with the results shown in Fig.10. The relationship between force and change in length is plotted for the stronger element B, the weaker element A, and the entire structure for all three modes.

The incremental changes of nodal forces due to small changes in nodal displacements are computed by Eq.3 as

$$\left\{d\ F_{eq,i}\right\} = [K]\left\{d\ u_i\right\}, \qquad [K] = \iiint_v B^T D B\ dv \tag{3}$$

where, u_i is nodal displacement and D is the tangent stiffness.

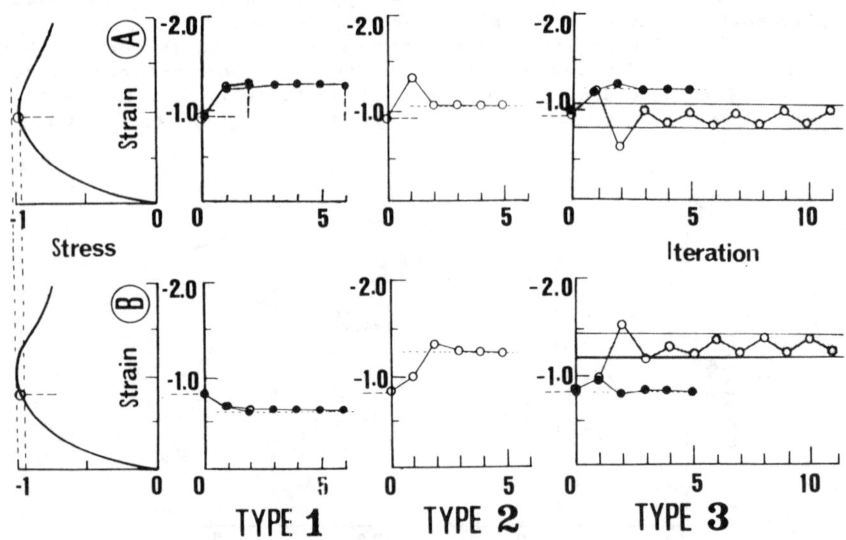

Fig.9 Variation of strains during iteration
O: tangent stiffness ●: non-negative stiffness

<u>Mode 1</u> solution in Fig.10 is the case where the weaker element is in the strain softening region and the stronger element is in the unloading region, which is correct answer.

<u>Mode 2</u> soution is the case where both elements are in the strain softening region.

<u>Mode 3</u> solution is the case where the stronger element is in the strain softening region and the weaker element is in the strain hardening region.

It can be seen from Fig.10 that there are regions for mode 2 and mode 3 where the displacement solution does not exist (jump region).

For the type 2 analysis, where the larger displacement increment is used, the first trial displacement value given is such that it causes strain softening in the two elements. The convergence procedure is started by using the tangent stiffness at the point corresponding to the assumed displacement. Thus, convergence is completed at the equilibrium point for the mode 2 which is the location nearest to the trial starting point.

For the type 3 analysis where an intermediate step size is used for the displacement increment, the stronger element is assumed to be in the strain softening region, and hence, the procedure is the one to search the mode 3 solution. However, as indicated in Fig.10, the assumed trial displacement falls in the region where the equilibrium solution does not exist. Hence, oscillation takes place. For the case such as this one, where the assumed mode is incorrect, conventional remedies of numerical procedure [7] have no effect.

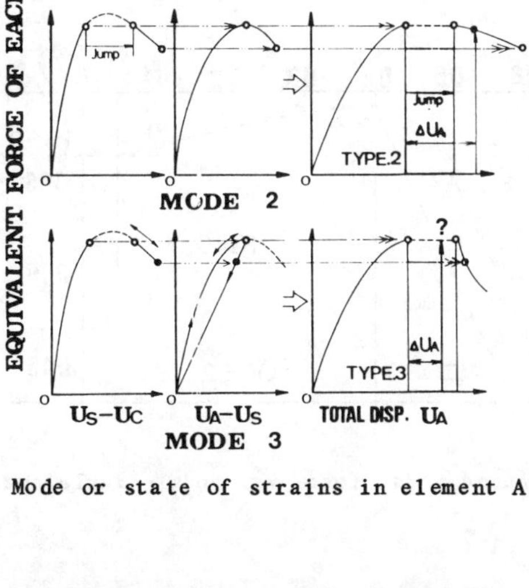

Fig.10 Mode or state of strains in element A and B

Next, a description will be made of the process of divergence for the type 3 solution. The equilibrium modification of the nodal displacement is given by Eq.4, where the structure tangent stiffness is K_s, the unbalanced nodal forces is ΔF_s, the correction increment of the nodal displacement is Δu_s and the tangent stiffness for element A and B are K_a and K_b respectively.

$$\Delta u_s = K_s \Delta F_s, \qquad K_s = K_a + K_b \tag{4}$$

The relationship between the nodal displacement $\bar{u}_s$ and the unbalanced nodal force ΔF_s for the internal node S can be obtained analytically, and is plotted in Fig.11. The displacement $\bar{u}_s$ is the movement from the equilibrium position for the preceding step. The solution for $\bar{u}_s$ which satisfies equilibrium is the point on the curve of Fig.11, where the ΔF_s coordinate is zero. The tangent stiffness K_s corresponds to the slope of the $\bar{u}_s$ – ΔF_s curve. On the figure the modes that tell the state of the two elements that are dictated by the value of $\bar{u}_s$ are also indicated. The positive and negative sign of the tangent stiffness K_s is also indicated.

In the convergence algorithm which uses the tangent stiffness, correction for the nodal displacement is always directed in the direction that decreases the value of ΔF_s. If the sign of K_s adopted was opposite to the one which is directing to the correct solution, the result is that the computed $\Delta \bar{u}_s$ is moved away from the correct solution. In the type 3 analysis the trial point for the first step is indicated by P_1 in Fig.11. This point is projected from the preceding converged solution using tangent stiffness and load increment at that stage. The point resides in mode 2 region. Locally the modification for the displacement is to move into the direction to reduce the unbalanced force. However, as is evident from Fig.11, the true solution is located in the direction opposite to the direction of the movement. Hence, the second trial value is moved to mode 3, giving rise to the unbalanced force. Then, if redistribution is executed using K_s at this point, the movement is then toward the true solution. Nevertheless, the displacement again falls inside mode 2 region, and the procedure oscillates.

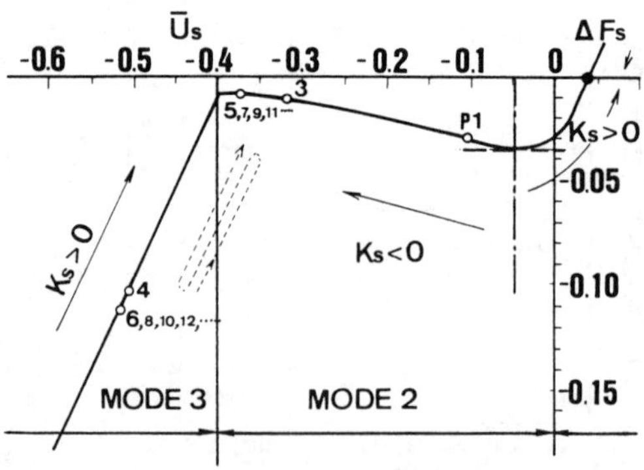

Fig.11 Unbalanced nodal force vs. nodal displacement (node S)

3.2 Convergence Process When Using Non-negative Convention for Negative Stiffness

The modified type 3 analysis is performed using the convention to set the element stiffness to zero in forming the global stiffness matrix even if the actual stiffness is negative. Then, for the global stiffness, which is to be denoted by $K_{s,NZ}$, each element of the stiffness matrix is positive. In the neighborhood of the trial displacement the direction of correction is to increase ΔF_s, but convergence is attained eventually (See Fig.9 ,11). In order for successful use of the negative stiffness, the starting (trial) value of displacement has to be positioned in the region of the correct mode (mode 1 for this case).

A question now would be whether or not there is generality in expecting convergence by using a non-negative convention for the element stiffness as was just mentioned above. For the type 2 analysis it is possible that two elements enter into strain softening region simultaneously. (It is frequently experienced that large unbalanced force is carried over from other elements even when the load increment is small, because of the stress released from cracked elements.) Then, if $K_{s,NZ}$ is zero or very small number, and hence by Eq.4, $\Delta\bar{u}_s$ should diverge. This is the mechanism that cracking is prescribed during solution process for the elements connected to the softening elements.

In the analysis of type 1 and convergent in mode 1 case, the number of repetitions increases when negative stiffness is ignored. This is because the process approaches the true solution in either way of treating the stiffness, since K_s and $K_{s,NZ}$ have the same sign owing to the fact that the first trial value for strain is set as mode 1 in Fig.11. In this case convergence is better with tangent stiffness.

Next, an analysis is performed for another material model which is to lose resistance very drastically after the peak strss as shown by broken lines in Fig.7. The relationship between the nodal displacements $\bar{u}_s$ and the unbalanced nodal forces ΔF_s is shown in Fig.12 for the type 2 case.

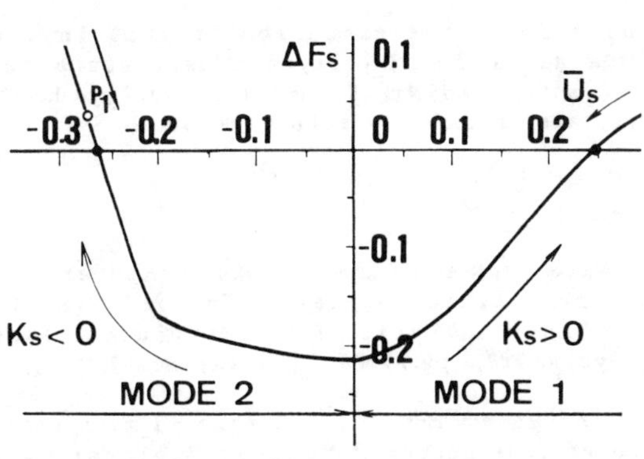

Fig.12 Unbalanced nodal force vs. nodal displacement with non-negative stiffness convention

When the tangent stiffness is used, immediate convergence is attained from point P_1 in Fig.12 to the solution of mode 2. Whereas, when positive stiffness is used, divergence occurs, because correction of the trial value moves away from the equilibrium solution owing to the fact that K_s and $K_{s,NZ}$ are of the same sign.

Convergence is fast when unbalanced nodal force is relaxed using tangent stiffness, when the direction of displacement to reduce the unbalanced force and the direction toward the equilibrium coincide on the trial displacement position. There is no problem if load increment is small. There occur cases where the direction toward true solution and the direction of correction by Eq.4 disagree. Then the stiffness to project the process toward the true solution is undecided.

4. CONCLUSION

The following characteristics of the behavior of the numerical solution process for the strain softening region were found from the investigation of example structures. A general solution process was not established.

For the non-linear solution process using Newton's iteration there can be cases where the direction of correction of trial solution projected by the tangent stiffness differs from the direction where the true solution exists.

Then, it is possible that divergence or oscillation occurs and that the true solution is not obtained. Further, there is no guarantee that there is the true solution in the direction opposite to the direction of modification that led to oscillation.

Acknowledgment

This research was supported by the Grant-in Aid for Scientific Research, Ministry of Education, No.58850100, under the supervision of Professor Hajime Okamura, University of Tokyo as the Principal Investigatior.

Contributions made to the execution of this investigation by Messrs. Masashi Odagawa and Li Baolu, graduate students of Nagaoka Technological University, and Mr. Kazumasa Ozawa, graduate student of University of Tokyo, are thankfully acknowledged.

References

[1] Maekawa, K., Niwa, J., and Okamura, H., 'Computer program COMM2 for analysing reinforced concrete,' Proceedings of JCI Second Colloquium on Shear Analysis of RC Structures, Japan Concrete Institute, Tokyo, 1983, pp.79-86, (in Japanese).

[2] Bathe, K. J., and Ramaswamy, S., 'On three-dimensional non-linear analysis of concrete structure,' Nuclear Engineering and Design, 1979, pp.385-409.

[3] Niwa, J., Maekawa, K., and Okamura, H., 'Non-linear finite element analysis of deep beams,' Final Report IABSE Colloquium, Delft, 1981.

[4] Hinton, E., Owen, D. R. J., and Taylor, C., 'Recent advances in non-linear computational mechanics,' Pineridge Press Ltd., Swansea.

[5] Okamura, H., and Maekawa, K., 'Evaluation of the finite element analytical modeling for reinforced concrete structures,' Proceedings of JCI Colloquium on Finite Element Analysis of RC Structures, Japan Concrete Institute, Tokyo, 1984, (in Japanese).

[6] Nishion, F., 'Continuum mechanics,' New Series in Civil Engineering, No. 5, Gihoudou Shuppan, 1984, (in Japanese).

[7] Washizu, K., et al., 'Handbook of finite element method of analysis,' Vol.1, Fundamentals, Baifuukan, 1981, (in Japanese).

[8] Maekawa, K., and Okamura, H., 'The deformational behavior and constitutive equation of concrete using the elasto-plastic and fracture model,' Journal of Faculty of Engineering, The University of Tokyo, (B), Vol.XXXII, No.2, 1983.

[9] Task Committee on Finite Element Analysis of Reinforced Concrete Structures of Structural Division, ASCE, 'State-of-the-art report on finite element analysis of reinforced concrete,' 1982.

[10] Niwa, J., 'Shear resisting mechanism of reinforced concrete deep beams,' Doctoral Dissertation, University of Tokyo, 1983, (in Japanese).

Microcomputer Challenge to Finite Element Analysis
- Possibilities for Application to Reinforced Concrete Structures -

Yoshiaki Yamada*

ABSTRACT

The behavior of reinforced concrete, either as a structural element or as a whole structure, is highly nonlinear and requires main frame computers with considerable memory and extreme speed for the analysis and/or design of actual structures. However, before using the finite element method in inappropriate and wasteful ways, particularly by inexperienced engineers, careful testing of the basic formulation as well as the performance of numerical procedures are mandatory. The advent of microcomputers has been fostering an environment in which scientists and engineers can develop, examine and verify the various facets of relevant computational routines. The present paper first gives an overview of recent microcomputer applications in the strucutral analysis field in order to demonstrate their usefulness. In the second part, which is retrospective in a sense, the paper surveys the development of nonlinear analysis procedures in Japan. They have been applied to metallic and composite structures and may hopefully find useful applications in the area of reinforced concrete.

INTRODUCTION

The engineering application of the finite element method dates back to the mid 1950's. Its power and usefulness have been well recognized in a variety of fields of engineering, and notable progress has been achieved in the past. However, the development has mainly relied on large computers, so that the updating of programs has not necessarily been an easy task, particularly when one is concerned with general purpose programs. It can be pointed out that many of the programs are based on obsolete numerical methods and the selection of elements is made from complicated, not modernized, libraries. Moreover, operation of many of such programs is confined to a batch data input mode.

The introduction of inexpensive and powerful microcomputers or personal computers seems to have completely changed the situation. Useful program

*Y. Yamada, Professor of Computational Mechanics
Ikutoku Technical University, 1030 Shimo-ogino, Atsugi City, Kanagawa 243-02, Japan
Professor Emeritus, University of Tokyo

segments can be easily developed on micros, based on the FORTRAN programming language. Programs can be operational both in batch and interactive modes. Furthermore, it should be pointed out that with appropriate procedures of opening files, the preparation and modification of data are facilitated to a great extent. Thus, softwares can be developed to operate effectively on both micro and main frame computer systems.

In this paper the possibilities of micros are examined, first with emphasis on their use for the development of fundamental program segments in the academic or scientific research environment. The experiences of the writer in this area are quite encouraging.

In the second part, the paper traces in a retrospective way the development of inelastic or nonlinear analysis methods in Japan. Unique solution procedures have been well established in the areas of both material and geometric nonlinearities and verified through studies of fundamental benchmark test problems. The future tasks are the implementation of the methods on microcomputers and their incorporation in novel analysis programs which are operational on both small and large computer systems.

It is important that programs are user-friendly, while the users are required to be familiar with the foundations and spirit of finite element analysis.

MICROCOMPUTER APPLICATION TO FINITE ELEMENT ANALYSIS

The most important advances in the use of microcomputers in structural analysis were made through the efforts of Wilson of University of California, who started the development in the late 1970's, following the completion of an earlier educational program CAL [12]. The software that evolved is the SAP-80 series of structural analysis programs, and version 1.0 of the user's manual appeared in 1980 [13]. In the beginning, SAP-80 was written specifically for small computer systems but has been modified to operate effectively on large computers as well. The most recent version features system-independent characteristics and can be rapidly converted to new computer hardware. It is worthwhile to trace the growth of SAP-80 by a series of articles [14,15,16,17] published in succession.

The following table from ref.[16] illustrates the hardware/software development cycle, in parallel with the birth and growth of the program series SAP-80:

Table 1. Hardware-Software Development Cycle

1977	CP/M 80 operating system verified
1978	First version of FORTRAN-80 released
1980	Bugs removed from FORTRAN-80
1981-83	Development of engineering software

Important new developments were the introduction of the Zilog Z-80 8 bit chip and recent Intel 8086 and 8088, Zilog Z-8000 and Motorola 68000 16 bit

CPUs and the Intel 8087 and 80827 floating point chip, with the reliable FORTRAN 77 compilers which have proved to be operational on the above novel CPUs. As for the OS (Operating System), the CP/M or CP/M 80 with 64K bytes of core storage is popular and versatile with 8 bit based microcomputers. However, it will be crucial to keep an eye on the innovation added by the introduction of MS DOS which is connected with 16 bit computer equipment, e.g. the IBM PC/XT and AT in the U.S., and the NEC PC 9801 in Japan. A promising alternative has been the Japanese TALOS 68K microcomputer, which is based on the Motorola 68000 chip and operaters under the CP/M 68K operating system.

It is interesting to note that the acronym CAL has been used in different ways, extending its meaning in the course of time. In the early days, CAL stood for Computer Analysis Language. The latest implication is connected with the development of the CAL/SAP program [6,18] which is based on the Computer Adaptive Language and also is intended principally for use in an educational environment for Computer Assisted Learning.

The kernel of the CAL/SAP program development consists of subroutines which are divided into the following three categories:

a series of free-field input routines,
in-core data management subroutines, and
an out-of-core data management system.

The FORTRAN source statements of the subroutines are given in three Appendices of ref.[18]. It is suggested that interested scientists and engineers familiarize themselves with the spirit of this development by a thorough study of the subroutines. The free-field input routines are indispensable for the compilation of software to be used on both small and large systems. In-core data management subroutines allow dynamic memory allocation with a minimum storage requirement, as well as programming effort. The out-of-core data management system allows data transfer from and to back-up storage at will.

It should be emphasized that the SAP-80 series programs have been written in FORTRAN, in contrast to others which were written in BASIC [3] exclusive for micro or personal computers. More importantly, the source statements of the code, e.g. of CAL 80-SIMPAL [2], have been open to the public. The source contains plenty of know-how which is believed to be quite beneficial for developers of engineering software.

In Japan, the use of microcomputers in structural analysis originated in the late 1970's almost at the same time as the launching of SAP-80 in the U.S. The hardware comprised the Z-80 CPU, and the first medium-sized program written was the program FEM4, which solved elastic plane stress, plane strain and axisymmetric problems [26]. The finite element adopted was an isoparametric element with 3 to 8 variable number nodes. It is pointed out that a pilot in-core data management routine similar to the one used in SAP-80 has been incorporated in FEM4. The extension of FEM4 to the nonlinear analysis regime has been program FEMN, which found heurisitc applications in the area of reinfoced concrete engineering.

In an earlier work [8] presented at 1981 Delft Colloquium, three useful routines were devised which take advantage of the welcome interactive characteristics of microcomputer systems. They addressed (1) restart of program execution, (2) node splitting for the simulation of crack formation along element boundaries, and (3) node relocation for the modification of element shapes. One application example was a block of concrete loaded symmetrically through a pair of top and bottom steel plates. A brief description of programs FEM4 and FEMN will be given in the following section.

Recent involvement of the writer in microcomputer applications has been directed towards the coordination of educational software. A series of programs for class instruction in mechanics of materials have been written: BAR1DM for one-demensional bar structures, FETRUS for plane truss, and FEBEAM for plane frame structures. Another fascinating task is the exploration of graphics capabilities, which form the basis for CAD/CAM applications. A recent product, PLOTBM, is a program that draws bending moment, shear force and axial force diagrams of plane frame structures together with a picture of the structure with the input loading condition. Combination of FEBEAM and PLOTBM enables the postprocessing of the results of finite element analyses. It has been found that the out-of core data management that opens the pertinent files in a systematic manner is quite useful for establishing efficient means of pre-and postprocessing of data associated with the finite element analyses. The microcomputer system on which the basic routines and application programs have been compiled and tested are TALOS-II and III.

PROCEDURES OF NONLINEAR ANALYSIS

The possibilities of the finite element method, as applied to reinforced concrete, have been reviewed extensively in the report [4], particularly with reference to the development in the U.S. from the earliest attempt made in 1967 to 1982, the time of publication of the report. Since then, the progress has been very rapid and quite a few examples of succcess will be reported in this Seminar as the result of enthusiastic pursuit of the applicability of the method. However, a systematic tracing of the progress as a whole may be rewarding, as useful techniques or strategies might be discovered even in publications which do not seemingly have direct connection with concrete engineering. The purpose of this section is to survey the methods that are used in Japan for the analysis of inelastic or nonlinear problems but have not necessarily been given frequent reference in the foreign literature.

The application of the finite element method in Japan to nonlinear problems started in 1967 when the elastic-plastic stress-strain matrix was established in a covenient closed form [19]. The constant stress and strain triangular element was used in an example solution and the computational procedure let one element yield at a time at each incremental step. This technique, which was later called the r-minimum method, has been extesively used in inelastic analysis programs in Japan. From the beginning, the tangent stiffness method has been popular in Japan, in contrast to the trend in Europe, particularly in England, where preference has been given for the initial stress method until recently.

The elastic-plastic stress-strain matrix described for the isotropic von Mises material in ref.[19] was soon extended to include Hill's orthotropic-anisotropic material and then to the Mohr-Coulomb or equivalently to Prager-Drucker soil-like material.

The next extension of the theory was for material exhibiting time-dependent behavior [20]. The mechanical model approach has been adopted, in which the material is simulated either by the multi-element generalized Voigt (Kelvin) or the Maxwell model. It should be noted that the mechanical model formulation embraces the equation-of-state approach by a procedure properly adjusting the property of its viscous component, e.g. of the generalized Maxwell model, so that it represents the relevant nonlinear behavior determined experimentally by the creep and/or relaxation test [21]. The finite element method, when applied to time-dependent materials, reduces the system of governing linear differential equations to the time integration of the discretized finite difference equations.

A series of studies described above which were made in the area of material nonlinearity have been summarized in a book [25]. It is worth remarking that the formulaiton given in [25] includes an anisotropic Prager-Drucker material, as well as Tresca material in a consistent manner, without introducing a complicated auxiliary angular parameter as used in the formulation in refs.[10,32]. The program which was written in the course of the above development has been the EPIC-IV.

The program EPIC-IV, though its coverage is limited to the Hill's anisotropic plastic material, is completely bug-free and has proved to give accurate solutions within its capacity for plane stress, plane strain, axisymmetric and generalized plane strain problems. Equations are solved using a conjugate gradient method, but a profile sky-line solver version is available as well. The monograph [27] which introduced the EPIC-IV program contains the complete source code and serves as the user's manual.

The EPIC-IV and other inelastic analysis programs based on the r-minimum method have been used for the benchmark testing, qualification and verification of structural analysis codes [28]. Though the augmentation of the scope of EPIC-IV is not contemplated, it can be used as a standard for the verification of the programs, both academic and commercial. In connection with the benchmark problem test, the action of the NAFEMS [7] is worth noticing. One of the original aims of NAFEMS is the establishment of standards and testing procedures in the development of error-free finite element codes.

Besides the material constitutive relations and failure theories, with which we are profoundly concerned when analysing reinforced concrete, modelling of reinforcement and representation of bond are other crucial subject areas. Ref.[22] introduced a procedure of incorporating bond or joint element into finite element analysis. Also the singularity or crack tip element has been the subject of ref.[23] and applied successfully to the evaluation of singular stress and strain fields adjacent to the notch under plane stress, plane strain and 3D stress conditions. In the formulation of the writer and his colleagues, a singularity transformation is applied to the mapping of geometric configuration rather than the inclusion of singular terms in the shape or interpolation functions.

A unified approach to the subject of singularity and infinite elements has been discussed in ref.[31] in the light of the concept of general parametric element. In this article a useful program segment has been given that eliminates ill-conditioning of the matrix equation connected with the very thin bond element. It has the effect of introducing relative displacements between neighboring nodes without any essential alteration of the source code.

Reinforced concrete can be treated as a composite material, characterized by anisotropic mechanical properties. Anisotropy poses problems that violate accepted principles, e.g. that the stress and strain singularity which is inversely proportional to the root of radial distance from the crack tip, or that the Saint Venant principle which is valid for isotropic material. Computational techniques for the evaluation of the eigenstates in composite materials have been discussed in ref.[30]. Large deformation or geometric nonlinearity may not be an important factor in the strength and failure of reinforced concrete, but the updated Lagrangian formulation summarized in ref.[29] could be the choice in future applications.

Finally a few remarks shall be made on software for the analysis of reinforced concrete. The report [4] contains an appendix with programs developed in America and Europe, up to 1982, which can be used for strength and failure analysis or simulation of concrete structure behavior. As for programs used in Japan, the outcome of a survey on finite element analysis programs is described in ref.[11].

There have been a lot of work and publications that demonstrate the power and possibilities of the finite element method for reinforced concrete applications, and a variety of in-house and proprietary nonlinear analysis software exists which meets the special and often urgent demand from the industry. However, programs in the public domain which have been used or have potential applicability to reinforced concrete are scarce, and the writer, to his knowledge as a nonspecialist in concrete engineering, can only refer to a series of programs published in ref.[10] in the United Kingdom, NONSAP [1] in the U.S. and COMPOSITE-III [24] and FEMN [9] in Japan. Source codes of these programs have been available to the public or at least to members of interested users groups.

Regarding the software developed in Japan, COMPOSITE-III is a prototypical general purpose nonlinear analysis program and plays the role of NONSAP, which is the forerunner for ADINA in the U.S. COMPOSITE-III was a product of the special research project on composite materials supported by a grant-in-aid for scientific research from the Japanese Ministry of Education. This program synthesizes time-independent and dependent materials in a unified expression of the constitutive equations. Program development spanned from 1973 to 1976.

Soon after, FEM4 was written and its complete source was published in a handbook [5], together with useful related routines for preprocessing, element mesh generation, and postprocessing of computation results, graphical display of results on plotters such as CALCOMP. FEMN is an extension of FEM4 and has been maintained by members of NAPRA (Nonlinear Analysis Program Research Association). The expansion from the primary Level-0 program amounts to four levels which feature, respectively:

Level-1 Optional input by a separate subroutine of the material constitutive relation and choice of computation scheme among the Newton-Raphson, modified Newton-Raphson and initial stress methods,
Level-2 Processing of smeared crack model and incorporation of bond element,
Level-3 Inclusion of temperature dependent material property,
Level-4 Complementary subprogram for the solution of heat conduction problem in concrete structures.

It is hoped that these programs will be continually updated and that microcomputer versions will become available in the future. In passing, it is noted that an extended commercial version of FEMN may be obtained from RCCM (Research Center of Computational Mechanics) Inc., Japan.

CONCLUSION

This paper summarizes the potential of the microcomputer for finite element analysis and reviewed the development of nonlinear analysis procedures in Japan, with the hope that they may find increased application in the area of reinforced concrete. Also a brief overview has been given of available programs, especially those that are in the public domain.

Due to the recent interests of the writer and his limited knowledge in the area of concrete engineering, this review was confined mainly to the work done by the writer and his colleagues. However, publications to which reference has been made contain summaries of achievements in the respective area through the detailed literature survey.

Finally, the writer acknowledges the efforts of the organizers of this Seminar and wishes to express his gratitude for having been given an opportunity to add his thought on the application of the finite element method to concrete engineering through the preparation of this article.

REFERENCES

[1] Bathe, K. J., Wilson, E. L. and Iding, R. H., NONSAP -- A Structural Analysis Program for Static and Dynamic Response of Nonlinear Systems, Structural Engineering Laboratory, University of California, Berkeley, Report No. UC SESM 74-3, February 1974.

[2] CAL 80 (SD), Computer Analysis Language for the Static and Dynamic Analysis of Structural Systems, Distributed by NISEE (National Information Service for Earthquake Engineering), University of California, Berkeley.

[3] Falk, H. and Beardsley, C. W., Finite Element Analysis Packages for Personal Computers, Mechanical Engineering, Vol. 107, January, 1985, pp.54-71.

[4] Finite Element Analysis of Reinforced Concrete, Committee on Concrete and Masonry Structures, Structural Division, ASCE, 1982.

[5] Handbook of Finite Element Method, Vol. 2, Baifukan, 1983, pp.845-1024. (in Japanese)

[6] Hoit, M. I., New Computer Programming Techniques for Structural Engineering, Ph. D. Dissertation, Department of Civil Engineering, University of California, Berkeley, 1983.

[7] Mair, W. M. and Morris, A. J., The National Agency for Finite Element Method and Standards, Engineering Computation, Vol. 1, September, 1984, pp. 280-284.

[8] Maekawa, k., Yamazaki, J. and Sakurai, T., Man Machine Interactive Analysis in Concrete Structures using Microcomputers, Advanced Mechanics of Reinforced Concrete, Colloquium, Delft, 1981, pp.335-350.

[9] Nonlinear Analysis Program FEMN, Nonlinear Analysis Program Research Association, User's Manual, Level-0, May, 1982. (in Japanese)

[10] Owen, D. R. J. and Hinton, E., Finite Elements in Plasticity: Theory and Practice, Pineridge Press, U.K., 1980.

[11] Proceedings of Colloquium on Finite Element Analysis of Concrete Structures, Japan Concrete Institute, December, 1984.

[12] Wilson, E. L., CAL -- A Computer Analysis Language for Teaching Structural Analysis, Computers and Structures, Vol. 10, 1978, pp.127-132.

[13] Wilson, E. L., The SAP-80 Series of Structural Analysis Program for CP/M Microcomputer Systems, Ver. 1.0, Jan. 1980/ Ver. 84.06, June 10, 1984.

[14] Wilson, E. L., SAP-80 Structural Analysis Programs for Large or Small Computer Systems, CEPA 1980 Fall Conference and Annual Meeting, Newport Beach, California, Oct. 13-15, 1980.

[15] Wilson, E. L., The Use of Microcomputers in Structural Analysis, Computers and Structures, Vol. 12, 1980, pp.695-698.

[16] Wilson, E. L., Finite Element Analysis on Microcomputers, paper presented at Symposium on Impact of New Computing System on Computational Mechanics (II), ASME WAM, Boston, Massachusetts, Nov. 14, 1983.

[17] Wilson, E. L., Development of Finite Element Software for Microcomputers, Computers in Engineering, Vol. 1, ASME, 1984, pp.392-396.

[18] Wilson, E. L. and Hoit, M. I., A Computer Adaptive Language for the Development of Structural Analysis Programs, Computers and Structures, Vol. 19, 1984, pp.321-338.

[19] Yamada, Y., Yoshimura, N. and Sakurai, T., Plastic Stress-Strain Matrix and Its Application for the Solution of Elastic-Plastic Problems by the Finite Element Method, Int. J. of Mechanical Sciences, Vol. 10, 1968, pp.343-354.

[20] Yamada, Y., Time Dependent Materials, Shock and Vibration Information Center, United States Department of Defence, 1975, pp.173-188.

[21] Yamada, Y., Constitutive Modelling of Inelastic Behavior and Numerical Solution of Nonlinear Problems by the Finite Element Method, Computers and Structures, Vol. 8, 1978, pp.533-543.

[22] Yamada, Y., Ezawa, Y. and Nishiguchi, I., Handy Incorporation of Bond and Singularity Elements in the Finite Element Solution Routine, Transactions of 5th SMiRT Conference, Berlin, Papaer M9/8, 1979.

[23] Yamada, Y., Ezawa, Y., Nishiguchi, I and Okabe, M., Reconsiderations on Singularity or Crack Tip Element, Int. J. for Numerical Methods in Engineering, Vol. 14, 1979, pp.1525-1544.

[24] Yamada, Y. and Sakurai, T., Composite-III User's Manual, Nonlinear Analysis Program Research Association, 1978. (In Japanese)

[25] Yamada, Y., Plasticity and Viscoelasticity, Baifucan, 1980. (in Japanese)

[26] Yamada, Y., Okumura, H. and Sakurai, T., Microcomputer System for Medium-Sized and Experimental Finite Elemnt Analysis, Research in Nonlinear Structural and Solid Mechanis, NASA Conference Pub. 2147, 1980, pp.277-289.

[27] Yamada, Y. and Yokouchi, Y., Programming for Elastic-Plastic Analysis by the Finite Element Method, Introducing Program EPIC-IV, Baifukan, 1981. (in Japanese)

[28] Yamada, Y. and Yokouchi, Y., Modelling and Accuracy of Solution in Inelastic Analysis -- Through Studies of Benchmark Problems, Nuclear Engineering and Design, Vol 67, 1981, pp.1-11.

[29] Yamada, Y., Nonlinear Matrices, Their Implications and Applications in Inelastic Large Deformation Analysis, Computer Methods in Applied Mechanics and Engineering, Vol. 33, 1982, pp.417-437.

[30] Yamada, Y. and Okumura, H., Finite Element Analysis of Stress and Strain Sigularity Eigenstates in Inhomogeneous Media or Composite Materials, Hybrid and Mixed Finite Element Methods, John Wiley, 1983, pp.325-342.

[31] Yamada, Y. and Nishiguchi, I., Concept of General Parametric Elements and Applications, Computers and Structures, Vol. 19, 1984, pp.277-284.

[32] Zienkiewicz, O. C., The Finite Element Method, Third Edition, McGraw-Hill (UK), Maidenhead, Berkshire, England, 1977.

VERIFICATION OF MODELING FOR REINFORCED CONCRETE FINITE ELEMENT

Hajime Okamura*, Koichi Maekawa** and S. Sivasubramaniyam***

ABSTRACT

The objective of this paper is to present an idea on how to evaluate finite element analytical models. The analytical models used in Japan are classified into three categories, micro, macro, and general, according to the material models used. A reinforced concrete finite element is modeled by using the recently developed material models in Japan and is verified as an example of a typical macro level model.

1. INTRODUCTION

Finite element analysis is applied to both practical design and the research to improve the understanding the behavior of reinforced concrete structures. The basic mechanical models used in an analysis, such as models of cracked concrete, bond between concrete and reinforcement and shear transfer across the cracked plane, are not always the same among many researchers. Therefore, it is important for engineers to evaluate the reliability of an analytical method. The object of this paper is to present an idea on how to evaluate finite element analytical models. However, it is extremely difficult to give a satisfactory answer to this problem. Accordingly, the authors have only tried to express their own opinions as to the evaluation process of an analytical model or a reinforced concrete finite element model.

2. ANALYTICAL MODELS USED IN JAPAN

Representative finite element analytical software developed in Japan is collected in Table 1. This table was prepared from the responses to a survey carried out in 1984 by the Japan Concrete Institute Committee on the Finite Element Method of Analysis for Reinforced Concrete. The modeling of uncracked and cracked concrete, bond expression, and the problems to which the method was applied are summarized. The programs are classified into three categories.

* Professor, Department of Civil Engineering, University of Tokyo, Hongo 7-3-1, Bunkyo-ku, Tokyo JAPAN

** Assistant Lecturer, Department of Civil Engineering, University of Tokyo, Hongo 7-3-1, Bunkyo-ku, Tokyo JAPAN

*** Graduate Student, Department of Civil Engineering, University of Tokyo, Hongo 7-3-1, Bunkyo-ku, Tokyo JAPAN

The programs classified as micro are developed for research or for understanding the behavior of reinforced concrete. For this category microscopic assumptions are the most important questions. A discrete crack model together with crack-link and bond-link elements are often used.

The programs categorized as macro mainly aim to analyze full scale reinforced concrete structures for practical use. The modeling of concrete cracking is limited to the smeared one. This is because investigation of the overall response to external actions is the main subject of these programs. Accordingly, the applications are concentrated on shell and wall types of structures for which the macroscopic expression of distributed cracking is suitable. Because of the degrees of freedom for shell structures, the majority of this category take three dimensional displacements into consideration. The dimensions of the stress field are reduced in many cases to a two dimensions, considering the direction of the stress vectors produced in the shell sections. The programs considering three dimensional stress fields and discussing the basic hypothesis are re-classified as in Table 1.

The analytical and logical flow of programs in the macro class start from the mathematical formulation of reinforced concrete finite elements. Bond of concrete to reinforcement between cracks is indirectly expressed by sophisticated tension stiffening or a residual stress normal to the crack faces in the cracked reinforced concrete constitutive laws. Tension stiffening of cracked concrete is usually formulated by using bond constitutive equations. Some coefficients are determined from limited experimental observations, such as from uni-axial pull-out tests.

However, we must recognize that there exists no reinforced concrete model subject to in-plane stresses which has been directly derived from experiments of reinforced concrete panels. In other words, the programs in this class do not have the reinforced concrete element model strictly based on the experimentally obtained behavior of elements. The components which cannot be described from the microscopic assumptions are supplemented by the macroscopic observations.

The programs classified under the level general have many options and are more universal in application. The original systems of analysis usually require the selection of their functions and modifications.

3. LEVELS OF MATERIAL MODELING

In a FEM analysis, it is necessary to use some material models which are derived directly from experimental results or indirectly from analyses based on microscopic material models. The microscopic models are also derived from experimental results or from analyses based on further microscopic material models. Finally, any analytical model must be derived from experimental results. And the material models or the experimental results govern the precision and the applicability of the programs.

Table 1 Programs used in Japan

	Program (Reporter)	Concrete model	Type of crack	Stiffness Tensile	Stiffness Shear	Bond model	Applied problems [Reference]
Micro	RC-FEM (Noguchi)	Darwin	Discrete	Crack-link		Bond-link	Beam-Column [1-4], Wall [5-7]
	-- (Morita)	Darwin	Optional	Crack-link		Bond-link	Column, Bond-Anchorage
	RCSAFE (Tsubaki)	Endochronic	Optional	Crack-link			Panel [8,9]
	FEMH.FORT (Hayashi)	Elastic	Discrete	Crack-link		Bond-link	Effect of bond [10-12]
	-- (Takiguchi)	Elasto-plastic	Smeared	Zero	Zero	Bond-link	Beam shear [13,14]
	NACOS (Shioya)	Isohata	Smeared	Considered	Considered		Beam, PC beam, Slab [15]
	COMM2NS (Maekawa)	Maekawa	Smeared	Zero	Yamada		Deep beam [16]
	RC-ELAPLA (Shirai)	Elasto-plastic	Smeared	Considered	Considered		Shear wall [17,18]
	CARC-SOLID (Inoue)	Elasto-plastic	Smeared	Zero	Considered	Bond-link	3D column [19,20]
Macro	NAPSS (Ouchi)	Elasto-plastic	Smeared	Zero	Considered		RC shell [21-23], Wall [24]
	URANUS (Imoto)	Elasto-plastic	Smeared	Zero	--		PC vessel [25], Wall [26]
	RC-SHELL (Ueda)	Elasto-plastic	Smeared	Considered	Considered		Shell [27], Wall [28]
	NACS (Fujita)	Isohata	Smeared	Zero	Considered		Cylindrical shell [29,30]
	CARC-PLAIN (Miyashita)	Elasto-plastic	Smeared	Zero	Considered	Bond-link	Wall [31], Column [32], Cylindrical shell [33, 34]
	RC-HYST2D (Imoto)	Elasto-plastic	Smeared	Zero	Considered	Bond-link	Beam-Column [35,36], Wall [37,38], Vessel [39,40]
	MN-1 (Aoyagi)	Kupfer	Smeared	Considered	Yamada	Joint element	Beam, Cylindrical wall
	FRACT-82 (Isohata)	Isohata	Smeared	Considered	Considered		Cylindrical shell [41]
	TERESA (Nomura)	Visco-elasto-plastic	Smeared	zero	Considered		Thermal loading on tank [42], Actual load [43]
	PROVIP (Kawasaki)	Elasto-plastic	Smeared	Zero	Considered	Joint element	Shell [15]
	K-NASS (Shiroishi)	Darwin	Smeared	Considered	Considered	Bond-link	Shell
	-- (Torigaya)	Elasto-plastic	Smeared	Considered	Considered		Shell
	N-FEAP (Shudo)	Maekawa	Smeared	Zero	Yamada		Shell
	FSTAN-7 (Mikame)	Elasto-plastic	Smeared	Considered	--		Vessel [44,45]
General	TOTAL-RC (Endo)	Kupfer	Optional	Optional	Optional	Joint element	Bond-Anchorage, 3D comp.
	PCM (Yoshikawa)	Optional	Optional	Optional	Optional	Joint element	In-plane shear
	RCCM MN-3 (Sakurai)	Optional	Optional	Optional	Optional	Joint element	Thermal loading on tank

Remarks :
Concrete : Endochronic [51], Kupfer [52], Isohata [53], Maekawa [54], Elasto-Plastic [27][55][56][57]
Cracked concrete : Crack-link [58][59], Yamada [60]
Bond : Bond-link [1][61]

It is quite natural that an analytical model, which depends on material models which are closer to the analysed behavior, will give more precise predictions. However, more extensive applicability can be expected from an analytical model which depends on microscopic material modeling. Apparent complicated behavior is usually due to the combination and mutual interaction of microscopic properties of concrete and steel.

The material models in FEM analysis for reinforced concrete may be divided into two levels, the microscopic level and the macroscopic level. For the analysis at the microscopic level, even more microscopic material models may be used.

The feature of the micro level modeling is to use a discrete type crack model together with a crack-link and a bond-link element or a joint element, and is applicable to the analysis of the local behavior in reinforced concrete. The mechanical properties of these elements should be verified by carefully executed experiments or analyses, because it is extremely difficult to carry out the experiments of the micro level applicable to the behavior of reinforced concrete.

The feature of the macro level modeling is to use a smeared type crack model, together with tension stiffening of cracked concrete to express the average bonding between concrete and steel in a reinforced concrete element. Accordingly, a reinforcing bar is modeled as a thin plate element attached on the concrete element as shown in Fig.1. In this sense, this level of modeling can be considered as reinforced concrete finite element modeling.

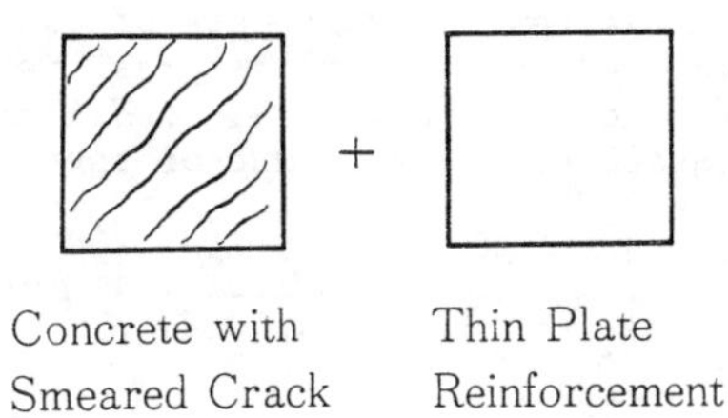

Fig.1 Macroscopic level of material modeling

4. MODELING OF A REINFORCED CONCRETE FINITE ELEMENT

A reinforced concrete finite element model is naturally useful for a finite element applicable to full scale analysis of actual structures. At present, it can be said that there exists no well verified reinforced concrete finite element model, and no accepted method of verification for an analytical model. Therefore, the authors have constructed a simple reinforced concrete finite element model and have verified it to demonstrate one verification procedure.

(1) Uncracked state

The So called Maekawa model [54] is used for the uncracked concrete, and the actual stress-strain relation is used for the steel bars. Then, the concrete and steel elements are attached to each other, making a reinforced concrete finite element.

(2) Cracking in concrete

When the calculated principal tensile stress violates the following cracking criteria, the principal stress is held constant, as shown in Fig.2. These equations are introduced by modifying Niwa's for compression-tension region [63] and Aoyagi-Yamada's [60] for tension-tension region.

$$\frac{\sigma_1}{f_t} < \sqrt[3]{1+(\frac{\sigma_2}{0.9f'_c})} \qquad \text{for } \sigma_2 \leq 0 \tag{1}$$

$$< 1 - 0.26(\frac{\sigma_2}{\sigma_1}) \qquad \text{for } \sigma_2 > 0 \tag{2}$$

where,

σ_1, σ_2 : principal stresses, $\sigma_1 \geq \sigma_2$
f_t : tensile strength of concrete
f'_c : compressive strength of concrete

Cracking in concrete is assumed to occur when the principal tensile strain reaches the cracking strain, which is set equal to two times the strain corresponding to the maximum tensile stress.

(3) Cracked state

The tension stiffness of concrete after cracking has usually been developed based on test results conducted under conditions where the direction of principal tensile stress coincides with that of the reinforcing bars. One of the typical examples is Morita-Kaku [62], where

$$\frac{\sigma_t}{f_t} = \frac{1}{4.5\text{x}10^3\varepsilon_t+0.84} + 0.22 \tag{3}$$

A simpler expression, having characteristics similar to these of Eq.3, is used here, because the characteristics of the tension stiffness could be varied easily by changing the constant c in the equation as shown in Fig.2.

$$\frac{\sigma_t}{f_t} = (\frac{\varepsilon_{tu}}{\varepsilon_t})^c \tag{4}$$

where,

σ_t : assumed tensile stress of cracked concrete
ε_t : tensile strain normal to cracks
ε_{tu} : cracking strain
c : constant, 1/5 was used for Collins-Vecchio's panels

The cracking direction is not usually normal to the direction of steel bars. This is not the condition under which the model has been developed. The factors which influence the bond behavior or the tension stiffness, such as concrete cover and the stress acting normal to the bars, are also different from those in the structures to be analyzed. Therefore, the tension stiffening model should be carefully verified by experimental tests of reinforced concrete panels.

Regarding the shear stiffness of concrete after cracking, the Aoyagi-Yamada model (see Fig.3) [60], which has been introduced from their reinforced concrete panel tests, is used as standard.

$$G_{cr} = \frac{3.6}{\varepsilon_t} \text{ in MPa} \qquad (5)$$

$$= \frac{500}{\varepsilon_t} \text{ in psi}$$

where,

G_{cr} : shear stiffness of cracked concrete

ε_t : tensile strain normal to the crack direction

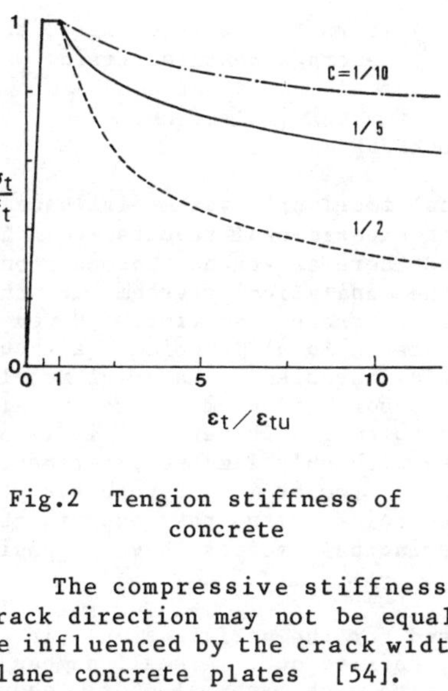

Fig.2 Tension stiffness of concrete

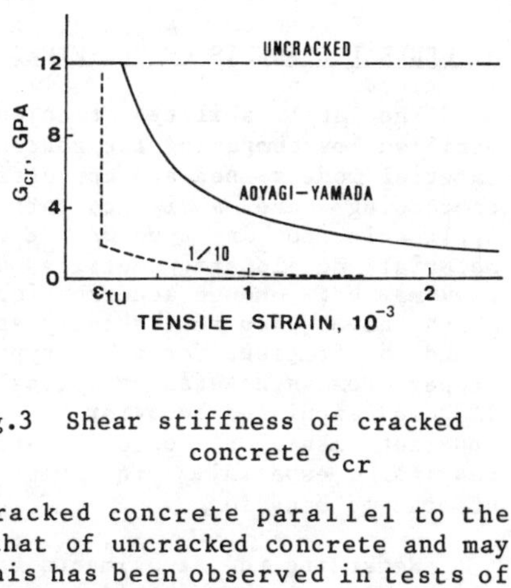

Fig.3 Shear stiffness of cracked concrete G_{cr}

The compressive stiffness of cracked concrete parallel to the crack direction may not be equal to that of uncracked concrete and may be influenced by the crack width. This has been observed in tests of plane concrete plates [54]. Maekawa has developed a constitutive model by modifying his basic elasto-plastic-fracture model as

$$\sigma_c = E_o K_c \varepsilon_e \qquad (6)$$

where,

E_o : initial tangent modulus of uncracked concrete

K_c : fracture parameter of cracked concrete

$= w K$

K : fracture parameter of uncracked concrete

w : parameter relating to the average tensile strain of element as illustrated in Fig. 4.

$$= 1 \qquad \text{for } \varepsilon_t \leq \varepsilon_{t1} \qquad (7)$$

$$= 1 - (\varepsilon_t - \varepsilon_{t1})(1 - w_o)/(\varepsilon_{t2} - \varepsilon_{t1}) \geq w_o \qquad \text{for } \varepsilon_t \geq \varepsilon_{t1}$$

w_o = 0.20, ε_{t1}=0.0012, ε_t=0.0076 were used as standard.

Reinforcing steel bars in any direction are modeled as a thin plate element which has a tension stiffness for one direction only and has the same cross sectional area of the bars. The dowel action is neglected, because this can be treated in the variation of shear transfer models. The reinforcing bars at cracked section will start to yield when the applied tensile forces become equal to the yield strength. Therefore, the apparent yield strength is temporarily modified. Concrete and steel elements with these properties are attached to each other, and make a reinforced concrete finite element.

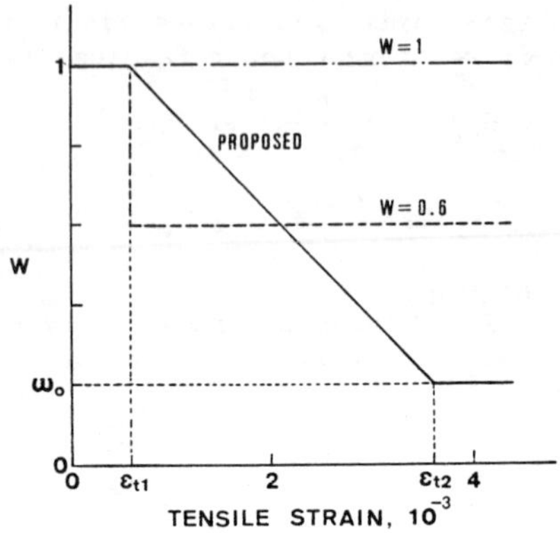

Fig.4 Parameter w relating to the average tensile strain

5. GENERAL CONCEPTS OF VERIFICATION PROCEDURE

The applicability of an analytical model or system is usually verified by comparing its response with experimental results. If the material models used are appropriate and there exists no inconsistency concerning the analytical flow, the analytical system is then applicable to any type of reinforced concrete behavior. However, material models are generally not applicable to all problems although they may have enough accuracy for a certain problem. Material models, which have been successfully applied to some problems, are usually found insufficient for other types of problems. Therefore, it is not proper to evaluate an analytical system with only limited experiments. If some calculated behavior is dependent on two or more hypotheses, the apparent behavior could sometimes coincide with the experimental results, especially in which two principal models have opposite effects on behavior.

Regarding the experiments to be used for the verification, it is essential that they should be carefully carried out. A small number of good experiments independently sensitive to each material model, should be selected, rather than a great number of tests having a similar nature. Furthermore, it is effective to use some extreme cases.

The authors want to discuss the sequence of verification as portrayed in Fig.5, where the process of the modeling and the verification for an analytical model is illustrated.

The analytical model for a reinforced concrete finite element is constructed by using the several macroscopic material models. The analytical results should be verified by comparison with the experimental results at the reinforced concrete element level. By this comparison, applicability of the macroscopic material models to this level of analysis can be indirectly verified while the material models may sometimes have to be modified. This level of verification is very important, because the conditions of experiments referred to the material level are usually not similar to conditions at the reinforced concrete element level. Comparisons with experiments at the reinforced concrete member level should also be carried out, because there exist some questions regarding the differences in conditions between the element level and the member level.

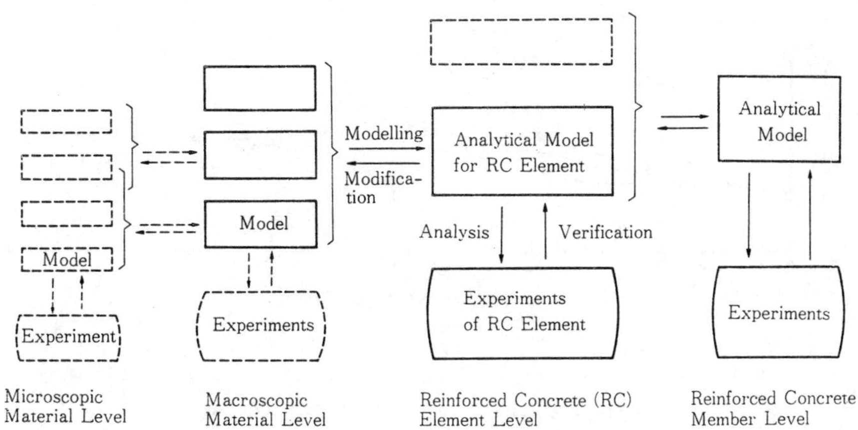

Fig.5 Process of the modeling and the verification

6. EXPERIMENTAL VERIFICATION FOR ELEMENT LEVEL

By using the reinforced concrete finite element model in an example the authors try to demonstrate the concepts mentioned above. The square reinforced concrete panels tested by Collins-Vecchio [49] are used for the verification. The analysis was executed on a personal computer using only one element.

Ten panels shown in Table 2 are selected for the verification of the analysis. The reinforcing bars are arranged in two orthogonal directions. Two panels, PV3 and PV4, are selected for verification of the tension stiffening model. These panels are isotropically and lightly reinforced and external load is pure shear. This geometrical and mechanical symmetry enforces the deformational mode also to be symmetric while no shear slip appears along the crack plane. This extreme deformational mode is never controlled by the shear transfer idealization. The concrete model along the crack has very little effect on the overall behavior of these panels as they are lightly reinforced. Therefore, tension stiffness and cracking criteria are the main parameters to be verified.

The shear stress strain relations of these panels calculated according to the standard model are shown in Fig.6 along with the experimental data. The standard modeling give a good estimation of the overall behavior of these panels. A sensitivity check of tension stiffening models is shown in the same figure. It appears that the tension stiffness has an influence on the overall deformation, and a little greater stiffness than the usual value (c=1/2) would be better. For these panels, welded wire fabric is used for reinforcement, differing from the usual deformed bars. Accordingly, some difference of the tension stiffness is expected and the parameter c in Eq.4 is taken as 1/5 for all the panels.

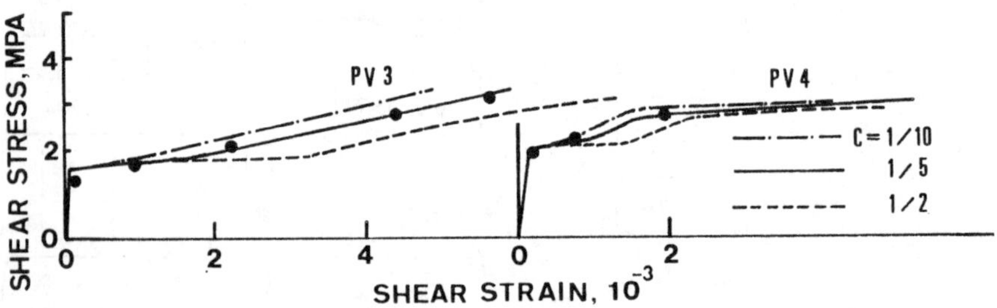

Fig.6 Effect of tension stiffening

The tensile strength of concrete in reinforced concrete panels, which is affected by many factors such as shrinkage stress, affects also the behavior of panels. However, it could not be specified. Accordingly, data fitting values are used as in Table 2.

Table 2 Properties of panels used for the verification

Specimen	Loading	Longituginal Steel		Pt	Concrete	
		Pl(%)	fyl(MPa)	Pl	f'c(MPa)	ft(MPa)
PV3	Shear	0.483	662	1	26.6	1.68
PV4	Shear	1.056	242	1	26.6	1.92
PV25	Shear+Comp.	1.785	466	1	19.2	2.71
PV23	Shear+Comp.	1.785	518	1	20.5	2.82
PV27	Shear	1.785	442	1	20.5	2.82
PV28	Shear+Tens.	1.785	483	1	19.0	2.30
PV22	Shear	1.785	458	0.85	19.6	2.54
PV21	Shear	1.785	458	0.73	19.5	2.54
PV20	Shear	1.785	460	0.50	19.6	2.34
PV19	Shear	1.785	458	0.40	19.0	2.07

Four panels, PV25, PV23, PV27 and PV28, are selected for verification of the tension stiffening model and the cracked concrete model parallel to the crack direction. These panels are isotropically and heavily reinforced. External load is pure shear or combined shear and biaxial compression or tension with equal magnitudes. Accordingly, the shear transfer model has no effect, and the behavior of these panels are governed mainly by the tension stiffening model under low stress level and by the compressive stiffness under higher stress.

A sensitivity check of tension stiffening models is shown in Fig.7. It appears that the tension stiffening model with c of 1/5 again gives satisfactory results for all these panels, while the model with c of 1/2 gives lower stiffness. This figure clearly shows the importance of tension stiffening even on the behavior of heavily reinforced concrete panels.

A sensitivity check of cracked concrete models is shown in Fig.8. It appears that the standard model gives satisfactory results for all those panels failing in shear. The strength of these panels is estimated to be higher than the tested value when the constant value is used as 1 for w, that is, no effect of cracking on the stiffness of concrete is considered. The constant value of 0.6 for w overestimates the effect of cracking under the low stress level.

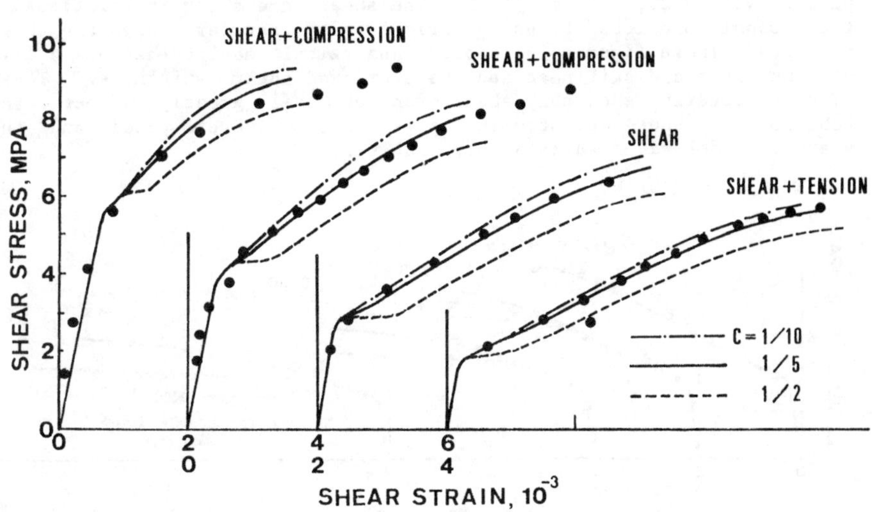

Fig.7 A sensitivity check of tension stiffening models

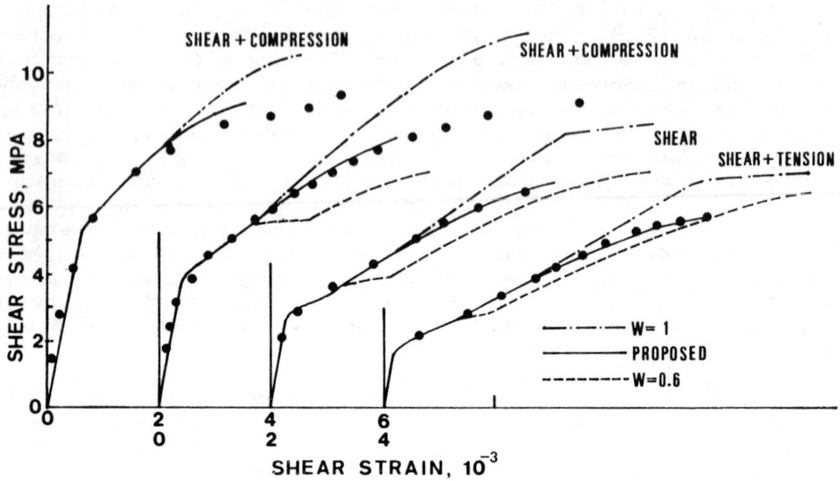

Fig.8 A sensitivity check of cracked concrete models

The shear transfer model after cracking is one of the factors that influence the behavior of the anisotropically reinforced concrete panels PV19, PV20, PV21 and PV22. The shear stress strain relations of these panels predicted by using various shear transfer models are shown in Fig.9. These figures include the analytical results that use a 1/10 of the standard stiffness and the uncracked shear stiffness. These figures clearly show that shear transfer model greatly affects the behavior of these anisotropically reinforced concrete panels and the standard model gives satisfactory results.

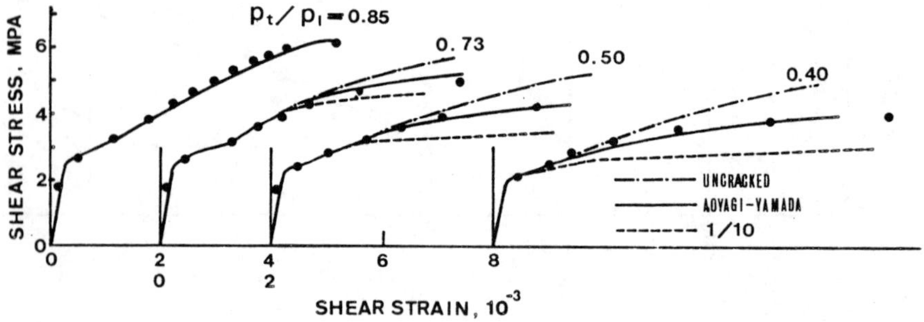

Fig.9 The effect of shear transfer models

The models discussed above can simulate the panel behavior used in the experimental verification very well. However, with regard to concrete model between cracks, further discussions are necessary relating to the effects of specimen size and bi-directional confinement by the reinforcement. The effect of concrete shrinkage on the cracking strength and tension stiffness should be investigated.

Even if the experimental verifications at the element level are satisfactory, the reinforced concrete finite element model should be verified by comparing it with tests at more macroscopic structural level. In the FEM analysis, the finite idealization of high ordered displacement fields is often used and the nodal force-displacement relations of structural elements are calculated with non-uniform strain fields. Therefore, it is necessary to analyze the representative response using the simplified geometry of a member level element. Furthermore, special attention should be paid to the post-peak behavior of the elements. Post-peak behavior will control the behavior of the member near the ultimate load. Because of the instability of correlations between the panel and the external loading system, it is difficult to verify analytical models for post-peak behavior in a uniform strain field. Accordingly, experimental verification at the member level will prove to be an indispensable technique for checking models.

7. CONCLUDING REMARKS

The procedures for the verification of an analytical model are demonstrated. A rational verification should be developed by a combination of experimental and analytical investigations as discussed in Section 5.

References

[1] Noguchi, H. and Maruta, M., 'Nonlinear Finite Element Analysis of Shear Resistance Mechanism of Reinforced Concrete Beams', JCI-C1, June 1982, pp.99-106 (in Japanese)
[2] Noguchi, H. and Iizuka, S., 'Nonlinear Analysis of the Effects of Bond Characteristics on the Mechanical Behavior of Reinforced Concrete Short Columns', Annual Conference of AIJ, Sept.1983, pp.2105-2106 (in Japanese)
[3] Noguchi, H., 'Non-Linear Finite Element Analysis of Reinforced Concrete Beam-Column Joints', IABSE Colloquium Delft, June 1981, pp.625-633
[4] Noguchi, H. and Naganuma, K., 'Nonlinear Analysis of Reinforced Concrete Beam-Column Joints under Reversed Cyclic Loading', JCI-C5, October 1983, pp.139-146 (in Japanese)
[5] Noguchi, H. and Naganuma, K., 'Nonlinear Parametric Analysis of Reinforced Concrete Panels', JCI-C5, October 1983, pp.147-154 (in Japanese)
[6] Noguchi, H. and Maruta, M., 'Nonlinear Analysis of Reinforcement for Openings of Reinforced Concrete Shear Walls Containing Openings', Annual Conference of AIJ, Sept. 1983, pp.2059-2060 (in Japanese)
[7] Noguchi, H. and Maruta, M., 'Nonlinear Analysis of Strength and Ductility of Infilled RC Shear Walls', Transactions of JCI, July 1984, pp.745-748 (in Japanese)
[8] Tsubaki, T., Yamanobe, S. and Ikeda, S., 'Analysis of Concrete Plates Subjected to In-Plate Forces--Finite Element Method and Boundary Element Method', JCI-C5, October 1983, pp.29-36 (in Japanese)
[9] Hashimoto, H., Tsubaki, T. and Ikeda, S., 'Parametric Analysis on Shear Transfer Mechanism of Crack Surfaces of Reinforced Concrete Plates', Annual Conference of JSCE, Division 5, Oct. 1984, pp.383-384 (in Japanese)
[10] Hayashi, S., Kokusho, S. and Osanai, T., 'Study of the Bending Behavior of Reinforced Concrete Columns Under Changing Axial Force', Transactions of AIJ, No.312, February 1982, pp.28-35 (in Japanese)
[11] Hayashi, S. and Kokusho, S., 'Fundamental Research on Bond Mechanism of Deformed Bars and Concrete', Annual Conference of AIJ, Oct.1982, pp.1329- 1330 (in Japanese)
[12] Hayashi, S., Kokusho, S. and Morimoto, T., 'Experimental study of the Bond Behavior of Reinforced Concrete Bars through the Interior Beam Column Joints', Report of the Research Laboratory of Engineering Materials, Tokyo Institute of Technology, No.9, 1984 (in Japanese)
[13] Takiguchi, K. and Kokusho, S., 'Experiments on Reinforced Concrete Beams with Emphasis on Strain Measurements', Transactions of AIJ, No.19 No.19494, April 1972, pp.1-8 (in Japanese)
[14] Kokusho, S. and Takiguchi, K., 'Two Dimensional Nonlinear Analysis of Reinforced Concrete Members by the Finite Element Method', Transactions of AIJ, No.200, Oct. 1972, pp.45-54 (in Japanese)
[15] Shioya, T., Kawasaki, H. and Kunita, J., 'Nonlinear Finite Element Analysis of Reinforced Concrete Circular Slabs', JCI-C5, October 1983, pp.183-190 (in Japanese)
[16] Niwa, J., 'Equation for Shear Strength of Reinforced Concrete Deep

Beams Based on FEM Analysis', JCI-C5, October 1983, pp.119-128 (in Japanese), Concrete Library International No.4, 1984 pp.283-296 (in English)

[17] Shirai, N. and Sato, T., 'Inelastic Analysis of Reinforced Concrete Shear Wall Structures', IABSE Colloquium Delft, June 1981

[18] Shirai, N. and Sato, T., 'Bond-Cracking Model for Reinforced Concrete', JCI-C5, October 1983 (in Japanese)

[19] Inoue, N., 'Experiment of Concrete Prisms with Hoop under Axial Force and their Analysis by 3-Dimensional Finite Element Method', Transaction of AIJ, No.336, February 1984, pp.1-10 (in Japanese)

[20] Inoue, N., '3-Dimensional Analysis of Reinforced Concrete Column Under Repeated Lateral Forces by Finite Element Method', Transactions of AIJ, No.341, July 1984, pp.1-10 (in Japanese)

[21] Ouchi, H., Koike, K., Ito, M. and Takeda, T. 'Experimental and Analytical Study on Inelastic Behavior of the RC Cooling Tower Model subjected to Horizontal Loads', Proceedings of JSCE, No.266, October 1977, pp.39-50 (in Japanese)

[22] Ohuchi, H. and Takeda, T., 'Nonlinear Analysis of Mechanical Behavior of LNG Underground Tanks Subjected to Horizontal Forces', Annual Conference of JSCE, Division 1, No.36, Oct. 1981, pp.99-100 (in Japanese)

[23] Ogaki, Y., Kobayashi, M., Takeda, T., Yamaguchi, T. and Yoshioka, K., 'Horizontal Loading Tests on Large-Scale Model of Prestressed Concrete Containment Vessel', 6th SMiRT, J4/2, August 1981

[24] Seya, Y., Takeda, T., Yamaguchi, T., Ohkuma, M. and Tsumura, H., 'A Proposal on the Thermal Stress Evaluation for the Reinforced Concrete Structures', 7th SMiRT, J4/6, August 1983

[25] Imoto, K., Yamaguchi, T., Takeda, T. and Takemoto, Y., 'Study on Prestressed Concrete Pressure Vessel of Nuclear Reactor', Annual Conference of AIJ, Oct. 1975, pp.1319-1320 (in Japanese)

[26] Imoto, K., Matsumoto, T. and Tomura, H. 'Nonlinear Analysis of Prestressed Concrete Containment Structure', Annual Conference of AIJ, Oct. 1982, pp.945-946 (in Japanese)

[27] Ueda, M., Seya, H. and Kei, T., 'Nonlinear Analysis of RC Cylinder and Box Type Wall subjected to Shear Force', JCI-C5, October 1983, pp.163-172 (in Japanese)

[28] Ueda, M., Seya, H. and Kei, T., 'Finite Element Analysis of Reinforced Concrete Shell Structure (No.1), (No.2)', Proc. Annual Conference of AIJ, Oct. 1983, pp.1699-1701 (in Japanese)

[29] Fujita, F., Isohata, O., Kawamata, S. and Tsuboi, Y., 'Elasto-Plastic Analysis of Reinforced Concrete Shell by Finite Element Method', Proc. Annual Conference of AIJ, 1980, pp.1021-1-22 (in Japanese)

[30] Fujita, F., Isohata, O., Kawamata, S. and Tsuboi, Y., 'Finite Element Elasto-Plastic Analysis of Thin Walled Structure of Reinforced Concrete as Applied to Reactor Facilities', 6th SMiRT, 1981

[31] Muto, K., Ohmori, N., Sugano, T., Miyashita, T. and Shimizu, H., 'Non-Linear Analysis of Reinforced Concrete Buildings', Tokyo Seminar of Finite Element Analysis, University of Tokyo Press, 1973

[32] Muto, K., Tsugawa, T., Miyashita, T. and Shimizu, H., 'Bearing Capacity and Deformation of Reinforced Concrete Column Subject to Various Axial Forces', Proc. Annual Conference of AIJ, Oct. 1972 (in Japanese)

[33] Morikawa, H. et al., 'Model Test of Reinforced Concrete Containment Structure for Nuclear Power Plant Subject to Simultaneously accidental Inner Pressure and Horizontal Force', Proc. Annual Conference of AIJ, Sept. 1978, pp.1831-1832 (in Japanese)

[34] Muto, K., Morikawa, H. and Aihara, S., et al., 'Experimental and Analysis of Reinforced Concrete Cylinder under Heat and Lateral Forces', Proc. Annual Conference of AIJ, Sept. 1980, pp.1877-1878 (in Japanese)

[35] Imoto, K. and Takeda, T., 'Elasto-Plastic Analysis of Reinforced Concrete Member by Finite Element Method (No.1)', Proc. 43 Annual Conference of AIJ (Kanto), 1972, p.209 (in Japanese)

[36] Imoto, K. and Takeda, T., 'Elasto-Plastic Analysis of Reinforced Concrete Member by Finite Element Method (No.2)', Proc. Annual Conference of AIJ, Oct. 1973, pp.505-506 (in Japanese)

[37] Imoto, K., Yamashita, N., Takeda, T. and Yoshioka, K., 'Experiment and Analysis of Prestressed Concrete Wall with Inside Brace (No.2)'. Proc. Annual Conference of AIJ, Oct. 1974, pp.1389-1390 (in Japanese)

[38] Koike, K., Takeda, T. and Imoto, K., 'Study on Shear Wall with Frame', Proc. Annual Conference of AIJ, December 1977, pp.1891-1892 (in Japanese)

[39] Tada, T., Imoto, K., Yamaguchi, T., Takeda, T. and Takemoto, Y., 'Study on Prestressed Concrete Pressure Vessel of Nuclear Reactor', Proc. Annual Conference of AIJ, Oct. 1974, pp.1217-1218 (in Japanese)

[40] Imoto, K. and Takeda, T., 'Elasto-Plastic Analysis of Reinforced Concrete Structure by Finite Element Method (No.3)', Proc. Annual Conference of AIJ, Oct. 1977, pp.1749-1750 (in Japanese)

[41] Isohata, O., 'On an Anisotropic Constitutive Model of Concrete for the FEM Nonlinear Stress Analysis', Trans. of AIJ, No.265, March 1978, pp.11-18 (in Japanese)

[42] Nomura, T., Ohuchi, H. and Takeda, T., 'Thermal Stress Analysis of Storage Tank of Liquefied Natural Gas (No.2)', Proc. Annual Conference of JSCE, Division 5, Oct. 1982, pp.419-420 (in Japanese)

[43] Nomura, T., Ohuchi, H. and Takeda, T., 'Thermal Stress Analysis of Storage Tank of Liquefied Natural Gas (No.3)', Proc. Annual Conference of FSCE, Division 5, Sept. 1983, pp.417-418 (in Japanese)

[44] Mikame et al., 'Mixed Finite Element Method for Nonlinear Analysis of Concrete Containment Vessels', 6th SMiRT, J3/6, 1981

[45] Mikame, A., 'Study on Concrete Containment Structure for Nuclear Reactor (Elasto-Plastic Analysis by Mixed Type of Finite Element Method)', Proc. Annual Conference of AIJ, Sept. 1981, p.1415 (in Japanese)

[46] Murazumi, Y., et al., 'Application of ADINA to Nonlinear Analysis of Reinforced Concrete Shear Walls with Opening', Computer and Structure, Vol.13, No.5-6, 1981, pp.727-736 (in Japanese)

[47] Murakado, Y., Sotomura, K., Yoshizaki, M. and Ezaki, T., 'Strength and Deformation of Reinforced Shear Wall with Many Small Openings'. Proc. Annual Conference of AIJ. September 1981. pp.1627-1628 (in Japanese)

[48] Vecchio F. and Collins, M. P., 'The Response of Reinforced

Concrete to In-Plane Shear and Normal Stresses', Publication No.82-03, University of Toronto, March 1982
[49] Collins, M.P., Vecchio, F. and Mehlhorn, G., 'An International Competition to Predict the Response of Reinforced Concrete Panels', Festschrift Prof. Dr. Bruno Thurlimann zum 60. Geburtstag, 1983, pp.471-491
[50] JCI Shear Committee, 'Collected Experimental Data of Specimens for Verification of Analytical Models', JCI-C6, October 1983 (in Japanese)
[51] Bazant, Z.P. and Shieh, C.L., 'Endochronic Model for Nonlinear Triaxial Behavior of Concrete', Nuclear Engineering and Design, Vol.47, 1978
[52] Kupfer, H. and Gerstle, K., 'Behavior of Concrete Under Biaxial Stress', Journal of Engineering Div., ASCE, Vol.99, No.EM4, Aug.1973, pp.852-866
[53] Isohata, O., 'On an Anisotropic Constitutive Model of Concrete for the FEM Nonlinear Stress Analysis', Transactions of AIJ, No.265, 1978, pp.11-18 (in Japanese)
[54] Maekawa, K. and Okamura, H., 'The Deformational Behavior and Constitutive Equation of Concrete Using Elasto-Plastic and Fracture Model', Journal of Faculty of Engineering, The University of Tokyo (B), Vol.37, No.2, 1983, pp.253-328 (in English), Concrete Library International No.4, 1984, pp.183-260 (in English)
[55] Takiguchi K. and Kokusho, S., 'Two Dimensional Nonlinear Analysis of Reinforced Concrete Members by the Finite Element Method (Part 1)', Transactions of AIJ, No.189, November 1971, pp.51-57 (in Japanese)
[56] Miyashita, T. and Hayami, Y., 'Stress-Strain relationship of 2 dimensional Concrete Derived from Plastic Theory', Proc. Annual Conference of AIJ, October 1982 (in Japanese)
[57] Mutoh, K., Sugano, T., Miyashita, T. and Inoue, N, '3-Dimensional Non-Linear Analysis of Reinforced Concrete Members by Finite Element Method', Trans. of AIJ, No.249, November 1976, pp.25-34 (in Japanese)
[58] Walraven, J. C., 'Fundamental Analysis of Aggregate Interlock,' ASCE, Vol.107, No.ST11, Nov.1981
[59] Tsubaki, T. and Ikeda, S., 'Elasto-Plastic Analysis of Concrete Members Containing Crack Surfaces', Proc. Annual Conference of JSCE, Division 5, October 1982 (in Japanese)
[60] Aoyagi, Y. and Yamada, K., 'Strength and Deformation Characteristics of Reinforced Concrete Shell Elements Subjected to In-plane Forces', Proc. of JSCE, No.331, March 1983, pp.167-180 (in English), Concrete Library International, No.4, 1984, pp.129-160 (in English)
[61] Miyashita, T. and Hayami, Y., 'Bond Analysis of Deformed Bar and Concrete by Finite Element Method', Proc. Annual Conference of AIJ, September 1983
[62] Morita, S. and Kaku, T., 'Experimental Study on the Deformation of Axially Reinforced Concrete Prisms Subjected to Tension and Drying', CAJ Review of the 18 General Meeting, 1964, pp.205-209
[63] Niwa, J. and Maekawa, K., 'A Proposal for the Failure Criterion of Concrete', Proc. Annual Conference of JSCE, September 1980, pp.273-274 (in Japanese)

Nonlinear Finite Element Analysis of Reinforced Concrete in Engineering Practice

Robert H. Iding,* M. ASCE

Abstract

To an increasing degree, nonlinear methods of analysis are finding their way into the structural engineering profession and are permitting the solution of problems considered intractable in the past. One of the most promising of these new methods is the finite element approach as applied to reinforced concrete. This paper discusses recent experience the author's consulting firm has had in the application of this method. General considerations for the use of finite elements in a commercial rather than academic context are given, especially with respect to stringent cost and time limitations. In addition, a case study is presented from recent work in the design of ice walls for offshore concrete gravity structures.

Introduction

Nonlinear finite element methods have been extensively used in research for many years, but only recently have they been finding their way into engineering practice. It is still true that the vast majority of analysis in engineering offices follows traditional patterns - linearized, simplified models which are appropriate for the degree of accuracy required in design or in the field. But, to an increasing degree, special problems which would have been handled by intuition or experiment in the past are being solved using nonlinear models of concrete material behavior coupled with finite element discretization. Use of finite element methods in an engineering office is quite different than in a research context, and some of these considerations are enumerated below. In addition, some typical practical problems which have been studied using nonlinear finite element methods are discussed.

General Considerations

The use of nonlinear finite element methods in a commercial context often differs from their use in an academic or research environment. Some of these considerations are the following:

1. Orientation is Toward Solving a Problem

There is almost always a very well defined objective for an analysis. It may be the need to choose design parameters, to evaluate the cause of

*Consultant, Wiss, Janney, Elstner Associates, Inc., 2200 Powell Street, Suite 925, Emeryville CA 94608-1836.

distress or collapse, or to address a point in litigation. Quite often, a client is not primarily interested in advancing the state of the art, but rather in applying the available information to the problem at hand. The scope of work should fit this objective.

2. Time Limitations

Analysis results are needed in weeks, or, at best, in months. Delay is usually unacceptable since the analysis may lie on the critical path of a much larger project. One must not be overoptimistic about what can be accomplished within the given time constraints and should plan the analysis model accordingly.

3. Cost Limitations

The budget for an analytical approach for solving a problem must be competitive with other possible approaches, such as a testing program, or, for design, perhaps just using a large factor of safety. Labor tends to be expensive and commercial computer charges can be high if not carefully controlled. This means that the approach selected for solving the problem may not always be able to utilize the latest research, but rather somewhat simpler or more approximate methods which can be implemented within the time and cost constraints.

4. Need for Verification

In order for analysis results to be accepted by practicing engineers, the methods upon which they are based must consistently give calculated values which reasonably agree with experimental test data. This is especially true for nonlinear approaches which may not have been used in practical applications before. The most effective analytical studies include a test program to calibrate the analytical model. However, this is usually not possible, in which case one must compare calculated results with test data in the literature.

Typical Problems Encountered in Practice

Numerous problems arise in any engineering office which cannot be solved using conventional analytical methods. Some of these problems must somehow be resolved without analysis due to time or budget limitations, but others can be dealt with using nonlinear finite element techniques. Some cases encountered in our consulting practice are the following:

1. Design of Ice Walls in Offshore Concrete Structures to optimize response to high-intensity, short-duration loading from floating sheet ice [1]. Such walls experience very high deflections and material deformation before peak load capacity is reached, and conventional analysis techniques are of limited usefulness. This particular application is discussed in detail in the next section.

2. Prediction of Amount of Cover required to prevent popout of expansive particles embedded in concrete. Expansive aggregate was accidentally added to a concrete mix and it was necessary to estimate the extent of

damage possible as these particles expanded during the hydration process [2].

3. Prediction of Fire Endurance of Concrete Structural Elements in order to evaluate adequacy of cover and other fire-related parameters. Of particular interest were the temperatures of embedded reinforcement and prestressed strands and the associated load capacity of the member. An analysis of this type has, in some cases, replaced furnace tests conventionally used to determine fire rating [3,4].

4. Prediction of Shrinkage Stresses in concrete structures through methods combining moisture diffusion theory and nonlinear structural analysis [5].

To illustrate the approach a practicing engineer might take to such applications, the following case study is offered.

Case Study - Ice Wall in Offshore Concrete Gravity Structure

The new generation of deepwater concrete gravity structures currently being developed must withstand very severe short duration loading from floating sheet ice. Consequently, most design concepts for such structures provide for an encircling ice wall which can resist high local ice pressure with adequate strength and energy-absorption properties. The most desirable response is a ductile, flexural failure mode permitting large deformation rather than a punching shear failure mode. A ductile failure mode would permit redistribution of ice pressure to the supports of each span of the icewall and would prevent puncturing of the external shell of the structure.

The response possibilities of a typical icewall span are summarized in Figure 1 where load capacity is shown for gradually increasing midspan deflection. A well designed ice wall will show only a moderate drop in ability to support load after peak capacity is reached. At much larger midspan deflections, load-carrying capacity can increase when circumferential reinforcing forms a catenary. Ultimate capacity is reached when this reinforcement fractures under large tensile strain. The key to such favorable response is to provide confinement for the concrete so that it can continue to carry some load even when severely cracked and crushed, to provide adequate shear reinforcement so that a punch-through shear failure does not occur before the ductile response can develop, and to accommodate end displacements associated with catenary action.

Analysis for Global Response to Ice Loading

Prediction of the response of a large concrete gravity structure to ice loading involves both a global analysis of the structure as a whole and a more detailed local analysis of the ice wall itself. The global analysis requires a three-dimensional model of the entire structure in order to study the spatial load path away from the point of ice impact. This information is needed to verify that interior walls supporting the ice wall have adequate strength and to determine support conditions for a local analysis of an isolated segment of ice wall.

Typically, a global analysis requires discretizing the entire structure into finite elements, as shown for a hypothetical concrete gravity structure in Figure 2. Components such as walls and horizontal diaphragms are modeled using plate elements while stiffeners and struts can be modeled using beam or truss elements. For most structures, flexure in the walls (especially the ice wall) can have an important effect on response, and the plate elements must contain rotational degrees of freedom and associated flexural stiffness terms. Due to the size and geometric complexity of the overall structure, a linearized model must suffice for the global three-dimensional analysis; i.e., reinforced concrete is treated as an elastic, isotropic material with steel reinforcing included in the evaluation of overall stiffness. This approach is suitable for determining forces and moments in interior and peripheral walls adjacent to the loaded ice wall span since these walls are proportioned so that they will be stressed considerably below capacity when failure in the loaded ice wall occurs. Response in the ice wall, of course, will be highly nonlinear and the local analysis of individual ice walls must be nonlinear.

There are a number of methods to model the ice loading. One approach that gives a good representation of ice behavior is to assume that the loading is represented by a uniform pressure in the range 1200-2000 psi which gradually spreads out vertically and laterally from the point of impact and midspan until the entire span is loaded. As the loaded area increases, the ice pressure intensity decreases to a range of 500-1200 psi. Such a load model requires a series of linear analyses for different stages in the load sequence to find critical forces and moments in the various components of the structure. For example, maximum forces in the interior walls tend to occur late in the load sequence, while maximum moments in the ice wall occur earlier in the load sequence.

Analysis for Local Response to Ice Loading

Prediction of response and failure modes in the ice wall itself requires a more detailed discretization and nonlinear material models. A portion of the ice wall is isolated from the global model, and in conjunction with appropriate boundary conditions, analyzed separately. The size and shape of this local model will depend on factors unique to each configuration. If there is adequate distance between horizontal diaphragms and the ice wall can be treated as a one-way slab, then a two-dimensional deep beam model of the type shown in Figure 3 can be used. That is, a one-meter high section of wall is isolated, forces in the vertical direction are assumed to be constant (i.e., the variation in local pressures is negligible), and deflections, moments, and forces in the horizontal plane are calculated. For some configurations, such a two-dimensional model will not be possible and a fully three-dimensional local model will be necessary. A two-dimensional model may include only one span of ice wall (or half a span if loads, reinforcing, and boundary conditions are symmetric) or several spans, and in some cases, portions of the interior supporting walls. Local models require careful characterization of boundary conditions since force must be applied not only to the outer edges of the two-dimensional mesh to model in-plane restraint, but also within the mesh to model the vertical dissipation of force due to shear in the walls.

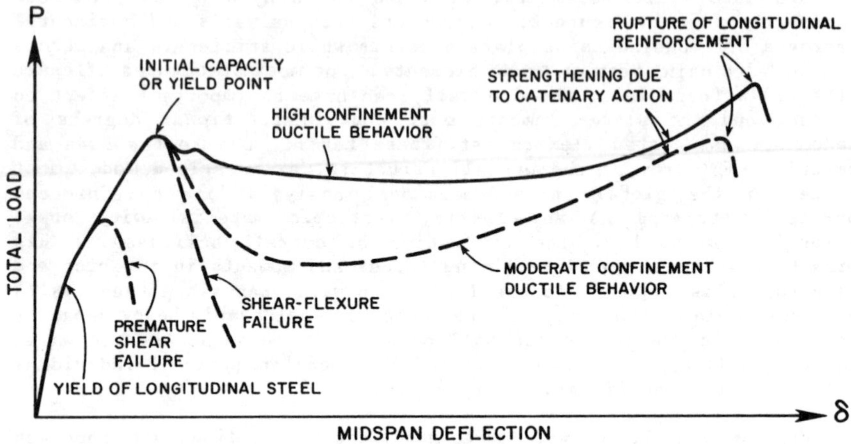

Figure 1 - Typical Icewall Midspan Deflection Response

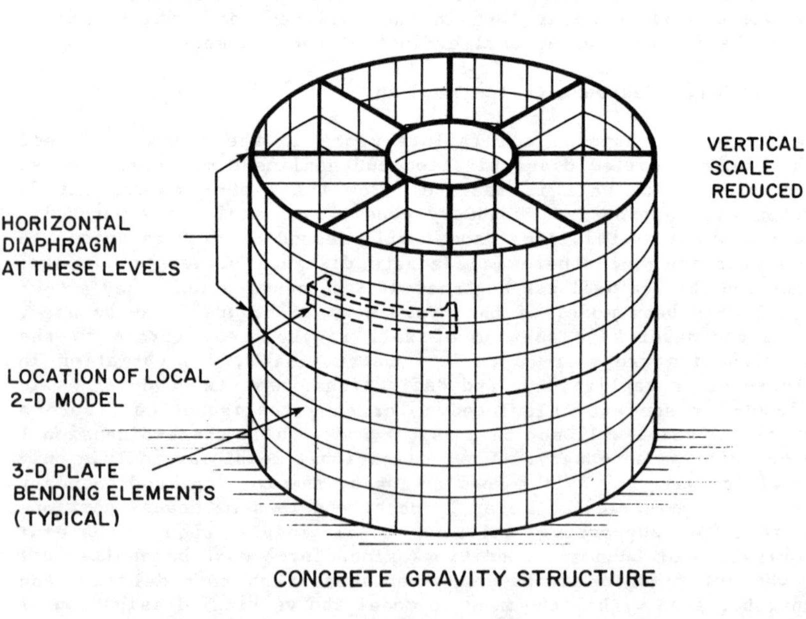

Figure 2 - Typical 3-D Global Finite Element Model

In many cases, a single span with a portion of its connection to the adjoining walls is all that is necessary for a local ice wall analysis, as shown in Figure 3. Boundary conditions for this particular 2-D model consist of the axial and rotational restraint of the adjacent ice wall and interior wall and can be derived from the 3-D global analysis. The ice wall is discretized into 2-D plane strain finite elements to represent the concrete and into one-dimensional elements to represent longitudinal steel reinforcing bars and shear reinforcement (stirrups, etc.). The mesh shown in Figure 4 would give good results without using excessive amounts of computer time. Concrete and reinforcing steel elements are connected to the same nodal mesh, which assumes composite action and no loss of bond between the two materials. However, reinforcing without surface deformations may be connected into the mesh only where locked into other reinforcing, as in the case of the stirrups shown in Figure 4.

Concrete and steel material behavior is modeled through nonlinear stress-strain curves. Typical stress-strain curves for concrete are shown in Figure 5 and depend on degree of confinement, which is parameterized by the intensity of lateral stresses σ_2 and σ_3 with respect to principal stress σ_1. These curves are parabolas defined by the material parameters f_{cp} (prism compressive strenth), E_o (initial tangent modulus), f_t (tensile strength), ε_f (strain at f_{cp}), ε_{ult} (ultimate strain), and f_{ult} (stress at ultimate strain). These parameters defining the stress-strain curve are modified for varying degrees of confinement using failure surfaces of the type shown in Figure 6 (for the case of compressive strength f_{cp}). Relationships of this type, based on experimental data on concrete specimens subjected to various combinations of triaxial stress [6], show that even moderate levels of confinement can double the strength and compressibility of concrete.

The above model describes the behavior of concrete in triaxial states of compression. However, under tensile stress, it can be assumed that concrete is essentially linear until cracking occurs at its tensile strength f_t. Experimental data show that for reinforced concrete tension is gradually released in an element of cracked concrete as tensile "strain" perpendicular to the cracks increases. This tension softening process can be modeled using a tensile stress-strain curve of the type shown in Figure 7a. Likewise, shear stiffness can be assumed to be essentially constant up to cracking and then degrades gradually rather than instantaneously due to aggregate interlock, dowel action, and other factors. This is modeled by a shear modulus <u>vs</u> tensile strain relationship, as shown in Figure 7b. Parameters which define the tension and shear softening models are calibrated using test data. Mesh size can affect the calibration of these parameters.

Steel reinforcing is subject to essentially uniaxial stress, so triaxial material laws are not necessary. The uniaxial stress-strain relationship shown in Figure 8, which models yielding, strain hardening and ultimate fracture, represents a typical reinforcing steel.

With these material laws and finite element discretization, the nonlinear solution technique follows the familar pattern:

1. Apply an increment of ice pressure as loading. Or, alternatively, apply an increment of displacement at the loaded point.

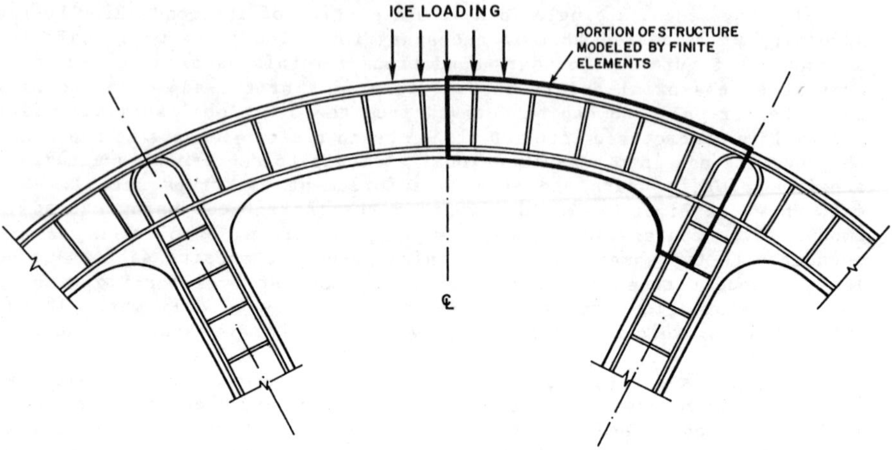

Figure 3 - 2-D Local Model of Typical Icewall

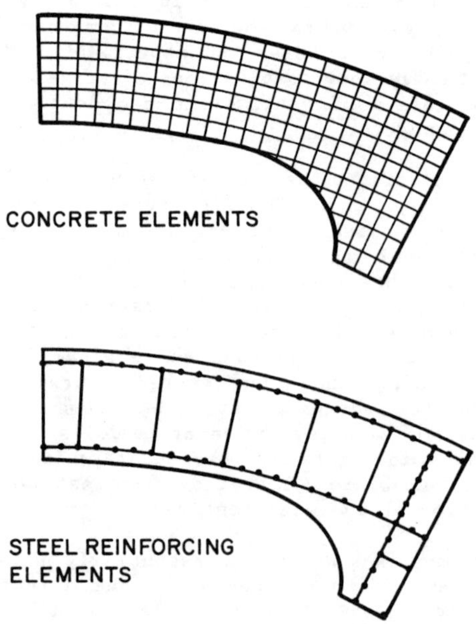

Figure 4 - Finite Element Mesh for 2-D Local Model

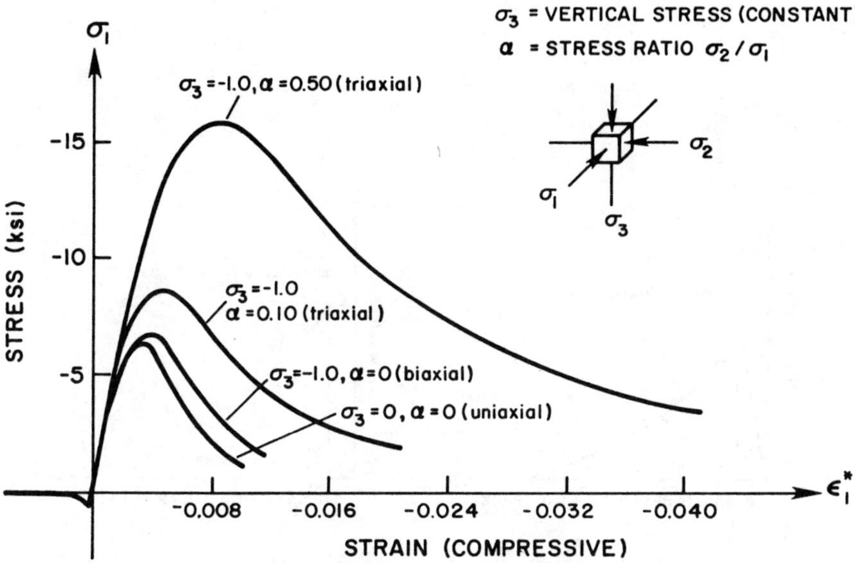

Figure 5 - Triaxial Stress-Strain Law for Concrete

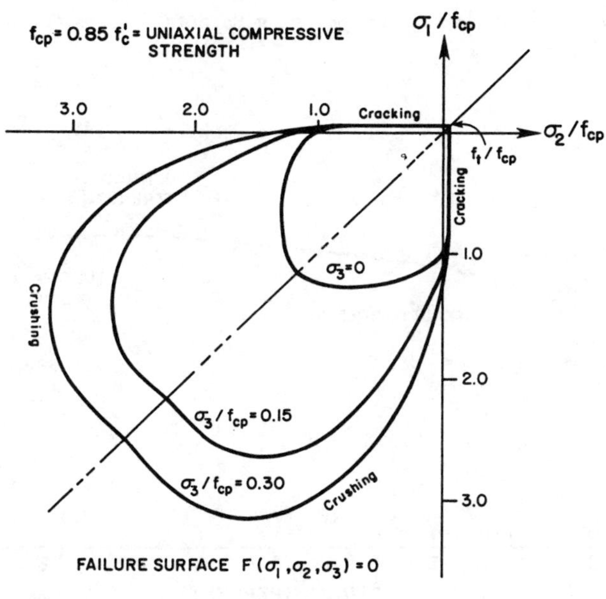

Figure 6 - Triaxial Law for Compressive Strength of Concrete

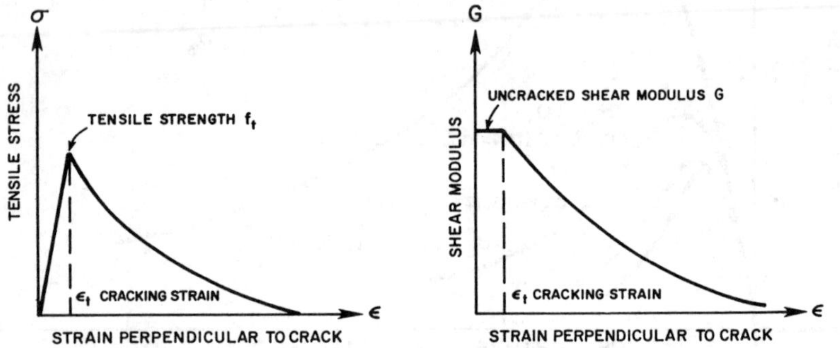

a) Stress-Strain Law for Concrete in Tension

b) Relationship Between Shear Modulus and Strain

Figure 7 - Tension and Shear Softening Material Models for Concrete

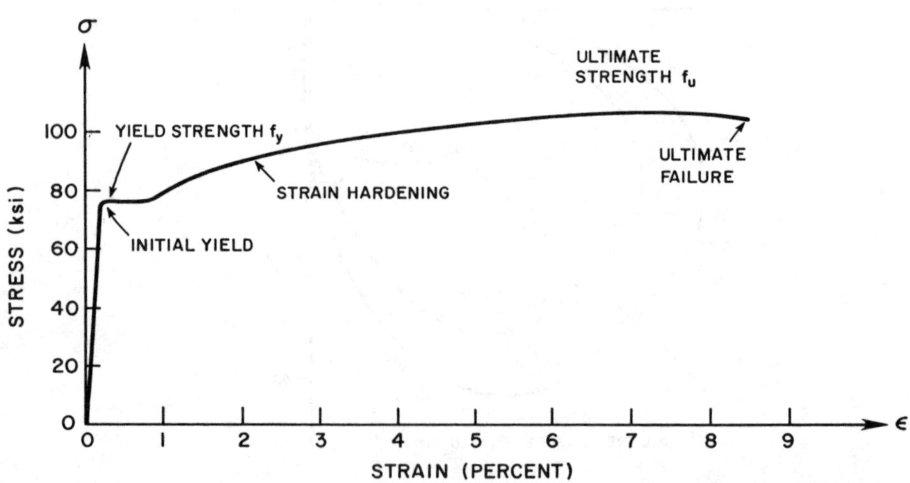

Figure 8 - Typical Stress-Strain Relationship for Steel Reinforcement

2. Calculate element tangent stiffnesses based on current deflections and strain levels at beginning of load step.

3. Assemble element stiffnesses into a global structural stiffness, including boundary support stiffnesses.

4. Solve for deflected shape of ice wall.

5. Calculate new element strains, and using the stress-strain law for concrete or steel, verify that in each element internal stresses and forces equilibrate external loads. It should be noted that this model permits each concrete element to have a different stress-strain curve depending on the amount of confinement in that element. That is, for each increasing level of ice loading, principal strains and current degree of confinement are calculated in each element and the corresponding principal stresses are found from that element's current stress-strain curve. If maximum principal stress is greater than the tensile strength, that element is considered to be cracked and tensile and shear stresses are released according to the descending branches of the curves shown in Figure 7. Likewise, when peak compressive strength is reached, compressive stress is released according to the descending branch of the stress-strain curves shown in Figure 5.

6. If internal forces do not equal external forces, calculate the out-of-balance forces, and, using the global structural tangent stiffness, solve for a corrected deflected shape. Continue this iterative process until convergence is obtained; i.e., until out-of-balance forces are very small.

Analysis results will give stress and strain history in each concrete element and steel reinforcing element, cracking patterns and displacement history of the icewall. For a typical icewall, cracking patterns as predicted by analysis are shown in Figure 9. This plot shows the extent of cracking, and using a smeared crack approach, the calculated direction of cracking in each element. The severity of cracking can be determined by a plot of concrete tensile strains (Figure 9). Concrete compressive stress patterns can also be plotted (Figure 9), and are useful in understanding load transfer through the ice wall. The very pronounced arching action in this short-span wall is readily apparent, as well as the stress concentration at the haunch. Stress and strain distribution in each reinforcing bar and stirrup can also be displayed graphically and can be used to identify zones of high stress and yielding in the steel reinforcing.

Analysis for Ductile Response to Ice Loading

As discussed earlier, typical icewall response consists of an initial peak load capacity followed by some drop in load through a transitional zone leading to catenary response at very large midspan deflections and ultimate rupture of the longitudinal steel reinforcement, as shown in Figure 1. The 2-D local analysis described above will predict this first peak both with respect to magnitude (total ice pressure) and the failure mechanism that causes the load to peak. Stress and strain plots in the concrete will identify areas where concrete begins to strain soften and ultimately crush. Locations of yielding steel reinforcement can be

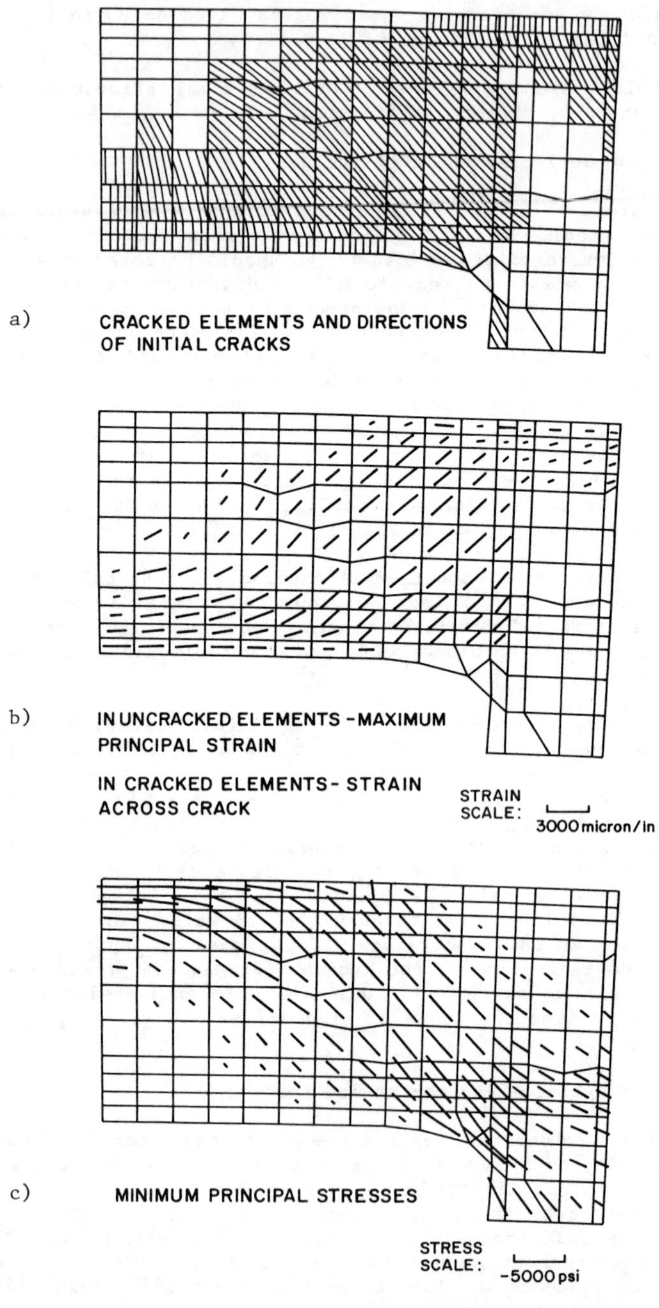

Figure 9 - Typical Stresses, Strains and Crack Pattern for 2-D Model identified. Data from experimental ice wall specimens have been compared with analytical predictions using these models and good correlation has been obtained.

The analysis can be continued beyond the peak load capacity. More and more of the concrete is pushed into the compressive strain softening range and total load capacity of the ice wall gradually decreases. However, conducting analyses through the entire transitional zone is hindered by uncertainties in accurately characterizing the post-yield behavior of concrete that is severely cracked and has undergone large deformations in the presence of heavy confinement. The descending branch of the stress-strain curve for concrete in this degraded state can show much more ductility than the curves discussed earlier (Figure 5) which were based on tests of plain or lightly reinforced concrete. Current experience indicates that the shape of the descending branch of the stress-strain curve to be used in a macroscopic computer model depends on the confinement system, its anchorages, etc. Perhaps a better approach than explicit analysis in the transitional zone is to formulate a set of requirements for confinement, shear reinforcement, and anchorage so that adequate ductility is obtained.

By pushing the analysis through the transitional zone and by including large-deflection theory in the computer model, it is possible to predict the second load peak of the ice wall response - the load at which longitudinal reinforcement ruptures due to catenary action. Since concrete in the shear span zone has lost most of its strength by this time, uncertainties in the descending branch of the stress-strain curve are not as important. Good correlation with test data has been obtained using models of this type.

Summary of Analysis Procedure

A complete analytical evaluation of a prototype ice wall configuration for response due to a particular ice impact scenario would include the following:

1. A 3-D global linear analysis to predict approximate forces and moments in interior walls.

2. A 2-D local nonlinear analysis (with boundary conditions from the 3-D global analysis) of an individual ice wall. This will predict load capacity (first peak) of the ice wall and the initial failure mode (crushing of concrete, yield of stirrups, etc.).

3. Verification that the confinement system will hold up under large displacements (welds, shear reinforcement, anchorage, etc.).

4. Catenary analysis to predict ultimate load capacity at rupture of circumferential reinforcing. This requires good identification of the rupture point of steel and the boundary conditions on the ice wall - particularly, the circumferential restraint on the steel reinforcement which forms a catenary. The predicted ultimate capacity as a catenary (and associated strains and deflections), when coupled with predicted initial load capacity, can serve as a rough measurement of the energy absorption possible in the ice wall.

A computer program (RECON 84) has been developed at WJE to implement the above solution procedure, and a number of different icewall designs have been investigated. Also, several test specimens have been constructed and loaded to failure and good correlation between calculated and measured response has been consistently observed.

Concluding Remarks

There is great potential for use of reinforced concrete finite element methods in engineering practice, and more and more this potential is being realized. Application to situations such as high-intensity ice loading or fire exposure have produced useful information not easily obtainable by more conventional methods of analysis. As research in material characterization and associated numerical techniques progresses and as practicing engineers become more familiar with the products of this research, this trend should accelerate.

Appendix I.—References

1. Iding, R. H., Bresler, B., and Gerwick, B. C., Jr., "Prediction of Failure Mechanisms in Concrete Structures Subjected to Ice Loads," Proceedings, ARCTIC'85 - National Conference on Civil Engineering in the Arctic Offshore, San Francisco, California, March 25-27,1984.
2. Iding, R. H., and Bresler, B., "Analytical Studies of Expansive Solids in Concrete," Proceedings, 4th Engineering Mechanics Division Specialty Conference ASCE/EMO-Purdue University, West Lafayette, Indiana, May 23-25, 1983.
3. Iding, R. H., and Bresler, B., "Prediction of Fire Response of Buildings Using Finite Element Methods," Proceedings, Third Conference on Computing in Civil Engineering CEPA/ASCE, April 3-6, 1984, San Diego, California.
4. Bresler, B., and Iding, R. H., "Fire Response of Prestressed Concrete Members," Chapter on 'Fire Safety of Concrete Structures', M. S. Abrams, Editor, ACE Special Publication No SP80, pp. 69-113, 1983.
5. Iding, R. H., and Bresler, B., "Prediction of Shrinkage Stresses and Deformations in Concrete," Proceedings, International Symposium on Fundamental Research on Creep and Shrinkage of Concrete, Lausanne, Switzerland, edited by F. H. Wittman, M. Nijhoff, pp. 341-352, 1982.
6. Cedolin, L., and Dei Poli, S., "Finite Element Nonlinear Plane Stress Analysis of Reinforced Concrete," Costruzioni in Cemento Armato - Studi e Rendiconti, Politechnico di Milano, 1976.

REALISATIONS AND RESTRICTIONS
Application of numerical models to concrete structures

J. Blaauwendraad

SYNOPSIS

Capabilities and limitations of two computer programs for the nonlinear analysis of reinforced concrete in the Netherlands are discussed. The MICRO program utilizes discrete cracks and is therefore advantageous in connection with research. The DIANA program is a general purpose package and has been structured according to the concept of smeared cracks. This program is of great value in research and also for professional use in structural design. The statements made are illustrated by analysis examples which have been assembled from several research associates and students. The final part of the paper raises the question how to make the newly acquired knowledge and techniques available to professional engineering practice.

Johan Blaauwendraad graduated from Delft University in 1962 and received the doctors degree in 1973. He started his career at the research institute TNO-IBBC, and moved to Rijkswaterstaat (State Public Works) in 1970. Until 1980 he was scientific director of the Data Processing Division and thereafter Head of the Structural Research Department. He has been professor of Civil Engineering at Delft University of Technology since 1979 and is project leader of "Betonmechanica". He is a member of IABSE working commission on "Safety, Reliability and Analysis".

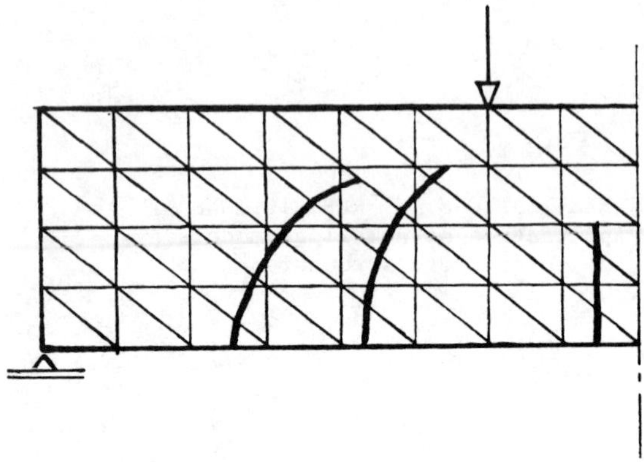

Figure 1 In the MICRO program discrete cracks can propagate curvilinearly and run across concrete elements. It differs from other programs for discrete cracks in that neither the mesh must be modified nor elements be disconnected from one another to allow for cracks.

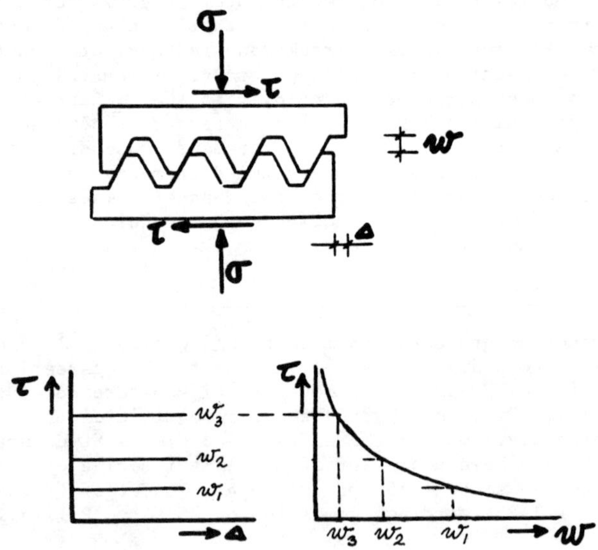

Figure 2 Initially a reasonable model for aggregate interlock was assumed in MICRO (Grootenboer model).

1. OUTLINE OF THE PAPER

At the IABSE colloquium on the mechanics and analysis of concrete in 1984, papers on the activities in the Netherlands were presented. Since then the Dutch project "Betonmechanica" (Concrete Mechanics) has been under way, both experimentally and analytically. The experimental investigations have been directed especially at three aspects:

- The mechanical behaviour (aggregate interlock) of a rough crack subjected to cyclic loading or sustained loading (creep). Executed at the Delft University of Technology (DUT).
- The mechanical behaviour of the contact layer (bond) between a reinforcing bar and the surrounding concrete; both for cyclic loading and sustained loading. Executed at the TNO Institute for Building Materials and Building Structures (TNO-IBBC).
- The influence of a marine environment on the creep rate and differences in this respect for several kinds of cement. Executed at the DUT.

The results of these investigations will be published elsewhere in due course.

This paper more particularly reviews developments which have meanwhile been achieved in the area of anlytical finite element models. In the Netherlands, analytical research proceeds along two distinct lines. On the one hand, the DIANA smeared crack program is continuously undergoing development and, on the other hand the MICRO program for discrete cracks is also being further developed. The name MICRO has nothing to do with microcomputers, but reflects the intention to use the program for very detailed analysis of structures. The DIANA program is a proprietary product of TNO-IBBC, and the MICRO program of Rijkswaterstaat (State Public Works). Delft University is a user of the programs but also contributes to designing adaptations to the software, within the framework of close cooperation between all parties concerned.

The DIANA program aims at being a general-purpose tool for use in consulting engineering practice and in research. The ultimate (indeed ambitious) goal is to feed in constantly newly generated algorithms, options and facilities in order to have at one's disposal a software package which reflects the most up-to-date state of the art in the field.

The aim for the MICRO progam is different. Though professional engineers certainly are encouraged to use this program to solve structural problems in practice, its main advantage will be found in research. Because of its discrete nature and the use of specific contact elements for both bond and aggregate interlock, we can profit from the MICRO program to study the distinct influence of different rough crack models and different bond characteristics on such aspects as the failure mechanism and the spacing and width of cracks.

This paper explains what progress has been made in developing both programs and also indicates which problems remain to be solved.

Section 2 is devoted to the research-oriented MICRO program and Section 3 to the DIANA general-purpose program. The broad outline presented in these sections will be illustrated with examples of analysis and the lessons to be drawn therefrom. In general, the activities reported here are carried out by research associates of TNO-IBBC and Rijkswaterstaat and by students and post-graduate researchers of Delft University.

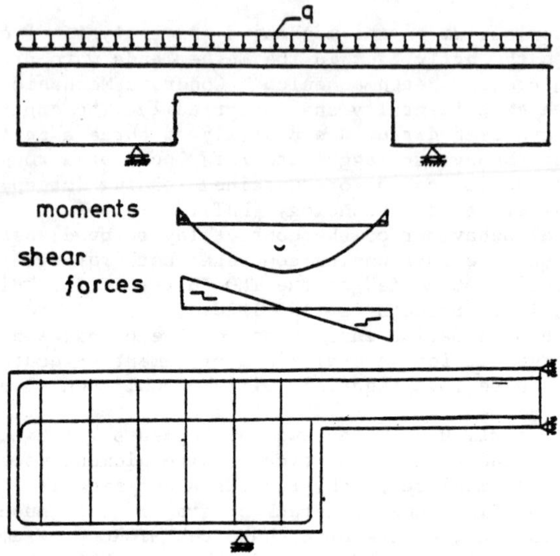

Figure 3 Partially clamped reinforced concrete beam without stirrups, subjected to uniformly distributed load. The failure due to shear was analysed before (IABSE, Delft 1981).

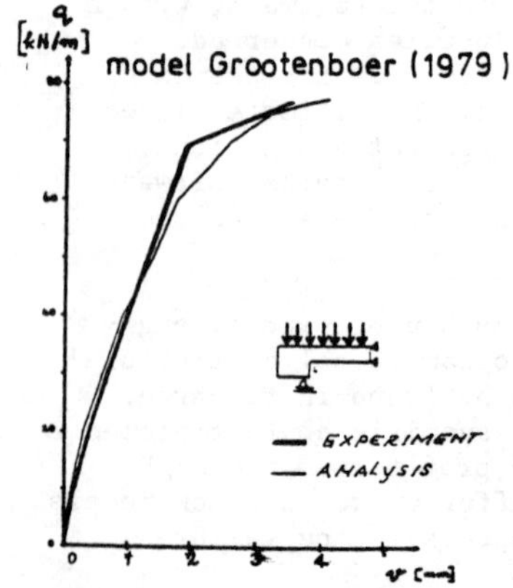

Figure 4
Load-deflection diagram for the beam of Fig. 3 using the Grootenboer model for aggregate interlock in MICRO.

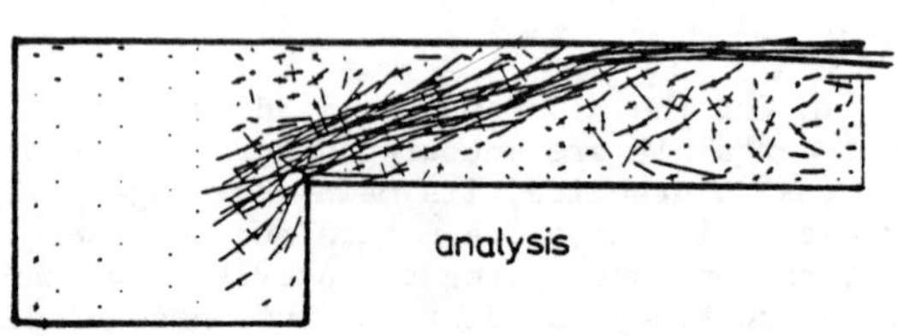

Figure 5
Principal stresses for the beam of Fig. 3 using the Grootenboer model for aggregate interlock in MICRO.

2. THE MICRO PROGRAM

2.1 The essentials

The MICRO program has been extensively described in |1|, |2| and |3|. From these references we will here recall the most relevant facts. The program has explicitly been designed on the basis of two-dimensional plane stress for incremental loading. Concrete elements are triangular and reinforcement elements are line elements. All reinforcement bars must be modelled by such elements. Essentially no smeared reinforcement is considered (hereafter one exception will be mentioned). Both reinforcement and concrete behave like nonlinear materials. For concrete the basic assuptions are due to Kupfer's well-known investigations. If a certain crack criterion (defined by stress) is met, the element in question splits into two separate parts (three in cases of crack branching). The mesh needs not be changed to allow for discrete cracks, nor need the elements be disconnected. Instead, cracks are permitted to run across the triangular elements and can propagate curvilinearly (actually as a polygon) when the load increases (see Fig. 1).
A contact element is used between each reinforcement element and the surrounding concrete elements to ensure proper modelling of bond. Likewise a post-cracking contact element is used between the parts of a cracked triangular concrete element to model accurately the aggregate interlock of the rough crack faces. The contact elements for bond and aggregate are intrinsically nonlinear.
The program has been based on an equilibrium mixed type of FEM, resulting in a set of nonlinear equations for the incremental element boundary displacements.

2.2 Realizations of MICRO

We shall now go into a number of aspects of progress made since the program was written.

Aggregate interlock: Noticeble progress has been achieved in modelling the aggregate interlock in rough cracks as may be clarified with the aid of an example. THe original author of MICRO, H.J. Grootenboer, had no experimental data at his disposal. At the time, he made a best possible guess for modelling aggregate interlock which accounted for the major effect that aggregate interlock disappears for large crack widths. The assumption was that only shear stresses τ, and no compressive stresses σ, would occur. For fixed values of the crack opening w, a rigid plastic relation between τ and the slip Δ of the crack faces was adopted. The plastic plateau for τ decreased for increasing crack opening w, see Fig. 2.
In 1981 results obtained with this rough crack model were reported for a partially clamped reinforced concrete beam without shear reinforcement subjected to uniform load (Fig. 3). The agreement between the experimental and the calculated load-deflection diagram was considered to be convincing (Fig. 4), even though the experiment revealed a sudden loss of stiffnes, whereas the analysis predicted merely a gradual detoriation of stiffness. At the time, little attention was paid to this difference for two reasons. First, the predicted failure load was in good agreement with the experimental investigation. Secondly, the

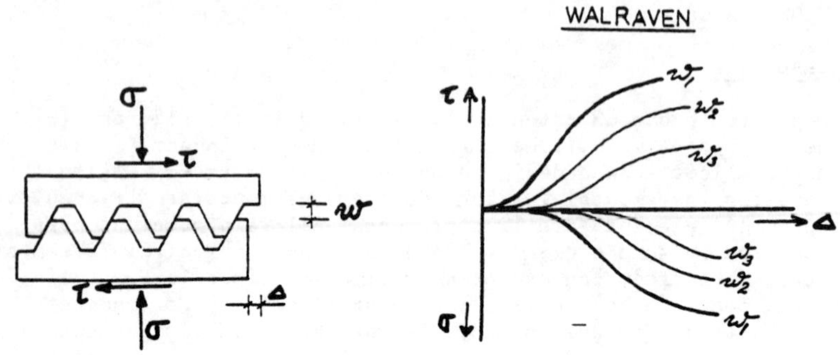

Figure 6 Schematic representation of Walraven's experimental aggregate interlock model. (Functions are given in |4|).

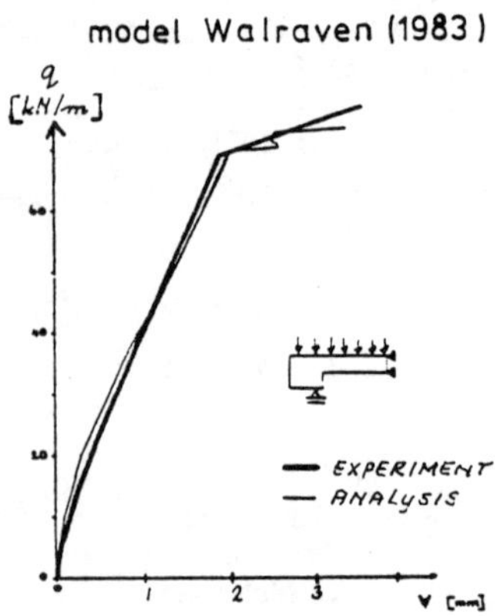

Figure 7
Load-deflection diagram for the beam of Fig. 3 using the Walraven model for aggregate interlock in MICRO.

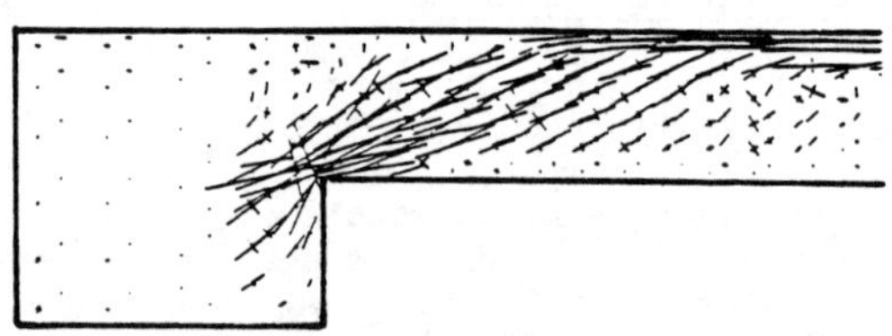

Figure 8
Principal stresses for the beam of Fig. 3 using the Walraven model for aggregate interlock in MICRO.

pattern of principal stresses (Fig. 5) showed typical arch action in the beam, accompanied by slipping of reinforcement near the supports, which to a great extent meets the expectations of an engineer. The analysis was stopped when excessively large strains (crushing) occured in the compressive zone at a position between the support and midspan of the beam.

Meanwhile we are now better informed about aggregate interlock, e.g. from Walraven's experiments |4|. The results are of the type as shown in Fig. 6. This new knowledge has been fed into the MICRO program, resulting in remarkably different analysis results, Fig. 7, 8 and |5|. Now the MICRO program does predict an abrupt type of failure, keeping the stiffness of the beam intact until the failure load is reached. Again the reinforcement starts to slip in this analysis, but now no sharply pronounced arch appears. The compressive zone does not yet fail in this analysis, but the program is not able to find any new equilibrium state, not even for very small load steps. In this analysis, failure is due to high diagonal tensile stresses in the beam, and the program reveals a wide opening of the diagonal crack which has developed. This is a better simulation of what happened in the experimental investigation. The diagonal crack really started at mid-depth of the beam and rather suddenly propagated into the tensile and compressive zone.

We conclude from this experience that it is much necessary to model the aggregate interlock correctly. Otherwise wrong conclusions may be drawn.

Fracture mechanics: In the MICRO program we implemented fracture mechanics options by making use of Hillerborg's fictituous crack model |6|. According to this approach the fracturing process zone which has a width of approximately 2 or 3 times the maximum aggregate size is modelled as one single discrete crack. This fits in perfectly with the set-up of MICRO. The softening branch can be of any shape, and unloading and reloading has been modelled for reversed and cyclic loading according to experimental investigations at the Delft University of Technology |7|. In this way a fatigue simulation can be made for the fracture propagation due to cyclic loading |8|.

It remains to be solved how a smooth transition is achieved from the Hillerborg model in which tension stresses occur (which force the crack to open) to the Walraven model for an already existing crack in which compressive stresses occur in the case of a shearing mode. The energy release for mixed modes in which tension and shear occur simultaneously also requires further research.

Bond: The bond model in MICRO only relates the shear stress τ to the associated relative displacement Δ. No attempt is made to include the effect of radial pressure on the bond characteristics. On the basis of a number of comparative analyses it has been found that we can use a standard τ - Δ diagram (nonlinear, including softening) for beams, columns and beam-column connections.

The results of creep tests can easily be fed into the program. It will, however, require drastic modifications to make the program suitable for reversed or cyclic bond behaviour.

It is planned to use the program to study the influence of low temperatures (LNG storage tanks; artic regions) on the spacing and width cracks, after experiments on low temperature effects on bond have been completed.

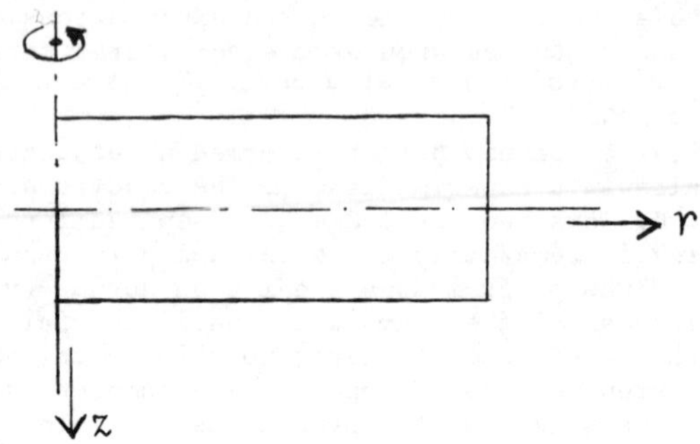

Figure 9
Schematic representation of a punching shear problem to be analysed with an axisymmetric version of MICRO.

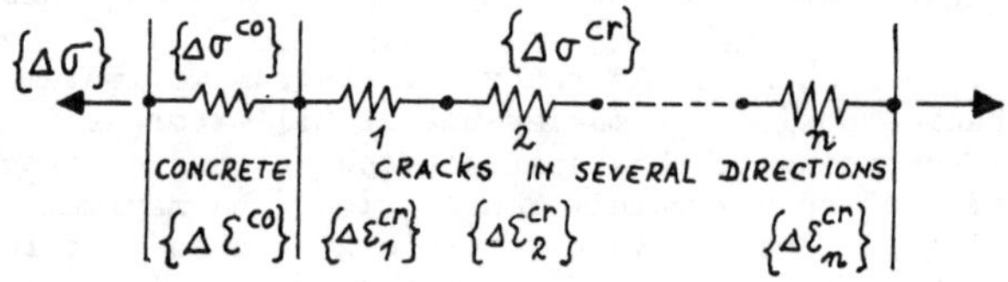

Figure 10
To handle cracks properly in the DIANA program the total strains are serially decomposed into a contribution of the plain concrete and contributions of the occuring crack directions (De Borst).

decomposition: $\Delta \underline{\varepsilon} = \Delta \underline{\varepsilon}^{co} + \Delta \underline{\varepsilon}^{cr}$

concrete: $\Delta \underline{\sigma}^{co} = D^{elastic} \, \Delta \underline{\varepsilon}^{co}$

crack: $\begin{bmatrix} \Delta\sigma^{cr}_{nn} \\ \Delta\sigma^{cr}_{nt} \end{bmatrix} \begin{bmatrix} D_c & 0 \\ 0 & G_c \end{bmatrix} \begin{bmatrix} \Delta\varepsilon^{cr}_{nn} \\ \Delta\gamma^{cr}_{nt} \end{bmatrix}$

D_c = SOFTENING MODULUS

G_c = SHEAR CRACK MODULUS

Mesh choice: In the case of mixed mode fracture mechanics the use of smeared cracks in analysis yields different results, depending on the choice of mesh, as results obtained by Rots |9| show, see Section 3.3. With discrete cracks this type of mesh dependence does not occur.

Axisymmetry: The MICRO program is currently being adapted for application to axisymmetric structures. We are trying to use the program for punching shear problems in reinforced concrete slabs which are subject to local loads and extreme temperatures (Fig. 9). Axisymmetry in finite element methods can be handled as a two-dimensional problem, but stresses and strains occur in three orthogonal directions. When tensile stresses occur in a circumferential direction, cracks may be formed in the radial direction. These crack planes coincide with the r-z plane of Fig. 9 and basically could disturb axisymmetry. Experiments, however, show that these radial cracks are densely distributed. Many such cracks occur, with relatively small crack widths. In the circumferential direction a smeared crack concept is therefore introduced in order to preserve axisymmetry and yet to account for the changes in the circumferential direction due to tension softening. The adaptation of the program is made in two steps. First, a linear-elastic axisymmetric version has been made by Valk |10|. From this a nonlinear version will be derived.
If we wish to change the plane stress version of MICRO into a plane strain version, we have to proceed along similar lines. In that case the third direction also plays a role and has to be handled with the smeared crack approach.

Mix of reinforcement: In section 3.3 it will be discussed that difficulties are still encountered in the smeared crack approach when a mix of several types of reinforcement is used. For instance, the treatment of a fine mesh of reinforcement basically differs from the treatment of a single main heavy reinforcing bar. In the smeared concept it is not yet clear how two such types can be combined. In the approach based on discrete cracks such difficulties are unknown by the very nature of the concept. Each bar is modelled separately and has its own contact element for bond. Admittedly, this advantage is gained at the price of adopting a very fine triangular element mesh for the concrete.

Support for smeared concept: The MICRO program which is based on the discrete crack concept requires in principle fewer assumptions than a program using the smeared crack approach, at least for a specified class of problems. From this point of view the MICRO program could be used to validate the results of smeared crack programs like DIANA. No such dedicated applications can be reported however. Still, it is believed that the results of MICRO did support further development of DIANA namely, in the sense that the DIANA team was faced with a challenge to achieve similar results in situations involving localized phenomena.

Supermicrocomputer: The MICRO program was originally developed for use on mainframe computers and minicomputers (VAX 11/750). Meanwhile it has been implemented on a TNO-designed Geminix supermicrocomputer, including the general engineering system GENESYS (systems software) which was used to develop MICRO. THe main processor of this computer is a 16 bit Motorola 68000 chip, supported by a hardware Floating Point Accelerator. As for the memory, we use 1.5 Mbyte core . Also a Winchester 19 Mbyte hard disk is available.

To provide graphic displays, the GINO library of plotting routines has been implemented on the Geminix as well. In this way it has been attempted to create a computer-aided engineering workstation for easy and inexpensive use of MICRO.

2.3 Limitations of MICRO

To balance the capabilities of the MICRO program mentiond in Section 2.2, a number of restrictions will now be indicated.

No 3D version: To generalize the discrete crack approach of MICRO to three dimensions is a complicated problem. Strictly speaking it is not impossible. However, we are not contemplating such a development for the time being. The 2D version and axisymmetric version are judged to be sufficient for existing research needs.
Besides, we are hopeful of achieving such improvements in smeared crack analysis facilities that duplication may be wasteful.

Time consuming: As stated earlier, in most cases we need a fine element mesh to account for real local effects. This demands much core capacity and long computer time on a VAX 11/750 and especially on a Geminix. Similar comments will be made with regard to DIANA in Section 3.3, to which the reader is accordingly referred.

No rotation of crack directions: The MICRO program has been designed such that it is actually valid only for proportional loading of the concrete. If stress paths in the concrete can change in a non-proportional manner, we know from experimental investigations that cracks which initially occurred may close again and that cracks in other directions start to develop. Such phenomena can be expected in plates with isotropic reinforcement meshes subjected to non-proportional external loading or in plates with orthotropic reinforcement meshes subjected to proportional loading. The MICRO program does not in general take account of the effect of rotations of principal stresses and the associated change in crack direction. The schedule does not provide for adaptions to take this phenomenon into account.

Local details only: Because of the fine mesh needed and the long computer time involved, only local details of structures can be analysed. Examples are: a beam failing in shear; a column subjected to combinations of normal force, bending moment and shear force; beam-column connections; connections between bottom slabs and cylindrical walls in bulk storage tanks for LNG; the effect of high and low temperatures on bond and cracks; and so on. As stated in Section 1, the MICRO program typically meets research requirements, but as such it also can be a tool for solving difficult problems which may become decisive in a particular professional design.

No cyclic loading: In the case of cyclic loading it is essential to have the capability that initially generated cracks can close again and that new cracks are allowed to open up in directions which cross the initially generated cracks. The MICRO program does not provide this option. Hence it cannot easily be extended to dynamic analysis. The capability of MICRO is restricted to reversed and repeated loading for structures in which crack rotations are of minor importance.

This is a severe limitation of the program. For reversed or repeated loadings the contact elements for bond and aggregate interlock should first be adapted to fit the experiments which have been mentioned in Section 1.

Numerical Stability problems: The MICRO program has a problem in common with all smeared crack programs. When a structure exhibits overall post-failure softening behaviour, the analysis becomes far more difficult. Depending on the load increments used, different descending branches may occur in the load-displacement diagram. So far, unrealistically small load steps have to be used to assure proper convergence. This subject is being examined in depth.

3. THE DIANA PROGRAM

3.1 Essentials

The DIANA software package has been extensively described elsewhere, so the reader is referred to the appropriate paper |11|. Essentially the program is based on the well-known stiffness method with incremental loading and allows for 1D elements, 2D elements and 3D elements, incuding axisymmetric ones. Numerical integration is applied using Gauss quadrature, but layers may be used in some specific elements. The nonlinear analysis of reinforced and prestressed concrete is only one application of DIANA, though an important one. The program is also applicable to the plastic analysis (including hardening) of metal structures and geotechnical problems and to nonlinear behaviour of rubber components. Moreover, the program is suited for time-dependent analysis for both creep and dynamic effects. This means that options for cyclic loading have basically been provided.

The development of DIANA has occurred along strict lines of sound software engineering in an environment with high standards for this profession. The program has been designed in a pronounced modular way and much attention has been paid to the structuring of data. In fact, an engineering database called FILOS (file organization system) has been developed in association with DIANA but can also be used separately for other purposes. It is a strict rule that at an integration point you can only use information which is available at that specific point.
A choice can be made from a library of several element types, and a separate library of constitutive models is available. Various combinations of elements and constitutive laws can be made.
Much attention has been paid to solution procedures for the nonlinear equations. The user has a number of equation solvers at his disposal, and it is a matter of experience to select the appropriate one |12|.
We finally mention an extension of DIANA for linear and nonlinear heat flow, which can be used for determining thermal stresses and strains |13|.

As for the subject of reinforced and prestressed concrete the DIANA package offers the following options. Reinforcement can be specified as line elements to represent single discrete bars and as smeared layer elements to represent reinforcement meshes. The reinforcement can have any position within the concrete elements.

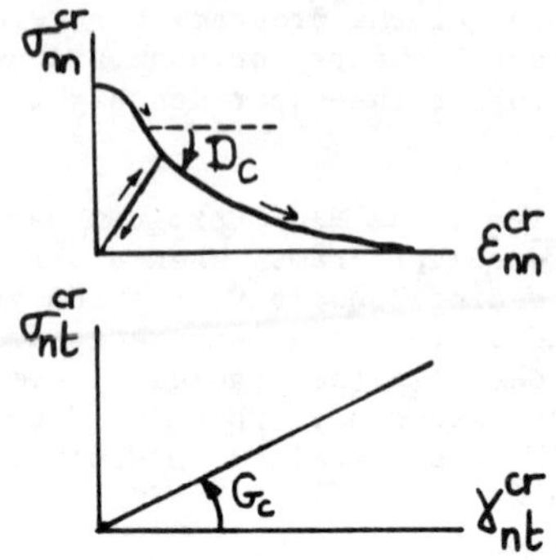

Figure 11 In DIANA a simple stiffness relation between the incremental crack stresses and strains has hitherto been used. D_c is the fracture softening modulus and G_c the crack shear modulus for aggregate interlock. Functions are defined in |9|.

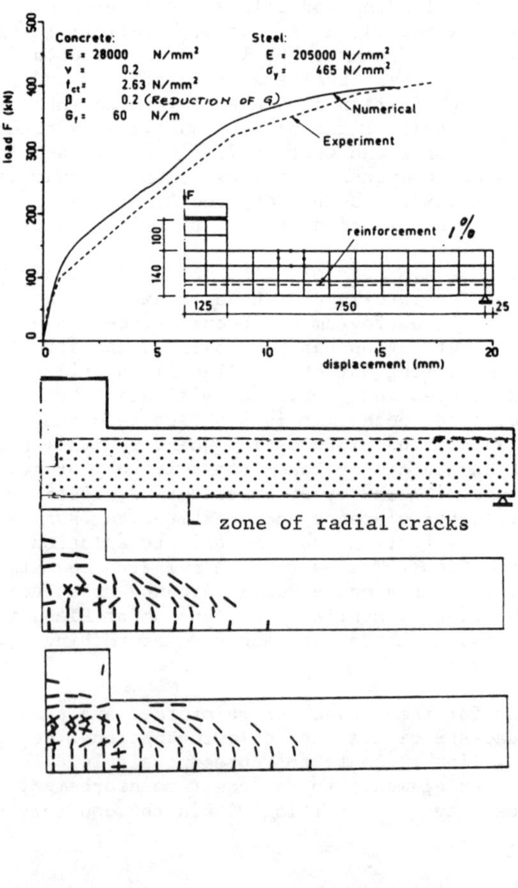

Figure 12
An axisymmetric punching shear analysis with DIANA shows the capability to consider more than one crack direction at an intgration point (De Borst).

The program fundamentally starts from the concept of smeared cracks. As a rule perfect bond is assumed. A departure from this rigid assumption does occur, but is not discussed here. The material law for the reinforcement can be linear-elastic or elastic-plastic with hardening. In the case of cyclic loading, due account is taken of the Bauschinger effect.
The concrete element library contains triangular and quadrilateral elements and they can be isoparametric. As already stated, DIANA uses elastic-plastic hardening types of material behaviour. In the domain of tensile stresses one can choose from a number of tension cut-off criteria.

3.2 Realizations of DIANA

In this section we shall concentrate on the capabilities of DIANA for applications to reinforced and prestressed concrete. Progress made in this respect is reported.

Smeared crack strain concept: The handling of cracks allows more than one crack direction at each Gaussian integration point. And, if necessary, cracks in any direction are able to close again and possibly reopen in the case of non-proportional loading and reversed or cyclic loading. For that purpose De Borst and Nauta |14| decomposed the total concrete strain increment $\{\Delta\varepsilon\}$ into two parts $\{\Delta\varepsilon^{co}\}$ and $\{\Delta\varepsilon^{cr}\}$. Here $\{\Delta\varepsilon^{co}\}$ is the incremental strain of the plain concrete and $\{\Delta\varepsilon^{cr}\}$ the incremental strain which represents the contribution of the crack widths. The latter component in its turn may consist of several contributions from cracks in different directions which have developed at an integration point. Actually, the concept envisages a serial chaining of the stiffness of the uncracked parts of the concrete and the stiffness of the cracks or the fracture processing zones, Fig. 10. The plain concrete between the cracks has to be handled with a suitably chosen constitutive law. If only low stress levels occur, the linear-elastic Hooke's law can be adopted.
For each crack direction two stress components (normal stress and shear stress) and two strain components (normal strain and shear strain) play a role. The incremental stiffness matrix has a rank of two, and in general four stiffness coefficients are involved.
In DIANA it has so far been assumed that the off-diagonal stiffness terms are zero. The term D_c relates the incremental normal crack strain to the incremental normal crack stress; it is the tension softening modulus in the range beyond maximum tensile strength. The diagonal stiffness term G_c connects the incremental shear strain and the incremental shear stress and accounts for aggregate interlock in the rough cracks.
Fig. 11 shows which types of D_c and G_c have been used in DIANA. The constant value of G_c actually means in terms of total stresses and strains that the uncracked shear modulus G has been reduced by a constant factor, independently of the crack widths (the normal crack strain).
Fig. 12 shows a typical punching shear example of De Borst in which the strain decomposition was applied. Some integration points have two or even three crack directions |14|.
Essentially Diana is now capable of handling rotating principal stresses, strains and the associated cracks.

Figure 13
An axisymmetric pull-out specimen which was analysed with DIANA to prove the capability of reproducing localization of fracture (Rots).

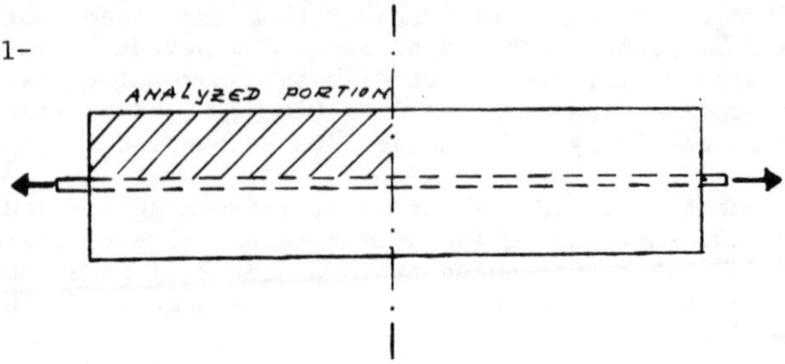

Figure 14
DIANA analysis results for the pull-out specimen of Fig. 13 (Rots). All cracks are shown (a) or only cracks which proceed to a given limit (b). Localization of incremental deformation is apparant from (c).

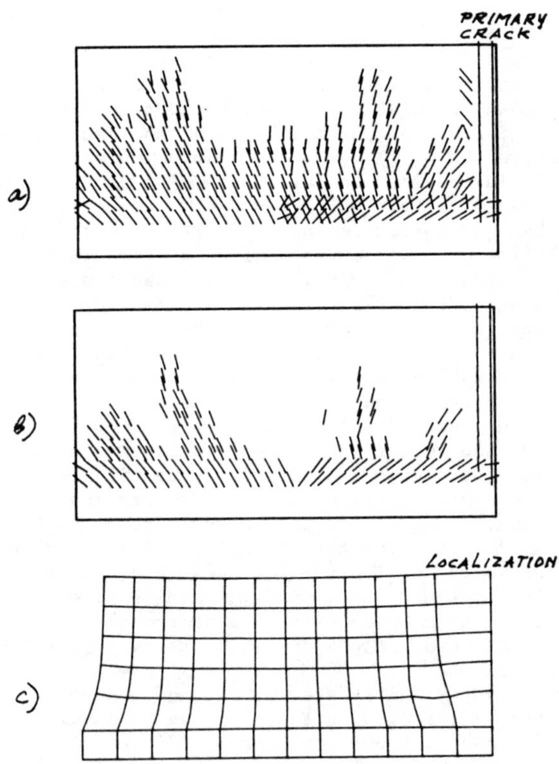

As yet, however, new crack directions are not generated continuously for each load increment. New directions are introduced only if the directional changes exceed a predefined minimum value.

Strain localization: Bazant and Oh |15| were first to point out how blunt crack bands can be objectively achieved within the concept of a smeared crack concept. Their approach is used in DIANA. To obtain strain localization it has proved to be very important not just to allow for softening, but also for unloading after the partial occurrence of softening. If this is not done the following happens. Suppose that at some location a single discrete crack starts to develop (i.e., the tensile strength has been reached), so that the stress level will decrease. Neighbouring elements may also have reached the tensile strength in the same load step. If the stress gradient is small, many elements descend along the tension softening branch and the crack strain increases for all elements and no marked localization of strains will develop. On the other hand, if the opportunity of an unloading branch is provided, the stresses in some elements can decrease in conjunction with decreasing crack strains, while in other elements they can do so for increasing crack strains. In this way the smeared crack concept can reproduce localization of strains and thus represent real local effects.
Rots |16| has convincingly analysed an example, simulating a symmetric pull-out test assuming perfect bond, Fig. 13. At some load level F a crack runs across the whole cross-section halfway along the specimen. This is a primary crack which actually splits the specimen into two new ones of half length. In earlier days, when DIANA had no fracture mechanics options (not to mention unloading in the softening domain), it was not possible to reproduce pull-out test results |17|. At that time only a very diffuse distribution of strains and stresses could be calculated, and the analysis had to be stopped upon reaching the tensile strength level. The phenomenon of the primary crack could not be simulated. Now DIANA can do this. In Fig. 14a all cracks are shown at the onset of the primary crack on the axis of symmetry. If you merely show cracks which exceed a given small crack strain, only a small number of them need to be included in the illustration, Fig. 14b. So the regions of localization for the primary crack and other secondary cracks are very nicely isolated. Apparently many initial cracks close again. In the interpretation of the results it is also very helpful to show incremental displacements. These indicate which changes occur in an important load step. In fig. 14c we see the localization happening in the central part of the specimen.
Rots also applied DIANA to the problem of a statically indeterminate clamped reinforced beam without stirrups which was subjected to uniformly distributed load q and failed in shear at load q_u |9|. Fig. 15 shows the crack pattern at 65% of the failure load. The prominently marked cracks represent those which reveal real crack widths. The thin cracks reveal fracturing zones which are still in the softening process; so no distinct individual crack has yet developed.

Non-associated flow: The examples of DIANA presented so far have been dominated by high tensile stresses and strains. When the critical stress combinations are in the compressive range, the multiaxial criterion for (so called) initial yielding and failure are far more important. We then need a proper description of the plastic strain

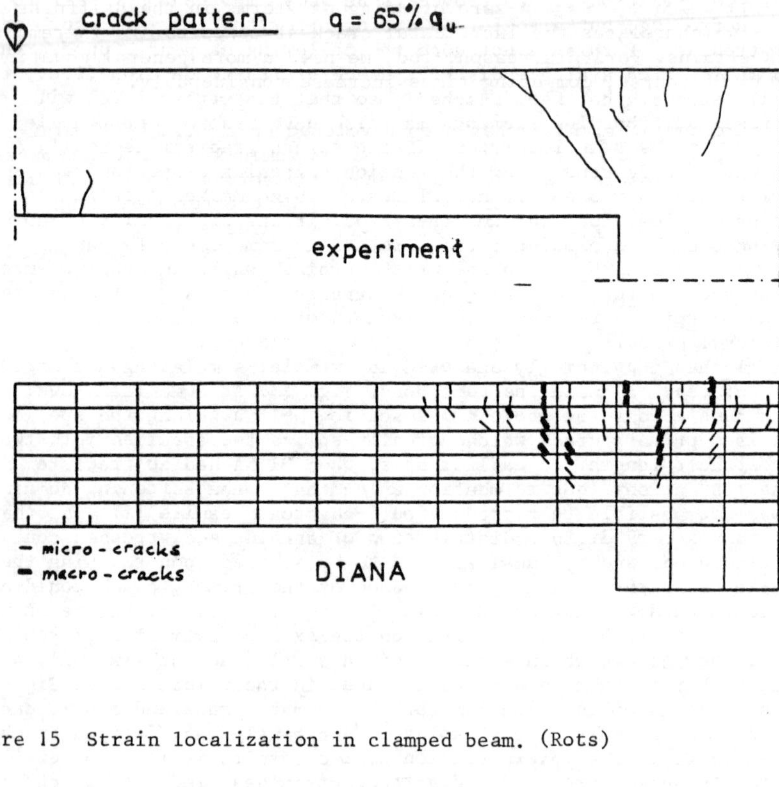

Figure 15 Strain localization in clamped beam. (Rots)

increments, and concrete does not obey the normality flow rule which follows from Drucker's postulate. The same holds for geomaterials. DIANA has been adapted for non-associated flow rules in order to better fit experimental data, as was reported by Vermeer and De Borst |18|. As a result the set of equations becomes asymmetric, so a new equation solver had to be developed.Incidentally, symmetry will also be lost when a correct aggregate interlock description is used (see Fig. 11) which yields non-zero off-diagonal terms in the stiffness matrix which relates the incremental crack stresses to the incremental crack strains. For this reason, too, we need a more general equation solver. Of course, computing times increase considerably.

Supermicrocomputer: DIANA has been developed on a VAX 11/780 mini-computer, but has also been implemented on Geminix superminicomputers. The main features are a Motorola 68000 processor, a hardware floating point accelerator, 1 to 2 Mbyte core and a Winchester hard disk ranging from 19 Mbyte to over 100 Mbyte. An investment of roughly 25,000 dollars can provide any consulting engineers' office with such powerful CAD facilities, except for the annual subscription for maintenance and continuous development of DIANA.

3.3 Limitations of DIANA

In this section a survey is given of aspects for which further improvements are desired.

Regional and local details only: Because of the long computing times in the nonlinear range the analysis of problems has to be restricted to regional or local details. This is even more applicable when time-dependent states are to be considered. Especially combinations of nonlinear behaviour and transient analysis are very time consuming and hence very expensive. Purchase of a faster computer ranging from an attached processor up to a supercomputer defintely will help for a certain class of problems. However, nonlinear shock analysis and similar time consuming tasks fundamentally require a different approach. The need to speed up the operations is so great that only parallel computing with arrays between 100 and 1000 processors would be really meaningful. Otherwise, continuous development may be blocked in the near future. This aspect of nonlinear analysis is, from now on, just as important as the mechanics of concrete itself.

Mesh sensitivity: As already stated in Section 3.2, we know how to obtain objective results with refining element meshes in fracture analysis |15|. It appears that this information is valid when the blunt crack bands are parallel to the edges of a rectangular mesh. Rots |9| has shown that objectivity is illusory in the case of mixed mode fracture analysis in which curvilinear cracks propagate from a notch. Refinement of a rectangular mesh results in several distinct crack bands, whereas the coarse mesh just produces one broad band. However, an inclined mesh using parallelogram-shaped elements yields a totally different analytical result.

Rough crack model: The model for aggregate interlock in Diana as explained in Section 3.2 can be improved. First of all, the crack shear modulus G_c should be made dependent on the normal crack strain

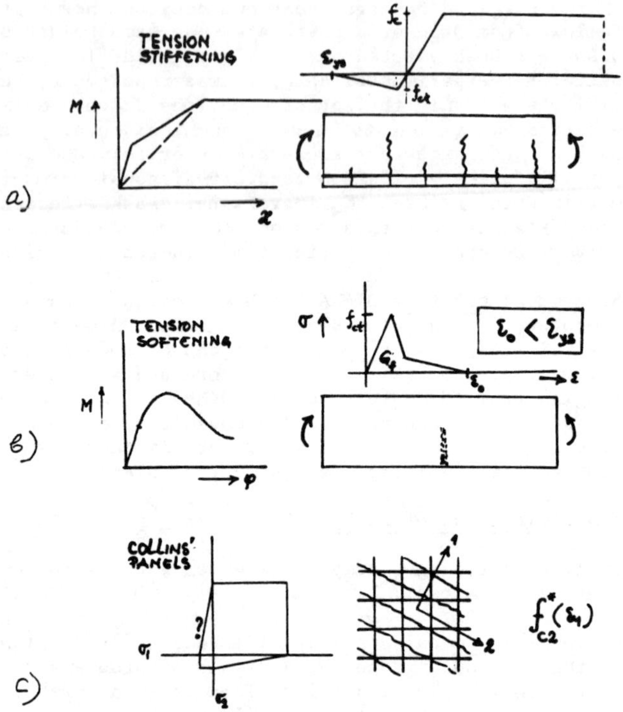

Figure 16
Schematic representation of three different and inconsistent models in cracked concrete. Tension stiffening (a), tension softening (b) and Collins' model (c).

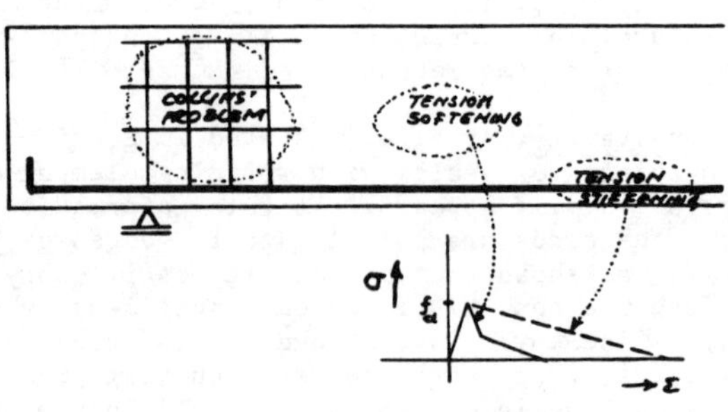

Figure 17
The three different and inconsistent models of cracking shown in Fig. 16 may occur in one structure, which gives rise to difficulties.

ε_{nn}^{cr} such that the crack shear modulus vanishes for high values of the normal crack strain. Besides, compressive stresses must also be generated for increasing shear crack strains, as shown by Walraven's experiments (Fig. 6). This will result in non-zero off-diagonal terms. Furthermore it must be born in mind that Walraven's model applies to a crack which has completely developed. As long as a crack is developing we have a fracture processing zone in which microcracks occur. It has not yet been elucidated how such fracture zones behave when the load path is changed such that combinations of tensile and shear stresses occur (mixed mode) in the fracture zone which has so far developed.

Inconsistent constitutive models: For dealing with cracked concrete we have learnt to use a number of different models, each quite suitable for a particular situation. One of them is tension stiffening in the neighbourhood of a main reinforcement bar, to account for the contribution of plain concrete between two cracks. It is called *stiffening* because the stiffness of the beam is greater than when no contribution of the concrete in the tensile zone is considered (as is done to analyse the ultimate strength of a reinforced cross-section). However, in the FEM analysis the effect is introduced by adding a fictitious *softening* branch in the tensile domain of the stress-strain diagram of the concrete, Fig. 16a. Tensile stresses are assumed to vanish when the tensile strains become equal to the strain at which the reinforcement bar starts to yield.

In the case of fracture in an unreinforced specimen we use another model. In this case we likewise bring in softening, but now realistically modelling the microcracking phenomenon in the fracture zone, Fig. 16b. The tensile strain at which the tensile stress vanishes is normally much smaller than in the case of tension stiffening. The value is now determined by the fracture energy release G_f and by the shape of the softening branch.

Yet another model has been proposed for reinforced panels in which reinforcement meshes are used. In this case it was apparent from Collins' test panels that not only tension stiffening had to be introduced, but also a reduction of the associated compressive strength of the concrete, depending on the tensile strains which occur in the transverse direction (Fig. 16c). This strain dependent reduction replaces the condition which is usually defined in the two-dimensional stress plane.

There arises the question what one should do in case where a structure consists of several parts which are of a different nature and refer respectively to the three models discussed, Fig. 17. Actually different models should be used for the respective regions, but how does the program know which model to apply to which element?

Besides, different models may even be linked within one region, for instance when a heavy main reinforcing bar is added to a smeared reinforcing mesh. We must admit that we then run into difficulties. We know quite a lot about each individual model, but it is in-depth knowledge of "bits and pieces" without a smooth transition from one to the other. To obtain a program suitable for ordinary users a generalized model is needed of which the models mentioned earlier, are special cases. This subject merits further effort to be devoted to it.

4. APPLICATION TO PROFESSIONAL USE

4.1 Experienced users

Nonlinear analysis so far has been restricted to qualified analysts because of the thorough knowledge that it requires of so many material mechanics aspects and problems of numerical nature. Admittedly, this was also said when finite element analysis was introduced for dealing with linear-elastic design. However, such design tasks are fairly easy to perform because in that case hardly any knowledge of the material is needed and normally no numerical instability problems arise. Actually, we need not only hardware and software for nonlinear computations but also brainware. We need a specially trained structural analyst who is familiar with the nonlinear program system and knows all its ins and outs. We call him a "model engineer". The training of such an engineer takes several months if indeed not a whole man-year. And to keep his experience up to date he has to make several analyses per year. Therefore it is questionable whether any particular firm would seek to cater independently for its own "nonlinear" requirements. Nevertheless we are working on the proliferation of nonlinear FEM analysis of reinforced concrete. The motivation is pressure from the construction industry rather than a conviction that is is well equipped for such tasks. The market for nonlinear analysis is really restricted compared to linear analysis, and not many contractors or consultants can afford to spend many man-years on keeping this capability up to standard. Yet this is what they will have to do if they want to be self-supporting for nonlinear studies.

4.2 Education

The introduction of nonlinear analysis into the structural engineering profession can be improved by means of adequate education. In the Netherlands such education is offered to students in the university of Delft and to professional engineers. Structural engineering students take a course in nonlinear FEM analysis for steel and reinforced concrete and execute a run with DIANA or MICRO on the university's Geminix supermicrocomputers. It is hoped thus to create a new generation of young engineers who at least are aware of the capabilities and limitations of the existing programs like DIANA and MICRO.
In this way it is hoped gradually to develop in the profession an appreciation of nonlinear analysis.
In addition to this education in the regular curriculum of the students, two courses for professional engineers are also planned. One of these courses is intended for designers who are in positions where decisions for nonlinear analyses have to be made. They need an overview of the capabilities and limitations of the programs and should be informed about the theory in a global sense. However, they themselves need not necesarily carry out such analyses. It is hoped that a relatively large number of engineers will attend this introductory course.
The other course will go in depth and is intended for the training of model engineers. The number of applicants for this course will be much smaller. These engineers are the structural analysts who will actually use the programs in research and design, and who will develop such programs themselves.

4.3 Utilization

Apart from the educational effort it is considered worthwhile to make another special effort for professional utilization in structural design. After having developed new program features in the Netherlands year after year, the decision was made to spend substantial sums of money on promoting the use of the DIANA program. This new activity constists of two phases. In a first short preliminary phase of about one year an "example book" will be compiled, showing how particular structural problems can be tackled by DIANA. For this phase about $ 100,000 have been made available, and an ad hoc committee of professional representatives has been invited to submit typical design problems relating to the construction of bridges, mass storage tanks, buildings and off-shore structures.
After this phase has been completed with (hopefully) positive results, it is expected to be allocated additional funds for a further and far longer phase, in which the practical use of the program will again be stressed and new problems solved. For instance the use of the program in developing a new code of practice could be encouraged. This phase should consist of clear and convincing demonstration projects. At the same time, during this phase, we intend to improve the program and eliminate (most of) the shortcommings listed in Section 3.3. The new phase will require substantial funds (say $ 2,000,000) and will extend over a number of years.

5. ACKNOWLEDGEMENTS

The program developments and use discussed in this paper were made possible by financial support from industry and the Ministry of Economic Affairs in the Netherlands. This money is allocated through CUR-VB, the Netherlands Committee for Research, Codes and Specifications for Concrete. Besides that the Ministry of Education and Science contributes support through STW, the Netherlands Technology Foundation. Sincere thanks are also due to TNO-IBBC, the Delft University of Technology and Rijkswaterstaat, who contributed valuable manpower and computer resources.

REFERENCES

1 Grootenboer, H.J., "Finite element analysis of two-dimensional reinforced concrete structures, taking account of nonlinear physical behaviour and the development of discrete cracks", Dissertation Delft University of Technology, 1979.

2 Blaauwendraad, J., Grootenboer, H.J., "Essentials for discrete crack analysis", Proceedings IABSE colloquium Advanced Mechanics of Reinforced Concrete, Delft 1981.

3 Grootenboer, H.J., Leyten, S.F.C.H., Blaauwendraad, J., "Concrete Mechanics, part C", Heron, Vol. 26, 1981, no 1c.

4 Walraven, J.C., "Aggregate interlock: A theoretical and experimental

analysis", Doctoral thesis, Delft University of Technology, 1980.

5 Souren, W.H., "Simulation of shear behaviour of reinforced concrete beam, using MICRO with discrete cracks" (in Dutch), graduate project, Delft University, 1984.

6 Hillerborg, A., "A model for fracture analysis", Coden Lutvdg/ (TVBM-3005)/1-8/1978.

7 Berg, F.J.M. van den, "Fictitious Crack Model", Rijkswaterstaat Bouwspeurwerk, Memo 84/08, 1984.

8 Moelands, J.M., "Modeling of fatigue behaviour of unreinforced concrete subjected to cyclic or repeated loading" (in Dutch), graduate project, Delft University, 1984.

9 Rots, J.G., Nauta, P., Kusters, G.M.A., Blaauwendraad, J., "Smeared crack approach and fracture localization in concrete", Heron, Vol. 30, 1985, no. 1.

10 Valk, J.Th., "Axisymmetric version of the program MICRO" (in Dutch), graduate project, Delft University of Technology, 1985.

11 Borst, R. de, Kusters, G.M.A., Nauta, P., De Witte, F.C., "DIANA - Comprehensive, but flexible finite element system", Proc. 4th Int. Seminar on F.E. Systems, Southampton England, 1983.

12 Borst, R. de, "Application of advanced solution techniques to concrete cracking and non-associated plasticity", Proc. Int. Conf. Univ. Pol. Barcelona, 1984

13 Jansen, J.C.M., "The solving of heat flow problems by the finite element method", graduate project, Delft University, 1985.

14 Borst, R. de, Nauta, P., "Smeared crack analysis of reinforced concrete beams and slabs failing in shear", Proc. Int. Conf. on Computer-Aided Analysis and Design of Concrete Structures, Split, 1984.

15 Bazant, Z.P., Oh, B.H., "Crack Band Theory for Fracture of Concrete", Materials and Structures (Rilem, Paris), Vol. 16, 1983.

16 Rots, J.G., "Bond-slip simulations using an advanced crack model and interface elements", Delft University, 1985.

17 Groot, A.K. de, Kusters, G.M.A., Monnier, Th., "Concrete Mechanics, part B", Heron, Vol. 26, 1981, no 1b.

18 Vermeer, P.A., Borst R. de, "Non-associated plasticity for soils, concrete and rock", Heron, Vol. 29, 1984, no. 3.

Collapse Analysis of RC Shells
Based on a New Fracture Criterion

Herbert A. Mang,[1] M. ASCE and Josef Eberhardsteiner[2]

Abstract

A new fracture criterion for concrete subjected to biaxial states of stress is presented and applied to ultimate load analysis of a built natural draft cooling tower shell made of reinforced concrete. It is based on the assumption that the biaxial tensile strength of concrete at a point of the shell increases with increasing stress gradient at this point. For the limiting case of a vanishing stress gradient, peak stresses from Kupfer's biaxial tests are taken as the biaxial tensile strength. For a non-vanishing stress gradient, a magnifying factor for Kupfer's biaxial tensile strength at the considered point is determined by means of a suitable hypothesis. It is shown that the significant increase of Kupfer's biaxial tensile strength at the integration points in the finite element model of the cooling tower does not have much influence on the structural response of this shell.

Introduction

The critical value of a physical quantity at a point in a given body frequently depends on the gradient of this quantity at the considered point. This applies, e.g., to critical compressive stresses of thin, perfect steel plates, signalling loss of stability of the undeformed configuration of the plates. Fig. 1(a) shows two steel plates with identical dimensions. For load case 2, the compressive stress at the bottom edge of the plate, σ_b, is assumed to be the same as the uniformly distributed compressive stress σ for load case 1. (For the problems of Fig. 1(a) compressive stresses are regarded as positive quantities.) It is well known (22) that $\sigma_{cr}^{(2)} > \sigma_{cr}^{(1)}$ where $\sigma_{cr}^{(1)} = \chi^{(1)}\sigma$ and $\sigma_{cr}^{(2)} = \chi^{(2)}\sigma > \chi^{(2)}\sigma_t$ are the critical compressive stresses for the two load cases, $\chi^{(1)}$ and $\chi^{(2)}$ are corresponding numerical factors and σ_t is the compressive stress at the top edge of the plate for load case 2. Similarly, for two concrete plates with identical dimensions (Fig. 1(b)), $\sigma_{cr}^{(2)} > \sigma_{cr}^{(1)}$, where $\sigma_{cr}^{(1)}$ and $\sigma_{cr}^{(2)}$ denote the tensile strength of concrete for the two load cases. $\sigma_{cr}^{(1)}$ is equal to the uniaxial (direct) tensile strength of concrete, representing the peak stress in the stress-strain diagram of a direct tension test. Disregarding the dependence of $\sigma_{cr}^{(1)}$ on the size of the test specimen, $\sigma_{cr}^{(1)}$ is considered to be a material property. Just as the modulus of rupture is not a material property, $\sigma_{cr}^{(2)}$ is not a material

[1] Prof. and Director of Institute of Strength of Materials, Technical University of Vienna, Vienna, Austria.

[2] University Assistant, Institute of Strength of Materials, Technical University of Vienna, Vienna, Austria.

property. Both examples show that σ_{cr} increases with increasing stress gradient.

As opposed to physical reality, conventional ultimate load analyses of the two concrete plates of Fig. 1(b) by the finite element method (FEM) on the basis of the classical strength criterion would result in $\sigma_{cr}^{(2)} = \sigma_{cr}^{(1)}$. According to this criterion, the principal stresses at discrete points of the individual elements (computed either only at the center of gravity or at all integration points) are compared to corresponding peak stresses obtained from biaxial tests such as, e.g., the ones conducted by Kupfer (14). These tests were carried out for biaxial states of uniformly distributed principal stresses. Consequently, the resulting peak stresses represent lower bounds to fracture stresses which would be obtained from corresponding tests for biaxial states of nonuniformly distributed principal stresses.

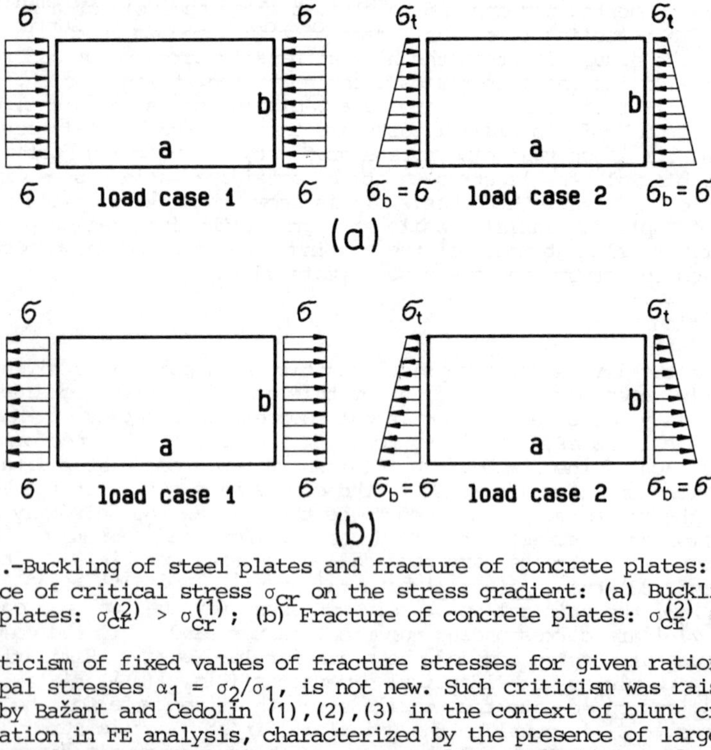

FIG. 1.-Buckling of steel plates and fracture of concrete plates: dependence of critical stress σ_{cr} on the stress gradient: (a) Buckling of steel plates: $\sigma_{cr}^{(2)} > \sigma_{cr}^{(1)}$; (b) Fracture of concrete plates: $\sigma_{cr}^{(2)} > \sigma_{cr}^{(1)}$

Criticism of fixed values of fracture stresses for given ratios of principal stresses $\alpha_1 = \sigma_2/\sigma_1$, is not new. Such criticism was raised first by Bažant and Cedolin (1),(2),(3) in the context of blunt crack propagation in FE analysis, characterized by the presence of large stress gradients. Restricting attention to linear elasticity, for a sharp crack, representing the limiting case of a blunt crack band, the stress gradient at the tip of the crack tends to infinity. A strength criterion based on tests carried out for biaxial states of uniformly distributed stresses is obviously not suitable for situations characterized by large stress gradients.

By adjusting the uniaxial tensile strength of concrete to the element size, Bažant and Cedolin (1) obtained objective results for the problem of a Mode I crack in an infinite solid and a finite strip, respectively, subjected at infinity to uniaxial stresses normal to the cracks. In other words, their results converge to the exact solutions as the FE grids are refined. The same applies to the approach of adopting the energy release rate as the controlling parameter for crack extension (1).

On the macroscopic level, to which the present work is restricted, it is useful to distinguish between the initiation of new cracks and the propagation of existing cracks. The problems of Fig. 1(b), e.g., refer to crack initiation. Consequently, methods for dealing with crack propagation such as the two aformentioned techniques proposed by Bažant and Cedolin are not applicable to those problems. On the other hand, the writers are not aware of tests carried out for biaxial states of nonuniformly distributed stresses, of which load case 2 of Fig. 1(b) is a uniaxial special case.

As opposed to the simple problems of crack initiation according to Fig. 1(b), on one hand, and to problems of propagation of discrete macro-cracks observed in *existing* reinforced concrete (RC) surface structures, on the other hand, for ultimate load analysis *at the time of design* of RC surface structures, the problem of crack initiation cannot be separated from the one of crack propagation. Such a separation would require the labelling of some of the integration points as "potential crack-initiation points" while labelling the remaining integration points associated with tributary domains of yet intact concrete as "potential crack-propagation points". For this purpose, a criterion for allocating "intact integration points" to one of the two categories of "potential crack-points" would be needed.

What instead is needed from the standpoint of the practical feasibility of ultimate load analysis, is a general fracture criterion which does not require distinguishing between crack initiation and crack propagation. On checking if the tributary domain of the considered integration point of a finite element remains intact after application of a certain load increment, it should be irrelevant for the algorithm whether or not this point was located in the proximity of already cracked subregions of the structure.

As follows from the preceding considerations, studies on concrete fracture restricted to the investigation of individual cracks "as though under a magnifying glass", to cite Rots et al. (20), only cover part of the problems encountered in ultimate load analysis of RC surface structures. Although such studies are important in their own right, it is usually not easy to adapt the employed methods to ultimate load analysis of real-life structures.

This statement, e.g., applies to consideration of the size effect in failure criteria for concrete. This effect was recently investigated by Bažant (4) who introduced a so-called "size reduced-strength" representing a characteristic of the entire structure rather than of the material. While the existance of the mentioned effect is undisputed, it is not obvious how to extend the concept of a global size reduced-strength from the relatively simple academic example of blunt crack band propagation to ultimate load analysis of RC structures with zones of large as well as small stress gradients.

The importance of the size effect when considering gradual tensile strain-softening within the framework of a constitutive model for tensile fracture of concrete was stressed by Rots et al. (20). These authors emphasized the need of adjusting the strain-softening modulus to the chosen size of finite elements, for otherwise the fracture energy release would be mesh-dependent. This argument is convincing.

However, apart from the difficulty of properly adjusting the strain-softening modulus in case of irregular FE meshes occurring in practical applications, it is not consistent to work with a variable strain-softening modulus while disregarding the dependence of the strength limit on the stress gradient as was done by the authors of (20).

The benefit of gaining a better insight in the mechanics of crack propagation by investigating cracks as though under a magnifying glass is beyond all doubt. However, such investigations should not result in the magnifying of problems which actually are of minor importance for ultimate load analysis of RC surface structures. As was recently shown by Dodds, Darwin and Leibengood (7) for RC beams, the classical stress-controlled smeared-crack representation adequately models the general softening that results from cracking in the beams. Employing a fixed value of the tensile strength of concrete, these authors obtained load-deflection curves exhibiting convergence with grid refinement. Moreover, they demonstrated that the tensile strength of concrete has a negligible influence on the response of the considered beams.

The purpose of the present paper is to report on the development and application of a new fracture criterion for concrete. Its range of application covers ultimate load analysis of thin RC plates and shells. The fracture criterion is based on the initially mentioned fact that the biaxial tensile strength of concrete increases with increasing stress gradient. For the limiting case of a vanishing stress-gradient, peak stresses from Kupfer's biaxial tests are taken as the biaxial tensile strength. The other limiting case is characterized by a stress gradient tending to infinity as is the case at the tip of a sharp crack if considerations are restricted to linear elasticity. In the context of smeared cracking, the second limiting case refers to the practically unattainable limit of finite discretizations.

The paper reports on the determination of a magnifying factor for Kupfer's biaxial tensile strength relating the "state of fracture" at the considered integration point of the FE model of the given surface structure to the "state of fracture" at the tensile fiber of a hypothetical concrete beam subjected to pure bending. The depth of the beam is determined by means of a suitable hypothesis of comparison between the considered integration point of the surface structure and the tensile fiber of the beam, based on the elastic part of the strain-energy density and on its gradient.

The presented fracture criterion is first verified for a concrete beam subjected to elementary states of stress. It is then applied to ultimate load analysis of a built RC natural draft cooling tower shell subjected to dead load and quasistatic wind load. The influence of the stress-gradient dependence of the biaxial tensile strength on the response of the shell is studied. The value obtained for the ultimate load is compared to values for buckling loads resulting from three different modes of physically linear instability analysis.

Fundamentals of ultimate load analysis of thin RC shells by the FEM

The new fracture criterion for concrete was incorporated in a computer program for ultimate load analysis of thin RC shells by the FEM. Mechanical fundamentals and analysis algorithms of this program are described in detail by Floegl (8). The following summary of the mechanical fundamentals is based on previous work by Mang and Trappel (17).

The kinematics of deformation is approximated by means of the Kirchhoff-Love assumptions. Fig. 2(a) refers to the first two of these three assumptions. The first one is the Kirchhoff normal-hypothesis. It can be formulated mathematically as $\gamma_{i3} = 0$, $i = 1,2$, where $\gamma_{13} \equiv \gamma_{\alpha\zeta}$ and $\gamma_{23} \equiv \gamma_{\beta\zeta}$ are the transverse engineering shear strains. The curvilinear coordinate system α,β,ζ is assumed to be orthogonal. The second Kirchhoff-Love assumption can be formulated mathematically as $w(\zeta) = w(\zeta = 0)$ where $w(\zeta = 0)$ and $w(\zeta)$ are the components of displacement vectors $\mathbf{u} = \overrightarrow{PP'}$ and $\mathbf{v} = \overrightarrow{QQ'}$, respectively, in the direction of the normal to the middle surface of the undeformed shell, S. The last Kirchhoff-Love assumption is concerned with neglecting the influence of the normal stress σ_ζ in the constitutive equations.

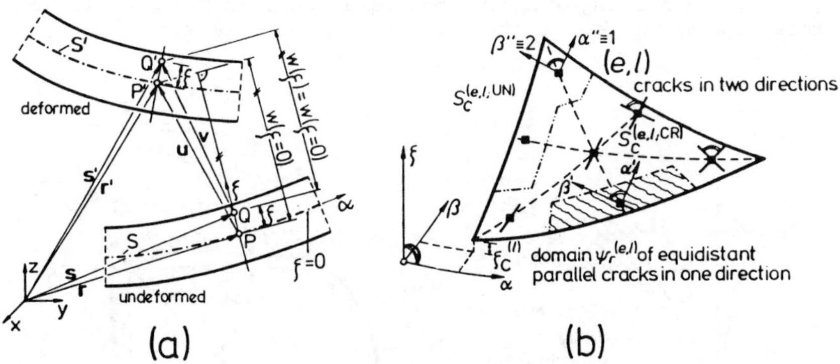

FIG. 2.-Analysis of thin RC shells by the FEM (17): (a) Kirchhoff-Love assumptions; (b) Middle surface of partially cracked concrete-layer l of finite element e

A popular shell theory based on the Kirchhoff-Love assumptions is the one by Koiter-Sanders (11), (5). This theory is used in the aforementioned computer program. Consideration of geometric nonlinearity is restricted to taking account of moderately large rotations. This is sufficient, as has been shown by Mang et al. (16) in the context of ultimate load analysis of RC cooling tower shells.

Curved, triangular, C^1-conforming, thin-shell elements (21) serve as the discretization tool. The elements are subdivided into sufficiently many thin layers such that approximately a plane state of stress may be assumed to exist in each layer. Fig. 2(b) shows the middle surface of layer l of finite element e. The layer is assumed to be partially cracked. $S_C^{(e,\, l;UN)}$ denotes the uncracked and $S_C^{(e,l;CR)}$ the cracked part of the middle surface of the layer. The state of stress is determined at seven integration points of the layer. Local, orthogonal directions α'',β'' at integration points (Fig. 2(b)) denote the principal directions of normal stresses in the uncracked concrete. The reason for this choice of local coordinate systems is that the employed constitutive equations for the intact concrete, proposed by Liu, Nilson and Slate (15), are referred to principal directions of normal stresses.

Let $\sigma_{C1} \equiv \sigma_1$ be the algebraically larger principal normal stress. Then, the constitutive equations for the intact concrete, based on

Kupfer's biaxial tests, can be cast in the following form (16):

$$\sigma_{Ci} \equiv \sigma_i = \frac{E_C^{(0)}}{1-\nu_C\alpha_i} \cdot \frac{1}{1+g_1(\alpha_2)\left[\frac{E_C^{(0)}}{\sigma_{iu}(1-\nu_C\alpha_i)} - \frac{1}{\varepsilon_{iu}}\right]\varepsilon_i - g_2(\alpha_2)(1-\frac{\varepsilon_i}{\varepsilon_{iu}})\frac{\varepsilon_i}{\varepsilon_{iu}}} \quad (1)$$

where $\alpha_1 = \sigma_2/\sigma_1$, $\alpha_2 = 1/\alpha_1$ are ratios of principal stresses, σ_{iu} and ε_{iu} are experimentally obtained values of peak stresses and corresponding strains (fracture stresses and strains) and $E_C^{(0)}/(1-\nu_C\alpha_i)$ are effective initial tangent moduli; $E_C^{(0)}$ and ν_C are the initial tangent modulus and Poisson's ratio, respectively, for uniaxial loading. Table 1 contains functions $g_i(\alpha_2)$ appearing in equation (1). Fig. 3(a) shows a comparison of typical stress-strain diagrams, based on equation (1), with corresponding diagrams from experiments conducted by Kupfer. In this figure, f_{cu} denotes the prism strength of concrete.

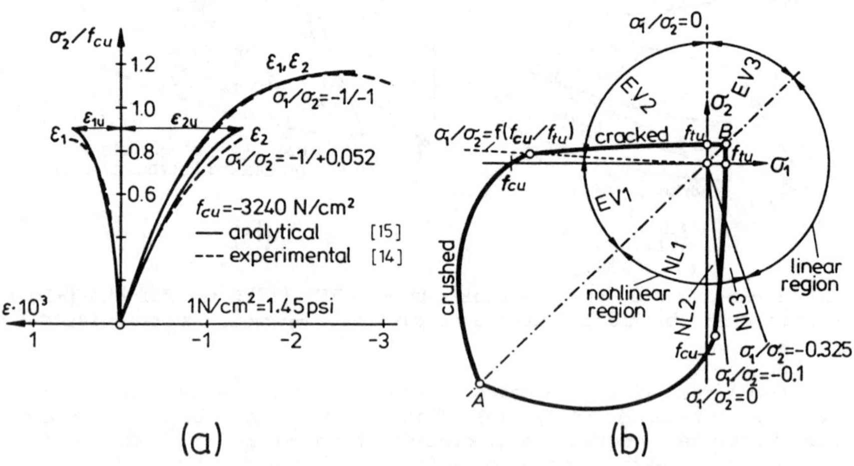

FIG. 3.-Constitutive modelling of plain concrete (17): (a) Typical stress-strain diagrams for plain concrete subjected to biaxial states of stress; (b) Fracture envelope for plain concrete subjected to biaxial states of stress (14)

The employed constitutive model is a so-called "equivalent uniaxial model" because of relating biaxial states of strain to biaxial states of stress by means of equivalent uniaxial stress-strain equations. Advantages and disadvantages of this constitutive model are discussed, e.g., by Chen and Ting (6). For thin shells analyzed for monotonically increasing, proportional loads, the chosen constitutive model is adequate.

For $|\sigma_i| \geq |\sigma_{iu}|$ concrete fails by fracturing. For $1 \leq \alpha_2 \leq \infty$ (biaxial-tension domain) and $-\infty \leq \alpha_2 \leq \alpha_{2L}$ (part of tension-compression domain) with $\alpha_{2L} = f(f_{cu}/f_{tu})$, where f_{tu} denotes the uniaxial (direct) tensile strength of concrete, it is assumed that fracture occurs by

TABLE 1.-Functions $g_1(\alpha_2)$, $g_2(\alpha_2)$ for equation (1)

	Ratio of principal stresses: $\alpha_2 = \sigma_1/\sigma_2$, $\sigma_1 \geq \sigma_2$				
	NL1	NL2	NL3	linear	
α_2	≤ 1 ≥ 0	≤ 0 ≥ -0.1	≤ -0.1 ≥ -0.325	≤ -0.325 $\geq -\infty$	$\leq \infty$ ≥ 1
$g_1(\alpha_2)$	1	1	1	0	
$g_2(\alpha_2)$	1	$1+10\alpha_2$	0	0	

cracking of concrete normal to the direction of σ_1. Cracked concrete retains its capacity to carry internal forces in the direction parallel to the cracks so long as $f_{cu} < \sigma_2 < f_{tu}$. Because of tension stiffening, concrete also retains part of this strength normal to the cracks. In the mentioned computer program for ultimate load analysis of thin RC shells by the FEM, tension stiffening is taken into account (9). For $\alpha_{2L} < \alpha_2 \leqq 1$ (part of tension-compression domain, biaxial-compression domain), it is assumed that concrete fails by crushing, thereby losing all of its strength.

Local orthogonal directions α',β' at integration points denote the directions of normals and parallels, respectively, to bands of equidistant parallel cracks distributed over tributary domains of these points such as, e.g., the tributary domain $\psi_r^{(e,l)}$ of integration point r in Fig. 2(b). Cracks in one or two directions are considered. The second direction is restricted to be orthogonal to the first one. Although this simplifying assumption is inconsistent with the state of strain, it is regarded as admissible (16). The described way of modelling the cracked concrete is the basis for the stress-controlled "smeared" technique used by the writers.

Fig. 3(b) illustrates the fracture envelope of concrete according to Kupfer (14). It is the geometric locus of biaxial states of fracture stresses σ_{1u}, σ_{2u}. Each one of the three symbols EV1, EV2 and EV3 denotes a certain part of the half of the fracture envelope above the symmetry line through points A and B. The three corresponding parts of the half of the fracture envelope below the symmetry line are described by the following analytical expressions (14):

$$\left(\frac{\sigma_{1u}}{f_{cu}} + \frac{\sigma_{2u}}{f_{cu}}\right)^2 + \frac{\sigma_{1u}}{f_{cu}} + 3.65\,\frac{\sigma_{2u}}{f_{cu}} = 0, \quad \alpha_{2L} \leqq \alpha_2 \leqq 1 \tag{2.1}$$

$$\frac{\sigma_{1u}}{f_{tu}} = \left(1+\frac{\sigma_{2u}}{f_{cu}}\right)^{\frac{1}{3}}, \quad -\infty \leqq \alpha_2 \leqq \alpha_{2L} \tag{2.2}$$

$$\sigma_{1u} = f_{tu}, \quad 1 \leqq \alpha_2 \leqq \infty. \tag{2.3}$$

Determination of Magnifying Factor for Kupfer's Biaxial Tensile Strength

Condition for Determination of Magnifying Factor.-Based on the fact that the fracture stress increases with increasing stress gradient, Kupfer's biaxial tensile strength of concrete, σ_{1u} ($\sigma_{1u} \geqq \sigma_{2u}$), valid for the special case of a vanishing stress gradient, represents a lower bound to the true biaxial tensile strength σ^*_{1u} ($\sigma^*_{1u} \geq \sigma^*_{2u}$). Setting $\sigma^*_{1u} = k\sigma_{1u}$, where k is a magnifying factor for Kupfer's biaxial tensile strength, the new criterion for failure of concrete by cracking is $\sigma_1 \geq k\sigma_{1u} \geq \sigma_{1u}$. For $-0.1 \leq \alpha_2 \leq 1$ (part of tension-compression domain, biaxial-compression domain) σ_{1u} is not increased, i.e., k is set equal to 1. For this range of values for α_2, with the exception of the very small subdomain $\alpha_{2L} \leq \alpha_2 \leq -0.1$, concrete fails by crushing. Reasons for setting k = 1 in case of crushing of concrete are (a) lack of experimentally corroborated knowledge about the dependence of concrete failure by crushing on the stress gradient and (b) the fact that global failure of thin RC shells is usually triggered by cracking rather than crushing of concrete.

The condition for determination of k is

$$\left.\frac{U^*_{0R}}{U_{0R}}\right|_{(r,l,e)} = \frac{U^{(b)}_{0R}\left(\frac{l}{3} < x < \frac{2l}{3},\ z = \frac{h}{2}\right)}{U^{(t)}_{0R}} \quad (3)$$

The left side of this equation refers to the considered integration point in the FE model of the investigated shell. The symbol $|(r,l,e)$ identifies this point as the integration point r of concrete layer l of finite element e (Fig. 2(b)). In the following, this symbol will only be used exeptionally. The right side of equation (3) is related to a hypothetical beam with rectangular cross-section of yet to be determined depth h, subjected, for $l/3 < x < 2l/3$, to pure bending (Fig. 4(a)), and to uniform axial tension (Fig. 4(c)), respectively. The argument of $U^{(b)}_{0R}$ refers to the center part of the tensile fiber of the beam shown in Fig. 4(a). In the following, the argument of $U^{(b)}_{0R}$ will only be written exceptionally.

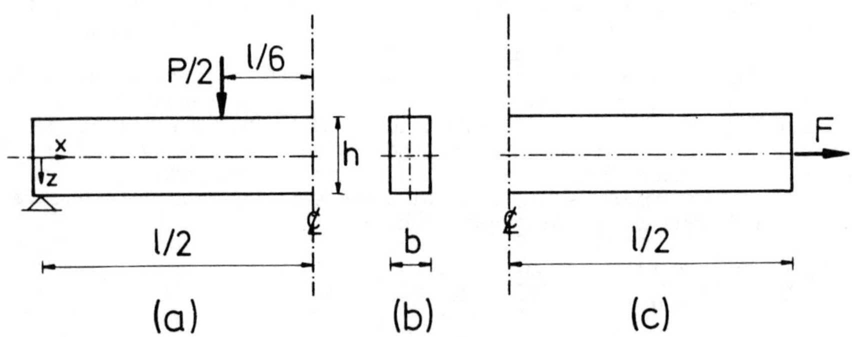

FIG. 4.-Hypothetical concrete beam with rectangular cross-section: (a) Pure bending for $l/3 < x < 2l/3$; (b) Cross-section; (c) Uniform axial tension

U^*_{0R} is the true strain-energy density released at the considered integration point when concrete fails by cracking. U_{0R} is the strain-energy density which would be released at this point when concrete fails by cracking, if the stress gradient was zero and if $\alpha_1 = \sigma_{2u}/\sigma_{1u} = \sigma^*_{2u}/\sigma^*_{1u}$, noting that Kupfer's tests were carried out for proportional loading. $U^{(b)}_{0R}$ $(U^{(t)}_{0R})$ is the strain-energy density released at incipient cracking of the hypothetical beam of Fig. 4(a), (Fig. 4(c)). (It is assumed that cracks start opening when the peak stress in the stress-strain diagram is reached). As will be shown in the following, equation (3) relates the *unknown* influence of the stress gradient on the biaxial strength of concrete at the considered point of the shell to the *known* influence of the stress gradient on the uniaxial tensile strength of concrete for a hypothetical beam of yet to be determined depth, subjected to pure bending.

U^*_{0R} is given as

$$U^*_{0R} = \hat{U}^*_{0I} - \hat{U}^*_{0II} \tag{4}$$

where $\hat{U}^*_{0I}$ and $\hat{U}^*_{0II}$ are elastic portions of the strain-energy density just before and after cracking, respectively. Similarly, U_{0R} is obtained as

$$U_{0R} = \hat{U}_{0I} - \hat{U}_{0II} \tag{5}$$

where $\hat{U}_{0I}$ $(\hat{U}_{0II})$ corresponds to $\hat{U}^*_{0I}$ $(\hat{U}^*_{0II})$. With reference to Fig. 5(a), $\hat{U}_{0I}$ and $\hat{U}_{0II}$ are obtained as

$$\hat{U}_{0I} = \frac{1}{2E_C^{(0)}}\left[(1-\nu_C\alpha_1)\,\sigma^2_{1u} + (1-\nu_C\alpha_2)\,\sigma^2_{2u}\right] \tag{6}$$

and

$$\hat{U}_{0II} = \frac{1}{2E_C^{(0)}}\,\sigma^2_{2u}\,. \tag{7}$$

Using the relationship $\alpha_1 = 1/\alpha_2 = \sigma_{2u}/\sigma_{1u}$ for eliminating σ_{2u} in equations (6) and (7) and substituting the results into equation (5), yields

$$U_{0R} = \frac{1-2\nu_C\alpha_1}{2E_C^{(0)}}\,\sigma^2_{1u}\,. \tag{8}$$

Formally, U^*_{0R} can be written as

$$U^*_{0R} = \frac{1-2\nu_C\alpha_1}{2E_C^{(0)*}}\,\sigma^{*2}_{1u} \tag{9}$$

where σ^*_{1u} and $E_C^{(0)*}$ are unknown quantities (Fig. 5(b)) which would have to be determined from suitable tests for nonuniform loading. This seems to be unfeasible because of the relatively complicated test set-up required for conducting such experiments and the large number of tests needed in consequence of the variation of the stress-gradient. (If σ^*_{1u} was known from tests, theoretical considerations for determining k would be superfluous).

Since Hooke's law holds at incipient cracking of a concrete beam subjected to pure bending, $U_{0R}^{(b)}$ is obtained as

$$U_{0R}^{(b)} = \frac{1}{2E_C^{(0)*}} (f_{btu}(h))^2 \tag{10}$$

where f_{btu} is the bending tensile strength (modulus of rupture) of concrete, depending on h and, thus, on the stress gradient. Similarly, $U_{0R}^{(t)}$ is obtained as

$$U_{0R}^{(t)} = \frac{1}{2E_C^{(0)}} f_{tu}^2 \ . \tag{11}$$

Strictly speaking, the use of $E_C^{(0)}$ of equation (8) in equation (11) and of $E_C^{(0)*}$ of equation (9) in equation (10) is an approximation disregarding the fact that Kupfer's biaxial tests were conducted for quadratic plates of 20 cm side length and 5 cm thickness whereas modulus-of-rupture tests are usually carried out for beams of 50 - 70 cm length with quadratic or rectangular cross-sections of 10 - 15 cm width and depth (10). In view of the considerable scattering of test results for f_{btu} and f_{tu}, the mentioned approximation is regarded as admissible.

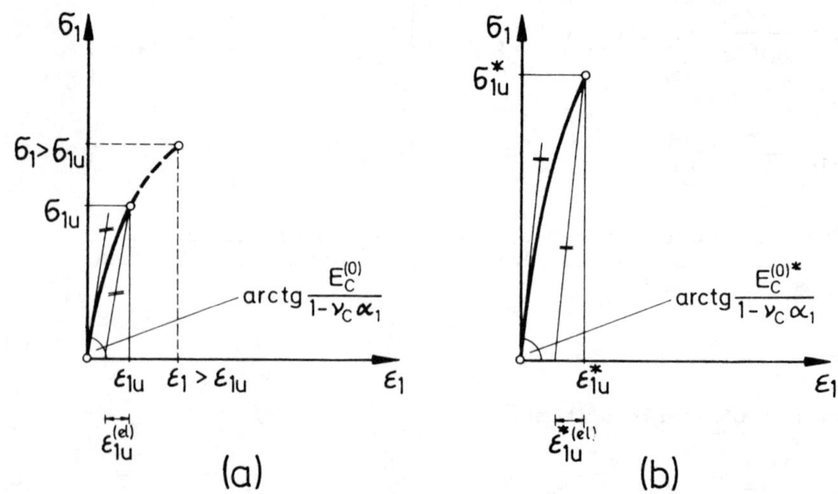

FIG. 5.-Magnification of Kupfer's biaxial tensile strength σ_{1u}: (a) $\sigma_1 - \varepsilon_1$ diagram for uniform loading, based on Kupfer's biaxial tests; (b) Corresponding $\sigma_1 - \varepsilon_1$ diagram for nonuniform loading

Substituting the expressions for U_{0R}, U_{0R}^{*}, $U_{0R}^{(b)}$ and $U_{0R}^{(t)}$ according to the equations (8) - (11) into the relationship (3), yields

$$\frac{\sigma_{1u}^*}{\sigma_{1u}} = k = \frac{f_{btu}(h)}{f_{tu}} \ . \tag{12}$$

Equation (12) states that the ratio of the true biaxial tensile strength of concrete, σ^*_{1u}, to Kupfers biaxial tensile strength σ_{1u}, at the considered point of the shell, is the same as the ratio of the bending tensile strength f_{btu} for a concrete beam of yet to be determined depth h, subjected to pure bending, to the direct tensile strength f_{tu}.

Determination of k on the Basis of Results from Tests for Beams.-If h was known, k(h) could be determined by using the results of modulus-of-rupture and direct-tensile-strength tests. For this purpose the expression for k (equation (12)) is written as

$$k = \frac{f_{btu}}{f_{tu}} = \frac{f_{btu}}{f_{btu}^{(10)}} \cdot \frac{f_{btu}^{(10)}}{f_{tu}} \tag{13}$$

where $f_{btu}^{(10)}$ is the bending tensile strength of a 10 cm deep beam.

According to Hellmann (10), $f_{btu}^{(10)}$ and f_{tu} can be expressed in terms of the cube strength of concrete, f'_{cu}, as follows:

$$f_{btu}^{(10)} = c_{bt}(f'_{cu})^{\frac{2}{3}} \tag{14.1}$$

$$f_{tu} = c_t(f'_{cu})^{\frac{2}{3}} \ . \tag{14.2}$$

Evaluating a large number of test results from the literature, Hellmann obtained average values of c_{bt} and c_t, reported as c_{bt} = 0.98 and c_t = o.52. Thus,

$$\frac{f_{btu}^{(10)}}{f_{tu}} = \frac{c_{bt}}{c_t} = 1.88 \tag{15}$$

resulting in

$$k = 1.88 \frac{f_{btu}}{f_{btu}^{(10)}} \ . \tag{16}$$

Evaluating a large number of test results from the literature, Mayer (18) presented a $f_{btu}/f_{btu}^{(10)}$ - h diagram (Fig. 6(a)). This diagram shows that the bending tensile strength increases with decreasing depth of the beam and, consequently, with increasing stress gradient. k is obtained by multiplying the ordinates of the diagram by 1.88. Fig. 6(b) shows the k - h diagram. The dashed part of the diagram is an extrapolation from the part of the curve which is based on experimental results. As will be demonstrated in the numerical investigation, the dashed part of the diagram is of no practical relevance. For the limiting cases $h \to 0$ and $h \to \infty$, k tends to ∞ and 1, respectively. Thus, $1 \leqq h \leqq \infty$.

Determination of h by means of a Hypothesis of Comparison.-The depth of the bending beam of Fig. 4(a) is determined by means of the following hypothesis of comparison, valid for uncracked concrete:

$$\left.\frac{|\nabla \hat{U}_0|}{\hat{U}_0}\right|_{(r,l,e)} = \left.\frac{\dfrac{\partial U_0^{(b)}}{\partial z}}{U_0^{(b)}}\right|_{\frac{l}{3} < x < \frac{2l}{3},\ z = \frac{h}{2}} \qquad (17)$$

where $\hat{U}_0 |_{(r,l,e)}$ is the elastic portion of the strain-energy density at the considered integration point in the FE model of the investigated shell, $|\nabla \hat{U}_0||_{(r,l,e)}$ is the absolute value of the gradient vector $\nabla \hat{U}_0$ at this point

$$U_0^{(b)}\left(\frac{l}{3} < x < \frac{2l}{3},\ z = \frac{h}{2}\right) = \frac{1}{2E_C^{(0)*}}\left(\frac{Pl}{bh^2}\right)^2 \qquad (18)$$

is the strain-energy density at a point in the center part of the tensile fiber of the beam of Fig. 4(a) and

$$\frac{\partial U_0^{(b)}}{\partial z}\left(\frac{l}{3} < x < \frac{2l}{3},\ z = \frac{h}{2}\right) = \frac{2}{E_C^{(0)*}h}\left(\frac{Pl}{bh^2}\right)^2 \qquad (19)$$

is the gradient of $U_0^{(b)}$ at the mentioned point. Noting that Hooke's law holds for the hypothetical bending beam, it appears logical to employ the elastic portions of the strain-energy density U_0 and of its gradient ∇U_0 for the comparison expressed by equation (17). This is also consistent with the determination of U_{0R}^* and U_{0R} according to equations (4) and (5).

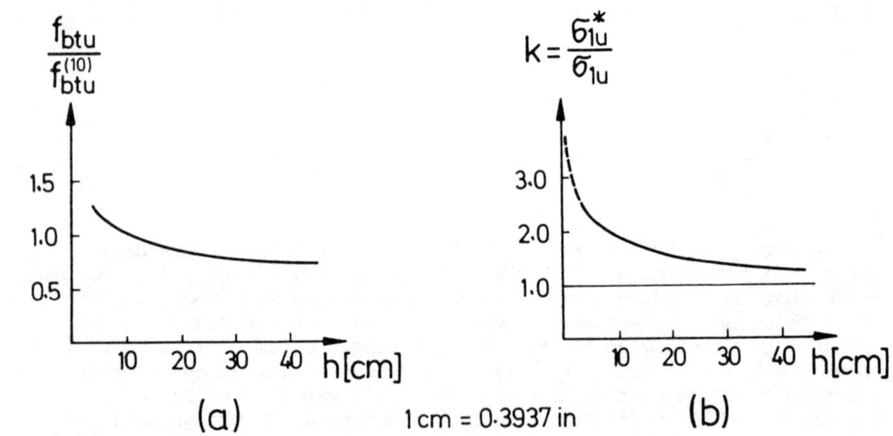

FIG. 6.-Determination of magnifying factor k for Kupfer's biaxial tensile strength: (a) Dependence of bending tensile strength of concrete on depth of beam (18); (b) Dependence of biaxial tensile strength of concrete on depth of hypothetical beam

With the help of the relations (18) and (19), h is obtained from equation (17) as

$$h = 4 \left. \frac{\hat{U}_0}{\left|\nabla \hat{U}_0\right|} \right|_{(r,l,e)} . \tag{20}$$

For the limiting cases $\left|\nabla \hat{U}_0\right|\Big|_{(r,l,e)} \to 0$ and $\left|\nabla \hat{U}_0\right|\Big|_{(r,l,e)} \to \infty$, h tends to ∞ and 0, respectively. The second limiting case, associated with $\hat{U}_0\Big|_{(r,l,e)} \to \infty$, is of no practical relevance. When h takes on a value outside the range of the depth of a typical beam of the previously mentioned length and width, this value should be regarded as an abstract analysis quantity related to the elastic portion of the strain-energy density and on its gradient at the considered integration point in the FE model of the shell.

Computation of $\hat{U}_0$ *and* $\left|\nabla \hat{U}_0\right|$.–Analogous to $\hat{U}_{0I}$ (equation (6)), $\hat{U}_0$ is obtained as

$$\hat{U}_0 = \frac{1}{2E_C^{(0)*}} \left[(1-\nu_C\alpha_1)\sigma_1^2 + (1-\nu_C\alpha_2)\sigma_2^2 \right]. \tag{21}$$

The absolute value of the gradient vector $\nabla \hat{U}_0$ is given as

$$\left|\nabla \hat{U}_0\right| = \left[\left(\frac{1}{A_1}\hat{U}_{0,1}\right)^2 + \left(\frac{1}{A_2}\hat{U}_{0,2}\right)^2 + \hat{U}_{0,3}^2 \right]^{\frac{1}{2}} \tag{22}$$

where $\hat{U}_{0,1} \equiv \partial\hat{U}_0/\partial\alpha$, $\hat{U}_{0,2} \equiv \partial\hat{U}_0/\partial\beta$, $\hat{U}_{0,3} \equiv \partial\hat{U}_0/\partial\zeta$ and A_1 and A_2 are the Lamé parameters which are related to the metric tensor. Since the wall thickness of the shell is assumed to be small in comparison to the minimum radius of curvature, A_1 and A_2 may be referred to the middle surface. From equation (21),

$$\hat{U}_{0,j} = \frac{1}{E_C^{(0)*}} \left[(\sigma_1-\nu_C\sigma_2)\sigma_{1,j} + (\sigma_2-\nu\sigma_1)\sigma_{2,j} \right]. \tag{23}$$

For the special case of a membrane state of stress, characterized by $\sigma_{1,3} = \sigma_{2,3} = 0$, $\hat{U}_{0,3} = 0$.

As mentioned previously, σ_{iu} is not increased for $-0.1 \leq \alpha_2 \leq 1$. Consequently, determination of h is restricted to the linear material region and to a comparatively small part of the nonlinear material region, termed as NL3 (Fig. 3(b)). According to Table 1, for these material regions, $g_2(\alpha_2) = 0$. Thus, analogous to equation (1),

$$\sigma_i = \frac{E_C^{(0)*}}{1 - \nu_C\alpha_i} \cdot \frac{\varepsilon_i}{1 + C_i^*\varepsilon_i} \tag{24}$$

where

$$C_i^* = \frac{E_C^{(0)*}}{\sigma_{iu}^*(1-\nu_C\alpha_i)} - \frac{1}{\varepsilon_{iu}^*} . \tag{25}$$

The quantities σ^*_{iu}, ε^*_{iu} and $E_C^{(0)*}$ would have to be determined from suitable tests for nonuniform loading. If these parameters were available, theoretical considerations for determining k would be superfluous.

For the linear material region,

$$C_i^* = \frac{E_C^{(0)*}}{\sigma^*_{iu}(1-\nu_C\alpha_i)} - \frac{1}{\varepsilon^*_{iu}} = 0. \tag{26}$$

It can be shown that for $C_i^* = 0$ the ratio $\hat{U}_0/|\nabla\hat{U}_0|$ does not depend on σ^*_{iu}, ε^*_{iu} and $E_C^{(0)*}$. Consequently, h does not depend on these unknown quantities. For the comparatively small part of the nonlinear material region, $C_i^* \neq 0$. Setting

$$C_i^* = C_i = \frac{E_C^{(0)}}{\sigma_{iu}(1-\nu_C\alpha_i)} - \frac{1}{\varepsilon_{iu}} \tag{27}$$

permits determination of h. There is no practical alternative to this rather insignificant simplification.

The quantities $\sigma_{1,j}$ and $\sigma_{2,j}$, $j = 1,2,3$, occurring in equation (23), are obtained as

$$\sigma_{1,j} = \frac{d_{1j}\,k_{22} - d_{2j}\,k_{12}}{k_{11}\,k_{22} - k_{12}\,k_{21}} \tag{28.1}$$

and

$$\sigma_{2,j} = \frac{d_{2j}\,k_{11} - d_{1j}\,k_{21}}{k_{11}\,k_{22} - k_{12}\,k_{21}} \tag{28.2}$$

where

$$k_{11} = \left[(1+\underline{C_1\varepsilon_1}) + \underline{(1-\nu_C\alpha_1)\frac{\partial C_1}{\partial \alpha_1}\alpha_1\varepsilon_1}\right]\frac{\varepsilon_1}{\sigma_1} \tag{29.1}$$

$$k_{12} = \left[-\nu_C(1+\underline{C_1\varepsilon_1}) + \underline{(1-\nu_C\alpha_1)\frac{\partial C_1}{\partial \alpha_1}\varepsilon_1}\right]\frac{\varepsilon_1}{\sigma_1} \tag{29.2}$$

and

$$d_{1j} = (1-\nu_C\alpha_1)\,\varepsilon_{i,j}. \tag{29.3}$$

The underlined terms vanish for the linear material region. k_{22}, k_{21} and d_{2j} are obtained by replacing the subscript "1" in the equations (29) by the subscript "2". From equation (27), $\partial C_i/\partial\alpha_i$ is obtained as

$$\frac{\partial C_i}{\partial \alpha_i} = -E_C^{(0)}\,\frac{\left[(1-\nu_C\alpha_i)\dfrac{\partial\sigma_{iu}}{\partial\alpha_i} - \nu_C\sigma_{iu}\right]}{[\sigma_{iu}(1-\nu_C\alpha_i)]^2} + \frac{1}{\varepsilon_{iu}^2}\frac{\partial\varepsilon_{iu}}{\partial\alpha_i}. \tag{30}$$

Equation (2.2) can be used to compute $\partial\sigma_{iu}/\partial\alpha_i$. Polynomial approximations to the part of experimentally obtained $\varepsilon_{iu} - \alpha_i$ diagrams (13) which refers to the material region NL3 (Fig. 3(b), Table 1) allow computation of $\partial\varepsilon_{iu}/\partial\alpha_i$. Principal strains occurring in the expressions for k_{11}, k_{22}, k_{12} and k_{21} and partial derivatives of principal strains occuring in the expressions for d_{1j} and d_{2j} are computed from the element displacement functions.

Equation (1) is used to compute the principal stress σ_1 which is then compared to $\sigma^*_{iu} = k\sigma_{1u}$. For $\varepsilon_1 > \varepsilon_{1u}$ this involves extrapolation from the part of the $\sigma_1 - \varepsilon_1$ diagram which is based on the results of Kupfer's biaxial tests (Fig. 5(a)).

Verification of the new fracture criterion for a concrete beam

The presented fracture criterion is verified for a 10 cm deep concrete beam. Three load cases are considered. They are: uniform tension (load case 1), non-uniform tension (load case 2) and pure bending (load case 3) (Fig. 7(a)). The reference stress at the bottom edge of the beam, σ, is assumed to be the same for the three load cases. Fig. 7(b) shows the FE mesh for treating the beam as a plate. The employed type of finite element represents a specialization of the type of curved, triangular, C^1-conforming thin-shell element in (21).

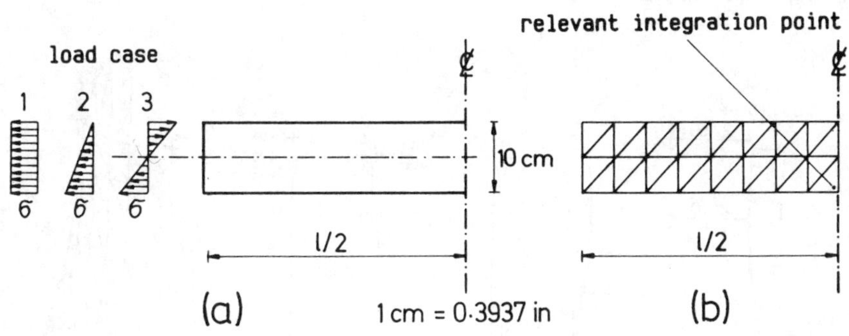

FIG. 7.-Concrete beam for verification of the new fracture criterion: (a) Concrete beam: 3 load cases; (b) FE mesh (solution as a plane stress problem)

Application of the classical strength criterion to the three load cases would result in the same load factor at incipient cracking of concrete, i.e., $\chi^{(1)} = \chi^{(2)} = \chi^{(3)}$. This is obviously not correct. Table 2 refers to the new fracture criterion. Noting that $\chi^{(1)} : \chi^{(2)} : \chi^{(3)} = k^{(1)} : k^{(2)} : k^{(3)}$, it follows from Table 2 that $\chi^{(1)} : \chi^{(2)} : \chi^{(3)} = 1 : 1.58 : 1.91$.

According to equation (20), for load case 2 (3) the depth of the hypothetical beam, h, referred to points of incipient cracking of the true beam, i.e., to the bottom edge of the latter, should be equal to 20 cm (10 cm). The reason for the deviations of FEM-based results for h from corresponding exact solutions is the distance of the integration points next to the bottom edge of the true beam from this edge. (In Fig.

7 (b) such a point is designated as "relevant integration point"). In consequence of these deviations, FEM-based results for the magnifying factor k differ slightly from corresponding values for k obtained from exact solutions for h.

TABLE 2.-Verification of the new fracture criterion for the beam of Fig.7

Load case	h [cm] [equation (20)]	k [1] [equation (12)]
1	∞	1.
2	19.00 (20)	1.58 (1.57)
3	9.27 (10)	1.91 (1.88)

( ) ... exact 1 cm = 0.3937 in

Collapse analysis of a built RC natural draft cooling tower

Fig. 8 shows characteristic dimensions, the thickness profile and the finite element mesh of a hyperbolic cooling tower of approximately 93 m height, erected in Voitsberg, Styria, Austria, designated as cooling tower Voitsberg III.

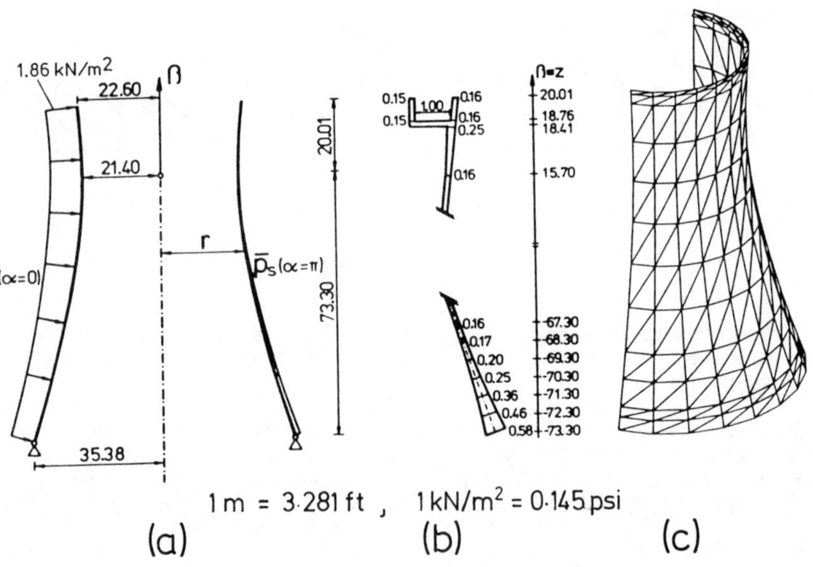

FIG. 8.- Cooling tower Voitsberg III, Styria, Austria (17): (a) Characteristic dimensions; (b) Thickness profile; (c) Finite element mesh

Table 3 contains coefficients A,B,...F for determination of the generatrix of the shell above and below the throat. They are to be substituted into the equation for a general curve of second order, given as

$$Az^2 + 2Brz + Cr^2 + 2Dz + 2Er + F = 0 \tag{31}$$

where r(z) is the radius of the middle surface of the shell at z ≡ ß.

TABLE 3.-Coefficients for determination of generatrix of shell

	Dimension	Hyperbola	
		$\beta \geq 0$	$\beta \leq 0$
A	[1]	-0.034131332	-0.186872088
B	[1]	-0.00783252	0.013830985
C	[1]	1.0	1.0
D	[m]	0.167616057	-0.295983075
E	[m]	-16.14902588	7.349552072
F	[m]	233.2183077	-772.5208287

1 m = 3.281 ft

The simplifying assumption of a rigid, hinged base was made. Recently, Mehl (19) has demonstrated that the influence of the flexibility of the supports on the buckling pressure of a cooling tower of larger dimensions than the Voitsberg III cooling tower is relatively small.

Material data are compiled in Tables 4 and 5. Quantities in Table 4 which were not defined previously are γ (specific weight), τ_{Bu} (bond strength), $E_S^{(0)}$ (modulus of elasticity of steel), $E_S^{(h)}$ (hardening modulus of steel), σ_{Sy} (yield stress of steel), and σ_{Su} (ultimate stress of steel). The quantities $E_S^{(0)}$, $E_S^{(h)}$, σ_{Sy} and σ_{Su} define a bilinear stress-strain diagram for steel.

TABLE 4.-Material properties of concrete and steel

Concrete		
$E_C^{(0)}$	3.00	[MN/cm^2]
ν_C	0.15	-
γ	24.25	[kN/m^3]
f_{cu}	-2.40	[kN/cm^2]
f_{tu}	0.222	[kN/cm^2]
τ_{Bu}	1.12	[kN/cm^2]
α_{2L}	-0.04347	-
Steel		
$E_S^{(0)}$	20.6	[MN/cm^2]
$E_S^{(h)}$	3.815	[MN/cm^2]
σ_{Sy}	50.0	[kN/cm^2]
σ_{Su}	56.0	[kN/cm^2]

1 kN/cm = 1450 psi

TABLE 5.-Location and thickness of layers of "smeared reinforcement"

No. of layer	$\beta \lessgtr$ [m]	$\lvert\zeta_S\rvert/h$ [1]	$\mu_\alpha h$ [cm]	$\mu_\beta h$ [cm]
	+20.01			
1	+19.21	0.450	0.276	0.083
2	+15.70	0.250	0.039	0.042
3	+4.20	0.250	0.039	0.043
4	-7.30	0.250	0.039	0.044
5	-18.80	0.250	0.039	0.043
6	-30.30	0.250	0.039	0.040
7	-41.80	0.250	0.039	0.037
8	-53.30	0.250	0.039	0.033
9	-60.30	0.250	0.039	0.061
10	-67.30	0.250	0.039	0.058
11	-69.90	0.300	0.039	0.078
12	-73.30	0.428	0.074	0.077

1 cm = 0.3937 in

Table 5 contains information on two orthogonal nets of reinforcement (one near the outer and one near the inner surface) varying with the height. The numbers in the first column of Table 5 identify layers of "smeared reinforcement". The second column refers to the boundaries β = const. of these layers. The third column contains absolute values of coordinates ζ_S, normalized with respect to the shell thickness h, indicating the distance of the steel layers from the middle surface of the shell. (A confusion of the shell thickness with the depth of the hypothetical beam for determination of the magnifying factor k for Kupfer's biaxial tensile strength, also designated as h, seems to be impossible). The concrete cover of the reinforcement is relatively large, resulting in relatively small values of $|\zeta_S|/h$. The last two columns of Table 5 contain the thicknesses of the steel layers; μ_α and μ_β are the percentages of reinforcement in the hoop and the meridional direction, respectively.

Because of symmetry of the quasistatic wind load with respect to the luff and the lee meridian, only one half of the shell needs to be analyzed. Fig. 9 shows the distribution of the wind-load function $\bar{p}_S$ according to (12).

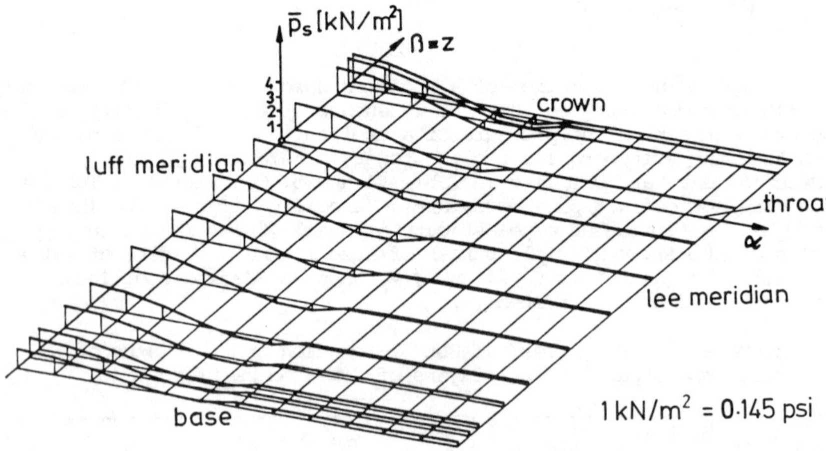

FIG. 9.-Distribution of wind-load function $\bar{p}_S$ (17)

Ultimate load analysis of the Voitsberg III cooling tower is based on the load combination $g + \chi\bar{p}_S$ where g is the dead load. Thus, only $\bar{p}_S$ is amplified in the course of the incremental-iterative FE analysis. The value of the load factor for which the analysis signals collapse of the shell is denoted as χ_{cr}.

Fig. 10 (a) shows two plots of the displacement component w, at a point on the luff meridian below the throat of the shell, versus the load factor χ. The solid (dashed) curve in this figure was obtained by means of the new (old) fracture criterion. The point A (B) divides the load-displacement diagram into two parts, characterized by intact concrete (state I) and cracked concrete (state II), respectively. The wind load associated with point A (B) in Fig. 10 (a) is termed "crack load".

It represents a lower bound to the ultimate load. From the standpoint of the analysis the ultimate load is reached if the equilibrium iteration fails to converge while the displacements increase rapidly. Fig. 10(b) shows two plots of the displacement component u at the point $\alpha = \pi/2$ of the crown of the shell. Figs. 10(a) and 10(b) illustrate that the influence of the increase of Kupfer's biaxial tensile strength on the response of the shell is insignificant. In this context ist should be mentioned that $k_{MAX}(\chi=2.35) = 1.84$ and $k_{MAX}(\chi=2.81) = 2.14$, indicating a considerable increase of Kupfer's biaxial tensile strength. These values are located in the solid part of the k - h diagram (Fig. 6(b)), based on experimental results.

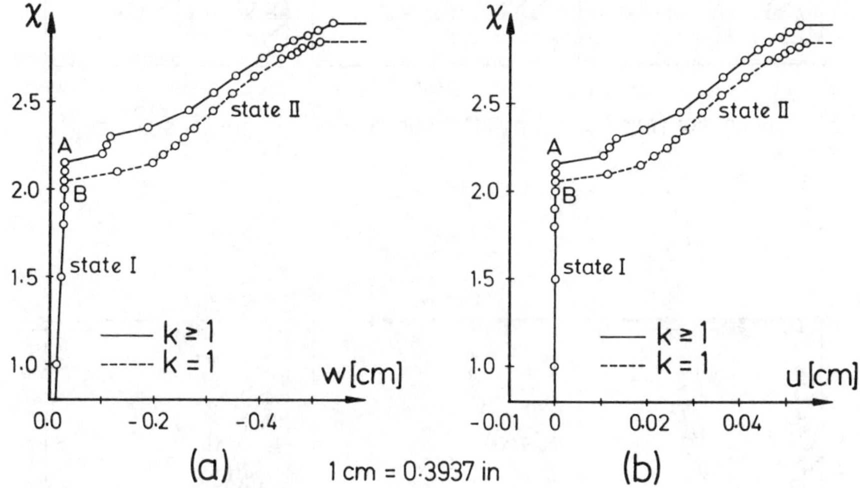

FIG. 10.–Load displacement diagrams: (a) w at $\alpha = 0$, $\beta = -18.8$ m; (b) u at $\alpha = \Pi/2$, $\beta = 20.0$ m

Fig. 11 shows plots of "smeared cracks" in three different concrete layers, at $\chi = 2.35$, for $k \geqq 1$ (new fracture criterion) as well as for $k = 1$ (old (classical) fracture criterion). The direction of the dashes indicates the direction of the cracks. The lengths of the dashes are proportional to the tributary areas of the respective integration points. In spite of the significantly larger amount of cracks for the case of $k = 1$, the overall response of the shell is not much softer than for the case of $k \geqq 1$. Consequently, the influence of the biaxial tensile strength of concrete on the response of the cooling tower is small. This corresponds to the initially mentioned result by Dodds, Darwin and Leibengood (7) according to which the tensile strength of concrete has a negligible influence on the response of the beams considered by these authors.

Table 6 contains a comparison of crack loads and ultimate loads with buckling loads resulting from three different modes of physically linear buckling analysis. (The letter symbol χ_{cr} is used in any case. The idealization of the crown of the shell, used for the classically (geo-

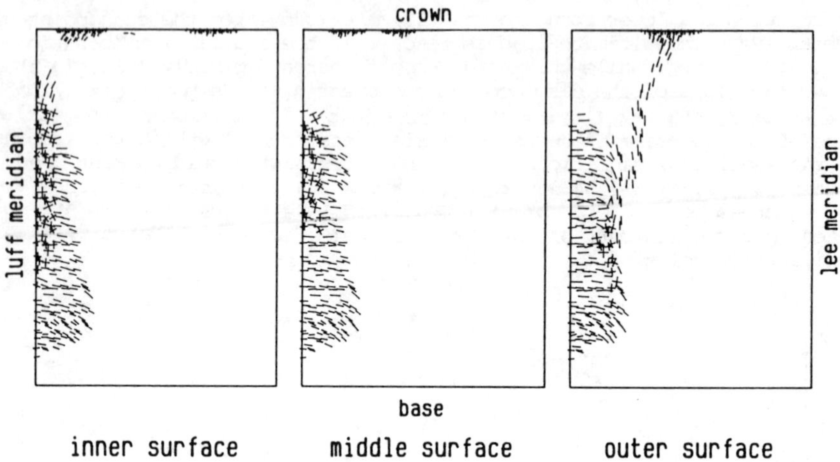

(a) k ≥ 1

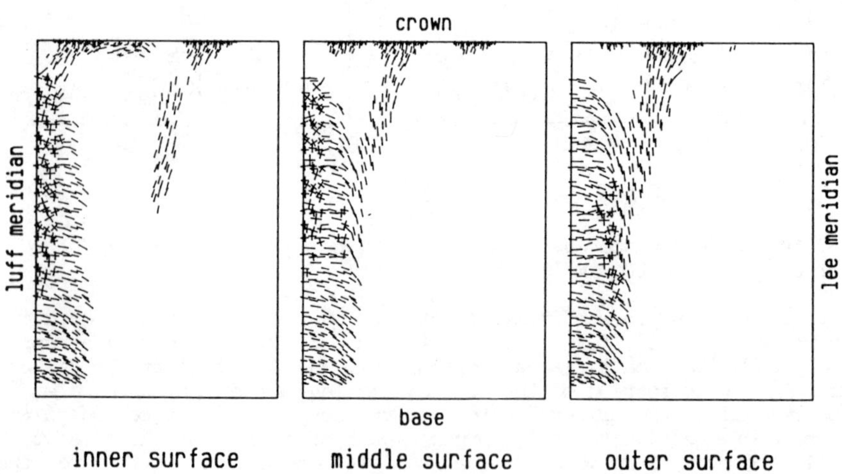

(b) k = 1

FIG. 11.-"Smeared cracks" in three different concrete layers, at χ = 2,35 metrically) linear buckling analysis, differs from the idealization of the crown employed for the other analyses.) It is seen that, whether or not geometric nonlinearity is considered, physically linear buckling analysis yields unrealistic results which are far on the unsafe side of the ultimate load (17).

TABLE 6.- Comparison of crack and ultimate loads with buckling loads from physically linear analysis

Mode of analysis (Dead load + χ_{cr}. wind load)		
Ultimate load analysis	crack load	ultimate load
$k \geq 1$ (new fracture criterion)	2.15	2.95
$k = 1$ (old fracture criterion)	2.05	2.85
Physically linear buckling analysis	buckling load	
initially (geometrically) linear	13.78 [23]	
classically (geometrically) linear	43.80* [23]	
geometrically nonlinear	59. [23]	

* ... for a different idealization of the crown of the shell

Concluding remarks

Application of the new fracture criterion for concrete to ultimate load analysis of a built RC natural draft cooling tower has shown that the significant increase of Kupfer's biaxial tensile strength σ_{1u} at the integration points in the FE model of the shell does not have much influence on the structural response of the shell. (The increase of σ_{1u} at these points is assumed to depend on the respective stress gradients.) Consequently, this response is insensitive to a variation of the biaxial tensile strength of concrete.

Based on the result presented in this paper and on a similar result for a groined vault, to be published in the near future, it seems that the classical strength criterion is adequate for ultimate load analysis of thin RC shells. This criterion ignores the strength-increasing influence of the stress gradient. Hence, slightly conservative results should be expected when applying the classical strength criterion.

The result of the numerical investigation confirms the fact that problems relevant to constitutive modeling of individual macro-cracks need not be relevant also to ultimate load analysis of thin RC surface structures. Identifying alleged problems of this complex mode of analysis as pseudo-problems should be one of the tasks of further pertinent research.

Acknowledgements

The writers are indebted to H. Walter for many stimulating discussions on the presented fracture criterion. His comments are based on earlier, unpublished work by him and the first writer. Financial support of the Fonds zur Förderung der wissenschaftlichen Forschung (Austrian Science Foundation) and the Hochschuljubiläumsstiftung der

Stadt Wien (University Jubilee Foundation of the City of Vienna) is gratefully acknowledged.

References

1. Bažant, Z. and Cedolin, L., "Blunt crack band propagation in finite element analysis", Journal of the Engineering Mechanics Division, ASCE, Vol.105, 1979, pp. 297 - 315.
2. Bažant, Z. and Cedolin, L., "Fracture mechanics of reinforced concrete", Journal of the Engineering Mechanics Division, ASCE, Vol.106, 1980, pp. 1287 - 1306.
3. Bažant, Z. and Cedolin, L., "Finite element modeling of crack band propagation", Journal of Structural Engineering, ASCE, Vol.109, 1983, pp. 69 - 92.
4. Bažant, Z., "Size effect in blunt fracture: concrete, rock, metal", Journal of Engineering Mechanics Division, ASCE, Vol.110, 1984, pp. 518 - 535.
5. Budiansky, B. and Sanders, L.J., On the "best" first order linear shell theory, Progress in Solid Mechanics: The Prager Anniversary Volume, Macmillan, New York, 1963.
6. Chen, W.F. and Ting, E.C., "Constutive models for concrete structures", Journal of the Engineering Mechanics Division, ASCE, Vol.106, 1980, pp. 1 - 19.
7. Dodds, R.H., Darwin, D. and Leibengood, L.O., "Stress controlled smeared cracking in RC beams", Journal of Structural Engineering, ASCE, Vol.110, 1984, pp. 1959 - 1976.
8. Floegl, H., Traglastermittlung dünner Stahlbetonschalen mittels der Methode der Finiten Elemente unter Berücksichtigung wirklichkeitsnahen Werkstoffverhaltens sowie geometrischer Nichtlinearität, Dr.techn.thesis, Technische Universität Wien, 1981.
9. Floegl, H. and Mang, H.A., "On tension stiffening in cracked reinforced concrete slabs and shells considering geometric and physical nonlinearity", Ingenieurarchiv, Vol.51, 1981, pp. 215 - 242.
10. Hellmann, H.G., "Beziehungen zwischen Zug- und Druckfestigkeit des Betons", beton, Vol.19, 1969, pp. 68 - 70.
11. Koiter, W.T., A consistent first approximation in the theory of thin elastic shells, Proceedings of the IUTAM Symposium on Theory of Thin Shells, North-Holland, Amsterdam, 1960.
12. Krätzig, W.P., Peters, H.L. and Zerna, W., "Naturzugkühltürme aus Stahlbeton - Derzeitiger Stand und Entwicklungsmöglichkeiten", Beton- und Stahlbetonbau, Vol. 73, 1978, pp. 37 - 42.
13. Kupfer, H., Hilsdorf, H.K. and Rüsch, H., "Behavior of concrete under biaxial stresses", ACI Journal, 1969, pp. 656 - 666.
14. Kupfer, H., "Das Verhalten des Betons unter mehrachsiger Kurzzeitbelastung unter besonderer Berücksichtigung der zweiachsigen Beanspruchung", Deutscher Ausschuß für Stahlbeton, Vol.229, Wilhelm Ernst, Berlin, 1973.
15. Liu, T.C.Y., Nilson, A.H. and Slate, F.O., "Biaxial stress-strain relations for concrete", Journal of the Structural Division, ASCE, Vol.98, 1972, pp. 1025 - 1034.
16. Mang, H.A., Floegl, H., Trappel, F. and Walter, H., "Wind-loaded reinforced concrete cooling towers: buckling or ultimate load?", Engineering Structures, Vol.5, 1983, pp. 163 - 180.
17. Mang, H.A. and Trappel, F., Physically linear buckling analysis of reinforced concrete cooling towers: design necessity or academic exercise?, Proceedings of the 2nd International Symposium on Natural-Draught Cooling Towers, Ruhr-Universität Bochum, Springer, Berlin, 1984. 1984.

18. Mayer, H., "Die Berechnung der Durchbiegung von Stahlbetonbauteilen", Deutscher Ausschuß für Stahlbeton, Vol.194, Wilhelm Ernst, Berlin, 1967.
19. Mehl, M., Über das Tragverhalten von Naturzugkühltürmen aus Stahlbeton, Dr.techn.thesis, Technische Universität Wien, 1982.
20. Rots, J.G., Nauta, P., Kusters, G.M.A. and Blaauwendraad, J, "Smeared crack approach and fracture localization in concrete", Heron, Vol.30, No.1, Stevin Laboratory, Delft University of Technology, The Netherlands, and Institute TNO for Building Materials and Building Structures, Rijswijk (ZH), The Netherlands, 1985.
21. Thomas, G.R. and Gallagher, R.H., A triangular thin shell finite element: linear analysis", NASA Contractor Report 2482, Washington D.C., 1975.
22. Timoshenko, St.P. and Gere, J.M., Theory of elastic stability, 2nd edn., Mc Graw-Hill, London, 1977.
23. Trappel, F., Vergleichende Beul- und Traglastuntersuchungen windbeanspruchter Kühlturmschalen aus Stahlbeton, Dipl.-Ing. thesis, Technische Universität Wien, in progress.

Estimate of Strength and Deformation Characteristics of Reinforced Concrete Shell Elements Subjected to In-Plane Forces

By Yukio AOYAGI*, Dr. of Eng.

Synopsis

It is of importance to understand the stress-strain behavior of reinforcement and concrete and to consider it properly in the sectional design of reinforced concrete (RC) shell elements subjected to in-plane forces. Although some theoretical research has been conducted so far, little experimental data are available to substantiate the validity of proposed theories. With this in mind an experimental study was carried out, in which 24 models of orthogonally reinforced concrete shell plate elements were loaded by in-plane forces inclined to the directions of reinforcements, simulating the boundary conditions prevailing in the actual structures. Also, 7 hollow cylindrical models reinforced with orthogonal, three-way and four-way reinforcing systems were tested in torsion. Based on the experimental evidence that the direction of cracks and the average shear rigidity across cracks are dependent on ratios of orthogonal principal stresses and on crack widths, respectively, a simplified analytical procedure, by which the accuracy of the estimate could be improved, is presented.

* Y. Aoyagi is a member of Japan Society of Civil Engineers, Japan Concrete Institute, ACI, etc. and is the head of material mechanics section at Civil Engineering Laboratory, Central Research Institute of Electric Power Industry, Chiba, Japan. He received his MS in 1964 and his Doctor of Engineering Degree in 1974 from the University of Tokyo, Japan. His research interests include design and construction of concrete containments for nuclear reactors, in-ground tanks for storage of LNG, prediction of cracking and deformational behaviors of reinforced concrete members under extremely low and elevated temperatures, creep, shrinkage and thermal effects on concrete structures, etc.

1. Introduction

Reinforced concrete (hereafter referred to as RC) shells have been applied to such structures as containments for nuclear power reactors, storage tanks for liquefied natural gas (LNG), water tanks, cooling towers, silos, etc. The field for application of RC shell structures is expected to increase also in the future. The design of RC shells has so far been based on sectional forces,which are induced by mechanical loadings such as internal pressure or earthquakes as well as restrained forces such as temperature effects. The internal forces are usually calculated by elastic analysis. The sectional forces are then utilized to determine the amount of reinforcement, taking into account the effect of cracking.

A typical critical stress condition occurs in an RC containment at the time of loss of coolant accident combined with an earthquake load. That is, the parts of the cylindrical wall parallel with the direction of earthquake force are likely to be subjected to combined stresses of N_x and N_y as well as N_{xy} (see Fig. -1).

This kind of problems was first studied in Germany in 1920s, and since the early part of 1970s it has become again one of the main themes of research interest not only in Germany but also in the USA, Japan, etc., because in connection with the construction of reinforced as well as prestressed concrete containments in high seismic regions, rationalization of design of an RC shell element against earthquake forces has become a problem requiring a solution. The present paper discusses the mechanical behavior of RC shell elements subjected to shearing stresses combined with in-plane membrane stresses, referring to the author's experimental results.

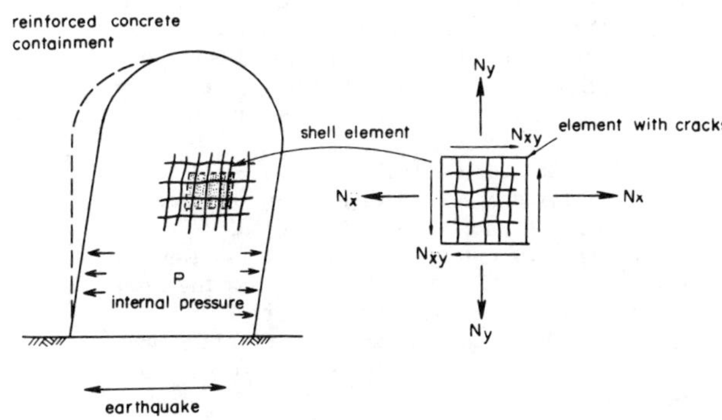

Fig.-1 Stress Conditions in Reinforced Concrete Containment Subjected to Earthquake with Presence of Internal Pressure

2. Specimens and Loading Method Employed by the Author

2.1 Test of Plate Specimens

Plate specimens with plan dimensions of test area 150 × 150 cm and a thickness of 10 cm (Fig. 2) were used for the test of orthogonally reinforced concrete shell elements. Around the periphery of specimen slitted thickened load introductory zones were attached to anchor the reinforcement and to facilitate uniform introduction of in-plane forces. The loads were applied by what is called a tournament scheme. 24 specimens were tested. Two percentages of steel were used and deviation angles of the principal forces with respect to the orthogonal reinforcement directions were varied from 0° to 45°. The disadvantages of this test method are the limited magnitude of shear forces applied to the specimens and the inability to introduce reversible shear stresses in the specimens. The main parameters of the specimens are listed in Table-1, together with the main test results.

2.2 Test of Cylindrical Models

To improve the controversial points of the above plate tests a reversal torsional loading test on RC hollow cylindrical models was conducted.

Configurations and dimensions of the test specimens are depicted in Fig.-3, together with loading scheme. Thickened stubs were monolithically connected to the cylindrical test area. Height, internal diameter and wall thickness were 200, 150 and 10 cm, respectively. Parameters varied in the specimens were the arrangements of the reinforcement as well as the ratio of reinforcement, and presence or absence of internal pressure. Main properties as well as concrete strengths of the models tested are listed in Table 2.

The inner face of the models, which were subjected to constant internal pressure, was lined with rubber sheet to keep the specimen watertight during the torsion test. After the internal pressure had attained the prescribed value, a torsion load was applied, sustaining the pressure up to failure by using a pressure, balancing nitrogen gas tank.

2:3 Materials used for Tests

The mix proportions of the concrete used for the experiment are shown in Table -3. Cylinder compressive strength of concrete at the time of test ranged from 200 to 280 kg/cm² for the cases of plate specimens. The yield point of the D-10 reinforcing bars used as the main reinforcing bars was 3780 kg/cm² and the tensile strength was 5510 kg/cm².

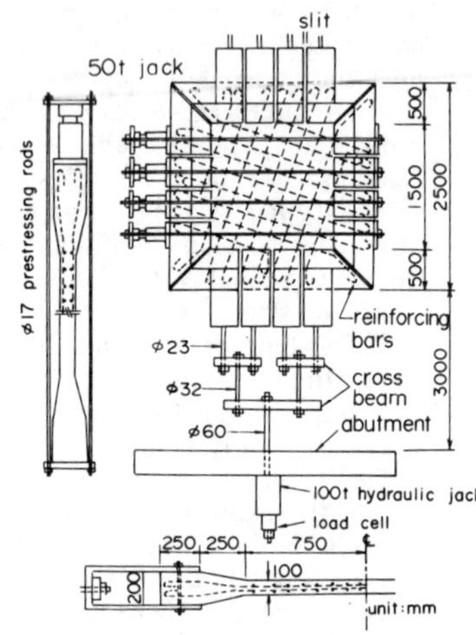

Fig.-2 Method of Applying In–plane Forces to Reinforced Concrete Plate Specimens Employed by the Author

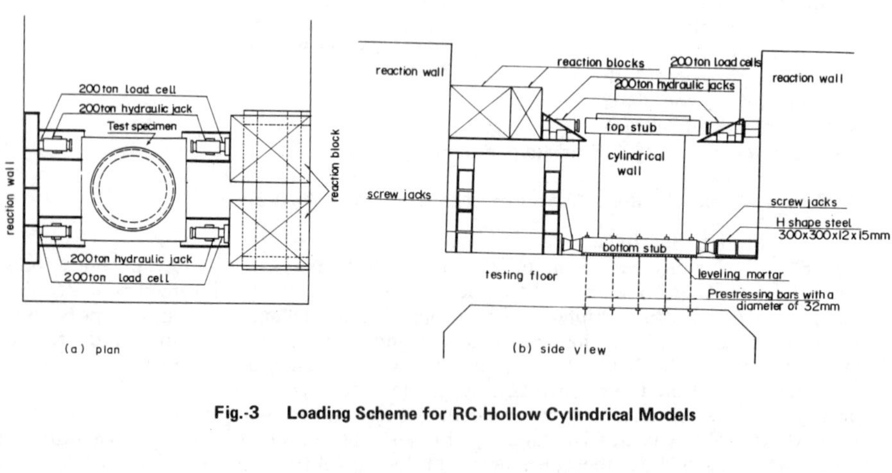

Fig.-3 Loading Scheme for RC Hollow Cylindrical Models

Table-1. Properties of Specimens and Comparisons of Calculated and Measured Yield Loads

No. of Specimens	Properties of specimen				N_1 Xyield	N_1 max	Leitz		Baumann		The Author				
	ρ (%)	α (°)	Presence of Crack	$k=\frac{N_2}{N_1}$			N_1 cal Xyield	Cal/ Mes.	N_1 cal Xyield	Cal/ Mes.	N_1 cal Xyield	Cal/ Mes.	N_1y, yield Cal	Mes	Cal/ Mes
20	0.761	0		0	44.0	63.5	43.2	1.02	43.2	1.02	43.2	1.02	-	-	-
6	0.761	0		0.5	40.0	64.0	43.2	0.93	43.2	0.93	43.2	0.93	-	-	-
11	0.713	12.5		0	44.0	56.0	35.7	1.23	37.2	1.18	38.6	1.14	56.0	40.5	1.38
12	0.713	12.5		1.0	38.0	46.0	40.5	0.94	40.5	0.94	40.5	0.94	42.0	40.5	1.03
7	0.713	22.5		0	36.0	50.0	33.5	1.07	36.1	1.00	35.6	1.01	44.0	40.5	1.09
8	0.713	22.5		0.5	36.0	48.0	36.7	0.98	37.4	0.96	35.0	1.03	41.0	41.3	1.00
9	0.713	22.5		1.0	36.0	44.0	40.5	0.89	40.5	0.89	40.5	0.89	44.0	40.5	1.09
13	0.713	22.5		-1.5	38.0	46.0	28.6	1.33	35.0	1.09	35.5	1.07	46.0	40.5	1.14
15	0.713	30		0.0	38.0	52.0	34.2	1.11	36.3	1.05	34.4	1.10	48.5	40.5	1.19
2	0.713	30		0	34.0	52.0	37.1	0.92	37.8	0.90	37.6	0.90	48.0	42.8	1.12
4	0.713	30		1.5	36.0	52.0	40.5	0.89	40.5	0.89	40.5	0.89	38.5	40.5	0.95
19	0.713	45		0.5	44.5	49.0	40.5	1.10	40.5	1.10	40.5	1.10	49.0	40.5	1.21
21	0.713	45		-1.0	42.0	51.5	40.5	1.04	40.5	1.04	40.5	1.04	51.5	40.5	1.27
3	0.713	30	ϕ=30	0.0	33.5	48.0	34.2	0.99	36.3	0.92	34.4	0.97	40.0	40.5	0.99
17	0.713	30	ϕ=57	0	47.5	56.0	37.1	1.28	37.8	1.26	37.6	1.26	47.0	42.8	1.10
10	0.713	22.5	ϕ=22.5	0	34.5	44.0	36.7	0.94	37.4	0.92	35.0	0.99	40.5	41.3	0.98
14	0.713	22.5	ϕ=22.5	-1.5	34.0	50.0	28.6	1.19	35.0	0.97	35.5	0.96	50.0	40.5	1.23
16	0.713	22.5	ϕ=49	-1	39.5	49.0	28.6	1.38	35.0	1.13	35.5	1.11	49.0	40.5	1.21
22	1.183	0		0	66.0	92.0	67.2	0.98	67.2	0.98	67.2	0.98	-	-	-
23	1.183	22.5		0	53.0	79.0	55.6	0.95	56.8	0.93	57.1	0.93	79.0	68.5	1.15
27	1.183	22.5		0	60.0	87.5	60.9	0.99	59.1	1.02	56.4	1.06	64.0	65.6	0.98
25	1.183	22.5	ϕ=22.5	0	51.0	82.5	55.6	0.92	56.8	0.90	57.1	0.89	82.5	68.5	1.20
26	1.183	45		0	61.5	82.0	67.2	0.92	67.2	0.92	67.2	0.92	69.0	67.2	1.03
24	x=1.183 y=0.592	22.5		0	51.0	65.0	55.6	0.92	52.5	0.97	54.2	0.94	58.5	57.0	1.03
	ϕ : Angle of the crack previously introduced (°)			average				1.04		1.00		1.00			1.11
				coefficient of variation				14.4%		9.6%		9.2%			10.1%

Table-2 Main Properties of the Cylindrical Specimens and Test Results

Mark of Specimens		T-1.18 - 0.0	T-118A - 0.0	T-180 - 00	T-1.18T-0.0	T-1.80 - 35	T-1.18 - 3.0	T01.18T - 3.0
Reinforcing System	Arrangement of Deformed bars ϕ10 mm	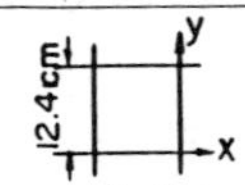 x.y : double	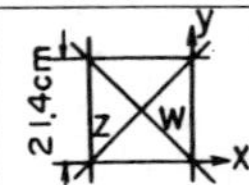 x.y : double w.z : single	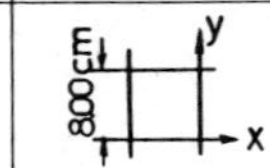 x.y : double	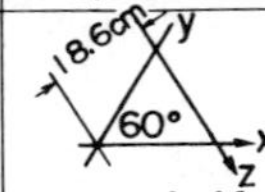 x.y.z : double	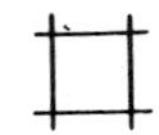 same as T-1.80-00	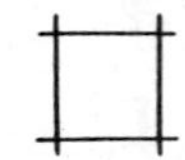 same as T.1.18-00	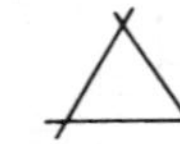 same as T.1.18T-00
Ratio (%)	x bars	1.15	0.67	1.79	0.77	1.79	1.15	0.77
	y bars	1.15	0.67	1.79	0.77	1.79	1.15	0.77
	w,z bars	0	0.47	0	0.77	0	0	0.77
	Volumetric	2.30	2.28	3.58	2.31	3.58	2.30	2.31
Internal pressure (kg/cm^2)		0	0	0	0	3.5	3.0	3.0
Conc. streng	Comp (kg/cm^2)	297	352	367	287	336	290	2.67
	Tens (")	25.8	26.6	32.1	24.1	300	23.7	22.0
Tensile cracking stress due to internal pres.		-	-	-	-	24.5	19.3	19.3
Torsional Shear Stress at kg/cm^2	Cracking	19.2	20.2	21.1	18.1	7.4*	7.4*	5.7*
	Yielding in x	36.8	42.4	59.4	40.8	31.9	14.3	19.7
	in y	46.6	46.6	72.7	35.4	40.2	27.0	22.9
	in w or z	-	32.7	-	42.2	-	-	-
	Maximum τ_u	57.4	60.6	76.0	50.7	61.8	49.8	59.2
	$\tau u/\sqrt{f'_c}$ **	3.32	3.22	3.96	2.99	3.37	2.92	3.62
Torsional deformation at τ_u (rad x 10^{-3})			23.7	83	20.8	2.38	37.6	48.8

* Shear stress at cracking due to torsion

** τ_u : Shear stress at ultimate (kg/cm^2) f'_c : concrete compressive strength (kg/cm^2)

Table-3. Mix. Proportion of Concrete used for Experiment

Mix Proportion

Water cement ratio (%)	Ratio of fine aggregate (%)	Slump (cm)	Air content (%)	Unit content (kg/m^3)					
				Water	Cement	Fine aggregate	Coarse aggregate 10 - 20 mm	Coarse aggregate 5 - 10 mm	Pczzolith No. 51
77	52	10±1	3±1	179	332	966	626	313	0.580

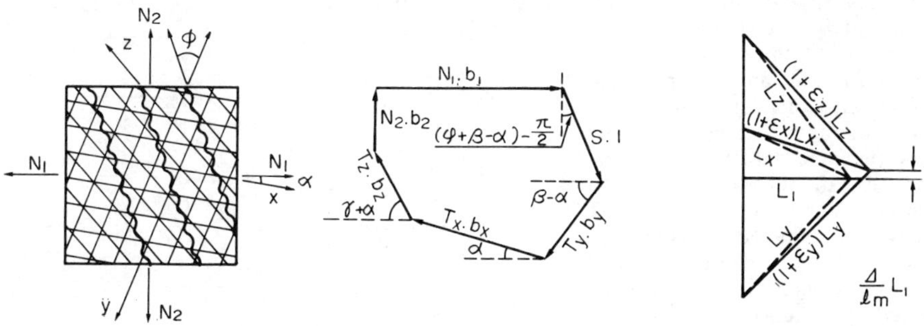

(a) Equilibrium (b) Strain Compatibity

Fig.-6 Equilibrium of Forces and Strain Compatibility in Three Way RC Shell Element

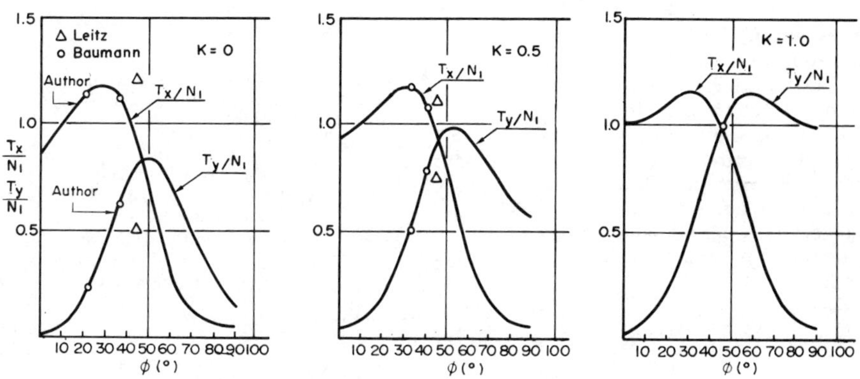

Fig.-7 Relationships between ϕ and T_x/N_1 as well as T_y/N_1 Obtained by Leitz, Baumann and the Author (α=22.5°, ρ_x=0.713%, N_1=30tons)

3. Method of Analysis for RC Shell Element

3.1 Orthogonal Reinforcing System[12],[13],[14]

According to Baumann,[4] [5] stresses in the reinforcement as well as compressive forces in concrete struts can be estimated based on the equilibrium of forces acting on a preassumed crack surface. Because of the indeterminate nature of the equilibrium equations the angle of the crack with respect to y–reinforcement as well as shear forces occurring along the crack must be determined based either on the compatibility conditions for deformations or on experimentally obtained evidence. A statistical analysis of the crack angles ϕ (Fig.-6) up to yielding in x-reinforcement obtained through the tests of RC shell elements [15] showed that the main influencing factors on ϕ are the deviation angle of bars α (Fig.-6) and the value κ, defined as the ratio between the applied principal forces in both directions. That is, it was found to be practically justifiable to assume that in the domain of uniaxial tension to compression-tension ($\kappa \leq 0$), ϕ may be assumed equal to α, while at the point of equal biaxial tension ($\kappa=1$), ϕ may be fixed at an angle of 45° regardless of the values of α. In the intermediate region ($0<\kappa<1$) ϕ can be linearly interpolated in terms of κ as illustrated in Fig.-4.

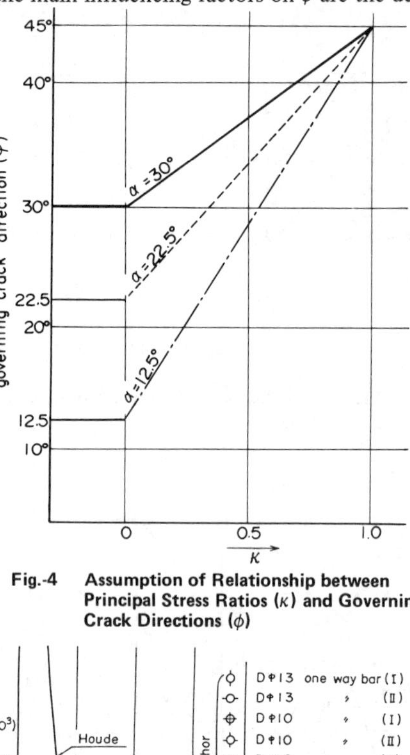

Fig.-4 Assumption of Relationship between Principal Stress Ratios (κ) and Governing Crack Directions (ϕ)

Unknown shear force transferred across the crack, S, is calculated in the same way as was proposed by Baumann.[4],[5] However, G_{cr}, which is defined as average shear rigidity across cracks, is experimentally determined. Referring to the results of the push-off tests obtained by the author et al. (see Fig.-5)[15] on precracked reinforced concrete blocks, G_{cr} could be approximated as a function of crack widths as follows:

$$G_{cr} = \frac{36}{w_m} \ell_m = \frac{36}{\epsilon_{\phi m}} \text{ (kg/cm}^2\text{)} \quad (1)$$

where, w_m : average crack width in cm. l_m : average crack spacing $\epsilon_{\phi m}$: average strain perpendicular to the direction of crack.

w_m is, on the other hand, a function of the average strains of shell element perpendicular to the direction of cracks. w_m is calculated by multiplying the average strain $\epsilon_{\phi m}$ by the average

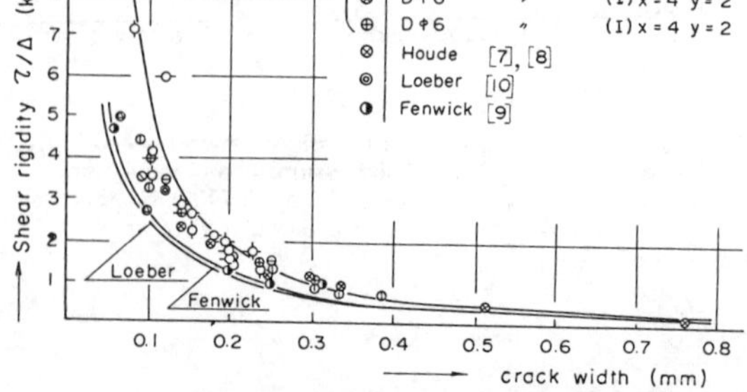

Fig.-5 Experimental Relationship between Widths of Cracks and Shear Rigidities

strain $\epsilon_{\phi m}$ by the average crack spacing ℓm. $\epsilon_{\phi m}$ can be predicted using the following equations (2), also developed by Baumann.[4],[5]

$$\epsilon_\phi = \frac{\epsilon_x}{\cos^2\phi} + \frac{\Delta}{\ell_m}\tan\phi = \frac{\epsilon_y}{\sin^2\phi} - \frac{\Delta}{\ell_m}\cot\phi$$

$$\epsilon_\phi = \epsilon_x + \epsilon_y$$

$$\Delta/\ell_m = \epsilon_y\cot\phi - \epsilon_x\tan\phi$$

$$\epsilon_1 = \epsilon_\phi\cos^2(\phi-\alpha) - \sin(\phi-\alpha)\cdot\cos(\phi-\alpha)\Delta/\ell m$$

$$\epsilon_2 = \epsilon_\phi\sin^2(\phi-\alpha) + \sin(\phi-\alpha)\cdot\cos(\phi-\alpha)\Delta/\ell m \qquad (2)$$

In the formula to relate the average strains in the reinforcement to those at a cracked section, trial fitting of the experimental data to the estimated[15] led to a modification of the original CEB formula. That is, in the modified equation the exponent of the second term of CEB formula was changed to 3 in lieu of the original 2. This may explain the fact that in a shell element, subject to membrane forces, degradation of bond between concrete and steel proceeds more rapidly than in ordinary RC beams or slabs, subject to flexural moment. The resulting equations adopted are as follows; [15]

$$\epsilon_{xm} = \epsilon_x\left[1 - \left(\frac{\sigma_{x,cr}}{\sigma_x}\right)^3\right]$$

$$\epsilon_{ym} = \epsilon_y\left[1 - \left(\frac{\sigma_{y,cr}}{\sigma_y}\right)^3\right] \qquad (3)$$

The crack spacing can be obtained by applying formulas such as provided in CEB Code. From the equilibrium of forces, the following equation is derived, which relates the sum of the forces in reinforcement in both directions to shear forces along the crack S.

$$T_x + T_y = (N_1+N_2) + (N_1-N_2)\sin\alpha\cdot\cos\alpha(\tan\phi) + S(\tan\phi-\cot\phi) \qquad (4)$$

As the crack angle ϕ is automatically given by the deviation angle α and external stress ratio κ, the problem is reduced to the determination of only S. The computation can be conducted by an iterative process.

3.2 Three Way Reinforcing System

Suppose a triangularly reinforced concrete shell element (Fig.-6) is loaded by biaxial principal membrane forces N_1 and N_2, the equilibrium of forces acting on the crack plane and the corresponding polygon of forces can be depicted as in Fig.-6(a). Also, the conditions of strain compatibility in x, y and z reinforcements at the cracked section can be drawn as in Fig.-6(b), neglecting the compressive strain in the concrete strut.

Referring to Fig.-6(b), the strain in z-reinforcement is expressed by the following equation.

$$\epsilon_z = [\epsilon_x\sin(\beta+\gamma)/\sin(\phi+\beta) - \epsilon_y\sin\gamma/\sin\phi] \times \sin(\phi+\beta+\gamma)/\sin\beta \qquad (5)$$

Based on the force equilibrium shown in Fig.-6(a), T_x, T_y and T_z, which denote the tensile forces in x, y and z reinforcements per unit width of shell element, respectively, can be expressed as follows:

$$
\begin{aligned}
T_z &= [\lambda_x T_x \sin(\beta+\gamma)/\sin(\phi+\beta) - \lambda y T_y \sin\gamma/\sin\phi] \cdot \sin(\phi+\beta+\gamma)/\sin\gamma \\
T_x &= (bN_1 + cN_2 + dS)/a \\
T_y &= (fN_1 + gN_2 + hS)/e
\end{aligned}
\tag{6}
$$

where $\lambda_x = \rho_z/\rho_x$, $\lambda_y = \rho_z/\rho_y$, ρ_x, ρ_y, ρ_z : reinforcement ratios in x, y and z directions respectively, α, β, γ : angle of the x, y and z reinforcements with respect to the principal force N_1, ϕ : angle between the directions of y–reinforcement and crack.

The coefficients appearing in the equation (6) are as follows:

$$
\begin{aligned}
a &= \sin(\phi+\beta) \cdot \sin^2\beta \cdot \sin\phi + \lambda_x \sin^2(\phi+\beta+\gamma) \cdot \sin^2(\beta+\gamma)\sin\phi / \\
&\quad \sin(\phi+\beta) + \lambda_y \sin^2(\phi+\beta+\gamma) \cdot \sin^2\gamma \cdot \sin(\phi+\beta)/\sin\phi \\
b &= \sin(\phi+\beta-\alpha) \cdot \sin(\beta-\alpha) \cdot \sin\beta \cdot \sin\phi + \lambda_y \sin^2(\phi+\beta+\gamma) \times \\
&\quad \sin(\phi+\beta-\alpha) \cdot \sin\gamma \cdot \sin(\alpha+\gamma)/\sin\phi \\
c &= \cos(\phi+\beta-\alpha) \cdot \cos(\beta-\alpha) \cdot \sin\phi \cdot \sin\beta - \lambda_y \cos(\phi+\beta-\alpha) \times \\
&\quad \sin^2(\phi+\beta+\gamma) \cdot \sin\gamma \cdot \cos(\alpha+\gamma)/\sin\phi \\
d &= \sin^2\phi \cdot \sin\beta + \lambda_y \sin^2(\phi+\beta+\gamma) \cdot \sin\gamma \cdot [\sin\phi \cdot \sin\gamma - \sin(\phi+\beta) \times \\
&\quad \sin(\beta+\gamma)]/\sin\phi \cdot \sin\beta \\
e &= \sin\phi \cdot \sin(\phi+\beta) \cdot \sin^2\beta + \lambda_x \sin^2(\phi+\beta+\gamma) \cdot \sin^2(\beta+\gamma) \cdot \sin\phi / \\
&\quad \sin(\phi+\beta) + \lambda_y \sin^2(\phi+\beta+\gamma) \cdot \sin_2\gamma \cdot \sin(\phi+\beta)/\sin\phi \\
f &= \lambda_x \sin^2(\phi+\beta+\gamma) \cdot \sin(\phi+\beta-\alpha) \cdot \sin(\beta+\gamma) \cdot \sin(\alpha+\gamma) / \\
&\quad \sin(\phi+\beta) + \sin(\phi+\beta-\alpha) \cdot \sin\alpha \cdot \sin(\phi+\beta) \cdot \sin\beta \\
g &= -\lambda_x \sin^2(\phi+\beta+\gamma) \cdot \cos(\phi+\beta-\alpha) \cdot \sin(\beta+\gamma) \cdot \cos(\alpha+\gamma) / \\
&\quad \sin(\phi+\beta) - \cos(\phi+\beta-\alpha) \cdot \cos\alpha \cdot \sin(\phi+\beta) \cdot \sin\beta \\
h &= \lambda_x \sin^2(\phi+\beta+\gamma) \cdot \sin(\beta+\gamma) \cdot [\sin\phi \cdot \sin\gamma - \sin(\phi+\beta) \cdot \sin(\beta+\gamma)] / \\
&\quad \sin(\phi+\beta) \cdot \sin\beta - \sin^2(\phi+\beta) \cdot \sin\beta
\end{aligned}
$$

The shear stress transferred across crack ,τ, , the shear force per unit width of shell element ,S, and the average shear rigidity of the shell element, G_{cr}, may be expressed as follows:

$$
\begin{aligned}
\tau &= G_{cr} \cdot \Delta/\ell_m \\
S &= \tau \cdot d \\
G_{cr} &= 36/\epsilon_{\phi m} \ (\text{kg/cm}^2)
\end{aligned}
\tag{7}
$$

where Δ: relative shear slip between edges of a crack, ℓ_m : average crack spacing, d: thickness of shell element.

Averaged normal strain in the direction perpendicular to cracks ($\epsilon_{\phi m}$) and relative shear slip divided by crack spacing (Δ/ℓ_m) are given as functions of ϵ_{xm} and ϵ_{ym}, which are average strains in x and y reinforcements, respectively.

$$\begin{aligned} \epsilon_{\phi m} &= \epsilon_{xm} \cos\phi/\sin(\phi+\beta)\cdot\sin\beta - \epsilon_{ym}\cos(\phi+\beta)/\sin\phi\cdot\sin\beta \\ \Delta/\ell_m &= -\epsilon_{xm}\sin\phi/\sin\beta\cdot\sin(\phi+\beta)+\epsilon_{ym}\sin(\phi+\beta)/\sin\phi\cdot\sin\beta \end{aligned} \qquad (8)$$

are defined independently as a function of principal stress ratio $\kappa = N_2/N_1$, as is described in the Section 3.1. Substituting the value of ϕ in the above equations, T_x, T_y, T_z and S are computed by solving interatively the simultaneous equations of (6) and (7).

The analytical procedure is also applicable to four way reinforcing system, when the reinforcement in one of the four directions is stressed in compression.

4. Verification of Analytical Method

4.1 An Example of Comparisons of Analytical Results Obtained by Various Authors

According to Flügge's formula,[2] if an RC shell element with a ratio of the reinforcements in the two directions $\lambda = \rho_x/\rho_y = 3$ is subjected to uniaxial tension with an angle of deviation $\alpha = 30°$, then the stresses in both reinforcements become identical, which also correspond to the minimum volume of reinforcements under the given loading conditions. To compare the analytical results among various proposed methods, T_x, T_y, R and S were calculated and are summarized in Table-4. In the table N_I is the value of principal stress in concrete occurring between the two adjacent cracks. R is the stress in the compressive strut.

Since in the author's as well as Tsubaki's formulas[6] absolute values of reinforcement ratios in both directions are required, ρ_x and ρ_y are assumed to be 1.18%, and 0.39%, respectively. The dimensions of the shell element was assumed to be the same as that of author's plate specimens, and the load considered for calculation was fixed as $N_1 = 30$ tons. The coefficients used in Tsubaki's formula are assumed as $\mu = 1.7$ and $\alpha_d = 1.0$.

Referring to the dimensionless values of forces in the x-reinforcement expressed by T_x/N_1, some differences can be recognized among the methods compared, the values being scattered in the range from the minimum of Flügge's 0.75 to the maximum of Baumann's 1.29. To the contrary, the values related to the y-reinforcement T_y/N_1 show an extremely wide difference such as that between Peter's[3] 0.15 and Tsubaki's 15.10. This is because Tsubaki assumes the cracks being nearly parallel to the y-reinforcement, increasing the steel stress extraordinarily due to friction, which, however, is inconceivable in practical situations. Even if disregarding Tsubaki's theory, considerable differences are observed in the stresses of the y-reinforcement among the researchers, ranging from Peter's 0.15 to Leitz's 0.68.

Looking at the stress in concrete strut R/N_1, somewhat higher values are obtained from Baumann and Leitz[1] than from Peter and the author. This implies that the direction of the assumed cracks plays an important role for the stress of concrete strut. As for shear force transferred along the cracked plane, Flügge, Peter and the author give approximately the same results. The fact that Flügge's value for N_I/N_1 is the largest one reflects his basic preposition that all the shear stresses are transmitted through aggregate interlock. Similar values are obtained by Peter and the author for all of the above parameters.

The above comparison led to the conclusion that although there are no significant differences among the methods hitherto proposed for the stresses in the x-reinforcement which has a smaller angle with the larger external principal tensile force, discrepancies are noted for the y-reinforcement.

Therefore, the author investigated the dependence of stresses in reinforcement on the crack inclinations. The calculations were conducted by the author's method for the

Table-4. Comparisons of Calculated Results from Various Analytical Methods (α=30°, $\kappa=N_2/N_1=0$, $\lambda=\rho_x/\rho_y=3$)

Analytical Methods	Assumption of Crack Formation	$\frac{T_x}{N_1}$	$\frac{T_y}{N_1}$	$\frac{R}{N_1}$	$\frac{S}{N_1}$	$\frac{N_I}{N_1}$
Flügge ; Orthogonal two-way cracking ϕ= 0 and ϕ= 90°	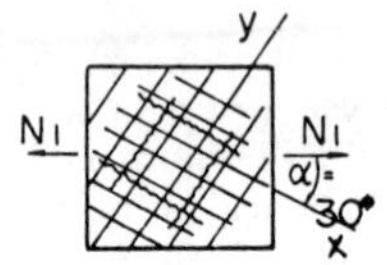	0.75	0.75	0.00	0.43	0.43
Leitz ; ϕ = 45° χ = ∞	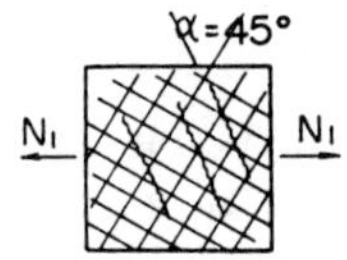	1.18	0.68	0.87	0	0
Peter ; Cracks perpendicular to principal force N_1 $\phi=\alpha$=30° χ=0	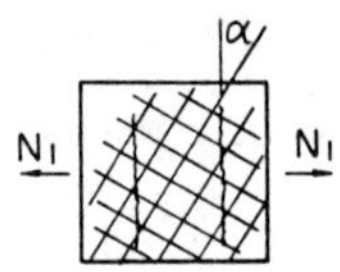	1.28	0.15	0.43	0.49	0.32
Baumann ; Principle of least deformation energy $\phi=\phi_1$ χ: variable	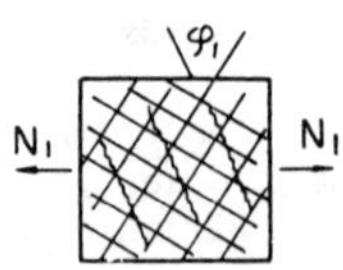	1.29	0.61	0.90	0	0
Tsubaki ; Shear friction theory ϕ=1° χ: variable	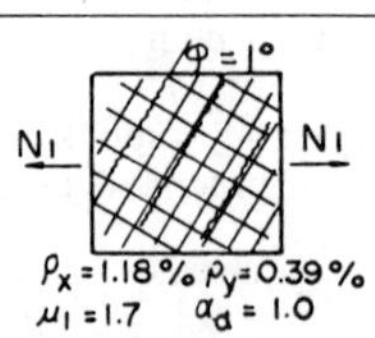	0.89	15.10	14.8	0.23	0
The author ; $\phi=\alpha$=30° χ: variable	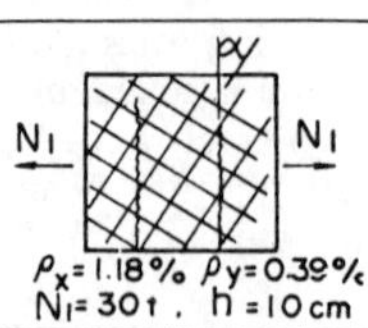	1.26	0.22	0.50	0.44	0.26

particular parameters of RC shell elements, that is, $\rho_x=\rho_y=0.713\%$, $\alpha=22.5°$ and $\kappa=0$, 0.5 and 1.0. The results are expressed in Fig. 7 as the relationship between ϕ and the dimensionless parameters T_x/N_1 and T_y/N_1, together with Baumann's and Leitz's estimations.

As can be seen from the figure, in the domain of angle ϕ from 20°~45°, for which practical designs are generally made, the inclinations of the curves for T_y/N_1 are so steep that the stresses in the y-reinforcement are far more sensitive to the direction of cracks assumed than those in the x-reinforcement. Coincidence of the calculated results by the three methods is only attained for the case of equal biaxial tension ($\kappa=1$). This again shows the importance of assigning the crack angles as true as possible to the actual conditions in the estimation of stresses in the y-reinforcement.

4.2 Comparisons of Plate Specimens

4.2.1 Average strains in reinforcements

In Fig.-8 the steel strain variation of the No.23 specimen averaged over a length spanning several cracks, versus the load is compared with those calculated by the methods proposed by Leitz, Baumann and the author. While the calculated strains in the x direction are practically identical among the three methods and provide a good fit to the measured, in the y direction the agreement seems distinctively better for the author's than the other two methods. Substantially the same results of comparisons were obtained for the cases of the unbalanced reinforced specimen (No. 24) subjected to uniaxial tension as well as the equally reinforced specimen (No. 27) subjected to biaxial tension.

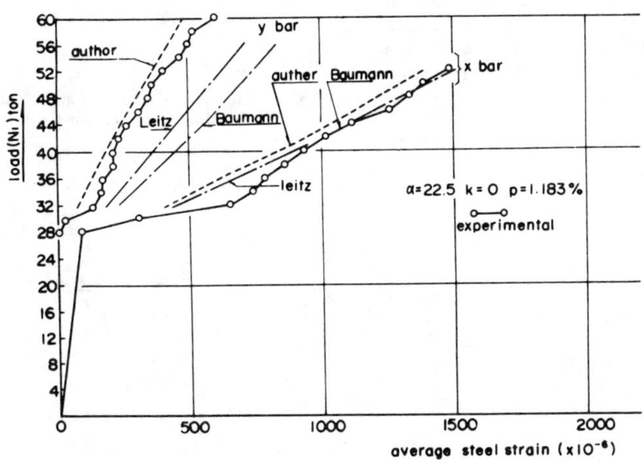

Fig.-8 Comparison of Measured Average Strains in Steels with Calculated

4.2.2 Yielding load in the x (weaker) direction.

The ratios of calculated and measured loads causing yielding in the x-bars are shown in Table-1 as well as in Fig.-9 in terms of principal stress ratios of κ. It can be seen that although there are no distinct differences between the accuracies of prediction obtained by Baumann and the author over the range of κ investigated, Leitz's theory has a tendency to underestimate the test results in the compression-tension region, especially at the value of κ equal to −1, that is, pure shear.

4.2.3 Deformations in the N_1 direction and shear slip

Fig.-10 shows comparisons of measured average post-cracked deformations in the direction of the principal tensile force N_1 with the results computed by the author for the cases of α=0°, 22.5° and 45°. The deformations, which become remarkably larger with increase in deviation angle ϕ, can also be predicted fairly well by the analytical procedure proposed by the author.

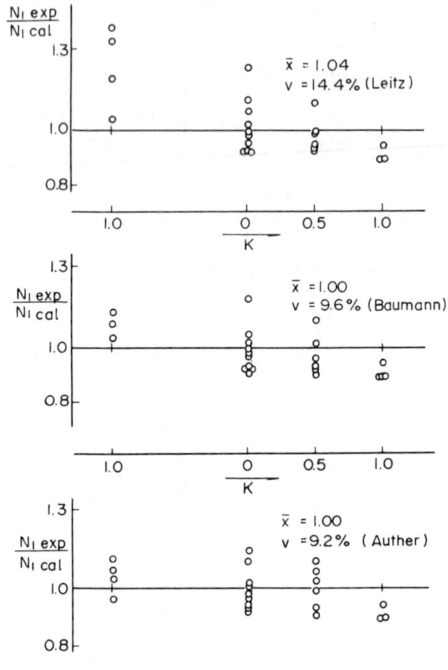

Fig.-9 Ratios of Calculated and Measured Loads Causing Yield in x Bars

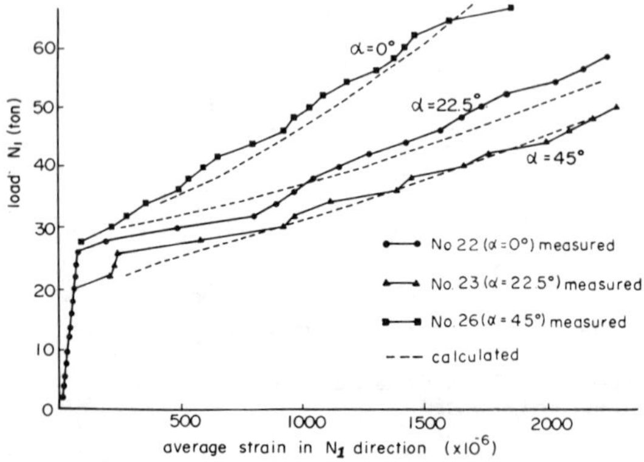

Fig.-10 Comparison of Measured Deformations in N_1 Direction with Calculated Deformations.

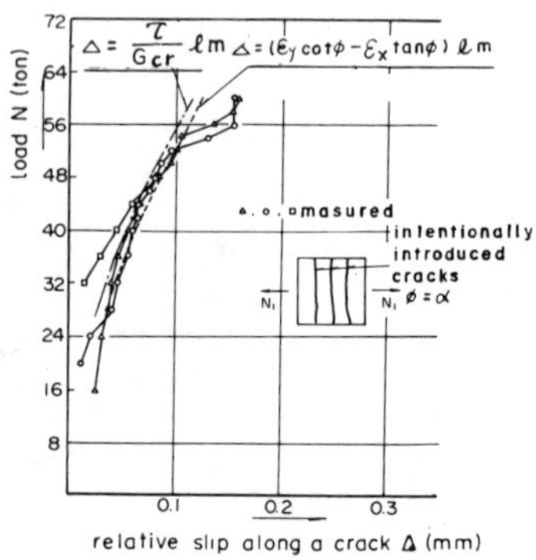

Fig.-11 Comparison of Measured Relative Slips along a Crack with Calculated No.25

In Fig.-11 relative shear slips along a crack are plotted against the applied principal force N_1,together with the calculated values based on the data of push-off tests as well as derived by the theoretically obtained strains in the reinforcement. It seems that the two estimates give similar results with a good fit of the experimental values. This seems to support the validity of the assumption adopted for the evaluation of the shear rigidities.

4.2.4 The load causing yielding in the y (stronger) reinforcement

The author derived a formula to predict the load causing yielding in the y reinforcement based on the theory mentioned before. In this analysis an assumption was made that after the x-reinforcement has yielded the load increment is to be resisted only by the stress increase in the y- reinforcement and the shear to be transferred across the crack S. The crack angle ϕ was assumed to retain its original value in the pre-yielding stage. The resulting formula to calculate the load increment required to cause yielding in the y reinforcement is obtained as follows: [15]

$$\Delta N_1 = \Delta T_y \tan^2\phi / [\cos^2\alpha \cdot (1 + \tan\alpha\tan\phi)^2 + \kappa \sin^2\alpha \cdot (1 - \cot\alpha\tan\phi)^2] \qquad (9)$$

where ΔN_1 : increase of the principal force N_1 per unit width of the plate element after yielding in the x-reinforcement.

$\Delta T_y = T_y^{yield,y} - T_y^{yield,x}$

$T_y^{yield,y}$: yield capacity per unit width in the y direction

$T_y^{yield,x}$: the y directional force per unit width when the x-reinforcement ment just yielded

The ratios between measured and calculated values of loads causing yielding in the y direction are listed in Table-1. An average ratio between experimental and calculated of 1.11 and a coefficient of variation of 10.1% suggest applicability of the proposed procedure.

4.3 Comparisons of Cylindrical Models

Effects of the reinforcing systems, reinforcement ratios and presence of internal pressure on the experimental torsional deformation behaviors can be seen on the skeleton curves for all the cylindrical models tested. (see Fig. 12)

In Figs-13 to 15 measured and calculated results for torsional deformations are shown for all of the tested specimens. Calculated torsional shear stresses at which first yielding of the steel occurs are listed in Table-2 together with the experimental results.

Fairly good agreements of the author's computed results with experimental results warrant the validity of the analytical method for the practical design purposes of RC shell elements subjected to in-plane shear combined with membrane tension. The range of applicability of the method should, however, be restricted up to the loading level, at which any of the reinforcements in the shell element starts to yield. Also, the preceding membrane cracks caused by internal pressure were found to affect markedly the deformational behaviours of shell elements, pointing out the necessity of further study directed to the behaviours of a shell element with pre-existing multi-directional cracks.

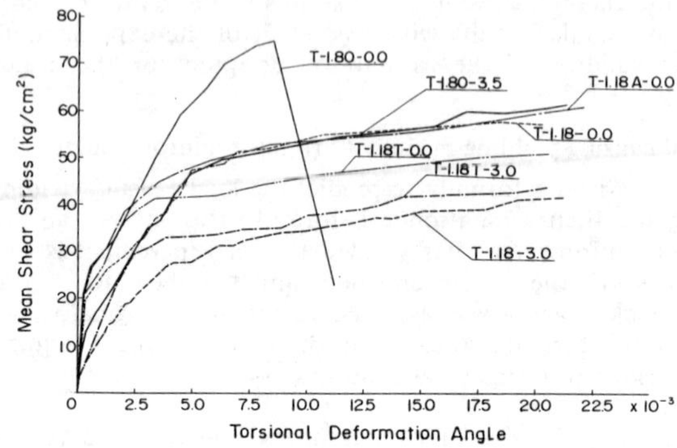

Fig.-12 Skeleton Curves of Torsional Deformations for all the Cylindrical Models Tested

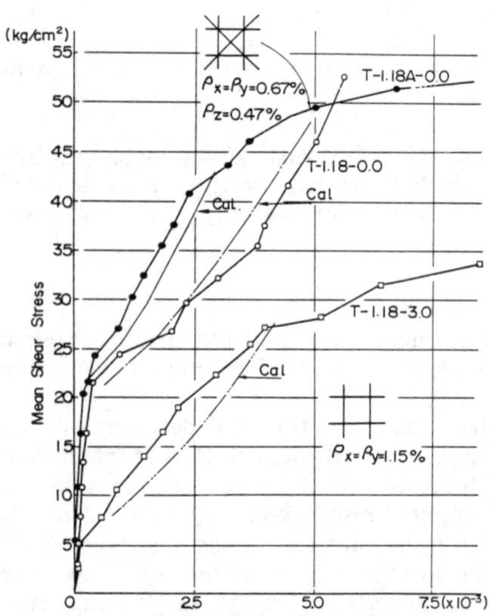

Fig.-13 Comparison of Torsional Deformation Angle as Measured and Calculated by the Author (Orthogonal and four way ρ_V. 2.3%) (ρ_V: volumetric percentage of steel)

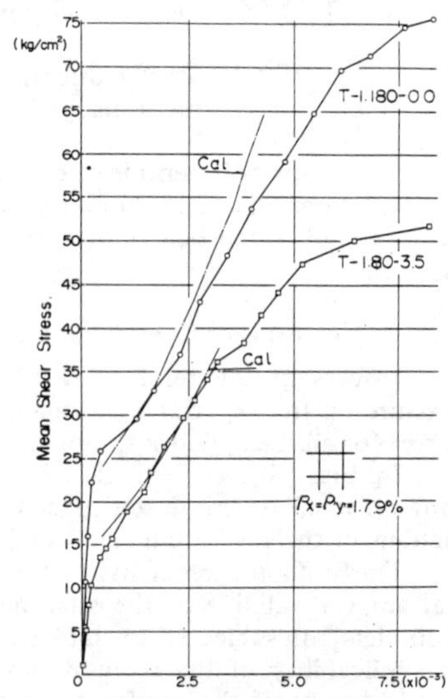

Fig.-14 Comparison of Torsional Deformation Angle between Measured and Calculated by the Author (Triangular ρ_V=2.31%)

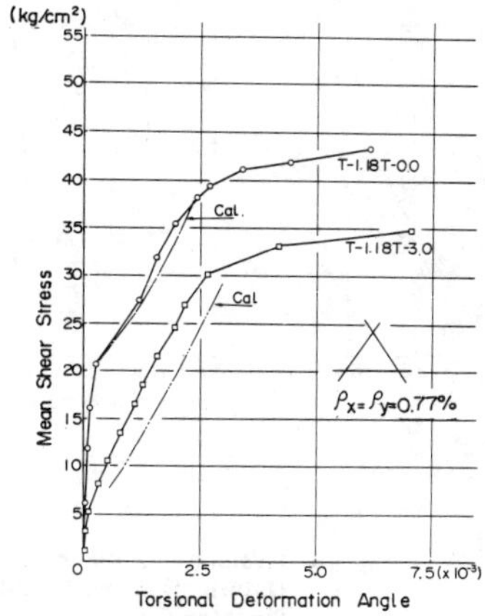

Fig.-15 Comparison of Torsional Deformation Angle as Measured and Calculated by the Author (Triangular ρ_v=2.31%)

5. Application of Finite Element Method

5.1 Outline of FEM Analysis

In the previous section the applicability of the author's practical method of analysis to the three as well as four way reinforcing systems was confirmed. In this section, mention of a trial use of a finite element plane stress crack analysis for estimating the mechanical behavior of the tested models is given.

Finite meshes of shell elements were selected so that the reinforcing bars are positioned just at the boundaries of adjacent elements. Concrete and reinforcing bars were modeled using 8 node isoparametric plane elements and 3 node isoparametric truss elements, respectively. Young's modulus, Poisson's ratio, yielding point and post yielding strain hardening rate were assumed as $Es=1.98\times10^6$ kg/cm², $\nu=0.3$, $\sigma_{sy}=3900$kg/cm² and $H'=E_s/100$, respectively. Kupfer's[11] constitutive equation for plane stress, which was modified by the author's test results, was utilized.

The basic assumptions concerning the effects of cracks are as follows;

(1) After the formation of the cracks the stress drops to zero, and tensile axial rigidities in the direction perpendicular to the crack also becomes zero.

(2) Cracks do not change the material properties determining the axial rigidities in the direction parallel to the crack.

(3) Average shear rigidities across cracks are dependent on the crack widths in the same manner as described in section 3.1, that is, $G_{cr}=36/\epsilon_{\phi m}$ (kg/cm²).

In the process of computation torsional loads were applied monotonically in steps with an increment of 10 tons, except for the loading stage immediately after cracking, when a negligibly small increment was added.

5.2 Comparisons of Test Results with the Computed Results.

Experimental hysteretic curves for torsional deformations and corresponding calculated skeleton curves are shown in Fig.-16 to 18, for the three specimens subjected to pure torsion without internal pressure. It may be seen that the analytical curves follow fairly well the experimental tendencies up to a larger torsional angle of nearly 10×10^{-3} rad. that is 2 to 3 times as large as that at the first yielding of the steel, indicating the applicability of the FEM even in the post-yielding range of shear deformations.

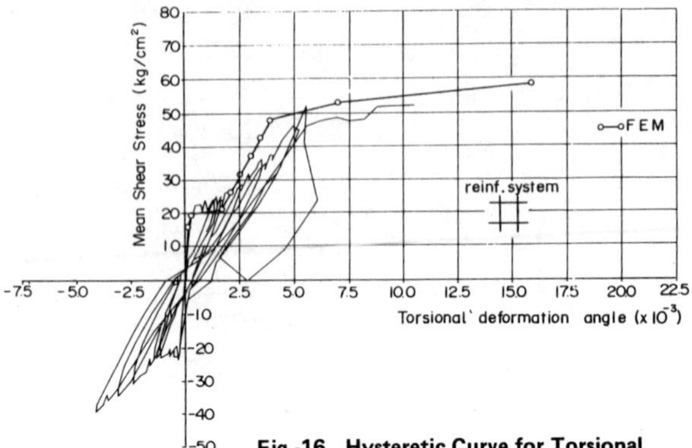

Fig.-16 Hysteretic Curve for Torsional Deformation and Calculated Skeleton Curve (T-1.18-0.0)

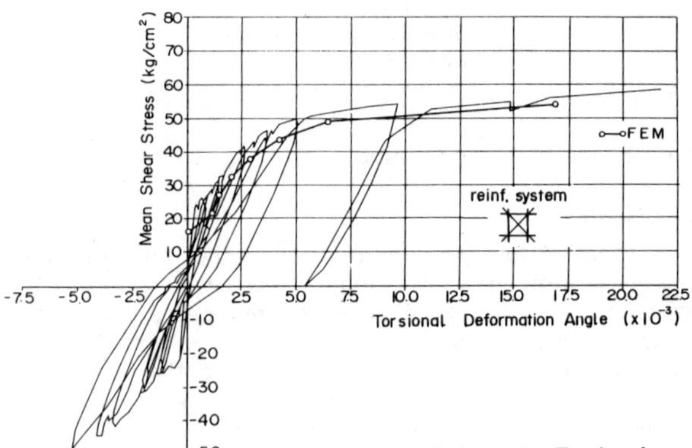

Fig.-17 Hysteretic Curve for Torsional Deformation and Calculated Skeleton Curve (T-1.18A-0.0)

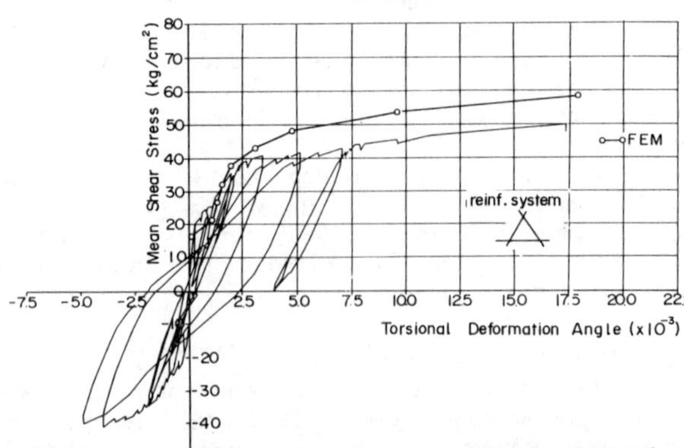

Fig.-18 Hysteretic Curve for Torsional Deformation and Calculated Skeleton Curve (T-1.18T-0.0)

6 Conclusions

Experimental results for the reinforced concrete plate specimens tested under in-plane force and hollow cylindrical models subjected to torsion were reviewed with particular reference to clarification of mechanical behavior as well as rationalization of the design method for RC shell elements.

The main conclusions obtained within the limitations of this study are:

(1) As a method for calculating the stress and deformation of RC shell elements, using the method of Baumann as the basis, an analysis procedure was proposed in which the stress dependence of cracking direction described previously, the crack width dependence of the shearing rigidity transfered through crack edges, and the contribution of the concrete portion between cracks to the tensile rigidity were taken into consideration. It was confirmed that the results of analysis by this procedure showed better agreement with the measured values than the calculated values of the stress, deformation and crack width by the procedures proposed by other authors. In particular, for the calculation in the stress of reinforcing bars which make a larger angle with the larger principal tensile stress (y-reinforcing bars), it was shown that the applicability of the equations proposed by the author was especially good.

(2) Experimental evidence showed that for the case of a constant volumetric reinforcement ratio, an orthogonal reinforcing system combined with diagonal bars and a triangular reinforcing system are advantageous for the loading conditions of pure shear and of shear plus membrane tension, respectively, from the view point of strength as well as deformational characteristics.

It was confirmed that the author's analytical method originally proposed for orthogonal bar networks is applicable to three as well as four way reinforcing arrangements with the limitation that the loading level considered should be lower than that causing yielding in the steel.

(3) A finite element analysis considering the effect of cracks was found to be effective to estimate the mechanical behavior of RC shell elements subjected to pure shear up to a state of shear deformation far beyond the yielding.

References

[1] Leitz, H. : "Eisenbewehrte Platten bei allgemeinen Biegungszustanden", Bautechnik, Heft 32, 1923.

[2] Flügge, W.: "Statik und Dynamik der Schalen", 3rd ed., Berlin, Springer-Verlag, 1962.

[3] Peter, J. : "Zur Bewehrung von Scheiben und Schalen für Hauptspannungen Schiefwinklig zur Bewehrungsrichtung", Die Bautechnik, Vol. 43, No. 5, pp. 149–154, 1966, and Vol. 43, No. 7, pp. 240–248, 1966.

[4] Baumann, T. : "Zur Frage der Netzbewehrung von Flachentragwerken", Der Bauingenieur, Vol. 47, Heft 10, pp. 367–372, 1972.

[5] Baumann, T. : "Tragwirkung Orthogonaler Bewehrungsnetze beliebiger Richtung in Flachentragwerken aus Stahlbeton", Deutcher Ausschuss für Stahlbeton, Heft 217, Berlin, 1972.

[6] Bazant, Z.P. and T. Tsubaki : "Concrete Reinforcing Net: Optimum Slip-Free Limit Design, Proc. of ASCE, Vol. 105, No. ST2, pp. 327–346, February 1979.

[7] Houde, J. : "Study of Force-Displacement Relationship for the Finite Element Analysis of Reinforced Concrete", Structural Concrete Series, No. 73-2, McGill University, Montreal, December 1973.

[8] Mirza, M.S. and J. Houde : "A Finite Element Analysis of Shear Strength of Reinforced Concrete Beams", ACI Special Publication SP 42-5, Vol. 1, "Shear in Reinforced Concrete", Detroit, Michigan, pp. 103–128, 1974.

[9] Fenwick, R.C. and T. Paulay : "Mechanisms of Shear Resistance of Concrete Beams", Journal of Structural Division, ASCE, Vol. 94, No. ST10, Proc. Paper 2325, pp. 2325–2350, October 1968.

[10] Loeber, P.J. : "Shear Transfer by Aggregate Interlock", M.E. Thesis, University of Canterbury, Christchurch, New Zealand, 1970.

[11] Kupfer, H. : "Das nicht lineare Verhalten des Betons bei zweiachsiger Beanspruchung", Beton und Stahlbetonbau, Heft 11, 1973.

[12] Aoyagi, Y. and K. Yamada : "An Experimental Approach to the Design of Network Reinforcement against In-plane Shear in Reinforced Concrete Containments", 5th SMIRT Conference, J4/7, Berlin, Germany, 1979.

[13] Aoyagi, Y., S. Ohmori and K. Yamada : "Strength and Deformational Characteristics of Orthogonally Reinforced Concrete Containment Models Subjected to Lateral Forces", 6th SMIRT Conference, J4/5, Paris, France, 1981.

[14] Aoyagi, Y. : "Analytical Approach to the Design of Orthogonally Reinforced Concrete Shell Elements Subjected to In-plane Shear", Proc. of JCI Colloquium on Shear Analysis of RC Structures, June 1982.

[15] Aoyagi, Y. and K. Yamada : "Strength and Deformational Characteristics of Reinforced Concrete Shell Elements Subjected to In-plane Forces", Concrete Library of JSCE, No. 4, Dec. 1984.

Aknowledgement

The author expresses his deepest appreciation to Dr. H. Okamura, Professor of University of Tokyo, and to Dr. K. Yamada, Chief Research Engineer, Maeda Construction Company, for their helpful suggestions and kind cooporation.

APPLICATION OF THE FINITE ELEMENT METHOD TO THE DESIGN OF CONCRETE STRUCTURES

by Shoji Ikeda[1] and Tatsuya Tsubaki[2]

ABSTRACT

A general method to apply the finite element method to the design of concrete structures is proposed in the first part. Then, in the second part, the effect of statistical variation of material properties and environmental conditions on the design of concrete structures is discussed. It is found that the statistical variation of strength has a strong influence on the mechanical behavior of RC structures such as the load-deflection relationship and the crack pattern. Also, it is confirmed that the statistical variation of creep and shrinkage and that of environmental conditions affect the long-term behavior of RC structures.

INTRODUCTION

Reinforced concrete is a structural system which is expected to behave monolithically under mechanical conditions even though it consists of different materials such as steel reinforcement and concrete. The theory of reinforced concrete has been developed to reproduce this behavior. Although each reinforcing bar could be treated independently, the theory is simplified to avoid the treatment of reinforcement as concentrated elements for main reinforcement or equally distributed elements for stirrups. The validity of this treatment is widely accepted because of experiments and experience. Obviously, some adjustment is included in the theory to consider the properties of concrete such as brittleness, viscoelasticity and plasticity.

When a new analysis method such as the finite element method (FEM) is applied to an actual reinforced concrete design, it is essential to simplify the complex properties of reinforced concrete such as in the case of the theory of reinforced concrete as mentioned above. In the actual design, the structures to be designed must fulfill the performance requirements. The major requirements are, in general,

---

[1]Professor, Dept. of Civil Engineering, Yokohama National University, Yokohama, Japan.

[2]Associate Professor, Dept. of Civil Engineering, Yokohama National University, Yokohama, Japan.

serviceability, safety, and durability. Serviceability requires elasticity and linearity so that allowable stress method may be suitable for this purpose, which can also indicate the nature of safety and durability. Safety is considered to correspond to the ultimate state which is based on the plastic behavior of materials. However, it must be noted that the ultimate state is not a safe state so that most of the actual structures would not experience this state in the life of the structure. Ductility is required in this state to avoid collapse of the structure. Hence, any new design method is desired to be able to calculate both linear and nonlinear behavior according to the serviceability state and ultimate state in a simplified manner.

The finite element method has already been applied to the design of actual concrete structures. Most of the procedure in application is such that the structures are assumed to consist of elastic isotropic materials and that the amount of reinforcement is compared to the tensile forces in the concrete which is calculated by the finite element method. This procedure is simple and inexpensive. However, it might be difficult to establish rational design rules for this simple procedure.

For the analysis of reinforced concrete, bar elements, bond-link elements and cracked elements are introduced to represent the actual behavior of reinforced concrete and analytical models are constructed to agree with the experimental results. On the other hand, practical design rules are based on the wide experience of actual designs and construction. In order to utilize the finite element method in the actual design of reinforced concrete, it is important to recognize the difference between analysis and design, that is, a practical design method must be simple and clear.

Considering the above, a simple procedure using FEM to the design of RC structures is outlined first in Chap.2. Then, in the subsequent chapters, the effect of statistical variation of material properties and environmental conditions on concrete structures is discussed, since the statistical aspects involved in concrete should be taken into account in the actual design so that RC structures can maintain adequate safety and reliability.

FINITE ELEMENT METHOD FOR REINFORCED CONCRETE DESIGN

Finite Element Method for Actual Design

The finite element method is the most powerful tool for the design calculations of a reinforced concrete structure where the conventional beam or plate theory cannot be adequately applied. In order to utilize the finite element method widely and to design more rationally, a specific design rule is necessary for proper application of finite element calculations.

The following is a proposal for design rules specifying the simplification of the material properties under required performance of reinforced concrete.

(1) Stress-strain relationship of concrete is assumed as shown in Fig.1(a).
(2) Stress-strain relationship of steel reinforcement is assumed as shown in Fig.1(b).
(3) Tensile strength of concrete is assumed to be $\sigma_{ck}/100$ for the serviceability state. Here, σ_{ck} is the specified compressive strength of concrete.
(4) Crack width under service load is restricted by the specified tensile strain in cracked elements, e.g., 0.001 for common circumstances.
(5) Multi-axial properties of concrete are assumed to be a simple combination of uniaxial properties, in principle. However, the following strength increase can be utilized for a high compression state.

$$\sigma_{1(max)} \leqq \sigma_{ck} + \sigma_{3(min)} \tag{1}$$

where

$\sigma_{1(max)}$ = maximum compressive stress in a principal direction;
$\sigma_{3(min)}$ = minimum compressive stress in other principal directions.

(6) Yielding and failure of materials are governed only by the normal stresses in principal directions.
(7) Tensile strength of concrete at ultimate state is assumed to be $\sigma_{ck}/50$.
(8) For design details, a minimum reinforcement ratio must be specified for every direction of cross section. The value of 0.001 is recommended for ordinary reinforced concrete.

For utilization of finite elements in a continuous body, concrete and steel reinforcement work together without any slippage both under service load and at ultimate state. This is the same assumption as in conventional reinforced concrete theory. However, degradation of stiffness due to cracks in concrete cannot be ignored. Therefore, a smeared crack is assumed in its principal direction. Reinforced concrete elements should be utilized to represent the above-mentioned behavior of reinforced concrete.

In the case of a plane stress condition, constitutive matrix $\underset{\sim}{D}$ is expressed by Eq.(2).

$$\underset{\sim}{D} = \frac{E_c}{1-\nu^2}\begin{bmatrix} 1 & \nu & 0 \\ \nu & 1 & 0 \\ 0 & 0 & \frac{1-\nu}{2} \end{bmatrix} + E_s \begin{bmatrix} p_x & 0 & 0 \\ 0 & p_y & 0 \\ 0 & 0 & \beta \end{bmatrix} \tag{2}$$

where

E_c, E_s = Young's modulus of concrete and reinforcement, respectively;
ν = Poisson's ratio of concrete;
p_x, p_y = Percentage of steel reinforcement in x and y direction, respectively;
β = Coefficient for dowel action. $\beta = 0$ for usual cases.

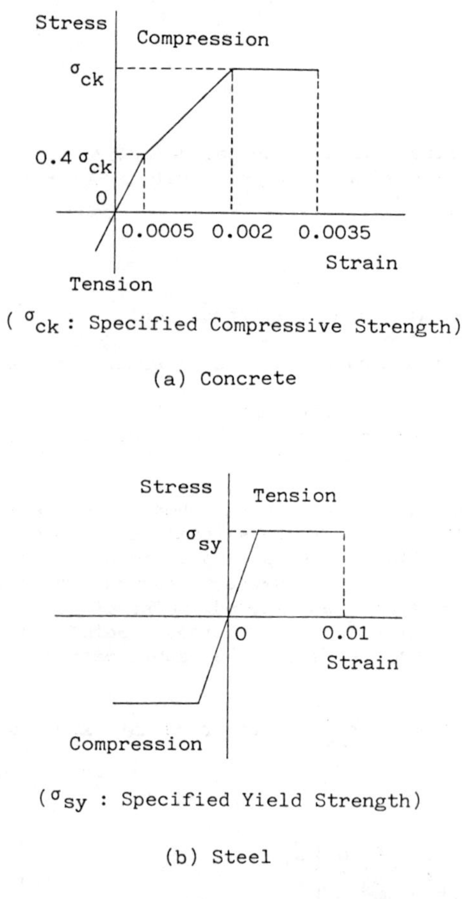

Fig.1 Stress-Strain Relationship for Designing RC Structures

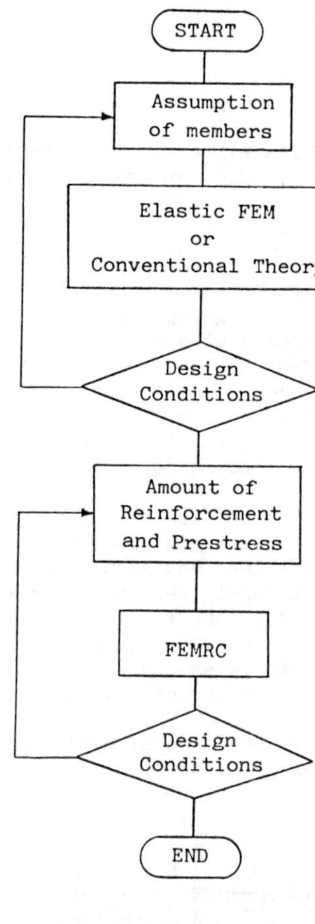

Fig.2 Flow of Designing RC Structures

For well distributed reinforcement such as stirrups in a web, all the elements of the web have uniform rigidity of the vertical reinforcement as expressed by Eq.(2), while the rigidity of concentrated reinforcement such as main tensile bars may be included in the nearby concrete elements.

The clarification and simplification as described above may be useful for practical application of the finite element method to the actual design of reinforced concrete. The method proposed here is named as the finite element method for reinforced concrete (FEMRC) hereafter.

Procedure for Actual Design

The flow of design work may be drawn as shown in Fig.2, where assumed structural members, obtained using the conventional theory or the elastic finite element method, are analyzed by FEMRC proposed in the previous section.

It is important for the design of reinforced concrete using the finite element method to check the input and output data, the constitutive laws of materials and the applicability of the program. A benchmark test is essential for this purpose(36). As the element sizes may affect the analysis results, discretization of mesh must be done with reference to previous works. Calculation errors due to this should be desirably within about five percent.

In the design of reinforced concrete, statistical variations must be considered in many cases. Simplification and statistical considerations make the design more rational for concrete structures. In addition to the statistical variation of strength and stiffness, it is also important to consider the statistical variation of the long-term material properties such as creep and shrinkage. To predict the scatter of these factors, a probabilistic method was applied in a recent work by Bazant, et al. (cf. Fig.3) (8), for example. The treatment of statistical considerations of these aspects is presented in the following chapters.

The boundary element method(BEM) and the combination solution method of FEM and BEM are quite effective for the calculation of reinforced concrete structures. This also is shown in the following chapter.

Finite Element Method for Actual Structures

In the following, some actual examples of the application of FEM to RC structures are briefly summarized.

The report of ASCE Committee shows the applicability of the finite element method to actual structures such as PCRVs(5). The report also lists the names of available computer programs using the finite element method.

JCI's proceedings on the finite element method colloquium includes some papers analyzing actual concrete structures(30). Nonlinear analysis of an actual cooling tower was done after the failure of the tower under wind load(34).

A recent application of the finite element method to an actual structure is the egg-shaped digestion tank having a diameter of 22.7m and height of 33.6m, now being constructed in Yokohama City (47). The structure has a sludge pocket at the upper part and is supported by precast concrete piles at the lower circumferential part. The structure was discretized by 25 axisymmetric shell elements and 48 solid elements supported by spring elements in three directions. The structure is unique because all design calculations were performed by the finite element method. The design process was such that the tensile forces calculated by an elastic finite element analysis were counteracted by prestressing forces.

Another example is the verification of design calculations for a complex bridge structure using the three-dimensional finite element method. The structure is a continuous prestressed concrete curved bridge having a branch in its span. The analysis model has 7616 isoparametric plate elements and 6663 nodal points(32). The analysis results were used to check the amount of reinforcement in detail.

The finite element method has been used widely in the design of massive concrete structures such as anchorage blocks of suspension bridges and thick footings of bridge piers.

Considering that the finite element method is used for various RC structures, as indicated above, a comprehensive and acceptable design philosophy for reinforced concrete using the finite element method as proposed here seems to be urgent for rational and economical construction of concrete structures.

STATISTICAL ASPECTS IN DESIGN OF RC STRUCTURES (STRENGTH AND STIFFNESS)

For a reinforced concrete member, the tensile strength of concrete is one of significant influence factors governing the ultimate strength and the failure mode. The tensile strength may be influenced by many factors such as the mix proportions, the curing conditions, and the casting direction of concrete. Each of these factors may cause a significant statistical variation of the tensile strength and stiffness of concrete, which consequently causes a scatter in the mechanical behavior of reinforced concrete structures.

In the following, the effect of the statistical variation of the strength and stiffness of concrete on the overall mechanical behavior of reinforced concrete structures is investigated so that the statistical aspects to be considered in the design procedure of RC structures described in the previous chapter and their consequences may be clarified.

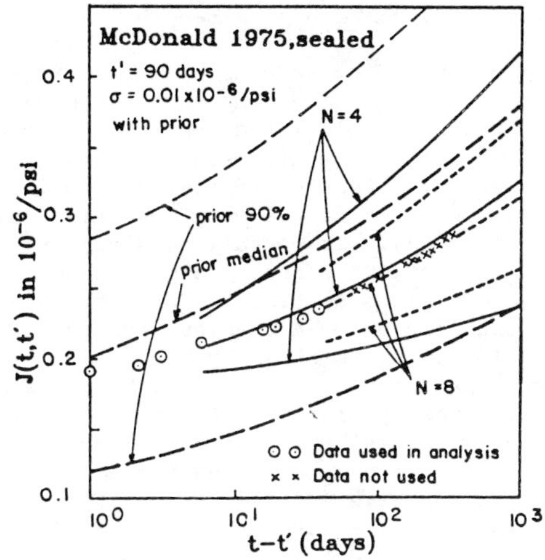

Fig.3 Example of Bayesian Prediction of Creep Strain for McDonald's Test Data (1975)
(From Ref.8) (1psi=6.89KPa)

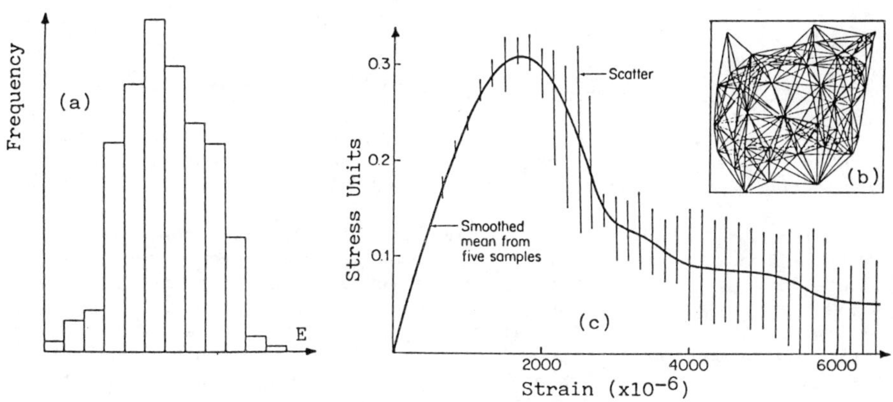

Fig.4 Effect of Statistical Variation of Stiffness and Strength (Uniaxial Extension) :
(a) Distribution of Member Modulus of Elasticity, E;
(b) Random Networks;
(c) Mean Stress-Strain Curve and Scatter for Five Samples of Random Material
(From Ref.20 and 27)

Effect of Statistical Variation of Strength in Uniaxial Extension

A model heterogeneous medium made up of a random network of linear elastic brittle members was introduced by Burt and Dougill (20). The network comprises a plane pin-jointed frame with either a random geometry or a repetitive lattice structure. The values of stiffness and strength for each member are allocated to give a normal distribution of member properties throughout the sample with specified mean and standard deviation (cf. Fig.4(a)) (27). The resulting structure is analyzed for a sequence of increments of the boundary displacement, at each stage reducing the stiffness of failed members to zero and using an iterative scheme to restore equilibrium. A typical random network and its response under uniaxial loading is shown in Fig.4(b) and (c). Because of the way the model is specified, it is possible to examine different samples of material, each having the same internal geometry and distribution of member properties, but in which the properties of individual members vary from sample to sample. The differences in behavior observed in testing five such macroscopically identical samples are shown in Fig.4(c). It is observed that the local variation of stiffness and strength has little effect for a low-stress state, while it has a significant influence in the range where the structure is failing. The size effect is also examined by changing the size of samples in a network. A greater scatter of results occurs when many members are failing and the material has passed the peak stress. During this period, there is a tendency for member failures to occur in local groupings. The occurrence of localization of failure implies that there will be significant differences in the final form of breakdown between different sized samples.

Effect of Statistical Variation of Strength in Bending

In the following, the effect of the statistical variation of tensile strength of concrete on a reinforced concrete beam is considered, which is based on a recent work of the authors.

The tensile strength of concrete is assumed to have the distribution shown in Fig.5. A reinforced concrete beam subject to two-point loading shown in Fig.6, with material properties summarized in Fig.7 is discretized by using finite elements as shown in Fig.8(a). Following the assumed distribution, the random tensile strength is allocated to each finite element (cf. Fig.8(b)). The crack distribution for each tensile strength distribution is shown in Fig.8(c). It is observed that the crack distribution is significantly influenced by the tensile strength distribution. Moreover, the load carrying capacity of the beam consequently is influenced by the crack distribution. The scatter of load-deflection curve for twelve samples of beams with random strength distribution is shown in Fig.9. For this topic, see also Refs. 9, 35, 39, 40, and 45.

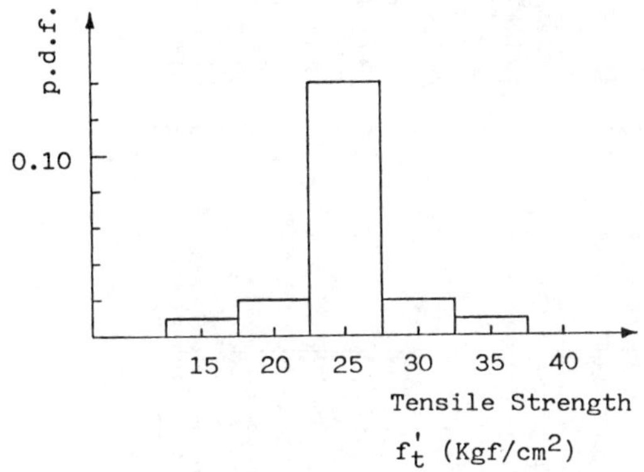

Fig.5 Probability Distribution Function of Tensile Strength ($1Kgf/cm^2$=98.1KPa=14.2psi)

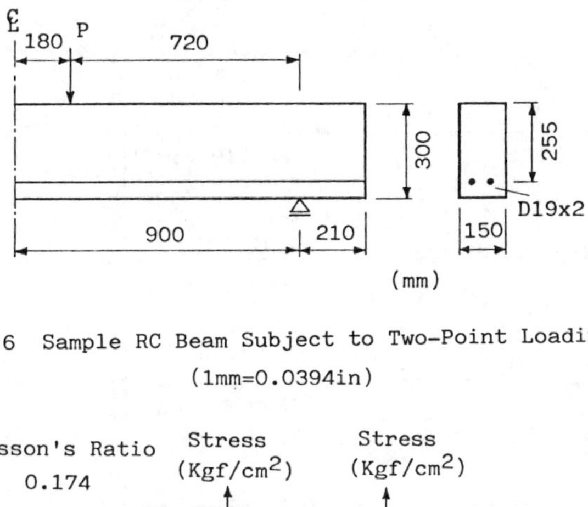

Fig.6 Sample RC Beam Subject to Two-Point Loading (1mm=0.0394in)

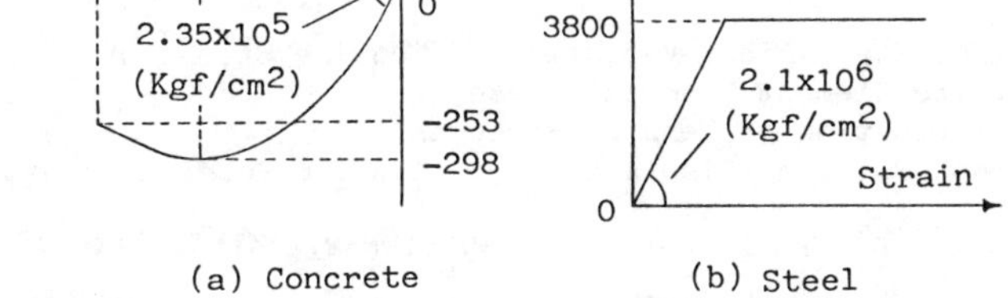

Fig.7 Stress-Strain Relationship for Concrete and Steel Reinforcement ($1Kgf/cm^2$=98.1KPa=14.2psi)

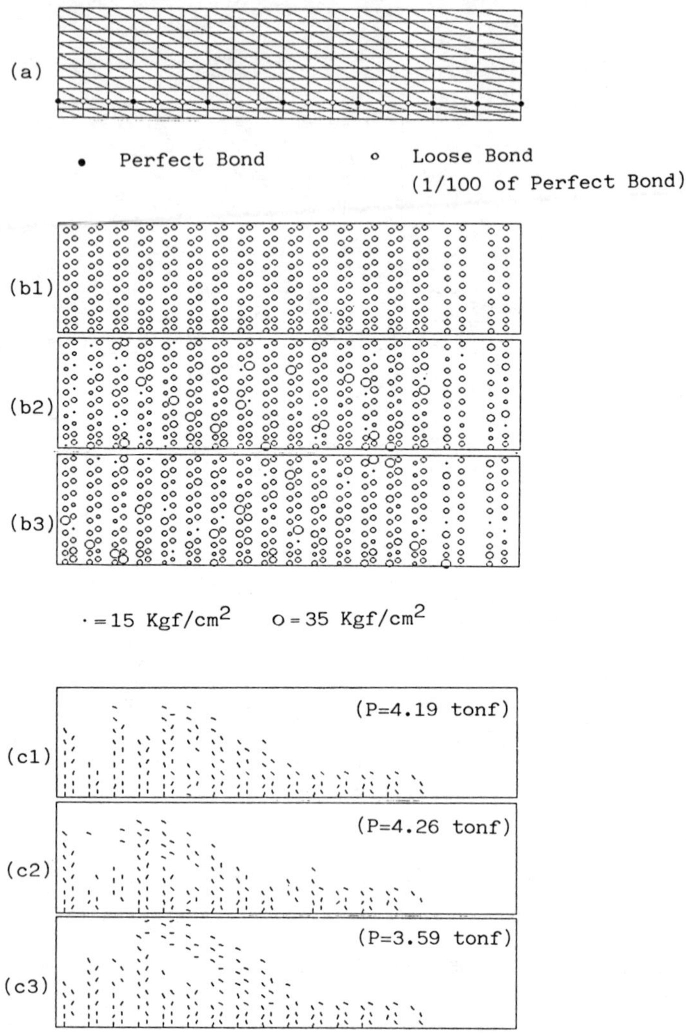

Fig.8 Effect of Statistical Variation of Tensile Strength (Bending);
(a) Finite Element Discretization;
(b) Distribution of Tensile Strength;
(c) Crack Pattern (Displacement at Loading Point u = 3 mm)

($1Kgf/cm^2$=98.1KPa=14.2psi, 1tonf=9.81KN=2.21Klbf, 1mm=0.0394in)

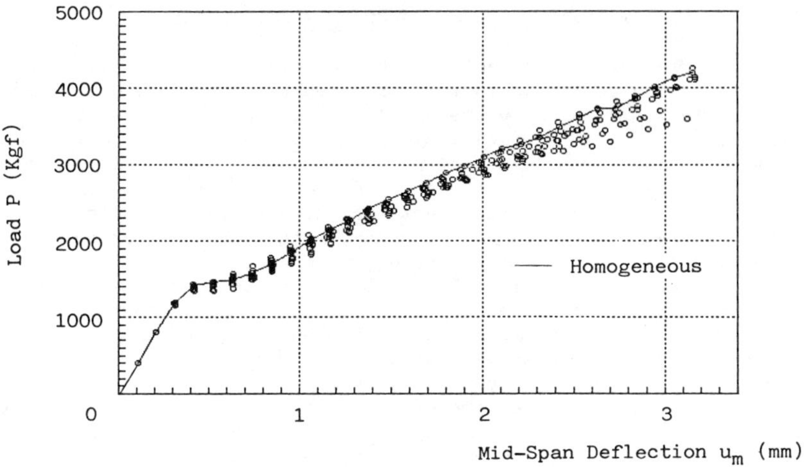

Fig.9 Scatter of Load-Deflection Relationship

(1Kgf=9.81N=2.20 lbf, 1mm=0.0394in)

STATISTICAL ASPECTS IN DESIGN OF RC STRUCTURES (CREEP, SHRINKAGE AND ENVIRONMENTAL CONDITIONS)

Creep and shrinkage are the main factors which influence the long-term mechanical behavior of concrete structures (12,14,17,46,48). In the following, the effect of the statistical variation of these properties on concrete structures is investigated. Since creep and shrinkage are long-term phenomena, the effect of environmental conditions on concrete structures is also considered subsequently.

Effect of Statistical Variation of Creep and Shrinkage

Creep and shrinkage of concrete are influenced by the statistical variation involved in materials and environmental conditions. Even in a laboratory experiment, it is possible that the variation of creep data due to such factors cannot completely be eliminated. Therefore, we might well realize that the creep data reported in the literature are influenced by those factors to some extent, and some statistical scatter is included. In the following, some experimental works to investigate the scatter in the data of creep and shrinkage are reviewed. Then, an overview of statistical methods to predict the scatter of creep and shrinkage is given.

Experimental Investigations of Creep and Shrinkage

A couple of experimental works to investigate the magnitude of scatter and the influence factors of creep and shrinkage have recently been reported. Alou and Wittmann(4) studied the shrinkage of concrete to estimate the final shrinkage by extrapolation. Cylindrical specimens between 100 and 500mm in diameter of two different concrete mixes were used. The specimens were set under 18°C and 65% RH for 4 months during measurement. It is reported that variation in the measured shrinkage was caused by the precision of the measuring technique during the first few days. After one month of drying, then, the scatter of the material properties became dominant. It is also reported that, the CEB-FIP code prediction is an underestimate. Since shrinkage of a concrete structure is influenced by the stochastic variables involved in the regional climate, such as temperature, wind speed, and humidity, the variation observed in a controlled laboratory test is a lower bound. Reinhardt, et al.(37) reported the variability of drying creep and shrinkage. Factors held constant were the specimen size and the mix proportions. Data were measured for 300 days. Load for creep was 20 and 40% of the compressive strength. Cornelissen(23) also reported the test result on the scatter of creep due to variations of concrete composition (water-cement ratio, aggregate-cement ratio). The effect of alternating temperature and relative humidity was also examined. Factors held constant were the size and shape of specimens, the type and grading of aggregate, the curing conditions, the load and age at loading, and the load duration.

Statistical Method for Creep Prediction

The variation of creep data and its influence on deterministic creep models were reported by Bazant and Panula(11). Test data for 80 different countries from various laboratories throughout the world, consisting of over 800 experimental curves and over 10,000 data points have been analyzed statistically. It has been shown that, if no measurements for a given concrete are made, the uncertainty of its creep and shrinkage prediction on the basis of the chosen mix parameters and the chosen design strength is enormous(11). For example, the prediction errors that are expected with a 10% probability (90% confidence limits) are 31% for the BP Model(11), 63% for the ACI-1971 Model(1,2), and 76% for the CEB-FIP-1978 Model(21). An important conclusion from the above works is that a creep prediction formula for a design code must indicate the magnitude of scatter expected from using the formula. As in the CEB-FIP model code, it is desirable for a design code to indicate the extreme values, or the scatter around the mean value, of various material parameters used in concrete. So far, only the statistical distribution of the strength of material is taken into account. The statistical distribution of other material properties, e.g., creep and shrinkage of concrete should also be taken into account.

A statistical comparison of several deterministic models for the prediction of creep and shrinkage of concrete with creep data available in the literature was also reported by Bazant and Zebich (18). The models were algebraically transformed into a linearized form, and statistical regression analysis was then carried out. The outline of the linear regression analysis is presented in the following. Prediction formulas for creep and shrinkage may be transformed into a linear form as follows.

$$y = a + bx + e \tag{3}$$

where x and y are the transformed independent and dependent variables, b is the slope, a is the y-intercept, and e is the error. Although the scatter is large for all models, it is reported that the BP Model(11) performs distinctly better than the ACI(1,2) and CEB-FIP(21) Models.

A Bayesian approach for the prediction of creep was presented by Bazant and Chern(8). The method is described briefly as follows. The "prior" information, i.e., the coefficient of variation of deviations from the creep law for concrete is combined, according to Bayes' theorem, with the probability of the creep values measured on a given concrete for a relatively short time to yield the "posterior" probability distribution of the creep values of the given concrete for any load duration and age at loading. The linearity of creep and the normal distribution of errors are assumed for the given concrete as well as the "prior" information. The Bayesian statistical prediction for basic creep of concrete is outlined in the following. Many creep prediction formulas may be transformed into a linearized form as follows.

$$J(t,t') = q_1 x + q_2 + e \tag{4}$$

where x may be called the reduced time, q_1, q_2 are material parameters, and e is the error. In particular, for the double power law, one may set

$$q_1 = \phi_1/E_0 \; ; \; q_2 = 1/E_0 \; ; \; x = (t'^{-m} + \alpha)(t - t')^n \tag{5}$$

where E_0, ϕ_1, m, n, and α are the material parameters included in the double power law. Parameters m, n, and α must be considered as fixed, while parameters E_0 and ϕ_1 are random and may be expressed from the random basic material parameters q_1 and q_2. According to the Bayes' theorem, the posterior probability density distribution $f''(q_1, q_2)$ is expressed by the prior probability density distribution $f'(q_1, q_2)$ as follows.

$$f''(q_1, q_2) = kP(J_j \mid q_i) f'(q_1, q_2) \tag{6}$$

$$k = \left(\int_{-\infty}^{\infty} \int_{-\infty}^{\infty} P(J_j \mid q_i) f'(q_1, q_2)\, dq_1 dq_2 \right)^{-1};$$

$$J_j = J(x_j) \quad (j = 1,2, \ldots ,N) \tag{7}$$

where $P(J_j \mid q_i)$ is the probability of observing values J_j under the condition that the parameter values are q_i. Assuming that J_j are statistically independent of each other and normal random variables for certain q_1, q_2, and x_j with the mean $q_1 x_j + q_2$ and the standard deviation σ, and that σ is independent of x, the posterior probability of J(x) for a given value $\hat{J}$ is obtained as follows.

$$P(J(x)<\hat{J}) = \int_{-\infty}^{\infty} \int_{-\infty}^{\infty} \phi(y(q_1,q_2)) f''(q_1,q_2)\, dq_1 dq_2 \tag{8}$$

$$\phi(y(q_1,q_2)) = P(J(x)<\hat{J} \mid q_1,q_2) = (1/(\sigma\sqrt{2\pi}))\int_{-\infty}^{y} e^{-z^2/2\sigma^2}\, dz \;;$$

$$f''(q_1,q_2) = a_0 \exp(-(1/2)\sum_{j=1}^{N} (y(q_1,q_2))^2) f'(q_1,q_2) \;;$$

$$a_0 = k(\sigma\sqrt{2\pi})^{-N};\; y(q_1,q_2) = (\hat{J} - q_1 x - q_2)/\sigma \tag{9}$$

If $f'(q_1, q_2)$ is a normal distribution, the posterior (updated) probability that the compliance J(x) at reduced time x will be less than some given value $\hat{J}$ may be calculated by the above equations. An example of the Bayesian prediction is shown in Fig.3. It is shown that this method gives good prediction of long-term creep, given the "prior" information and short-term creep data. The above Bayesian approach may be applied to predict the shrinkage strain of concrete. It has been confirmed by the authors that the method works for shrinkage if the linear transformation of the deterministic prediction formula is done adequately. Also, a method to generalize the above method to a problem with more than two variables has been developed recently by the authors. For the prediction formula of creep, see also Ref.24.

Probabilistic Analysis of Concrete Structures

In order to take into account the uncertainty of creep of concrete in the analysis of RC structures, various methods of predicting the long-term deflection of RC beams have been introduced by, for example, Benjamin et al.(19), Zundelevich and Benjamin(49), and Ditlevsen(26)(See also Refs. 3 and 28). For the analysis of a reinforced concrete beam-column, the following assumptions are employed in Ditlevsen's work: 1) the curvature of the beam axis at x and time t is given by C(x,t) (Bernoulli modeling hypothesis (1)); 2) the stress field σ (x,y,z,t) varies linearly with y and z in any cross section (Bernoulli modeling hypothesis (2)); 3) the uncracked concrete section only carries longitudinal stresses (the effective section assumption); 4) the reinforcement force N(x,t) is non-random and determined by a strain value E(B(x,t)-hC(x,t)) in the reinforcement. Here, h is the distance between the centers of gravity of the effective concrete section and the reinforcement section.

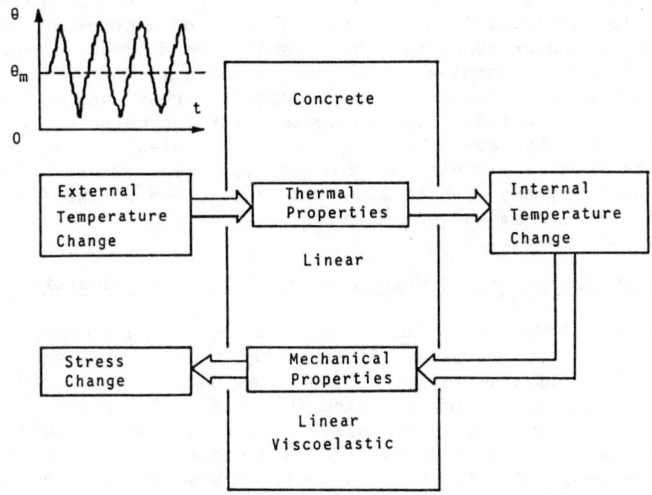

Fig.10 Thermal Stresses in Concrete Subject to Periodic Temperature Change

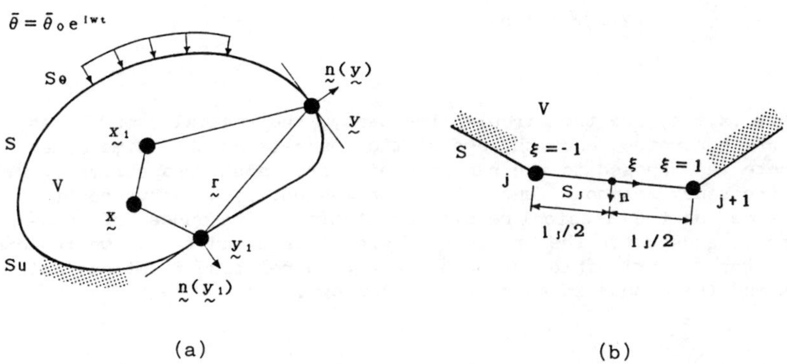

Fig.11 Spectral Method with the Boundary Element Method :
(a) Domain and Boundary Subject to a Periodic Input;
(b) Linear Boundary Element
(From Ref.44)

For a structure with an arbitrary shape, the finite element method with Monte Carlo simulation and Latin hypercube sampling may be useful. Latin hypercube sampling, in general, needs a small number of samples relative to random sampling such as that used in the Monte Carlo method. The method is particularly efficient (in terms of a decrease in number of necessary computer runs) for establishing trends from large finite element analyses of complex structures. An example of a reliability analysis for a prestressed simply supported beam under transverse loading is shown in Ref.6. The parameters for the material properties of concrete were assumed to be uniformly distributed, while the random load variable P was a normally distributed variable. Then, by the Latin hypercube sampling, the probability density function of the fiber strain at the base of the beam was estimated.

Effect of Statistical Variation of Environmental Conditions

The effect of the variation of environmental conditions on concrete structures is, in most cases, larger than that of the variation of internal factors such as material properties and mix proportions (38). Tsubaki and Bazant(43) investigated the effect of random change of ambient humidity on the shrinkage and stresses of a viscoelastic concrete hollow cylinder. Bazant (7) used a different method to solve the above problem. In his work, a modified Wiener-Khintchine relation obtained by shifting the time scale was used, instead of the impulse response function method used in Ref.(43), to take into account the nonlinearity in time (i.e., nonstationarity) due to aging of concrete. This method is more efficient in calculation than the method using an impulse response function. The relationship between the power spectral density S_f of input f and the power spectral density S_g of response g can be expressed as follows.

$$S_g(\omega,x,t,t_0) = |Y(\omega,x,t,t_0)|^2 \, S_f(\omega) \tag{10}$$

where ω ,x,t,t_0 are the circular frequency, the spatial coordinates, the age of concrete, and the age of the concrete at the time when the concrete was exposed to the random stationary input, respectively. Y is the frequency response function of the system. The above equation is the same as the classical result for stationary response of non-aging systems, except for the presence of arguments t and t_0 . On the other hand, the relationship between the autocorrelation functions of the input and the output is expressed as follows.

$$R_g(\lambda,x,t,t_0) = \int_{-\infty}^{\infty}\int_{-\infty}^{\infty} R_f(\xi-\eta+\lambda)\,y(x,t,t_0,\xi)\,y(x,t,t_0,\eta)\,d\xi\, d\eta \tag{11}$$

$$R_f(\lambda) = \lim_{T\to\infty} (1/(2T))\int_{-T}^{T} f(\theta)f(\theta+\lambda)\,d\theta \;;$$

$$y(x,t,t_0,\xi) = (1/(2\pi))\int_{-\infty}^{\infty} Y(\omega,x,t_0+\xi,t_0)\, e^{i\omega\xi}\, d\omega \tag{12}$$

y is called the impulse response function of the system and θ is the time. Note that Eq.(10) is algebraic, while Eq.(11) is given by a double integral. From a computational point of view, Eq.(10) is more convenient for the spectral analysis of aging systems. The frequency response function for a stationary heat conduction problem can be obtained from the following matrix equation for temperature $\underset{\sim}{\theta}$ and flux $\underset{\sim}{q}$.

$$(\kappa \underset{\sim}{K}_2 + i\omega \underset{\sim}{K}_1)\underset{\sim}{\theta} = \underset{\sim}{q} \tag{13}$$

$$\underset{\sim}{K}_1 = \sum_{el} \int_V \underset{\sim}{N}^T \underset{\sim}{N}\, dV \quad ; \quad \underset{\sim}{K}_2 = \sum_{el} \int_V \underset{\sim}{N}^T_{,i}\, \underset{\sim}{N}_{,i}\, dV \tag{14}$$

where $i = \sqrt{-1}$, κ =thermal diffusivity, ω =circular frequency, and $\underset{\sim}{N}$= interpolation function. Then, the frequency response functions of stresses are obtained, solving the following matrix equation for displacements $\underset{\sim}{u}$.

$$\underset{\sim}{K}\,\underset{\sim}{u} = \underset{\sim}{F}'' \tag{15}$$

$$\underset{\sim}{K} = \sum_{el} \int_V \underset{\sim}{B}^T \underset{\sim}{D}\,\underset{\sim}{B}\, dV \quad ; \quad \underset{\sim}{F}'' = \sum_{el} \int_V \underset{\sim}{B}^T \underset{\sim}{D}\,\underset{\sim}{\varepsilon}''\, dV \quad ;$$

$$\underset{\sim}{\sigma} = \underset{\sim}{D}\,(\underset{\sim}{\varepsilon} - \underset{\sim}{\varepsilon}'') \quad ; \quad \underset{\sim}{\varepsilon} = \underset{\sim}{B}\,\underset{\sim}{u} \tag{16}$$

where $\underset{\sim}{D}$ is the tangent modulus matrix, and $\underset{\sim}{\varepsilon}''$ is the strain vector due to thermal expansion. For a stationary problem, $\underset{\sim}{D}$ is obtained by replacing Young's modulus by the complex modulus calculated from the relaxation modulus E_R . The application of this method is shown by Bazant and Wang(15,16).

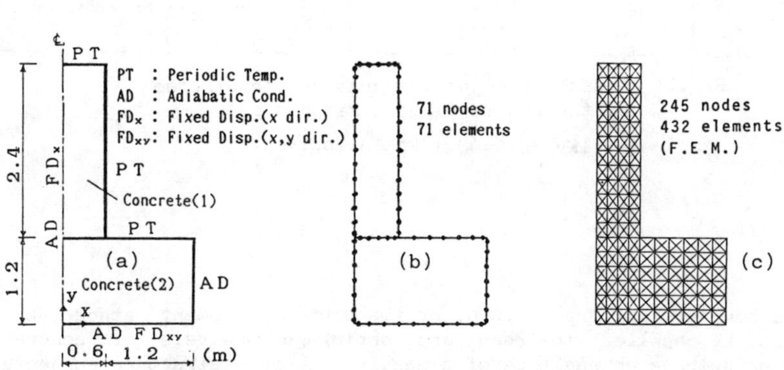

Fig.12 Massive Concrete Structure Subject to a Periodic Temperature Input :
(a) Boundary Conditions; (b) Boundary Element Discretization; (c) Finite Element Discretization

(From Ref.44) (1m=3.28ft)

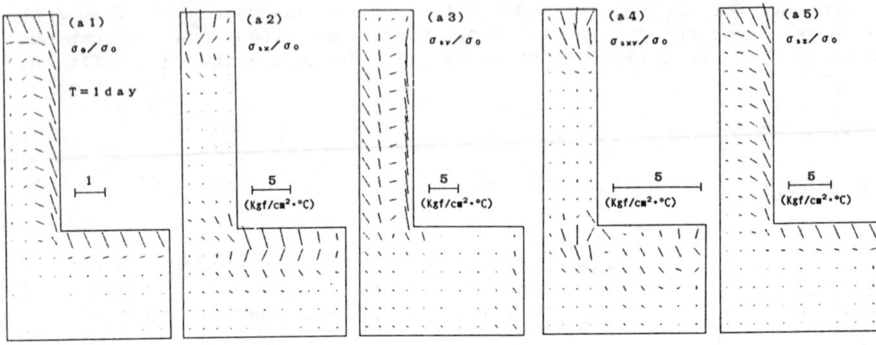

Fig.13 Distribution of Temperature and Stresses by the Boundary Element Method ($1Kgf/cm^2$=98.1KPa=14.2psi)

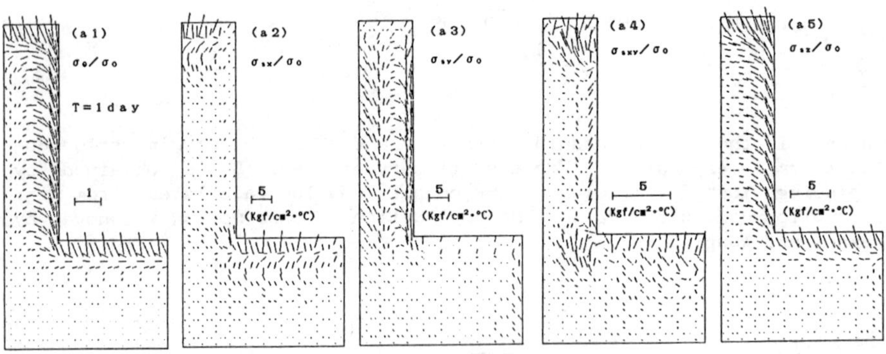

Fig.14 Distribution of Temperature and Stresses by the Finite Element Method ($1Kgf/cm^2$=98.1KPa=14.2psi)

The boundary integral method, or the boundary element method (BEM), can also be applied to creep and shrinkage analysis of concrete structures such as an analysis of a massive concrete structure subjected to a stationary random periodic temperature change. Since the boundary element method needs less elements, in general, than the finite element method, it is particularly effective for a three-dimensional concrete structure. The problem dealt with above by the finite element method can be solved using the boundary element method. For a stationary two-dimensional heat conduction problem, the frequency response function of temperature θ is obtained by the following integral equation(cf. Figs.10,11).

$$c(\underset{\sim}{y})\theta(\underset{\sim}{y}) + \int_S q^*(\underset{\sim}{y},\underset{\sim}{y}_1)\theta(\underset{\sim}{y}_1)\,dS(\underset{\sim}{y}_1)$$

$$=\int_S \theta^*(\underset{\sim}{y},\underset{\sim}{y}_1)q(\underset{\sim}{y}_1)\,dS(\underset{\sim}{y}_1) \tag{17}$$

$$q^*(\underset{\sim}{y},\underset{\sim}{y}_1) = \partial\theta^*(\underset{\sim}{y},\underset{\sim}{y}_1)/\partial n \quad ; \quad \theta^*(\underset{\sim}{y},\underset{\sim}{y}_1) = (1/(4i))H_0^{(2)}(\beta r) \quad ;$$

$$\beta = (1 - i)\sqrt{(\rho c\omega)/(2K)} \quad ; \quad r = |\underset{\sim}{y} - \underset{\sim}{y}_1| \quad ;$$

$$q(\underset{\sim}{y}_1) = \partial\theta(\underset{\sim}{y}_1)/\partial n \quad ; \quad c(\underset{\sim}{y}) = 1/2 \tag{18}$$

where K, ρ, and c, respectively, stand for the heat conductivity, the density, and the specific heat of the structural material. $\partial/\partial n$ is the directional derivative in the direction of the outer normal $\underset{\sim}{n}$ on the boundary surface S. Then, the frequency response functions of the stresses are obtained by the following integral equation for displacement u_i and traction f_i .

$$C_{Ki}(\underset{\sim}{y})u_i(\underset{\sim}{y}) + \int_S f^*_{Ki}(\underset{\sim}{y},\underset{\sim}{y}_1)u_i(\underset{\sim}{y}_1)\,dS(\underset{\sim}{y}_1)$$

$$= \int_S u^*_{Ki}(\underset{\sim}{y},\underset{\sim}{y}_1)f_i(\underset{\sim}{y}_1)\,dS(\underset{\sim}{y}_1) + b_K(\underset{\sim}{y}) \tag{19}$$

$$C_{Ki}(\underset{\sim}{y}) = (1/2)\,\delta_{Ki} \tag{20}$$

where u^*_{Ki} is the fundamental solution of the two-dimensional elastic body, and f^*_{Ki} is the traction corresponding to u^*_{Ki} . b_K is the body force term due to the temperature distribution $\theta(x)$, and δ_{Ki} is the Kronecker's delta. The details of the analysis method are given in Ref.44. An example of an analysis using the above analysis method is shown in Figs.12,13, and 14. Figs.13 and 14 show the distributions of the temperature and stresses of a concrete structure subject to a periodic temperature change (the period T=1 day) (cf. Fig.12). The length and slope of a line segment stand for the magnitude and the phase angle of the temperature and stresses, respectively. It is observed that the result obtained by BEM is in good agreement with that by FEM.

Madsen and Bazant(33) also reported the effects of variations in environmental humidity and temperature, together with those of variations in shape and size of the structure and other parameters related to mix composition and strength, on concrete structures. The effects of variations of these parameters at different points of the same concrete structure on the total uncertainty were investigated. The model uncertainties related to choosing one specific set of shrinkage and creep formulas were also discussed. Model uncertainty was accounted for by applying a random factor or a model uncertainty factor to each term: shrinkage, basic creep, and drying creep. Although any deterministic creep and shrinkage formula can serve as the basis for this approach, the BP formulas are chosen in their work. The formula for shrinkage is then:

$$\varepsilon_{sh}(t,t_o) = \Psi_1 \varepsilon_{sh\infty} k_h \; S(t,t_o) \tag{21}$$

Also, the formula for creep is

$$J(t,t') = \Psi_2((1/E_0) + C_0(t,t')) + \Psi_3(C_d(t,t',t_o) - C_p(t,t',t_o)) \tag{22}$$

Here, ε_{sh} =shrinkage strain; t_0 =age of concrete when drying begins; $J(t,t')$=compliance function(creep function)=strain at age t caused by a uniaxial unit stress acting since concrete age t'; E_0 =asymptotic modulus ($\simeq 1.5E_c$ where E_c =conventional elastic modulus of concrete); C_0, C_d, C_p =functions describing basic creep (creep at constant temperature and humidity), increase of creep during drying, and decrease of creep after drying; $C_0(t,t') = (\phi_1/E_0)\,(t' + \alpha)(t-t')$ (double power law) where ϕ_1, m, n, and α =parameters depending on the type of concrete; C_d =function similar to S; and C_p=function similar to C_0; k_h=function of relative humidity h, defined as $(1-h^3)$ if $h \leqslant 0.98$, or else, -0.2; and S= empirical function of the variable $(t-t_o)/\tau_{sh}$, where τ_{sh} =shrinkage-square half time, proportional to the square of the dimension of the cross section, $S(t,t_0)=(1+\tau_{sh}/(t-t_0))^{-1/2}$ where τ_{sh} =const. x $(k_s D)^2$, D=2 x volume-to-surface ratio, and k_s=factor taking into account the shape of cross section. For $\Psi_1 = \Psi_2 = \Psi_3 = 1$, Eqs.(21) and (22) yield the ordinary BP formulas. From the test data, the mean values and coefficients of variation of the Ψ -factors can be determined. The following values are used in the example.

$$\begin{array}{lll} \text{Shrinkage} & E(\Psi_1)=1 \;;\; V_{\Psi_1}=0.14 & \\ \text{Basic Creep} & E(\Psi_2)=1 \;;\; V_{\Psi_2}=0.23 & (23) \\ \text{Drying Creep} & E(\Psi_3)=1 \;;\; V_{\Psi_3}=0.13 & \end{array}$$

Then, calculation of the response of concrete structures with the above random creep and shrinkage properties is carried out by using the matrix formulation which allows a convenient algebraic treatment of linearly viscoelastic structures. For the computation of the statistics of the structural response, the method of point estimates for probability moments is used. In the simplest form of this approach, only two points $\mu \pm s$ need to be considered for each random variable(μ =mean of the variable, and s=standard deviation).

The spectral analysis of structural response is no doubt sophisticated for regular design applications. The Bayesian approach to creep prediction is also intended for special structures and would not be used in regular design. In this sense, the use of model uncertainty factors is a relatively simple method for an analysis of creep-sensitive structures. The above problem was also solved by using Latin hypercube sampling in the work of Bazant, et al.(10).

CONCLUSIONS

In the first part, a general method to apply the finite element method to the design of concrete structures is proposed. In the second part, the effect of statistical variation of material properties and environmental conditions on reinforced concrete structures is discussed in order to clarify the influence of statistical aspects on the design of RC structures. It is found that the statistical variation of strength has a significant influence on the mechanical behavior of RC structures. Also, for long-term behavior of RC structures, recent developments in Bayesian prediction of creep parameters of a given concrete and Bayesian extrapolation of short-term creep data are introduced. In order to predict the scatter of creep and shrinkage of concrete, the methods of analyzing the effects of stochastic material properties and stochastic changes of environmental conditions and loads on concrete structures are discussed.

For simple beam-like structures, it is possible to use a quasi-elastic solution (e.g., 13,22,25,29,31) to deal with random environmental changes. For general concrete structures subjected to random environmental changes, it is desirable to formulate the problem by a numerical method such as the finite element method and the boundary element method, as used with deterministic creep models. For a stationary problem, it is possible to use the conventional finite element method for creep analysis (6,41,42).

REFERENCES

1. ACI Committee 209/II, "Prediction of Creep, Shrinkage and Temperature Effects in Concrete Structures," ACI SP-27, Designing for Effects of Creep, Shrinkage and Temperature, American Concrete Institute, Detroit, Mich., 1971, pp.51-93.
2. ACI Committee 209, "Prediction of Creep, Shrinkage, and Temperature Effects in Concrete Structures," ACI SP-76, Designing for Creep and Shrinkage in Concrete Structures, American Concrete Institute, Detroit, Mich., 1982, pp.193-301.
3. ACI Committee 435, "Variability of Deflection of Simply Supported Reinforced Concrete Beams," American Concrete Institute Journal, Vol.69, No.1, 1972, pp.29-35.
4. Alou, F., and Wittman, F. H., "Experimental Study of the Variability of Shrinkage in Concrete," Fundamental Research on Creep and Shrinkage of Concrete, Edited by F. H. Wittmann, Martinus Nijhoff Publishers, The Hague, 1982, pp.75-92.
5. American Society of Civil Engineers, "State-of-the-Art Report on Finite Element Analysis of Reinforced Concrete," 1982, New York.
6. Anderson, C. A., "Numerical Creep Analysis of Structures," Creep and Shrinkage in Concrete Structures, Chap. 8, Edited by Z. P. Bazant and F. H. Wittmann, John Wiley, New York, 1982, pp.259-303.
7. Bazant, Z. P., "Response of Aging Linear Systems to Random Input," Report No.82-12/665r, Center for Concrete and Geomaterials, Northwestern University, Evanston, IL., December, 1982.

8. Bazant, Z. P., and Chern, J. C., "Bayesian Statistical Prediction of Concrete Creep and Shrinkage," ACI Journal, Vol.81, No.4, July/August, 1984, pp.319-330.
9. Bazant, Z. P., and Kim, J.-K., "Size Effect in Shear Failure of Longitudinally Reinforced Beams," ACI Journal, Vol.81, No.5, 1984, pp.456-468.
10. Bazant, Z. P., and Liu, K.-L., "Random Creep and Shrinkage in Structures : Sampling," Report No.84-5/679r, Center for Concrete and Geomaterials, The Technological Institute, Northwestern Univ., Evanston, IL., May, 1984.
11. Bazant, Z. P., and Panula, L., "Practical Prediction of Time-Dependent Deformations of Concrete," Materials and Structures, Parts I and II : Vol.11, No.65, 1978, pp.307-328, Parts III and IV : Vol.11, No.66, 1978, pp.415-434, Parts V and VI : Vol.12, No.69, 1979, pp.169-183.
12. Bazant, Z. P., and Tsubaki, T., "Weakly Singular Integral for Creep Rate of Concrete," Mechanics Research Communications, Vol.7, No.5, 1980, pp.335-340.
13. Bazant, Z. P., and Tsubaki, T., "Nonlinear Creep Buckling of Reinforced Concrete Columns," Journal of the Structural Division, ASCE, Vol.106, No.ST11, November, 1980, pp.2235-2257.
14. Bazant, Z. P., Tsubaki, T., and Celep, Z., "Singular History Integral for Creep Rate of Concrete," Journal of the Engineering Mechanics Division, ASCE, Vol.109, 1983.
15. Bazant, Z. P., and Wang, T. S., "Spectral Analysis of Random Shrinkage Stresses in Concrete," Journal of the Engineering Mechanics Division, ASCE, Vol.110, No.EM2, February, 1984, pp.173-186.
16. Bazant, Z. P., and Wang, T. S., "Algorithm for Aging Viscoelastic Structures under Periodic Loading," Journal of Engineering Mechanics, ASCE, Vol.110, No.6, June, 1984, pp.972-984.
17. Bazant, Z. P., and Wu, S. T., "Rate-Type Creep Law of Aging Concrete Based on Maxwell Chain," Materials and Structures, Vol.7, No.37, 1974, pp.45-60.
18. Bazant, Z. P., and Zebich, S., "Statistical Linear Regression Analysis of Prediction Models for Creep and Shrinkage," Cement and Concrete Research, Vol.13, No.6, November, 1983.
19. Benjamin, J. R., Cornell, C. A., and Gabrielsen, B. I., "A Stochastic Model for the Creep Deflection of Reinforced Concrete Beams," Proceedings, International Symposium on the Flexural Mechanics of Reinforced Concrete, American Concrete Institute, SP-12 , Detroit, Mich., 1965, pp.557-580.
20. Burt, N.J., and Dougill, J.W., "Progressive Failure in a Model Heterogeneous Medium," J. of Engineering Mechanics Division, ASCE, Vol.103, No.EM3, pp.365-376.
21. CEB-FIP, "Model Code for Concrete Structures," Comite Eurointernational du Beton and Federation Internationale de la Precontrainte, CEB Bulletin No.124/125-E, Paris, 1978.
22. Chiorino, M.A., and Mola, F., "Analysis of Linear Visco-Elastic Structures Subjected to Delayed Restraints," Fundamental Research on Creep and Shrinkage of Concrete, Edited by F. H. Wittmann, Martinus Nijhoff Publishers, The Hague, 1982, pp.485-508.

23. Cornelissen, H. A. W., "Creep of Concrete - A Stochastic Quantity," Fundamental Research on Creep and Shrinkage of Concrete, Edited by F. H. Wittmann, Martinus Nijhoff Publishers, The Hague, 1982, pp.109-124.
24. Diamantidis, D., Madsen, H. O., and Rackwitz, R., "On the Variability of the Creep Coefficient of Structural Concrete," Materials and Structures, Vol.17, No.100, 1984, pp.321-328.
25. Dilger, W. H., "Methods of Structural Creep Analysis," Creep and Shrinkage in Concrete Structures," Chap. 9, Edited by Z. P. Bazant and F. H. Wittmann, John Wiley, New York, 1982, pp.305-339.
26. Ditlevsen, O., "Stochastic Visco-Elastic Strain Modeled as a Second Movement White Noise Process," International Journal of Solids and Structures, Vol.18, No.1, 1982, pp.23-35.
27. Dougill, J. W., "Mechanics of Concrete Systems: Current Approaches to Assessing Material Behaviour and Some Possible Extensions," Creep and Shrinkage in Concrete Structures, Chap.2, Edited by Z. P. Bazant and F. H. Wittmann, John Wiley, New York, 1982, pp.23-49.
28. Grant, L. H., Mirza, S. A., and MacGregor, J. G., "Monte Carlo Study of Strength of Concrete Columns," ACI Journal, Vol.75, No.8, August, 1978, pp.348-358.
29. Haas, W., "Numerical Analysis of Creep and Shrinkage in Concrete Structures," Fundamental Research on Creep and Shrinkage of Concrete, Edited by F. H. Wittmann, Martinus Nijhoff Publishers, The Hague, 1982, pp.369-379.
30. Japan Concrete Institute, "Proceedings of JCI Colloquium on Finite Element Analysis of RC Structures, 1984, Tokyo.
31. Jonasson, J. E., "Analysis of Shrinkage in Concrete," Fundamental Research on Creep and Shrinkage of Concrete, Edited by F. H. Wittmann, Martinus Nijhoff Publishers, The Hague, 1982, pp.259-268.
32. Kajima Corporation, "Detailed Design of Okaya Bridge of Japan Highway Public Corporation, 1985. (in Japanese, unpublished).
33. Madsen, H. O., and Bazant, Z. P., "Uncertainty Analysis of Creep and Shrinkage Effects in Concrete Structures," American Concrete Institute Journal, Vol.80, No.2, March/April, 1983, pp.116-127.
34. Milford, R.V., and Schnobrich, W.C., "Nonlinear Behavior of Reinforced Concrete Cooling Towers," Civil Engineering Studies, Structural Research Series No.514, University of Illinois at Urbana-Champaign, 1984.
35. Mirza, S.A., Hatzinikolas, M., and MacGregor, J.G., "Statistical Descriptions of Strength of Concrete," Journal of the Structural Division, ASCE, Vol.105, No.ST6, 1979.
36. National Agency for Finite Element Method and Standard, "Guidelines to Finite Element Practice," August, 1984, Scotland.
37. Reinhardt, H. W., Pat, M. G. M., and Wittmann, F. H., "Variability of Creep and Shrinkage of Concrete," Fundamental Research on Creep and Shrinkage of Concrete, Edited by F. H. Wittmann, Martinus Nijhoff Publishers, The Hague, 1982, pp.93-108.
38. Russell, H. G., Meyers, B. L., and Daye, M. A., "Observations on Structures," Creep and Shrinkage in Concrete Structures, Chap. 10, Edited by Z. P. Bazant and F. H. Wittmann, John Wiley, New York, 1982, pp.341-358.

39. Shinozuka, M., "Probabilistic Modeling of Concrete Structures," J. of the Engineering Mechanics Division, ASCE, Vol.98, No.EM6, December, 1972, pp.1433-1451.
40. Shinozuka, M., and Lenoe, E., "A probabilistic model for Spatial Distribution of Material Properties," Engineering Fracture Mechanics, Vol.8, 1976, pp.217-227.
41. Smith, P. D., and Anderson, C. A., "NONSAP-C: A Nonlinear Stress Analysis Program for Concrete Containments under Static, Dynamic, and Long-Term Loadings," Los Alamos Sci. Lab. Report LA-7496-MS, 1978.
42. The Task Committee on Finite Element Analysis of Reinforced Concrete Structures of the Structural Division Committee on Concrete and Masonry Structures, "Time-Dependent Effects," State-of-the-Art Report on Finite Element Analysis of Reinforced Concrete, Chap. 6, American Society of Civil Engineers, New York, N.Y., 1982, pp.309-400.
43. Tsubaki, T., and Bazant, Z. P., "Random Shrinkage Stresses in Aging Viscoelastic Vessel," Journal of the Engineering Mechanics Division, ASCE, Vol.108, No.EM3, June, 1982, pp.527-545.
44. Tsubaki, T., "Thermal Stresses in Concrete Subject to Periodic Temperature Change," Trans. of JCI, Japan Concrete Institute, Vol.6, 1984, pp.347-354.
45. Vanmarcke, E., and Grigoriu, M., "Stochastic Finite Element Analysis of Simple Beams," J. of Engineering Mechanics Division, ASCE, Vol.109, No.EM5, 1983, pp.1203-1214.
46. Wittmann, F. H., "Creep and Shrinkage Mechanisms," Creep and Shrinkage in Concrete Structures, Chap. 6, Edited by Z. P. Bazant and F. H. Wittmann, John Wiley, New York, 1982, pp.129-161.
47. Yoshida, S., Saito, K., Kubota, M., et al., "Design of Egg-shaped Digestion Tank in Yokohama," Journal of Prestressed Concrete, Japan, Vol.26, No.5, Sept., 1984. (in Japanese)
48. Young, J. F., "The Microstructure of Hardened Portland Cement Paste," Creep and Shrinkage in Concrete Structures, Chap. 1, Edited by Z. P. Bazant and F. H. Wittmann, John Wiley, New York, 1982, pp.3-22.
49. Zundelevich, S., and Benjamin, J. K., "Probabilistic Analysis of Deflections of Reinforced Concrete Beams," Probabilistic Design of Reinforced Concrete Buildings, ACI Publication SP-31, American Concrete Institute, Detroit, Mich., 1972, pp.223-246.

Analytical Study of Ultimate Behavior of Underground LNG Storage Tanks Subjected to Both Thermal and Seismic Earth Pressure Load

Hajime Ohuchi* Toshikazu Takeda**

Abstract

It is necessary for safety and for proper functioning of LNG storage tanks to predict nonlinear behavior up to the ultimate state when seismic earth pressure is the governing load for structural design. Failure behavior and thermal stresses are predicted by static finite element analyses for shell structures subjected to uniform earth pressure exerted on one side or both sides of the structure and to stationary axisymmetric loads including thermal load. The following results were obtained from this study: (1) Ultimate capacity reaches about three times the design load for seismic earth pressure and when the temperature dependence of material properties is considered, the safety factor of three tends to be increased. (2) Thermal stresses can be reduced for practical design by the occurrence of cracks. (3) The circumferential thermal bending moment $\dot{M}_{\theta}$ on pressurized sides at the top of the tank, thought to be greatly concerned with failure, disappears. In conclusion, a design method using a nonlinear response analysis computer program developed for the types of structures considered in these studies is discussed.

1. Introduction

Underground LNG storage tanks as shown in Fig-1, because of their structural characteristic of being built in the ground, are thought to be subject at all times to actual loads such as earth and water pressures, and thermal loads due to the very low-temperature liquid stored in them. It may be said that the present situation with this type of structure is that elastic design is conducted using combined loads including earthquake loads. Increased earth pressure, which is usually considered to be the governing earthquake load, is studied in this paper. Results of static FEM nonlinear analyses are reported for typical LNG storage tanks subjected to both seismic and permanently acting loads.

When considering the nonlinear material characteristics of concrete represented by cracking, thermal stresses which are self-restraining are subject to the influence of actual loads, and when

*Research Engineer, Technical Research Institute of Structural Mechanics of Ohbayashi Corp., Tokyo, Japan

**Deputy director, Technical Research Institute of Ohbayashi Corp., Tokyo, Japan combined with earthquake loads which tend to govern in design, it is expected that thermal stresses will vary greatly. Consequently, it is thought it would be reasonable, especially with regard to factors such as thermal stress, not to think in fixed terms, but to consider combinations with other loads, and to carry out design after first gaining a proper grasp of the behavior of the structure.

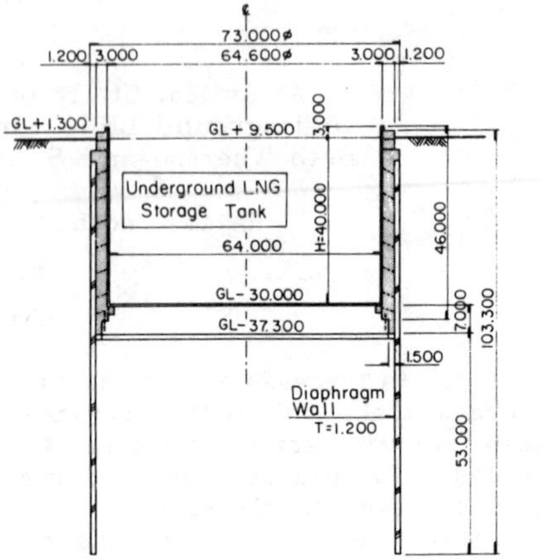

Fig-1 Underground LNG Storage Tank

2. Analysis Technique

This analysis method consists of a finite element nonlinear analysis of a thin plate and shell structure approximated as in-plane and out-of-plane quadrilateral folded plate element assemblages [1].

2.1 Basic Assumptions

The major items hypothesized in this analysis are listed below.

(1) Strain distribution in the direction of plate thickness within the element is determined by assuming that cross sections of the plate remain plane during bending, as shown in Fig-2. Consequently, reinforcing steel and concrete are assumed to be perfectly bonded.

(2) It is assumed that the inside of an element is in a state of plane stress.

(3) Deformations are assumed to be infinitesimal and, therefore, geometric nonlinearity is not considered.

2.2 Uniaxial Stress-Strain Relationship

The uniaxial stress-strain relationships of concrete and reinforcing steel are shown in Fig.-3. With regard to concrete, the extension

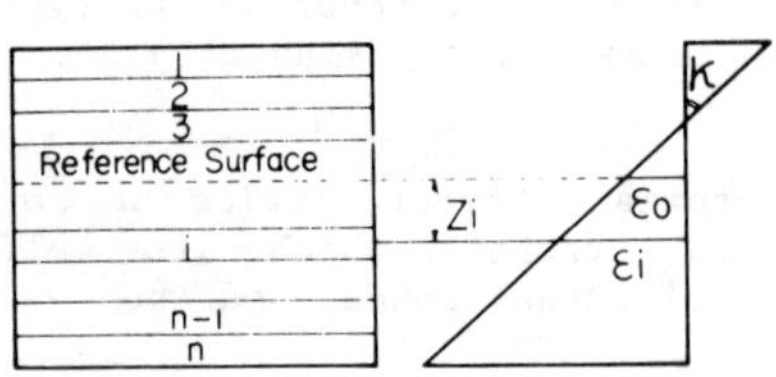

Fig-2 Layerd Element and Strain Distribution

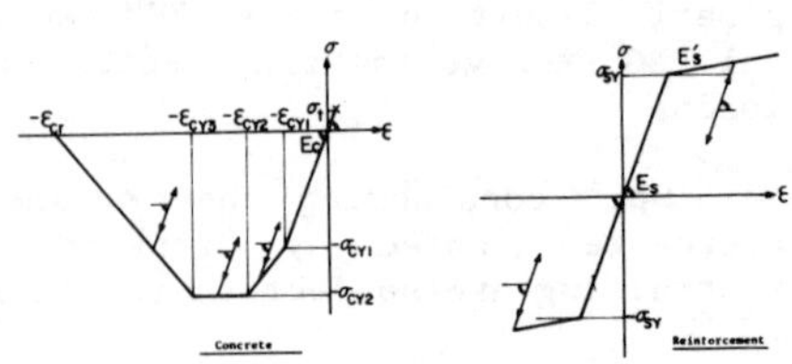

Fig-3 Uniaxial Stress-strain Relationship

to a biaxial plastic stress state is accomplished employing the concepts of equivalent stress and equivalent plastic strain.

2.3 Failure Conditions

For failure condition equations, Kupfer et al. [2] have given biaxial failure test values, and the following equations approximating those test values are employed [3].

$$\tau_{oct}+\sqrt{2}\frac{1-\alpha}{1+\alpha}\sigma_{oct}-\frac{2\sqrt{2}}{3}\frac{\alpha}{1+\alpha}\sigma_{cY}=0 \quad \text{(biaxial tension and tension - compression zone)} \qquad (1)$$

$$\tau_{oct}+\sqrt{2}\frac{\beta-1}{2\beta-1}\sigma_{oct}-\frac{\sqrt{2}}{3}\frac{\beta}{2\beta-1}\sigma_{cY}=0 \quad \text{(biaxial compression zone)} \qquad (2)$$

Cracking failure is considered using an orthogonal anisotropic body with cracks produced orthogonal to the direction of maximum principal stress, after which stiffness remains in the direction of cracking only. For the biaxial compression zone, behavior is modeled using an isotropic hardening body where the yielding surface can move to the $f\text{-}\sigma_{CY2}=0$ plane when the elastic limit plane $f\text{-}\sigma_{CY1}=0$ is exceeded. For the strain softening zones the yielding surface is isotropically shrunk by a equivalent plastic strain.

2.4 Stress-Strain Relationships

Stresses are generally expressed by the differences between actual strain ε and initial and plastic strain (εo: thermal strain, εp: plastic strain)

$$\{\sigma\} = [D_e]\{\varepsilon-\varepsilon_p-\varepsilon_o\} \quad \ldots\ldots (3)$$

which when incrementally expressed, generally will become following:

$$d\{\sigma\} = [D_e]\{d\varepsilon-d\varepsilon_p-d\varepsilon_o\}+d[D_e]\{\varepsilon-\varepsilon_p-\varepsilon_o\} \quad \ldots\ldots (4)$$

Thereupon, regarding the plastic strain increment, when the incremental plasticity theory based on the associated flow rule is employed, Eq.(4) will be formularized as follows:

$$\{\Delta\sigma\} = [D]\{\Delta\varepsilon-\Delta\varepsilon_o\}+[\Delta D_e]\{\varepsilon-\varepsilon_p-\varepsilon_o\} \quad \ldots\ldots (5)$$

where, $[D]=[De]-[Dp]$

$$[D_p] = [D_e] - \frac{[De]\{\partial f/\partial\sigma\}\{\partial f/\partial\sigma\}^T[De]}{g+\{\partial f/\partial\sigma\}^T[D_e]\{\partial f/\partial\sigma\}} \quad \ldots\ldots (6)$$

2.5 Equilibrium Equation and Analytical Method

The equivalent nodal force-nodal displacement relationships of elements derived employing the principle of virtual work is the following:

$$[K_e]\{\Delta\delta_e\} = \{\Delta F_e\}+\{\Delta F_{e1}\}+\{\Delta F_{e2}\}+\{\Delta F_{e3}\} \quad \ldots\ldots (7)$$

where,

$$\{\Delta F_1\} = \int[B]^T[D]\{\Delta\varepsilon_o\}dV$$

$$\{\Delta F_2\} = -\int[B]\ [\Delta D_e]\{\varepsilon-\varepsilon_p-\varepsilon_o\}dV$$

$$\{\Delta F_3\} = \int[B]^T\{\sigma_{ex}\}$$

σ_{ex}: unbalanced stress, [B]: nodal displacement-strain matrix

Incorporating the rotational stiffness term zero about Z axis in the direction perpendicular to the plane in the element, the stiffness matrix is expressed in the local coordinate system. Multiplying by the coordinate transformation matrix [L], the element matrix in the global coordinate system is derived. By superposing these element stiffnesses one on another, the global equilibrium equations are derived as follows:

$$[K]\{\Delta\delta\} = \{\Delta F\}+\{\Delta F_1\}+\{\Delta F_2\}+\{\Delta F_3\} \quad \text{..... (8)}$$

where $[K] = \Sigma[L]^T[Ke][L]$, $\{\Delta F\}$=external force vector, $\{\Delta F_1\}=\Sigma[L]\{\Delta Fe_1\}$, $[\Delta F_2]=-[L]\{\Delta Fe_2\}$, and $\{\Delta F_3\}=\Sigma[L]\{\Delta Fe_3\}$.

2.6 Temperature Dependence of Material Properties

For each element and each layer, coefficient of thermal expansion, modulus of elasticity, and Poisson's ratio are considered to be temperature-dependent and are modeled using a multilinear relationship. These temperature-dependent values are used to calculate stiffness [D] and initial strain $\{\varepsilon o\}$. Other concrete properties are compressive strength and tensile strength. Reinforcing steel properties considered are yield stress, thermal expansion coefficient, and modulus of elasticity [5].

3. Nonlinear Analysis of Underground LNG Tanks

The side walls of the underground LNG tank are taken as the structure to be considered here.

3.1 Basic Treatment of Structure Model

(1) Load: As permanently acting loads, dead weight, earth and water pressure, internal liquid pressure, and roof load are considered. The steady-state temperature distribution obtained by heat transfer analysis is used as thermal load (inner surface at middle part of side wall -30°C, outer surface -9°C), but in the direction of plate thickness it is to be a straight-line distribution. Next, increased earth pressure during earthquake is subject to several factors, but it is taken here to be uniformly distributed as shown in Fig-4, where uniformly distributed rectangular loads act from two opposing sides [4] or from one side based on the static seismic coefficient method, and the respective behavior is observed by analyses for the case of loading from one side and both sides.

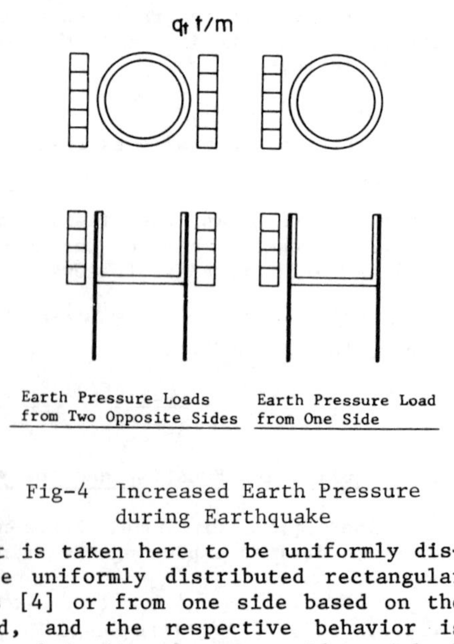

Fig-4 Increased Earth Pressure during Earthquake

(2) Structural Forms: Diaphragm walls are considered here as being temporary, and the structure studied is limited to the side walls

(180-deg. portion in circumferential direction divided into 12 equal parts, vertical direction 11 parts). Reinforcement Ratio at central-height-section are followings: Circumferential Pin=0.214% (2-D29@200), Pout=0.085% (D25@200), Vertical Pin=0.51% (2-D38@150), Pout=0.127% (D38@300), Wall Thickness=300 cm. The embedded portion is considered to be a beam on an elastic foundation, and is modeled by a radial spring at the base.

(3) Surrounding Ground: The influence of the ground as a medium restraining deflection of the sid wall will not be considered.

As the load histories in performing calculations, permanently acting actual loads such as earth and water pressures are applied as nodal forces. Then the permanent thermal loads are applied, and finally, the increased earth pressure during earthquake is applied by gradually increasing the load.

3.2 Stresses for Permanent Axisymmetric Loads

At the inner surface toward the bottom of the side wall, the vertical stress σz reaches a maximum of 45 kg/cm², but a state of cracking is not reached because of increase in tensile strength due to low temperature, and it may be said that under this loading condition an elastic state is maintained.

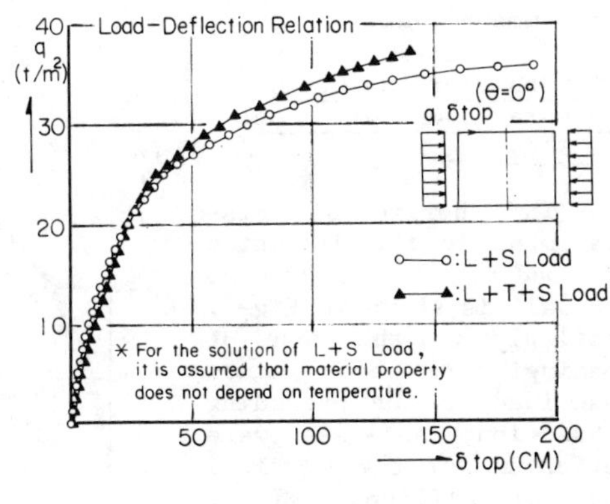

Fig-5 Load-Horizontal Displacement at Top

3.3 Behavior for Earth Pressure Loads Applied on Two Opposing Sides

The relationship between pressure load during earthquake and the horizontal displacement at the top in the direction of loading is shown in Fig-5. Here, L expresses permanently acting actual loads such as earth and water pressures, T expresses thermal loads, and S signifies increased earth pressure load during earthquake. The two cases of (L+S) and (L+T+S) were considered for the purpose of investigating the influence of thermal stress. In the case of (L+T+S) in

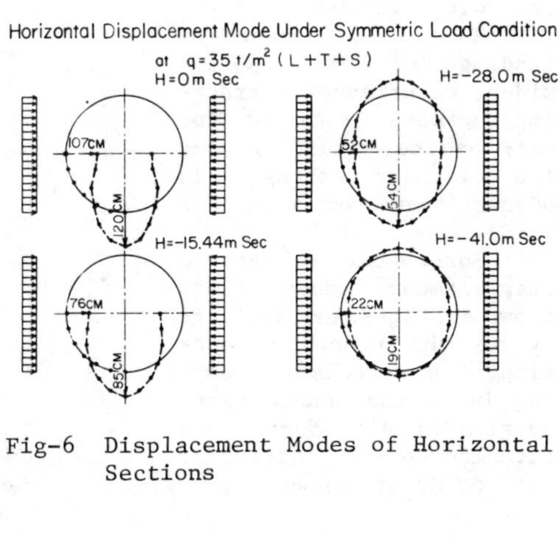

Fig-6 Displacement Modes of Horizontal Sections

the figure, so long as a history of thermal load is sustained, occurrence of cracking is more prominent than in the case of (L+S) from the beginning, and a slightly lower gradient is indicated. However, with regard to ultimate strength, the case with thermal load acting shows a higher value, and the effect of material strength increasing with temperature is apparent.

Fig.-6 shows the displacement modes of the various horizontal sections (H expresses a vertical distance from the top), where large effects of flexural deflection appear at the top and pushing in of the loaded faces and bulging of the sides are prominently produced. Circumferential reinforcing bars show yielding along their inner surface at the loaded surface sides, while at the web sides (θ=45°-135°), yielding along the outer surface of the bars is seen, so that bending moment M_θ in the circumferential direction corresponds with pushing in and bulging out of the walls (side surface passive earth pressure springs are not considered here). For vertical reinforcing bars at θ=0-45° at the bottom of the side walls, both the outer and inner bars yielded in tension and shows the influence of heaving up with the θ=90° line as the symmetric axis.

3.4 Behavior for Earch Pressure Load Applied on One Side

The horizontal displacement in the direction of loading at the top of the wall is shown in Fig-7. Similarly to the case of loading from both sides described in the preceding subsection, analyses were performed for the two load cases of (L+T) and (L+T+S). Load type symbols L, T, and S are the same as in the preceding section.

As in the case of loading from two opposing sides, occurrence of cracking is more prominent for the load case (L+T+S), but the ultimate strength is higher than for the case (L+S).

Horizontal section displacement modes are shown in Fig-8 and similarly to the preceding section, the effects of pushing in at the loaded face, bulging at side faces, and circumferential direction bending M_θ are observed.

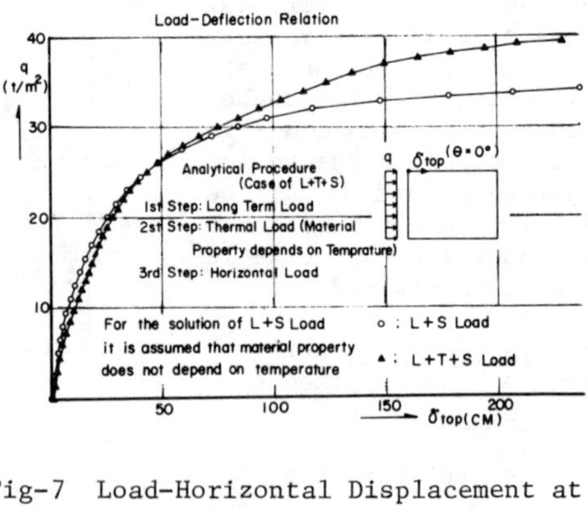

Fig-7 Load-Horizontal Displacement at Top

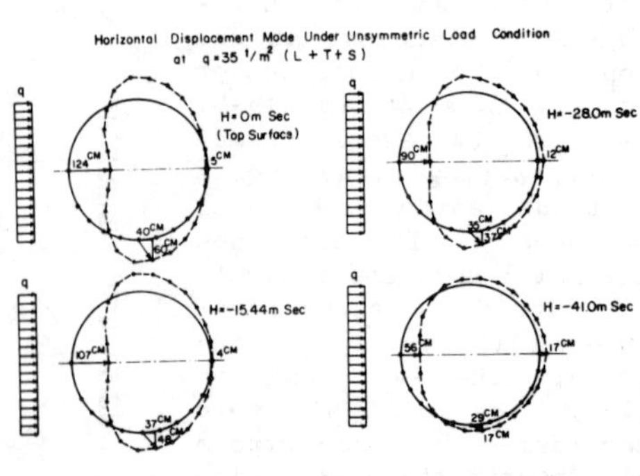

Fig-8 Displacement Modes of Horizontal Sections

However, deflections are concentrated especially at the loaded face side in this case and the nonloaded sides bulge out influenced by this so that displacement in the reverse direction is observed at the face opposite from the one loaded.

Circumferential bars as in the case of loading from two sides, yield at the inside at the loaded side and at the outside at the web sides, and appear to correspond with the distribution at Mθ.

Furthermore, at the top on the opposite side from the loaded surface, inside bars yield in flexurel, while at the base from around 90 to 180°, compressive yielding is observed. As for reinforcing bars in the vertical direction, yielding due to predominance of membrane tension against overturning moment is chiefly seen at the 0 to 60° part at the base, while at 90 to 180°, only the outside bars yield. Regarding the latter, this behavior is thought to be due to the following reasons in addition to the compressive yielding phenomena at the same locations as the circumferential bars. The case of loading of one side is different from the case of loading at two sides and the reaction to total horizontal force will result in base spring forces which represent the embedment effect of underground diaphragm walls.

Radial-direction spring reaction is shown in Fig-9, and values for locations of θ=90 to 180° are especially large. It is thought this force causes circumferential-direction compressive membrane force and vertical-direction bending moment to predominate.

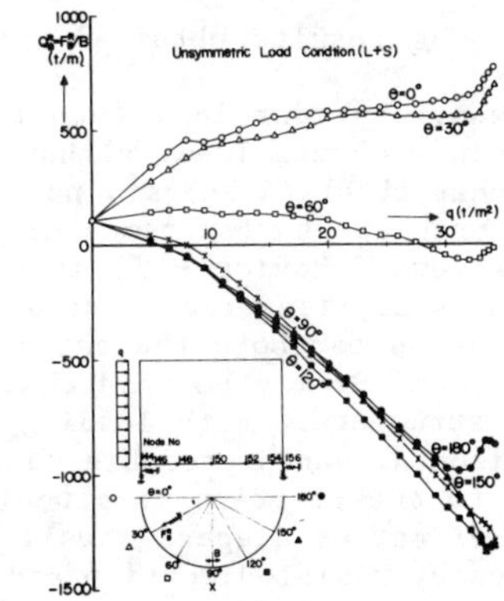

Fig-9 Radial-Direction Spring Reaction

3.5 Thermal Stress

In both cases, loading at two opposing sides and loading on one side, circumferential bending moment predominates in particular, and this is thought to be greatly involved with the failure properties of the structure. Therefore, the resultant stresses Mθ and Nθ at this part were forcused on and the variations under increased earth pressures during earthquake were observed. Fig-10 shows both cases, (L+T+S) and (L+S), including and not including thermal load history, and it is attempted to evaluate the difference between these as thermal stress. It is thought that bending moment at first ($q=0$ t/m^2) is almost all due to thermal stress, and through the growth of

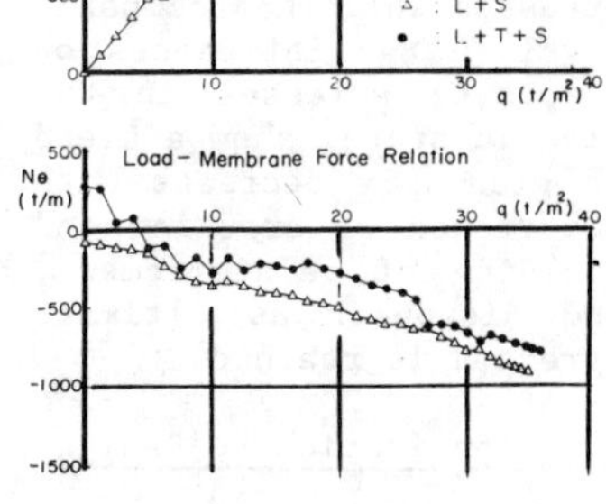

Fig-10 Load-Circumferential Resultant Stress

cracks the difference between the two is gradually narrowed, and it may be said these will more or less coincide in the end. This indicates that thermal stress disappears as cracks grow and reinforcing bars yield. For membrane force N_θ, although a difference in behavior is seen intermediately, the difference practically disappears at maximum load. These facts suggest that when ultimate strength is considered by assuming a failure mechanism, the influence of thermal stress can be ignored.

4. Conclusion

4.1 Summary of Analytical Results

The results obtained through these analyses are itemized as follows:

. Behavior under long term load - Under a permanently acting load including thermal load, although maximum stress σz of concrete at bottom reaches about 45 kg/cm^2, no cracks occur because of increase in tensile strength due to low temperature.

. Failure Behavior - It is observed that deformation at the top where load is applied predominates in accordance with circumferential bending moment M_θ for both the cases of loading from two sides and loading one side, and it may be said that this is greatly involved with failure of the structure. With loading from one side where the reaction to total horizontal earth pressure must be taken up by a diaphragm wall, failure due to the reaction is conceivable at the 90 to 180° part of the base. The effect of diaphragm wall embedment was replace in the present analyses by radial-direction and vertical-direction springs. Consequently, reaction to bulging was most apparent at the sides of the base, and it may be said that reaction to total horizontal force was concentrated at a radial-direction spring at the face opposite to that which was loaded.

. Strength - It was learned that ultimate strength against increased earth pressure during earthquake based on the seismic coefficient method reaches a level approximately three times greater than design load and that ultimate strength increases when variations in physical properties according to temperature are considered.

. Thermal Stress - Thermal stresses which had comparatively large influences on composite stresses in the elastic state, show a trend to gradually decrease with occurrence of cracking and yielding of reinforcement, and disappear as ultimate strength is reached.

4.2 Application to Design

A series of studies had been made using the ap-

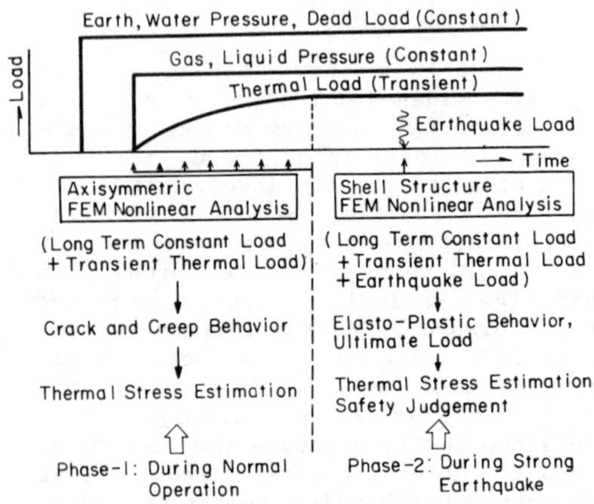

Fig-11 Analytical Study Approach on LNG Storage Tank

proach of Phase 1 and Phase 2 shown in Fig-11. The report of Ref [6] corresponds to Phase 1, which represents a load condition assuming a period of normal operation. This report corresponds to Phase 2 and is an analysis of a shell structure assuming action of a strong earthquake in a condition when permanent loads are also acting.

Self-restraining type stresses, of which thermal stress is representative, may be dependent on the load combination when the behavior of a structure in the nonlinear range is considered. Particularly, when combination with an earthquake load which is governing in design is considered, it is expected that thermal stress will be greatly affected, and therefore, considering this as fixed will not fit the actual situation. Permanently acting loads including thermal load act as long-term loads, and how the behavior of a structure is grasped and how the structure is designed against seismic load in such a situation is a matter of great importance.

With a current design method in which design force is determined based on elastic analysis values, it is difficult to rigorously grasp a stiffness-dependent stress such as thermal stress. It is also difficult to clearly determine safety with regard to the strength of a structure. These matters become more troublesome the more that the indeterminate degree of the analytical structure is increased, and the more that load combination becomes complex.

The current allowable stress design method which makes examinations based on standardized safety factors taking into consideration various indefinite factors may itself be a well-balanced design method. Today, however, when it can be expected that computer technology will become more and more highly advanced, it is thought that it has now become possible for designing to be done using a nonlinear analysis program such as described in this report. It would be rational to design based on a better understanding of the behavior of structures.

It is thus that a design method using the nonlinear analysis program for shell structures or axisymmetric structures described in this report was conceived as one measure of the design method shown in Fig-12. The procedure of this design method is as given below and is predicted on the assumption that the storage tank will never become completely empty of LNG liquid, and in effect, the temperature of the atmosphere inside the tank will be constant.

(1) Primary Design

Reinforcement arrangement calculations are made for combined design stresses (elastic stresses) including thermal stress estimated by a suitable stiffness evaluation method.

(2) Secondary Design

Nonlinear analysis is carried out on the above cross section, and then the cross section is changed based on the result obtained. In this case, it is necessary, for example, for studies according to load condition to be made as follows:

[During Normal Operation]: Axisymmetric rotation analyses are carried out on various conceivable load combinations and appropriate safety for serviceability with regard to response value is obtained.

[During Medium and Weak Earthquakes]: A condition combining seismic force of about the degree of design load according to the current seismic design method and long term loads is assumed. Shell structure analysis as adapted in this report is performed and appropriate safety from the standpoint of serviceability and ultimate strength with regard to response values is secured.

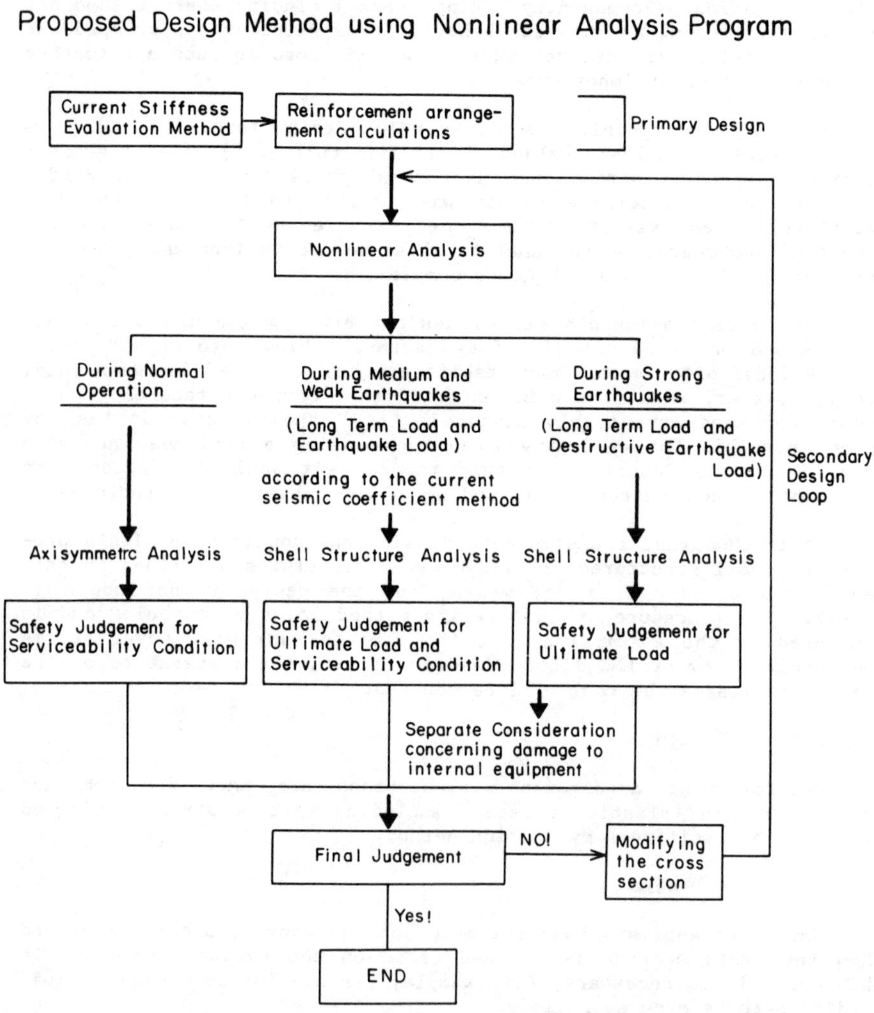

Fig-12 Design Flow using Nonlinear Analysis Program

[During Strong Earthquakes]: A similar shell analysis is carried out for a destructive earthquake load and appropriate safety from the standpoint of ultimate strength is secured with regard to response values. In this case, there will be a necessity for separate consideration to be given espetially to damage to internal equipment such as membranes and insulation.

In the event appropriate safety cannot be obtained for the various structural parts in the above process, it will be necessary to obtain a more rational cross section by modifying the cross section and by making calculations within the secondary design loop.

References:

[1] H. Ohuchi et al.: Experimental and Analytical Study of Inelastic Behavior of a Cooling Tower Model Subjected to Horizontal Load, Proc. of JSCE, No.266, Oct. 1977, pp.38-50 (in Japanese)

[2] H. Kupfer, H.K. Hilsdrof and H. Rusch: Behavior of Concrete under Biaxial Stresses, ACI Journal, pp.656-666, Aug. 1969.

[3] M.J. Mikkola and W.C. Schnobrich: Material Behavior Characteristics for Reinforced Concrete Shells Stressed Beyond the Elastic Range, Civil Engineering Studies, SRS, No.367, University of Illinois, Aug. 1970.

[4] Japan Gas Association: Recommended Practice for Underground LNG Storage Tanks (Draft), Feb. 1979 (in Japanese).

[5] K. Okada et al.: Bending Characteristics of Concrete Members at Low Temperatures, Concrete Journal, Vol.15, No.11, 1977, pp.9-20 (in Japanese).

[6] T. Nomura, H. Ohuchi, T. Takeda: Thermal Stress Analysis of Reinforced Concrete Low Temperature Storage Tank, Colloquium on Finite Element Analysis of RC Structures, Dec. 1984 (in Japanese).

PAST, PRESENT AND FUTURE DEVELOPMENTS

by Alex C. Scordelis[1], F. ASCE

ABSTRACT

The past history, present status, and future developments of the finite element analysis of reinforced and prestressed concrete structures are reviewed. Past history is divided into two periods; 1967 to 1976 and 1977 to 1985. Early (1972) recognized complicating factors, needed experimental research and future studies are summarized and important events and publications are described. The present status is reviewed and the questions and uncertainties on various topics in this research area, which need to be answered before the results are widely accepted for design applications are discussed. Finally, needed future studies and developments are discussed in terms of specific recommendations for implementation into design and identification of new research areas.

1. INTRODUCTION

Reinforced and prestressed concrete designs have been widely used for buildings, bridges, offshore oil platforms, nuclear containment vessels, and many other structures. The design of these structures must satisfy the requirements of safety and serviceability. While this can be accomplished in most cases by following approximate or empirical procedures prescribed in codes or recommended practices, it is desirable to have refined analytical models and methods available, which can trace the structural response of these structures throughout their service load history and under increasing loads through their elastic, cracking, inelastic and ultimate ranges.

Such refined analytical methods, after having been verified by selected experimental results, may be used to study the effects of important parameters in a systematic way to provide a firmer basis for the codes and specifications on which usual designs are based, or they may be used directly in the analysis and design of unusual and complex structures.

For the purposes of structural analysis, a reinforced or prestressed concrete structure can be analytically modelled as a system of joints (nodal points) interconnected by discrete structural elements made up of concrete, reinforcing steel and prestressing steel. The objective of the analysis may be stated simply: given the joint loading, prestressing and environmental history, the geometry of the structure (location of the joints) and the stiffness properties of the

---

[1]Prof. of Civil Engrg., Univ. of California, Berkeley, Calif.

structural elements; find the resulting joint displacements and the internal stresses in the concrete, reinforcing steel and prestressing steel within each structural element at any time up to ultimate failure. In its broadest sense, this is the definition of the finite element analysis of structures. Depending upon whether the objective is a global, regional or local analysis, the discrete structural elements may be large, such as beam, column, or panel type elements taken from an assumed rigid frame structure (Fig. 1a, b), or small, such as plane stress, plate bending, shell or three dimensional solid elements (Fig. 1c, d, e, f, g), taken from an assumed continuum structure.

The purpose of this paper is to present the past history, present status and future developments for the finite element analysis of reinforced and prestressed concrete structures. No attempt will be made to cite the hundreds of papers that have been written on this subject. Instead, only the author's views on the subject will be presented with respect to developments in North America and Europe.

2. PAST HISTORY

2.1 Period from 1967 to 1976

The earliest important applications of the finite element method to concrete structures was by the Berkeley group during the late 1950's and early 1960's in the analysis of large mass concrete dams being designed and built in the United States at that time. Cracking, time dependent, thermal and sequence of construction effects were traced in the analysis of these plain concrete structures. However, the earliest published application of the method to reinforced concrete structures was by Ngo and Scordelis [1] in March 1967. Beams with predefined "discrete cracks" (Figs. 2, 3) were studied using a plane stress analysis in which the concrete and steel reinforcement were represented by two-dimensional triangular finite elements or axial bar elements, and special linkage elements (Fig. 3) were used to connect the steel to the concrete. Concepts of fracture mechanics to predict stress concentration and extensions at the crack-tip were ignored in this early work.

Instead of using predefined discrete cracks, as in Figs. 2, 3, later studies traced the discrete cracks individually by progressively splitting nodes or elements and thence defining a new topology for the next analysis step. Concepts of fracture mechanics were added at a later date.

Because of the complications involved in using discrete crack modelling, the "smeared crack" concept was then introduced in which progressive cracking is assumed to be distributed over an entire element or at integration points within an element. This has the advantage of permitting the use of the same structural nodal point topology throughout the entire nonlinear solution. In general, the reinforcement is assumed to be distributed throughout the concrete element to form a composite stiffness. Full bond is generally assumed between this reinforcement and the concrete. In some cases the concept

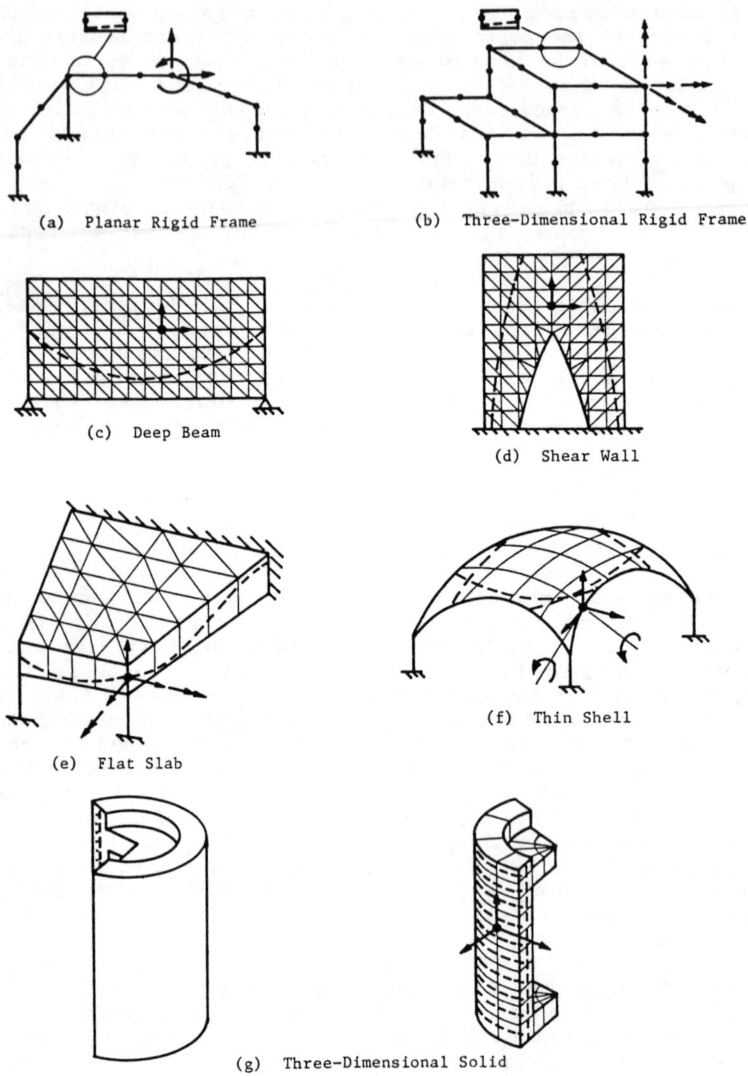

Fig. 1 Reinforced and Prestressed Concrete Finite Element Systems

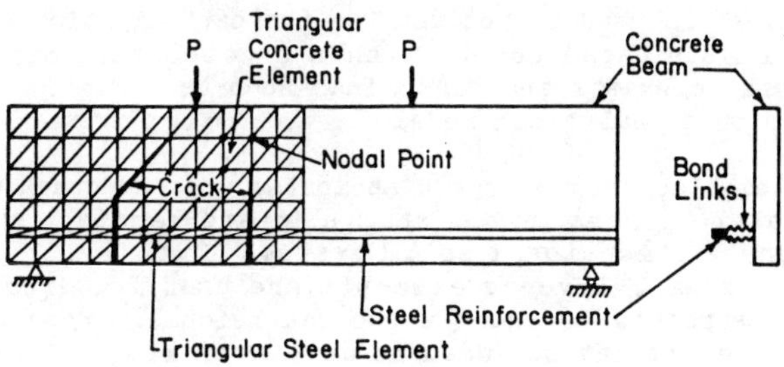

Fig. 2 Original Analytical Model for a Beam [1](1967)

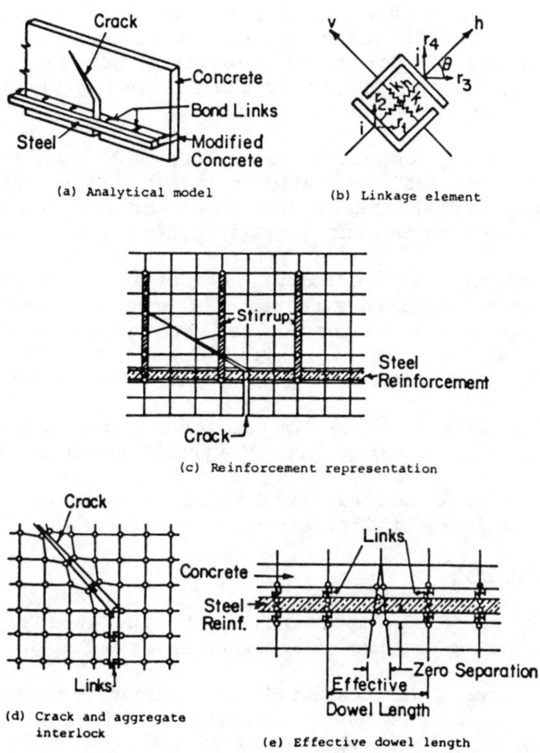

Fig. 3 Analytical Models and Linkage Elements [1](1967)

of "tension stiffening" is used to account for the participation of the concrete between the distributed cracks within an element. Separate reinforcing bar finite elements and bond links such as shown in Fig. 3 may also be added to this analytical model.

Over the past eighteen years, the distributed smeared crack model has been used for plane stress, plane strain, plate bending, shell, axisymmetric and three dimensional solid systems (Fig. 1). For plate bending and shell systems, layered elements are used in which each concrete layer is assumed to be under a two dimensional stress state. Distributed reinforcement may be added to any layer also. Cracking progresses through the thickness, layer by layer. An extensive family of elements has also been developed for axisymmetric and three dimensional solids in connection with the analysis and design of reinforced and prestressed concrete nuclear reactor vessels (PCRV), containment vessels and offshore oil concrete platforms.

State of the art papers which review much of research done in North America and Europe from 1967 to 1976 have been published by Scordelis [2], Schnobrich [3, 4] and Wegner [5]. It is interesting to note that Scordelis [2] in his state of the art report on "Finite Element Analysis of Reinforced Concrete" presented in June 1972 at a conference in Montreal, Canada, presented the following summary of important items at the time (1972) on three subjects.

Complicating Factors: (1) composite material; (2) cracking of concrete; (3) nonlinear constitutive relationship; (4) failure criteria for concrete; (5) creep and shrinkage; (6) load and environmental history; and (7) bond, dowel action and aggregate interlock.

Needed Experimental Research: (1) constitutive relations and failure criteria for concrete under combined stresses; (2) creep and shrinkage under combined stresses and various temperature and humidity conditions; (3) bond stress-slip relationship; (4) tension stiffening in cracked concrete; (5) aggregate interlock; and (6) dowel shear.

Future Studies: (1) modelling by three dimensional solid elements; (2) static, cyclic and reversible loading; and (3) dynamic response.

Much research effort has been expended to study the above items since 1972, but many questions still remain to be resolved.

2.2 Period from 1977 to 1985

From 1977 to 1985 several important events and publications occurred, which gave a strong impetus to research in this field.

In August 1977 a proposal was submitted to the Executive Committee of the Structural Division of the American Society of Civil Engineers requesting the formation of a Task Committee on Finite Element Analysis of Reinforced Concrete Structures. A Committee with twenty members under the chairmanship of Arthur Nilson was appointed with the stated purpose to review, correlate, develop and promote information and research on the finite element analysis of reinforced concrete and masonry structures, and to study proposed analytical models, theories, and experimental data, so as to make appropriate recommendations for their application in design.

A five year intensive effort by the Committee resulted in a comprehensive 545 page state-of-the-art report published in May 1982 [6] with the following Chapters: (1) Introduction; (2) Constitutive Relations and Failure Theories; (3) Modelling of Reinforcement and Representation of Bond; (4) Concrete Cracking; (5) Shear Transfer; (6) Time Dependent Effects; (7) Dynamic Analyses; (8) Numerical Examples and Applications; and an Appendix Compilation of Finite Element Programs for Nonlinear Analysis of Reinforced Concrete Structures.

In addition to the above report, the proceedings [7, 8, 9] of the following three international conferences held in Europe contain many recent papers on the subject:

1. International Symposium on Nonlinear Behavior of Reinforced Concrete Spatial Structures, July 3-7, 1978, Darmstadt, Federal Republic of Germany.

2. International Association for Bridge and Structural Engineering Colloquium on Advanced Mechanics of Reinforced Concrete, June 2-6, 1981, Delft, The Netherlands.

3. International Conference on Computer Aided Analysis and Design of Concrete Structures, September 17-21, 1984, Split, Yugoslavia.

3. PRESENT STATUS

3.1 General Remarks

While extensive research on the finite element analysis of reinforced concrete structures has been conducted over the past eighteen years, its implementation in the design of actual structures has been disappointing. Present designs are still generally based on determining internal forces and moments, by elastic analyses in which the reinforced or prestressed concrete system is assumed to be uncracked, homogeneous, isotropic and linearly elastic. Concrete and steel requirements for these elastic internal forces and moments are then designed on the basis of ultimate strength. Where deflections due to cracking or time dependent effects are considered at all the deflections are often taken as simple multiples of the elastic deflections. Even such complex structures as large offshore concrete platforms, thin concrete shell roofs or hyperbolic cooling towers, nuclear containment vessels, or segmentally erected prestressed concrete bridges are primarily analyzed and designed in this manner, although in some cases more refined nonlinear analyses are used for special problems in these structures. Computer programs for the nonlinear finite element analysis of reinforced or prestressed concrete structures have not been used widely for a variety of reasons: (1) there is a lack of dependable and available general programs for this purpose; (2) the computer analyses are time consuming, expensive, difficult to apply and to interpret the results in most cases; (3) there is a lack of consensus among the experts as to the correct input to model the many nonlinear phenomena indicated earlier; and (4) because of these uncertainties and the fact that experimental verification of the analyses have been obtained only for simple structures and loadings, there is a lack of confidence that they can accurately predict behavior and quantitative results for the complex problems that they are especially suited for.

Future research and development in this field must address the above problems if the method is to become a useful part of the design of structures, rather than simply a research tool for studying basic phenomena and interesting fundamental, but essentially academic problems.

3.2 Questions and Uncertainties

At present there are a number of questions and uncertainties that need to be answered before full confidence in the results of the nonlinear solutions can be achieved. Researchers must study and clearly answer these so as to come to an agreement on how best to implement their findings in design applications. An extensive discussion of many of these questions is given in Reference [6].

Some of these and additional questions are listed below under various topic headings.

3.3 Analytical Models and Finite Elements

1. A variety of these have been proposed. What are the important factors in making an appropriate selection for a particular type of structure or problem?

2. For global analyses, can larger elements or subsystems be used which capture the dominant nonlinearities in special classes of structures with a greatly reduced number of degrees of freedom?

3. What is the effect of size, shape and aspect ratio of the elements on the numerical results obtained? Does the solution converge to the same result, with refinement of mesh irrespective of element shape and layout and, if not, what criteria should be specified to get the best results?

3.4 Constitutive Relations and Failure Theories

1. Many of these have been proposed for plain concrete. Do they give significantly different results when implemented into the analysis of a reinforced concrete system? If so, why? If not, which is the simplest to use for various types of structures or problems?

2. Can the sum of the concrete and steel contributions taken independently model the response of the true composite system?

3. What is the effect of complex stress histories, cyclic loading, nonproportional loading, changing orthotropic directions and confinement of the concrete on the final significant results?

3.5 Cracking, Tension Stiffening, Aggregate Interlock, Dowel Action and Bond

1. Concrete cracking can be modelled using smeared cracks or discrete cracks with or without fracture mechanics concepts. How should a choice be made for application to a particular problem?

2. Tension stiffening of concrete between cracks has been accounted for after cracking either in the concrete, by a descending portion in the tensile stress-strain relationship, or in the steel, by using an increased equivalent steel modulus of elasticity. Which of the two should be used and why?

3. An additional complication after cracking is the problem of defining the shear resistance parallel to the crack due to aggregate interlock which is commonly accounted for in smeared crack models by using a reduced shear modulus for the concrete as a constant or some function of the crack width. Is this adequate? What is the effect on shear transfer of dowel tension and shear for a bar crossing the crack?

4. Where individual reinforcing bars are modelled as discrete elements, bond transfer is accomplished through linkage elements, between the steel and concrete, to which stiffnesses must be assigned. What is the simplest realistic constitutive relationship for these bond link stiffnesses?

5. All of the above effects are undoubtedly interrelated in an actual structure. When the effect of cyclic or reversed loading, opening and closing of cracks, etc., are added, the complexities multiply. However, in a nonlinear analysis, these effects must be considered wherever they have a significance on the prediction of results and, thus, recommendations on how this is to be done must be made.

3.6 Time Dependent Effects

1. Time dependent effects include loading, prestressing and environmental history and the time dependent relationships for creep, shrinkage, aging and temperature in the concrete and relaxation of the prestressing steel.

2. Large reinforced or prestressed concrete structures are built sequentially as cast in place or precast structures, so all of these effects occur. Often they are neglected or accounted for by very approximate methods in an analysis. While methods have been developed for including them in nonlinear finite element computer analyses, verification with experimental results is meager and the effect of upper and lower bounds of input material parameters on behavior and final results is lacking.

4. FUTURE STUDIES AND DEVELOPMENTS

4.1 General Remarks

It is evident from the preceding section that while much progress has been made on many of the questions identified back in 1972 [2], extensive research has raised other uncertainties regarding these questions. There is certainly a need in the future to continue basic analytical and experimental research on these questions to gain a better understanding of these phenomena. However, if this research is to be implemented and result in improved designs, then the researchers who are most knowledgeable in this field must develop recommendations for this implementation. This should be a major goal over the next few years. In addition, researchers in this field should identify new areas of research not yet studied which may have a direct impact on improved design of reinforced and prestressed concrete structures.

Thus future studies and developments may be summarized as follows: (1) continue basic analytical and experimental research on questions previously identified; (2) develop recommendations for implementing presently existing knowledge and methods of analysis into design applications; and (3) identify new areas of research which may have an impact on improved designs.

Item (1) has already been discussed. A few specific recommendations are given below for items (2) and (3) in the hope that participants at the U.S. - Japan Joint Seminar on Finite Element Analysis of Reinforced Concrete will discuss and improve them and contribute other recommendations.

4.2 Recommendation for Implementation into Design

1. Funds should be sought from the National Science Foundation for a cooperative research project for the purpose of securing or developing a general purpose computer program for the nonlinear finite element analysis of reinforced concrete structures. The proposal for the research project should be developed by one or more individuals knowledgeable in the subject and in computer programs, who are willing to act as Co-Principal Investigators. The program should be written in FORTRAN language in a modular fashion so that implementation of any of the analytical models, elements, material laws, etc. discussed in this paper would be possible. Specifications for the program should be developed, by the Co-Principal Investigators and reviewed by the ASCE Task Committee on Finite Element Analysis of Reinforced Concrete Structures. Every effort should be made to utilize existing appropriate computer software. Upon completion of the program and complete documentation, it should be publicized and made available to any interested party at the nominal cost of reproduction, handling and mailing.

2. A series of carefully selected numerical examples should be secured or developed which could be used as test cases for verifying and comparing numerical results obtained in the program using different analytical models. Two types of examples should be sought: (1) simple examples which test various parts of the program independently; and (2) comprehensive design examples, which test the ability of the analysis to predict the behavior of total structures, involving many parts of the program.

3. Existing experimental data should be sought or new experiments

proposed for verification of the two types of problems cited in paragraph 2.

4. Examples of existing or proposed structures for which the nonlinear analysis capability developed would be an important design aid should be sought, studied, described and publicized in engineering journals and magazines.

4.3 New Areas of Research

While the following areas have received some attention, additional applications of nonlinear finite element analysis methods for reinforced and prestressed concrete structures (FEARC) to study them should be considered for future study.

1. Dynamic, shock and impact loadings. Many important and costly concrete structures must be designed for these loadings. Analytical models and material input parameters are needed to predict their response by FEARC.

2. High strength concrete and effects of confinement. Compressive strengths of over 20,000 psi with ultimate strains of 8% have been obtained experimentally. Can FEARC be used to predict these experimental results and if so it should be used to make parameter studies for recommendations in design.

3. Combination of nonlinear material, geometric and time dependent effects. Additional analytical and experimental studies of structural systems such as hyperbolic cooling towers and thin shell roofs are needed to verify FEARC prediction of their structural response up to failure under these effects.

4. Nonlinear global, regional and local analyses of segmentally erected reinforced and prestressed concrete structures. Most large cast-in-place or pre-cast structures such as buildings, bridges, offshore concrete platforms, cooling towers, etc. are built segmentally over an extended time period. A unified segmental method of analysis, which could incorporate nonlinear material, geometric and time dependent effects using different types and sizes of finite elements for the global, regional and local analyses of the different types of structures would be a powerful design aid.

5. CONCLUSIONS

The past history, present status and needed future studies and developments of the finite element analysis of reinforced and prestressed concrete structures have been reviewed. Much progress has been made in research, but questions and uncertainties still remain. While basic research on these questions and in new areas should be continued, a strong future effort should be made to implement existing knowledge into design applications.

REFERENCES

[1] D. Ngo, A. C. Scordelis, "Finite Element Analysis of Reinforced Concrete Beams", ACI Journal, V.64, No.3, March, 1967.
[2] A. C. Scordelis, "Finite Element Analysis of Reinforced Concrete Structures", Proceedings of the Specialty Conference on Finite Element Method in Civil Engineering, Montreal, Canada, June 1-2, 1972.
[3] W. C. Schnobrich, "Finite Element Determination of Nonlinear Behavior of Reinforced Concrete Plates and Shells", Proceedings of the Symposium on Nonlinear Techniques and Behavior in Structural Analysis, Transport and Road Research Laboratory, United Kingdom, 1974.
[4] W. C. Schnobrich, "Behavior of Reinforced Concrete Predicted by Finite Element Method", Proceedings of Second National Symposium on Computerized Structural Analysis and Design, George Washington University, Washington, D.C., March, 1976.
[5] R. Wegner, "Finite Element Models for Reinforced Concrete", Proceedings, U.S. - Germany Symposium on Formulations and Computational Methods in Finite Element Analysis, M.I.T., Boston, August 8-13, 1976.
[6] ASCE Task Committee on Finite Element Analysis of Reinforced Concrete Structures, State-of-the-Art Report on "Finite Element Analysis of Reinforced Concrete," ASCE Special Publications, 1982.
[7] Proceedings of IASS Symposium on Nonlinear Behavior of Reinforced Concrete Spatial Structures, Vols. 1, 2, 3, Darmstadt, Federal Republic of Germany, July 3-7, 1978.
[8] Final Report of IABSE Colloquium on Advanced Mechanics of Reinforced Concrete, Vol-Band 34, Delft, The Netherlands, June 2-6, 1981.
[9] Proceedings of International Conference on Computer Aided Analysis and Design of Concrete Structures, Parts 1, 2, Split, Yugoslavia, September 17-21, 1984.

FUTURE PROSPECTS FOR FINITE ELEMENT ANALYSIS OF REINFORCED CONCRETE STRUCTURES

by

Hiroyuki Aoyama[1], M.ASCE and Hiroshi Noguchi[2], M.ASCE

ABSTRACT: Researchers and designers' views of the future prospects for finite element analysis of RC structures are summarized by compiling the replies submitted to the JCI-RCFEM Committee questionnaire. First, the replies concerning what we aim at in FEM analysis and the problems to be solved and future objectives of development of FEM computer programs are introduced. As for the relationship between FEM analysis and experiments, some strategies in case of disagreement of analytical results with experimental results and some opinions on coordinating FEM analytical and experimental programs are presented. Furthermore, suggested future courses for FEM analysis development are discussed, both for research and design applications. Finally, suggestions for information exchange on FEM analysis of RC structures are made.

1. INTRODUCTION

Eighteen years have passed since Ngo and Scordelis first analyzed reinforced concrete (RC) beams with the finite element method (FEM) in 1967. Studies on finite element analysis of RC structures are now entering a new stage in which the analytical method itself is reconsidered with a view to developing design equations rather than the earlier emphasis on aiming for agreement with experimental results. In these circumstances, the recent studies are going to be reported in ten sessions in this joint seminar. It is beyond the authors' capabilities to compile all these studies and mention the future studies and applications to the design problems. But by chance results of a questionnaire on the problems to be solved and future prospects for finite element analysis of RC structures which was conducted by the authors under the auspices of Japan Concrete Institute Committee on Finite Element Analysis of RC Structures (JCI-RCFEM Committee) in August, 1984 are now available. Though it had a descriptive style and very abstract contents, valuable replies were submitted not only by members of the JCI Committee but also by experimental researchers.

---

[1] Prof. of Struct. Engr., Dept. of Arch., Univ. of Tokyo, 7-3-1 Hongo, Bunkyo-ku, Tokyo, 113 Japan.

[2] Assoc. Prof. of Struct. Engr., Dept. of Arch. Engr., Chiba Univ., 1-33 Yayoi-cho, Chiba, 260 Japan.

In this paper, the views of researchers and designers on future prospects for finite element analysis of RC structures are compiled and reported.

2. WHAT DO WE AIM AT IN FINITE ELEMENT ANALYSIS OF RC STRUCTURES?

It is very important to consider what we aim at in finite element analysis of RC structures. Our aims will obviously strongly influence our future prospects.

The replies on that question submitted to another questionnaire on the investigation of FEM computer programs in Japan by Okamura and Maekawa under the auspices of JCI-RCFEM Committee in April, 1984 are arranged as follows:

	Number of Replies
Development of rational design methods and design equations	 11
*To deepen understanding from both aspects of experiments and analyses and reflect it in the macro models.	
*To understand the real behavior of an overall structural system under complicated load history and load combination is important in order to aid rational design. FEM analysis is a useful tool for this purpose.	
*To discuss the confinement effect of shear reinforcement of columns under cyclic loading and contribute to the development of rational design methods.	
*To grasp the restoring force characteristics and stress distribution.	
*To use for structural analysis in the design of arbitrarily shaped RC structures.	
*To use for experimental programs for shear walls and shell structures.	
It is possible to grasp the force flow and local stress intensity which are not given by experimental data	 6
Shear failure mechanism	 3
Development of simple mechanism models	 2
Development of RC finite elements: reinforcement + concrete	 1

Simulation up to ultimate strength under monotonic loading and analytical determination of behavior under cyclic loading 1

To use FEM Analysis as the estimating method of experiments and connect with the proposal of rational design methods 1

Bond characteristics in members (Bond deterioration near cracks) 1

To investigate the characteristics of FEM as compared with other discrete models (B.E.M.) 1

In the questionnaire on the problems and future prospects, a more concrete view was requested and given as follows:

Okamura mentioned (1) Improvement of macro model for shear failure, etc., and (2) Pursuit of mechanism of bond stress-slip relations.

Maekawa classified analytical programming methods into the following two large groups:

1) Limiting the subject of analysis, parametric analyses are carried out using efficient and simple models, and the results are used for the design.

2) The macro model is made by estimating the mechanism of RC structures synthetically and correctly by the refined micro model.

In the latter method, he indicated that it is important not to make the main assumption halfheartedly and to face the phenomenon straightforwardly.

Ichinose mentioned the following three items as the objective of FEM programs:

1) **Research-directed type:** The failure mechanism of members is made clear. The results are reflected in the macro models in order to deduce the design equation. (Yamada's opinion is the same. This is similar to Maekawa's opinion 2).

2) **Development-directed type:** Development of design method of reinforcement arrangement by numerical experiments.

3) **Design-directed type:** Required amount of reinforcement is obtained for complicated structures or a portion of them.

It seems that the researchers on FEM analysis have a tendency to aim at type 1, and the designers aim at types 1 and 3. But the development-directed type 2 is also important for reinforced concrete structures. As the reinforcement details have a considerable effect not only on the strength of members but also on the ductility, it is necessary to clarify the relationship between the ductility and the reinforcement details for the development of the ultimate strength design method. The studies on the development of reinforcement arrangement have often depended on experiments, but analytical studies such as Kokusho's [7] and Noguchi's [8] seem likely to increase in the future.

As for the others, Toritani and Endo mentioned that after the verification of the analytical model based on the feedback from the experiment to the analysis, preliminary parametric analysis is carried out in advance of the verification test for the aseismic safety.

Ueda and Sotomura mentioned the development of the design method by interpolating the experimental data analytically.

Shirai and Inoue mentioned the interpretation of the plasticity analysis and the physical meaning of previous experimental equations.

Endo mentioned that the materials for the structural design are made from the analysis of the behavior of large complicated structures, which is not given by the experiments.

It seems that there is a little difference in viewpoint between researchers and designers. The major points are understanding of the micro mechanism for the researchers and parametric study for the designers.

3. PROBLEMS TO BE SOLVED AND FUTURE OBJECTIVES IN THE DEVELOPMENT OF FEM COMPUTER PROGRAMS OF RC STRUCTURES

Replies to the questionnaire on the problems to be solved and the future development objectives for FEM computer programs are summarized as follows:

	Number of Replies
Cyclic loading (Basic information on the modelling of the restoring force characteristics)	 12
Development and improvement of crack models	 11
*Cracking criterion under strain gradients	
*Reconsideration of shear stiffness	
*Investigation of cracking algorithm	
*Initiation of the second crack from the shear transfer and closing of the first crack	
*Separation of dowel action from aggregate interlock	
Investigation of failure mode	 5
*Shear failure mechanism of shear walls with high reinforcement ratio	

*Shear sliding failure

*Introduction of fracture mechanics

Confirmation of ultimate strength by the displacement incremental method (Analysis of strain-softening behavior) 4

Bond 3

Reconsideration of material models 3

*It is indefinite how to decide the material properties of concrete in FEM model

*Development of RC element

Three-dimensional analysis (Analysis of the effect of confinement reinforcement on ductility) 2

Massive concrete (Consideration of properties at early age) 1

Nonlinear analysis of a structure including foundation and ground 1

Improvement of accuracy of current analytical models 1

Solution of large-scale structures by analytical models with fewer degrees of freedom 1

The most numerous replies for problems to be solved are on cyclic loading and development and improvement of a crack model. As a crack model is also very important for the opening and closing of cracks in cyclic loading, it is the most important problem to be solved to deal with the concrete element after cracking in FEM analysis of RC structures.

There are indications that the problems under cyclic loading are matters of primary concern in order to contribute to the development of the rational aseismic design method in our country.

There are some opinions on the introduction of fracture mechanics into the cracking criterion and other factors. For example, Shirai indicated that the concept of fracture mechanics should be introduced and the equivalent strength criterion can be obtained from the investigation of failure behavior. He also mentioned that it will be difficult to get further new information superior to the experimental data, as long as the previous strength criterion is used in the analysis. This opinion is very interesting, because the introduction of fracture mechanics will force reconsideration of previous analytical models and will be useful for deepening the understanding of structural behavior.

4. FEM ANALYSIS AND EXPERIMENTS

Most of the researchers and design engineers indicated that the relationship between analysis and experiments is important in FEM analysis of RC structures, as described in Chapter 2. There are many varieties of the relation between analysis and experiments. For example, in the preliminary analysis for making an experimental program and the analysis for the interpretation of the experimental results, the main stress is laid upon the experiment. When experiments on a few specimens are conducted for the verification of the analytical results, the stress is laid upon the analysis. There are also some cases which give equal emphasis to both. As it is necessary for FEM analysis and experiments to make up for each other's deficiencies, the relationship between analysis and experiments will be discussed.

4.1 On Strategies to be Used in Case of Disagreement between Analytical Results and Experimental Results

Researchers often try to re-analyze and obtain a solution closer to the experimental data, when the analytical results do not agree with the experimental results. In this case, only the final solution closer to the experimental results is usually reported in the technical paper. Therefore, it is unknown to the readers in what manner the good agreement has been obtained. But this process is very important for analytical researchers. The experience on this process gives a great help in understanding both the insufficient points of the analytical models and the behavior of the test specimens more clearly. When this process is stopped halfway, the analysis will not be given life.

According to the results of the questionnaire, almost all of the researchers investigate the material properties of the analytical models, the boundary conditions, finite element idealization, the constitutive laws of the materials and others together with a check of the numerical computational method.

In this investigation, Okamura mentioned that at first he tries to examine if the concrete element model accurately describes the real state after cracking. Inoue also mentioned that he investigates the dominant factors on the characteristics of the analytical object and re-examines the corresponding assumptions and criteria. Ueda indicated the following four items: (1) tension stiffening effect, (2) shear transfer across cracks, (3) constitutive laws of cracked concrete, and (4) limit strain at shear failure, as the dominant material characteristics of shear walls from his own experience in the analysis.

The estimation of the dominant factors in each analysis is based mainly on the researchers' experience in the analysis. If the investigation process is not published, it is feared that the recognition of the factors is limited to each researcher. It will be necessary for us to pool our own analytical experiences on the dominant factors on the behavior of each structural member and discuss actively and try to determine quantitatively the factors on which most researchers have the same view to a certain extent, as Tanaka mentioned.

4.2 Coordination and Balance between Analytical and Experimental Programs

Researchers often compare the predictions from a FEM analysis for the load deflection curves, strains in reinforcing bars and concrete, bond stress and slip distribution of reinforcing bars and crack patterns with the corresponding experimental results. In previous studies, much emphasis was placed on how well the analytical calculations agreed with the experimental results. Because of this emphasis on comparisons, investigations of analytical results dealing with changes in the principal stress flow, yielding of reinforcing bars and compression failure of concrete, which are difficult to compare with experimental results, have been somewhat neglected.

Many of the replies to the questionnaire stress the necessity to improve analytical accuracy near ultimate strength as an important problem on the analytical side. For this purpose, it is necessary to be able to analyze even in the strain-softening region beyond the maximum strength with some accuracy and to investigate the definition of failure of concrete elements, especially shear and shear sliding failures and the relationship between local failure and the overall failure of the member.

With regard to experimental results and macro model analytical studies, Takiguchi and Ichinose indicated that it is important to consider what information is available from the experimental side for comparisons with the FEM analytical results, and what is desired in the discussion of the FEM analytical results from the macro model analytical side. They propose to obtain the detailed deflection mode from the experimental results as a result of considerations on the deflection measurement method. They mentioned that it is possible to examine quantitatively the verification of the strain and stress distributions in the macro model. This method will be applicable to the strain and stress distribution of FEM analytical results. Hereafter we hope that such a quantitative investigation with the experiment and macro model analysis will be deepened in order to examine efficiently the large amount of information obtained from a FEM analysis.

In the JCI Colloquium on Shear Analysis of RC Structures (1982)[9], Kaku indicated in the discussion that it is important in the development of a well-balanced study for the analytical researchers to carry out the basic experiments by themselves on the unsolved points in the analysis, because the analytical researchers have a tendency to extend the scope of the application of the analytical model without a full understanding of the limited scope of application of the experiment which was a basis for the analytical model. Blaauwendraad also indicated in IABSE Colloquium (1981)[2] the necessity to fill a gap between researchers on concrete (experiment) and mechanics (analysis).

Most of those who answered the questionnaire agree on that point. It is proposed that an experimental researcher and an analytical researcher form a pair in each group and exchange information to carry out cooperative research and clarify what data is necessary in the experiment for the analysis and the prior condition, assumption and the scope of application in the experiment in the technical report. Okamura, Maekawa and Shirai indicated the necessity of FEM analysis in basic research, especially in the university, and the preparation of the circumstances where the basic experimental study, corresponding to the development of FEM analysis, is appraised high. It seems that there are not enough experimental data on the element which form the basis of modelling of FEM analysis in our country.

In contrast with that, Endo indicated the uncertainty of the reproducibility in the experiment and mentioned that it is not necessary to wrestle with the comparison with the experimental results and we should lay emphasis on the analysis. It may be necessary for us to take this opinion into consideration.

Many Japanese FEM analytical researchers, especially the university researchers, pursue experimental studies together with analytical studies. As for the relationships between FEM analytical studies and experimental experiences, there are many opinions that good analytical models are not made if the analytical researcher does not know the experiment, but it is not always necessary to have direct experience. In this case, the analytical researcher should participate actively in visiting laboratories and carefully observing experimental phenomena. In contrast with this, as a minority opinion, Yamada indicated that if the researcher has too much experience, he risks being particular about secondary phenomena and overlooking the essential mechanisms of the phenomena.

There are many unsolved problems, both analytical and experimental, such as the tensile strength of concrete in cracked RC members. Shirai mentioned that we must investigate by making up for each other as the solution only from analysis or experiment is difficult. For this purpose, it will be necessary for FEM analytical researchers to always pay attention to the relationships between analysis and experiment and investigate what kind of information is necessary from the experiment about the unsolved analytical problem. It is then necessary for them to carry out the basic experiment by themselves for the solution and solve the problem systematically keeping close communication with the experimental researchers.

5. ON OUR COURSE FOR THE FUTURE IN RESEARCH ASPECTS OF FEM ANALYSIS

The results to the questionnaire on the future prospects in research aspects of FEM analysis of RC structures are compiled and discussed in two parts: the short term of the next few years, and the longer term of 5 to 10 years.

5.1 The Next Few Years

The main items of investigation for the short term in research aspects are as follows:

	Number of Replies
Rationalization and verification of analytical models (especially cracked concrete elements). Investigation with emphasis on the important part of analytical models	6

To deepen the understanding of the behavior of RC structures by analytical models and connect the results to the macro design model (plastic analysis and others)	 5
To clarify the limitations of analytical models	 2
Accumulation of analytical examples of structural members	 2
Analysis of the behavior in the unstable region after maximum strength. Convergence of the solution in high stress levels	 2
Presentation of the necessary basic experiments for the development of FEM analytical models	 1
Verification of FEM analytical models by macro models	 1
Analysis of the structure level.	 1

The rationalization and verification of analytical models and the development of macro models through analytical investigations by FEM models are the predominant items.

As for the rationalization of analytical models, we should place emphasis on the factors which have a great influence on the analytical results (Maekawa). For example, are we treating the actual behavior accurately by using the smeared crack model? Is it appropriate that the failure criterion of the discrete crack model is similar to that of the smeared crack model? We hope to deal effectively with the shear problem, where only a few cracks occur near ultimate strength, by the discrete crack model. Is it possible to solve analytically the behavior at the unstable region after the maximum strength with good accuracy?

As for the development of macro models, it will be an effective method to deepen our understanding of the behavior of RC structures by comparison with experiments at the same time as we make efforts to rationalize the analytical model. Subsequently the verification and the range of application of previous macro models and experimental design equations may be investigated and rational macro design models may be proposed by the parametric analysis.

When we try to understand the behavior of RC structures, it is important to clarify the limitations of the analytical models. This is connected to the presentation of the basic experiments which are necessary for overcoming the analytical limitations, and this will more and more be the driving force to develop FEM analysis.

Those connected with design work indicated the necessity of the verification of micro models by macro models and the necessity to emphasize the analysis of structures rather than members. There is some divergence in viewpoints between researchers and designers on these points.

5.2 Long Term Objectives

The main items of investigation for the next 5 to 10 years are as follows:

	Number of Replies
Pursuit of deterioration phenomena under reversed cyclic loading	 4
*Accumulation of basic experimental data for modelling	
*Application of static analytical method to dynamic analysis	
Scale effect	 3
*Investigation for the essential pursuit of the behavior of real structures	
*Analysis of large complicated structures for which experiments are impossible	
Development of general computer programs which are not written for special members	 1
Theoretical study	 1
Does the existence of the discontinuity of cracks and the stability of the whole system guarantee the convergence of the solution?	
Understanding and formulation of material characteristics	 1
Analysis at the structure level by using the simple analytical model of the member level which is developed from FEM analysis and experiments	 1
Development of design equations by the macro model limit analysis from the numerical investigation of stress flows and failure mechanisms	 1
Pursuit of the behavior of members after strength decay	 1
Solution of a new phenomenon by the introduction of fracture mechanics.	 1

As for the long term, besides the item related to the development of macro models for the short term, the analyses under reversed cyclic loading and on the scale effect are mentioned by many researchers. As these two items are still more difficult problems, there are indications that researchers have decided to solve these problems after they have rationalized the analytical models based on the short term experiments.

The problems of predicting behavior under reversed cyclic loading are important for us to solve because we must consider seismic loading in our country and clarify the differences between strength, ductility and failure modes observed under cyclic loading and those under monotonic loading. We must also connect the static analytical method to the dynamic analytical method.

As full scale tests on large scale structures such as nuclear reactor buildings are difficult, scaled down models are now used for these tests. Hence the scale effect problem is an important subject in order to fully understand the behavior of large scale structures. Though some interesting papers on scale effect have been reported recently, the basic experiments on the material characteristics and their rational modelling will be necessary in the future.

6. ON OUR COURSE FOR THE FUTURE IN DESIGN APPLICATION ASPECTS OF FEM ANALYSIS

As Hardy Cross said over fifty years ago, "A method of analysis has value if it is ultimately useful to the designer; not otherwise." (1930) [5].

As the ultimate objective of FEM analytical studies of RC structures is to apply the research findings in actual design, this chapter's theme is the most important. According to the questionnaire results on the application to design, there is not so sharp a distinction between the short term and the long term as compared with that in research. This indicates that there are not so many study cases on the application to design and there are many problems to be solved in the longer term future.

Takiguchi indicated that as conditions before and after cracking are more important in the design than the ultimate condition, even the present FEM analytical model is useful to the designer in investigating behavior a little beyond the elastic limit.

The items for the short term which are indicated by many replies are as follows:

* Improvement of macro models

* Improvement of the reliability in the analysis of large scale structures

* To investigate the previous design equations and macro models thoroughly by FEM analysis and clarify their physical meaning and problems.

As for the improvement of macro models, we should investigate the previous macro models by the rationalization of the micro analytical models and develop RC elements which are easily applicable to design and apply them to the design of members. Subsequently based on those results, it will be necessary to aim at the simplified macro model on the member level which is applicable to the whole structure.

As for the improvement of the reliability in the analysis of large scale structures, it is necessary to improve the reliability on the analytical accuracy of each member and obtain an agreement in views on the efficiency of FEM analysis at first. (Shirai).

There are not many design equations which were derived from FEM analysis and they have not attracted special interest. Therefore, it is necessary to place emphasis also on the development of design equations in order to improve the reliability of FEM analysis. Hereupon, it is important to re-examine the previous design equations and macro models by FEM analysis at first in order to develop new rational design equations by extending the capabilities of macro and micro models, as indicated by Kanoh, Minami and Takiguchi in JCI-RCFEM Colloquium (1984)[6].

The items for the long term are given as follows:

*Analysis of large scale structures with limited degrees of freedom and reduction of number of experimental specimens for the verification of safety.

Development of a design program for RC plane element structures by smeared crack models.

Explanation of shear failure phenomenon by discrete crack models and application to design.

*Development of design equations by parametric analysis.

Investigation of behavior under combined loading such as temperature, internal pressure and lateral forces.

Development of the design method by nonlinear analytical method.

Development and publication of an accurate analytical program for special structures.

*Application of FEM to partial structures, such as a member which is critical in the design, and the design of detail.

Development of the reliability design method.

Computation of the potential strength of whole nuclear reactor buildings.

Thus there are various points of view and ways of thinking. The marked items with an asterisk may be pointed out as major and characteristic themes.

7. ON A PLACE FOR EXCHANGE OF INFORMATION ON FEM ANALYSIS OF RC STRUCTURES

Most of those who replied to the questionnaire agreed to make a regular place for the exchange of information at home and abroad on various topics from the research to the design application on FEM analysis of RC structures.

Arranging the reply results of the questionnaire, it seems to be ideal to hold a domestic colloquium once every two to three years in the future and to publish the proceedings in English as the periodical publications and to hold an international seminar by rotation once every four to five years in order to exchange information between national groups. Under these circumstances, the researchers will be able to have sufficient time for the basic investigations which contribute to the development of FEM analysis of RC structures, and we can have an opportunity to introduce abroad the information in our country which is characterized by many analytical examples of the application to structural members. It will be also important to make up a gap of the critical mind, even in a minor way, between researchers and designers on the place for the exchange of information as mentioned by Sotomura.

In conclusion, the technical paper which is usually published in our country is too short in content for detailed reference. In America, the circulation of the report such as the doctoral thesis has long been established. Tanaka proposed that JCI execute the circulation of the report until a similar system is established in our country.

8. CONCLUDING REMARKS

Future prospects for FEM analysis of RC structures were reported by compiling the reply results submitted to the JCI-RCFEM Committee questionnaire. As the number of replies was not very large, we cannot claim that the contents of this discussion include all points of view of researchers and engineers who participate in design work in our country. But every reply was written with serious consideration of the future in FEM analysis despite the claims of a busy schedule. Therefore, the authors hope that this paper will be a good guide to advance the research of FEM analysis and its application to design in the future.

The main points of this paper are as follows:

Ch. 2. We aim at the understanding of the micro mechanism and the development of design equations by parametric analysis using FEM analysis.

Ch. 3. The main problems to be solved and the future objective for FEM computer programs are the several problems concerning cyclic loading and the improvement of crack models.

Ch. 4. It is important for FEM analysis and experimental work to make up for each other's deficiencies and to deepen our understanding of the actual behavior of RC structures and put the results into practice.

Ch. 5. As for the subjects that confront us in research on FEM analysis, the rationalization and verification of analytical models and the development and improvement of macro models are important. For the longer term, it is desirable to solve the problems under cyclic loading and to account for scale effects.

Ch. 6. As for the application of FEM analysis to design, it is desirable at first to re-examine the previous design equations and macro models with the aid of FEM analysis and to develop design equations by the improvement of the macro models using parametric studies. We hope that an efficient element like a RC element with less degrees of freedom will be developed and that large scale structures can be analyzed with high accuracy and that the number of experimental specimens required for the verification of safety will be reduced.

ACKNOWLEDGEMENTS

The authors express their deepest appreciation to the following people who replied to the JCI-RCFEM Committee's questionnaire despite the claims of their busy research and design work.

H. Okamura	The University of Tokyo
N. Inoue	Muto Institute of Structural Mechanics, Kajima Corporation
K. Maekawa	Technological University of Nagaoka
N. Shirai	Nihon University
K. Sotomura Y. Murazumi	Taisei Corporation
K. Takiguchi	Nagoya Institute of Technology
M. Ueda	Takenaka Komuten Corporation
Y. Arai	Muroran Institute of Technology
A. Endo	Kajima Corporation
T. Ichinose	Nagoya Institute of Technology
R. Tanaka	Tohoku Institute of Technology
T. Toritani	Toda Corporation
K. Yamada	Maeda Corporation

APPENDIX. - REFERENCES

1. "Application of the Finite Element Method to Two-Dimensional Reinforced Concrete Structures", CEB Bulletin D'Information No. 159, June 1983, pp. 89.
2. Blaauwendraad, J., "Bridging the Gaps: an Intermezzo", Final Report, IABSE Colloquium, Delft, June 1981, pp. 717-724.
3. "Concrete under Multiaxial States of Stress Constitutive Equations for Practical Design", CEB Bulletin D'Information No. 156, 1983, pp. 27.
4. "Finite Element Analysis of Reinforced Concrete Structures", Task Committee on Finite Element Analysis of Reinforced Concrete Structures", ASCE, New York, 1982, pp. 545.
5. Hardy Cross, "Analysis of Continuous Frames by Distributing Fixed-End Moments", Transactions, ASCE, Paper No. 1793, May 1930, pp.1-10.
6. Kanoh, Y., Minami, K. and Takiguchi, K., "Micro and Macro Models in the Analysis of R/C Member", Proc., JCI Colloquium on Finite Element Analysis of RC Structures, Dec. 1984, pp. 119-122,(in Japanese).

7. Kokusho, S. and Hayashi, S. et al, "Experimental Study on RC Beams with Opening", Proc., Annual Conv., AIJ, Sept. 1981, pp. 1505-1506, (in Japanese).
8. Noguchi, H. and Maruta, M., "Nonlinear Analysis of Reinforcing Method of RC Shear Walls with Opening", Proc., Annual Conv., AIJ, Sept. 1983, pp. 2059-2060, (in Japanese).
9. "Proc. of JCI Colloquium on Shear Analysis of RC Structures", Introductory Report, JCI Research Committee on Shear Strength of RC Structures, JCI-C4E, Sept. 1983, pp. 140.
10. "Proc. of JCI 2nd Colloquium on Shear Analysis of RC Structures", JCI Research Committee on Shear Strength of RC Structures, JCI-C5, Oct. 1983, pp. 190, (in Japanese).
11. "Proc. of JCI 2nd Colloquium on Shear Analysis of RC Structures--Collected Experimental Data of Specimens for Verification of Analytical Models--", JCI Research Committee on Shear Strength of RC Structures, JCI-C6, Oct. 1983, pp. 62, (in Japanese).
12. "Proc. of JCI Colloquium on Finite Element Analysis of RC Structures", JCI Committee on Finite Element Analysis of RC Structures, JCI-C8, Dec. 1984, pp. 200, (in Japanese).

SUBJECT INDEX

Page number refers to first page of paper.

AUTHOR INDEX

Page number refers to first page of paper.